reproduction
in
farm animals

reproduction
in
farm animals

Edited by

E. S. E. Hafez

School of Medicine
Wayne State University
Detroit, Michigan

4th Edition

LEA & FEBIGER   *Philadelphia* • 1980

Library of Congress Cataloging in Publication Data

Hafez, E S E 1922–   ed.
    Reproduction in farm animals.

    Includes bibliographical references and index.
    1. Livestock—Reproduction   2. Veterinary
physiology.   I. Title. [DNLM: 1. Animals, Domestic
—Physiology. 2. Reproduction. SF768 R426]
SF768.H2 1980 636.089′26 80-10489
ISBN 0-8121-0697-0

First Edition, 1962
    Japanese translation, 1965
    Spanish translation, 1967
Second Edition, 1968
    Japanese translation, 1971
Third Edition, 1974

Published in Great Britain by Balliere Tindall, London

PRINTED IN THE UNITED STATES OF AMERICA

Print Number: 4 3 2 1

Preface

The objective of the first edition, appearing in 1962, was to present the basic and comparative aspects of reproductive physiology in a simplified manner that would meet the needs of students in reproductive biology, veterinary medicine and animal sciences. This aim has not changed.

The last five years have been characterized by a gratifying accumulation of data, due to important advances in methodology such as radioimmunoassay, scanning electron microscopy, and enzymology. The fourth edition is a complete revision of the third edition in order to include recent progress in gamete ultrastructure and transport, cytogenetics and biochemistry of reproduction, and hypothalamic control of reproduction and sexual behavior. There have been numerous deletions from the third edition, as well as integration of new references and modern concepts. Several new chapters have been added dealing with functional histology of reproduction, seminal plasma, folliculogenesis, egg maturation, ovulation, and estrus synchronization. The chapter on egg transfer was expanded to include recent techniques of superovulation and nonsurgical collection and transfer of embryos. The chapter dealing with reproduction in laboratory animals will be helpful in staging demonstrations for students of reproduction. Much of the information on laboratory animals presented in the third edition has been replaced by recent data on farm animals. No attempt was made to provide a detailed bibliography, but a selected number of classic papers and review articles are listed at the end of each chapter.

The material in the book is arranged into five parts. Part I deals with the functional anatomy of male and female reproduction, with emphasis on functional histology and physiologic mechanisms. Part II deals with the neuroendocrinology and endocrinology of reproduction; the reproductive life cycle; folliculogenesis, egg maturation, and ovulation; spermatozoa; the biochemistry of seminal plasma; the transport of ova and spermatozoa; fertilization, cleavage, and implantation; pregnancy, prenatal development, parturition; lactation; and reproductive and neonatal behavior. Part III includes the species-specific aspects of the reproductive cycles

v

of farm animals and laboratory animals. Part IV deals with reproductive failure in the male and female, intersexuality, and reproductive infections. Part V includes techniques for improving reproductive efficiency, such as artificial insemination, induction and synchronization of estrus, pregnancy diagnosis, and egg transfer.

Nine scanning electron micrographs were kindly provided by Dr. Duane Kraemer of College Station, Texas; Dr. Larry Johnson of Dallas, Texas; Dr. C. J. Connel of San Francisco, California; Dr. Sayako Makabe of Tokyo, Japan; Dr. J. E. Flechon of Jouy-en-Josas, France; and Dr. William P. Wergin of U.S.D.A., Beltsville, Maryland. The sincere thanks of the editor are extended to all contributors and to the staff of Lea & Febiger for their excellent co-operation during the preparation and production of the volume.

Detroit, Michigan   E.S.E. Hafez

Contributors

Alexander, G.: CSIRO, Division of Animal Physiology, Ian Clunies Ross Animal Research Laboratory, Prospect, NSW, Australia.

Anderson, L.L.: Department of Animal Science, 11 Kildee Hall, Iowa State University, Ames, Iowa 50010.

Ashdown, R.R.: Department of Veterinary Anatomy, The Royal Veterinary College, University of London, Royal College Street, London N.W. 1, OTU, England.

Bellows, R.A.: United States Department of Agriculture, Livestock and Range Research Station, Miles City, Montana 59301.

Britt, J.H.: Department of Animal Science, North Carolina State University, Raleigh, North Carolina 27650.

Buttle, H.L.: National Institute for Research in Dairying, Shinfield, Reading, RG 2 9AT, England.

Cowie, A.T.: National Institute for Research in Dairying, Shinfield, Reading, RG2 9AT, England.

Dunn, T.G.: Department of Animal Sciences, University of Wyoming, Laramie, Wyoming 82071

Foote, R.H.: Department of Animal Sciences, Cornell University, Ithaca, New York 14853.

Garner, D.L.: Department of Physiological Sciences, College of Veterinary Sciences, Oklahoma State University, Stillwater, Oklahoma 74074.

Gilbert, A.B.: Poultry Research Centre, West Mains Road, Edinburgh EH9 355, Scotland.

Hafez, E.S.E.: Departments of Gynecology-Obstetrics and Physiology, Wayne State University School of Medicine, Detroit, Michigan 48201.

Hancock, J.L.: Department of Veterinary Anatomy, The Royal Veterinary College, University of London, Royal College Street, London N.W.1, OTU, England.

Hawk, H.W.: United States Department of Agriculture, Animal Science Institute, Reproduction Laboratory, Beltsville, Maryland 20705.

Howarth, J.A.: Department of Epidemiology and Preventive Medicine, School of Veterinary Medicine, University of California, Davis, California 95616.

Hulet, C.V.: United States Department of Agriculture, US Sheep Experiment Station, Dubois, Idaho 83423.

Jainudeen, M.R.: Faculty of Veterinary Medicine, Department of Clinical Studies, Universiti Pertanian Malaysia, Serdang, Selangor, West Malaysia.

Kaltenbach, C.: Department of Animal Sciences, University of Wyoming, Laramie, Wyoming 82071.

Kendrick, J.W.: Department of Reproduction, School of Veterinary Medicine, University of California Davis, California, 95616.

Kressly, L.R.: New Bolton Center, School of Veterinary Medicine, University of Pennsylvania, R.D.1, Kennett Square, Pennsylvania, 19348.

Levasseur, M-C: Station de Recherches de Physiologie Animale, INRA, 78350 Jouy-en-Josas, France.

McFeely, R.A.: New Bolton Center, School of Veterinary Medicine, University of Pennsylvania, R.D.1, Kennett Square, Pennsylvania, 19348.

McLaren, A.: M.R.C. Mammalian Development Unit, University College London, Wolfson House, 4 Stephenson Way, London, NW1, 2HE, England.

Reeves, J.J.: Department of Animal Sciences, Washington State University, Pullman, Washington 99163.

Roche, J.F.: The Agriculture Institute, Grange, Dunsany, County Meath, Irish Republic.

Seidel, G.E., Jr.: Animal Reproduction Laboratory, Colorado State University, Fort Collins, Colorado 80523.

Signoret, J.P.: Station de Physiologie de la Reproduction, Centre de Recherches de Tours, B.P. 1 37-Nouzilly, France.

Shelton, M.: San Angelo Research and Extension Center, RFD 1, Box 950, San Angelo, Texas 76901.

Sugie, T.: National Institute of Animal Industry, Chiba, Japan.

Thibault, C.: Université Pierre et Marie Curie, Laboratoire de Physiologie de la Reproduction, 78350 Jouy-en-Josas, France.

White, I.G.: Department of Veterinary Physiology, The University of Sydney, Sydney, NSW, Australia.

Contents

APPENDICES

1

Introduction to Animal Reproduction

E.S.E. HAFEZ

During the course of mammalian evolution, there have been notable anatomic, endocrinologic and physiologic changes. Among the more obvious are: economy in the production of gametes, reduction in the size of the egg, internal fertilization, development of the corpus luteum as a temporary endocrine organ, development of the placenta as a nutritive, excretory, endocrine, and protective organ and finally, birth at the time appropriate to the species when the young can survive in the environment provided for them. The main effect of these changes was to ensure continuation of the species.

Birds (*Aves*), fish (*Pisces*), and amphibians (*Amphibia*) are "oviparous." Their eggs are large, produced in abundance, surrounded by an abundant yolk, and in many cases, fertilized outside the body of the female whose external genitalia are poorly developed. Reptiles (*Reptilia*) are "ovoviviparous"; their eggs are covered with a protective shell and have an abundant yolk. Furthermore, the larvae hatch inside the body of the female. Mammals, on the other hand, with the exception of *Monotremata*, are "viviparous." They produce fewer eggs which contain a scant yolk, fertilization is internal, fetal development is completed in the uterus, and the external genitalia are well-developed. The echidna (*Tachyglossus aculeata*) and the platypus (*Ornithorhynchus anatinus*) are the only mammals that lay eggs. The eggs of these mammals are relatively small but do contain enough nutriment to support development up to an advanced stage, though not to the stage of hatching.

There are several thousand mammalian species, but reproductive biology has been extensively studied in less than 25: namely, rodents (*Rodentia*), rabbits (*Lagomorpha*), primates (including man), farm animals and a few marsupials. Some of these species are characterized by peculiar reproductive phenomena, such as restricted sexual season, absence of estrus, presence of menstruation, dissociation of ovulation and estrus, nonspontaneous ovulation, spontaneous multiple ovulation with limited implantation, delayed implantation, and ovulation during pregnancy.

The animals which man has domesticated over the centuries to meet his own needs for food, clothing, power, or companionship include cattle, sheep, goats, pigs *(Artiodactyla)*; horses and asses *(Perissodactyla)*; cats and dogs *(Carnivora)*; and poultry *(Galliformes)*. These animals vary with respect to sexual season, sexual cycle, gestation period, type of placentation, litter size, lactation period, and susceptibility to reproductive diseases. For example, cattle, pigs and chickens breed throughout the year, horses and asses in the spring, and most sheep and goats in the fall. These seasonal variations are not so evident in tropical zones compared to temperate and frigid zones, where periodicity is seasonal. Furthermore, these variations are less evident in domesticated than in wild species.

The activity of the gonads and the accessory glands are influenced directly or indirectly by hereditary factors, ambient temperature, photoperiod, and nutrition. The reproductive cycle is regulated by interactions between the central nervous system, the hypothalamus, the pituitary, and the gonads. The hypothalamus controls the secretion of gonadotropins by releasing a regulatory substance into the portal blood flowing from the median eminence region of the tuber cinereum into the pars distalis of the pituitary gland.

Natural and synthetic LH-releasing hormones (LH-RH) have similar biologic and physiochemical properties. These hormones greatly enhance the release of both LH and FSH in vivo and from pituitaries incubated in vitro in doses of fractions of a nanogram.

Prostaglandins (PG), a group of chemically related 20-carbon chain hydroxy fatty acids, are widely distributed in mammalian tissue. They have been used to induce luteolysis (destruction of the corpus luteum) in early stages of pregnancy and to synchronize estrus. The potent oxytocic effect of prostaglandins has been used to induce abortion and labor in women.

Recent advances in reproductive endocrinology are primarily due to the development of radioimmunoassay, a standard and highly sensitive method used for assay of releasing factors, gonadotropins, steroids and prostaglandins.

There are remarkable species differences in the degree to which the scrotum is held near or away from the abdomen. Optimal temperature for spermatogenesis and storage of epididymal spermatozoa seems to be related to their survival time in the female reproductive tract. In cattle and sheep the scrotum is pendulous, and the life span of spermatozoa in the female reproductive tract is only about 30 hours. In horses the scrotum is close to the abdominal wall, and the life span of spermatozoa in the female tract extends to three days. The avian testes is located in the abdominal cavity, and the epididymal spermatozoa survive for about 30 days. Turkey spermatozoa survive in the female tract for prolonged periods and eggs can be fertilized for up to 30 days after a single mating. Sperm survive for several months in the female tract of hibernating bats of the families *Vespertilionidae* and *Phinolophidae*. Species differences in sperm survival are due to differences in the anatomy and physiology of the female reproductive tract, sperm concentration and motility, rate of sperm transport, the female's endocrine state at copulation and her inflammatory response to it, and the dilution of sperm by luminal fluids secreted in the female tract.

The mammalian spermatozoon of most species, before being able to penetrate the zona pellucida, undergoes final maturation in the female reproductive tract (capacitation). This phenomenon is followed by the acrosome reaction, involving multiple fusions between the plasma and outer acrosomal membranes, with subsequent vesiculation.

The striking differences that occur in the estrous cycles most likely reflect differences in the way in which exteroceptive stimuli are translated into hormone production and release. The rat, for example, ovulates spontaneously toward the end of estrous periods that recur at four- to five-day intervals. The rabbit is an induced ovulator and ovulates only 10 to 11 hours after copulation. All the large domestic animals except the Camelidae are spontaneous ovulators like the rat, but their corpora lutea function in the absence of any obvious additional stimulation as in the rabbit. Members of the camel family are induced ovulators.

Cleavage and blastulation of the embryo depend on critical oviductal and uterine factors. Quantitative and qualitative changes in the uterine fluids are related to some reproductive process or to specific needs of the embryo during particular stages during pre-implantation development. In some species the implantation of the embryo is delayed, increasing the duration of pregnancy.

Early pregnancy is characterized by active secretion of progesterone from the corpus luteum. In marsupials, such as opossum and kangaroo, the life span of the corpus luteum is similar in pregnant and nonpregnant females, and the length of gestation period is similar to that of the estrous cycle. In farm mammals, the life span of the corpus luteum is prolonged during pregnancy and ovulation is suppressed except in the mare. In general, the duration of gestation increases with the size of the species and with the stage of development at which the young are born.

The neonate in placental mammals is very immature, develops slowly, and depends on maternal care. The stages of development at birth vary greatly in different species and determine the extent to which parental care is required. In the rat and rabbit, neonates are born blind, naked, and with a poorly developed thermoregulatory system; thus, they require a warm maternal nest. In ungulates, the young are born in an advanced stage of development and can fend for themselves in a few days. The extent of mother-young social interactions also varies widely and is necessary for the full development of the physical and behavioral characteristics of the species.

The efficiency of reproduction in a given species depends on the length of the sexual season, frequency of estrus, number of ovulations, duration of pregnancy, litter size, suckling period, puberty age, and duration of the reproductive period in the animal's life. In general, the age at which puberty is attained is earlier in smaller sized species than in large ones, as well as in females compared to males. There is no definite age at which reproductive functions cease abruptly during life, constituting menopause or climacteric in man. Many other female mammals, however, die before arrest of reproductive functions occurs.

The efficiency of reproduction may decline as a result of seasonal, genetic, nutritional, anatomic, hormonal, neural, immunologic, humoral, or pathologic factors. These factors may result in partial or complete reproductive failure. Those concerned with farm animals have the continuous objective of preventing such failure. Several methods have been used to control and enhance fertility in an attempt to keep some balance between the supplies and demands of an exploding population.

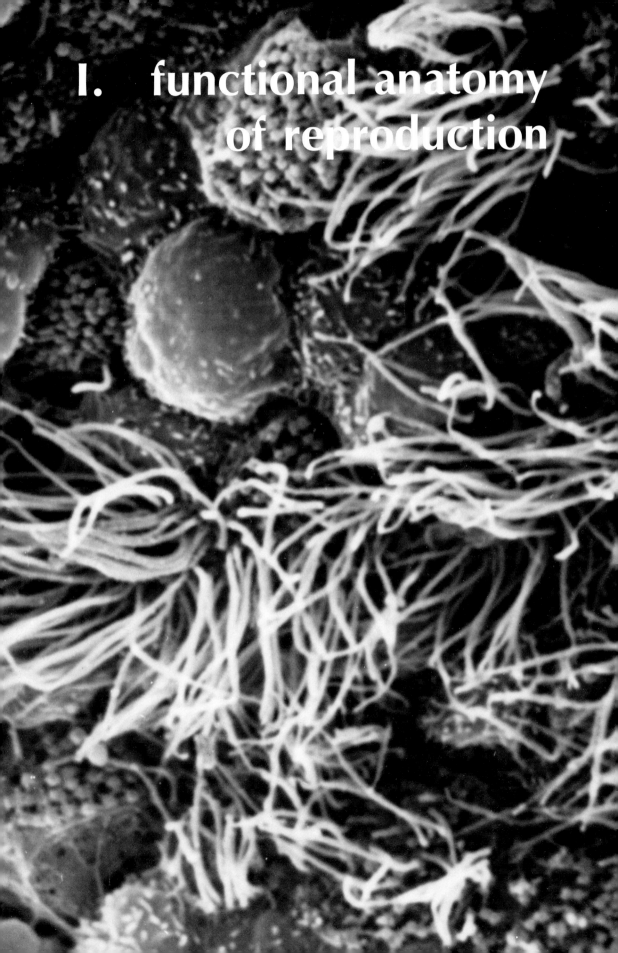

I. functional anatomy of reproduction

2

Functional Anatomy of Male Reproduction

R.R. ASHDOWN AND J.L. HANCOCK

The male gonads, the testes, lie outside the abdomen within the scrotum, which is a purselike structure derived from the skin and fascia of the abdominal wall. Each testis lies within the vaginal process, a separate extension of the peritoneum, which passes through the abdominal wall at the inguinal canal. The internal and external inguinal rings are the deep and superficial openings of the inguinal canal. Besides permitting the passage of the vaginal process and its contents, the inguinal canal also gives passage to important vessels and nerves supplying the external genitalia. Blood vessels and nerves reach the testis in the spermatic cord, which lies within the vaginal process; the *ductus deferens*, which at first accompanies the vessels, leaves them at the orifice of the vaginal process to join the urethra.

The spermatozoa produced by the testis leave by way of a number of efferent ductules that lead into the coiled duct of the epididymis, which becomes the straight ductus deferens. Sets of accessory glands discharge their contents either into the ductus deferens or near its termination in the pelvic portion of the urethra.

The urethra originates at the neck of the bladder. Its pelvic portion, which is enclosed by the striated urethral muscle and receives secretions from various glands at the pelvic outlet, leads into a second penile portion where it is joined by two more cavernous bodies to make up the body of the penis, which lies beneath the skin of the body wall. Throughout its length the urethra is surrounded by cavernous vascular tissue. A number of muscles grouped around the pelvic outlet contribute to the root of the penis. The apex or free end of the penis is covered by modified skin—the penile integument; in the resting condition it is enclosed within the prepuce. The topographic features of the organs of the important farm species are shown in Figure 2–1; their dimensions are listed in Table 2–3. Detailed descriptions of the organs are given by Nickel and associates (1973).

The testis is supplied with blood from the testicular artery, which originates from the dorsal aorta near the embryonic site of the testes. The internal pudendal artery supplies the pelvic genitalia and branches leave the pelvis at the ischial

7

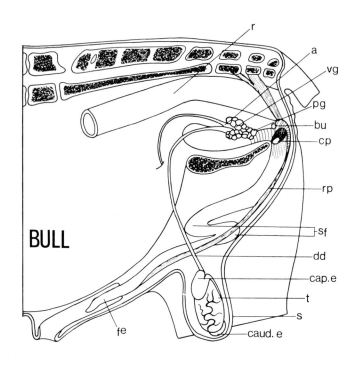

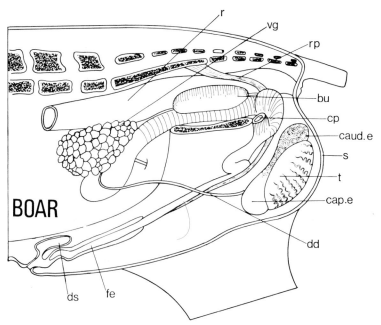

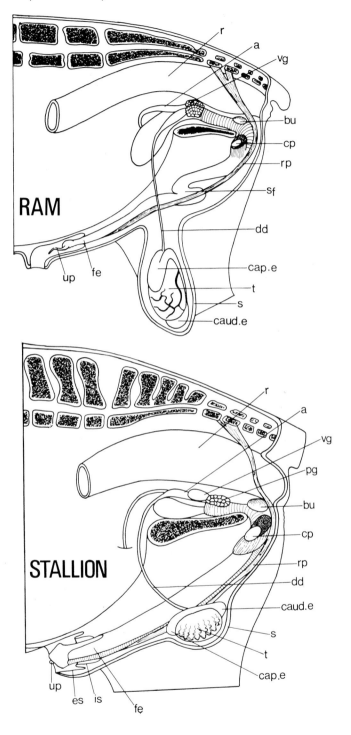

Fig. 2–1. Diagram of the male reproductive tracts as seen in left lateral dissections.
a, Ampulla; *bu*, bulbourethral gland; *cap.* e, caput epididymidis; *caud.* e, cauda epididymidis; *cp*, left crus of penis, severed from the left ischium; *dd*, ductus deferens; *ds*, dorsal diverticulum of prepuce; *es*, external prepuce; *fe*, free end of penis; *is*, internal prepuce; *pg*, prostate gland; *r*, rectum; *rp*, retractor penis muscle; *s*, scrotum; *sf*, sigmoid flexure; *t*, testis; *up*, urethral process; *vg*, vesicular gland. *(Adapted from Popesko, 1968. Atlas der topographischen Anatomie der Haustiere. Vol. 3, Jena, Fischer)*

arch to supply the penis. The external pudendal artery leaves the abdominal cavity via the inguinal canal to supply the scrotum and prepuce.

Afferent and efferent (sympathetic) nerves accompany the testicular artery to the testis. The pelvic plexus supplies autonomic (sympathetic and parasympathetic) fibers to the pelvic genitalia and to the penis. Sacral nerves supply motor fibers to the muscles of the penis and sensory fibers to the free end of the penis. Afferent fibers from the scrotum and prepuce travel to the spinal cord, mainly in the genitofemoral nerve (Larson and Kitchell, 1958; Hodson, 1970).

DEVELOPMENT

Prenatal Development

The testes develop in the gonadal ridge, which lies medial to the embryonic kidney *(mesonephros)*. The gonads of the male embryo differentiate following the arrival of the primordial germ cells; these migrate to the gonadal ridge about day 26 in the bull fetus (Gier and Marion, 1970).

The primordial germ cells, carried into the medulla by the primary sex cords, formed from the coelomic epithelium, provide the element from which the germinal epithelium of the seminiferous tubule is formed. The *rete testis* develops as a separate mass of cords that establishes connection with the mesonephric (kidney) tubules on one hand and with the future seminiferous tubules on the other; as a result, the mesonephric duct becomes the excurrent duct of the testis (Fig. 2–2). In the course of development in the bull, boar, ram and stallion, the rete comes to lie centrally in the testis and not peripherally as in man. The mesonephric tubules, which become connected to the rete, form the efferent ductules located in the head of the epididymis. The rest of the epididymis is formed from the first part of the mesonephric duct; the remainder of the duct forms the ductus deferens and a terminal diverticulum forms the vesicular gland. The paramesonephric duct, which is the primordium of the female duct system, degenerates in the male.

Two active agents that are produced by the fetal testis are responsible for differ-

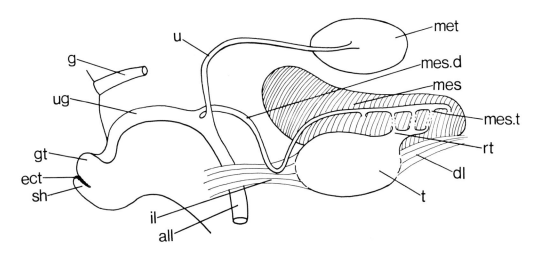

Fig. 2–2. Diagram to show the origins of the male reproductive organs in a mammal.
all, Allantois; *dl*, diaphragmatic ligament; *ect*, ectodermal lamella; *gt*, genital tubercle; *g*, termination of hindgut; *il*, inguinal ligament; *mes*, mesonephros; *mes. d*, mesonephric duct; *mes. t*, mesonephric tubules; *met*, metanephros; *rt*, rete testis; *sh*, developing penile prepuce; *t*, testis; *u*, ureter; *ug*, urogenital sinus. *(Adapted from Gier and Marion, 1969. Biol. Reprod. 1, 1)*

entiation and development of the duct system. Fetal androgen produced by the testis causes development of the male reproductive tract. The second agent is called "Müllerian inhibiting substance" because it is responsible for suppression of the paramesonephric (Müllerian) ducts from which the uterus and vagina develop in the female.

Early in fetal life the urogenital sinus, into which the mesonephric ducts open, is separated from the termination of the gut; the male urethra forms from the urogenital sinus. At the urogenital orifice the genital tubercle forms and within it the penile part of the urethra is developed. A separate fold of skin grows distally over the genital tubercle to form the penile prepuce (Fig. 2–2); it remains fused to the penis until after birth. Abnormalities in differentiation and development of gonads and ducts can result in varying degrees of intersexuality.

Descent of the Testis

Testicular descent (Fig. 2–3) involves abdominal migration to the internal inguinal ring, inguinal migration through the canal, and finally migration within the scrotum. Descent into the scrotum is preceded by formation of the vaginal process, a peritoneal sac extending toward the scrotum and enclosing the inguinal ligament of the testis, which together with the diaphragmatic ligament and the mesorchium suspends the fetal testis. The inguinal ligament connects the gonad and the mesonephric duct; distal to this point it is often called the *gubernaculum testis* and it terminates in the region of the scrotal rudiments. Before descent is accomplished, the gubernaculum enlarges greatly; after descent the gubernaculum regresses. Both processes are thought to be important in the mechanics of testicular descent. The time of descent varies according to the species (Table 2–1). In the horse the epididymis commonly en-

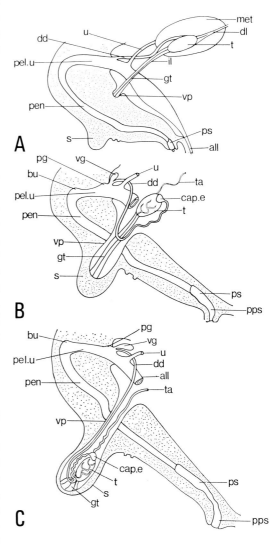

Fig. 2–3. Diagrams to show descent of the testis and development of the reproductive tract in the bovine fetus. *A* at 62 days, *B* at 102 days, *C* at 140 days. *all*, Allantois; *bu*, bulbourethral gland; *cap. e*, caput epididymidis; *dd*, ductus deferens; *dl*, diaphragmatic ligament; *gt*, gubernaculum testis; *il*, inguinal ligament; *met*, metanephros; *pel. u*, pelvic urethra; *pen*, penis; *pg*, prostate gland; *ps*, penile prepuce; *pps*, prepenile prepuce; *s*, scrotum; *t*, testis; *ta*, testicular artery; *u*, ureter; *vg*, vesicular gland; *vp*, vaginal process. *(Adapted from Gier and Marion, 1970. Development of the Mammalian Testis. In: The Testis. Vol. 1. Johnson, Gomes and VanDemark [eds], New York, Academic Press)*

Table 2–1. Chronology of Development of the Male Reproductive Tract in Farm Animals*

	Bull	Ram	Boar	Stallion
Testicular descent	Enters scrotum half-way through fetal life	Enters scrotum half-way through fetal life	Enters scrotum in last quarter of fetal life	Enters scrotum just before or just after birth
Primary spermatocytes in seminiferous tubules	24 weeks	12 weeks	10 weeks	Variable throughout seminiferous tubules of each testis
Spermatozoa in seminiferous tubules	32 weeks	16 weeks	20 weeks	56 weeks (variable)
Spermatozoa in cauda epididymidis	40 weeks	16 weeks	20 weeks	60 weeks (variable)
Spermatozoa in the ejaculate	42 weeks	18 weeks	22 weeks	—
Completion of separation between penis and penile part of prepuce	32 weeks	>10 weeks	20 weeks	4 weeks
Age at which animal can be considered sexually "mature"	150 weeks	>24 weeks	30 weeks	90–150 weeks (variable)

*Compiled from various sources.

ters the inguinal canal before the testis and that part of the inguinal ligament connecting testis and epididymis (proper ligament of testis) remains extensive until after birth.

Sometimes the testis fails to enter the scrotum. Recent work has failed to show any morphologic abnormalities in the fetal retained testes immediately after the period of normal descent (Van Straaten and Wensing, 1977) and the failure may be related to the anatomy of the vaginal process and gubernaculum (Smith, 1975). In this condition (cryptorchidism), the special thermal needs of the testis are not met and normal spermatogenic function is impossible, although the endocrine function of the testis is unimpaired. Cryptorchid males therefore show more or less normal sexual desire but are sterile. Occasionally some of the abdominal viscera pass through the orifice of the vaginal process and enter the scrotum; scrotal hernia is particularly common in pigs.

Postnatal Development

Each component of the reproductive tracts of all farm animals grows in size relative to overall body size and undergoes histologic differentiation (Abdel-Raouf, 1960; Nishikawa, 1959), but functional competence is not achieved simultaneously in all components of the reproductive system. Thus in the bull the capacity for erection of the penis precedes the appearance of spermatozoa in the ejaculate by several months. At puberty all the components of the male reproductive system have reached a sufficiently advanced stage of development for the system as a whole to be functional. The period of rapid development that precedes puberty is known as the prepubertal period, although this period is itself sometimes referred to as "puberty" (Donovan and Werff ten Bosch, 1965; Skinner and Rowson, 1968). During the postpubertal period, development continues and the reproductive tract reaches full sexual

Table 2–2. Growth of the Reproductive Tract in Holstein and Holstein-cross Bulls During the Postpubertal Period*

	Age in Months (Mean)			
Items	37	59	80	133
Number of bulls	7	20	4	7
Body weight (lb)	1864	2081	2046	2006
Testis weight (gm)	259	335	359	395
Epididymis weight (gm)	27	35	38	40
Vesicular gland weight (gm)	55	78	79	81
Bulbourethral gland weight (gm)	5.2	6.5	7.1	6.0
Penis length (cm)	95	97	103	106

*From Almquist and Amann, 1961. *J. Dairy Sci.*, 44, 1668.

maturity months or even years after the age of puberty. Some important anatomic changes that occur during postnatal development are summarized in Tables 2–1 and 2–2, and are illustrated in Figures 2–4 to 2–7.

TESTIS AND SCROTUM

Structure

The testis is secured to the wall of the vaginal process along the line of its epididymal attachment (Fig. 2–8). The position in the scrotum and the direction of the long axis of the testis relative to the body differs with the species (Fig. 2–1). The epididymis is closely apposed to the surface of the testis and the point of origin of the efferent ducts from the rete testis lies under the flattened expanded head of the epididymis (Dym, 1976). The surface of the testis is covered by an extension of the parietal peritoneum of the abdominal cavity. Beneath this lies a tough fibromuscular *tunica albuginea* from which, at the epididymal attachment, extensions penetrate the parenchyma of the organ to join the *mediastinum*, a cord of connective tissue running through the testis (Fig. 2–8). Fibrous septa divide the parenchyma into lobules of coiled seminiferous

tubules, which lead into the rete testis by way of straight tubules.

Measurements of testicular dimensions made in intact live animals have been shown to closely correlate with measurements of the organ after removal. Land and Carr (1975) have found a high correlation between testis diameter and organ weight for three breeds of sheep in two age groups. Differences between breeds in testis diameter were related to differences in female prolificacy between these breeds.

Endocrine Function

The interstitial (Leydig) cells which lie between the tubules are the source of the male hormone (testosterone). The epithelium of the tubule consists of spermatogenic cells and supporting sustentacular (Sertoli) cells. The basement membrane contains contractile "myoid" cells. Spermatozoa are produced by differentiation of the last of several generations of cells which result from the division of peripherally situated spermatogonia.

The two important functional roles of the testes are governed by the gonadotropic hormones of the pituitary gland. Follicle stimulating hormone (FSH) is closely connected with initiating activity

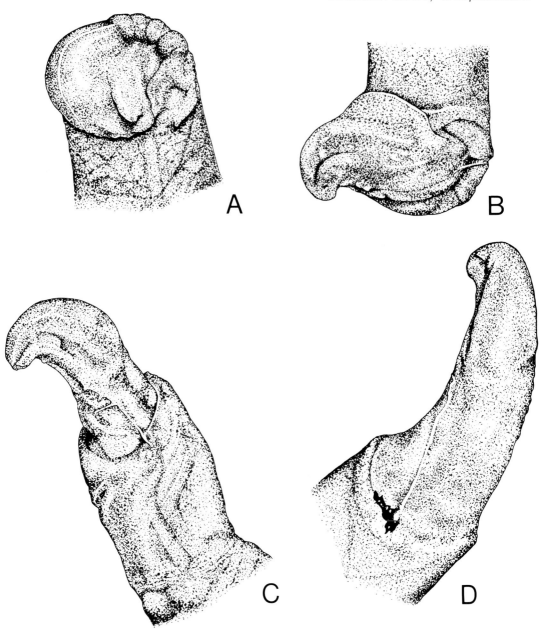

Fig. 2–4.   Penis of a crossbred Galloway calf, protruded to show separation of the free end of the penis from its penile prepuce at four different ages. *A*, 278 days; *B*, 306 days; *C*, 319 days; *D*, 326 days. *(Redrawn from Ashdown, 1960. J. Agric. Sci. 54, 348)*

Fig. 2–5.   Transverse sections through the adherent penis and prepuce of a week-old calf. *A*, low-power view; *B*, higher-power view of the ectodermal lamella; *C*, higher-power view of the ectodermal lamella to show the keratinizing changes that split the lamella and separate penis from prepuce. *c*, cavity of prepuce; *ccp*, corpus cavernosum penis; *ccu*, corpus spongiosum penis; *e*, "epithelial pearl" formed by keratinization; *f*, frenulum; *l*, ectodermal lamella; *oz, mz, iz*, outer, middle and inner zones of the ectodermal lamella; *p*, penile tissues; *s*, prepuce; *u*, urethra.   →

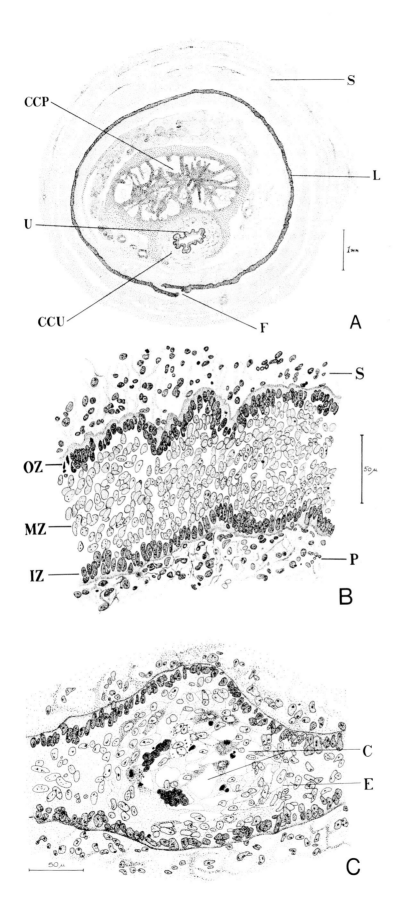

15

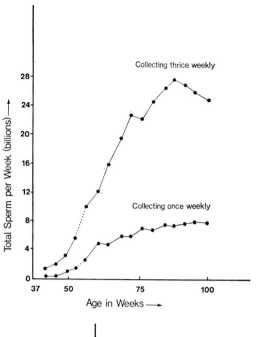

Fig. 2–6. Average sperm output of Holstein bulls during the postpubertal period, up to 100 weeks of age. *(Simplified from Almquist, Amann and Hale, 1963.* **Ann. Meet. Amer. Soc. Anim. Sci., Morgantown, West** *Virginia)*

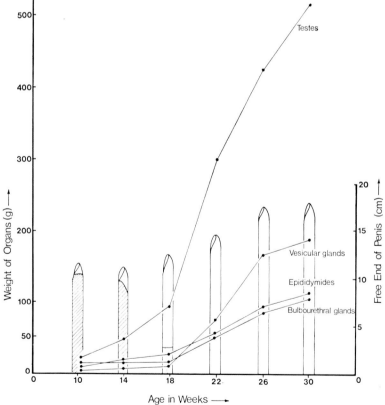

Fig. 2–7. Development of the reproductive tracts in six littermate Landrace boars from 10 to 30 weeks of age. Organ weight is the sum of left and right components. Length of the free end of the penis is shown; the shaded area represents the area still adherent to the penile prepuce.

Fig. 2–8. Diagrammatic horizontal section through the scrotum of the bull to show the relationships of the organs. The inset diagram shows more clearly the layers of the scrotal wall and the vaginal process.

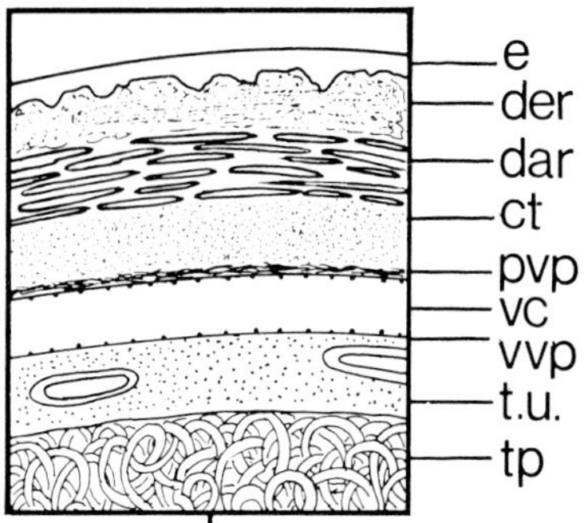

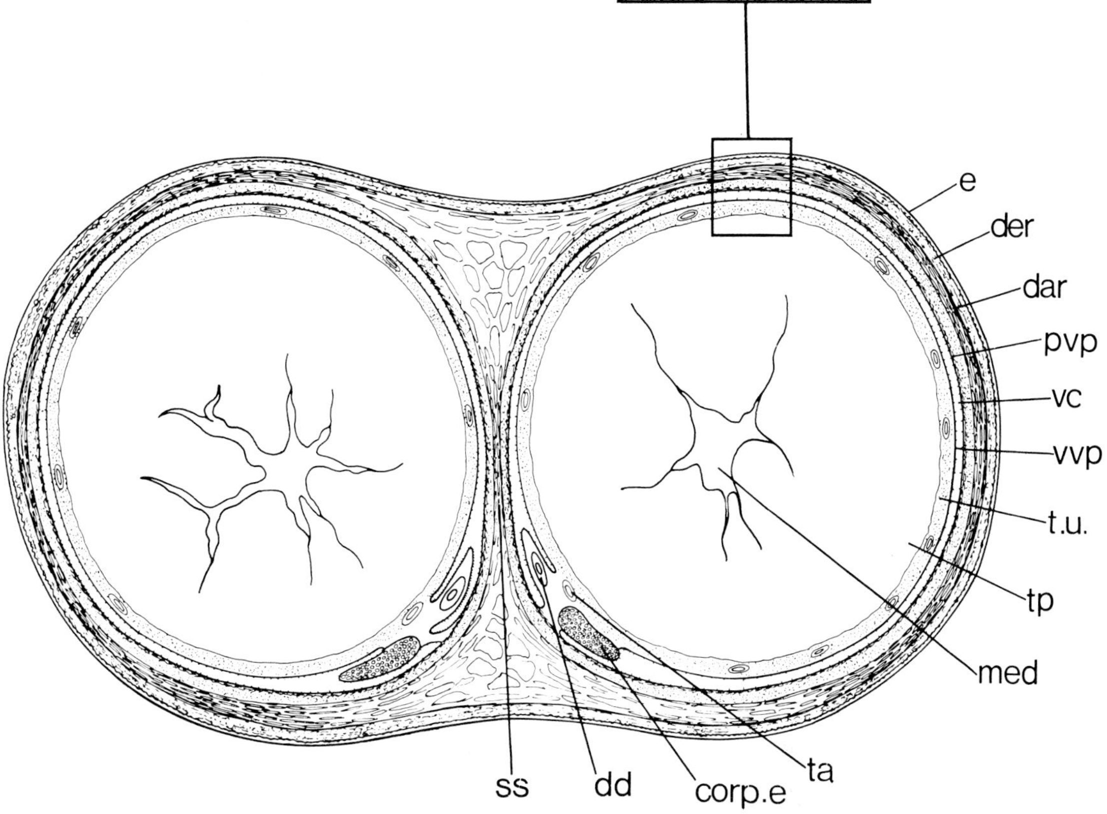

corp. e, Corpus epididymidis; *ct*, loose connective tissue lying between the vaginal process and the wall of the scrotum; *dar*, tunica dartos; *dd*, ductus deferens; *der*, dermis; *e*, epidermis; *med*, mediastinum testis; *pvp*, parietal layer of vaginal process; *ss*, scrotal septum; *t.u.*, tunica albuginea of testis; *ta*, testicular artery; *tp*, testicular parenchyma; *vc*, cavity of vaginal process; *vvp*, visceral layer of vaginal process. *(Adapted from Blom and Christensen, 1947. Skand. Vet. Tidskrift 37, 1)*

in the seminiferous tubules. Luteinizing or interstitial cell stimulating hormone (LH or ICSH) controls the endocrine activity of the cells. Male sex hormone produced by the interstitial cells supports the action of FSH on spermatogenesis and is responsible for development of secondary sexual characteristics and for the growth and functional integrity of the male reproductive tract as a whole. Castration of prepubertal males results in suppression of sexual development (Fig. 2–12). Regressive changes in behavior and structure take place following castration of adult males. Castration is a standard procedure in animal husbandry to modify aggressive male behavior and to eliminate undesirable carcass qualities, e.g., boar taint. Testosterone exerts some sparing effects on protein metabolism and there is now a trend toward the use of intact males for meat production.

Exocrine Function

The spermatozoa leave the testis in an important fluid secretion (Setchell, 1970). This "rete fluid" differs greatly in composition from blood plasma and lymph. Thus, there is an important blood-testis barrier which effectively separates the seminiferous epithelium from the general circulation. This barrier seems to be formed by special cells of the basement membrane of the tubule and by some special features of the sustentacular cells (Dym and Fawcett, 1970). The barrier effectively divides two compartments of the testis, with the parent spermatogonia separated by the sustentacular cells from their progeny. The integrity of the blood-testis barrier is believed to be important for normal testicular function. It serves to separate immunologically the germinal epithelium from the rest of the body tissues, and injuries that impair the effectiveness of the barrier result in immunologic damage to the testes; the barrier is weakest at the rete testis (Neaves,

1977). The damaging effects of certain heavy metals (e.g., cadmium) on testicular function are believed to be due to their effect on this barrier.

Sperm Production

Daily sperm production per gram of testis is lower for the bull (13 to 19 × 10^6) than for the ram (24 to 27 × 10^6) and boar (24 to 31 × 10^6) (Amann, 1970) or for the horse (19.3 to 22.3 × 10^6) (Swierstra et al,1975a). Sperm production in the bull increases with age up to seven years. Production is usually greater than output as measured by the number of spermatozoa recovered at ejaculation.

Thermoregulation of the Testis

For effective functioning, the mammalian testes must be maintained at a temperature lower than that of the body. Anatomic features of the testis and scrotum permit the regulation of testicular temperature. The scrotal skin is noticeably lacking in subcutaneous fat and is richly endowed with sweat glands, and its muscular *(dartos)* component enables it to alter the thickness of the scrotum and its surface area and to vary the closeness of the contact of the testes with the body wall. In the horse this action may be supported by the internal cremaster muscle within the spermatic cord which can lower or raise the testis. In cold conditions the *cremaster* and dartos muscles contract, elevating the testes and wrinkling and thickening the scrotal wall. In hot conditions the muscles relax, lowering the testes within the thin-walled pendulous scrotum. The advantages offered by these mechanisms are enhanced by the special relationship of the veins and arteries.

In all farm animals, the testicular artery is a convoluted structure in the form of a cone, the base of which rests on the cranial or dorsal pole of the testis (Fig. 2–9).

These arterial coils are enmeshed by the so-called pampiniform plexus of testicular veins. This arrangement provides an effective countercurrent mechanism by which arterial blood entering the testis is cooled by the venous blood leaving the testis. In the ram the temperature of the blood in the testicular artery falls 4° C in its course from the external inguinal ring to the surface of the testis; the blood temperature in the veins rises by a similar amount between the testis and the external inguinal ring. The position of the arteries and veins close to the surface of the testis tends to increase direct loss of heat from the testis (Waites and Setchell, 1969). Evidence of transfer of steroid hormones from venous to arterial blood has been obtained in several species (Free, 1977); its occurrence is presumably related to the anatomy of the testicular artery and the pampiniform plexus.

Temperature receptors in the scrotal skin of sheep can elicit responses that tend to lower *whole* body temperature.

The position of the testis relative to the heart, in a region where little muscular activity occurs, makes a rather unusual demand on the mechanism for the return of blood; venous return is promoted by the close contact between arteries and veins. A particular feature of the blood supply to the testis that is more difficult to explain is the absence of a pulsatile flow.

EPIDIDYMIS

Structure

Three anatomic parts of the epididymis are recognized (Fig. 2–10). The *caput epididymidis* (head), in which a variable number of efferent ductules (6 to 20) join the duct of the epididymis, forms a flattened structure applied to one pole of the testis. It is continued as the narrow *corpus epididymidis* (body) which terminates at the opposite pole in the expanded *cauda epididymidis* (tail). The contour of the cauda epididymidis is a visible feature in the live animal. The caput, corpus and cauda epididymidis are less clearly differentiated in the stallion than in other farm species, and in the foal the attachment to the testis is very loose.

The wall of the duct of the epididymis has a prominent layer of circular muscle fibers and a pseudostratified epithelium of columnar cells. Three regions of the duct of the epididymis can be distinguished histologically; these do not coincide with the gross anatomic regions (Nicander, 1957). The initial segment is characterized by a high epithelium with long straight stereocilia that almost obliterate the lumen. In the middle segment the

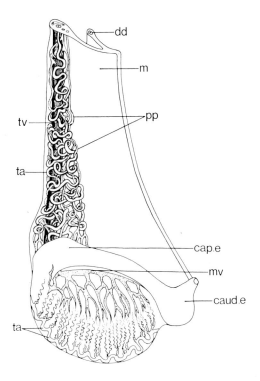

Fig. 2–9. Diagrammatic lateral view of the left testis of a stallion to show arrangement of arteries and veins. *cap. e*, Caput epididymidis; *caud. e*, cauda epididymidis; *dd*, ductus deferens; *m*, mesorchium; *mv*, marginal vein of testis; *pp*, pampiniform plexus of veins; *ta*, testicular artery; *tv*, testicular vein. *(Adapted from Tagand and Barone, 1956. Anatomie des Équidés Domestiques. 2, iii, Lyons, École Nat. Vet.)*

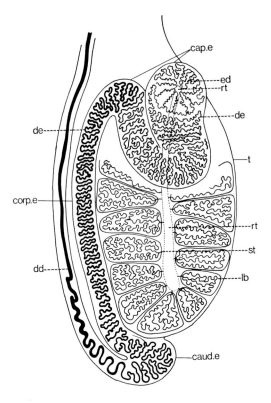

Fig. 2–10. Schematic drawing of the tubular system of the testis and epididymis in the bull (for clarity the duct system of the rete testis is omitted).
cap. e, Caput epididymidis; *caud. e,* cauda epididymidis; *corp. e,* corpus epididymidis; *dd,* ductus deferens; *de,* duct of the epididymis; *ed,* efferent ductule; *lb,* lobule with seminiferous tubules; *rt,* rete testis; *st,* straight tubule; *t,* testis. *(Simplified from Blom and Christensen, 1960. Nord. Vet. Med. 12, 453)*

stereocilia are not so straight and the lumen of the duct is wide. In the terminal segment stereocilia are short; the lumen is very wide and packed with spermatozoa.

Function

The cytologic features suggest that absorption is an important function of the initial and middle segments of the epididymis but not of the terminal segment. The effects of ligation of the epididymis at different levels show

clearly that active absorption of fluid occurs in the caput epididymidis.

By cannulating the efferent ductules it has been shown that, in the ram, up to 60 ml fluid leave the testis daily although total ejaculate volume averages about 1.0 ml. Hourly flow rate per 100 gm testis (0.4 ml) is less in the bull than in the ram and goat. The epithelium of the efferent ductules is also capable of removing particulate matter, including spermatozoa, from the lumen.

Secretory activity is a feature of the epithelium of the duct of the epididymis that is suppressed by castration; the secretions may maintain the viability of spermatozoa during storage. The role of the epididymis in the maturation of spermatozoa is less certain. Migration of the cytoplasmic bead from the neck of the spermatozoa to the terminal part of the midpiece normally occurs in the course of the journey through the epididymis. This change in morphology is associated with important physical and cytochemical changes (Hamilton, 1971) and with an increase in their capacity for motility and their fertilizing ability (Orgebin-Crist, 1969). However, there is evidence from laboratory species that these maturation changes still occur even when spermatozoa are confined to the caput epididymidis by ligatures. Experimental evidence indicates that boar spermatozoa from the corpus epididymidis have a lower fertilizing capacity than spermatozoa from the cauda epididymidis (Hunter et al, 1976).

Spermatozoa in dilute suspension are transported through the efferent ductules by the action of the ciliated epithelium, supported by contraction of the musculature of the duct wall and by the action of the smooth muscle cells in the tunica albuginea and the myoid cells in the walls of the seminiferous tubules. The time required for the transport of spermatozoa through the epididymis is remarkably constant for individual species. By incorporating a radioactive label into spermatocytes and by sampling different levels of the epididymis at varying intervals, it has been found that the duration of the epididymal journey for bull, ram and boar is 10 days, 13 to 15 days and 9 to 12 days respectively (Ortavant et al, 1961; Swierstra, 1968). In the horse, the minimum time required for passage of spermatozoa through the epididymis is three days, but the average time required is 8 to 11 days (Swierstra et al, 1975b). The epididymis is an important storage organ; although fertilizing capacity of spermatozoa lasts only a few hours outside the body at the temperature of the epididymis, it lasts for several weeks in the isolated epididymis. The life of the spermatozoa in the epididymis is shortened by castration.

The two epididymides of a mature bull can accommodate up to 74.1 $\times$ 10^9 spermatozoa, equal to 3.6 days' production by the testes. Of the total extragonadal sperm reserves in the ram, 15% are found in the caput epididymidis, 4% in the corpus epididymidis, and 68% in the cauda epididymidis. Depletion of these reserves occurs following repeated ejaculation, but depletion fails to alter significantly the characteristics of the sperm population of the cauda epididymidis. The speed of transport of spermatozoa through the epididymis is altered only slightly by exhaustive depletion.

The ductus deferens leaves the cauda epididymidis and is supported in a separate fold of peritoneum; it is readily separable from the rest of the spermatic cord. It has a thick muscular wall, and its terminal portion is furnished with branched tubular glands. It enters the pelvic urethra at the *colliculus seminalis*. In some species (Table 2–3) this portion forms a distinct ampulla. The ampullae contain only a small fraction (less than 1%) of the total extragonadal sperm reserves in the ram. The ampullae have muscular walls, which expel the semen from the ductus deferens

into the urethra; this process of emission is only one component of the process of ejaculation.

ACCESSORY GLANDS

The prostate, vesicular and bulbourethral glands pour their secretions into the urethra where, at the time of ejaculation, they are mixed with the fluid suspension of spermatozoa and ampullary secretions from the ductus deferens (Fig. 2–11). All the accessory glands are essentially lobular branched tubular glands with smooth muscle prominent in the interstitial tissue. Anatomic differences between farm species are summarized in Table 2–3 and Figure 2–12.

Comparative Anatomy

The Vesicular Glands. These lie lateral to the terminal parts of each ductus deferens. In ruminants and swine they are compact lobulated glands; in the stallion they are true vesicles consisting of large pyriform glandular sacs. The duct of the vesicular gland and the ductus deferens may share a common ejaculatory orifice into the urethra.

Prostate Gland. Two components are distinguished: a distinct lobulated external part or body which lies outside the thick urethral muscle surrounding the urethra, and a second internal or disseminated part distributed along the length of the pelvic urethra below the urethral muscle. The body of the prostate is small in the bull and large in the boar, while in the ram no body is visible. In the stallion the prostate gland is wholly external and consists of two lateral lobes joined by an isthmus.

The Bulbourethral Glands. These are paired bodies lying dorsal to the urethra near the termination of its pelvic portion. In the bull they are almost hidden by the *bulbospongiosus muscle* and in all species they are covered by a thick layer of

Table 2–3. Dimensions and Weights of the Components of the Male Reproductive Tract in Farm Animals*

Organ		Bull	Ram	Boar	Stallion
Testis	Length (cm)	13	10	13	10
	Diameter (cm)	7	6**	7	5
	Weight (gm)	350	275	360	200
Epididymis	Length of duct (m)	40	50	18	75
	Weight (gm)	36	–	85	40
Ductus deferens	Length (cm)	102	24	–	70
Ampulla	Length (cm)	15	7	Scattered lobules of gland tissue at termination of the duct	25
	Diameter (cm)	1.2	0.6		2
Vesicular gland	Length (cm)	13	4	13	15
	Breadth (cm)	3	2	7	5
	Thickness (cm)	2	1.5	4	5
	Weight (gm)	75	5	200	–
Prostate gland	Body (cm)	3 × 1 × 1	Scattered lobules of gland tissue	3 × 3 × 1 (20 gm)	Isthmus 2 × 3 × 0.5
	Disseminate part (cm)	12 × 1.5 × 1	–	17 × 1 × 1	Lobe 7 × 4 × 1
Bulbourethral gland	Length (cm)	3	1.5	16	15
	Breadth (cm)	2	1	4	2.5
	Thickness (cm)	1.5	1	4	2.5
	Weight (gm)	6	3	85	–
Penis	Total length (cm)	102	40	55	50
	Length of free end (cm)	9.5	4	18	20
	Urethral process (cm)	0.2	4	Not present	3
Prepuce	Length (cm)	30	11	23 (Preputial diverticulum volume about 100 ml)	External 25 Internal 15

*Average values are given from various sources. It is especially difficult to give figures for the horse because of breed variation in size.
**In the Soay breed, Lincoln (1978) showed that testicular diameter is increased by reducing the daily light period from 16 to 8 hours.

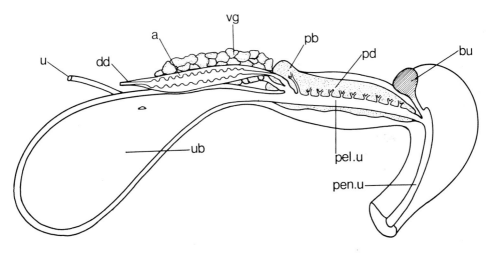

Fig. 2–11. **Diagram to show the disposition of the glands which discharge into the pelvic urethra of the bull.** *a*, Ampulla; *bu*, bulbourethral gland; *dd*, ductus deferens; *pb*, body of prostate gland; *pd*, disseminate part of prostate gland; *pel. u*, pelvic urethra; *pen. u*, penile urethra; *u*, ureter; *ub*, urinary bladder; *vg*, vesicular gland.

striated muscle. They are especially large in the boar and contribute the prominent gel-like component of boar semen.

Urethral Glands. The bull lacks urethral glands comparable with those found in man: earlier conclusions about their contribution to bull semen need to be re-examined (Kainer et al, 1969). Glands of this name in the horse have been considered comparable to the disseminated prostate of ruminants, but in the boar the disseminate prostate and the urethral glands are histologically distinct (McKenzie et al, 1938).

Function

Apart from providing a liquid vehicle for the transport of spermatozoa, the function of the accessory glands is obscure although much is known about the specific chemical agents contributed by the glands to the ejaculate (Mann, 1964).

These agents serve as markers of the contribution made to the semen by individual glands and as indicators of gland function. Fructose and citric acid are important components of vesicular gland secretions of domestic ruminants. Citric acid alone is found in stallion vesicular glands; boar vesicular glands also contain little fructose and are characterized by a high content of ergothioneine and inositol. Glyceryl phosphoryl choline is a distinctive component of the epididymal secretion. Ergothioneine is found in the ampullary glands of the horse and the jackass.

Spermatozoa from the cauda epididymidis are capable of fertilization when inseminated without the addition of accessory gland secretions. The gel-like fraction of the boar ejaculate forms a plug in the vagina of mated females but in commercial insemination practice this fraction is removed from the semen by filtration.

In large animals it is possible to palpate per rectum some of the accessory glands. The positions of these glands relative to the bony pelvis are shown in Figure 2–12. The vesicular glands of the bull are readily detectable, lying one on each side of the ampullae, and the body of the prostate gland is detectable as a hard smooth prominence caudal to the neck of the

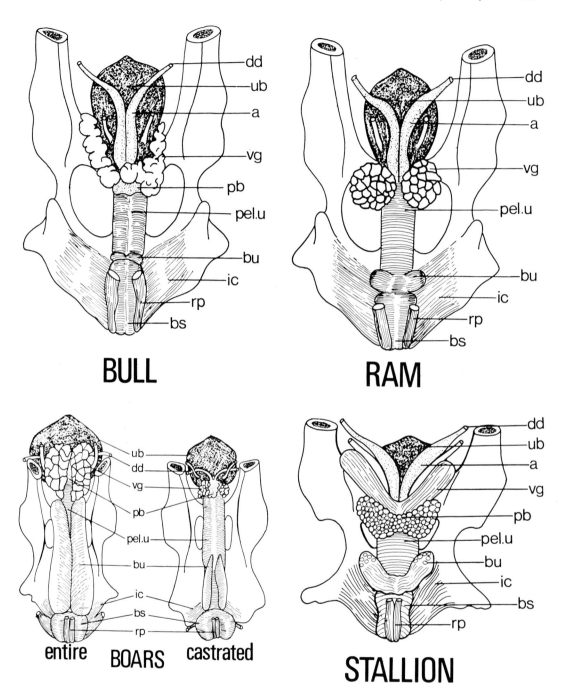

Fig. 2–12.   Diagrams of the pelvic genitalia, within the pelvic bones, as seen from a dorsal view. (The cranial parts of the ilium have been removed.)

a, Ampulla; *bs*, bulbospongiosus muscle; *bu*, bulbourethral gland; *dd*, ductus deferens; *ic*, ischiocavernosus muscle; *pb*, body of prostate gland; *pel. u*, pelvic urethra; *rp*, retractor penis muscle; *ub*, urinary bladder; *vg*, vesicular gland. *(Diagrams of bull, boar and stallion redrawn from Nickel, 1954. Tierärztl. Umschau 9, 386)*

bladder. The bulbourethral glands, because of their covering of muscle, are not identifiable in the bull.

Rectal palpation of the bulbourethral glands of the boar can be used to identify abdominal cryptorchid animals; the glands are much smaller in castrated males than in cryptorchid males (Fig. 2–12). Pathologic changes occur in the accessory glands of castrated male sheep that graze certain species of clover. This is due to estrogenic components of the clover which induce feminizing changes in the glandular epithelium.

PENIS AND PREPUCE

Structure of the Penis

In the mammalian penis, three cavernous bodies are aggregated around the penile urethra (Fig. 2–13). The *corpus spongiosum penis* which surrounds the urethra is enlarged at the ischial arch to form the penile bulb. This bulb is covered by the striated *bulbospongiosus* muscle. The *corpus cavernosum penis* arises as a pair of *crura* from the ischial arch under cover of the striated *ischiocavernosus*

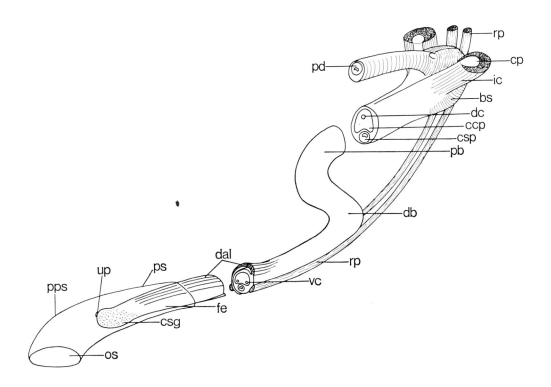

Fig. 2–13. Diagram to show the anatomy of penis and prepuce in the bull.
bs, Bulbospongiosus muscle; *ccp,* corpus cavernosum penis; *csp,* corpus spongiosum penis, surrounding the penile urethra; *cp,* left crus penis; *csg,* corpus spongiosum glandis (a thin mantle of cavernous tissue covers the fibrous tissue of the enlargement at the tip of the penis); *dal,* dorsal apical ligament; *dc,* dorsal erection canal; *db,* distal bend of the sigmoid flexure; *fe,* free end of the penis, covered by penile integument; *ic,* ischiocavernosus muscle; *os,* orifice of the prepuce; *pd,* disseminate part of the prostate gland; *pps,* prepenile prepuce; *ps,* penile prepuce; *pb,* proximal bend of sigmoid flexure; *rp,* left retractor penis muscle; *up,* urethral process; *vc,* left ventrolateral erection canal.

muscle. The corpus cavernosum penis continues to the apex of the penis as a more or less paired dorsal cavernous body. A thick collagenous covering (tunica albuginea) covers the cavernous bodies and from it numerous trabeculae enter the corpus cavernosum penis to support its cavernous tissue. A pair of

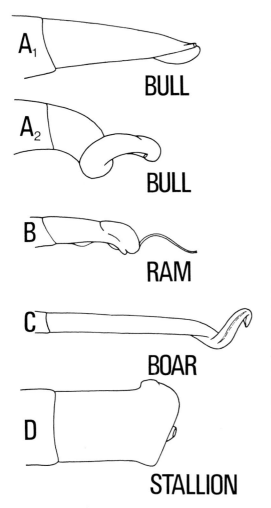

Fig. 2–14. Diagrams to show the shape of the free end of the penis. A_1 shows the shape of the penis just before intromission and A_2 shows the shape after intromission when spiral deviation has occurred. *B* shows the shape of the penis during natural service. *C* does not show the full degree of spiralling that occurs during service. *D* was drawn after injection and shows enlargement of the erectile bodies *(A_1, A_2 and B from photographs. C and D from fixed specimens. Not drawn to scale.)*

smooth *retractor penis* muscles arise from the sacral or coccygeal regions of the vertebral column and are especially large in ruminants and swine. In these animals the retractor penis muscles are able to control the effective length of the penis by their action on the sigmoid flexure.

In the stallion the cavernous bodies contain large cavernous spaces; during erection considerable increases in size result from accumulation of blood in these spaces. In bull, ram and boar, the cavernous spaces of the corpus cavernosum penis are small, except in the crura and at the distal bend of the sigmoid flexure. The cavernous spaces of the corpus spongiosum penis are large, but distention is limited by the tunica albuginea. In ruminants and swine, erection results from the inflow of a relatively small volume of blood.

The subcutaneous tissues of the free end of the penis in some species form a well-developed cavernous body, the *corpus spongiosum glandis*. It is poorly developed in the bull and indistinct in the boar. Figure 2–14 illustrates the features of the free end of the penis of bull, ram, boar and stallion. In the stallion a prominent urethral process (Fig. 2–14D) is encircled by a shallow groove, the *fossa glandis* (Getty, 1975). A dorsal diverticulum from the fossa, the *sinus urethralis*, is an important site of infection in stallions carrying the organism causing contagious equine metritis (CEM).

Structure of the Prepuce

In ruminants and swine the orifice of the prepuce is controlled by a special striated muscle (cranial muscle of the prepuce); a second (caudal) muscle may also be present. The prepuce can be divided into penile and prepenile parts. The epithelia of the penile part of the prepuce and the penile integument are derived from a single ectodermal lamella; during postnatal development this lamella is split by keratinization and the free end of the penis is liberated from the penile prepuce (Figs. 2–4 and 2–5, Table 2–1). In the boar, the prepenile prepuce consists of a short vestibule with a narrow external orifice and there is a large dorsal diverticulum in which urine and epithelial debris accumulate. The penile (inner) prepuce of the stallion is enclosed in a voluminous prepenile (outer) prepuce. Eversion of the lining of the prepuce can expose the epithelium to injury and infection; in cattle of European origin, this occurs most commonly in polled beef breeds (Long, 1969). Prolapse of the prepuce is an important disorder in some cattle breeds of Asian origin.

Erection and Protrusion

Sexual stimulation produces dilatation of the arteries supplying the cavernous bodies of the penis (especially the crura) (Gilanpour, 1972). Stiffening of the penis in ruminants is mainly brought about by the ischiocavernosus muscle, which pumps blood from the cavernous spaces of the crura into the rest of the corpus cavernosum penis by way of special longitudinal cavernous spaces (erection canals) (Watson, 1964). High peak pressures have been recorded from this cavernous body during erection in bull, stallion and goat (Beckett et al, 1974). In the normal bull there are no veins draining the distal levels of the corpus cavernosum penis, and this facilitates development of the erection pressure in the organ (Ashdown and Gilanpour, 1974). Rising pressure in the corpus cavernosum penis produces considerable elongation of the bovine penis, with little dilatation (Majeed, 1976). Distention of the cavernous spaces, especially those at the distal bend of the sigmoid flexure, eliminates the flexure as the retractor muscles of the penis relax (Ashdown, 1970). When the penis of the bull is protruded, penile and prepenile parts of the prepuce are everted over the

protruded organ. The spiral arrangement of the fibrous architecture of the penile integument causes the penis to spiral when the integument is stretched; the urethral orifice turns in a counterclockwise direction through 300 degrees as ejaculation occurs (Ashdown and Smith, 1969; Seidel and Foote, 1969).

Intromission in the bull lasts for about two seconds and straightening of the penis after withdrawal often occurs abruptly as the dorsal apical ligament reasserts its action in keeping the penis straight. Withdrawal into the prepuce follows as the pressure in the cavernous spaces subsides. The fibrous architecture of the corpus cavernosum penis in the region of the sigmoid flexure tends to re-form the flexure; this is assisted by shortening of the retractor penis muscle. The terminal 5 cm or so of the penis of the boar are spiralled like a corkscrew and during erection the whole visible length of the free end of the penis becomes spiralled around its long axis as a rope is twisted. Intromission lasts for up to seven minutes, during which time a large volume of semen is ejaculated. Spiral deviation does not occur in the horse and intromission lasts for several minutes.

Emission and Ejaculation

The fully integrated sequence of erection, emission and ejaculation is under complex nervous control; electrical stimulation of ejaculation in farm animals is a crude imitation of the natural mechanisms.

Passage of semen along the ductus deferens is continual during sexual inactivity. Tischner (1972) found that 1.5 to 6.0 $\times$ 10^9 spermatozoa pass through the ductus deferens daily in the sexually inactive ram. Sexual excitement and ejaculation increase the rate of flow but this is followed by a reduction in rate of flow for 10 to 20 hours after ejaculation so that, overall, the numbers of spermatozoa passing through the ductus deferens are not increased by sexual activity.

Muscular contraction of the wall of the duct is controlled by sympathetic autonomic nerves of the pelvic plexus derived from the hypogastric nerves. Emission of semen from the duct into the urethra is accompanied by muscular contraction of the walls of the ampullae of the ductus deferens. The stored secretions of some, if not all, of the accessory glands are probably released by muscular contraction controlled by autonomic nerves; secretory activity of the glandular epithelium may be under similar control. Ejaculation involves contraction of striated muscles that are innervated by sacral nerves. One of these, the bulbospongiosus muscle, compresses the penile bulb during ejaculation and so pumps blood from the penile bulb into the remainder of the corpus spongiosum penis (Gilanpour, 1972). Unlike the corpus cavernosum penis, this cavernous body is normally drained by distal veins; peak pressures recorded during ejaculation are only one-tenth of those in the corpus cavernosum penis (Beckett et al, 1975). The waves of pressure passing down the penile urethra may help to transport the ejaculate (Watson, 1964).

Reflux of semen into the bladder has been shown radiographically to occur in rams during electrical stimulation (Hovell et al, 1969). Spermatozoa can be recovered from the urine of sexually rested rams, and studies on rams with cannulated urinary bladders indicate that small quantities of spermatozoa (7 to 200 × 10^6) are pushed into the urinary bladder during natural ejaculation (Tischner, 1972).

OTHER SPECIES

Anatomic features of the male reproductive tracts have been described for the camel (Elwishy et al, 1972), the llama (Casas Peréz et al, 1967), the buffalo (Joshi et al, 1967), the goat (Yao and Eaton, 1954)

and the jackass (Tagand and Barone, 1956).

REFERENCES

Abdel-Raouf, M. (1960). The postnatal development of the reproductive organs in bulls with especial reference to puberty. Acta Endocrinol. (Kbh.) Suppl. 49, 1.

Amann, R.P. (1970). Sperm production rates. In The Testis. Vol. 1. A.D. Johnson, W.R. Gomes and N.L. VanDemark (eds), New York, Academic Press.

Ashdown, R.R. and Smith, J.A. (1969). The anatomy of the corpus cavernosum penis of the bull and its relationship to spiral deviation of the penis. J. Anat. 104, 153.

Ashdown, R.R. (1970). Angioarchitecture of the sigmoid flexure of the bovine corpus cavernosum penis and its significance in erection. J. Anat. 106, 403.

Ashdown, R.R. and Gilanpour, H. (1974). Venous drainage of the corpus cavernosum penis in impotent and normal bulls. J. Anat. 117, 159.

Beckett, S.D., Hudson, R.S., Walker, D.F., Vachon, R.I. and Reynolds, T.M. (1972). Corpus cavernosum penis pressure and external penile muscle activity during erection in the goat. Biol. Reprod. 7, 359.

Beckett, S.D., Walker, D.F., Hudson, R.S., Reynolds, T.M. and Vachon, R.I. (1974). Corpus cavernosum penis pressure and penile muscle activity in the bull during coitus. Am. J. Vet. Res. 35, 761.

Beckett, S.D., Purohit, R.C. and Reynolds, T.M. (1975). The corpus spongiosum penis pressure and external penile muscle activity in the goat during coitus. Biol. Reprod. 12, 289.

Casas Peréz, J.H., San Martin, M. and Copaira, M. (1967). Histology of the testis in the alpaca (Lama pacos). Rev. Fac. Méd. Vét. Univ. Nac. Lima, 18–20 (1963/1966), 223.

Donahoe, P.K., Ito, Y., Price, J.M. and Hendren, W.H.H. (1977). Müllerian inhibiting substance activity in bovine foetal, newborn and prepubertal testes. Biol. Reprod. 16, 238.

Donovan, B.T. and van der Werff ten Bosch, J.J. (1965). Physiology of Puberty. London, Arnold.

Dym, M. and Fawcett, D.W. (1970). The blood testis barrier in the rat and the physiological compartmentation of the seminiferous epithelium. Biol. Reprod. 3, 308.

Dym, M. (1976). The mammalian rete testis—a morphological examination. Anat. Rec. 186, 493.

Elwishy, A.B., Mobarak, A.M. and Fouad, S.M. (1972). The accessory genital organs of the one humped male camel (Camelus dromedarius). Anat. Anz. 131, 1.

Free, M.J. (1977). Blood supply to the testis and its role in local exchange and transport of hormones. In The Testis. Vol. 4. A.D. Johnson and W.R. Gomes (eds), New York, Academic Press.

Gier, H.T. and Marion, G.B. (1970). Development of the mammalian testis. *In* The Testis. Vol. I. A.D. Johnson, W.R. Gomes and N.L. VanDemark (eds), New York, Academic Press.

Getty, R. (1975). Sisson and Grossman's The Anatomy of Domestic Animals, 5th ed. Philadelphia, W.B. Saunders.

Gilanpour, H. (1972). Angioarchitecture and Functional Anatomy of the Penis in Ruminants. Ph.D. Thesis, University of London.

Hamilton, D.W. (1971). The mammalian epididymis. *In* Reproductive Biology. H. Balm and S. Glasser (eds), Basel, S. Karger, Excerpta Med. Fdn.

Hodson, N.P. (1970). The nerves of the testis, epididymis and scrotum. *In* The Testis. Vol. 1. A.D. Johnson, W.R. Gomes and N.L. VanDemark (eds), New York, Academic Press.

Hovell, G.J.R., Ardran, G.M., Essenhigh, D.M. and Smith, J.C. (1969). Radiological observations on electrically induced ejaculation in the ram. J. Reprod. Fertil. *20*, 383.

Hunter, R.H.F., Holtz, W. and Henfrey, P.J. (1976). Epididymal function in the boar in relation to the fertilizing power of spermatozoa. J. Reprod. Fertil. *46*, 463.

Joshi, N.H., Luktuke, S.N. and Chatterjee, S.N. (1967). Studies on the biometry of the reproductive tract and some endocrine glands of the buffalo male. Indian Vet. J. *44*, 137.

Kainer, R.A., Faulkner, L.C. and Abdel-Raouf, M. (1969). Glands associated with the urethra of the bull. Am. J. Vet. Res. *30*, 963.

Land, R.B. and Carr, W.R. (1975). Testis growth and plasma LH concentration following hemicastration and its relationship with female prolificacy in sheep. J. Reprod. Fertil. *45*, 495.

Larson, L.L. and Kitchell, R.L. (1958). Neural mechanisms in sexual behavior. Am. J. Vet. Res. *19*, 853.

Lincoln, G.A. (1978). Induction of testicular growth and sexual activity in rams by a "skeleton" short day photoperiod. J. Reprod. Fertil. *52*, 178.

Long, S.E. (1969). Eversion of the preputial epithelium in bulls at artificial insemination centres. Vet. Rec. *84*, 495.

Majeed, Z.Z. (1976). Biometrical and Functional Studies on the Penis in Ruminants. Ph.D. Thesis, University of London.

Mann, T. (1964). Biochemistry of Semen and of the Male Reproductive Tract. London, Methuen.

McKenzie, F.F., Miller, J.C. and Bauguess, L.C. (1938). The reproductive organs and semen of the boar. Res. Bull. Mo. Agric. Exp. Sta., No. 279.

Neaves, W.B. (1977). The blood testis barrier. *In* The Testis. Vol. 4. A.D. Johnson and W.R. Gomes (eds), New York, Academic Press.

Nicander, L. (1957). Studies on the regional histology and cytochemistry of the ductus epididymidis in stallions, rams and bulls. Act. Morphol. Neerl. Scand. *1*, 337.

Nickel, R., Schummer, A. and Seiferle, E. (1973). The Viscera of the Domestic Mammals. Transl. and rev., W.O. Sack, Berlin, Parey.

Nishikawa, Y. (1959). Studies on Reproduction in Horses. Tokyo, Japan Racing Association.

Orgebin-Crist, M.-C. (1969). Studies on the function of the epididymis. Biol. Reprod. Suppl. 1, *1*, 155.

Ortavant, R., Orgebin, M.C. and Singh, G. (1961). Étude Comparative de la Durée des Phénomènes Spermatogénétiques chez les Animaux Domestiques. *In* The Use of Radioisotopes in Animal Biology and the Medical Sciences. (Symposium, Mexico City.) New York, Academic Press.

Seidel, G.E., Jr. and Foote, R.H. (1969). Motion picture analysis of ejaculation in the bull. J. Reprod. Fertil. *20*, 313.

Setchell, B.P. (1970). Testicular blood supply, lymphatic drainage and the secretion of fluid. *In* The Testis. Vol. 1. A.D. Johnson, W.R. Gomes and N.L. VanDemark (eds), New York, Academic Press.

Skinner, J.D. and Rowson, L.E.A. (1968). Puberty in Suffolk and crossbred rams. J. Reprod. Fertil. *16*, 479.

Smith, J.A. (1975). The development and descent of the testes in the horse. Vet. Ann., 15th issue, 156.

Swierstra, E.E. (1968). Cytology and duration of the cycle of the seminiferous epithelium of the boar. Duration of spermatozoan transit through the epididymis. Anat. Rec. *161*, 171.

Swierstra, E.E., Gebaner, M.R. and Pickett, B.W. (1975a). The relationship between daily sperm production as determined by quantitative testicular histology and daily sperm output in the stallion. J. Reprod. Fertil. Suppl. 23, 35.

Swierstra, E.E., Pickett, B.W. and Gebaner, M.R. (1975b). Spermatogenesis and duration of transit of spermatozoa through the excurrent ducts of stallions. J. Reprod. Fertil. Suppl. 23, 53.

Tagand, R. and Barone, R. (1956). Anatomie des Équidés Domestiques. 2, iii. Lyons, École Nat. Vét.

Tischner, M. (1972). The role of the vasa deferentia and the urethra in the transport of semen in rams. Acta agraria silvestra *12*, 77.

Van Straaten, H.W.M. and Wensing, C.J.G. (1977). Histomorphometric aspects of testicular morphogenesis in the naturally cryptorchid pig. Biol. Reprod. *17*, 473.

Watson, J.W. (1964). Mechanisms of erection and ejaculation in the bull and ram. Nature (Lond.) *204*, 95.

Waites, G.M.H. and Setchell, B.P. (1969). Physiology of the testis, epididymis and scrotum. *In* Advances in Reproductive Physiology. Vol. 4. A. McLaren (ed), London, Logos, pp. 1–21.

Yao, T.S. and Eaton, O.N. (1954). Postnatal growth and histological development of reproductive organs in male goats. Am. J. Anat. *95*, 401.

3

Functional Anatomy of Female Reproduction

E.S.E. HAFEZ

The female reproductive organs are composed of ovaries, oviducts, uterus, cervix uteri, vagina and external genitalia. The internal genital organs (the first of four components) are supported by the broad ligament. This ligament consists of the mesovarium, which supports the ovary; the mesosalpinx, which supports the oviduct; and the mesometrium, which supports the uterus. In cattle and sheep, the attachment of the broad ligament is dorsolateral in the region of the ilium, so that the uterus is arranged like a ram's horns, with the convexity dorsal and the ovaries located near the pelvis (Fig. 3–1). The ovary, oviduct and uterus are supplied primarily by autonomic nerves. The pudic nerve supplies sensory fibers and parasympathetic fibers to the vagina, vulva and clitoris.

This chapter deals with the embryology, morphology, anatomy, physiology and biochemistry of the female reproductive organs in farm mammals.

EMBRYOLOGY

The fetal reproductive system consists of two sexually nondifferentiated gonads, two pairs of ducts, a urogenital sinus, a genital tubercle and vestibular folds (Fig. 3–2). This system arises primarily from two germinal ridges on the dorsal side of the abdominal cavity and it can differentiate into a male or a female system, a condition referred to as embryonic bisexuality. The developmental fates of the sexual rudiments in the male and in the female fetus are shown in Table 3–1.

The sex of the fetus depends on inherited genes, gonadogenesis, and the formation and maturation of accessory reproductive organs. The expression of the genetic sex is a developmental process that depends on the functioning of the fetal gonads and, occasionally, on the functioning of the adrenal cortex.

Estrogen and androgen can cause sex reversal in male and female embryos respectively during only a brief period early

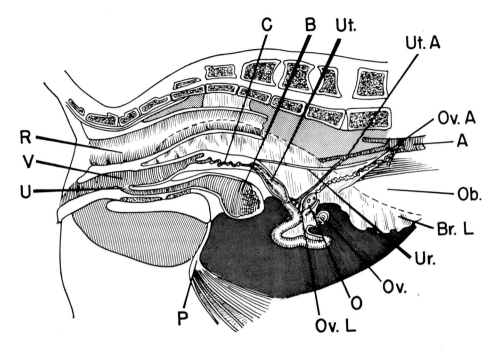

Fig. 3–1. Sagittal section through pelvic region (view of left side) showing attachments of rectum and urogenital tract, the pouches of the pelvic peritoneum, and attachments of the abdominal muscles to the prepubic tendon in the ewe.
A, Aorta; *B*, bladder; *Br.L*, broad ligament of uterus; *C*, cervix; *O*, ovary; *Ob.*, internal oblique muscle; *Ov.*, oviduct; *Ov. A*, ovarian artery; *Ov. L*, ovarian ligament; *P*, prepubic tendon; *R*, rectum; *U*, urethra; *Ur.*, ureter; *Ut. A*, uterine artery; *V*, vagina. *(Redrawn from Bassett, 1965. Aust. J. Zool. 13, 201)*

in sexual differentiation. In contrast, the accessory reproductive organs remain sexually labile for a much longer time, and hormone treatment late in development can induce considerable reversal in these structures. The age at which this bisexual potential is completely lost varies with the species.

Gonads

The gonads are formed from a group of large granulated yolk sac cells that invade the germinal ridges. Two invasions occur in the female. The initial one is abortive, but the second results in the formation of sex cords which later spread upward into primordial germ cells (oogonia). The sex cords of the female are called medullary cords; those of the male are the seminiferous tubules (Fig. 3–3).

The testis develops predominantly from the medulla of the sexually undifferentiated gonad, whereas the ovary arises primarily from its cortex.

Reproductive Ducts

Wolffian and Müllerian ducts are both present in the sexually undifferentiated embryo. In the female, the Müllerian ducts develop into a gonaductal system and the Wolffian ducts atrophy. The opposite is true in the male. The female Müllerian ducts fuse caudally to form a uterus, a cervix and the anterior part of a

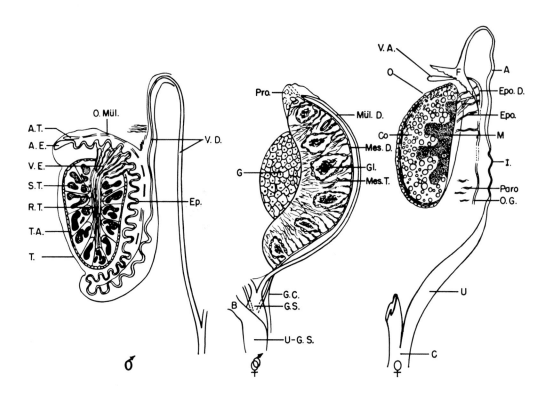

Fig. 3–2. Diagram representing embryonic differentiation of male and female genital systems. *Center:* the undifferentiated system with its large mesonephros, mesonephric duct, Müllerian duct and undifferentiated gonad. Note that the Müllerian and mesonephric ducts cross before they enter the genital cord. *Right:* the female system, in which the ovary and Müllerian ducts differentiate while the remnants of the mesonephros and mesonephric ducts atrophy into the epoophoron, paroophoron and Gartner's duct. *Left:* the male system in which the testes and mesonephric (Wolffian) ducts differentiate; the sole remnants of the Müllerian ducts are the testicular appendix and prostatic utricle (vagina masculinus). *(Moustafa and Hafez, unpublished data, 1972)*

A Ampulla	*G.C.* Genital cord	*Paro* Paroophoron
A.E. Appendage of epididymis	*Gl.* Glomerulus	*Pro.* Pronephros
A.T. Appendage of testis	*G.S.* Genital sinus	*R.T.* Rete tubules
B Bladder	*I.* Isthmus	*S.T.* Seminiferous tubules
C Cervix	*M* Ovarian medulla	*T.* Testis
Co Ovarian cortex	*Mes. D.* Mesonephric duct	*T.A.* Testicle artery
Ep. Epididymis	*Mes. T.* Mesonephric tubules	*U* Uterus
Epo. Epoophoron	*Mul. D.* Müllerian duct	*U-G.S* Urogenital sinus
Epo. D. Duct of epoophoron	*O* Ovary	*V.A.* Vesicular appendage
F Fimbriae	*O.G.* Obliterated Gartner's duct	*V.D.* Vas deferens
G Gonad (undifferentiated)	*O. Mul.* Obliterated Müllerian duct	*V.E.* Vasa efferentia

Table 3–1. Developmental Fate of the Sexual Rudiments in the Male and the Female Mammalian Fetus

Sexual Rudiment	Male	Female
Gonad		
Cortex	Regresses	Ovary
Medulla	Testis	Regresses
Müllerian ducts	Vestiges	Uterus, oviducts, parts of vagina
Wolffian ducts	Epididymis, vas deferens	Vestiges
Urogenital sinus	Urethra, prostate, bulbourethral glands	Part of vagina, urethra
Genital tubercle (phallus)	Penis	Clitoris
Vestibular folds	Scrotum	Labia

(Frye, 1967. *Hormonal Control in Vertebrates.* New York, Macmillan.)

vagina. The oviduct becomes coiled and acquires differentiated epithelia and fimbriae just before birth.

In the male fetus, testicular androgen plays a role in the persistence and development of the Wolffian ducts and the atrophy of the Müllerian ducts. However, the growth of the female Müllerian ducts beyond the ambisexual stage is apparently hormonally independent and the duct is capable of autonomous growth, coiling and epithelial differentiation.

Urogenital Sinus

The urogenital sinus gives rise to the vestibule. The folds of skin that border the sinus form the lips of the vulva. The

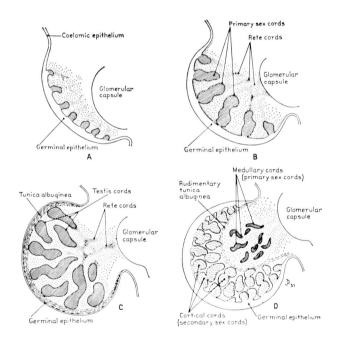

Fig. 3–3. Differentiation of the undifferentiated gonad of higher vertebrates into testis and ovary. *A*, The primary sex cords arise from the germinal epithelium. *B*, Primary sex cords have developed, but the gonad is still undifferentiated. *C*, Differentiation of the testis is taking place: the primary sex cords continue to proliferate, while the germinal epithelium diminishes in size; the tunica albuginea also develops. *D*, Differentiation into an ovary involves development of secondary sex cords from the cortex and reduction of the primary sex cords and tunica albuginea. *(Burns, 1961. In: Sex and Internal Secretions, Vol. 1, 3rd ed. W. C. Young [ed.], Baltimore, Williams & Wilkins)*

female phallus, or clitoris, homologous to the male penis, grows little in size.

OVARY

Anatomy

The ovary, unlike the testis, remains in the abdominal cavity. It performs both an exocrine (egg release) and endocrine function (steroidogenesis).

Prenatal Development

The predominant tissue of the ovary is the cortex. The primordial germ cells arise extragonadally and migrate through the yolk sac mesentery to the genital ridges. During fetal development the oogonia are produced by mitotic multiplication. This is followed by the first meiotic division to form several million oocytes, a process that is arrested in the prophase. Subsequent atresia reduces the number of oocytes at the time of birth, a further reduction occurs at puberty and only a few hundred are present during reproductive senescence. All ova arise from the original germ cell population of the genital ridge. The follicular cells, which surround the primary oocyte, develop from the germinal epithelium by downgrowth, whereas the endocrine cells—the theca and interstitial cells—originate from the ovarian medulla. With few exceptions, primordial follicles are not formed during postnatal life.

At birth, a layer of follicular cells surrounds the primary oocytes in the ovary to form the primordial follicles. At first these are scattered throughout the ovary, but during early neonatal life they become localized in the peripheral cortical zone, beneath the tunica albuginea and surrounding the vascular medulla.

The shape and size of the ovary vary both with the species and the stage of the estrous cycle (Table 3–2). In cattle and sheep the ovary is almond-shaped,

whereas in the horse it is bean-shaped owing to the presence of a definite ovulation fossa, an indentation in the attached border of the ovary. The porcine ovary resembles a cluster of grapes because the protruding follicles and corpora lutea obscure the underlying ovarian tissue.

The part of the ovary that is not attached to the mesovarium is exposed and bulges into the abdominal cavity. The ovary, composed of the medulla and cortex, is surrounded by the superficial epithelium, commonly known as germinal epithelium (Fig. 3–4). The ovarian medulla consists of irregularly arranged fibroelastic connective tissue and extensive nervous and vascular systems that reach the ovary through the hilus (attachment between ovary and mesovarium). The arteries are arranged in a definite spiral shape. The ovarian cortex contains ovarian follicles and/or corpora lutea at various stages of development or regression. The connective tissue of the cortex contains many fibroblasts, some collagen and reticular fibers, blood vessels, lymphatic vessels, nerves and smooth muscle fibers.

Ovarian Blood Flow. The vascular pattern of the ovary changes with different hormonal states. Variations in the architecture of the vessels allow adaptation of the blood supply to the needs of the organ. The relative distribution of the blood among the various compartments of the ovary is altered without affecting the total ovarian blood supply.

Niswender et al (1976) used Doppler ultrasonic blood flow transducers to measure the velocity of the blood flowing in the ovarian artery throughout the estrous cycle. Blood flow transducers were implanted around the ovarian arteries. The intraovarian distribution of blood undergoes remarkable changes during the preovulatory period. In sheep, ovarian venous flow drops from about 8 ml/min 73 days before ovulation to 2 ml/min at estrus. A corresponding drop in ovarian arterial blood flow per minute is paral-

Table 3–2. Comparative Anatomy of the Ovary in the Adult Female of Farm Mammals

Organ	Animal			
	Cow	Ewe	Sow	Mare
Ovary				
Shape	Almond-shaped	Almond-shaped	Berry-shaped (cluster of grapes)	Kidney-shaped; with ovulation fossa
Weight of one ovary (gm)	10–20	3–4	3–7	40–80
Mature graafian follicles				
Number	1–2	1–4	10–25	1–2
Diameter of follicle (mm)	12–19	5–10	8–12	25–70
Diameter of egg without zona pellucida (μ)	120–160	140–185	120–170	120–180
Mature corpus luteum				
Shape	Spheroid or ovoid	Spheroid or ovoid	Spheroid or ovoid	Pear-shaped
Diameter (mm)	20–25	9	10–15	10–25
Maximum size attained (days from ovulation)	10	7–9	14	14
Regression starts (days from ovulation)	14–15	12–14	13	17

The measurements included in this table vary with age, breed, parity, plane of nutrition and reproductive cycle.

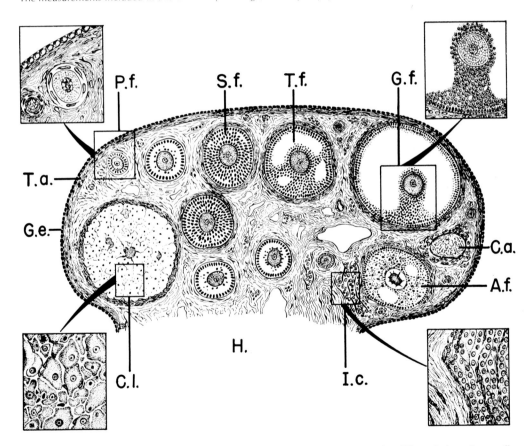

Fig. 3–4. A composite diagram of the mammalian ovary. Progressive stages in the differentiation of a graafian follicle are indicated *(upper left to upper right)*. The mature follicle may become atretic *(lower right)* or ovulate and form a corpus luteum *(lower left)*.
A.f., Atretic follicle; *C.a.,* corpus albicans; *C.l.,* corpus luteum; *G.e.,* germinal epithelium; *G.f.,* graafian follicle; *H,* hilus; *I.c.,* interstitial cells; *P.f.,* primary follicle; *S.f.,* secondary follicle; *T.a.,* tunica albuginea; *T.f.,* tertiary follicle. *(Partly adapted from Turner, 1948. General Endocrinology. Philadelphia, W. B. Saunders Co.)*

35

leled by a sharp decline in the flow of lymph from the ovary (Moor et al, 1975).

Arterial blood flow to the ovary varies in proportion to luteal activity. Hemodynamic changes seem to be important in regulating corpus luteum (CL) function and lifespan. Thus, changes in blood flow precede the decline in progesterone secretion, whereas restriction of ovarian blood flow causes premature CL regression. At the time of luteolysis in ewes there is a reduction in ovarian blood flow as well as an increase in arteriole-venule shunting within the ovary (Pharriss, 1970; Niswender et al, 1976).

Ovarian Follicles

Development. The development of ovarian follicles is often classified according to (a) size; (b) number of layers of granulosa cells; (c) development of the theca; (d) position of the oocyte within its surrounding cumulus oophorus; and (e) presence of an antrum (fluid-filled space). Primordial follicles are oocytes (potential eggs) surrounded by one layer of flattened cells; these cells increase in number by mitosis and become cuboidal, and the follicle is then called a secondary follicle. Vesicular follicles are then formed by accumulation of fluids in the antrum between the epithelial cells; the membrana granulosa has two, three and then four layers. Capillaries invade the fibrous layer of cells surrounding the follicle and form a vascular layer, the theca interna. The granulosa cells within the wall of the graafian follicle are deprived of a blood supply by the basement membrane.

With the progressive development of the cortical blood vessels in the neighborhood of the growing follicle and the formation of the two layers of the theca, a basket-like vascular meshwork develops around the follicle, particularly in the theca interna.

The granulosa cells cease dividing a few days before ovulation and just before rupture the membrana granulosa changes from a compact to an open, lace-like layer. The walls of the developing follicle contain smooth muscle cells that show contractile responses during all stages of the estrous cycle.

Number of Follicles That Develop. The number of graafian follicles that develop per estrous cycle depends on hereditary and environmental factors. In cattle and horses, one follicle usually develops more rapidly than the others, so that at each estrus only one egg is released. The remaining follicles then regress and become atrophied. In swine, 10 to 25 follicles ripen at each estrus. In sheep, one to three follicles may reach maturity depending on the breed, age and stage of sexual season.

In normally cycling heifers, the follicle that is largest three days before the onset of estrus is the one that ovulates, whereas there is no pattern for ovulation of the second largest follicle. Following unilateral ovariectomy there is an altered size distribution of ovarian follicles.

For caloric intake to influence ovulation rate in gilts, a minimal interval of time immediately prior to estrus is required. Full-feeding beginning on day 15 for as short a period as the last five to six days of the estrous cycle increased the number of large follicles at estrus, whereas fullfeeding for three to four days beginning on day 16 failed to affect significantly the number of large follicles at estrus (Dailey et al, 1976).

Hormonal Mechanism. Both the rate of follicular development and the number that ripen per estrus depend on the pituitary gonadotropins. In the adult female, a limited amount of hormones is available to the ovary. During each estrous cycle, as a result of the release of sufficient FSH by the pituitary gland, a crop of "growing" follicles is stimulated to undergo further growth and maturation.

Many follicles grow during the first stages of the estrous cycle, but few mature completely. Perhaps less FSH is required to initiate the growth of small follicles than to maintain larger follicles and bring them to ovulatory size, since the number of follicles that mature can be greatly increased (super-ovulation) by injecting animals with large doses of gonadotropins. Why then do so many follicles begin to grow? They may supply substances, like estrogen, that are essential for the ovulation of other, larger follicles.

Compensatory changes in ovarian function following unilateral ovariectomy have been studied in several laboratory animals. In unilaterally ovariectomized rodents, the remaining ovary ovulates as many viable ova as the combined ovaries of intact animals. In cattle and sheep a single ovulation occurs whether or not one ovary has been removed. The loss of one ovary may decrease competition for circulatory levels of gonadotropins. Alternatively, decreased estrogen feedback to the pituitary may permit elevation of serum gonadotropin levels. This increase in gonadotropin levels may act as an endogenous follicle-priming mechanism.

Steroidogenesis. Steroidogenesis by granulosa cells is one of the unique features of follicular development. All cell types in the ovary have the capacity to make steroid hormones. The steroid hormones secreted by a particular cell type are determined by the stage of the estrous cycle. The ovaries of farm animals lack the relatively large amounts of steroid-secreting interstitial tissue that are so prominent in the ovaries of rodents and rabbits. The theca cells of farm animals appear to differentiate almost completely during the atretric processes, and thus these animals lack an important source of certain steroids present in those animals that have large amounts of interstitial tissue (Hansel et al, 1973). Estrogens are secreted largely by the theca cells of the follicle. Some progesterone produced by the granulosa cells may be used by the theca cells for the synthesis of androgens

and estrogens (Baker, 1972). Following ovulation, the luteinized granulosa cells become vascularized and secrete increased amounts of progesterone in the ovarian vein.

Follicular Fluid. The follicular fluid, a transudate of blood plasma, provides a suitable environment for the developing oocyte. It is possible that a blood-follicle barrier may exclude certain substances from the follicular fluid.

The follicular fluid plays a role in the differentiation and steroidogenic capacity of ovarian cells. The granulosa cells and the oocyte are in contact with follicular fluid and are separated from blood plasma by a limiting membrane. In nonpregnant heifers, follicular fluid estrogen concentration is greater in the ovary with the corpus luteum than in the contralateral ovary (England et al, 1973). The estradiol-17β content of follicular fluid in the remaining ovary of gilts that have been unilaterally ovariectomized on day 4 or 15 is more than double the content of a single ovary of intact gilts (Rexroad and Casida, 1976). The increased content is due to increased follicle size, as unilateral ovariectomy does not affect estradiol-17β concentration in follicular fluid.

Follicular fluid exerts several effects on spermatozoa, such as stimulation of sperm respiration, increased rate of motility and amplitude of the flagellar beat, and alterations in the acrosome (Brewer and Wells, 1977). Little is known about the physiologic significance of follicular fluid in sperm capacitation and fertilization.

Rupture of Follicles (Ovulation). At ovulation, the egg is released while embedded in a solid mass of follicular cells, the cumulus oophorus, which protrudes into the fluid-filled antrum. The cumulus oophorus is usually attached to the granulosa cells opposite the side that will eventually rupture.

The rupture per se occurs at the apex of the follicle. The outermost layer is the first to part. The inner layers protrude through the gap to form a papilla or stigma. The stigma thins out, bulges on the surface of the ovary, and becomes completely avascular. The bulging stigma soon ruptures, releasing some of the thin follicular fluid. After a short interval the egg mass moves toward the rupture site, becoming elongated as it progresses. More fluid moves through the opening, carrying the egg.

Several theories have been advanced to explain why the follicle wall ruptures, but the precise mechanism involved remains uncertain. For example, it may be facilitated by muscular contraction of the fibers encircling the follicle.

Ovulation occurs in response to an LH surge. The stimulation of prostaglandin F (PGF) synthesis by LH provides evidence that the preovulatory release of LH causes increased PGF synthesis in the ovary, leading to increased steroidogenesis and ovulation. The follicle has two reactions in response to the LH-receptor complex: (a) stimulation of steroid synthesis by LH

and (b) activation of an ovulatory enzyme by the steroid released. The sequence of events leading to steroidogenesis involves the activation of adenyl cyclase by an LH-receptor complex perhaps through the local action of a prostaglandin, which results in increased tissue levels of cyclic AMP (O'Shea and Phillips, 1974). Cyclic AMP then induces a protein synthetic process essential to increased synthesis of progesterone.

Association of Follicle and Corpora Lutea. There is a physiologic association between follicles and functional corpora lutea because follicular development is enhanced in the ovary containing a corpus luteum. Follicular fluid weight and the number of large follicles are greater in the ovary bearing a corpus luteum than in the other ovary.

Progesterone may act both systemically and locally to alter time-dependent changes in follicle size, thus making it possible for some follicles to grow while

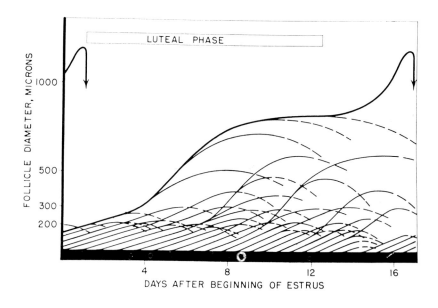

Fig. 3–5. A schematic representation of the follicular cycle in the guinea pig. The heavy solid line indicates the average diameter of the largest follicles. Ovulation occurs at the arrow. The other solid and broken lines represent the concomitant growth and atresia, respectively, of other groups of follicles that do not normally ovulate. *(From Evertt, 1961. In: Sex and Internal Secretions, Vol. 1, 3rd ed. W. C. Young [ed.], Baltimore, Williams & Wilkins)*

others undergo atresia. The effect of progesterone on follicular growth rates does not appear to be mediated through induction or alteration of ovarian estradiol-17β content (Rexroad and Casida, 1977). In the ewe, follicles in ovaries that contain corpora lutea grow faster than follicles on the opposite ovaries without corpora lutea. Such an effect may be mediated through differences in local concentrations of progesterone in the two ovaries. Estrogens also affect follicles directly, both preventing atresia and stimulating growth.

Atresia and Degeneration. The estimated number of oocytes present in both ovaries at the time of birth varies from 60,000 to 100,000 according to the species and breed. However, not all of these oocytes develop to the mature stage; for every egg that matures and is ovulated, several start to develop but never reach maturity (Fig. 3–5). Consequently every normal ovary contains some degenerating oocytes in follicles that fail to rupture (atretic follicles).

Once an ovarian follicle begins to grow it is destined to either ovulate or become atretic. In atretic follicles, mitotic activity in the granulosa cells ceases, pyknotic nuclei appear in the antrum, the cumulus cells undergo necrosis and meiotic spindles are often formed within the ovum. The degenerating oocyte is characterized by hyalinization, thickening of the zona pellucida and/or fragmentation of the cytoplasm. It is engulfed by the ovarian fibrocytes by phagocytosis and eventually disappears into a scar.

In swine there are three forms of follicular atresia: hemorrhagic, cystic and milky. Milky follicular degeneration is inversely related to the number of ovulatory follicles. Genetic differences in the number of ovulatory follicles are directly related to milky degeneration and inversely related to hemorrhagic degeneration.

The number of ovarian oocytes decreases progressively with advancing age. In 14- to 15-year-old cows, the average number of ovarian oocytes is reduced to 24,000 (Erickson et al, 1976).

The Egg (Ovum)

The egg, a highly differentiated cell, is capable of being fertilized and subsequently undergoing embryonic development. Mammalian eggs were first recognized by de Graaf in 1672, and they were described and identified by Cruickshank in 1797 and von Baer in 1827.

Development. The precursor cells of either male or female gametes, called gonocytes, originate probably from extra-embryonic endodermal tissue (extragonadally). They migrate to the presumptive intra-embryonic gonadal zone where they differentiate into either oogonia or spermatogonia. In the female fetus, the germinal epithelium forms into clusters in which one gonocyte differentiates into an oogonium containing typical cell constituents (eg, Golgi apparatus, mitochondria, nucleus and one or more nucleoli). The oogonia then proliferate prior to or shortly after birth, so that the fetal ovaries contain the sole reservoir of all future ova called oocytes.

The growth of the oocyte is characterized by: (a) the enlargement of the cytoplasm by accumulation of different sizes of granules of deutoplasm (yolk); (b) the development of an egg membrane (zona pellucida); and (c) the mitotic proliferation of follicular epithelium and adjacent tissue. These follicular cells may serve as nurse cells by providing the deutoplasm of the oocyte. By maturity, the egg has accumulated reserves of material to provide an energy source for subsequent development. Factors determining which of the ovarian oocytes are destined to begin their growth or to complete their growth during a reproductive cycle are unknown.

There are two stages in the growth of the oocyte (Fig. 3–6). During the first phase, growth is rapid and intimately associated with the development of the ovarian follicle. Attainment of its mature size occurs about the time antrum formation begins in the follicle. During the second phase, the oocyte does not grow in size while the ovarian follicle, responding to pituitary hormones, increases rapidly in diameter. This growth is confined primarily to follicles in which the egg has attained its full dimensions.

During the latter phase of follicular growth, the oocyte matures. The nucleus, which had entered into the prophase of the meiotic division during the growth of

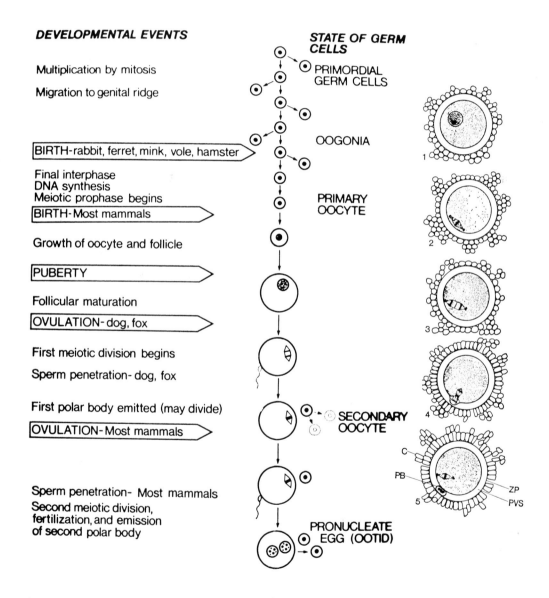

Fig. 3–6. *Left,* Diagrammatic representation of development events in the life cycle of the female germ cell. *Right,* Preovulatory maturation of the egg. *C,* Corona radiata; *PB,* polar body; *PVS,* perivitelline space, *ZP,* zona pellucida. *(Baker, 1972. In: Reproduction in Mammals. C. R. Austin and R. V. Short [eds.], Cambridge, Cambridge University Press)*

the oocyte, prepares to undergo reductive divisions. The nucleoli and the nuclear membrane disappear and the chromosomes condense into a compact form. The centrosome (a specialized area of dense cytoplasm) then divides into two centrioles around which asters appear (groups of radiations at both poles of the oocyte). These asters move apart and a spindle is formed between them. The chromosomes in diploid pairs are set free in the cytoplasm and become arranged on the equatorial plate of the spindle (metaphase I). The primary oocyte now undergoes two meiotic divisions. In the first division two daughter cells arise, each containing one-half of the chromosome complement (2n). However, unlike the divisions occurring in spermatogenesis, the egg acquires almost all of the cytoplasm. This cell is known as the secondary oocyte; the other, much smaller cell is known as the first polar body. At the second maturation division, the secondary oocyte divides into the ootid (n) and a second polar body (n). The two polar bodies, containing little cytoplasm, are entrapped within the zona pellucida of the oocyte, and here they degenerate. The first polar body may also divide; thus the zona pellucida may contain one, two or three polar bodies.

The time at which the two reductive divisions occur is not necessarily coincidental with the time of ovulation. The oocyte is usually in the pachytene or diplotene stage of prophase I during diestrus. Shortly before ovulation the oocyte may undergo the first meiotic division. The second division begins but is not completed until or unless fertilization takes place. Thus, the second polar body and the female pronucleus are formed at fertilization. The ova of cattle, sheep and swine contain one polar body at ovulation, whereas the ova of the horse, dog and fox are in the first maturation division at ovulation.

The secondary oocyte is liberated at ovulation (primary oocyte in the case of the horse). The oocyte continues the process of maturation until fertilization, when it becomes a "zygote." In the process of oogenesis, one primary oocyte gives rise to one egg; in spermatogenesis, one primary spermatocyte gives rise to four spermatozoa.

Structure. Differences in egg size depend almost entirely on the amount of accumulated deutoplasm. With the exception of the eggs of monotremes, the diameter of the intrazonal vitellus at the time of ovulation ranges between 80 and 200 μ. In farm mammals, it is usually less than 185 μ. Even though the egg is larger than most of the somatic cells, there is little relationship between the size of the egg and that of the adult animal.

Corona Radiata. Before ovulation, the egg lies at one side of the ovarian follicle, embedded in a solid mass of follicular cells called the cumulus oophorus. Recently ovulated eggs are usually surrounded by a variable number of granulosa cell layers (corona radiata) and a matrix of follicular fluid. The connection of the eggs with the granulosa cells is loosened by the development of new, liquid-filled, intercellular spaces in the cumulus. Both cumulus and corona cells present on cattle and sheep eggs persist for only a few hours after ovulation. Protoplasmic extensions of these cells penetrate the zona pellucida in oblique or irregular directions, and intertwine with thin projections (microvilli) present in the oocyte itself. However, soon after ovulation these projections are withdrawn. *In vitro* exposure of recently ovulated eggs to oviductal fluids containing fibrinolytic enzymes also results in retraction and degeneration of these projections, followed by regression of the main part of the cells. The ensuing necrosis leads directly to the denudation of the egg, which occurs more slowly with the eggs of pigs and rabbits than with those of cattle, sheep and horses.

Egg Membranes. The egg has two distinct membranes: the vitelline membrane and the zona pellucida. The vitelline membrane is a cortical differentiation of the oocyte and has essentially the same structure and properties of the plasma membrane of somatic cells (diffusion and active transport). The zona pellucida is a homogeneous and apparently semipermeable membrane composed of a conjugated protein that can be dissolved by proteolytic enzymes, such as trypsin and chymotrypsin.

In certain species, another "membrane" is present; such membranes accumulate during the passage of the eggs through the oviduct, which secretes a great variety of materials for the protection and nutrition of the eggs. For example, fish and amphibian eggs are provided with jelly envelopes, membranes and shells. A layer of mucin is deposited by the oviductal epithelium around the zona pellucida of rabbit eggs. However, cattle, sheep and swine oviductal eggs are not surrounded by such a coat.

The egg membranes probably are important for the protection of the egg as well as for the selective absorption of inorganic ions and metabolic substances, as exemplified by the physiochemical changes that occur at the time of ovulation, fertilization, cleavage and expansion of the blastocyst.

Vitellus. The vitellus comprises most of the volume within the zona pellucida at the time of ovulation. After fertilization, it shrinks and a perivitelline space is formed between the zona pellucida and the vitelline membrane in which the polar bodies are situated. The structure of the vitellus is species-dependent, mainly owing to varying amounts of yolk and fat droplets. In goat and rabbit eggs, the yolk granules are finely divided and uniformly distributed; thus, the various nuclear changes that occur during meiosis and fertilization are readily visible. However, horse and cow eggs are filled with fatty and highly refractile droplets, so that the nucleus is obscured by the dark mass of vitellus. If eggs fail to become fertilized, the vitellus fragments into units of unequal size, each containing one or more abortive nuclei.

Structural Abnormalities. Structural abnormalities include small or giant eggs, oval or flattened eggs, a ruptured zona pellucida, or the presence of vacuoles within the vitellus. Such abnormalities may be the result of faulty or incomplete maturation of the oocyte, genetic factors or environmental stress.

Corpus Luteum

The corpus luteum develops after the collapse of the follicle at ovulation. The inner wall of the follicle develops into macro- and microscopic folds that penetrate the central cavity. These folds consist of a central core of stromal tissue and large blood vessels, which become distended. The cells develop a few days before ovulation. They regress quickly and within 24 hours after ovulation all remaining thecal cells are in an advanced stage of degeneration. Hypertrophy and luteinization of the granulosa cells commence after ovulation. The luteal tissue enlarges mainly through hypertrophy of the lutein cells.

Progesterone is secreted by the lutein cells as granules. In the ewe this process appears to be maximal at day 10 of the cycle and begins to taper off noticeably at day 12. The secretory activity declines gradually until day 14, but up to this time the fine structures of other cytoplasmic organelles in the luteal cell show no remarkable changes, except for a gradual increase in the number of autophagocytic bodies which first begin to appear at day 12 (Gemmell et al, 1976).

In aged animals the functions of the corpus luteum decline as a result of (a) inability of follicular cells (granulosa and theca interna) to respond fully to hormonal stimuli, (b) changes in the quantity and/or quality of hormone secretion, and (c) reduced stimulus for hormone secretion (Erickson et al, 1976).

Development. The increase in the weight of the corpus luteum is initially rapid. In general the period of growth is slightly longer than half the estrous cycle. In the cow, the weight and progesterone content of the corpus luteum increase rapidly between days 3 and 12 of the cycle (Fig. 3–7) and remain relatively constant until day 16, when regression begins. In the ewe and sow, corpora lutea increase rapidly in weight and progesterone content from day 2 to day 8, and remain

relatively constant until day 15, when regression begins (Erb et al, 1971). The diameter of the mature corpus luteum is larger than that of a mature graafian follicle except in the mare, in which it is smaller (Table 3–2).

The corpus luteum has a possible stimulatory effect on follicular development and ovulation through a local intraovarian mechanism. For example, the presence of a previously formed corpus luteum increases the efficacy of pregnant mare serum gonadotropin (PMSG) in inducing ovulation in sheep.

Regression. If fertilization does not occur, the corpus luteum regresses, allowing other larger ovarian follicles to mature. As these cells degenerate, the whole organ decreases in size, becomes white or pale brown, and is known as the corpus albicans. Regressive changes include thickening of the walls of the arteries in the corpus luteum, a decrease in cytoplasmic granulation, a rounding of the cell outline, and peripheral vacuolation of the large luteal cells (Fig. 3–8). After two or

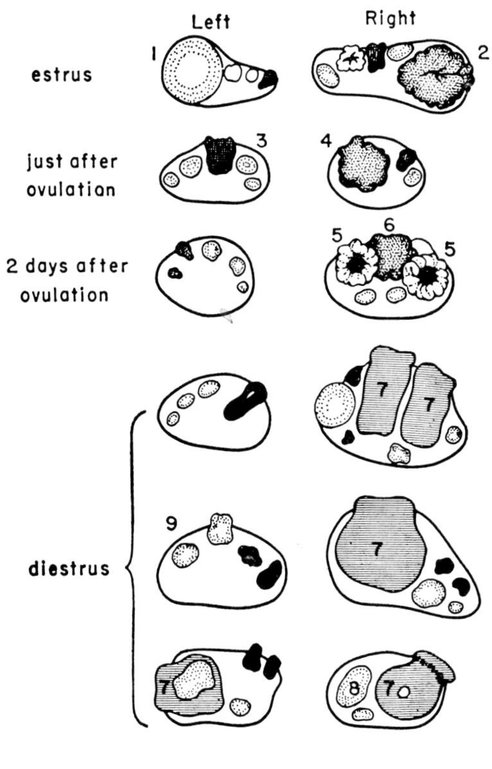

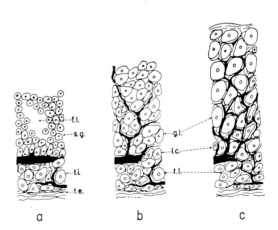

Fig. 3–7. Diagrammatic illustration of the morphologic changes in the estrous cycle of bovine ovary. *1*, Ripe follicle; *2*, regressing corpus luteum (brick brown); *3*, collapsed follicle—surface wrinkled and walls bloodstained; *4*, regressing corpus luteum (bright yellow); *5*, twin corpora lutea—some hemorrhage; *6*, regressing corpus luteum (bright yellow); *7*, corpus luteum of diestrus; *8*, largest follicle; *9*, corpus albicans. *(Redrawn from Arthur, 1964. Wright's Obstetrics. London, Bailliere, Tindall & Cox)*

Fig. 3–8. Diagram showing the organization of cells in the corpus luteum of the ewe. *a,* Corpus haemorrhagicum; *b,* corpus luteum of the second day following estrus; *c,* corpus luteum of the fourth day following estrus. Blood vessels are shown with heavy black lines. *f.l.,* Lake of follicular fluid; *g.l.,* lutein cells from the membrana granulosa; *t.i.,* theca interna; *t.e.,* theca externa. *(Adapted from Warbritton, 1934. J. Morphol. 56, 181)*

three cycles a barely visible scar of connective tissue remains. Remnants of the bovine corpus albicans persist during several successive cycles. The bovine corpus luteum of the estrous cycle begins to regress 14 to 15 days after estrus, and its size may be halved within 36 hours.

Luteolysis. Estrogen-induced luteolysis, probably mediated through stimulation of uterine prostaglandin $F_2\alpha$ ($PGF_2\alpha$) during the estrous cycle of the ewe, is responsible for the normal regression of the corpus luteum. An embryo must be in the uterus of ewes on days 12 and 13 after mating in order for the corpus luteum to be maintained. Apparently this time represents the state at which the uterus initiates steps leading to luteolysis.

Prostaglandin $F_2\alpha$ ($PGF_2\alpha$) is a potent, naturally-occurring luteolysin in sheep. Various doses and routes of administration of exogenous $PGF_2\alpha$ are effective in producing complete luteal regression. Injection of $PGF_2\alpha$ into a large follicle in the ovary with CL is one of the more effective routes. Indomethacin [1-(p-chlorobenzoyl)-5-methoxy-2-methylindole-3-acetic acid] blocks the estrogen-induced release of $PGF_2\alpha$ from the ovine uterus. Intrauterine injections of indomethacin prevent estrogen-induced luteal regression in cattle and sheep.

The main uterine vein and the ovarian artery are the proximal and distal components of a local veno-arterial pathway involved in the luteolytic and antiluteolytic effects. Hysterectomy abolishes the luteolytic effect and causes persistence of the corpus luteum.

Changes in Blood Flow to Corpus Luteum. The presence of a functional corpus luteum greatly increases blood flow to the ovary (Niswender et al, 1976); the regional blood flow within the corpus luteum uses 15μ radioactive microspheres. In the ewe, blood flow to the luteal ovary increases from less than 1 ml/min to 3 to 7 ml/min as the corpus

luteum develops (Niswender et al, 1976). During regression, blood flow to the luteal ovary declines sharply.

The changes in blood flow to the luteal ovary can be attributed to changes in flow to the corpus luteum, which receives most of the blood suppy. It seems that blood flow to the corpus luteum plays a role in the regulation of this gland and in regulating the activity of the gonadotropic hormones at the luteal cell level. A secondary action of LH may be to increase blood flow to the corpus luteum. $PGF_2\alpha$ affects the vascular component of the corpus luteum. Administration of $PGF_2\alpha$ to ewes reduces blood flow to the ovary with a corpus luteum and reduces secretion of progesterone.

Corpus Luteum and Pregnancy. Progestogens are necessary for the maintenance of pregnancy. They act in part by altering the ionic permeability across the myometrial muscle cell membrane, resulting in an increased resting membrane potential and lowered cellular conduction and excitability. Some progestogens serve as immediate precursors to other steroids that are also necessary during pregnancy.

Except in the mare, there is an obligatory requirement for continued secretory activity of the corpus luteum throughout pregnancy because the placenta does not secrete progesterone in these species. Ovariectomy of the gilt at any time during pregnancy results in abortion within two to three days. After removing one ovary or the corpora lutea from each ovary on day 40 of gestation, a minimum of five corpora lutea is needed to maintain gestation.

The corpus luteum of pregnancy is known as the corpus luteum verum and may be larger than the corpus luteum spurium (false yellow body) of the estrous cycle. In cattle it increases in size for two to three months of gestation, regresses for four to six months and thereafter remains relatively constant until calving, when it degenerates within one week postpartum.

THE OVIDUCT

An intimate anatomic relationship exists between the ovary and the oviduct. In farm mammals, the ovary lies in an open ovarian bursa in contrast to some species (eg, rat, mouse) in which it lies in a closed sac. This bursa in farm animals is a pouch consisting of a thin peritoneal fold of mesosalpinx, which is attached to a suspended loop at the upper portion of the oviduct (Fig. 3–9). In cattle and sheep the ovarian bursa is wide and open. In swine it is well-developed and, although open, it largely encloses the ovary. In horses it is narrow and cleft-like and encloses only the ovulation fossa.

Anatomy

The oviducts are suspended in the mesosalpinx, a peritoneal fold that is derived from the lateral layer of the broad ligament. The length and degree of coiling of the oviduct vary in farm mammals.

The oviduct may be divided into four functional segments: the fringe-like "fimbriae"; the funnel-shaped abdominal opening near the ovary—the "infundibulum"; the more distal dilated "ampulla"; and, the narrow proximal portion of the oviduct connecting the oviduct with the uterine lumen—the "isthmus."

The size of the infundibulum varies with the species and age of the animal; the surface area is 6 cm² to 10 cm² in sheep, and 20 cm² to 30 cm² in cattle. The opening of the infundibulum, the ostium abdominale, lies in the center of a fringe of irregular processes that form the extremity of the oviduct, the fimbriae. The fimbriae are unattached except for one point at the upper pole of the ovary. This assures close approximation of the fimbriae and the ovarian surface.

The ampulla, comprising about half of the oviductal length, merges with the constricted section known as the isthmus. The anatomic and physiologic significance of this ampullary-isthmic junction

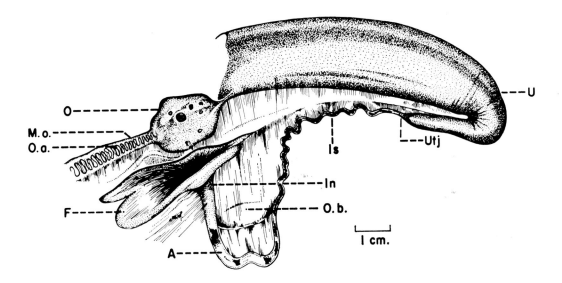

Fig. 3–9. Anatomic relationship between the ovary and the oviduct in the ewe.
A, Ampulla; *F*, fimbriae; *In*, infundibulum; *Is*, isthmus; *M.o.*, mesovarium; *O*, ovary; *O.a.*, ovarian artery; *O.b.*, ovarian bursa; *U*, uterus; *Utj*, uterotubal junction. Note the suspended loop to which the ovarian bursa is attached. The oviduct in the ewe is pigmented.

is still unknown. The isthmus is connected directly to the uterus (it enters the horn in the form of a small papilla in the mare). No well-defined sphincter muscle is present at this point, the uterotubal junction. However, in the sow this junction is guarded by long finger-like mucosal processes. In the cow and ewe there is a flexure at the uterotubal junction, especially during estrus.

Muscle also extends from these layers into the connective tissue of the mucosal folds, permitting coordinated contractions of the entire wall. The thickness of the musculature increases as one advances from the ovarian to the uterine end of the oviduct (Fig. 3–10,*A*).

Vasculature. The vasculature of the oviduct is derived from the uterine and ovarian arteries, which together supply arcades of vessels along the length of the oviduct. The remarkable increase in the prominence of the vasculature, largely regulated by ovarian estrogens, is partly associated with the enhanced secretory function of the oviduct.

Lymphatic drainage occurs by way of the para-aortic or lumbar nodes.

Innervation. As with other segments of the female reproductive tract, the oviduct is partially supplied by "short" adrenergic neurons, which are physiopharmacologically different from those of the more common "long" adrenergic neurons. In addition to receiving nerves from pre- and paravertebral ganglia (long adrenergic neurons), the oviduct receives a portion from ganglionic formations in the uterovaginal area (short adrenergic neurons). In the rabbit, adrenergic innervation of the oviduct is derived from postganglionic fibers that originate in the inferior mesenteric ganglia (long adrenergic neurons), and others that originate in ganglia located near the cervix at the level of the cervicovaginal junction (short adrenergic neurons) (Pauerstein et al, 1974). The two types of neurons seem to differ anatomically and functionally.

The degree of innervation varies in the different muscle layers and in different regions of the oviduct. Adrenergic innervation is particularly prominent in the circular musculature of the isthmus and at the ampullary-isthmic junction, where adrenergic nerve terminals are in close contact with individual smooth muscle cells. The dense adrenergic innervation permits the isthmus to act as a physiologic sphincter, which may be important for regulating egg transport.

In the ampulla and infundibulum the adrenergic nerves are limited primarily to the smooth muscle of the blood vessel walls.

Function of the Oviduct

Several in vivo and in vitro techniques have been used to study the function of the oviduct. The oviduct has the unique function of conveying the eggs and spermatozoa in opposite directions, almost simultaneously. The structure of the oviduct is well adapted to its multiple functions. The fringe-like fimbriae transport ovulated eggs from the ovarian surface to the infundibulum. The eggs are transported through the mucosal folds to the ampulla where fertilization and early cleavage of fertilized eggs take place. The embryos remain in the oviduct for three days before they are transported to the uterus. The mesosalpinx and oviductal musculature coordinate ovarian hormones, estrogen and progesterone. The uterotubal junction controls, in part, the transport of sperm from the uterus to the oviducts.

Oviductal Fluid. The oviductal fluid provides a suitable environment for fertilization and cleavage of fertilized eggs. The rate of accumulation of oviductal fluid is regulated by ovarian hormones. By using different methods of cannulation for continuous collection of oviductal secretions, it has been shown that the volume of fluid secreted by both oviducts

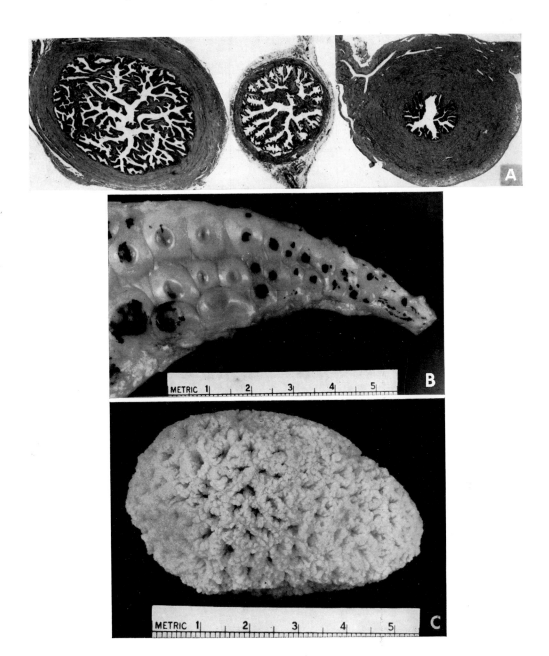

Fig. 3–10.   *A*, Cross section of ampulla in the sow *(left)* and cow *(center)*. Note differences in the diameter of the lumen and complexity of mucosal primary and secondary folds. Cross section in isthmus in the sow *(right)*; note the thickness of muscular coat ($\times$ 4.5). *(Photographs by E. Schilling)*
B, Mucosa of uterine horn of the nonpregnant ewe. Note caruncles and pigmentation of the endometrium.
C, Maternal caruncle from a pregnant cow. Note the spongy-like crypts to which the chorionic villi were embedded.

varies during the estrous cycle. The volume is low during the luteal phase, increases at the onset of estrus, reaches a maximum a day later and then declines to characteristic luteal phase levels (Fig. 3–11).

The direction of flow of a large part of the oviductal fluid is toward the ovary, since the isthmus blocks or partially blocks the flow of fluids into the uterus.

Several physiologic factors may be involved in creating currents and countercurrents: (a) quantitative and qualitative changes in oviductal secretions throughout the menstrual cycle and in response to contraceptives, (b) beat of kinocilia in oviductal compartments which vary in size and shape, (c) constant change in the diameter of the oviductal lumen in different segments as a result of muscle contraction and reorientation of mucosal folds.

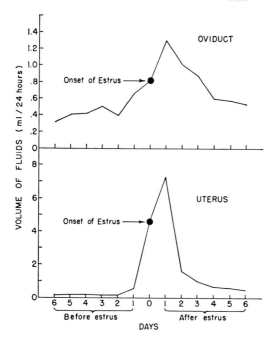

Fig. 3–11. Fluctuations in the volume of secretions of the oviduct and the uterus during the estrous cycle of sheep. Maximal secretion rates occur one day following the onset of estrus; this period coincides with the time of ovulation and reception of the ovum by the fimbriae. During estrus the volume of uterine fluids exceeds that of the oviduct, whereas the reverse is true during the luteal phase. *(Perkins et al., 1965. J. Anim. Sci. 24, 383)*

The directional movement of oviductal secretions may contribute to ovum transport to the uterus. In sheep most of the oviductal secretions pass out of the oviductal ostium early in the estrous cycle (Belve and McDonald, 1968). However, on day 4, when ova usually enter the uterus, fluid flow through the uterotubal junction increases remarkably.

Several protein components are common to oviductal fluid and serum. However, some of these are present in different proportions in these two body fluids; for example, the quantity of transferrin and prealbumin in oviductal fluid is far greater relative to albumin than is the quantity of these proteins in the serum. Many serum proteins have no counterparts in oviductal fluid and conversely, several proteins are unique to oviductal fluid. The presence of unique proteins in oviductal fluid implies that oviductal cells may be engaged in specific secretory activity.

In some species, the oviductal fluid has a direct influence on the respiration of spermatozoa and early embryonic development. Following ovulation in the rabbit, significant changes occur in the activity of a heat-labile oviductal component that enhance the developmental capacity of the preimplantation embryo.

Oviductal Musculature and Related Ligaments

Oviductal contractions facilitate mixing of oviductal contents, help to denude the ova, promote fertilization by increasing egg-sperm contact, and partly regulate egg transport. Unlike intestinal peristalsis, oviductal peristalsis tends to delay slightly the progression of the ovum instead of transporting it.

Patterns of Oviductal Contractions. The oviductal musculature undergoes various types of complex contractions: localized peristalsis-like contractions originating in isolated segments or loops and traveling only a short distance; segmental contractions; and worm like writhings of the entire oviduct. Contractions in an obovarian direction are more common than those in an adovarian direction. In general the ampulla is less active than the isthmus.

Because longitudinal muscle fibers, which cause shortening, and circular muscle fibers, which cause annular constriction, are constantly activated, the contractile pattern of the oviduct is complex. Additional complicating factors are the contractile activities of the mesosalpinx, the myometrium and the supporting ligaments, and ciliary movement.

Oviductal muscular contractions are stimulated by contractions of two major membranes that contain smooth musculature and are attached to the fimbriae, ampulla, and ovary: the mesosalpinx and the mesotubarium superius. The frequency and amplitude of spontaneous contractions vary with the phase of the estrous cycle. Before ovulation, contractions are gentle with some individual variations in the rate and pattern of contractility. At ovulation, contractions become most vigorous. At this time the mesosalpinx and mesotubarium superius contract vigorously, independently and intermittently; the mesotubarium contracts more vigorously than the mesosalpinx. These contractions draw the oviduct into the form of a crescent, slide the fimbriae over the surface of the ovary, and cause continuous change in the contour of the oviduct. At ovulation, the fringelike folds contract rhythmically and "massage" the ovarian surface.

The pattern and amplitude of contraction vary in different segments of the oviduct. In the isthmus, peristaltic and antiperistaltic contractions are segmental, vigorous and almost continuous. In the ampulla strong peristaltic waves move in a segmented fashion toward the midportion of the oviduct. The varying patterns of oviductal contraction may be associated with cyclic changes in glycogen content of oviductal musculature. Glycogen in the oviduct is more abundant in the inner circular musculature than in the outer longitudinal musculature. The amount of glycogen is low after menstruation, higher after ovulation, and highest during pregnancy. Glycogen content in human myometrium fluctuates similarly.

The excitability and contractile properties of the oviductal musculature are regulated by ovarian steroids and prostaglandins.

Prostaglandins and Oviductal Contractility. PGE_1 and PGE_2 exert a characteristic effect on the longitudinal musculature of the oviduct: an increase in tonus of the proximal part and relaxation of the rest of the organ. However, PGE_3 relaxes the whole oviduct. On the other hand, PGF_1 and PGF_2 act as stimulators, the strongest effect being exerted by PGF_2 with no apparent change in sensitivity or action throughout the menstrual cycle. PGF_2 has a relaxing effect on the whole oviduct.

Following intravenous injection of PGE_1 and PGE_2 in rabbits, the spontaneous motility of the oviduct is suppressed with a large decrease in tone and frequency of contraction. Injection of PGF_1 and PGF_2 evokes a sustained increase in oviductal muscular tone. Thus it would appear that the prostaglandins are involved in the regulation of the oviductal function.

Utero-Ovarian and Related Ligaments. The utero-ovarian ligament contains smooth muscle cells arranged primarily in longitudinal bundles, which continue into the myometrium but not into the ovarian stroma. The smooth muscles in the mesovaria and the various ligaments of the mesenteries attached to the ovaries and the fimbriae contract intermittently. These rhythmic muscular contractions assure that the fimbriae remain in a constant position relative to the surface of the ovaries. These spontaneous contractions vary with the endocrine state of the female. The response of these ligaments to

Table 3–3. Comparative Anatomy of the Reproductive Tract in the Adult Nonpregnant Female of Farm Mammals

Organ	Animal			
	Cow	Ewe	Sow	Mare
Oviduct				
Length (cm)	25	15–19	15–30	20–30
Uterus				
Type	Bipartite	Bipartite	Bicornuate	Bipartite
Length of horn (cm)	35–40	10–12	40–65	15–25
Length of body (cm)	2–4	1–2	5	15–20
Surface of lining of endometrium	70–120 caruncles	88–96 caruncles	Slight longitudinal folds	Conspicuous longitudinal folds
Cervix				
Length (cm)	8–10	4–10	10	7–8
Outside diameter (cm)	3–4	2–3	2–3	3.5–4
Cervical lumen				
Shape	2–5 annular rings	Several annular rings	Corkscrew-like	Conspicuous folds
Os uteri				
Shape	Small and protruding	Small and protruding	Ill-defined	Clearly-defined
Anterior vagina				
Length (cm)	25–30	10–14	10–15	20–35
Hymen	Ill-defined	Well-developed	Ill-defined	Well-developed
Vestibule				
Length (cm)	10–12	2.5–3	6–8	10–12

The dimensions included in this table vary with age, breed, parity and plane of nutrition.

prostaglandins also varies with hormonal conditions.

THE UTERUS

The uterus consists of two uterine horns (cornua), a body and a cervix (neck). The relative proportions of each, as well as the shape and arrangement of the horns, vary from species to species (Table 3–3). In swine, the uterus is of the bicornuate type (uterus bicornis). The horns are folded or convoluted and may be as long as four to five feet, while the body of the uterus is short (Fig. 3–12). This length is an anatomic adaptation for successful litter bearing. In cattle, sheep and horses, the uterus is of the bipartite type (uterus bipartitus). These animals have a septum that separates the two horns and a prominent uterine body (the horse has the largest). Superficially the body of the uterus in cattle and sheep appears larger than it actually is because the caudal parts of the horns are bound together by the intercornual ligament.

Both sides of the uterus are attached to the pelvic and abdominal walls by the broad ligament. In multiparous animals the uterine ligaments stretch, allowing the uterus to drop into the pelvic cavity. In the mare this may hinder the removal of endometrial fluids or even allow small amounts of urine to flow through the cervix during estrus, resulting in mild catarrhal inflammation.

Like most other hollow internal organs, the walls of the uterus consist of a mucous membrane lining, an intermediate smooth muscle layer and an outer serous layer, the peritoneum. From the physiologic standpoint, only two layers are recognized—the endometrium and the myometrium. The sympathetic nerves are

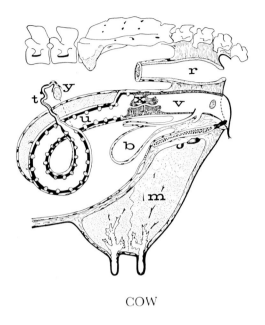

COW

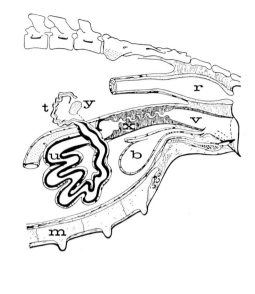

SOW

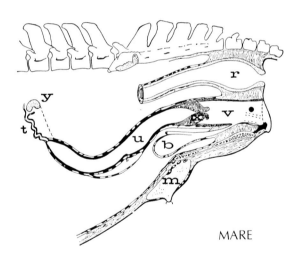

MARE

Fig. 3–12. Comparative anatomy of the reproductive organs in the female.
b, Bladder; *m*, mammary gland; *r*, rectum; *t*, oviduct; *u*, uterus; *v*, vagina; *x*, cervix; *y*, ovary. Note species differences in anatomy of cervix, uterus and mammary gland. *(Redrawn from Ellenberger and Baum, 1943. Handbuch der vergleichenden Anatomi der Haustiere, 18th ed. Zietszchmann, Ackernecht and Grau [eds.], Berlin, Springer)*

supplied through the uterine and pelvic plexuses, terminating partly in the muscle fibers and partly in the mucosa.

Vasculature of the Uterus

The uterus receives its blood and nerve supply through the broad ligament (Fig. 3–13). The blood vessels are numerous, thick-walled and tortuous. The middle uterine artery, a branch of either the internal iliac artery or the external iliac artery, provides the chief blood supply to the uterus in the region of the developing fetus, and thus it enlarges greatly with advancing gestation. The cranial uterine artery, a branch of the utero-ovarian artery, supplies blood to the ovary by the ovarian artery and to the anterior extremity of the uterine horn by the cranial uterine artery. The utero-ovarian artery supplies blood to the oviduct. The caudal uterine artery is a forward continuation of a branch of the internal pubic artery, which supplies blood to the vagina, vulva and anus.

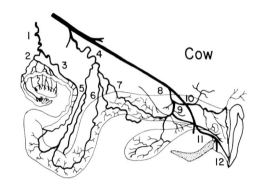

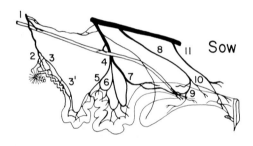

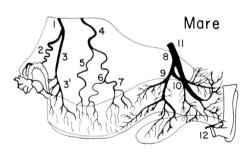

Fig. 3–13. Diagrammatic illustration of the arterial blood supply in the female reproductive tract.
1, utero-ovarian artery; *2,* ovarian artery; *3,* uterotubal artery; *3′,* anterior cornual artery or anterior artery of uterine horn; *4,* uterine artery; *5,* middle cornual artery; *6,* posterior cornual artery; *7,* artery of corpus uterii; *8,* vaginal artery; *9,* cervicouterine artery; *10,* vagino-rectal artery; *11,* internal pudic artery; *12,* artery of the clitoris. *(Barone and Pavaux, 1962. Bull. Soc. Sci. Vet. Lyon No. 1, 33–52)*

The utero-ovarian artery runs closely along the surface of the corresponding vein and shows only a minor degree of tortuosity. This artery terminates by giving rise to a small branch to the tip of the uterine horn and oviduct, one or more highly convoluted ovarian branches and a branch to the broad ligament. The utero-ovarian vein is formed by the fusion of one or more large uterine branches, the plexiform ovarian veins and a branch from the broad ligament. An extensive perivascular plexus of small veins is present between and around the utero-ovarian vessels, concentrated mainly around the arteries.

There is prominent anastomosis between the uterine and ovarian arterial systems. The uterine artery ipsilateral to an ovary bearing a corpus luteum (CL) in nonpregnant heifers responds *in vitro* to periarterial nerve stimulation with greater constriction than the contralateral uterine artery (Ford et al, 1976). The differential response of the uterine arteries to nerve stimulation may be due to exposure of the artery adjacent to the ovary with the CL to increased local ovarian lymph concentrations of progesterone or estrogen.

Radioactive microspheres have been used to measure regional blood flow to the reproductive organs. There is considerable variation among species in the relative amount of arterial blood that reaches the uterus and ovaries by the uterine versus the ovarian artery. The ovarian artery appears to contribute to the uterine arterial supply through the uterine branch of the ovarian artery. In cattle, on the side ipsilateral to the CL, the uterine artery may contribute to the ovarian supply.

Constriction of the uterine artery with a resultant reduction in blood flow is attributed to the response of vascular smooth muscle to norepinephrine liberated from innervating adrenergic nerves. The regulatory actions of ovarian steroids on the ovine uterine artery may also be potentiated by prostaglandin $F_2\alpha$ ($PGF_2\alpha$), which is synthesized and released by the uterus.

Endometrial Glands and Uterine Fluid

The endometrial glands are branched, coiled, tubular structures lined with columnar epithelium. They open onto the endometrial surface, except in the caruncular areas (in ruminants). The glands are relatively straight at the time of estrus; they grow, secrete, and become more coiled and complex as the level of progesterone produced by the developing corpus luteum rises. They begin to regress when the first signs of luteal regression are also noted. The endometrial surface epithelial cells are relatively tall during estrus; following a period of active secretion during estrus they become low and cuboidal at two days postestrus.

The volume and biochemical composition of the uterine fluid show consistent variation during the estrous cycle (Fig. 3-11). In sheep the volume of the fluid in the uterus exceeds that of the oviduct during estrus, whereas during the luteal phase the reverse is true (Perkins et al, 1965).

The endometrial fluid contains mainly serum proteins but also small amounts of uterine-specific proteins. The ratio and amounts of these proteins vary according to the reproductive cycle. Differences in concentration as well as distribution of components in the uterine fluids compared to the blood serum provide evidence that secretion as well as transudation occurs. In the rabbit, a protein named blastokinin (uteroglobulin) can influence blastocyst formation from morulae, whereas uterine fluid in the mouse contains a factor that initiates implantation. Uterine-specific proteins can be detected in the pig at a similar stage of blastocyst development, and at least one of these is a cathode-migrating protein on disc electrophoresis. In the pig, the appearance of these uterine-specific proteins is pro-

gesterone-dependent. In ruminants, the situation is less clear, for although specific proteins have been identified, they are not present in amounts comparable to that of blastokinin in the rabbit.

The uterine fluid has two important functions, namely, to provide a favorable environment for sperm capacitation and to provide nutrition for the blastocyst until implantation is completed. In the cow, the embryo lies free in the medium for approximately 30 days, during which extensive embryonic differentiation takes place before the conceptus becomes firmly attached to the endometrium.

Uterine Contraction

The contraction of the uterus is coordinated with the rhythmic movements of the oviduct and ovary. There is considerable variation in the origin, direction, amplitude degree and frequency of contractions in the reproductive tract. In the estrous rabbit, contractions in the uterus are continuations of contractions of the oviducts and move along the uterine horn from the uterotubal junction toward the cervix. In general the greatest number of uterine contractions move toward the oviducts during early estrus but toward the cervix after the end of estrus.

During early estrus in the ewe, peristaltic contractions of the uterus originate predominantly near the uterine body and move toward the oviducts. During late estrus, peristaltic contractions originate near the uterotubal junction and move toward the cervix. In the ewe, anterior segments of the uterus have stronger contractions, but the duration of contractions is greater in posterior segments. Endogenous estrogen initiates the type of uterine motility during early estrus, and declining estrogen secretion is responsible for the change in direction of contractions during estrus. The frequency of contractions increases in response to both norepinephrine and epinephrine in both early and late estrus (Rexroad and Barb, 1978).

As ovulation occurs, ovarian changes cause a modification of these patterns. Thus long before the egg reaches the uterine lumen, the uterine muscles have become quiescent; moreover, they remain so throughout pregnancy. This sort of myometrial activity can also be induced experimentally by the injection of estrogen and progesterone.

Uterine Metabolism

The endometrium metabolizes carbohydrates, lipids and proteins to supply the necessary requirements for cell nutrition, rapid proliferation of the uterine tissue and development of the conceptus. Cyclic metabolic variations in this tissue consist of changes in the rate of nucleic acid synthesis, the availability of glucose, and the amount of glycogen reserves. These reactions depend on four phenomena: (a) the enzymatic reactions involved in glucose metabolism; (b) the increase in circulation through the spiral arterioles; (c) the morphologic changes that occur in the endometrium and myometrium; and (d) the stimulating action of the ovarian and other hormones.

Two compounds that are of special significance in endometrial metabolism are glucose and glycogen. Glucose is converted to glucose-6-phosphate by hexokinase, perhaps as it passes through the cell wall. Once in the form of glucose-6-phosphate it may undergo further conversions.

Ovarian hormones play a substantial role in regulating uterine metabolism. Growth of the uterus (both protein synthesis and cell division) is induced by estrogen; in the process it utilizes a large amount of energy in the form of ATP. Estrogen causes hyperemia followed by an increase in amino acid incorporation, nucleic acid synthesis and nitrogen reten-

tion. This hormone also stimulates phosphorus incorporation, oxidative metabolism, aerobic and anaerobic glycolysis and glycogen deposition.

Progestational responses in the endometrium involve major growth, a striking increase in DNA and RNA and a loss of water. A rapid change occurs in the metabolism of the endometrium about the time the egg passes through the uterotubal junction.

Function of the Uterus

The uterus serves a number of functions. The endometrium and its fluids play a major role in the reproductive process: (a) sperm transport from the site of ejaculation to the site of fertilization in the oviduct; (b) regulation of the function of the corpus luteum; and (c) initiation of implantation, pregnancy and parturition.

Sperm Transport. At mating, the contraction of the myometrium is essential for the transport of spermatozoa from the site of ejaculation to the site of fertilization. Large numbers of spermatozoa aggregate in the endometrial glands; the physiologic and immunologic significance of this phenomenon is not known. As spermatozoa are transported through the uterine lumen to the oviducts they undergo "capacitation" in endometrial secretions.

Luteolytic Mechanisms. There is a local utero-ovarian cycle whereby the corpus luteum stimulates the uterus to produce a substance which in turn destroys the corpus luteum. The uterus plays an important role in regulating the function of the corpus luteum. Corpora lutea are maintained in a functional state for long periods following hysterectomy of cattle, sheep and swine. If small amounts of uterine tissue remain in situ, luteal regression occurs and cycles are resumed after variable periods. Following unilateral hysterectomy, corpora lutea adjacent to the excised uterine horn are usually better maintained than those adjacent to the remaining horn (Figs. 3–14, 3–15).

It appears that the uterus produces or participates in the production of some luteolytic substance, and this uterine luteolysin may be selectively transferred from the utero-ovarian vein to the closely adherent ovarian artery, and thus reaches the ovary in much greater concentrations than that in peripheral blood.

Intramuscular or intrauterine administration of prostaglandin causes complete luteal regression in the cow and ewe. Infusions of PGF$_2\alpha$ into the utero-ovarian vein in ewes also causes rapid regression of the corpora lutea and a decline in plasma progesterone levels. Prostaglandin F$_2\alpha$ (PGF$_2\alpha$) appears to be the uterine luteolysin in the ewe, transmitted by way of a veno-arterial pathway directly from the uterus to the corpus luteum, where it causes luteal regression. Evidence that this mechanism occurs in the cow has not been well established. However, administration of PGF$_2\alpha$ is useful in artificial

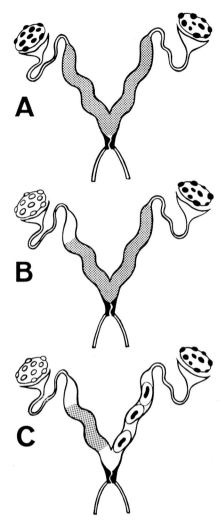

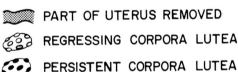

PART OF UTERUS REMOVED

REGRESSING CORPORA LUTEA

PERSISTENT CORPORA LUTEA

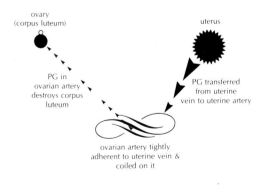

Fig. 3–14. Postulated route by which prostaglandin manufactured by the progesterone-primed uterus is able to enter the ovarian artery and destroy the corpus luteum.

ovary
(corpus luteum)

uterus

PG in ovarian artery destroys corpus luteum

PG transferred from uterine vein to uterine artery

ovarian artery tightly adherent to uterine vein & coiled on it

Fig. 3–15. Diagram showing the effect of partial hysterectomy on persistence of corpora lutea in the pig. *A*, Total hysterectomy during the luteal phase causes retention of the corpus luteum for a period similar to gestation. *B*, Unilateral hysterectomy and partial removal of the other horn, retaining only a fragment 20 cm in length, causes asymmetric functioning of the two ovaries. *C*, The corpora lutea on the intact side are normally maintained whereas those on the other side regress before the 22nd day of gestation. *(Data from Du Mesnil du Buisson, 1961. Ann. Biol. Anim. Bioch. Biophys. 1, 105)*

regulation of the estrous cycle in cattle. Intrauterine doses of 25 to 30 mg cause a rapid decline in plasma progesterone levels and a return to estrus in two to three days in most treated cows.

The gravid uterine horn exerts an antiluteolytic effect at the level of the adjacent ovary. This effect is exerted through a local utero-ovarian veno-arterial pathway.

In the nonpregnant ewe the uterus causes regression of the corpus luteum through a direct or local pathway between the uterine horn and the adjacent ovary. The local pathway between a uterine horn and adjacent ovary in sheep is veno-arterial in nature, involving the main uterine vein (uterine branch of ovarian vein) as the uterine component and the ovarian branch of the ovarian artery as the ovarian component.

Uterine dilation and irritation inhibit the normal development and function of the cyclic corpora lutea of cows and ewes. This mechanism may play a role in reproductive failure. Infusions of large numbers of nonspecific bacteria into the uterus and insemination of heifers with semen containing a virus cause luteal inhibition and induce precocious estrus.

Implantation and Gestation. The uterus is a highly specialized organ that is adapted to accept and nourish the products of conception from the time of implantation until parturition. A well-described though obscurely defined uterine "differentiation" occurs, governed by the ovarian steroid hormones. This process must evolve to some critical stage when the uterus is prepared to selectively accept the blastocyst. Unless such differentiation occurs, the uterus is unsuited to permit implantation.

After implantation, the embryo depends upon an adequate vascular supply within the endometrium for its development. Throughout gestation the physiologic properties of the endometrium and its blood supply are important for the survival and development of the fetus. The uterus is capable of undergoing tremen-

dous changes in size, structure and position to accommodate the needs of the growing conceptus.

Parturition and Postpartum Involution. The contractile response of the uterus remains dormant until the time of parturition, when it plays the major role in fetal expulsion. Following parturition, the uterus almost regains its former size and condition by a process called involution (Fig. 3–16). In the sow the uterus continuously declines in both weight and length for 28 days after parturition; thereafter it remains relatively unchanged during the lactation period. However, immediately after the young are weaned, the uterus increases in both weight and length for four days.

During the postpartum interval the destruction of endometrial tissue is accompanied by the presence of large numbers of leukocytes and the reduction of the endometrial vascular bed. The cells of the myometrium are reduced in number and size. These rapid and disproportional changes in the uterine tissue are a possi-

Fig. 3–16. Diagrammatic illustration of the changes taking place in size and shape of the ruminant uterus during pregnancy. Three uteri are shown in the diagram; the inner one represents a nonpregnant uterus; the outer one represents a gravid uterus prior to delivery, and the middle one represents a uterus after delivery in the process of involution.

ble cause of low postpartum conception rate. Neither the presence of suckling calves nor anemia delays uterine involution. Caruncular tissues are sloughed off and expelled from the uterus 12 days after calving. Regeneration of the surface epithelium over the caruncles occurs by growth from the surrounding tissue and is completed 30 days after calving.

Effects of Foreign Bodies and IUDs. The stimulation of the uterus during the early stages of the estrous cycle hastens regression of the corpus luteum and causes precocious estrus. Uterine stimulation can be initiated by placing a small foreign body in the lumen. The subsequent estrous cycle will be either shortened or prolonged depending on when the foreign body was inserted and on the nature and size of the material introduced. That the nervous system is responsible for this effect is implicated by the fact that the estrous cycle is unaffected when the uterine segment containing the foreign body is denervated.

Although intrauterine devices (IUDs) have an antifertility effect in several domestic animals, their apparent mode of action varies widely. The major antifertility effect of IUDs seems to be exerted between the time the embryo enters the uterus and the time of implantation. For example, the insertion of large-diameter (uterus-distending) IUDs in sheep and cattle results in an alteration of the estrous cycle by shortening the functional lifespan of the corpus luteum. In sheep, large diameter IUDs inhibit sperm transport and fertilization.

CERVIX UTERI

The cervix is a sphincter-like structure that projects caudally into the vagina.

Anatomy

The cervix is characterized by a thick wall and constricted lumen. Although the structure of the cervix differs in detail among farm mammals, the cervical canal is characterized by various prominences. In ruminants these are in the form of transverse or spirally interlocking ridges known as annular rings, which develop to varying degrees in the different species (Fig. 3–17). They are especially prominent in the cow (usually four rings) and in

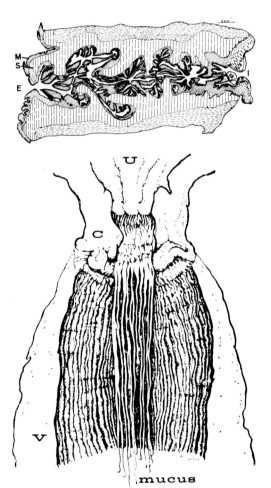

Fig. 3–17. *(Top)* Tracing of a longitudinal section of the bovine cervix showing the complexity of the cervical crypts which attract massive numbers of spermatozoa. *E,* external, or *I,* internal, or *M,* mucus-secreting mucosa; *S,* cervical stroma.
(Bottom) Diagrammatic illustration showing how the strands of cervical mucus flow from the crypts of the cervix *(C)* to the epithelium of the vagina *(V).* The biophysical characteristics of cervical mucus and arrangement of the macromolecules of mucus facilitate sperm transport from the vagina to the uterus *(U).*

Table 3–4. In Vivo and In Vitro Techniques Used to Study Functions of the Oviduct

Function Under Study	Techniques
Structure and ultrastructure of epithelium, secretory activity and cilia	Scanning and transmission electron microscopy
	Culture of fragments of oviductal mucosa and transfer of growth to rose chambers
	Histochemical observations of frozen section
Identification of adrenergic or cholinergic receptors	Fluorescence histochemical technique
	Physiopharmacology of oviductal contractility (eg, response to autonomic drugs)
Contractility of oviductal musculature	Visual observation of oviduct through abdominal wall or abdominal window
	Direct kymography
	Intraluminal catheter or microballoon placed in lumen to record intratubular pressure of oviduct
	Cineradiography
Biochemical composition of oviductal fluid	Extra-abdominal or intra-abdominal device to collect fluid
	Ligature of oviduct at the uterotubal junction and fimbriae to collect fluid
	Radioisotopes
Detection of protein uptake in oviductal epithelium	Immunofluorescence
Egg transport in oviduct	Pharmacology and neuropharmacology
	Effects of prostaglandins, steroid hormones and oral contraceptives
	Segmental flushing of oviduct after salpingectomy
	Use of surrogate eggs
	Recovery of eggs from uterus *in vivo*
Sperm transport in oviduct	Segmental flushing of oviduct after salpingectomy at intervals following previous artificial insemination
	Flushing of oviduct from fimbriae, by laparoscopy, at intervals following breeding or AI
Kinetics of cilia beat	High speed cinematography
	Photo probe to measure ciliary beat optically
	Laser photo correlation spectroscopy to study cilia beat in cultured ciliated cells

the ewe, where they fit into each other to close the cervix securely. In the sow, the rings are in a corkscrew arrangement which is adapted to the spiral twisting of the tip of the boar's penis. Distinguishing features of the mare's cervix are the conspicuous folds in the mucosa and the projecting folds into the vagina.

The cervix is tightly closed except during estrus, at which time it relaxes slightly, permitting sperm to enter the uterus. Mucus discharged from the cervix is expelled from the vulva.

Cervical Mucus

Cervical mucus consists of macromolecules of mucin of epithelial origin, which are composed of glycoproteins (particularly of the sialomucin type) that contain about 25% amino acids and 75% carbohydrates. The mucin is composed of a long, continuous polypeptide chain with numerous oligosaccharide side chains. The carbohydrate portion is made of galactose, glucosamine, fucose and sialic acid. The proteins of cervical mucus include prealbumin, lipoprotein, albumin and beta and gamma globulins. The cervical mucus contains several enzymes, including glucuronidase, amylase, phosphorylase, esterase and phosphatases. Cervical mucus that accumulates in the vaginal pool may also contain endometrial, oviductal, follicular and peritoneal fluids, as well as leukocytes and cellular debris from uterine, cervical and vaginal epithelia.

Owing to its unique biophysical characteristics, the cervical mucus has several rheologic properties, such as ferning, elasticity, viscosity, thixotropy and tack (stickiness).

The cervical mucus during estrus shows a fern pattern of crystallization upon drying on a glass slide. This fern pattern, associated with the high chloride content of the mucus, does not occur with drying of mucus obtained at stages of the cycle when progesterone levels are high or during pregnancy. The phenomenon may have some value, when combined with other observations, for early pregnancy diagnosis. The secretion of cervical mucus is stimulated by ovarian estrogen and inhibited by progesterone. Cyclic qualitative changes in the cervical mucus throughout the estrous cycle, and cyclic variations in the arrangement and viscosity of these macromolecules, cause periodic changes in the penetrability of spermatozoa in the cervical canal. Optimal changes of cervical mucus properties, such as an increase in quantity, viscosity, ferning and pH, and decrease in viscosity and cell content, occur during estrus and ovulation, and these are reversed during the luteal phase when sperm penetration in the cervix is inhibited. Under the influence of estrogens the macromolecules of glycoprotein of the mucus are oriented so that the spaces between them measure 2μ to 5μ. In the luteal phase the spaces of the meshwork of macromolecules become increasingly smaller. Thus, at the time of estrus and ovulation the large size of the meshes allows the transport of spermatozoa through the meshwork of filaments and through the cervical canal.

Functions

The cervix plays several roles in the reproductive process: (a) it facilitates sperm transport through the cervical mucus to the uterine lumen, (b) it acts as sperm reservoir (see Chapter 11), and (c) it may play a role in the selection of viable sperm, thus preventing the transport of nonviable and defective sperm.

Sperm Transport. Upon ejaculation, spermatozoa are oriented toward the internal os. As the flagellum beats and vibrates the sperm head is propelled forward in the channels of least resistance. The macro- and microrheologic properties of cervical mucus play a major role in sperm migration. Sperm penetrability increases with the cleanliness of mucus, since cellular debris and leukocytes delay sperm migration. The aqueous spaces between the micelles permit the passage of sperm as well as diffusion of soluble substances. Proteolytic enzymes may hydrolyze the backbone protein or some of the crosslinkages of the mucin and reduce the network to a less resistant mesh with more open channels for the migration of sperm. When cervical mucus and semen are placed in apposition *in vitro*, phase lines immediately occur between the two fluids. Sperm phalanges soon appear and develop high degrees of arborization, the terminal aspects of which consist of channels through which one or two spermatozoa can pass.

After mating or artificial insemination, massive numbers of spermatozoa are lodged in the complicated cervical crypts. The cervix might act as a reservoir for spermatozoa, thus providing the upper reproductive tract with subsequent releases of sperm. It is also possible that spermatozoa that are trapped in the cervical crypts are never released, thus preventing excessive numbers of spermatozoa from reaching the site of fertilization.

In ruminants prolonged survival of sperm in the cervix relative to other parts of the reproductive tract suggests that the cervix acts as a sperm reservoir. In the cervices of cattle and goats, most sperm were not randomly distributed but were located between cervical crypts. Penetration of sperm to these sites in the cervix depends upon sperm viability and upon the structure and consequently, the rheologic properties of the cervical mucus.

The Cervix During Pregnancy. During pregnancy, a highly viscid, nonferning, thick and turbid mucus occludes the cervical canal, acting as an effective barrier against sperm transport and invasion of bacteria in the uterine lumen, thus preventing uterine infections. The only other time the cervix is open is prior to parturition. At this time the cervical plug liquifies and the cervix dilates to permit the expulsion of the fetus and fetal membranes.

VAGINA

The vaginal wall consists of surface epithelium, muscular coat and serosa.

Anatomy

The muscular coat of the vagina is not as well-developed as the outer parts of the uterus. It consists of a thick inner circular layer and a thin outer longitudinal layer; the latter continues for some distance into the uterus. The muscularis is well-supplied with blood vessels, nerve bundles, groups of nerve cells and loose and dense connective tissue. The cow is unique in possessing an anterior sphincter muscle in addition to the posterior sphincter (at the junction of the vagina and vestibule) found in the other farm mammals.

The surface epithelium is composed of glandless, stratified, squamous epithelial cells, except in the cow, in which some mucous cells are present in the cranial part next to the cervix and the epithelial surface fails to cornify, probably because of low levels of circulating estrogens.

There are species differences in vaginal changes during the estrous cycle. These

differences probably reflect different secretion rates for estrogen and progesterone and ultimately for the gonadotropins. However, vaginal smears are not useful in diagnosing the stage of the cycle or hormonal abnormalities. In the ewe, growth of the vaginal epithelium is accelerated during estrus, and desquamation occurs in late estrus and early metestrus. The incidence of leukocytes is greatest during diestrus, when progesterone secretion is maximal. In the sow the vaginal epithelium increases in height to a maximum at estrus and decreases to a low point at days 12 to 16. The superficial layers of vaginal epithelium slough away between days 4 and 12. During diestrus and anestrus the vaginal epithelium of the mare is covered by a sticky, grayish secretion.

Physiologic Responses

Vaginal Contractions. Vaginal contractility plays a major role in psychosexual responses and possibly sperm transport. The contraction of the vagina, uterus and oviducts is activated by fluid secreted into the vagina during precoital stimulation.

Immunologic Responses. The vagina appears to be one of the major sites for sperm antigen-antibody reaction since the vagina is more exposed to sperm antigen than are the uterus and oviduct. Local production of antibodies to sperm antigens may occur within the vaginal tissues.

Immature and mature plasma cells, located beneath the epithelium, seem to be under endocrine control since these cells increase in number during the luteal phase, following ovariectomy, and during the postmenopausal stage. These plasma cells seem to be involved in the secretion of immunoglobulins A and G, which seem to prevent bacterial infection and produce antibodies against spermatozoa.

Vaginal Fluid. The vaginal fluid is composed primarily of transudate through the vaginal wall, mixed with vulvar secretions from sebaceous glands and sweat glands and contaminated with cervical mucus, endometrial and oviductal fluids and exfoliated cells of the vaginal epithelium. As estrus approaches, the vascularity of the vaginal wall increases and the vaginal fluid becomes thinner.

A specific and distinct odor is present in the urogenital tract of cows during estrus. This odor apparently disappears or is greatly attenuated during diestrus. Results indicate that dogs can be trained to detect and respond to the odors associated with estrus in cattle (Kiddy et al, 1978).

Microbiologic Flora. The vaginal flora is made of a dynamic mixture of aerobic, facultatively anaerobic, and strictly anaerobic microorganisms with new strains constantly being introduced. The flora of microorganisms varies throughout the life cycle. The various populations of microorganisms are equipped enzymatically to survive and replicate under a given vaginal environment. Only those most suited to replicate and compete for nutrients can become established and join the flora or even replace other microorganisms. During periods of high glycogen content, acidophilic organisms predominate but other organisms are present among the heterogeneous group making up the normal flora.

Functions of the Vagina

The vagina has multiple functions in reproduction. It is a copulatory organ in which semen is deposited and coagulated until spermatozoa are transported through the macromolecules of the cervical mucus column. The dilated bulbous vagina provides a postcoital semen pool to supply spermatozoa for cervical reservoirs. The rugae vaginales and the fence-like, rhomboid-shaped arrangement of the musculature allow distention of the vagina during mating and parturition. Although the vagina contains no glands, its walls are moistened by transudates through the vaginal epithelium (incorrectly called mucosa), by cervical mucus and by endometrial secretions.

Following ejaculation, the seminal plasma is not transported into the uterus; most of it is expelled or absorbed through the vaginal walls. Some of the biochemical components of the seminal plasma, when absorbed in the vagina, exert physiologic responses in other parts of the female reproductive tract.

The pH of the vaginal secretion is unfavorable to spermatozoa. A complex interaction of the cervical mucus, vaginal secretion and seminal plasma induces a buffering system that protects spermatozoa until they are transported through the micelles of cervical mucus. Pathologic conditions resulting in insufficient buffering of the seminal pool (such as low volume of ejaculate, scanty amounts of thick cervical mucus and leakage of semen) may cause rapid immobilization of spermatozoa.

The narrow canalis vaginae and the biochemical and microbiologic milieu of the vagina protect the upper reproductive duct from invading microorganisms. The vagina serves as an excretory duct for secretions of the cervix, endometrium and oviduct; it also serves as the birth canal during parturition. These functions are accomplished through various physiologic characteristics, namely, contraction, expansion, involution, secretion and absorption.

THE EXTERNAL GENITALIA

The vestibule, the labia majora, the labia minora, the clitoris and the vestibular glands comprise the external genitalia.

Vestibule. The junction of the vagina and vestibule is marked by the external urethral orifice and frequently by a ridge (the vestigial hymen). In some cattle the humen may be so prominent that it interferes with copulation.

The vestibule of the cow extends inward for approximately 10 cm, where the external urethral orifice opens into its ventral surface. Just posterior to this opening lies the suburethral diverticulum, a blind sac (Fig. 3–12). Gartner's tubes (remnants of the Wolffian ducts) open into the vestibule posteriorly and laterally to Gartner's ducts. The glands of Bartholin, which secrete a viscid fluid, most actively at estrus, have a tuboalveolar structure similar to the bulbourethral glands in the male.

Labia Majora and Labia Minora. The integument of the labia majora is richly endowed with sebaceous and tubular glands. It contains fat deposits, elastic tissue and a thin layer of smooth muscle, and has the same outer surface structure as the external skin. The labia minora have a core of spongy connective tissue. The surface contains many large sebaceous glands.

Clitoris. The ventral commissure of the vestibule conceals the clitoris, which has the same embryonic origin as the male penis. It is composed of erectile tissue covered by stratified squamous epithelium and it is well-supplied with sensory nerve endings. In the cow, the greater part of the clitoris is buried in the mucosa of the vestibule. However, in the mare it is well-developed, and in the sow it is long and sinuous, terminating in a small point or cone.

REFERENCES

Baker, T. (1972). Oogenesis and ovulation. *In* Reproduction in Mammals. C.R. Austin and R.V. Short (eds), Cambridge, University Press.

Belve, A.R. and McDonald, M.F. (1968). Directional flow of fallopian tube secretion in the Romney ewe. J. Reprod. Fertil. *15*, 357.

Brewer, D.J. and Wells, M.E. (1977). Effects of in vitro incubation of bovine spermatozoa in bovine follicular fluid. J. Anim. Sci. *44*, 262.

Dailey, R.A., Clark, J.R., Staigmiller, R.B., First, N.L., Chapman, A.B. and Casida, L.E. (1976). Growth of new follicles following electrocautery in four genetic groups of swine. J. Anim. Sci. *43*, 175.

England, B.G., Karavolas, H.J., Hauser, E.R. and Casida, L.E. (1973). Ovarian follicular estrogens in Angus heifers. J. Anim. Sci. *37*, 1176.

Erb, R.E., Randel, R.D., and Callahan, C.J. (1971). Female sex steroid changes during the reproductive cycle. J. Anim. Sci. Suppl. 1, *32*, 80. (IX Biennial Symposium on Animal Reproduction)

Erickson, B.H., Reynolds, R.A. and Murphree, R.L. (1976). Ovarian characteristics and reproductive performance of the aged cow. Biol. Reprod. *15*, 555.

Ford, S.P., Wever, L.J. and Stormshak, F. (1976). In vitro response of ovine and bovine uterine arteries to prostaglandin F_2 and periarterial sympathetic nerve stimulation. Biol. Reprod. *15*, 58.

Gemmell, R.T., Stacy, B.D. and Thorburn, G.D. (1976). Morphology of the regressing corpus luteum in the ewe. Biol. Reprod. *14*, 270.

Hansel, W., Concannon, P.W. and Lukaszewska, J.H. (1973). Corpora lutea of the large domestic animals. Biol. Reprod. *8*, 222.

Kiddy, C.A., Mitchell, D.S., Bolt, D.J. and Hawk, H.W. (1978). Detection of estrus-related odors in cows by trained dogs. Biol. Reprod. *19*, 389.

Moor, R.M., Hay, M.R. and Seamark, R.F. (1975). The sheep ovary: regulation of steroidogenic, haemodynamic and structural changes in the largest follicle and adjacent tissue before ovulation. J. Reprod. Fertil. *45*, 595.

Niswender, G.D., Reimers, T.J., Diekman, M.A. and Nett, T.M. (1976). Blood flow: A mediator of ovarian function. Biol. Reprod. *13*, 381.

O'Shea, J.D. and Phillips, R.E. (1974). Contractility *in vitro* of ovarian follicles from sheep and the effects of drugs. Biol. Reprod. *10*, 370.

Perkins, J.L., Goode, L., Wilder, W.A., Jr. and Henson, D.B. (1965). Collection of secretions from the oviduct and uterus of the ewe. J. Anim. Sci. *24*, 383.

Pharriss, B.B. (1970). The possible vascular regulation of luteal function. Perspect. Biol. Med. *13*, 434.

Rexroad, C.E., Jr. and Barb, C.R. (1978). Contractile response of the uterus of the estrous ewe to adrenergic stimulation. Biol. Reprod. *19*, 297.

Rexroad, C.E., Jr. and Casida, L.E. (1976). Ovarian follicular atresia and follicular estradiol-17β after unilateral ovariectomy in pregnant gilts. J. Anim. Sci. *43*, 802.

Rexroad, C.E., Jr. and Casida, L.E. (1977). Effect of injection of progesterone into one ovary of PMSG-treated anestrous ewes on follicle growth and ovarian estradiol-17β. J. Anim. Sci. *44*, 84.

4

Functional Histology of Reproduction

E.S.E. HAFEZ

FEMALE REPRODUCTIVE ORGANS

Ovary

The ovary consists of an outer zone, the "cortex," and an internal zone, the "medulla" (Baker, 1972). The germinal epithelium on the ovarian surface is cuboidal (Fig. 4–1, A). The ovarian follicles are in different stages of development (Table 4–1).

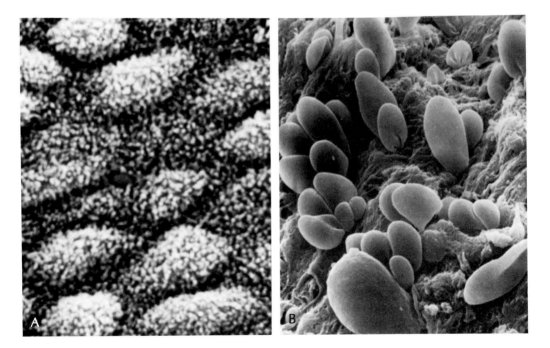

Fig. 4–1. *A*, Scanning electron micrographs of superficial (germinal) epithelium of the ovary. Note cobblestone arrangements and abundance of microvilli. *B*, Corpus hemorrhagicum. Note red blood cells protruding from the surface (5,040×. Courtsey Dr. S. Makabe)

Table 4–1. Functional Histology of the Mammalian Ovary

Anatomical Functional Unit	Histologic Characteristics
Superficial epithelium	A surface layer of flattened epithelium (commonly and incorrectly known as germinal epithelium) with abundant microvilli.
Tunica albuginea	Dense, fibrous connective tissue covering the whole ovary just beneath the superficial epithelium. Connective tissue cells near the surface are arranged roughly parallel to the ovarian surface and are somewhat denser than the cells lying toward the medulla.
Ovarian cortex	Contains several primary follicles (with oocytes in a quiescent state) and a few large follicles. During each estrous cycle, a variable number of follicles undergo rapid growth and development, culminating in the process of ovulation.
Ovarian medulla	Loose connective tissue that contains nerves, lymphatics, and tortuous thin-walled blood vessels, collagen and elastic fibers, fibroblasts and scattered smooth muscle cells.
Ovarian stroma	Poorly differentiated, embryonal-mesenchymal-like cells capable of undergoing complex morphologic alterations during the reproductive life; stromal cells can give rise to theca interna cells or interstitial cells.
Smooth muscle	Smooth muscle cells are present throughout the ovary, especially in the cortical stroma where they intermingle with the cells of the theca. Ovarian myoid cells are similar to smooth muscle cells of other tissues. These cells contain large numbers of microfilaments arranged in characteristic bundles and also micropinocytotic vesicles located just beneath the plasmalemma. The cholinergic receptors on smooth muscle cells may mediate the contraction of the graafian follicle. Thus, smooth muscle cells and neural elements may be directly involved in ovulation. The presence of smooth muscle cells, especially in the perifollicular regions, may be involved in "squeezing the follicle" during ovulation.
Ovarian Follicles Primary follicle	An oocyte enclosed by a single layer of flattened or cuboidal follicular cells and surrounded by interstitial tissue.
Growing follicle	Oocyte with increased diameter and increased number of layers of follicular cells; zona pellucida is present around the oocyte.
Secondary follicle	Flattened granulosa cells of the primordial or unilaminar follicle proliferate, assuming an irregular, polyhedral appearance.
Tertiary (vesicular) follicle	Under the influence of pituitary gonadotropins, the granulosa cells of multilayered follicles secrete a fluid, the liquor folliculi, which accumulates in the intercellular spaces. The continued secretion and accumulation of liquor folliculi result in the dissociation of granulosa cells, which causes the formation of a large, fluid-filled cavity—the antrum. The zona pellucida is surrounded by a solid mass of radiating follicular cells, forming the corona radiata.
Graafian follicle	Follicular cells increase in size; the antrum is filled with follicular fluid (liquor folliculi). The oocyte is pressed to one side, surrounded by an accumulation of follicular cells (cumulus oophorus); elsewhere in the follicular cavity an epithelium of fairly uniform thickness called the membrana granulosa has formed. The theca interna and theca externa have formed.
Preovulatory follicle	A blister-like structure protruding from the ovarian surface due to rapid accumulation of follicular fluid and thinning of the granulosa layer. The viscous liquor folliculi is formed from the secretions of granulosa cells and plasma proteins transported into the follicle by transudation. The cumulus oophorus is detached from the thinned and extensively dissociated stratum granulosum. The oocyte lies free in the liquor folliculi, surrounded by an irregular mass of cells. Dramatic changes are noticeable at the subcellular level, particularly in the Golgi complex, which is involved in the formation of the zona pellucida and cortical granules. The oocyte, in the prophase of meiosis (dictyate stage), resumes meiosis several hours before ovulation. The first meiotic (maturational) division is associated with extrusion of the first polar body, which may briefly remain attached to the oocyte by a cytoplasmic bridge.
Corpus hemorrhagicum	The follicular cavity is filled with lymph and blood from broken thecal vessels, blood from follicular fluids and blood from small vessels that hemorrhaged at ovulation. It acts as a "stopper," sealing the residual cavity after discharge of oocyte. Intact vessels and connective tissue cells from the surrounding theca begin to proliferate.
Corpus luteum	The transformation of a ruptured follicle into a corpus luteum involves characteristic folding of the granulosa layer toward the central portion of the residual cavity; luteinization of the granulosa cells occurs under the influence of LH. Lutein cells, polyhedral in form or without definite cell walls, are arranged in irregular masses. The cytoplasm may be clear or granular according to the secretory and functional activity.
Corpus albicans	White fibrous tissue that forms from the corpus luteum of previous ovulations.

Follicular Cells. In order to remain alive and viable, the germ cells must be enclosed by follicle cells. Contact between the oocyte and the follicle cells is maintained within the developing ovary even

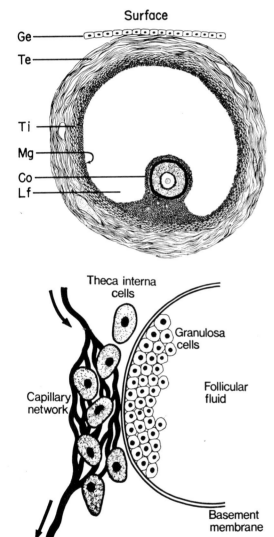

Fig. 4–2. *(Top)* Illustration of a graafian follicle. *Co,* Cumulus oophorus; *Ge,* germinal epithelium; *Lf,* liquor folliculi; *Mg,* membrana granulosa; *Te,* theca externa; *Ti,* theca interna. *(Bottom)* The structure of the wall of the graafian follicle showing how the granulosa cells are deprived of a blood supply by the basement membrane. *(Baird, 1972. In: Reproduction of Mammals. C.R. Austin and R.V. Short [eds.], Cambridge, Cambridge University Press.)*

though the deposition of the zona pellucida induces a topographic separation (Fig. 4–2). The follicle cell and the oocyte are connected by cytoplasmic processes that originate from the follicle cell and contain mitochondria, lipid droplets and pleomorphic bodies reminiscent of lysosomes. These projections terminate in either the perivitelline space or at the oolemma (Fig. 4–3). While the projections within the perivitelline space degenerate, the other projections become surrounded by oocyte microvilli and are connected to the oocyte by desmosomes.

Although the association of the oocyte with follicle cells is essential for the maintenance of the oocyte, follicle cell contact does not ensure the survival of the oocyte, because oocytes within cystic follicles are enclosed by follicle cells but are atretic. In addition, grossly degenerative oocytes are found within apparently "healthy" antral follicles, which suggests that the oocyte deteriorates prior to the follicle.

The functions of granulosa cells and the maturational process of the oocyte are coordinated by intercommunication between cells within the membrana granulosa and between granulosa cells and the oocyte. The integrity of the junctional complexes may be estrogen-dependent, thus indicating a possible intrafollicular relationship between the steroidogenic function of the thecal cells and the maturation of the oocyte.

Degeneration of Oogonia. Oogonia have three phases of degeneration: phase 1—chromatin condenses at the periphery of the nucleus with no effect on the cytoplasmic organelles; phase 2—a dense ring of chromatin develops at the margin of the nucleus and the cytoplasmic organelles swell and rupture, resulting in cytoplasmic vacuolization; phase 3—granulosa cells often invade and subsequently phagocytize the atretic oogonia.

Corpus Luteum. After ovulation, the corpus luteum (Fig. 4–1, *B*) differentiates fully into a typical endocrine gland, se-

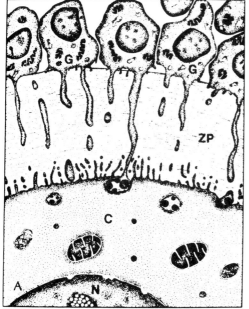

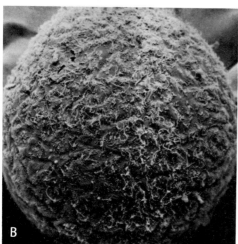

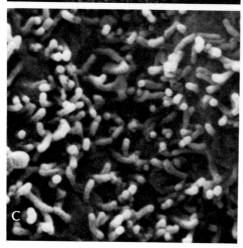

creting progesterone. The steroid hormone secreted by the corpus luteum has several functions: (a) it prepares the endometrium for implantation, (b) it enables the blastocyst to implant, and (c) it maintains pregnancy after implantation (Hansel et al, 1973). The functional lifespan of the corpus luteum, which varies with the species, depends primarily upon the fate of the embryo. Following implantation of the blastocyst, the corpus luteum undergoes extensive enlargement and continues to synthesize and secrete steroid hormones (the corpus luteum of pregnancy).

Female Reproductive Tract

Remarkable differences exist in the pattern of tissue organization of surface epithelium of different segments of the reproductive organs (Hafez, 1978 a, b). In most organs, the cells are uniform in size, closely packed, and resemble cobblestones. In some instances, the surfaces of the cells are ill-defined and covered with short microvilli and residual mucus.

The oviduct, uterus and cervix are composed basically of secretory cells and ciliated cells. The secretory cells secrete the luminal fluids necessary for the transport and nutrition of spermatozoa and eggs and for cleavage of fertilized eggs and development of preimplantation blastocysts.

Ciliated cells are found singly or in groups arranged in rows or a mosaic pattern. Cilia first grow at the periphery of the cell and then fill in the central area. Cilia appear first as stubby, short cylin-

Fig. 4–3. *A,* Structure of fully formed zona pellucida *(ZP)* around an oocyte in a graafian follicle. Microvilli arising from the oocyte interdigitate with processes from the granulosa cells *(G).* These processes penetrate into the cytoplasm of the oocyte *(C)* and may provide nutrients and maternal protein; *(N)* oocyte nucleus. *(Baker, 1972. In: Reproduction in Mammals. C.R. Austin and R.V. Short [eds.], Cambridge, Cambridge University Press.)*
B, Egg after removal of zona pellucida.
C, Microvilli shown at a higher magnification.

ders on those cells that have large numbers of microvilli on their surfaces.

In the female reproductive tract, kinocilia play an important role in the transport of particles and in directing the flow of luminal fluids. Two types of ciliary motility have been recognized: an effective stroke and a recovery stroke. In the effective stroke the cilia sweep rapidly in a still and slightly wavy motion. In the recovery stroke the cilia bend near the basal body and the degree of bending proceeds as a slow wave toward the tip. The recovery stroke carries the cilia back to the effective stroke position. Cilia beat approximately 1200 times/minute. The kinocilia in the female reproductive tract beat rhythmically toward the vagina creating a directional flow of luminal fluids.

Oviduct

The wall of the oviduct is composed of mucosa, muscularis and serosa (cf. Hafez and Blandan, 1969).

Oviductal Mucosa. The oviductal mucosa is made of primary, secondary and tertiary folds (Fig. 4–4). The mucosa in the ampulla is thrown into high, branched folds that decrease in height toward the isthmus, and become low ridges in the uterotubal junction. The complex arrangement of these mucosal folds in the ampulla almost completely fills the lumen so that there is only a potential space. Fluid is at a minimum; thus, the cumulus mass is the intimate contact with the ciliated mucosa.

The mucosa consists of one layer of columnar epithelial cells. The underlying submucosa of smooth muscle fibers and connective tissue are permeated by fine blood and lymph vessels. The epithelium contains ciliated and nonciliated cells, together with "peg cells" (Stiftzellen or intercalary cells), which are presumably depleted secretory cells (Fig. 4–5). The peg cells are long, rod-like slender cells with a dark compressed nucleus and luminal surface indentations with practically

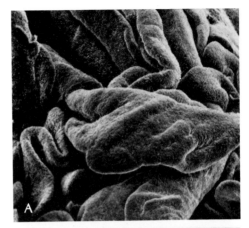

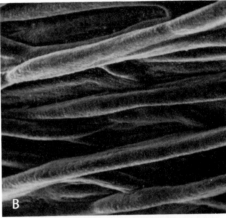

Fig. 4–4. Scanning electron micrographs of the oviductal epithelium.
A, Mucosal folds that protrude in the lumen of ampulla (37 ×).
B, Mucosal folds that protrude in the lumen of the isthmus (40 ×).
C, Ciliated and secretory cells near the ampullary isthmic junction (1,460 ×).

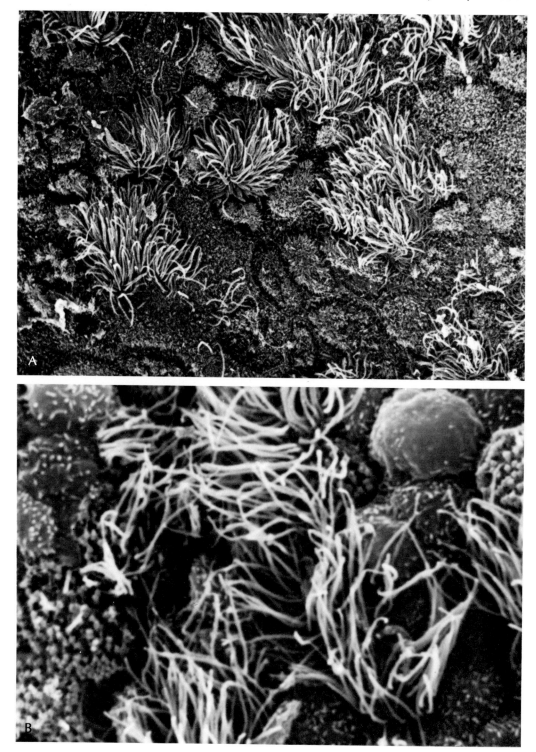

Fig. 4–5. Scanning electron micrograph of the oviductal epithelium in the isthmus (*top*, 6,000×) and uterotubal junction (*bottom*, 10,000×). Note the cilia protruding over the surface of secretory cells undergoing a cell cycle (*bottom*) with varying number of microvilli.

no cytoplasm. They are usually located in the basal portions of the folds near secretory cells, and are less common at the uterine end of the oviduct. The oviductal epithelium undergoes morphologic and cytologic changes during the estrous cycle.

Ciliated Cells. The ciliated cells of the oviductal mucosa have a slender motile cilia (kinocilia) that extend into the lumen (Fig. 4–6). The cilia conform to the standard structural plan of kinocilia, with nine pairs of fibrils arranged concentrically around another central pair. Ciliated cells are characterized by a fine granular cytoplasm containing vesiculated endoplasmic reticulum, large mitochondria and cytoplasmic droplets. Large, centrally placed, round or oval nuclei are surrounded by a perinuclear halo of clear cytoplasm.

The apical surfaces of both ciliated and secretory cells are covered by microvilli that are uniformly distributed. Microvilli found on secretory cells tend to be long and slender; on ciliated cells, these surface modifications are usually hidden by cilia in many histologic secretions because of the angle of the secretion.

The rate of beat of cilia is affected by the levels of ovarian hormones, activity being maximal at ovulation or shortly afterward when the stroke of the cilia in the fimbriated portion of the oviducts is closely synchronized and direction-oriented toward the ostium. The action of ciliary beat seems to enable the egg within the surrounding cumulus cells to be stripped from the surface of the collapsing follicles toward the ostium of the oviduct. The percentage of ciliated cells decreases gradually in the ampulla toward the isthmus, and reaches a maximum in the fimbriae and infundibulum. Ciliated cells are noted in large numbers at the apices of the mucosal folds. Variations in the percentage of ciliated and secretory cells along the length of the oviduct have some functional significance. Ciliated cells are most abundant where the egg is picked up from

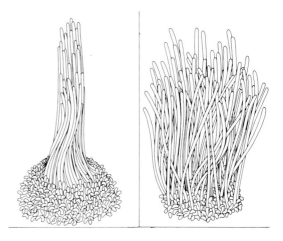

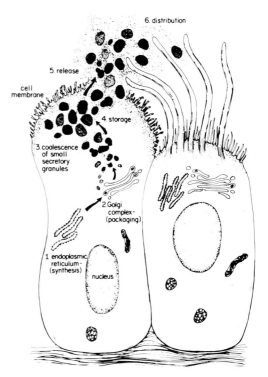

Fig. 4–6. Line drawings of ciliated and secretory cells, the main cellular components of the oviductal and uterine epithelium.
(Top) Ciliated cells from the oviduct (right) and uterus (left). Note presence of microvilli on the apical surface of cell.
(Bottom) Secretory cell showing the biosynthesis, packaging, storage, release and distribution of secretory material, which is the main component of the luminal fluid in the oviduct and uterus. It is believed that the action of kinocilia facilitates the release of secretory granules from the surface of secretory cells. *(Hafez et al., 1976, Am. J. Obstet. Gynecol., and Hafez et al., 1977, Int. J. Fertil.).*

the ovarian surface, whereas secretory cells are abundant where luminal fluids are needed as a medium for the interaction of eggs and spermatozoa.

Although the ciliated cells in the ampulla appear to arise from crypt-like crevices because of the apical protrusion of the nonciliated cells, the ciliary tips have free access of movement into the lumen. In the isthmus there are a few ciliated cells located deep between the tall secretory ones. The tips of their cilia hardly reach the oviductal lumen and thus exert no effect on egg transport. The cilia beat toward the uterus. Their activity, coupled with oviductal contractions, keeps oviductal eggs in constant rotation, which is essential for bringing egg and sperm together (fertilization) and preventing oviductal implantation.

Ciliation of the oviduct is hormonally controlled in the rhesus monkey: cilia disappear almost completely after hypophysectomy and develop in response to administration of exogenous estrogens. The oviducts atrophy and deciliate during anestrus, hypertrophy and become reciliated during proestrus and estrus, and atrophy and deciliate during pregnancy.

Infections of the female reproductive tract are associated with dramatic changes in cell morphology. Infection is usually associated with the loss of ciliated cells. Inflamed oviductal epithelium has ciliated cells devoid of cilia that retain a certain number of ciliary basal bodies, the precursors of ciliary shafts. A decrease in the number of cilia may lead to the accumulation of tubal fluid and inflammatory exudate, which contributes to the agglutination of tubal plicae and subsequent development of salpingitis.

Secretory Cells and Fluid. The secretory cells of the oviductal mucosa are nonciliated and characteristically contain secretory granules, the size and number of which vary widely among species and during different phases of the estrous cycle. The apical surface of the nonciliated

cells is covered with numerous microvilli (Figs. 4–6, 4–7). The structure of the secretory cells, the epithelial height and the secretory activity change throughout the estrous cycle, with maximal changes occurring near the time of ovulation. Secretory granules accumulated in epithelial cells during the follicular phase of the cycle are released into the lumen after ovulation, causing a reduction in epithelial height. These variations generally become less pronounced moving from the infundibulum to the isthmus. There are quantitative and qualitative variations in the secretory granules in the same cell during different stages of the reproductive cycle. The release of secretory granules from the nonciliated cells of the oviducts is programmed by gonadal hormones.

Definite cytologic changes during the follicular phase precede secretory activity. The endoplasmic reticulum of the secretory cells spreads out, the mitochondria swell as their matrices fill with a granular substance, and the Golgi apparatus becomes well developed. During the luteal phase, numerous secretory droplets appear, the endoplasmic reticulum dilates, mitochondria become fewer and the Golgi apparatus expands.

The metabolic activity of the epithelium of the oviduct appears to be most pronounced in the vicinity of the egg. In the rabbit, for example, the uptake of radioactive sulfate is always highest in the segment that contains the egg as it passes along toward the uterus.

By means of extra- and intraabdominal devices that are used to cannulate the oviduct and collect its fluids, it has been possible to show that ovarian hormones also regulate the secretory activity of the epithelium of the oviduct (Perkins et al, 1965). The secretions are composed predominantly of mucoproteins and mucopolysaccharides.

Muscularis and Serosa. The tunica muscularis consists of an inner circular layer and an outer longitudinal layer of

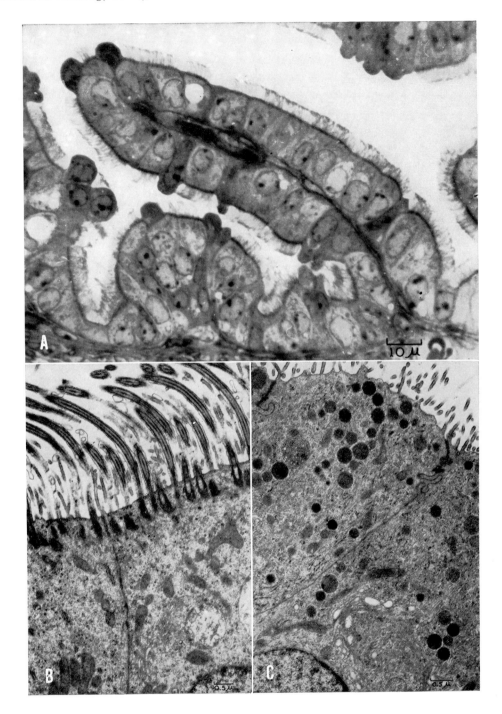

Fig. 4–7.   Structure and ultrastructure of mammalian oviduct. (Scale in microns).
A, Infundibulum of mouse oviduct, day 1 of pregnancy. Ciliary cells dominate and among them secretory cells ("peg cells") can be seen. Toluidine blue staining of a plastic section.
B, Ciliary cells in the infundibulum of mouse oviduct, day 1 of pregnancy (time of egg passage).
C, Secretory cells in the isthmus of mouse oviduct, day 2 of pregnancy (time of egg passage). Secretory granules of varying density are observed in the apical part of the cells. *(Photographs by S. Reinius).*

smooth (unstriated) muscle (Fig. 4–8). The contractility of oviductal musculature, under hormonal control, regulates in part the transport of eggs and spermatozoa.

Uterus

The uterus consists of a thin outer layer, the "perimetrium," a thick "myometrium" composed of inner circular and outer longitudinal smooth muscle layers, and an inner layer, the "endometrium." The changes that occur in the endometrium during the estrous cycle prepare the uterus to receive the blastocyst and therefore play a major role in the reproductive process.

Endometrium. The endometrium is a highly glandular structure consisting of an epithelial lining of the lumen, a glandular layer and connective tissue. It varies in thickness and vascularity with changes in ovarian hormones throughout the estrous cycle and pregnancy.

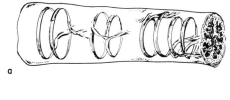

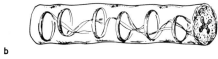

Fig. 4–8. Diagrammatic illustration of the musculature in the oviduct of ungulates.
A, Ampulla: the musculature consists of spiral fibers arranged almost circularly. B, Isthmus: note differences in morphology of muscle fibers. C, Uterotubal junction: note the longitudinal muscle coat of uterine origin, as well as peritoneal fibers. *(Schilling, 1962. Zentralbl. Veterinaermed. 9, 805)*

The columnar epithelial cells that line the uterine lumen play an important role in the interaction between the blastocyst and endometrium during the initial stages of pregnancy. To facilitate the attachment of the blastocyst, the luminal epithelial cells undergo structural changes that are regulated by progestins and estrogens.

The endometrium is characterized by the presence of numerous openings of the endometrial glands (Figs. 4–8, 4–9). Ciliated cells are less abundant in the endometrium than the oviductal epithelium. Large cytoplasmic projections are noted at the apical membrane of the nonciliated cells. The abundance, length, shape and interbranching of apical microvilli vary throughout the cycle. The development of apical microvilli, the synthesis, storage and release of endometrial secretory granules and ciliogenesis are hormone-dependent. Degenerated cells are found at random at different stages of the estrous cycle. Following the release of secretory material from the protruding cell membrane, the cell collapses and becomes wrinkled and devoid of microvilli.

Caruncles. The inner surface of the uterus in ruminants contains nonglandular projections, the *caruncles*. They are arranged in four rows, extending from the uterine body into the two uterine horns, and consist of connective tissue comparable to that found in the cortical stroma of the ovary. The deeper areas between the prominences are rich in blood vessels but contain no glands.

The uterus of the nonpregnant cow has 70 to 120 caruncles, each measuring approximately 15 mm in diameter. During pregnancy, they may attain a diameter of 10 cm and appear spongy due to the numerous crypts that receive the placental chorionic villi. These villi develop in localized areas, the *cotyledons*, which invade the caruncles. The cotyledons and caruncles are referred to together as the *placentome*.

In horses and swine, the uterus has no

caruncles. Instead, the mucosa is characterized by conspicuous longitudinal folds that pass into the cervix to form the internal and external orifices.

Endometrial Glands. The endometrial glands are simple branched tubular glands that are more or less coiled, especially toward their ends. The glands possess a lamellar connective tissue and are lined by simple or ciliated columnar epithelium.

These glands are scattered throughout the endometrium with the exception of the caruncles. They are branched, tubular and coiled, especially toward the ends. Their density varies with the species, breed, parity and estrous cycle. The variation in proximity to one another during the estrous cycle is largely a result of changes in diameter and amount of stromal ground substance. The number of glands is higher in the horns than in the

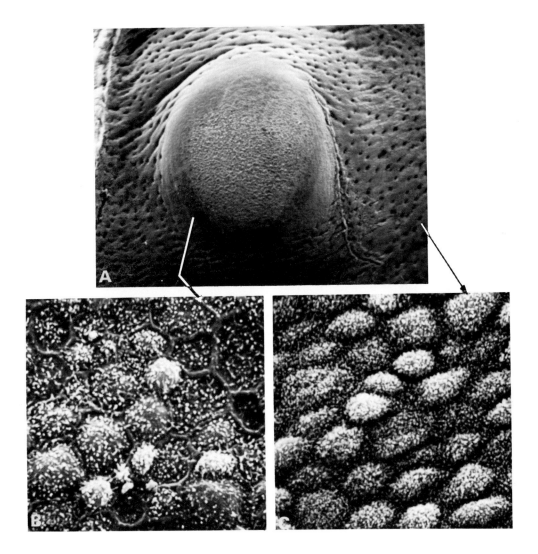

Fig. 4–9. Scanning electron micrographs of the endometrium in the ewe. *A*, Caruncle surrounded by openings of endometrial glands. *B*, Endometrial cells (2,000×) from the perimeter of the caruncles have distinct borders, and are twice as large as endometrial cells around the openings of endometrial glands. *C*, (5,000×).

mucosa adjacent to the cervix. The glands may be rapidly increased by budding and outward growth from the basal zone. Buds that do not reach the surface become dilated and cystic.

The cyclic shedding of the endometrial surface epithelium followed by regeneration causes no extensive bleeding, such as that observed in primates in which a substantial portion of the endometrium is shed. During "metestrus bleeding" in the cow, the caruncles show a pronounced capillary distention, but the epithelium remains intact.

Myometrium. The myometrium is the muscular portion of the uterine wall. It consists of two layers of smooth muscle: a thick inner circular layer and a thinner outer longitudinal layer. Between them lies a vascular layer made of blood and lymph vessels, nerves and connective tis-

sue. This layer is not distinct in the sow and mare, and may even occur within the circular layer in ruminants. During pregnancy the amount of muscle tissue in the uterine wall increases, both by cell enlargement and cell number.

The serosa (perimetrium), the stratum musculare and the layer of longitudinal muscle are all continuations of the broad ligament, which invests and suspends the uterus.

Cervix

The cervix is a sphincter-like structure that projects caudally into the vagina. It is characterized by a thick wall, a constricted lumen, and transverse or spirally interlocking ridges that develop to varying degrees in the different species. These ridges are especially prominent in rumin-

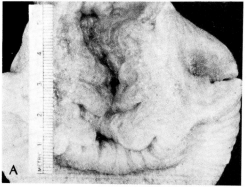

Fig. 4–10. *A,* Cervix (cut open) of a heifer 4 days after estrus. Note the right annular folds around the cervical canal.
B, Bovine cervical canal when the cervix was cut open. Note the complexity of cervical crypts. (59×).
C, Bovine cervical epithelium showing the kinocilia overlapping on secretory cells. (600×).

ants, in which they fit into each other to close the cervix securely.

Cervical Mucosa. The cervix does not contain any glands, as is assumed erroneously. The cervical mucosa is thrown into primary and secondary mucosal crypts (Fig. 4–10) that provide an extensive secretory surface. These intricacies give the cervix its typical "fern leaf" microscopic appearance. The cervical mucosa is made of two types of columnar epithelial cells: ciliated cells with kinocilia and nonciliated secretory cells (Fig. 4–11). The kinocilia beat toward the vagina and the nonciliated cells contain massive numbers of secretory granules. The greatest secretory activity of these cells occurs at estrus (Fig. 4–12).

The cervical crypts are more developed in the cow than in other farm animals. The mucosa lies in large folds, which are tallest and thickest in the cow and thinnest in the ewe. The folds vary in height, particularly within the cow and goat cervix. In all three species, the cervical folds are taller, more numerous and less regular

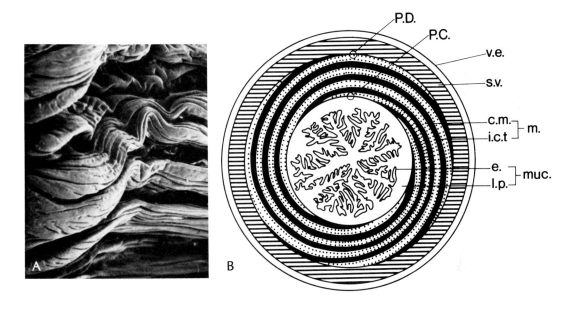

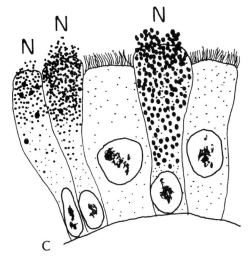

Fig. 4–11. *A*, Scanning electron micrograph showing complex arrangement of cervical folds in the ewe. (12×).

B, Cross section in the cervix showing the complex arrangement of several layers of circular musculature *(c.m.)* with interconnective tissue *(i.c.t.)*. Note the interwoven layers of tissues. P.D., P.C., v.e. and s.v., various histologic layers.

C, Diagrammatic illustration of epithelial cells of the cervix showing two types of cells—ciliated cells with kinocilia and nonciliated cells with abundant secretory granules (N). *(Hafez et al, 1972. Acta Biochim. 39, 195).*

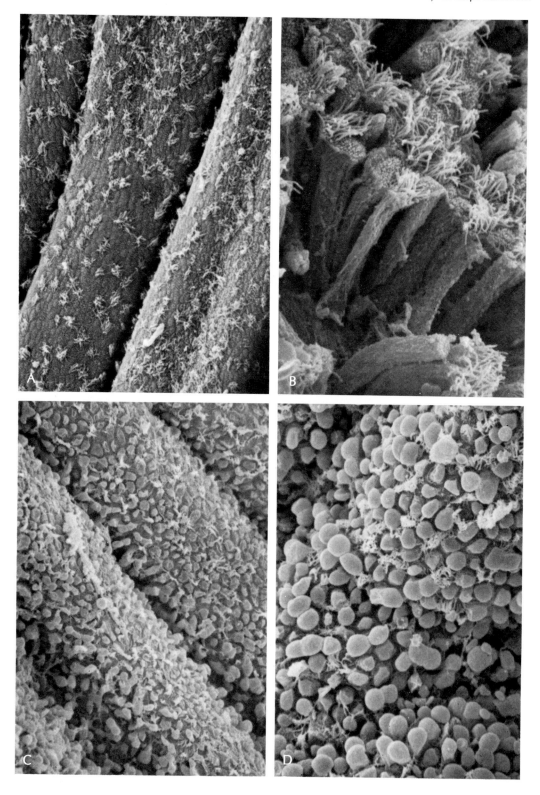

in size in the cranial part of the cervix than in the caudal part.

At estrus, the cells in the bases of the folds and the secondary indentations act as if they are more responsive to estrogenic stimulation than the cells near the lumen (Heydon and Adams, 1979). This has implications for sperm transport, because the mucus coming out of cells is believed to guide spermatozoa along the "lines of strain" down to the mucosa near its origin. A greater mucous response to estrogen in the bases of the folds and secondary indentations will result in more sperm being guided specifically into these areas, thus aiding in the physical entrapment of spermatozoa. There is evidence that the biophysical characteristics of cervical mucus vary in various parts of the cervical crypts. Using histochemical techniques, sulfomucins occur in cells of the cervical folds closest to the lumen, while in the cells in the crypts or the secondary indentations mucus is more abundant and stains for sialomucins (Heydon and Adams, 1979). The epithelium of the cervix of ovariectomized females contains only a small amount of mucus, which stains for sulfomucin. Treatment with estrogen produces large amounts of mucus in the cells, and this stains as sialomucin.

Cervical Wall. The cervical wall consists primarily of fibrous, elastic and collagenous tissue and a small amount of muscle. The connective tissue is made of fibrous constituents, cellular components and ground substance. Collagen, the principal fibrous element, has a high tensile strength. Collagenase, a specific degenerative enzyme for collagen, may be involved in the process of labor. The reticular network is an intimate part of the collagen system. The mass of the cervix increases as gestation advances and, although involution occurs after parturition, part of the gain in mass is maintained.

Vagina

The normal squamous epithelium of the lower vagina is smooth, with little undulation of the surface architecture. The cells are flat, polygonal, with demonstrable but thin-edged interdigitating borders. The cells, in multiple layers, overlie each other irregularly, much like haphazardly layered shingles on a roof. The edges of these cells roll back, giving the characteristic picture of exfoliation as they lift up from their borders. The exfoliated cells appear wrinkled as if dried up, with lost or obscured surface microridges.

The surface of the vaginal cell is made of numerous microridges that run longitudinally or in circles. In this multilayered stratified epithelium, the cells are wedged one upon the other by interlocking opposed microridges, thus form-

Fig. 4–12. Scanning electron micrographs of the cervix in the ewe. *A,* Anterior portion of the cervix illustrating the highly convoluted structure of this organ. Although the exposed surface of the lumen can be examined, the folds or crypts cannot be observed unless this tissue is further segmented. (500×)
B, Portion of the anterior cervix that was mechanically damaged during preparation. As a consequence of the damage, which resulted in the loss and separation of cells, the shape and structure of individual columnar epithelial cells can be observed more clearly. These cells, which are attached to a basement membrane, are elongate but normally have only the external membrane surfaces of their apices exposed to the lumen of the cervix. (2,000×)
C, Portion of the cervix from a ewe. The apical portions of the secretory cells, which were present during the luteal-phase, have become dilated. This development appears to be synchronous but, at any one time, is confined to distinct regions that may encompass several folds. (600×)
D, The dilated apical portions of the secretory cells frequently exceed the height of adjacent ciliated cells, which comprise about 30% of the surface area at estrus. (1,000×) (Courtesy Dr. W.P. Wergin, Reproduction Laboratory, U.S. Department of Agriculture, Beltsville, Maryland)

Table 4–2.   Functional Histology of the Testis

Segment	Histologic Characteristics
Tunica albuginea	A thick, white capsule of connective tissue surrounding the testis; made primarily of interlacing series of collagenous fibers with elongated and flattened nuclei of the fibroblasts.
Seminiferous tubules	Appear as large isolated structures, round or oblong in outline; varying appearance due to the complex coiling of the tubules at many different angles and levels. Between the tubules are blood vessels of variable diameter (smaller than tubules) packed with erythrocytes and embedded in small masses of interstitial (Leydig) cells, which produce the male sex hormones. Strands of connective tissue form a thin layer around each tubule.
Spermatogonium	Lie in the outermost region of the tubule; round nuclei appear as an irregular layer just within surrounding connective tissue. Nuclei may be recognized by small size and dark stain due to presence of large numbers of chromatin granules.
Primary spermatocytes	Located just inside an irregular layer of spermatogonia and Sertoli cells; nuclei are noticeably larger than those of the spermatogonia and stain lighter, although they still contain a considerable amount of granular chromatin.
Secondary spermatocytes	Maturation divisions and secondary spermatocytes are not seen in the average tubule owing to the short duration of these stages.
Spermatid	Located internally to primary spermatocytes. Cells are small and round with light-staining nuclei. Layer of spermatids may be several cells in thickness. Spermatozoa lie along the border of the lumen; their long tails extending into the cavity as filamentous structures, their heads appearing as dark-staining small points or short lines. The sperm heads are lodged in deep indentations of the surface of the Sertoli cell, but never actually within the cytoplasm.
Sertoli cells	Large and relatively clear except for the prominent, dark-staining nucleolus. Cytoplasm is diffuse and its limits are indefinite.

ing a firm surface. The morphology and pattern of these microridges, which affect the firmness of the epithelium, vary throughout the reproductive cycle. The microridges exhibit a regular pattern during pregnancy, while pores appear within the microridges of the cells during the estrous cycle.

Sequential changes in the cell types present in the vaginal smears are unreliable as the sole means of detecting ovulation in ewes.

Vulva

The vulva, made of the labia majora and minora, the clitoris and the vaginal introitus, is covered by stratified squamous epithelium similar to all other skin areas, with some follicles and sebaceous and sweat glands. The surfaces of the cells show the usual interlacing microridges. The edges of these cells are thin in con-

trast to the slightly raised terminal bars on the squamous cells of the upper vagina.

MALE REPRODUCTIVE ORGANS

Extensive investigations have been conducted on the anatomy, physiology and biochemistry of the mammalian testis (Johnson et al, 1970), epididymis (Martan, 1969) and male accessory organs (Spring-Mills and Hafez, 1979, 1980).

The histologic and cytologic characteristics of the cellular components of the seminiferous tubules are summarized in Table 4–2 and illustrated in Figures 4–13 and 4–14. While in the epididymis, the spermatozoa, helped by the secretions of the epithelium, approach full maturity and increase their motility. They move through the duct from the caput epididymidis to the cauda epididymidis where they are transported to the vas deferens.

The histologic characteristics of the

male accessory organs are illustrated in Figures 4–14 and 4–15. Secretions are stored in the lumen of the male accessory glands. Following the emptying of the glands, there is a continuous accumulation of secretion in the lumen. The accessory glands are continuously active in synthesizing and secreting protein, the secretion either being used during copulation or lost by spontaneous ejaculation.

Epididymis

The efferent ducts connect the rete testis to the epididymal duct, which follows a continuous but highly convoluted course. The epididymis, more than 120 feet long in the bull and about 180 feet long in the boar, is divided macroscopically into three gross regions: head (caput or initial), body (corpus or middle) and tail (cauda or terminal). There is a progressive decrease in the height of the epithelium and stereocilia and a widening of the lumen throughout the three segments. The first two segments are concerned with sperm maturation whereas the terminal segment is for sperm storage. The epithelial lining of the epididymis consists of different kinds of cells: princi-

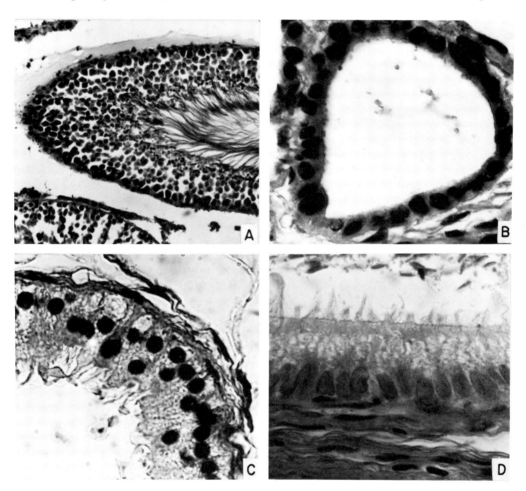

Fig. 4–13. *A,* Seminiferous tubule of a mature rat (115×). *B,* Epithelium of a rete tubule of the rat (776×). *C,* Epithelium of a ductulus efferens of the rat. (776×). *D,* Wall of a ductulus efferens of the rabbit (776×). *(S. McKeever, 1970. In: Reproduction and Breeding for Laboratory Animals, E.S.E. Hafez, [ed.], Philadelphia, Lea & Febiger)*

pal columnar cells, small basal cells on basal membrane and some lymphocytes. The lumen of the epididymal tubules is lined with epithelium made of a basal layer of small cells and a surface layer of tall columnar ciliated cells. Masses of sperm are often found in the lumen. Spaces between tubules are filled with a loose connective tissue.

Vas Deferens

The mucosa is thrown into longitudinal folds. Near the epididymal end of the vas the epithelium resembles that of the epididymis: the nonciliated cells have little secretory activity. The lumen is lined with pseudostratified epithelium which may or may not be stereociliated. The

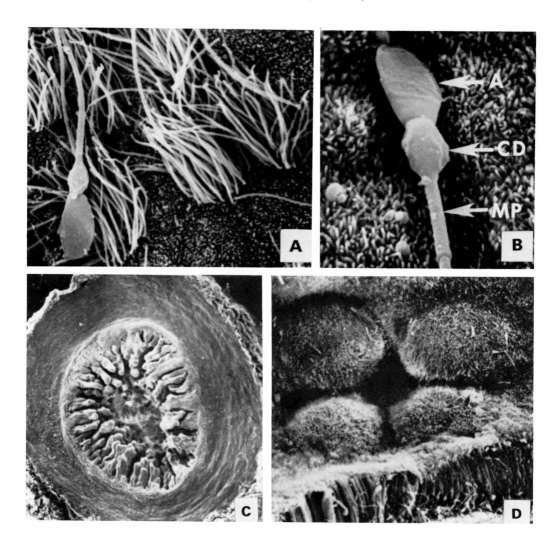

Fig. 4–14. *A*, Luminal surface of an efferent duct with ciliated and nonciliated cells and a spermatozoon.
B, Short microvilli on the luminal surface of nonciliated cells in the *efferent ducts*. The spermatozoal cytoplasmic droplet (CD), acrosome (A), and middle piece (MP) are distinguishable by SEM (6,500×).
C, Cross section of the distal cauda epididymidis. Thick layers of smooth muscle surround the highly-infolded epithelium (25×).
D, Sagittal section of the corpus epididymidis. Columnar cells are covered with stereocilia. Undulations are due to differences in cell height (276×). (Johnson, L. et al., 1978. Am. J. Vet. Res. Courtesy of Dr. Larry Johnson)

height of the epithelium is lower than that in the epididymis. The musculature consists of an inner and outer longitudinal layer and a thicker middle circular layer. The three layers are not easily recognizable because muscle fibers of the layers are interwoven. There are species differences in the thickness of the various muscle layers, the length of the duct and the presence or absence of the ampulla.

Male Accessory Organs

Ampullary Gland (Ampulla Ductus Deferentis). These glands open into the urethra. The vas deferens forms a thickening called the ampulla; this thickening is due to an increase in the size of the mucosa, which possesses branched tubular glands with sac-like dilations. The gland consists of tubules enclosed by a

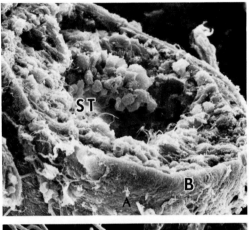

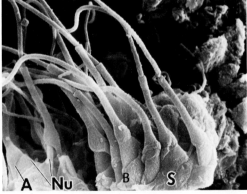

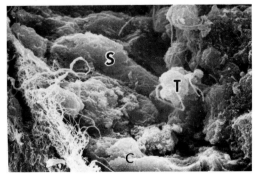

Fig. 4–15. Scanning electron micrographs of testicular components.
A, Cross section of a seminiferous tubule (ST). Note several "stages" of spermatogenesis, encased in a muscular boundary tissue (B) composed of myoid cells, collagen fibers and glycoprotein matrix (438×).
B, Clone of mid-late elongated spermatids from a canine testis. Note their insertion into the cytoplasm of Sertoli cells (S). The spermatids on the left side of the micrograph clearly show the relationship of the nucleus (Nu) and the cytoplasmic attachment (A) of the spermatid to the Sertoli cell (2,640×).
C, Conical Sertoli cells (S) extend from the basement membrane, where they are obscured by collagen fibers, to the lumen of the tubule; round spermatids with flagellae are shown at the right (T). (930×). (C.J. Connell, 1978. Spermatogenesis, Chap. 3 In: Scanning Electron Microscopy of Human Reproduction, E.S.E. Hafez [ed.], Ann Arbor Science Pubs., Ann Arbor, Michigan).

capsule of connective tissue and smooth muscle. Strands of connective tissue extend from the capsule to form septa between the tubules. Smooth muscle cells appear in the anterior septa, but disappear as the duct approaches the urethra. The ampullae, absent in the boar, are highly developed in the stallion and contribute ergothioneine—a sulfur-containing nitrogenous base—to the ejaculate.

Seminal Vesicles (Vesicula Seminalis). The seminal vesicle originates as an evagination of the vas deferens. Each consists of an elongated simple or branched tubule that is distended with fluid during the breeding season. The gland is composed of an inner secretory epithelium, a middle layer of smooth muscle, and an external layer of collagenous and elastic connective tissue. The epithelium is pseudostratified or columnar. Folds of mucous membrane unite to form a reticulated surface. Vesicles may contain a granular secretion that stains strongly with eosin (red). The muscular walls consist chiefly of circular muscles. The seminal vesicles of the bull have numerous lobes. In the stallion, elongated pear-shaped glands contribute the gel to the ejaculate. In the boar, the well-developed seminal vesicles are filled with a milky, highly viscous fluid. The dog and cat have no seminal vesicles.

Prostate Glands (Glandula Prostata). The prostate gland is present in two forms: (1) as a mass of tissue located at the vas deferens-urethral junction; this type is encapsulated and lobulated to varying degrees; it penetrates the urethral muscularis and opens into the urethra by means of one pair or more ducts; (2) as a disseminate type, which extends along the entire prostatic and most of the membranous urethra.

In the ram and the goat, the prostate is a primitive disseminate type and consists of glands that do not penetrate muscle surrounding the pelvic part of the urethra. In the bull and boar, the prostate resembles both the lobed and disseminate type. In

general, the prostate is a tubulo-alveolar gland with epithelial folds extending into the lumen. Spherical secretory granules are released from the free surface, particularly during periods of high sexual activity.

Bulbourethral Glands (Cowper's Glands, Glandula Bulbourethralis). These glands are round, compact bodies except in the boar, in which they are large, almost cylindrical and filled with a viscous, rubberlike, white secretion that forms a gel in the ejaculate. They are located dorsally to the bulb of the penis, partially or completely embedded in the bulbocavernous muscle. The compound tubulo-alveolar glands resemble mucous glands and are surrounded by a capsule of connective tissue and striated muscle. Connective tissue forms septa and trabeculae that subdivide the glands.

REFERENCES

Baker, T. (1972). Oogenesis and ovulation. *In* Reproduction in Mammals. C.R. Austin and R.V. Short (eds), Cambridge, Cambridge University Press.

Hafez, E.S.E. (ed) (1978a). Scanning Electron Microscopy of Human Reproduction, Vol. 4. Ann Arbor, Mich., Ann Arbor Science Pubs.

Hafez, E.S.E. (ed) (1978b). Human Reproductive Physiology, Vol. 5. Ann Arbor, Mich., Ann Arbor Science Pubs.

Hafez, E.S.E. and Blandau, R.J. (eds) (1969). The Mammalian Oviduct. Chicago, Chicago University Press.

Hansel, W., Concannon, P.W. and Lukaszewska, J.H. (1973). Corpora lutea of large domestic animals. Biol. Reprod. 8, 222.

Heydon, R.A. and Adams, N.R. (1979). Comparative morphology and mucus histochemistry of the ruminant cervix; differences between crypt and surface epithelium. Biol. Reprod. (In press)

Johnson, A.D., Gomes, W.R., and Van Demark, N.L. (1970). The Testis, Vol. 1. New York, Academic Press.

Martan, J. (1969). Epididymal histochemistry and physiology. Biol. Reprod. Suppl. 1, 134.

Perkins, J.L., Goode, L., Welder, W.A., Jr. and Henson, D.B. (1965). Collection of secretions from the oviduct and uterus of the ewe. J. Anim. Sci. 24, 383.

Spring-Mills, E. and Hafez, E.S.E. (eds) (1979). Accessory Glands of the Male Reproductive Tract. Ann Arbor, Mich., Ann Arbor Science Pubs.

Spring-Mills, E. and Hafez, E.S.E. (eds) (1980). Male Accessory Organs. New York, Elsevier.

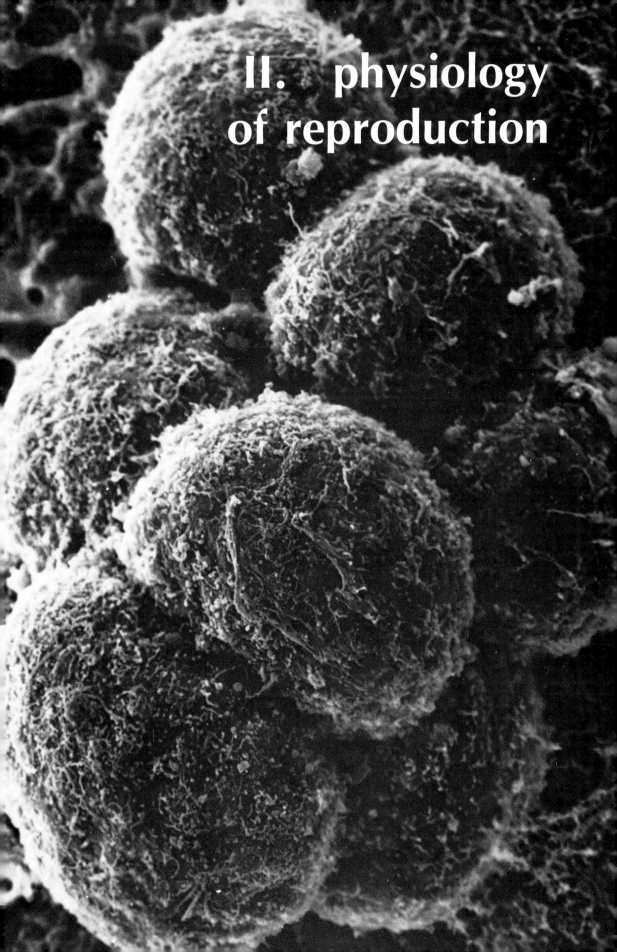

II. physiology of reproduction

5

Endocrinology of Reproduction

C.C. KALTENBACH
and
T.G. DUNN

✳   Hormones are defined classically as substances produced by endocrine glands in one part of the body that are subsequently carried by either the blood or lymph to another part of the body where they modify the activity of specific target organs. There are many hormones and these have a wide range of activities. Hormones that regulate reproductive processes are derived primarily from the hypothalamus, pituitary, gonads and placenta. Although all hormones are highly specific and selective in their action, a particular response to any hormone at a given time is often modified by the presence or absence of other hormones.

HYPOPHYSEAL HORMONES

The importance of the pituitary gland to reproductive processes was discovered early in this century (for a review of early experiments, see Greep, 1974). The "breakthrough" discoveries came when two groups of scientists found that implants of fresh anterior pituitary tissue would cause an immediate growth and maturation of the ovaries and uterus (Smith, 1926; Zondek and Aschheim, 1926). These studies were then extended to show that the gonads of immature hypophysectomized rats failed to develop (Smith and Engle, 1927). The gonads of mature animals rapidly regressed to an immature state following hypophysectomy. These findings clearly indicated that something from the pituitary gland was essential for proper function of the gonads (ie, reproduction), for growth and for resistance to stress.

Gonadotropins

Several investigators began to fractionate extracts of anterior pituitary glands and soon learned that not one but two gonad-stimulating hormones were present in extracts of the anterior pituitary (Fevold et al, 1931). These two hormones were subsequently named for their principal biologic effects. The hormone that caused ovulation and transformation of the ovarian follicle to a corpus luteum was called *luteinizing hormone* (LH). Because this hormone also stimulated the Leydig cells or interstitial cells of the testis, it was also called *interstitial cell-stimulating hormone* (ICSH). The other pituitary gonad-stimulating hormone was called *follicle-stimulating hormone* (FSH) because this fraction promoted the development of graafian follicles in the ovary.

Chemistry. Major progress in the isolation and chemical characterization of the gonadotropins did not happen until the late 1950s when several new techniques (eg, gel filtration, isoelectric focusing, polyacrylamide gel electrophoresis) became available to protein biochemists. Pure preparations of ovine (o), bovine (b), porcine (p), equine (e), rat (m), rabbit (r), and human (h) LH are available. Purification of FSH has not proceeded as rapidly as for LH; apparently, the FSH molecule does not withstand the rigors of purification as well as LH. Human FSH seems to be an exception and fairly pure preparations of this gonadotropin are available. High-potency ovine and equine FSH have been prepared, but both porcine and bovine FSH appear to have less potency.

The highly pure preparations of LH from different species and pure hFSH have allowed determination of some aspects of the chemical structure of gonadotropins. These hormones are *glycoproteins*, which means that the hormones are composed of chains of amino acids linked together by peptide bonds and chains of carbohydrates linked to the polypeptides; hence the name glycopro-

teins. Both LH and FSH consist of two polypeptide chains or subunits that are linked together (Fig. 5–1). These subunits have been designated as α and β. The β subunit of oLH consists of 96 amino acid residues and two carbohydrate chains. The α subunit contains 119 amino acid residues and a single carbohydrate chain (Papkoff et al, 1977). When the two subunits are chemically dissociated, neither subunit retains biologic activity. When the chains are recombined, biologic activity is restored, sometimes to the level of native hormone (Papkoff et al, 1973).

Scientists were surprised to learn that the amino acid sequence of oLHα was remarkably similar to that of another pituitary glycoprotein hormone, bovine thyroid-stimulating hormone (bTSH) (Liao and Pierce, 1970). As more information on the chemical nature of the pituitary glycoprotein hormones became available, it became apparent that the α subunits of the hormones were quite similar not only between hormones within a species, but also between species (Ward et al, 1973; Sairam and Papkoff, 1974; Papkoff et al, 1977). Ovine and bovine LHα are identical. Porcine LHα is about 95% similar and hLH is about 75% similar to oLHα. On the other hand, great diversity exists for the β subunit. The β subunit imparts both hormone and species specificity; hence this subunit is also called the hormone specific subunit. The hormone specific subunits for oLH and bLH are identical but are only about 65% similar to hLHβ.

The primary structure (amino acid sequence) of FSH is known only for hFSH (Papkoff et al, 1977). Only recently have highly purified preparations of ovine and bovine FSH become available (Grimek and McShan, 1974; Grimek et al, 1979). The current information shows that ovine and bovine FSH, like hFSH, are composed of two dissociable subunits, a common α subunit and a hormone-species specific β subunit.

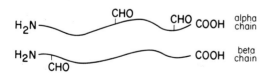

Fig. 5–1. Diagrammatic representation of the α and β subunits of luteinizing hormone showing the approximate position of the polysaccharide units.

The structures of the carbohydrate chains of the gonadotropins are not known. The carbohydrate content and makeup vary considerably between hormones and within hormones between species. For example, the carbohydrate content of LH varies from 12% for bLH to 24% for eLH (Sherwood and McShan, 1977). The carbohydrate content of FSH is greater than LH, being close to 25% for ovine, equine, human and bovine FSH (Sherwood and McShan, 1977; Grimek et al, 1979). The carbohydrate chains are composed of the monosaccharides (mannose, galactose, fucose, N-acetylglucosamine, N-acetylgalactosamine, and sialic acid) linked together to form complex polysaccharide units. The polysaccharide unit is attached to the polypeptide chain by a linkage between N-acetylglucosamine and asparagine.

Sialic acid plays an important role in the biologic activity of the gonadotropins. Apparently, the sialic acid moiety extends the circulatory half-life of hormones (Fig. 5–2, Niswender et al, 1974). The biologic activity of oFSH is much greater than that of bFSH and this has been related to the lower sialic acid content of bFSH (Grimek et al, 1979).

Estimated molecular weights of gonadotropins vary among species and between species. Molecular weights for ovine, bovine, porcine and equine LH respectively are: 28,000 to 32,500; 25,200 to 30,000; 27,000 to 34,000; and 33,500 (Sherwood and McShan, 1977). Molecular weights for ovine and equine FSH are 32,700 to 33,800 and 33,200 (Sherwood and McShan, 1977) and 37,300 for bFSH (Grimek et al, 1979). Why so much diversity in molecular weights of supposedly pure, homogeneous compounds? The reason is that the compounds often are not homogeneous when subjected to polyacrylamide gel electrophoresis. The electrophoretic heterogeneity happens because slight structural changes occur during extraction and isolation of the hormones (Sairam and Papkoff, 1974). It also seems likely that genetic diversity may exist in either the polypeptide chains or the polysaccharide units. Pituitary glands utilized for extraction are usually collected from slaughter houses; thus, depending upon the kinds of animals killed, pituitary glands consist of a heterogeneous lot of material. Glands from all kinds of animals may be present in one batch of material, for example, glands from sexually immature, sexually mature, intact, gonadectomized, pregnant and nonpregnant animals. Peckham et al. (1973) demonstrated that FSH extracted from pituitary glands from ovariectomized monkeys was different from that of intact monkeys.

Much progress has been made in understanding the chemical nature of the gonadotropins. While this progress is essential for understanding structure-function relationships, much more work remains to be done before gonadotropins can be synthesized chemically.

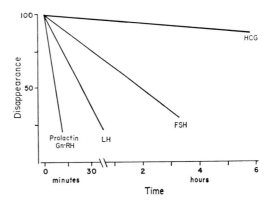

Fig. 5–2. The relative disappearance rates of some protein (prolactin, Gn-RH) and glycoprotein hormones (LH, FSH and human chorionic gonadotropin [HCG]). Prolactin and Gn-RH have no sialic acid residues, whereas the LH, FSH and HCG molecules have 1–2, 5 and 12 sialic residues, respectively. If one defines half life (T½) as the time required to reduce blood levels of a hormone by 50%, assuming no further secretion, then T½ for Gn-RH and prolactin is approximately seven minutes. The T½ for LH, FSH and HCG is approximately 0.5, 2 and 24 hours respectively. *(From Niswender, et al, 1974. In: Reproduction of Farm Animals, 3rd ed. Hafez, E.S.E. [ed], Philadelphia, Lea & Febiger)*

Table 5–1. Hormones of Reproduction

Source	Hormone	Function
Hypothalamus	Releasing hormones (Gn-RH, TRH)	Causes release of FSH, LH and TSH (from anterior pituitary)
	Somatostatin	Inhibits release of growth hormone
	Prolactin inhibiting factor (PIF)	Inhibits prolactin release
	Oxytocin (stored in posterior pituitary)	Stimulates uterine contractions, parturition, sperm and egg transport, milk ejection
Anterior pituitary	Follicle stimulating hormone (FSH)	Stimulates follicular growth, spermatogenesis, estrogen secretion
	Luteinizing hormone (LH)	Stimulates ovulation, corpus luteum function, secretion of progesterone, estrogen and androgen
	Prolactin	Promotes lactation; stimulates corpus luteum function and progesterone secretion in some species; may inhibit estrogen secretion
	Growth hormone (GH or STH)	Promotes tissue and bone growth
	Thyroid stimulating hormone (TSH)	Stimulates thyroxin secretion from thyroid gland
	Adrenocorticotropin (ACTH)	Stimulates adrenal cortical hormone secretion
Placenta	Human chorionic gonadotropin (HCG) (primates only)	Demonstrates LH activity
	Pregnant mare serum gonadotropin (PMSG)	Demonstrates FSH activity; stimulates formation of accessory CL
	Placental lactogen	Has GH activity
	Placental luteotropin (rodents)	Maintains CL
	Estrogens and progesterone	See ovary
Ovary	Estrogens	Promotes female sex behavior; stimulates secondary sex characteristics, growth of reproductive tract, uterine contractions, mammary duct growth; controls gonadotropin release; stimulates calcium uptake in bones; has anabolic effects
	Progesterone	Acts synergistically with estrogen in promoting estrous behavior and preparing reproductive tract for implantation; stimulates endometrial secretion; maintains pregnancy; stimulates mammary alveolar growth; controls gonadotropin secretion
Testes	Androgens	Develops and maintains accessory sex glands; stimulates secondary sexual characteristics, sexual behavior, spermatogenesis; has anabolic effects
	Inhibin (may also be secreted by the ovary)	Inhibits FSH release

Function. As the name implies, the function of gonadotropins is to stimulate the gonads (Table 5–1). In the female, gonadal (ovarian) stimulation produces the following results: growth of ovarian follicles, maturation of the oocytes within the ovarian follicles, secretion of estrogen by the cellular components of the ovarian follicles, ovulation, development of the corpus luteum and finally, secretion of progesterone by the corpus luteum. Most of the information on the functions of gonadotropins has been derived from experiments with laboratory animals because these animals are easily hypophysectomized. The hormones, in various combinations and sequences, can then be administered to these animals to elucidate the functions of the hormones in the normal animal.

Growth of the ovarian follicle proceeds in an ordered sequence. Once a follicle is recruited from the pool of nongrowing follicles and growth begins, it has one of two fates: atresia or ovulation. The gonadotropins determine which of the two fates will occur, although they do not cause recruitment of follicles from the pool of nongrowing follicles. FSH acts early in follicular development and certainly is required for formation of the antrum of the follicle. Perhaps more importantly, FSH, acting synergistically with estrogen, causes the formation of both FSH and LH receptors in the granulosa cells of the follicle (Richards and Midgley, 1976). Both gonadotropins act upon the ovarian follicle to stimulate estrogen secretion (see p. 103).

About three days prior to ovulation, circulating levels of estradiol increase. This increase of estradiol causes release of a surge of LH from the pituitary gland (Goding et al, 1969). This surge of LH causes the mature graafian follicle to ovulate. In addition to ovulation, events that are still not understood cause resumption of meiosis in the oocytes of mature follicles.

The granulosa cells of the ovulated follicle are destined to become the luteal cells of the corpus luteum. Although this transformation is a function of LH, the prerequisites for the development of a successful corpus luteum are established during the follicular phase. These prerequisites are: adequate granulosa cells (action of FSH and estrogen), ability to respond to LH (development of LH receptors by the action of FSH) and the ability to secrete progesterone (probably the action of LH) (McNatty, 1978). The continued function of the corpus luteum is maintained in some laboratory species by a luteotropic complex (Hilliard, 1973); LH appears to be the principal component of this complex in the ewe (Schroff et al, 1971).

In addition to its actions on steroidogenesis and oocyte maturation, there is some evidence that LH regulates the blood supply to the ovary. Niswender et al (1976) have shown that the blood flow to the ovary with a corpus luteum in ewes is correlated with peripheral serum progesterone concentrations. Moreover, administration of anti-LH serum reduced both progesterone concentration and blood flow, while administration of LH increased progesterone concentration and blood flow.

In the male the actions of gonadotropins are stimulation of steroid hormone secretion, spermatogenesis, and synthesis of specific androgen-binding protein (ABP). LH acts upon the Leydig cells of the testis to cause secretion of testosterone. Testosterone, in turn, plays an important role in the maintenance of spermatogenesis.

The role of FSH in spermatogenesis is not well defined because of the ability of testosterone to maintain spermatogenesis in hypophysectomized rats. In rats, FSH seems to be necessary for initiation of the spermatogenic sequence during puberty (Steinberger, 1975).

The Sertoli cells in the seminiferous tubules are target cells for FSH (Means et al, 1976). One of the actions of FSH on the Sertoli cell is to stimulate secretion of an androgen-binding protein (ABP). This protein binds testosterone, thereby maintaining high levels of testosterone in the lumen of the seminiferous tubule. FSH may also play a role in release of the spermatid from the syncytium of spermatids surrounded by Sertoli cells.

The ability of gonadotropins to stimulate such diverse processes as steroidogenesis and gametogenesis suggests a unique system for the molecular action of gonadotropins. Stimulation of cells to produce steroids is not the result of modification of the gonadotropins, but is a modification of the steroidogenic cell itself.

Mechanism of Action. The ability of gonadotropins to increase steroidogenesis is preceded by the binding of a gonadotropin to a specific receptor located in the cell membrane of the steroid-producing cell (Catt and Dufau, 1976; Catt and Pierce, 1978). Gonadotropin binding to the receptor activates the membrane-bound enzyme, adenylate cyclase. This enzyme catalyzes the conversion of adenosine triphosphate (ATP) to cyclic adenosine monophosphate (cAMP). Cyclic AMP has been designated the "second messenger" and apparently plays a role in the actions of several hormones (Sutherland and Rall, 1960). One of the known actions of cAMP is the activation of protein kinase enzymes. These enzymes phosphorylate other enzymes—usually a prerequisite for enzyme activation. The substrates for the cAMP-dependent protein kinase enzymes in steroid secreting cells have not yet been identified, but probably involve the enzymes that catalyze rate-limiting reactions in steroidogenesis. The sequence of events associated with gonadotropic stimulation of their target cells is shown in Figure 5–3.

The function of cAMP as the second messenger has recently been questioned because isolated Leydig cells were stimulated to maximal testosterone secretion by a dose of LH that did not increase levels of cAMP (Mendelson et al, 1975). Although this lack of correlated response makes one question the second-messenger hypothesis, recent experiments indicate that doses of LH that increase steroidogenesis also increase binding of cAMP to its receptor(s) without increasing cellular levels of cAMP (Dufau et al, 1977; Fletcher, 1978).

Steroidogenic cells (both in the ovary and the testis) seem to have an abundance of gonadotropin receptors. The total number of LH receptors isolated from membranes of luteal cells increases about 40-fold from day 2 to day 10 of the estrous cycle in the ewe (Diekman et al, 1978). The receptor population remains high on days 12 and 14 and declines on day 16 (the estrous cycle length of the ewe is about 16 days). The number of occupied receptors show the same pattern, but amount to only 0.6% of the total receptors. Thus, it appears that only a small portion of the LH receptors need be occupied for normal function of the corpus luteum. Apparently, the same holds true for the Leydig cell (Mendelson et al, 1975).

There is some evidence that once the gonadotropin binds to the membrane receptor, the entire hormone-receptor complex is taken into the cell and degraded (Chen et al, 1978; Ascoli and Puett, 1978). An abundance of receptors may, therefore, be necessary for continued function of the cell even though only a small proportion of the total number is occupied at any one time.

The mechanism of action by which gonadotropins stimulate target cell response seems to occur in an ordered fashion: (1) binding of the gonadotropin to a membrane-bound receptor of the target cell, (2) activation of adenylate cyclase to produce cAMP, (3) binding of cAMP to specific, intracellular receptors (regulatory subunit of protein kinase?), (4) activation of protein kinase enzymes and (5) activation, by yet unknown mechanisms, of enzymes involved in target cell function.

Prolactin

The hormone prolactin is a misfit among gonadotropins because it does not have a common role in animals. Prolactin was discovered as a hormone that induced milk secretion in rabbits and subsequently one that caused crop "milk" production in pigeons. Since that time, ovine, bovine, porcine, and rat prolactin have been obtained in highly purified form and the chemical properties determined.

Chemistry. Ovine prolactin consists of 198 amino acid residues in a known sequence and it has a molecular weight of 24,000 (Li, 1974). The molecule does not contain carbohydrate and is a single chain (ie, does not have subunits like LH and FSH) formed into loops by three disulfide bonds.

Prolactin molecules from swine, cattle, and sheep are similar (Sherwood and McShan, 1977). Ovine, bovine and porcine prolactin are also similar to human prolactin, which in turn is similar to human growth hormone and human placental lactogen (Friesen, 1977).

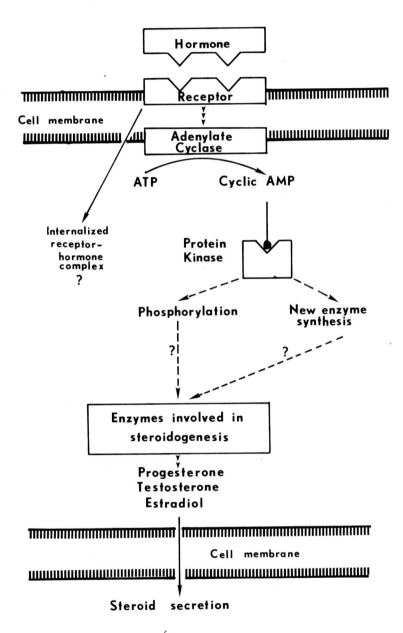

Fig. 5–3. The sequence of cellular events that occur following binding of gonadotropic hormones to a receptor in the membrane of a target cell. This illustration shows most of the known actions following the binding of LH to luteal or Leydig cells. The reactions shown by solid lines represent known actions while the dashed lines and question marks indicate speculated actions.

Function. The broad diversity of functions of prolactin within the animal kingdom classify the hormone more as a metabolic hormone than as a gonadotropin. Nicoll (1974) listed 134 functions of prolactin in five categories: reproduction, growth promotion, water and electrolyte balance, actions on ectodermal structures, and actions involving synergism with steroid hormones. Certainly one of the major functions of prolactin involves the ability of the female mammal to nurture her young (Table 5–1). This aspect is covered in Chapter 14.

The most controversial aspect of the function of prolactin is its luteotropic properties among various mammalian species. There is abundant evidence that prolactin is part of the luteotropic complex of rats, mice, and hamsters (Hilliard, 1973). The luteotropic function of prolactin in large domestic animals is questionable (Hansel et al, 1973). The two groups that have worked with hypophysectomized sheep are not in agreement on the luteotropic role of prolactin (Denamur et al, 1973; Kaltenbach et al, 1968; Schroff et al, 1971; Kiser et al, 1973).

More recently, the role of prolactin in reproductive malfunction is being questioned. A galactorrhea-amenorrhea syndrome has been described in women in whom suppression of menses is associated with inappropriate lactation and elevated prolactin levels (Yen, 1978). Administration of drugs such as ergocryptine derivatives suppress prolactin levels and result in return to normal menstrual cycles. Attempts to shorten the extended postpartum interval in suckled cows with administration of similar drugs as well as antiprolactin serum have usually met with failure.

Other Pituitary Hormones

Three other anterior pituitary hormones indirectly influence reproduction. Growth hormone (GH) or somatotropin stimulates growth of all body tissues and influences carbohydrate, lipid and protein metabolism. Growth hormone is a single polypeptide chain of 191 amino acid residues (Wilhelmi, 1974). The primary structure of human, bovine and ovine GH is known. A polypeptide of 188 amino acid residues similar to human GH has been synthesized.

Thyroid-stimulating hormone (TSH) stimulates the thyroid gland to secrete thyroxine and triiodothyronine. These hormones regulate the basal metabolic rate of the animal and thus are essential to reproduction. The thyroid hormones appear to be especially important during fetal development because severe abnormalities are observed in thyroidectomized ovine fetuses. Thyroid-stimulating hormone is a glycoprotein with α and β subunits (Pierce, 1974). The α subunit consists of 96 amino acid residues and two carbohydrate chains, similar to LHα and FSHα. The hormone-specific subunit consists of 113 amino acid residues and a single carbohydrate chain. The primary structure of only bovine TSH is known.

Adrenocorticotropin (ACTH) stimulates the adrenal cortex to synthesize and release the glucocorticoids and mineralocorticoids. These two classes of steroid hormones are involved in regulation of glucose metabolism and osmotic balance. In addition, ACTH plays an important role in initiation of parturition. (This function is discussed further on page 107.) ACTH is one of the smaller pituitary hormones consisting of 39 amino acid residues (Hofmann, 1974). It contains neither carbohydrate nor disulfide bonds. The amino acid sequences of porcine, ovine, bovine, and human ACTH are known and porcine and human ACTH have been synthesized.

GONADAL STEROID HORMONES

The gonadal hormones are lipid compounds known as steroids. All steroids have a common cyclopentanoperhydrophenanthrene nucleus composed of three six-membered phenanthene rings designated A, B and C and one five-membered cyclopentane ring designated D (Fig. 5–4). The basic nucleus contains 17 carbon atoms. Carbons 18 and 19, called angular methyl groups, project from carbons 13 and 10, respectively. Carbons 20 and 21, if present, project as a side chain from carbon 17.

Steroidogenesis

The immediate precursor for all steroids is pregnenolone which is derived from cholesterol. Cholesterol, in turn, is synthesized from acetyl-CoA which contains only two carbon atoms. Three molecules of acetyl-CoA are condensed and reduced to mevalonic acid, a six-carbon compound. Mevalonic acid is phosphorylated and decarboxylated to yield isopentenyl pyrophosphate, a five-carbon structure that serves as the immediate precursor of all sterols. Three units of isopentenyl pyrophosphate are condensed to form farnesyl pyrophosphate. Squalene, which is subsequently cyclized to give lanosterol, is formed by the union of two molecules of farnesyl pyrophosphate. The loss of three methyl groups from lanosterol

yields cholesterol. Synthesis up to this point is thought to occur primarily in the smooth endoplasmic reticulum of the cell.

Cholesterol is transported to the mitochondria where cleavage of the side chain between carbons 20 and 22 occurs resulting in pregnenolone (Fig. 5–5). Pregnenolone is subsequently converted to progesterone in the endoplasmic reticulum. Hydroxylation and decarboxylation of progesterone which leads to androgen formation occurs in the cytoplasm. The enzymes involved in steroid secretion are thus located in various subcellular fractions. This scheme provides a simplistic view of a highly organized, complicated process that requires multiple enzyme systems and produces a set of concerted reactions.

The ultimate product of a steroid-producing endocrine gland is determined by several factors including the relative proportion of cell types, the anatomic relationship of these cells, the concentration of precursors and cofactors as well as the presence of trophic hormones. The importance of these relationships can be illustrated by a review of follicular estrogen secretion.

Estrogens are derived from androgens by elimination of the C19 methyl group and aromatization of the A ring. LH stimulates thecal cells of follicles to secrete testosterone. The testosterone is subsequently aromatized to estradiol in granulosa cells under the influence of FSH stimulation (Fortune and Armstrong, 1978). This so called two-cell two-gonadotropin model is similar to that previously demonstrated in testicular tissue, in which LH stimulates testosterone production in the Leydig cells and FSH stimulates aromatization of the testosterone to estrogen in the Sertoli cells of the seminiferous tubules (Dorrington et al, 1978, Fig. 5–6).

Secretion of steroids into the circulation has always been considered a process of passive diffusion controlled by a concen-

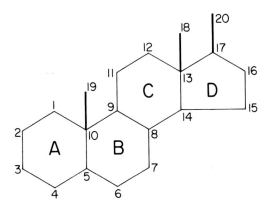

Fig. 5–4. Conventional representation of the steroid ring system with letters used to designate the rings and numbers to identify the carbon atoms.

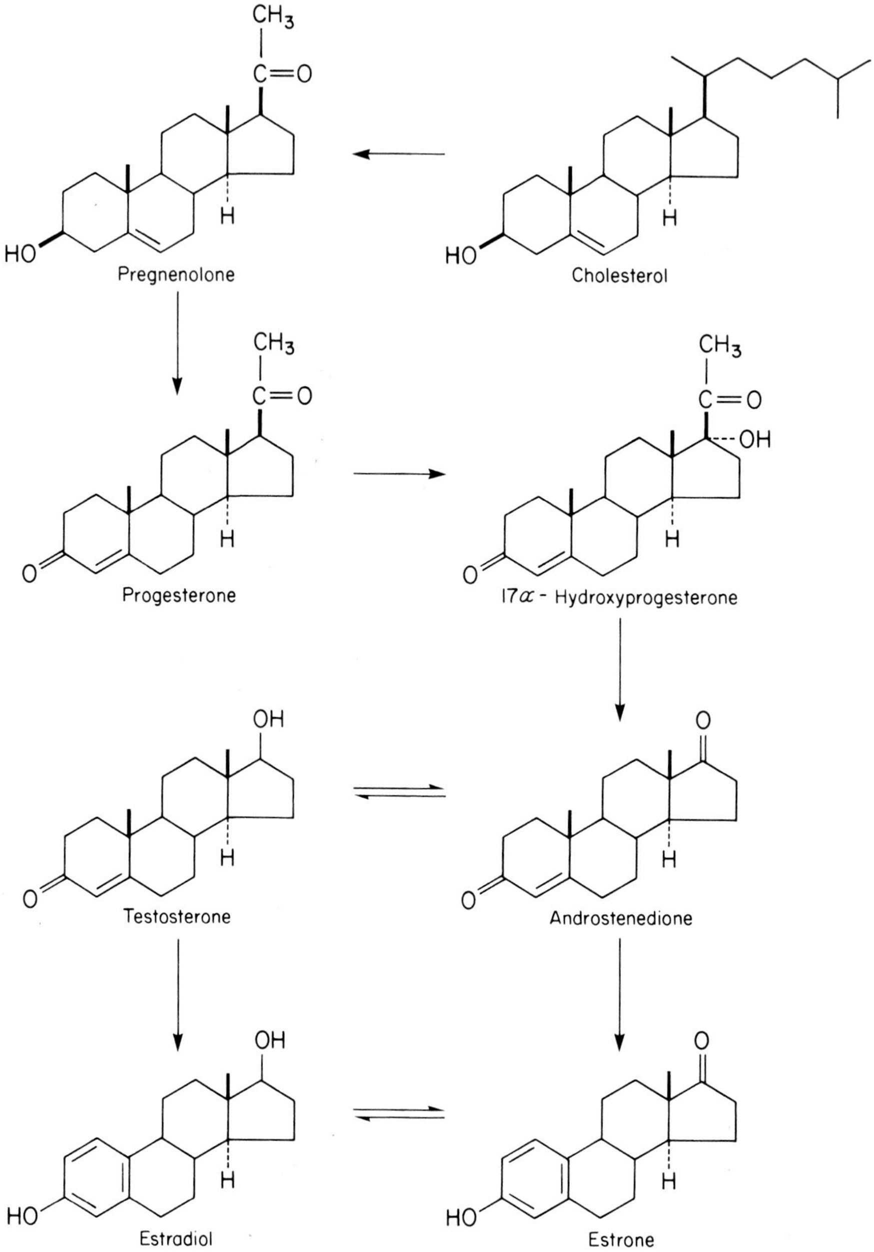

Fig. 5–5. The biosynthesis of steroid hormones from cholesterol.

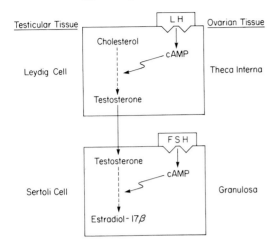

Fig. 5–6. Model of the two-cell two-gonadotropin hypothesis. *(Modified from Dorrington et al, 1978).*

tration gradient of the diffusible molecules. However, recent evidence indicates that protein secretory granules produced at the level of the Golgi apparatus may actually accumulate steroids from tubular elements of smooth endoplasmic reticulum. These carrier proteins migrate toward the plasma membrane where they release the steroid by means of exocytosis (Sawyer et al, 1977).

Once secreted into the circulation, most steroids appear to be bound to and carried by specific binding proteins. These binding globulins assist in transporting the steroids which are relatively insoluble in aqueous medium. Among their other functions, these binding proteins may also regulate the rate at which steroids are inactivated by the liver. Most steroids are inactivated by conjugation with either a sulfate or glucuronic acid residue. Such conjugates are water-soluble and therefore excreted as waste products in the urine or feces.

Function

Estrogens. Of all steroids, estrogens have the widest range of physiologic functions. Estrogen is required for the

psychologic manifestations of heat or estrus. Such behavior usually can be induced with estrogen alone; however, small amounts of progesterone are necessary for estrus in some species, and in most species studied, less estrogen is required if the female is pretreated or primed with progesterone. Glandular duct growth of the uterine endometrium, histologic changes in the vaginal epithelium during the estrous cycle and duct growth in the mammary gland during lactogenesis are attributable to estrogen. Other effects of estrogens with regard to reproduction include their ability to control the release of pituitary hormones, potentiate the effects of oxytocin and prostaglandins on uterine contractions and assist in the process of implantation.

Nonreproductive effects of estrogens include stimulation of calcium uptake and ossification of bones. The protein anabolic effect of estrogen has been capitalized upon by the livestock industry as administration of estrogenic compounds increases weight gains and feed efficiency in ruminants.

Progesterone. Progesterone acts synergistically with estrogen in several physiologic functions including the growth of uterine and mammary glands. Progesterone inhibits uterine contractions and stimulates the endometrial glands to secrete endometrial fluid which is necessary for nourishment of the preimplantation blastocyst. Progesterone is also required for the continuous maintenance of pregnancy in most mammals, at least during the first one-third of gestation.

High levels of progesterone inhibit estrus and the ovulatory surge of LH, thus establishing the importance of this hormone in regulation of the estrous cycle.

Androgens. Maintenance of secondary sex characteristics, ie, the physical and anatomic features that are unique to males of a species, is the most visible function of androgens. Sex drive is totally androgen dependent and standing in the social or pecking order is also related to androgen production. Physiologically, androgens are required for normal spermatogenesis and these steroids are necessary to maintain structure and function of the accessory sex glands including the seminal vesicles, prostate and bulbourethral glands. Androgens also have protein anabolic effects, which promote nitrogen retention and increase the number and thickness of muscle fibers.

Mechanism of Action

Steroids are thought to pass through cell membranes by simple diffusion although limited data suggest the possibility of protein-mediated transport. Once inside the cell, steroids are specifically bound to cytoplasmic proteins called receptors. Binding of the steroids induces conformational changes in the receptor which allows translocation of the bound complex from the cytoplasm to the nucleus. The mechanism by which translocation occurs is unknown, although the rate of translocation is directly proportional to the concentration of steroid-receptor complex in the cytoplasm. Similarly, it is not known whether the receptor complex binds to a variety of nuclear fractions (DNA, histone, ribonucleoprotein) or a single component. Regardless, entry into the nucleus results in synthesis of specific mRNA molecules. The resultant mRNA is translocated to the cytoplasm where it directs synthesis of specific proteins. The newly synthesized protein accounts for changes observed in target tissues following exposure to steroids.

PLACENTAL HORMONES

Placental hormones include pregnant mare serum gonadotropin, human chorionic gonadotropin and placental lactogens.

Pregnant Mare Serum Gonadotropin

Pregnant mare serum gonadotropin (PMSG) was discovered when blood serum from pregnant mares produced sexual maturity in immature rats (Cole and Hart, 1930). PMSG is a glycoprotein hormone with a high content of sialic acid (Sherwood and McShan, 1977). It can be dissociated into α and β subunits and molecular weight estimates vary from 28,000 to 53,000.

This gonadotropin is secreted by endometrial cups in the equine uterus. These structures are formed by specialized trophoblastic cells that invade the maternal endometrium and are thus of fetal, not maternal, origin. Fetal genotype greatly influences the amount of PMSG secreted by the endometrial cups. Mares carrying mule fetuses have lower levels than mares carrying horse fetuses (Clegg et al, 1962).

The uterine cups are formed about the sixth week of pregnancy and persist until about the twentieth week. PMSG, which appears in the circulation coincident with the development of the endometrial cups, stimulates excessive follicular development. Several of these follicles ovulate and develop into accessory corpora lutea. Contrary to claims made in earlier literature, the original corpus luteum does not regress prior to formation of the accessory corpora lutea.

PMSG is isolated from the urine of pregnant mares and consequently, it was one of the first commercially available gonadotropic materials. It displays both FSH-like and LH-like activity. PMSG has frequently been used to promote extensive follicular development prior to superovulation for embryo transfers. It has also been used to promote follicular development during the anestrous period of ewes. The high sialic acid content increases the circulatory half-life of PMSG and thereby makes it more effective than FSH.

Human Chorionic Gonadotropin

Human chorionic gonadotropin (hCG) is a gonadotropin excreted in the urine of pregnant women. It is synthesized by the syncytiotrophoblastic cells of the placenta. In humans, hCG is detectable about eight days after ovulation, which is only one day following implantation (Jaffe, 1978). Human chorionic gonadotropin converts the corpus luteum of the menstrual cycle to the corpus luteum of pregnancy; thus, hCG provides the signal for maintenance of the corpus luteum and establishment of early pregnancy in humans.

The chemical structure of hCG is well-defined (Bahl, 1977). It is a glycoprotein consisting of α and β subunits with a molecular weight of 40,000. The α subunit has 92 amino acid residues and two carbohydrate chains. The α subunit of hCG is similar to the α subunits of hLH, pLH, oLH and bLH. The β subunit has 145 amino acid residues and five carbohydrate chains. The amino acid sequence and the saccharide sequence of both polypeptide chains and the seven carbohydrate chains is known. The high carbohydrate content of hCG is one of the distinguishing features of this hormone. Removal of sialic acid greatly reduces the biologic activity of hCG.

Human chorionic gonadotropin has LH-like functions. As with PMSG, it is a commercially available source of luteinizing activity. As such, it was used in treatment of cystic ovaries of dairy cows and is still used in many situations to induce ovulation. One of the more recent uses of hCG is in the study of LH receptors, since both hormones compete for the same receptor. Human chorionic gonadotropin often is used in lieu of LH because it can be radioiodinated with ^{125}I without losing biologic activity. Since hCG appears early in human pregnancy, detection of hCG in the urine by immunologic methods is the basis for early pregnancy diagnosis.

Placental Lactogens

Placental lactogenic (PL) hormones have been demonstrated in a number of species, including the human, rat, goat, sheep and cow. These hormones have chemical and biologic properties similar to those of both GH and prolactin and they are extracted from placental tissue (Fellows et al, 1976; Chan et al, 1976). Ovine PL has an estimated molecular weight of 22,000 to 23,000 and 192 amino acid residues, making it similar to both oGH (191 amino acid residues) and ovine prolactin (198 amino acid residues). When evaluated in hypophysectomized rats, oPL induced weight gain and epiphyseal growth—functions similar to GH. On the other hand, oPL also promoted casein synthesis in rabbit mammary tissue—a function of prolactin.

The concentrations of oPL in maternal serum are low during the first two trimesters of pregnancy and rise dramatically during the last trimester (Handwerger et al, 1977; Kelly et al, 1974). The function of the placental lactogenic hormones is unknown. A study of specific binding sites for oPL showed that the greatest binding occurred in the livers of nonpregnant ewes, followed by adipose tissues, ovaries and uteri and finally, in the livers of fetuses (Chan et al, 1978). This suggests that oPL may mediate important metabolic events in the ewe (ie, act more like GH instead of prolactin). Infusions of arginine into pregnant ewes caused dramatically increased levels of oPL, GH and prolactin (Handwerger et al, 1978), suggesting that the mechanisms that control the release of oPL are similar to those that control the release of GH and prolactin.

Placental lactogens may be important regulators of maternal metabolism to ensure the availability of adequate nutrients to the developing fetus. The maternal levels of oPL are greatest during the last trimester of gestation, when the most rapid fetal growth occurs. It will be interesting to learn the effect (if any) that maternal malnutrition during the last trimester of gestation has on PL levels.

The role of PL in milk production cannot be ignored. Levels of bovine PL were nearly twice as high in dairy cows (high milk production) as in beef cows (low milk production). In addition, bPL levels in high-milk-producing cows tended to be higher than in low-milk-producing cows (Bolander et al, 1976).

The physiologic roles of PL hormones offer some exciting new areas of research to develop methods for predicting and improving milk production and growth.

PROSTAGLANDINS

Prostaglandins were initially discovered in the 1930s in extracts of human semen as a vasopressor material that stimulated smooth muscle. The chemical structure of these compounds was not established until about 20 years later. Unlike other humoral agents, prostaglandins are not localized in any particular tissue. Prostaglandins appear to act locally at the site of their production in most instances, and therefore they do not conform to the classic definition of a hormone that is transported through the blood or lymph.

Prostaglandins exist in the forms of at least six parent compounds and numerous metabolites that exhibit a wide variety of pharmacologic effects. Prostaglandins have been shown to affect blood pressure, lipolysis, gastric secretion, blood clotting and other general physiologic processes including renal and respiratory function. Although the pharmacologic effects of prostaglandins are mostly known, their physiologic role is not well delineated.

Blood levels of most prostaglandins are generally very low although they appear to be elevated under certain conditions, such as parturition. Prostaglandins are rapidly metabolized and degraded, which probably accounts for their transient

pharmacologic activity and low blood levels.

All prostaglandins are 20-carbon unsaturated hydroxy fatty acids with a cyclopentane ring at C8-C12. Arachidonic acid, an essential fatty acid, is the precursor for prostaglandins that are most closely associated with reproductive processes, namely prostaglandin $F_{2\alpha}$ ($PGF_{2\alpha}$) and prostaglandin E_2 (PGE_2) (Fig. 5–7).

Prostaglandins have been implicated in gonadotropin release, ovulation, regression of corpora lutea, uterine motility, parturition and sperm transport. Even though prostaglandin involvement in these processes is thoroughly discussed in the specific chapters, a brief summary is appropriate.

PGE_2 causes release of LH in vitro and in vivo, while $PGF_{2\alpha}$ releases LH in vivo in several laboratory and farm animals. However, these are all pharmacologic responses and the physiologic roles that prostaglandins play in gonadotropin release are yet to be determined.

Evidence that prostaglandins are involved in ovulation has been obtained almost exclusively with rabbits, in which the administration of indomethacin, an inhibitor of prostaglandin synthesis, blocks ovulation. LH release was not affected in these animals and the effect of indomethacin was reversed at least partially by simultaneous or subsequent treatment with PGE_2 and $PGF_{2\alpha}$.

The primary emphasis in reproduction studies has centered on the luteolytic effects of $PGF_{2\alpha}$ and the oxytocic effects of both $PGF_{2\alpha}$ and PGE_2. Initial studies that demonstrated that $PGF_{2\alpha}$ is luteolytic in the rat have been extended to include the cow, ewe, sow and mare. The exact mechanism by which $PGF_{2\alpha}$ exerts its action has not been determined; however, one of several postulates contends that the veno-constrictive effect of $PGF_{2\alpha}$ may induce hypoxia, which in turn leads to luteolysis (Niswender et al, 1976). Indirect support for this hypothesis was recently obtained in an experiment demonstrating that PGE$_2$, a potent vasodilator, prevents estrogen-induced luteolysis in the ewe (Colcord et al, 1978).

It is well established that in all farm animals the uterus is the primary source of the luteolytic activity that controls the corpus luteum. Whether the luteolytic substance is PGF$_{2\alpha}$, its precursor, arachidonic acid, or some other substance has been the subject of controversy, as has the mode by which this substance reaches the ovary. Further details relevant to this area are given in the chapters on reproductive cycles.

The uterine source of prostaglandins also plays an important role in the process of parturition. At term in ewes there is a parallel increase in estrogen, PGF$_{2\alpha}$ and uterine activity. In the cow, estrogen increases linearly over the last 30 days of pregnancy. Progesterone drops rather rapidly during the last two or three days. The withdrawal of progesterone is followed by a major release of PGF$_{2\alpha}$ into the uterine venous drainage. The primary effect of the rise in prostaglandins appears to be an increase in myometrial contraction of the uterus. Prostaglandins also appear to increase the release of oxytocin, which plays an important role in uterine contractions during the second stage of parturition.

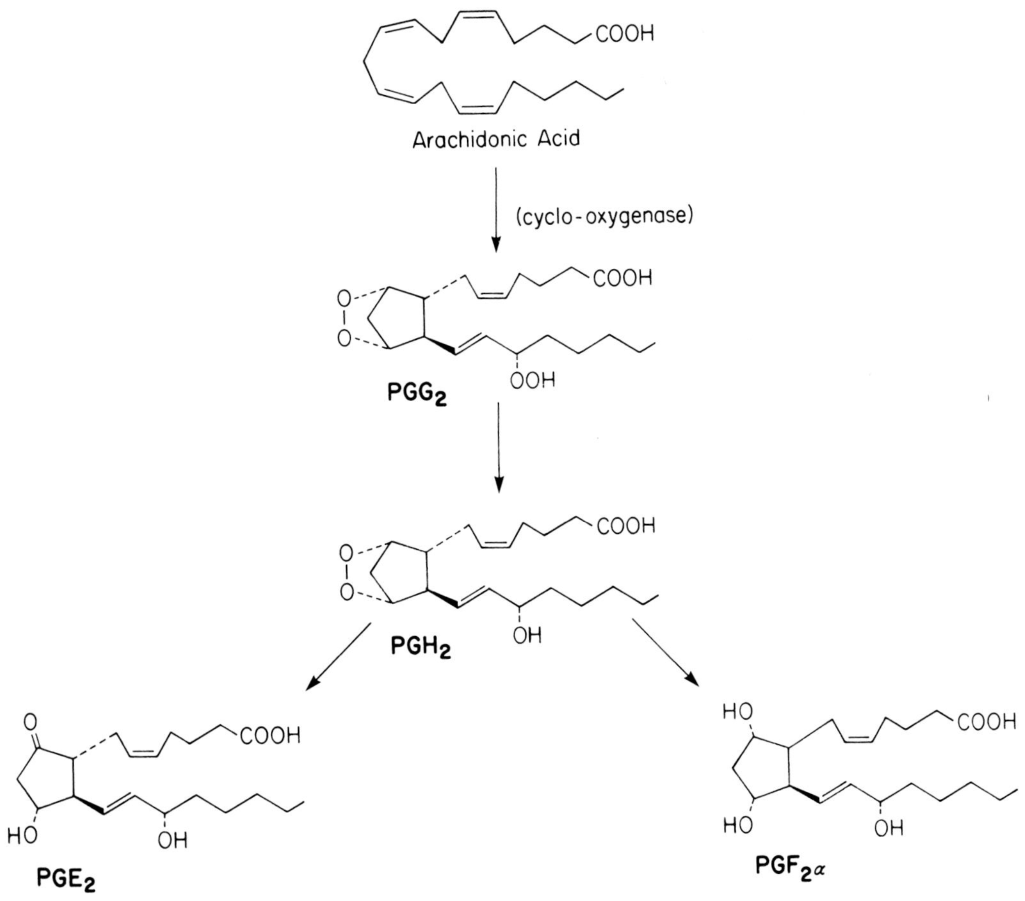

Fig. 5–7.   The biosynthesis of prostaglandins from arachidonic acid.

SYNTHETIC HORMONES

Synthetic hormones have been used widely in the livestock industry, primarily as growth promotants and for synchronization of estrus and control of ovulation. The exogenous administration of natural hormones has little value in most instances because of their relatively short half-life. Modification of the basic structure of a compound, such as the addition of an ester in the case of steroids, often delays absorption from an injection site. Delayed absorption extends the half-life and thus increases the potency. These compounds are also active when taken orally as they resist destruction by microflora in the gut which rapidly destroy natural hormones.

Diethylstilbestrol (DES) is a classic example of a synthetic hormone that has been used since the 1950s for growth promotion in cattle and sheep (Fig. 5–8). This compound, which is anabolic and promotes nitrogen retention, increases feed efficiency and weight gain. Although DES does not have the basic steroid structure, it competes with estradiol for the same tissue receptor protein and thus possesses estrogenic activity.

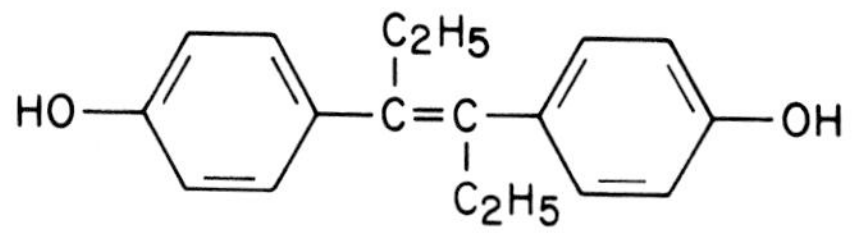

Diethylstilbestrol

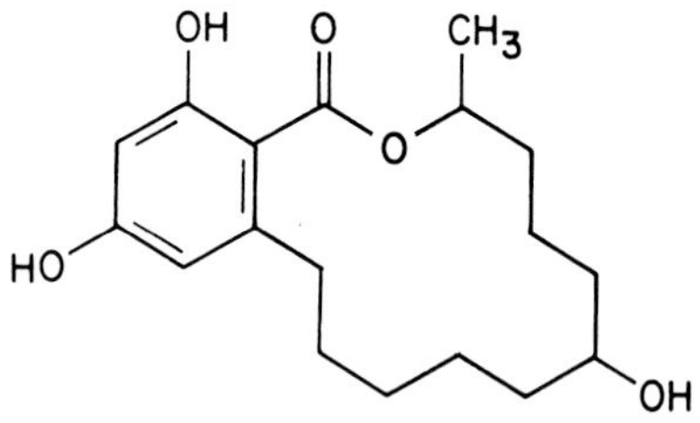

Zeranol

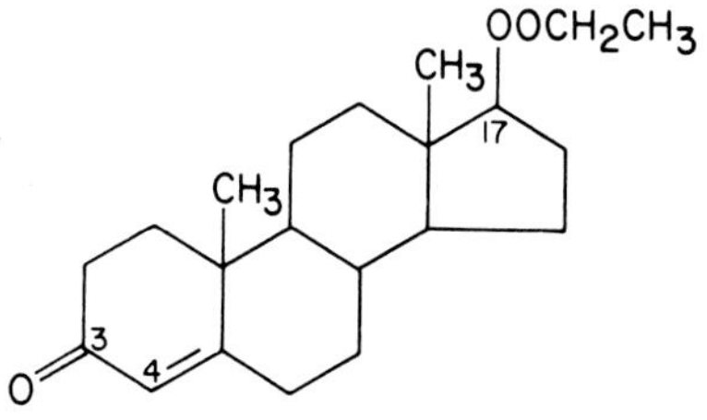

Testosterone Propionate

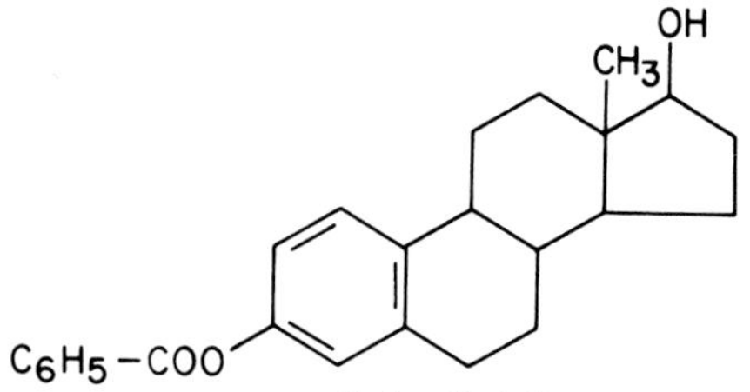

Estradiol Benzoate

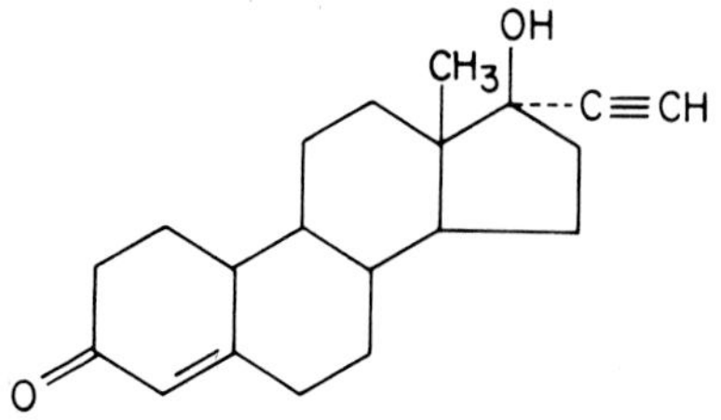

Norgestomet

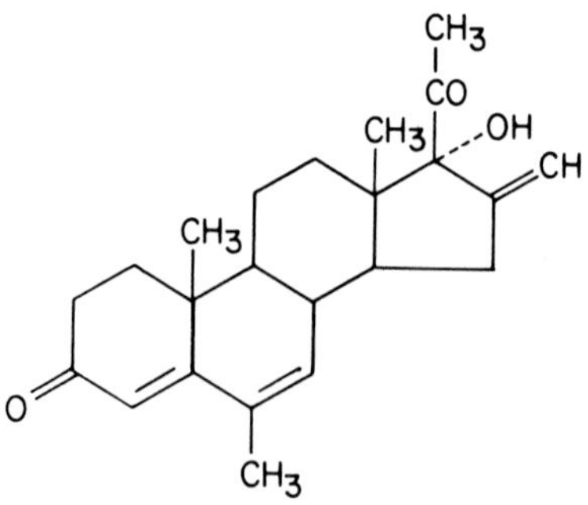

Melengestrol Acetate

Fig. 5–8. Synthetic hormones.

Androgens are also anabolic and esters such as testosterone propionate have been used in combination with derivatives of estrogen (estradiol benzoate, for example) as growth promotants (Fig. 5–8).

Synthetic progestogens such as melengestrol acetate (MGA) have also been used for growth promotion, although their mechanism of action is different. MGA is effective only in heifers with intact ovaries. Continuous administration of MGA appears to inhibit gonadotropin release, which prevents ovulation but allows development of ovarian follicles. Estrogen produced by these follicles promotes nitrogen retention and increases feed efficiency.

Zeranol, which is marketed under the trade name of Ralgro, is one of the newer compounds approved for use as a growth promotant. The exact mechanism of action for this plant extract has not been determined; however, it does compete with estradiol for the same tissue receptor protein which probably accounts for its anabolic action.

Control of ovarian function for synchronization of estrus and ovulation can also be accomplished with synthetic hormones. Continuous administration of synthetic progestogens such as MGA or norgestomet inhibits gonadotropin secretion and prevents ovulation. Removal of the progesterone block following natural regression of corpora lutea allows simultaneous development of follicular activity and ovulation in treated animals. Similar results can be obtained with $PGF_{2\alpha}$ and its derivatives. These compounds cause immediate luteolysis of existing corpora lutea, which initiates follicular development and ovulation.

HORMONE INTERRELATIONSHIPS

Puberty

Puberty is defined as the age at which an animal becomes capable of reproducing sexually. Females of all species reach puberty at an earlier age than males. Numerous factors including breed, growth rate, ambient temperature and season of birth affect age at puberty.

Endocrine patterns associated with the onset of puberty vary with the species studied. In ewes, pulsatile secretion of LH with peaks of high amplitude occur throughout the period of sexual maturation and continue through the first breeding season (Fig. 5–9). A similar pattern of LH secretion has been observed in prepubertal heifers, except that a definite decrease in the magnitude of the LH peaks commences approximately one week prior to first ovulation (Gonzalez-Padilla et al, 1975). A decrease in LH secretion prior to puberty also occurs in the rat. In contrast, LH increases linearly through puberty in bulls (Lunstra et al, 1978) and human females.

There is no distinct correlation between serum levels of FSH, estrogen or Gn-RH and the onset of puberty; however, the first ovulatory surge of LH in heifers is associated with an elevation in progesterone. The origin of this elevation in progesterone is unknown, as it could arise from either the ovaries or the adrenals.

The onset of puberty is not characterized by any deficiency in circulating levels of gonadotropins nor does ovarian sensitivity appear limiting, as prepubertal pigs, lambs and calves will all ovulate if given exogenous gonadotropins. Continuous high levels of estrogen suppress LH and FSH secretion, and a single injection of estrogen will induce LH release in prepubertal animals, suggesting that both negative and positive feedback systems are operational prior to puberty. The exact mechanism controlling the onset of puberty is thus unknown. Even though the hypothalamus and ovaries appear competent, these have been tested only with large quantities of exogenous hormone. It is thus possible that puberty requires maturation of the hypothalamic mechanism controlling the surge mode of LH secretion. Increases in ovarian sensitivity

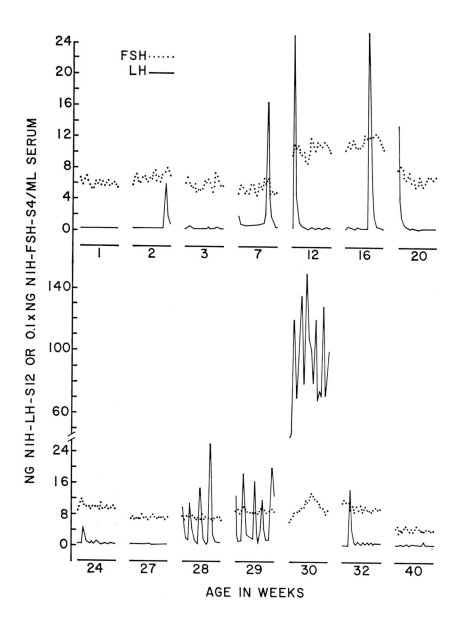

Fig. 5–9. Blood levels of LH and FSH throughout the life of a sheep. In this study 20 samples were obtained over a six-hour period once each week beginning at birth and continuing until the ewe became pregnant. First estrus and ovulation occurred during week 30 in this ewe and the samples were collected during the ovulatory LH peak. This ewe was bred during week 32 and was approximately eight weeks pregnant by week 40. *(From Niswender, et al, 1974. In: Reproduction of Farm Animals, 3rd ed. Hafez, E.S.E. [ed], Philadelphia, Lea & Febiger)*

to physiologic levels of gonadotropins may also be involved.

Estrous Cycle

The estrous cycle is controlled by the interaction of FSH, LH, estrogen and progesterone (Fig. 5–10). These hormones are common to most domestic animals; however, their secretory patterns and their relative effects vary among the different species. These differences lead to variations in length of luteal and follicular phases of the cycle as well as to differences in duration of estrus. Therefore, an explanation of the sequence of events and hormonal interactions for one species does not necessarily apply to all species, although these can be typified by events in the ewe.

A stylized version of the relative blood levels of gonadotropic and steroid hormones during the various stages of the estrous cycle of the ewe is shown in Figure 5–11. The follicular phase of the cycle is characterized by rapidly decreasing levels of progesterone and a peak in blood levels of estradiol. This decline in the level of progesterone followed by the rapid rise in estrogen is an essential requirement for the onset of behavioral estrus. Blood levels of LH increase approximately twofold during the follicular phase (Fig. 5–12). This may result from removal of the negative feedback influence of progesterone. Prior to the development of assays that are capable of quan-

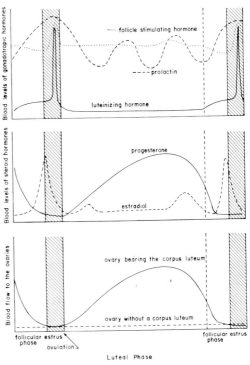

Fig. 5–11. Relative blood levels of gonadotropic and steroid hormones throughout the estrous cycle of the ewe. The data represent mean daily levels from several animals. The relative blood flow to both ovaries was determined using Doppler ultrasonic blood flow equipment. *(From Niswender, et al, 1974. In: Reproduction of Farm Animals, 3rd ed. Hafez, E.S.E. [ed], Philadelphia, Lea & Febiger)*

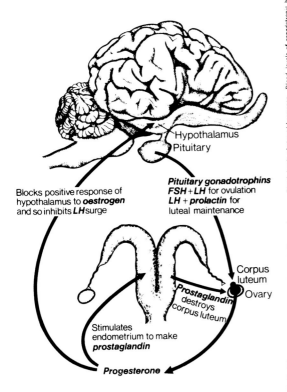

Fig. 5–10. Interrelationships between the pituitary, the corpus luteum, the uterus and the hypothalamus in sheep *(Short, 1972. In: Reproduction in Mammals, Book 3. Austin and Short [eds], Cambridge, Cambridge University Press).*

tifying blood levels of FSH, it was thought that this hormone also increased during the follicular phase. This does not occur, although it is assumed that FSH is needed for the rapid increase in follicular development that occurs during this phase.

The peak of estrogen during the follicular phase exerts a positive feedback influence on the hypothalamo-hypophyseal axis resulting in the ovulatory surge of LH, which occurs approximately 12 hours after the onset of estrus. Peak levels of

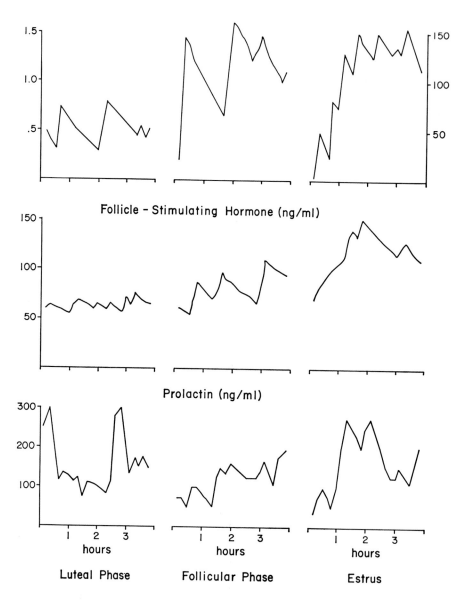

Fig. 5–12. Patterns of blood gonadotropins at different stages of the estrous cycle. In all cases blood was collected every six minutes for four hours and assayed for all three gonadotropic hormones. Note the similarity of the pattern of LH release at different stages of the cycle even though the absolute quantity of LH on the day of estrus was 100 times higher than during the luteal or follicular phases of the cycle. As can be seen, FSH and prolactin are also released in bursts at all times during the cycle. *(From Niswender, et al, 1974. In: Reproduction of Farm Animals, 3rd ed. Hafez, E.S.E. [ed], Philadelphia, Lea & Febiger)*

FSH and prolactin occur simultaneously with the ovulatory surge of LH. The increase in LH and FSH is caused by hypothalamic release of Gn-RH, whereas prolactin secretion is due to inhibition of PIF by the preceding peak of estrogen.

The precise role of LH in the ovulatory process has not been determined, although it appears to stimulate proteolytic enzymes including collagenase, which degrades connective tissues in the wall of the follicle. This process requires approximately 24 hours. Ovulation does not appear to be a mechanical process of rupturing due to excessive internal pressure. In most species the ovulatory follicle becomes flaccid several hours prior to ovulation, which results in a slow release of the follicular fluid following rupture of the follicular wall.

In addition to the stimulation of proteolytic activity, the ovulatory surge of LH also increases follicular synthesis of prostaglandins, especially prostaglandin E_2. Injections of indomethacin, an inhibitor of prostaglandin synthesis, concurrently with the ovulatory surge of LH prevent follicular rupture. This indicates that prostaglandins have an essential but undefined role in the mechanism by which LH causes ovulation.

Following ovulation, the wall of the follicle gradually thickens due to hypertrophy and hyperplasia of the granulosa cells. The rapidly proliferating cells fill the remaining cavity and begin to secrete progesterone. The resulting corpus luteum continues to increase in size and weight and obtains full growth and function seven to nine days after ovulation in the ewe. The size of the corpus luteum is highly correlated with its ability to secrete progesterone. The secretory capacity of the corpus luteum is also highly correlated with blood flow to the ovary (Fig. 5–11).

The secretion of progesterone depends on continuous support of luteotropic hormones from the pituitary gland. Pro-gesterone secretion ceases and the corpus luteum regresses within a few days following hypophysectomy. Partial maintenance of the corpus luteum occurs if LH is administered continually following hypophysectomy. Full function of the corpus luteum appears dependent on prolactin in addition to LH, indicating that a complex of hormones rather than any single hormone is necessary for luteotropic support. Some controversy exists regarding this aspect, and there appear to be species differences in hormonal requirements for maintenance of luteal function.

The corpus luteum is maintained throughout pregnancy in the ewe but regresses abruptly 13 to 15 days following ovulation in the nonpregnant animal. The rapid decline in progesterone due to regression of the corpus luteum is the trigger for the whole sequence of events leading to the next estrus and ovulation. Removal of one uterine horn in the ewe and other domestic animals prevents normal regression and prolongs life of the corpus luteum on that side only. Corpora lutea on the side of the intact uterine horn regress normally, indicating that the uterus exerts its influence in a local manner.

The mechanism for local control remained a mystery for many years until a unique experiment performed by J. R. Goding in Australia provided an explanation (Goding et al, 1971–72). Goding meticulously separated the ovarian artery and ovarian vein, which are closely adherent (Fig. 5–13). Ewes treated in this manner failed to show luteolysis. This led to the theory that a countercurrent transfer mechanism exists whereby a luteolytic substance from the uterus passes directly from the utero-ovarian vein to the ovarian artery. Subsequent studies indicated that the uterine luteolytic substance is most probably prostaglandin $F_{2\alpha}$.

More recent studies have failed to demonstrate differences in endometrial and uterine venous blood levels of $PGF_{2\alpha}$ be-

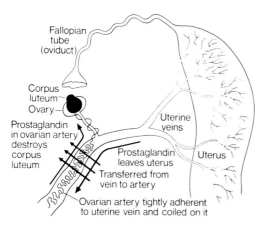

Fallopian
tube
(oviduct)

Corpus
luteum
Ovary

Prostaglandin
in ovarian artery
destroys
corpus
luteum

Uterine
veins

Prostaglandin
leaves uterus
Transferred from
vein to artery

Uterus

Ovarian artery tightly adherent
to uterine vein and coiled on it

Fig. 5–13. Postulated route by which prostaglandin manufactured by the progesterone-primed uterus is able to enter the ovarian artery and destroy the corpus luteum in sheep. *(Short, 1972. In: Reproduction in Mammals, Book 3. Austin and Short [eds], Cambridge, Cambridge University Press)*

tween pregnant and nonpregnant ewes. Such findings have prevented universal acceptance of the countercurrent theory. Although $PGF_{2\alpha}$ does not vary between pregnant and nonpregnant ewes, there is increased uptake and transport of PGE_2 to the ovary of the early pregnant ewe (Lewis et al, 1978). PGE_2 may function as an antiluteolysin that counteracts the luteolytic effects of $PGF_{2\alpha}$. In addition, the role of arachidonic acid as it relates to luteolysis, particularly in the cow, is yet to be resolved.

It is thus evident that the exact mechanism whereby the uterus influences the corpus luteum remains obscure.

Pregnancy and Parturition

The various aspects of pregnancy, including gamete transport, implantation, development of the conceptus and expulsion of the fetus, are all controlled by modulation of several hormones from the pituitary, ovary and placenta. The source, level and function of these hormones varies among species.

Endocrine levels immediately following conception are the same as those observed during the estrous cycle. The increasing levels of progesterone and relatively low levels of estrogen are essential for sperm transport and subsequent passage of the fertilized ovum into the uterus. Alteration of estrogen-progesterone ratios during this time results either in fertilization failure or interruption of normal placentation.

The first observable change in hormonal patterns during pregnancy is associated with maintenance of the corpus luteum and its continued secretion of progesterone. The secretion of pituitary hormones is essential for maintenance of the corpus luteum, as hypophysectomy during early pregnancy inevitably results in luteal regression. However, maternal recognition of pregnancy appears to involve blockage of uterine induced luteolysis. In most species, the mechanism by which the embryo transmits this message is unknown. The appearance of increased levels of PGE_2 from unknown sources which act as an antiluteolysin may be the responsible mechanism in the ewe (Lewis et al, 1978).

Progesterone is essential for the maintenance of pregnancy and it serves many functions during this period. These include increased coiling and thickening of the endometrial glands and the production of uterine secretions that are essential for the nutrition of the early developing embryo. Progesterone levels in the cow reach a peak approximately two weeks following ovulation and remain at this level until shortly before parturition. In contrast, progesterone levels in the ewe gradually rise throughout the duration of gestation. This rise is attributed to placental production of progesterone. Animals with this ability generally tolerate ovariectomy during early pregnancy, whereas animals such as the cow, pig, rabbit and rat will abort if the ovaries are removed.

The most dramatic changes in the levels of hormones during pregnancy occur shortly before parturition. The relationship between fetal and maternal hormones during this period in the cow is shown in Fig. 5–14. The first significant change appears to be an increase in fetal levels of corticoids. Such an increase in fetal adrenal activity appears to trigger initiation of parturition in both the ewe and the cow. Although maternal levels of corticoids do not change during this period, there is a concomitant decrease in maternal levels of progesterone.

Utero-ovarian venous levels of estrogen begin to increase in the cow three weeks prepartum. There is a rapid increase in estrogen during the last ten days that reaches a peak 1 to 4 days prior to parturition. Utero-ovarian levels of PGF remain constant until 48 hours before calving, when there is an exponential increase that peaks at parturition. The final stage of labor is accompanied by a surge in oxytocin, which probably acts in combination with PGF to cause expulsion of the fetus.

The precise role of each hormone in the complex series of events leading to parturition remains to be determined. Progesterone blocks myometrial activity and the continuous administration of progesterone delays parturition. Removal of progesterone is therefore generally considered essential.

Estrogen increases the uterine response to oxytocin and appears to stimulate PGF production. For these reasons estrogen has usually been considered to have an essential and positive role in the process of parturition. This belief has been brought into question by the results of two recent experiments in ewes. ACTH-induced parturition of the hypophysectomized ovine fetus occurs without an increase in maternal plasma concentrations of estrogen (Kendall et al, 1977). Furthermore, immunization of ewes against total unconjugated estrogens has no effect on parturition of intact fetuses

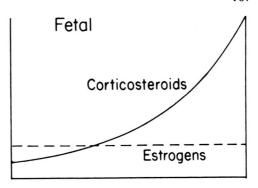

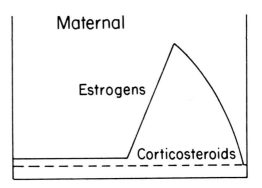

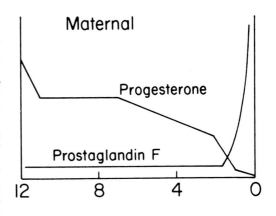

Days before calving

Fig. 5–14. Mean changes in bovine fetal and maternal hormones preceding parturition. Fetal hormones were quantified in plasma from the vena cava. The maternal utero-ovarian vein was sampled for estrogen and prostaglandins. Maternal corticosteroid and progesterone levels were determined on jugular plasma. *(Adapted from Hunter et al, 1977. Acta Endocrinol. 84, 653)*

(Rawlings et al, 1978). These findings may require a reassessment of the role of estrogen as well as that of other hormones in the cascade of events that culminate in parturition.

Postpartum Period

Studies on endocrine profiles during the postpartum period are limited largely to the cow. The onset of the first postpartum estrus is influenced by the level of nutrition before and after calving, body condition at calving, lactation, dystocia, breed and age of the cow. Postpartum anestrus varies in length from three to 17 weeks or longer depending on the foregoing factors. The economic necessity to have cows calve at a minimum of 12-month intervals has stimulated considerable research in this area.

Postpartum endocrine profiles have been established for LH, prolactin, progesterone, cortisol, estradiol-17β and estrone (Humphrey, 1977). Secretory patterns of prolactin are extremely variable. Onset of postpartum estrus cannot be related to frequency, magnitude or duration of secretory peaks or to mean levels of prolactin. It is thus difficult to ascribe a particular role to prolactin regarding onset of the postpartum estrus. A similar conclusion has been reached for cortisol and estradiol-17β, as there are no appreciable changes in either of these steroids during the postpartum intervals.

In contrast, LH appears to be very important in the onset of estrus. Following parturition, peripheral levels of LH are very low. This pattern is followed by a period of intense secretory activity that lasts two to three weeks. An abrupt decrease in the frequency and magnitude of LH peaks occurs approximately six days prior to estrus and ovulation. This decrease in LH secretory activity immediately precedes a transitory elevation in progesterone that lasts for three to four days.

Preliminary evidence indicates that the intense secretory activity of LH may be initiated by surges of estrone. These observations have to be confirmed. The importance of LH as it relates to postpartum estrus has been confirmed in a recent study (Walters et al, 1979). In this instance, removal of calves at day 21 postpartum caused an increase in frequency and magnitude of LH peaks and increased the number of unbound LH receptors in ovarian follicular tissue at 96 hours after removal of the suckling stimulus compared to suckled controls.

A complete understanding of the various complexities relating to duration of the postpartum anestrous period awaits further investigation.

Spermatogenesis

The process of spermatogenesis consists of a complex series of cell division and morphologic changes. The process can be viewed as three steps: mitosis of spermatogonia to yield primary spermatocytes; meiosis of primary spermatocytes to yield spermatids and morphologic changes of the spermatids to yield spermatozoa.

Spermatogenesis ceases in hypophysectomized mammals, indicating that pituitary support is essential. Specifically, hypophysectomy of rats causes a loss of type A spermatogonia (one of the early spermatogonial cell types). Whether this loss is due to blockage of the mitotic process or simply to cell degeneration is not known. Some type B spermatogonia form primary spermatocytes and begin subsequent meiotic divisions. Only a few complete meiosis and these spermatids do not survive to form spermatozoa (Steinberger, 1971).

The hormonal requirements for spermatogenesis still are not defined clearly. Evidently, the hormonal requirements to *initiate* spermatogenesis in the immature animal differ from the hormonal require-

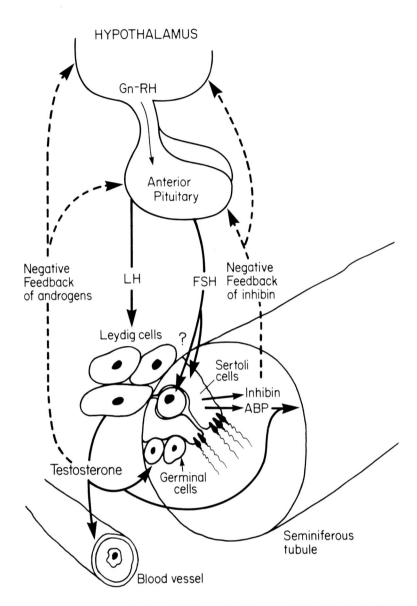

Fig. 5–15. A diagram illustrating the hypothalamic and pituitary control of testis function. Gonadotropin hormone-releasing hormone (Gn-RH) from the hypothalamus stimulates secretion of FSH and LH from the anterior pituitary. FSH acts upon Sertoli cells to cause production of androgen-binding protein (ABP), conversion of testosterone to estrogen, completion of sperm release and postulated secretion of inhibin. FSH may also act upon the germinal cells during the process of spermatocytogenesis. LH acts upon the Leydig cells (interstitial cells) of the testes causing them to secrete testosterone. The postulated action of inhibin is negative feedback regulation of FSH secretion. Testosterone, in turn, acts upon the body in general to cause development of secondary sex characteristics, causes development and maintenance of the male reproductive tract, and acts upon the germinal cells in such a way as to maintain spermatogenesis. Testosterone is bound in the seminiferous tubule by ABP.

ments to *maintain* spermatogenesis in the hypophysectomized animal (Fawcett, 1976). Other discrepancies in such studies have appeared because of the ages and species of animals studied, contamination of FSH preparations with LH, the type of model used (eg, treatment of immature rats with estradiol to suppress gonadotropin to undetectable levels versus the use of hypophysectomized animals), and use of qualitative methods versus quantitative methods for evaluating the effects of different hormonal treatments (Steinberger and Steinberger, 1974).

For maintenance of spermatogenesis in the rat, Steinberger and Steinberger (1974) have proposed the following: mitosis of spermatogonia and the initial stages of meiosis are hormone-independent; the final reduction division of meiosis requires testosterone; and the final morphologic differentiation of spermatids to spermatozoa requires FSH and perhaps testosterone. The initiation of spermatogenesis in the immature animal as well as reinitiation of spermatogenesis following posthypophysectomy testicular regression probably requires FSH (Ortavant et al, 1977).

The role of FSH on the spermatogenic process is not at all understood. The Sertoli cell is a target for FSH and upon stimulation it secretes androgen-binding protein (ABP) (Means et al, 1976). ABP probably ensures that high levels of androgens (testosterone or dihydrotestosterone) are in the lumen of the seminiferous tubule, thereby exposing the germinal cells to high androgen levels. Androgen-binding protein may also be responsible for transporting testicular androgens into the epididymis. Unfortunately, the requirement for FSH is not that clearcut because Sertoli cells have an intracellular androgen receptor as well as membrane receptors for FSH (Means et al, 1976). Moreover, testosterone treatment of Sertoli cell cultures causes secretion of androgen-binding protein.

In addition to secreting ABP and probably seminiferous tubule fluid, Sertoli cells also secrete inhibin (Steinberger and Steinberger, 1977). This hormone appears to exert a negative feedback effect on the secretion of FSH from the pituitary gland. The complex hormone interrelationships between the hypothalamus, the pituitary and the testis are shown in Figure 5–15.

REFERENCES

Ascoli, M. and Puett, D. (1978). Fate of the receptor-bound human chorioniogonadotropin in Leydig tumor cells. Abstr. 338, 60th Annual Meeting, Endocrine Society, Miami, FL.

Bahl, O.P. (1977). The chemistry and biology of human chorionic gonadotropin and its subunits. *In* Frontiers in Reproduction and Fertility Control. Greep, R.O. and Koblinsky, M.A. (eds), Cambridge, Mass., MIT Press, pp. 11–24.

Bolander, F.F., Jr., Ulberg, L.C. and Fellows, R.E. (1976). Circulating placental lactogen levels in dairy and beef cattle. Endocrinology 99, 1273.

Catt, K.J. and Dufau, M.L. (1976). Basic concepts of the mechanism of action of peptide hormones. Biol. Reprod. 14, 1.

Catt, K.J. and Pierce, J.G. (1978). Gonadotropic hormones of the adenohypophysis (FSH, LH and prolactin). *In* Reproductive Endocrinology. Yen, S.S.C. and Jaffe, R.B. (eds), Philadelphia, W.B. Saunders, pp. 34–62.

Chan, J.S.D., Robertson, H.A. and Friesen, H.G. (1976). The purification and characterization of ovine placental lactogen. Endocrinology 98, 65.

Chan, J.S.D., Robertson, H.A. and Friesen, H.G. (1978). Distribution of binding sites for ovine placental lactogen in the sheep. Endocrinology 102, 632.

Chen, T.T., Endres, D.B., Abel, J.H., Jr. and Niswender, G.D. (1978). Binding and degradation of human chorionic gonadotropin by ovine luteal cells. Abstr. 337, 60th Annual Meeting, Endocrine Society, Miami, FL.

Clegg, M.T., Cole, H.H., Howard, C.B. and Pigon, H. (1962). The influence of foetal genotype on equine gonadotropin secretion. J. Endocrinol. 25, 245.

Colcord, M.L., Hoyer, G.L. and Weems, C.W. (1978). Effect of prostaglandin E$_2$ (PGE$_2$) as an antiluteolysin on estrogen-induced luteolysis in ewes. J. Anim. Sci., Suppl. 1 (Abstr), 47, 352.

Cole, H.H. and Hart, G.H. (1930). Potency of blood serum of mares in progressive stages of pregnancy in affecting sexual maturity of immature rats. Am. J. Physiol. 93, 57.

Denamur, R., Martinet, J. and Short, R.V. (1973). Pituitary control of the ovine corpus luteum. J. Reprod. Fertil. 32, 207.

Diekman, M.A., O'Callaghan, P., Nett, T.M. and Niswender, G.D. (1978). Validation of methods and quantification of luteal receptors for LH throughout the estrous cycle and early pregnancy in ewes. Biol. Reprod. *19*, 999.

Dorrington, J.H., Fritz, I.B. and Armstrong, D.T. (1978). Control of testicular estrogen synthesis. Biol. Reprod. *18*, 55.

Dufau, M.L., Tsuruhara, T., Horner, K., Podesta, E., and Catt, K.J. (1977). Evidence for the intermediate action of cyclic AMP and protein kinase in the steroidogenic action of gonadotropin in isolated Leydig cells. Proc. Natl. Acad. Sci. *74*, 3419.

Fawcett, D.W. (1976). The male reproductive system. *In* Reproduction and Human Welfare: A Challenge to Research. Greep, R.O., Koblinsky, M.A. and Jaffe, F.S. (eds), Cambridge, Mass., MIT Press, pp. 165–277.

Fellows, R.E., Bolander, F.F., Hurley, T.W. and Handwerger, S. (1976). Isolation and characterization of bovine and ovine placental lactogen. *In* Growth Hormone and Related Peptides. Pecile, E. and Muller, E.E. (eds), Amsterdam, Excerpta Medica, pp. 315–326.

Fevold, H.L., Hisaw, F.L. and Leonard, S.L. (1931). The gonad-stimulating and luteinizing hormones of the anterior lobe of the hypophysis. Am. J. Physiol. *97*, 291.

Fletcher, P.W. (1978). A study of adenosine 3'–5' cyclic monophosphate binding sites in the ovine corpus luteum using 8-azidoadenosine 3'–5' cyclic monophosphate. Ph.D. Thesis, Laramie, University of Wyoming.

Fortune, J.E. and Armstrong, D.T. (1978). Hormonal control of 17β-estradiol biosynthesis in proestrous rat follicles: estradiol production by isolated theca versus granulosa. Endocrinology *102*, 227.

Friesen, H.G. (1977). Prolactin. *In* Frontiers in Reproduction and Fertility Control. Greep, R.O. and Koblinsky, M.A. (eds), Cambridge, Mass., MIT Press, pp. 25–32.

Goding, J.R., Catt, K.J., Brown, J.M., Kaltenback, C.C., Cumming, I.A. and Mole, B.J. (1969). Radioimmunoassay for ovine luteinizing hormone: secretion of luteinizing hormone during estrus and following estrogen administration in the sheep. Endocrinology *85*, 133.

Goding, J.R., Cumming, I.A., Chamley, W.A., Brown, J.M., Cain, M.D., Cerini, J.C., Cerini, M.E.D., Findlay, J.K., O'Shea, J.D. and Pemberton, D.H. (1971/72). Prostaglandin F$_{2\alpha}$, "the" luteolysin in the mammal? Hormones and Antagonists. Gynec. Invest. *2*, 73.

Gonzalez-Padilla, E., Wiltbank, J.N. and Niswender, G.D., (1975). Puberty in beef heifers. I. The interrelationship between pituitary, hypothalamic and ovarian hormones. J. Anim. Sci. *40*, 1091.

Greep, P.O. (1974). History of research on anterior hypophysial hormones. *In* Handbook of Physiology. Knobil, E. and Sawyer, W.H. (vol eds), sec. 7: Endocrinology, vol. IV: The Pituitary Gland and its Neuroendocrine Control, part 2. Baltimore, American Physiological Society (distributed by Williams & Wilkins Co.) pp. 1–27.

Grimek, H.J., Gorski, J. and Wentworth, B.C. (1979). Purification and characterization of bovine follicle-stimulating hormone: comparison with ovine follicle-stimulating hormone. Endocrinology *104*, 140.

Grimek, H.J. and McShan, W.H. (1974). Isolation and characterization of the subunits of highly purified ovine follicle-stimulating hormone. J. Biol. Chem. *249*, 5725.

Handwerger, S., Crenshaw, C., Maurer, W., Barrett, J., Hurley, T.W., Golander, A. and Fellows, R.E. (1977). Studies on ovine placental lactogen secretion by homologous radioimmunoassay. J. Endocrinol. *72*, 27.

Handwerger, S., Crenshaw, M.C., Lansing, A., Golander, A., Hurley, T.W. and Fellows, R.E. (1978). Stimulation of ovine placental lactogen secretion by arginine infusion. Endocrinology *103*, 1752.

Hansel, W., Concannon, P.W. and Lukaszewska, J.H. (1973). Corpora lutea of the large domestic animals. Biol. Reprod. *8*, 222.

Hilliard, J. (1973). Corpus luteum function in guinea pigs, hamsters, rats, mice and rabbits. Biol. Reprod. *8*, 203.

Hofmann, K. (1974). Relations between chemical structure and function of adrenocorticotropin and melanocyte-stimulating hormones. *In* Handbook of Physiology. Knobil, E. and Sawyer, W.H. (vol eds), sec. 7: Endocrinology, vol. IV: The Pituitary Gland and its Neuroendocrine Control, part 2. Baltimore, American Physiological Society (distributed by Williams & Wilkins Co.), pp. 29–58.

Humphrey, W.D. (1977). Characterization of hormone patterns in the postpartum anestrous beef cow. Ph.D. Thesis, Laramie, University of Wyoming.

Jaffe, R.B. (1978). The endocrinology of pregnancy. *In* Reproductive Physiology. Yen, S.S.C. and Jaffe, R.B. (eds), Philadelphia, W.B. Saunders Co., pp. 521–536.

Kaltenbach, C.C., Graber, J.W., Niswender, G.D. and Nalbandov, A.V. (1968). Luteotropic properties of some pituitary hormones in nonpregnant or pregnant hypophysectomized ewes. Endocrinology *82*, 818.

Kelly, P.A., Robertson, H.A. and Friesen, H.G. (1974). Temporal pattern of placental lactogen and progesterone secretion in the sheep. Nature *248*, 435.

Kendall, J.Z., Challis, J.R.G., Harts, I.C., Jones, C.T., Mitchell, M.D., Ritchie, J.W.K., Robinson, J.S. and Thorburn, C.D. (1977). Steroid and prostaglandin concentrations in the plasma of pregnant ewes during infusion of adrenocorticotrophin or dexamethasone to intact or hypophysectomized foetuses. J. Endocrinol. *75*, 59.

Kiser, T., Dunn, T.G., Corah, L.R. and Kaltenbach, C.C. (1973). CL maintenance in hypophysectomized, pregnant ewes. J. Anim. Sci. (Abstr) *37*, 318.

Lewis, G.S., Jenkins, P.E., Fogwell, R.L. and Inskeep, E.K. (1978). Concentrations of prostaglandins E$_2$ and F$_{2\alpha}$ and their relationship to luteal function in early pregnant ewes. J. Anim. Sci. *47*, 1314.

Li, C.H. (1974). Chemistry of ovine prolactin. *In* Handbook of Physiology. Knobil, E. and Sawyer, W.H. (vol eds), sec. 7: Endocrinology, vol. IV: The Pituitary Gland and its Neuroendocrine Control, part 2. Baltimore, American Physiological Society (distributed by Williams & Wilkins Co.), pp. 103–110.

Liao, T.H. and Pierce, J.G. (1970). The presence of a common type of subunit in bovine thyroid-stimulating and luteinizing hormone. J. Biol. Chem. 245, 3275.

Lunstra, D.D., Ford, J.J. and Echternkamp, S.E. (1978). Puberty in beef bulls: Hormone concentrations, growth, testicular development, sperm production and sexual aggressiveness in bulls of different breeds. J. Anim. Sci. 46, 1054.

McNatty, K.P. (1978). Follicular determinants of corpus luteum function in the human ovary. Proc. Workshop on Ovarian Follicular and Corpus Luteum Function, Americana Hotel, Miami Beach, June 11–13, 1978.

Means, A.R., Fakunding, J.L., Huckins, C., Tindall, D.J. and Vitale, R. (1976). Follicle-stimulating hormone, the Sertoli cell and spermatogenesis. Recent Prog. Horm. Res. 32, 477.

Mendelson, C., Dufau, M.L., and Catt, K.J. (1975). Gonadotropin binding and stimulation of cyclic AMP and testosterone production in isolated Leydig cells. J. Biol. Chem. 250, 8818.

Niswender, G.D., Nett, T.M. and Akbar, A.M. (1974). The hormones of reproduction. *In* Reproduction in Farm Animals, 3rd ed. Hafez, E.S.E. (ed), Philadelphia, Lea & Febiger, pp. 57–81.

Niswender, G.D., Reimers, T.J., Diekman, M.A. and Nett, T.M. (1976). Blood flow: a mediator of ovarian function. Biol. Reprod. 14, 64.

Nicoll, C.S. (1974). Physiological actions of prolactin. *In* Handbook of Physiology. Knobil, E. and Sawyer, W.H. (vol eds), sec. 7: Endocrinology, vol. IV: The Pituitary Gland and its Neuroendocrine Control, part 2. Baltimore, American Physiological Society (distributed by Williams & Wilkins Co.), pp. 253–292.

Ortavant, R., Courot, M. and Hochereau de Reviers, M.T. (1977). Spermatogenesis in domestic animals. *In* Reproduction in Domestic Animals, 3rd ed. Cole, H.H. and Cupps, P.T. (eds), New York, Academic Press, pp. 203–228.

Papkoff, H., Sairam, M.R., Farmer, S.W. and Li, C.H. (1973). Studies on the structure and function of interstitial cell-stimulating hormone. Recent Prog. Horm. Res. 29, 563.

Papkoff, H., Ryan, R.J. and Ward, D.N. (1977). The gonadotropic hormones, LH (ICSH) and FSH. *In* Frontiers in Reproduction and Fertility Control. Greep, R.O. and Koblinsky, M.A. (eds), Cambridge, Mass., The MIT Press, pp. 1–10.

Peckham, W.D., Yamaji, T., Dierschke, D.J. and Knobil, E. (1973). Gonadal function and the biological and physicochemical properties of follicle-stimulating hormone. Endocrinology 92, 1660.

Pierce, J.G. (1974). Chemistry of thyroid-stimulating hormone. *In* Handbook of Physiology. Knobil, E. and Sawyer, W.H. (vol eds), sec. 7: Endocrinology, vol. IV: The Pituitary Gland and its

Neuroendocrine Control, part 2. Baltimore, American Physiological Society (distributed by Williams & Wilkins Co.), pp. 79–101.

Rawlings, N.C., Pant, H.C. and Ward, W.R. (1978). The effect of passive immunization against oestrogens on the onset of parturition in the ewe. J. Reprod. Fertil. 54, 363.

Richards, J.S. and Midgley, A.R., Jr. (1976). Protein hormone action: a key to understanding ovarian follicular and luteal cell development. Biol. Reprod. 14, 82.

Sairam, M.R. and Papkoff, H. (1974). Chemistry of pituitary gonadotrophins. *In* Handbook of Physiology. Knobil, E. and Sawyer, W.H. (vol eds), sec. 7: Endocrinology, vol. IV: The Pituitary Gland and its Neuroendocrine Control, part 2. Baltimore, American Physiological Society (distributed by Williams & Wilkins Co.), pp. 111–131.

Sawyer, H.R., Abel, J.H., Jr., McClellan, M.C. and Chen, T.T. (1977). Mechanism of progesterone secretion from ovine corpora lutea *in vitro*. Proceedings Tenth Annual Meeting, Society for the Study of Reproduction, University of Texas, Austin.

Schroff, C., Kaltenbach, C.C., Graber, J.W. and Niswender, G.D. (1971). Maintenance of corpora lutea in hypophysectomized ewes. J. Anim. Sci. 33, 268 (Abstr).

Sherwood, O.D. and McShan, W.H. (1977). Gonadotropins. *In* Reproduction in Domestic Animals, 3rd ed. Cole, H.H. and Cupps, P.T. (eds), New York, Academic Press, pp. 17–47.

Smith, P.E. (1926). Ablation and transplantation of the hypophyses in the rat. Anat. Rec. 32, 221.

Smith, P.E. and Engle, E.T. (1927). Experimental evidence regarding the role of the anterior pituitary in the development and regulation of the genital system. Am. J. Anat. 40, 159.

Steinberger, A. and Steinberger, E. (1977). The Sertoli cells. *In* The Testis, vol. IV. Johnson, A.D. and Gomes, W.R. (eds), New York, Academic Press, pp. 371–399.

Steinberger, E. (1971). Hormonal control of mammalian spermatogenesis. Physiol. Rev. 51, 1.

Steinberger, E. (1975). Hormonal regulation of the seminiferous tubule function. *In* Current Topics in Molecular Endocrinology, Hormonal Regulation of Spermatogenesis. French, F.S., Hansson, V., Ritzen, E.M. and Nayfeh, S.N. (eds), New York, Plenum Press, p. 337.

Steinberger, E. and Steinberger, A. (1974). Hormonal control of testicular function in mammals. *In* Handbook of Physiology. Knobil, E. and Sawyer, W.H. (vol eds), sec. 7: Endocrinology, vol. IV: The Pituitary Gland and its Neuroendocrine Control, part 2. Baltimore, American Physiological Society (distributed by Williams & Wilkins Co.), pp. 325–345.

Sutherland, E.W. and Rall, T.W. (1960). The relation of adenosine-3′, 5′-phosphate and phosphorylase to the actions of catecholamines and other hormones. Pharm. Rev. 12, 265.

Walters, D.L., Dunn, T.G. and Kaltenbach, C.C. (1979). Unpublished observations.

Ward, D.N., Reichert, L.E., Jr., Liu, W-K, Nahm, H.S., Hsia, J., Lamkin, W.M. and Jones, N.S. (1973). Chemical studies of luteinizing hormone from human and ovine pituitaries. Recent Prog. Horm. Res. *29*, 533.

Wilhelmi, A.E. (1974). Chemistry of growth hormone. *In* Handbook of Physiology. Knobil, E. and Sawyer, W.H. (vol eds), sec. 7: Endocrinology, vol. IV: The Pituitary Gland and its Neuroendocrine Control, part 2. Baltimore, American Physiological Society (distributed by Williams & Wilkins Co.), pp. 59–78.

Yen, S.S.C. (1978). Chronic anovulation due to CNS-hypothalamic-pituitary dysfunction. *In* Reproductive Endocrinology. Yen, S.S.C. and Jaffe, R.B. (eds), Philadelphia, W.B. Saunders Co., pp. 341–372.

Zondek, B. and Aschheim, S. (1926). Uber die Funktion des Ovariums. Z. Geburtshilfe Gynaekol. *90*, 372.

6

Neuroendocrinology of Reproduction

J.J. REEVES

The changes in breeding and anestrous periods of many species are correlated with seasonal variations and climatic changes. For example, the ewe exhibits a breeding season during the part of the year when daylight is shortened and an anestrous season during long days, an observation that is logically explained by the nervous system's being an intermediary between the external environment and the endocrine system. In other species such as the cat, rabbit and mink, ovulation occurs only after copulation, which again is explained by the existence of a neural connection between stimulus and response.

Neuroendocrinology involves the interactions of the nervous system with endocrine glands. Three types of cells mediate communication between organs: neurons, neuroendocrine cells and endocrine cells. The neuroendocrine cell converts a neuronal input to an endocrine output. The neuroendocrine cell releases a neurosecretory product (hormone) that is transported in the circulation, in contrast to a neuron, which releases a neurohumor that diffuses only a short distance across the synaptic cleft. The basic difference between the transmission of signals by neurohumors and neurohormones involves the anatomic mode of transport. The neurohumor comes in contact with only a limited number of cells after being released from the presynaptic neuron, while a neurohormone is distributed by way of the circulatory system to many cells in the body.

Neuroendocrine reactions play a major role in the regulation of reproductive, metabolic and behavioral functions of the body. The neural structures that serve an endocrine function may be termed neuroendocrine glands. The neuroendocrine glands consist of the hypothalamus, the neurohypophysis, the pineal gland and the adrenal medulla.

HYPOTHALAMUS

The hypothalamus lies at the base of the brain; it is bordered anteriorly by the optic chiasma, posteriorly by the mammillary bodies, dorsally by the thalamus and ventrally by the sphenoid bone. Its entire size is about 1/300 that of the brain. The

114

hypothalamus is composed of many bilaterally paired nuclei, some of which are schematically described in Figure 6–1 to indicate their relative positions. The medial portion of the hypothalamus is known as the third ventricle of the brain which separates most of the paired nuclei.

A unique vascular connection exists between the hypothalamus and the anterior pituitary. Arterial blood comes into the pituitary by way of the superior hypophyseal artery and inferior hypophyseal artery. The superior hypophyseal artery forms capillary loops at the median eminence and pars nervosa. From these capillaries the blood flows into the hypothalamo-hypophyseal portal vessels, which pass down the pituitary stalk and terminate in capillaries in the anterior pituitary. A portal system begins and ends in capillaries without going through the heart. The hypothalamo-hypophyseal portal system is the vascular pathway that transports hypothalamic hormones to the anterior pituitary (Fig. 6–2). The inferior hypophyseal artery transports blood to the anterior and posterior pituitary. Not only does blood flow from the hypothalamus to the pituitary, but part of the venous outflow of the anterior pituitary is by way of a retrograde flow back to the hypothalamus (Oliver et al, 1977; Bergland and Page, 1978). Thus, the hypothalamus is exposed to high concentrations of pituitary hormones in blood passing in a retrograde direction. The physiologic importance of these findings can be appreciated because they are associated with a negative feedback regulation of the hypothalamus by the pituitary hormones. This type of feedback has been termed the short-loop feedback.

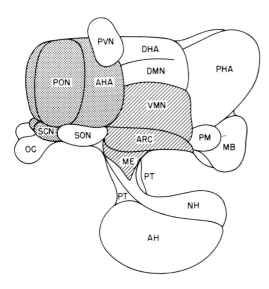

☳ control centers of preovulatory LH and FSH

▨ control centers of tonic LH and FSH secretion

Fig. 6–1. Schematic drawing of hypothalamic nuclei and pituitary. *AH*, Adenohypophysis; *ARC*, arcuate nucleus; *AHA*, anterior hypothalamic area; *DHA*, dorsal hypothalamic area; *DMN*, dorsal medial nucleus; *ME*, median eminence; *NH*, neurohypophysis; *MB*, mammillary body; *PM*, premammillary nucleus; *OC*, optic chiasm; *PVN*, paraventricular nuclei; *PON*, preoptic nuclei; *PHA*, posterior hypothalamic area; *PT*, pars tuberalis; *SCN*, suprachiasmatic nucleus; *SON*, supraoptic nuclei; *VMN*, ventromedial nucleus. The diagonal lines show nuclei that control tonic LH and FSH release from the pituitary. The dotted areas are nuclei that control the preovulatory surge of LH and FSH from the pituitary.

Luteinizing Hormone Releasing Hormone (LHRH)

The theory of neurohumoral control of the anterior pituitary was formulated by Green and Harris (1947) on the basis of anatomic and physiologic data that hypothalamic nerve fibers liberate hormonal substances into the hypothalamo-hypophyseal portal vessels. Thirty years later Schally and Guillemin (1977) shared the Nobel prize for their independent work on determining the chemical structures of various hormones of the hypothalamus that control pituitary function.

Chemical Nature. Substances of the hypothalamus that control the release of pituitary hormones were initially called releasing factors. However, as their chemical structures have become known they have been termed releasing hormones

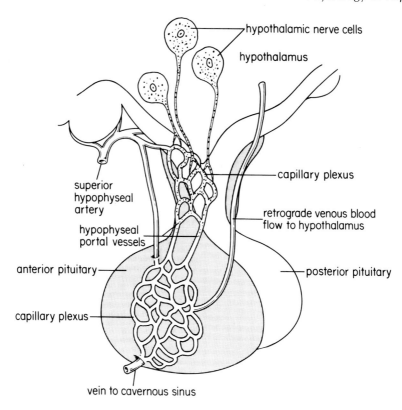

Fig. 6–2. Schematic representation of hypothalamic nerve cells releasing neurohormones into the portal vessels for transport to the anterior pituitary via the hypothalamo-hypophyseal portal vessels. Solid particles in nerve cell represent neurohormone. Blood is also transported by the retrograde venous system back to the hypothalamus.

(abbreviated RH), which could also signify regulating hormones, since some hypothalamic hormones seem to affect the synthesis as well as the release of respective anterior pituitary hormones. The chemical structures of some hypothalamic substances that control anterior pituitary function are not yet known and are still considered hypothalamic releasing factors.

LHRH is a decapeptide (10 amino acids) with a molecular weight of 1183 (Fig. 6–3,*A*). The synthesis of a large number of structural analogs of LHRH has been important in establishing the structure-activity relationship of this hormone. Two basic types of analogs to LHRH have been synthesized. The antagonistic analogs appear to bind at the receptor site on the pituitary but do not induce the release of luteinizing hormone (LH) or follicle stimulating hormone (FSH) and block the action of the natural hormone. Stimulatory analogs that induce release of more LH and FSH than the natural hormone have also been synthesized. Because of the increase in LH and FSH release caused by the stimulatory analogs, they are more effective in inducing ovulation, as demonstrated in anestrous ewes by Kinder et al (1976). Table 6–1 illustrates structural changes in LHRH that alter the biologic activity of the molecule.

The C-terminal portion of this molecule is necessary for attaching to the receptor and the first three amino acids are necessary for activating LH and FSH release. The reason for the increased biologic ac-

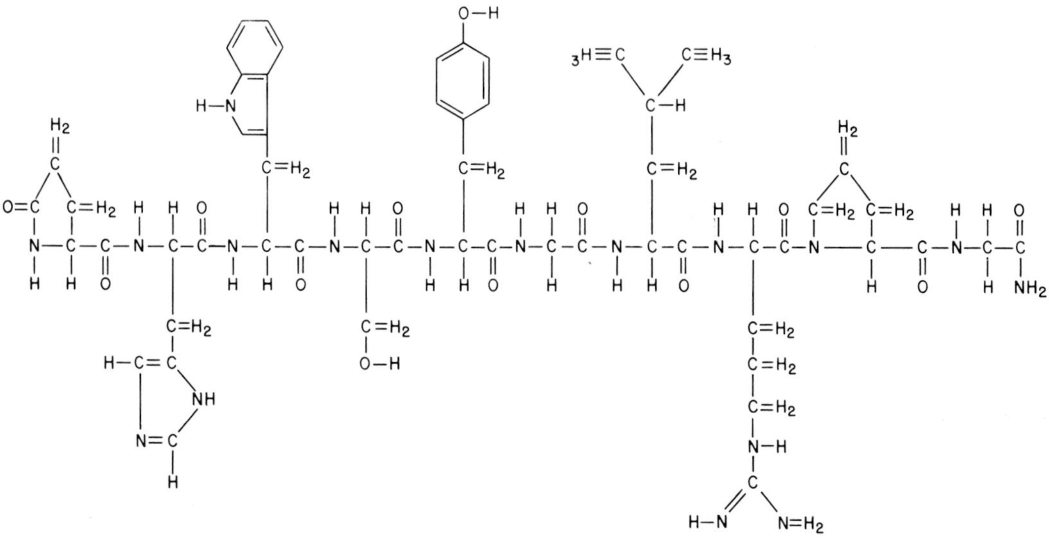

A   PYROGLU-HIS-TRP-SER-TYR-GLY-LEU-ARG-PRO-GLY-NH₂

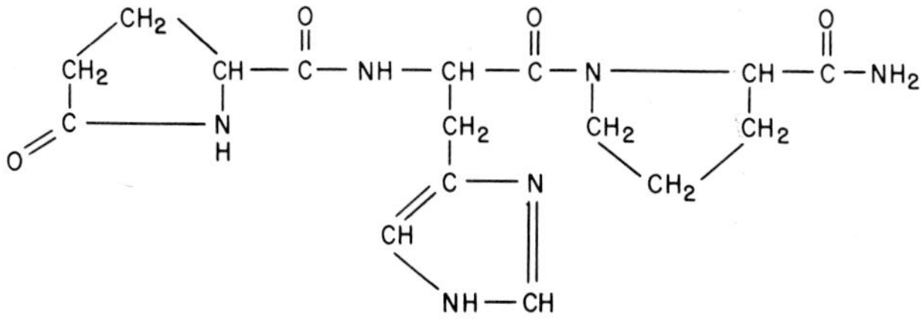

B   PYROGLU-HIS-PRO-NH₂

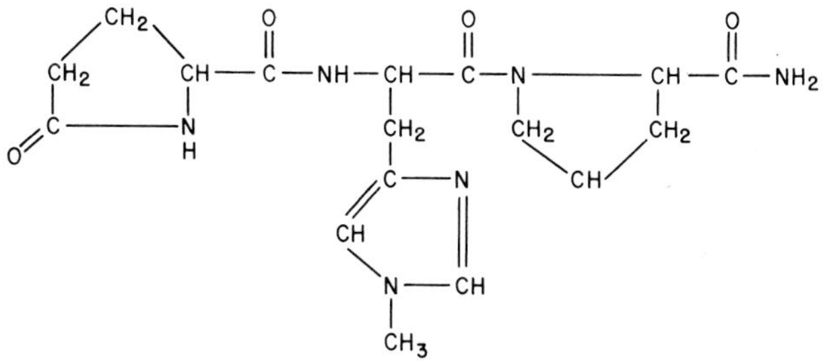

C   PYROGLU-N^{3im}Methyl-HIS-PRO-NH₃

Fig. 6–3. *A,* Molecular structure of LHRH, a decapeptide consisting of 10 amino acids, indicating that one hormone is responsible for release of both LH and FSH from the anterior pituitary. *B,* Molecular structure of TRH, a tripeptide consisting of three amino acids. *C,* Molecular structure of TRH analog, which has eight times more biologic potency than native TRH. Note the extra methyl group on the 3 position nitrogen of the imidazole ring of histidine.

Table 6–1.   Structural Relationships of LHRH and Some Stimulatory and Antagonistic Analogs

Name	Structure										Biologic Activity[a]	Species
LHRH	Pyro-glu	His	Trp	Ser	Tyr	Gly	Leu	Arg	Pro	Gly NH_2	1	Sheep
desglyNH_2^{10}-LHRH-ethylamide	1[b]	2	3	4	5	6	7	8	9	$NHCH_2CH_3$	2	Sheep
D-Leu[6]-desgly NH_2^{10}-LHRH-ethylamide	1	2	3	4	5	DLeu	7	8	9	$NHCH_2CH_3$	10	Sheep
D-Phe[2]-Phe[3]-DPhe[6]-LHRH-ethylamide	1	D-Phe	Phe	4	5	DPhe	7	8	9	$NHCH_2CH_3$	Inhibits Ovula-tion	Rat

[a] Biologic activities are compared on equal weight basis to LHRH. Activity of LHRH was assigned 1.
[b] Number denotes amino acid associated with natural LHRH.

tivity of D-Leu[6]-des-gly-NH_2^{10}-LHRH-ethylamide is due to its ability to stay bound to the pituitary longer than the natural hormone (Reeves et al, 1977). The half-life of this hormone in circulation is identical to that of the natural hormone and therefore does not contribute to its increased biologic activity (Dias and Reeves, 1978). The half-life of LHRH is seven minutes after a single injection of the hormone into a sheep (Nett et al, 1973) (Fig. 6–4,*A*). However, when sheep are infused for an extended period with LHRH, the half-life upon withdrawal of treatment is three hours (Fig. 6–4,*B*). As would be expected, clearance of the hormone from the blood after a pulse injection is faster than clearance when all tissues plus blood are saturated with the hormone, as would be the case after infusion for an extended period of time. In either case, the short-half life of LHRH has made its application to the livestock industry difficult.

The nomenclature of LH and FSH releasing hormone is summarized in Table 6–2.

Control of LH and FSH Release. The release of both LH and FSH was found to be controlled by one hypothalamic hor-mone. Figure 6–5 shows the release of both LH and FSH in a ewe injected with synthetic LHRH. The LHRH molecule has been given many names and abbreviations, which are summarized in Table 6-2. There is no agreement among researchers regarding which of these terms should be the universal name for the LH and FSH releasing hormone, but the student must recognize that all of these different names appear in the literature and refer to the same hormone. The term luteinizing hormone releasing hormone (LHRH) is used throughout this text as the preferred name.

Tonic LH and FSH Release. Serum LH and FSH are released in a tonic or basal fashion in both the female and male. Tonic levels of LH and FSH are controlled by negative feedback from the gonads. The tonic level of LH is not stationary, but shows circhoral (L. circa = about, L. horal = hour) oscillation (approximately hourly changes), which are most evident in the ovariectomized ewe (Fig. 6–6). Tonic serum LHRH concentrations are also elevated in the ovariectomized ewe. The circhoral oscillation of LH and the increased concentration after ovariectomy is due to the lack of a negative feedback

Table 6–2.   Nomenclature of LH and FSH Releasing Hormone Appearing in Literature

Name	Abbreviation
Luteinizing hormone-releasing hormone	LHRH or LRH
Luteinizing hormone-releasing hormone/follicle stimulating hormone-releasing hormone	LHRH/FSHRH
Gonadotropin releasing hormone	GnRH
Luteinizing hormone releasing factor	LHRF or LRF
Luliberin	LHRH
Gonadoliberin	GnRH

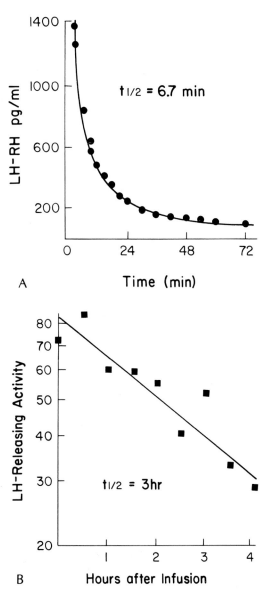

Fig. 6–4. A, LHRH levels in sheep serum after intracarotid injection of 100 μg. LHRH measured by radioimmunoassay. (*Modified from Nett et al, 1973. J. Clin. Endocrinol. Metab. 36:886*).
B, Serum LHRH levels in sheep after infusion of 1 mg/hr for 16.5 hr. LHRH measured by bioassay. (*Modified from Dias and Reeves, 1978. J. Anim. Sci. 46, 1707*)

from the ovarian steroids on the tonic LH control center in the hypothalamus.

Preovulatory LH and FSH Release. A second type of LH and FSH release, called the preovulatory surge of LH and FSH, is evident in the female prior to ovulation. The preovulatory surge of LH and FSH is responsible for ovulation and it lasts from

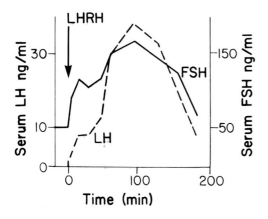

Fig. 6–5. Concurrent rise in serum LH and FSH concentration following an injection of synthetic LHRH in an anestrous ewe. *(Adapted from Reeves et al, 1972. J. Anim. Sci. 35, 84).*

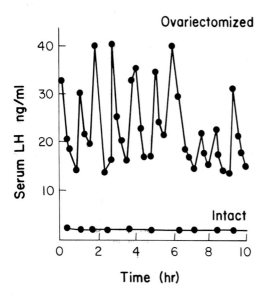

Fig. 6–6. Tonic serum LH in ovariectomized and intact ewes. Note the higher LH level in the absence of negative feedback from the ovary. Also note the circhoral fluctuation or episodic release of LH, which is easily noted in the ovariectomized ewe.

6 to 12 hours in most species. The preovulatory surge of LH is initiated by an increase in the circulating estrogen concentration, which has a positive effect on the hypothalamic-pituitary axis in inducing the surge of LH and FSH release.

In the anestrous ewe treated with estradiol-17β, a surge of LH and FSH occurs within 15 or 16 hours after treatment (Beck and Reeves, 1973; Reeves et al, 1974). The site of estrogen activity is at the anterior hypothalamus, at the preoptic and suprachiasmatic areas, where it increases LHRH release, and at the pituitary, where it increases the sensitivity of the pituitary to LHRH, resulting in an increased release of LH and FSH.

Halasz (1969) developed a knife that can separate parts of the hypothalamus from the rest of the brain. Thus was demonstrated that the preoptic and anterior hypothalamic nuclei control the preovulatory surge of LH and FSH while the arcuate nucleus, ventromedial nucleus and median eminence control tonic release of LH and FSH (Halasz, 1969; Jackson et al, 1978). With two centers in the hypothalamus controlling LH and FSH release (Fig. 6–1), estrogen can stimulate preovulatory LH and FSH release by positive feedback on the preoptic anterior hypothalamic area and decrease tonic LH and FSH release by negative feedback on the arcuate nucleus, ventromedial nucleus and median eminence of the hypothalamus.

Karsch, Foster, Legan, and Hauger (1976) have proposed a logical explanation of the anestrous season in the sheep. During the last estrous cycle of the breeding season, estrogen exerts a negative feedback on the tonic LH control center in the hypothalamus. Because of this negative feedback and decreased LH release, estradiol secretion also decreases. Therefore the estrogen-threshold required to initiate the preovulatory LH surge cannot be attained and ovulation will not occur. The resulting anestrous condition persists until the tonic LH control center in the hypothalamus is not inhibited by estrogen. The decrease in day length allows the tonic LH control center of the hypothalamus to become refractory to the negative feedback of estrogen and results in increased tonic LH release. The increase in circulating tonic LH results in increased estradiol release from the ovarian follicle, which stimulates the preovulatory LH control center in the hypothalamus. The breeding season would begin and progesterone rather than estradiol would be the major inhibitory reg-

ulator of tonic LH secretion. Figure 6–7 shows a proposed model for the hypothalamic involvement in the anestrous and breeding seasons of the ewe.

Extra-pituitary Sites of LHRH Action. *Brain.* In the female rat a clear relationship exists between the ovarian cycle and rhythms of sexual receptivity. Removal of the ovaries results in complete cessation of behavioral estrus. The heat response can be reinitiated by treatment with daily doses of exogenous estrogen or with relatively small doses of estrogen followed by progesterone. LHRH will also induce lor-

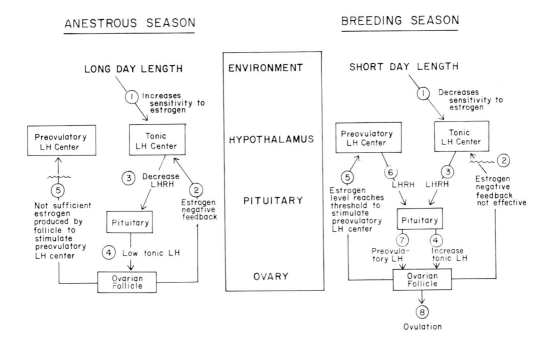

Fig. 6–7. The four major components to consider in the model for hypothalamic control of the anestrous season in the ewe are external environment (day length), hypothalamus, pituitary and ovary. During the anestrous season or long days, the ewe does not show reproductive cyclic activity. The length of daylight by some mechanism, probably through the optic nerve, sensitizes the tonic LH center in the hypothalamus to become sensitive to circulating estrogen concentrations (1). Estrogen (2) then negatively feeds back on the tonic LH control center and less LHRH is released (3). Less tonic LH is released into circulation (4), resulting in lower estrogen production(5) from the ovarian follicle. The circulating estrogen concentrations cannot reach a threshold concentration to stimulate the preovulatory LH center in the hypothalamus, resulting in the lack of ovulation during the anestrous season.

During the breeding season of the ewe, the day length decreases (1) and the tonic LH control center in the hypothalamus becomes refractory to circulating estrogen (2). Increasing levels of estrogen do not negatively affect the tonic LH control center at this time, resulting in increased LHRH release (3), which induces increased tonic LH release from the pituitary (4). The increased tonic LH release induces sufficient estrogen levels (5) in circulation to stimulate the preovulatory LH center of the hypothalamus, resulting in increased LHRH release (6). The resulting preovulatory surge of LH (7) induces ovulation (8).

dosis (female mating characteristic in rat) in ovariectomized rats treated with estrogen in dosages too low to provoke mating (Moss and McCann, 1973). Two hours after LHRH treatment, components of female sexual behavior appear in estrogen-primed rats.

Interestingly, the primary region concerned with mating behavior in the female rat is the preoptic nuclei of the hypothalamus, because lesions in this area abolish mating behavior. It is therefore possible that LHRH might be released from terminals of LHRH-containing neurons prior to estrus and interact with cells associated in the mediation of sexual behavior. LHRH may also reach cells associated in the mediation of sexual behavior by the retrograde blood flowing from the pituitary back to the hypothalamus. Good evidence supports the existence of a role for LHRH in producing behavioral estrus in the rat; however, no data are available at this time to indicate that LHRH acts on the central nervous system of domestic species.

Ovaries. D-Leu[6]-des-gly-NH$_2$[10]-LHRH-ethylamide, an analogue of LHRH, can decrease human chorionic gonadotropin-induced ovarian growth in hypophysectomized rats (Ripple and Johnston, 1976). These data suggest that the LHRH molecule may have some direct inhibitory actions on the ovary. A direct inhibitory action of LHRH on the ovary might be important to prevent the formation of cystic follicles in the normal female. However, this assumption has not been confirmed.

Potential and Existing Uses of LHRH. LHRH has been used in the treatment of cystic ovaries and ovulation in anestrus ewes, and has other potential uses.

Cystic Ovaries. LHRH is effective in overcoming cystic follicles in cows (Kittock et al, 1973). In this instance, 100 μg of LHRH induces the release of substantial amounts of endogenous LH to induce rupture of the cystic follicle, which allows the

cow to resume normal cyclic activity. Abbott Laboratories has placed LHRH on the market for therapeutic control of ovarian cysts; their trade name for this compound is Cystorelin.

Ovulation in Anestrous Ewes. D-Leu[6]-des-gly-NH$_2$[10]-LHRH-ethylamide is effective in releasing enough LH and FSH to induce ovulation in anestrous ewes (Kinder et al, 1976). Although functional corpora lutea are formed as a result of ovulation (Frandle et al, 1977), anestrous ewes induced to ovulate with LHRH or its analogues do not conceive when artificially inseminated with viable ram semen. The reason for this lack of conception in LHRH-treated ewes is not understood, but it may be associated with the need for presensitization of the reproductive tract with steroids from a previous estrous cycle. At the present time LHRH or its analogues are not effective for inducing out-of-season breeding in the ewe. However, the potential for this hormone to be used in anestrous breeding of ewes still exists. LHRH has not been effective in inducing superovulation in sheep or cattle.

Clinical Applications. LHRH inhibitory analogues developed in the future may be effective for nonsteroid contraception in humans and synchronization of estrus in domestic species. Inhibitory LHRH analogues can block the normal endogenous LH surge in rats and consequently block ovulation. These LHRH inhibitory analogues are being tested at this time as nasal sprays for human contraception.

Thyrotropin Releasing Hormone (TRH)

Thyrotropin releasing hormone (TRH), also called thyrolibrin, was the first hypothalamic hormone to be isolated. The structure was determined by Nair et al, 1970, and Burgus et al, 1970. In both pigs and sheep this hormone was found to be a tripeptide (Fig. 6–3B). Of the dozens of compounds that have been synthesized and tested for TRH activity, only one has been found that exceeds native TRH activity (Vale et al, 1971) (Fig. 6–3C).

TRH also induces prolactin release in cattle (Davis et al, 1977). Although one cannot conclude that TRH is the prolactin releasing hormone, it is not unusual for one hypothalamic hormone to be associated with the release of more than one anterior pituitary hormone. Figure 6–8 shows the effect of TRH on TSH and prolactin release in dairy heifers.

Prolactin Release Inhibiting Factor (PRIF)

Prolactin release inhibiting factor, a substance in the hypothalamus of pigs that inhibits prolactin release, was found to be a gamma-amino butyric acid (GABA) by Schally et al. (1977). GABA appears to act directly on the pituitary to inhibit prolactin release; however, it is not resolved at this time whether this is the physiologic messenger controlling prolactin release.

Presently the potential uses of either TRH or PRIF for increasing livestock production are difficult to predict. PRIF may be useful at weaning to inhibit lactation in the nursing dam. Although not directly associated with reproduction, TRH has the potential for increasing thyroid activity, which has been associated with increased growth and milk production.

NEUROHYPOPHYSIS

Oxytocin

Oxytocin and vasopressin were first identified in extracts prepared from pituitary glands of cattle and swine in 1953 by Du Vigneaud. These were the first hormonal peptides to be synthesized. Du Vigneaud received a Nobel prize for this accomplishment in 1955. The mammalian neurohypophysis is the source of three principal hormones: oxytocin, arginine vasopressin and lysine vasopressin (Table 6–3).

Oxytocin is found in all mammals. Arginine vasopressin, also called antidiuretic hormone (ADH), has been identified in marsupials, which suggests that it appeared early in the evolution of mammals and it is the most common vasopressin in mammals. Lysine vasopressin has been identified in domestic pigs. However, both types of vasopressin are found in the warthog, peccary and hippopotamus. Vasotocin is the neurohypophyseal hormone found in the bird and has both oxytocin and vasopressin-like actions.

Chemical Nature. Oxytocin and ADH are synthesized in the supraoptic and paraventricular nuclei of the hypo-

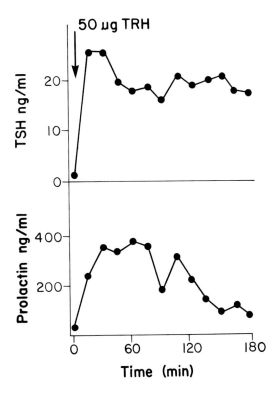

Fig. 6–8. Plasma concentrations of TSH and prolactin in dairy heifers injected with 50 µg TRH. Note that both plasma TSH and prolactin were elevated after injection with synthetic TRH. *(Adapted from Davis et al, 1977. Endocrinology 100, 1394)*

Table 6–3. Chemical Structure of Neurohypophyseal Hormones

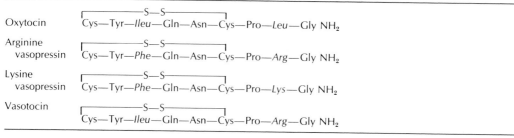

Oxytocin
Cys—Tyr—*Ileu*—Gln—Asn—Cys—Pro—*Leu*—Gly NH$_2$

Arginine
vasopressin
Cys—Tyr—*Phe*—Gln—Asn—Cys—Pro—*Arg*—Gly NH$_2$

Lysine
vasopressin
Cys—Tyr—*Phe*—Gln—Asn—Cys—Pro—*Lys*—Gly NH$_2$

Vasotocin
Cys—Tyr—*Ileu*—Gln—Asn—Cys—Pro—*Arg*—Gly NH$_2$

thalamus and are only stored or released from the neurohypophysis. These neurohypophyseal hormones are synthesized together with the carrier proteins called neurophysins. The complex of neurophysin I and oxytocin can be considered a prohormone for oxytocin. Like other neurosecretions, oxytocin and ADH are transported in small vesicles enclosed by a membrane. These secretory vesicles flow down the hypothalamic hypophyseal nerve axons by axoplasmic streaming and are stored at the nerve endings next to the capillary beds in the neurohypophysis until their release into the circulation (Fig. 6–9).

In-vitro studies with hypothalami have shown that labeled tyrosine can be incorporated into oxytocin or vasopressin but cultured neurohypophyses cannot incorporate labeled tyrosine into oxytocin or vasopressin. These results are proof that the neurohypophyseal hormones are synthesized only in the hypothalamus and not in the neurohypophysis.

Function. Vasopressin acts upon the epithelial cells of the distal portion of the renal tubule to cause reabsorption of water. Oxytocin in Greek means "rapid birth," thus describing one of its physiologic functions, which is contraction of uterine muscle. Oxytocin also causes increased contraction frequency in the oviduct and may thus be involved in the transport of both female and male gametes in the oviduct. How oxytocin directly affects the uterine and oviduct contractions

is not known; however, estrogen enhances the responsiveness of smooth muscle to oxytocin. In the bird and reptile vasotocin appears to be important in causing contraction of the shell gland and vagina to induce oviposition.

The milk ejection reflex or milk letdown is an example of a neuroendocrine reflex. This is one of the best established functions of oxytocin. The lactating female becomes conditioned to visual and tactile stimuli associated with suckling or milking; this conditioning induces the

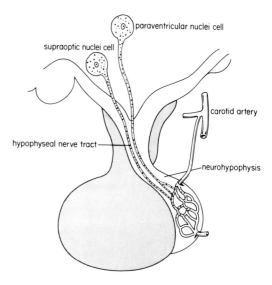

Fig. 6–9. Schematic drawing of transport of neurohypophyseal hormones down nerve tract from hypothalamus to neurohypophysis. Oxytocin is synthesized in the PVN and SON and stored and released from the neurohypophysis.

release of oxytocin into the circulation. Oxytocin then acts on the myoepithelial cells (smooth muscle cells) that surround the alveoli in the mammary gland. The contraction of the myoepithelial cells puts pressure on the alveoli, which displaces milk into the duct system of the mammary gland, resulting in milk letdown (Fig. 6–10).

Application. Oxytocin is used in the livestock industry to induce female animals to let down milk after parturition if a problem exists and to induce expulsion of retained placenta. It is also used to aid delivery of young animals when the female has been in labor for an extended period.

PINEAL GLAND

Anatomy

The pineal gland originates as a neuroepithelial evagination protruding from the roof of the diencephalon. This organ is conical in shape and lies within the posterior border of the corpus callosum and between the superior colliculi of the brain. The pineal gland has undergone great changes as vertebrates have evolved from amphibians to mammals. The amphibian pineal gland is a photoreceptor that sends information to the brain. The mammalian pineal is an endocrine organ which, despite its derivation

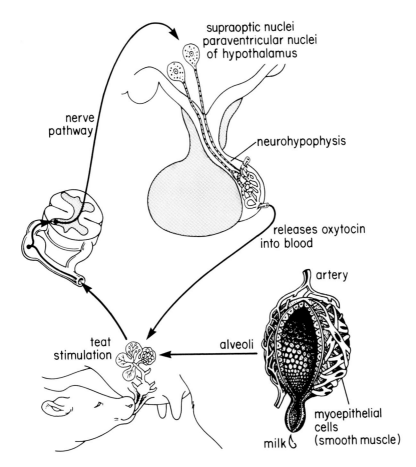

Fig. 6–10. Milk letdown may be considered a neuroendocrine reflex. The stimulation of the teat induces a neural signal to the hypothalamus to release oxytocin from the neurohypophysis, which stimulates the myoepithelial cells to constrict the alveoli resulting in milk secretion.

from the diencephalic roof, has no direct connection to the central nervous system. The metabolism of the pineal is controlled by environmental lighting by an indirect pathway involving sympathetic nerves. This pathway involves the eye and the optic nerve. It has not yet been possible to trace the path taken by the photic input to the pineal gland between the optic tract and the spinal cord. The neurotransmitter released from the postganglionic sympathetic axion is norepinephrine.

Melatonin

Chemical Nature. Pineal cells called pinealocytes take up the amino acid, tryptophan, from the circulation, and an enzyme catalyzes its hydroxylation and a second enzyme decarboxylates it to serotonin. Two steps in the metabolism of serotonin are under neural control. These

are the conversion of serotonin to N-acetylserotonin by the enzyme serotonin-N-acetyltransferase. This enzyme shows a circadian rhythm in its activity (Klein and Weller, 1970) and appears to be the rate-limiting step in melatonin formation. The second step involves the melatonin-forming enzyme, hydroxyindole-O-methyltransferase (HIOMT), which does not show a significant rhythm in its activity. A general outline of the indolamine metabolism of the pineal gland is presented in Figure 6–11.

Function. The skin of frogs and tadpoles blanches rapidly when the animals are fed extracts of bovine pineal glands. This skin-lightening principle in the pineal extracts is melatonin. Melatonin has no effect on the melanocytes responsible for pigmentation of mammalian skin. In place of dermatologic actions, melatonin has come to exert effects on

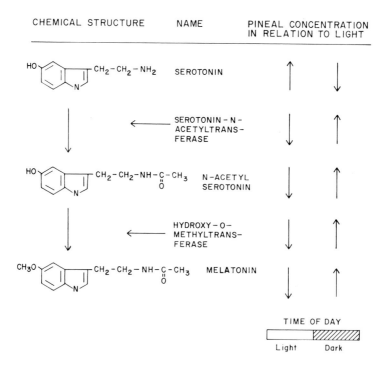

Fig. 6–11.   A model of indolamine metabolism in the rodent pineal gland. The metabolic pathway from serotonin to melatonin is shown on the left. The variations in pineal concentrations of metabolites and enzymes in relation to light/dark phases is shown on the right.

Table 6–4. **Summary of Origin and Function of Neurohormones Involved in Reproduction**

Gland	Hormone	Function
Hypothalamus	PRIF	Inhibits prolactin release
	PRF	Stimulates prolactin release
Anterior hypothalamic area Preoptic nuclei Suprachiasmatic nucleus	LHRH	Stimulates preovulatory surge of LH and FSH
Ventromedial nuclei Arcuate nucleus Median eminence	LHRH	Stimulates tonic release of LH and FSH
Dorsal medial nuclei Preoptic nuclei	TRH	Stimulates release of TSH and prolactin
Paraventricular nuclei Supraoptic nuclei	Oxytocin	Induces uterine and oviductal contractions and milk letdown and facilitates gamete transport
Pineal	Melatonin	Antigonadotropic in rodent; function in domestic animals unknown

another neuroectodermal derivative, the brain.

It is not surprising that many animals rely on the changing photoperiod to induce their breeding season. The seasonal changes in photoperiodic length are absolutely reproducible and probably have been so for eons. Animals have thus evolved mechanisms whereby this highly regular cycle could be used to signal periods of reproductive competence and incompetence. It is not sufficient for animals merely to recognize a given season; rather they must actually anticipate the

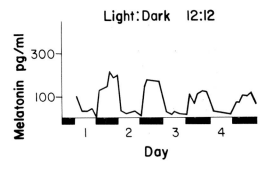

Fig. 6–12. Serum concentrations of melatonin from ewes exposed to a light regime of 12 hours light:12 hours dark. Darkness is indicated by solid bars. The sheep, like other species, releases melatonin during darkness. *(Adapted from Rollag and Niswender, 1976. Endocrinology 98:432)*

changing environmental conditions so they can mate at a time that will allow the delivery of the young in the spring. Mammals have developed a variety of means to ensure spring parturition. For example, sheep have long gestation periods (short-day breeders) whereas in some mammals like the hamster this interval is short (long-day breeders). Others breed in the fall but delay implantation of the ova until late winter (spotted skunk).

In hamsters, sheep and presumably other mammals, exposure to darkness stimulates melatonin synthesis and release while light suppresses it. The changes in melatonin serum levels in ewes exposed to light and dark periods have been evaluated using a radioimmunoassay specific for melatonin and are shown in Figure 6–12. Melatonin levels are high during the dark periods and low during light periods (Rollag and Niswender, 1976).

In a long-day breeder such as the hamster, continuous darkness will induce atrophy of the gonads; this effect is blocked by removal of the pineal. Thus melatonin appears to be antigonadotropic in the long-day breeder. Most research has been conducted on long-day breeders and it indicates that melatonin is inhibitory to reproduction. Little information is available on melatonin's effects on short-day breeders, such as sheep; it may be postulated that melatonin could possibly stimulate the reproductive process in the short-day breeder. However, the relationship of melatonin and reproduction in the short-day breeder remains to be determined.

The origin and functions of neurohormones involved in reproduction are summarized in Table 6–4.

REFERENCES

Beck, T.W. and Reeves, J.J. (1973). Serum luteinizing hormone (LH) in ewes treated with various dosages of 17β-estradiol at three stages of the anestrous season. J. Anim. Sci. 36, 566.

Bergland, R.M. and Page, R.G. (1978). Can the pituitary secrete directly to the brain? (affirmative anatomical evidence). Endocrinology 102, 1325.

Burgus, R., Dunn, T.F., Desiderio, D., Ward, D.N., Vale W. and Guillemin, R. (1970). Characterization of ovine hypothalimine hypophysiotropic TSH-releasing factor. Nature 226, 321.

Davis, S.L., Sasser, R.G., Thacker, D.L. and Ross, R.H. (1977). Growth rate and secretion of pituitary hormones in relation to age and chronic treatment with thyrotropin-releasing hormone in prepubertal dairy heifers. Endocrinology 100, 1394.

Dias, J.A. and Reeves, J.J. (1978). Disappearance of LHRH and D-Leu[6]-des gly NH$_2$[10]-LHRH ethylamide in ewes. J. Anim. Sci. 46, 1707.

du Vigneaud, V., Lawler, H. and Popenol, E.A. (1953). Enzymatic cleavage of glycinamine from vasopressin and a proposed structure for this pressor-antidiuretic hormone of the posterior pituitary. J. Am. Chem. Soc., 75, 4880.

Frandle, K.A., Kinder,J.E., Coy, D.H., Schally, A.V., Reeves, J.J. and Estergreen, V.L. (1977). Plasma progestins in anestrous ewes treated with D-Leu[6]-des gly NH$_2$[10]-LHRH ethylamide. J. Anim. Sci. 45, 486.

Green, J.D. and Harris, G.W. (1947). The neurovascular link between the neurohypophysis and adenohypophysis. J. Endocrinol. 5, 136.

Halasz, B. (1969). In Frontiers in Neuroendocrinology. W.F. Ganong and L. Martini (eds), p. 307. New York, Oxford University Press.

Jackson, G.L., Kuehl, D., McDowell, K. and Zaleski, A. (1978). Effect of hypothalomic deafferentation on secretion of luteinizing hormone in the ewe. Biol. Reprod. 18, 808.

Karsch, F., Foster, D.L., Legan, S.J. and Hauger, R.L.

(1976). On the control of tonic LH secretion in sheep: A new concept for regulation of the estrous cycle and breeding season. In Excerpta Medica International Congress Series Endocrinology. V.H.T. James (ed), Amsterdam, Excerpta Medica, 402, 192.

Kinder, J.E., Adams, T.E., Nett, T.M., Coy, D.H., Schally, A.V. and Reeves, J.J. (1976). Serum gonadotropin concentrations and ovarian response in ewes treated with analogs to LH-RH/FSH-RH. J. Anim. Sci. 42, 1220.

Kittock, R.J., Britt, J.H., and Convey, E.M. (1973). Endocrine response after GnRH in luteal phase cows and cows with ovarian follicular cysts. J. Anim. Sci. 37, 985.

Klein, D.C. and Weller, J.L. (1970). Indole metabolism in the pineal gland: a circadian rhythm in N-acetyltransferase. Science 169, 1093.

Moss, R.L. and McCann, S.M. (1973). Induction of mating behavior in rats by luteinizing hormone-releasing factor. Science 181, 177.

Nair, R.M.G., Barrett, J.F., Bowers, C.Y. and Schally, A.V. (1970). Structure of porcine thyrotropin releasing hormone. Biochemistry 9, 1103.

Nett, T.M., Akbar, A.M., Niswender, G.D., Hedlund, M.T. and White, W.F. (1973). A radioimmunoassay for gonadotropin releasing hormone (GnRH) in serum. J. Clin. Endocrinol. Metab. 36, 880.

Oliver, C., Mical R. and Porter, J.C. (1977). Hypothalamic-pituitary vasculature: Evidence for retrograde blood flow in the pituitary stalk. Endocrinology, 101, 598.

Reeves, J.J., O'Donnel, D.A. and Denorscio, F. (1972) Effect of ovariectomy on serum luteinizing hormone (LH) concentrations in the anestrous ewe. J. Anim. Sci., 35, 73.

Reeves, J.J., Arimura, A., Schally, A.V., Kragt, C.L., Beck, T.W. and Casey, J.M. (1972). Effects of synthetic luteinizing hormone-releasing hormone (LH-RH/FSH-RH) on serum LH, serum FSH and ovulation in anestrous ewes. J. Anim. Sci. 35, 84.

Reeves, J.J., Beck, T.W. and Nett, T.M. (1974). Serum FSH in anestrous ewes treated with 17β-estradiol. J. Anim. Sci. 38, 374.

Reeves, J.J., Tarnavsky, G.K., Becker, S.R., Coy, D. and Schally, A.V. (1977). Uptake of iodinated luteinizing hormone releasing hormone analogs in the pituitary. Endocrinology 101, 540.

Rippel, R.H. and Johnson, E.S. (1976). Inhibition of HCG-induced ovarian and uterine weight augmentation in the immature rat by analogs of GnRH. Proc. Soc. Exp. Biol. Med. 152, 432.

Rollag, M.D. and Niswender, G.D. (1976). Radioimmunoassay of serum concentrations of melatonin in sheep exposed to different lighting regimens. Endocrinology 98, 482.

Schally, A.V., Redding, T.W., Bowers, C.Y. and Barrett, J.F. (1969). Isolation and properties of porcine thyrotropin releasing hormone. J. Biol. Chem., 244, 4077.

Schally, A.V., Redding, T.W., Arimura, A., Dupont, A. and Linthicum, G.L. (1977). Isolation of gamma-amino butyric acid from pig hypothalami and demonstration of its prolactin release-inhibiting (PIF) activity in vivo and in vitro. Endocrinology *100*, 681.

Vale, W., Rivier, J. and Burgus, R. (1971). Synthetic TRF (thyrotropin releasing factor) Analogues: II pGlu-N^{31m}Me-His-Pro-NH$_2$: A synthetic analogue with specific activity greater than that of TRF. Endocrinology *89*, 1485.

7

Reproductive Life Cycles

MARIE-CLAIRE LEVASSEUR
and
C. THIBAULT

The life cycle may be divided into three distinct phases: (1) a preparatory period during fetal and neonatal life when gonadotropic function and gonads differentiate; (2) a puberal period during which the adjustment of the mechanisms acting as regulators between the gonadotropic function and the gonads gradually leads to full maturity; (3) a fertile period characterized by succeeding ovarian cycles (fertile or not) and seasonal variation of sexual activity. After puberty, fertility progressively increases, becomes maximal and then is curtailed, finally to cease completely. This gradual reduction of reproductive efficiency is related to the aging process. However, farm animals are usually slaughtered well before the decrease in fertility and the aging process has not been studied adequately in these species (Fig. 7–1).

FETAL AND NEONATAL LIFE

Gonadotropin Function

Secretion of both the gonadotropins FSH and LH and their hypothalamic releasing factor, GnRH, always begins during fetal life. In the ewe and cow it starts early, shortly after sex differentiation (month 1 or 2 of pregnancy) and in the sow only toward the end of fetal life (about 1.5 months after gonadal sex differentiation). Although the vascular relationship between the hypophysis and hypothalamus is rudimentary, the secretion of gonadotropins basically obeys the same regulatory mechanisms as in the adult, and judging by the blood levels of this hormone at this primitive stage, its secretion must be as efficient as in the adult.

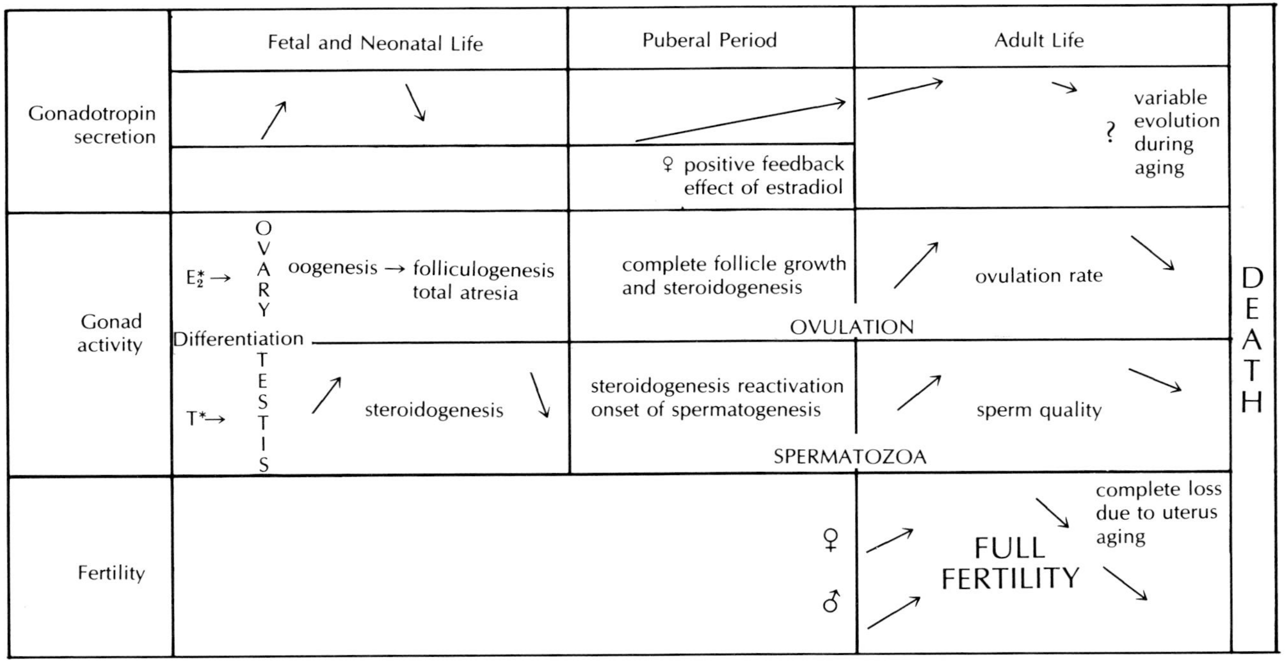
Fetal and Neonatal Life
Puberal Period
Adult Life
Gonadotropin secretion
♀ positive feedback effect of estradiol
variable evolution during aging
?
Gonad activity
E_2^* →
OVARY
oogenesis → folliculogenesis
total atresia
Differentiation
T* →
TESTIS
steroidogenesis
complete follicle growth and steroidogenesis
OVULATION
ovulation rate
steroidogenesis reactivation onset of spermatogenesis
SPERMATOZOA
sperm quality
Fertility
♀
♂
FULL FERTILITY
complete loss due to uterus aging
DEATH

*E_2 = estradiol, T = testosterone

Fig. 7–1. The three phases of sexual life

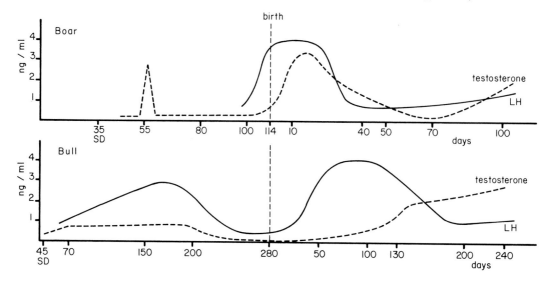

Fig. 7–2.   Plasma testosterone and LH levels in boar and bull from gonadal sexual differentiation (S.D.) during fetal life up to the beginning of puberty. *(Data for boar from Meusy-Dessolle, 1974. C.R. Acad. Sci., Paris, D, 278, 1257, and 1975, C.R. Acad. Sci., Paris D, 281, 1975; Colenbrander et al., 1977. Biol. Reprod., 17, 506. Data for bull from Challis et al., 1974. J. Endocrinol. 60, 107; Lacroix et al., 1977. Ann. Biol. Anim. Biochim. Biophys. 17, 1013)*

This secretion temporarily regresses; it is slightly reduced two months before birth in cattle, near term in sheep and one month after birth in pigs (Fig. 7–2). The damping of gonadotropin secretion must be related to the maturation of the central nervous system. It occurs when superior brain structures take charge of hypothalamic activity (Levasseur, 1977). The gonadotropin levels remain low up to the onset of puberty. The duration of this period of "infancy" is highly variable. It lasts a few days in the rat, one month in sheep and pig, three months in cattle and six to seven years in humans.

At the onset of the puberal period gonadotropin secretion rises. This process occurs in normal animals as well as in animals that had been castrated early, in which the process is clearer owing to the absence of negative feedback from the gonadal steroid (Fig. 7–3). Gonadotropin rise results in the removal of the inhibitory control of the central nervous system when the body's development progressively attains a level compatible with re-

production. The maintenance of an active role of the central nervous system on reproductive activity is evidenced in sheep by seasonal variations of gonadotropin secretion and sexual activity correlated with the daylight ratio.

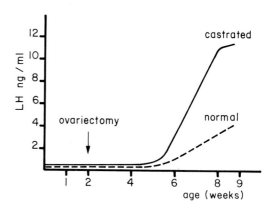

Fig. 7–3.   Comparative evolution of plasma LH levels in normal and castrated ewe lambs. The puberal rise occurs at the same age in both animals, but the increase is more pronounced in the castrate. *(From Foster et al., 1975, Endocrinology, 96, 15)*

The Gonads

During fetal and neonatal life gametogenesis and steroidogenesis seem independent, while at the onset of puberty they become closely related.

The Testis. The basic structure of the testis (seminiferous cords and interstitial tissue) remains unchanged from gonadal sex differentiation at the beginning of fetal life to the onset of puberty. The seminiferous cords are lined with supporting cells, whereas undifferentiated germ cells or gonocytes occupy the central part. The only modification during this period is a slow increase in the relative number of gonocytes. This increase is not gonadotropin-dependent; in the hypophysectomized sheep fetus, the increase in gonocyte number is similar to that of the control (Courot, 1971).

Interstitial tissue that fills the space between the sex cords is composed of elongated conjunctive-type cells and steroidogenic cells, recognized by an abundant smooth reticulum and the presence of mitochondria with tubular cristae. The Leydig cells secrete androgens as soon as the testis differentiates and before the gonadotropic function is triggered. However, the Leydig cells are gonadotropin-sensitive and their continued steroidogenic activity closely depends on gonadotropic secretion. In swine a transitory testosterone secretion occurs about day 55 when the Leydig cells differentiate; this secretion then drops until the fetus begins to secrete LH shortly before birth. The further decrease of gonadotropins one month after birth causes testosterone secretion to regress again during the short period of "infancy." In cattle and sheep gonadotropin secretion begins earlier; the fetal Leydig cells are quickly stimulated by LH and testosterone is secreted until gonadotropic function regresses (Fig. 7–2).

At the onset of puberty gonadotropin secretion resumes and Leydig cells are reactivated. In swine the Leydig cells, which were active during fetal and neonatal life, occupy large areas between the tubules, while after puberty the peritubular cells are more active (Van Straaten and Wensing, 1978). This observation tends to support the long-debated hypothesis that there are two Leydig cell populations, one fetal and the other puberal.

The Ovary. Initial ovarian structure is not fundamentally different from that of the testis. Sex cords, formed by somatic and germ cells, are present at the beginning of ovarian and testicular differentiation. While these structures remain basically unchanged in the testis as seminiferous cords or tubules, in the ovary germ cells actively divide, sex cords vanish and finally each oocyte is wrapped by a few somatic cells to form the primordial follicle. At the end of oogenesis, the ovary encloses millions of primordial follicles within a framework of interstitial tissue and is lined with ovarian epithelium erroneously called germinal epithelium. Oogonia and oocytes are formed during the first half of fetal life in the ewe and the cow. Oocyte formation also begins early during fetal life in the sow; however, oogenesis is completed only during the first weeks after birth (Fig. 7–4).

Gonadotropins are not involved in oogonial multiplication or in meiotic prophase; oogenesis progresses normally in a sheep fetus that is hypophysectomized before the onset of meiosis (Mauléon, 1973).

The appearance of meiotic prophase early in life is one of the main differences between ovarian and testicular germ cell evolution (Fig. 7–5). Moreover, as oogonia completely disappear, the oocytes formed during the fetal and neonatal period are the only source of oocytes available during the entire sexual life. As soon as the primordial follicle reserve is constituted, it rapidly diminishes by atresia. A cow fetus that has

2,700,000 oocytes at day 110 of gestation has only 70,000 oocytes at birth. From the end of the period of oogenesis, some primordial follicles continuously begin growth, but up to puberty all disappear owing to atresia (Fig. 7–5).

The ovary also contains interstitial tissue. When the gonad sexually differentiates this primary interstitial tissue may secrete estradiol early, before oocyte formation (18- to 30-day rabbit fetus: Milewich et al., 1977; 30- to 50-day ewe fetus; Mauléon et al, 1977; 60-day cow fetus: Shemesh et al, 1978). This steroidogenic activity definitely disappears.

In certain mammals such as rabbits, a secondary interstitial tissue may gradually form from the thecae of atretic follicles. This tissue can secrete progestins (20α-OH progesterone) and perhaps androgens, but never estrogens. In the puberal female estradiol can only be secreted when antral follicles have developed. Their theca interna, formed from intersti-tial tissue, mainly secretes testosterone, which is converted into estradiol by granulosa cells. The theca can thus be compared to peritubular Leydig cells in the testis.

PUBERTY

From a practical point of view, a male or female animal has reached puberty when it is able to release gametes and to manifest complete sexual behavior sequences. Puberty is basically the result of a gradual adjustment between increasing gonadotropic activity and the ability of the gonads to simultaneously assume steroidogenesis and gametogenesis.

Gonadotropic Function and Steroidogenesis

At the beginning of the puberal period, gonadotropin secretion levels rise due to the increase of both the amplitude and the

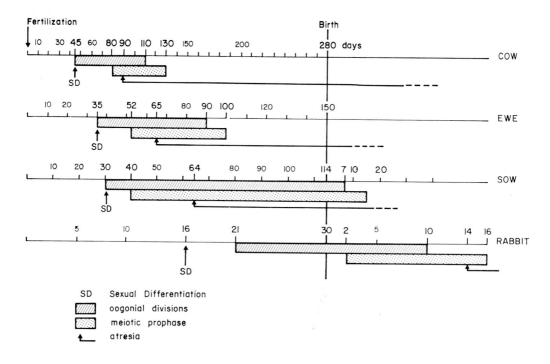

Fig. 7–4. Oogenesis in some mammals.

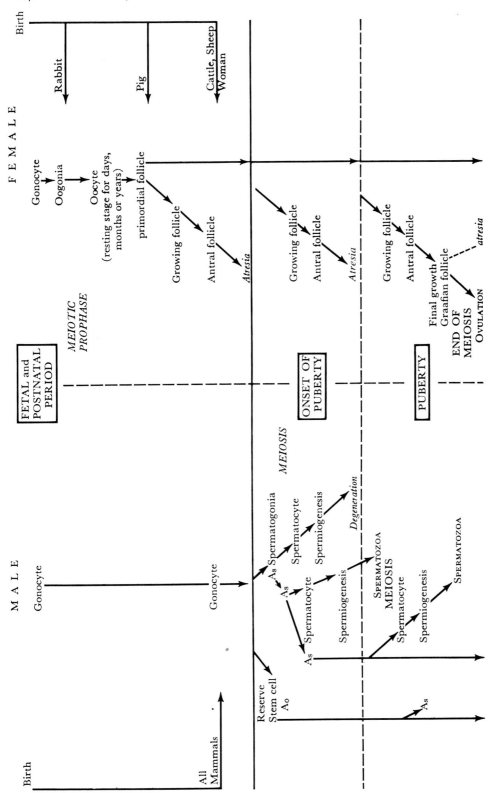

Fig. 7–5. Comparison of gametogenesis in male and female mammals from fetal life to active sexual life. A_o = reserve stem cells, A_s = stem cell spermatogonia (see text).

frequency of the periodic pulses of these hormones (ewe/lamb: Foster et al, 1975; ram lamb: Foster et al, 1978; bull calf: Lacroix et al, 1977). In two- to eight-week old lambs pulse frequencies increase from one to five in a six-hour period (Fig. 7–6); the magnitude of the surge is augmented threefold during that time.

In the male in response to gonadotropin secretion, testosterone progressively rises from very low to adult levels. Every LH pulse is followed at a one-hour interval by a transitory rise of testosterone secretion. The extent of the testosterone secretion increases as puberty advances, and finally the average testosterone level remains definitively high. The increase of blood testosterone level eventually causes a decrease in gonadotropin secretion by a negative feedback effect (Fig. 7–7). In the female, estrogen secretion gradually increases in response to the puberal gonadotropin rise as long as antral follicle formation has begun. This is the case in the ewe and the cow. On the other hand, estrogen level only rises in the gilt toward week 11 after birth, when the first antral follicles appear (Schlenker et al, 1973), while puberal gonadotropic secretion begins three

weeks earlier at eight weeks of age (Colenbrander et al, 1977).

Moreover, ovulation necessitates high estradiol levels, which cause a gonadotropin surge (positive estradiol feedback). While all castration experiments, or those using estradiol overloading, show that negative estradiol feedback on gonadotropin operates in fetal life, positive feedback is gradually established only during the puberal period. It is the last mechanism to be triggered in the puberal female. Thus in the ewe lamb the first LH surge after estradiol injection is noted at 4 weeks of age, but a surge equivalent to that seen during the estrous cycle is obtained only at 27 weeks (Land et al., 1970; Foster and Karsch, 1975) (Fig. 7–8).

Gametogenesis

Spermatogenesis. At the onset of puberty in male swine, cattle and sheep, the

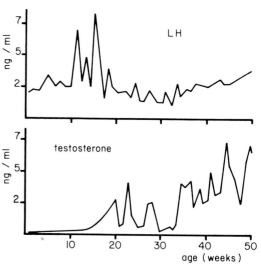

Fig. 7–7. LH and testosterone levels in a bull calf determined by weekly blood samples. LH increase precedes testosterone elevation. Their weekly variation is due to a pulsatile secretion pattern. When pulse frequencies and amplitude increase, there is more chance of obtaining high-level samples, e.g. between weeks 10 and 20 for LH. *(From Lacroix et al, 1977, Ann. Biol. Anim. Biochim. Biophys., 17, 1013)*

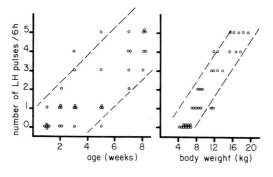

Fig. 7–6. Increase of LH pulse frequencies in six ram lambs at the onset of puberty. There is a better correlation with body weight (r = 0.90) than with age (r = 0.76). The onset of puberty is characterized by the appearance of pulsatile discharges of gonadotropins, which increase in frequency during the attainment of sexual maturity. From these graphs it is clear that the increase of pulse frequencies is more closely correlated with body weight than with age. *(From Foster et al., 1978, Endocrinology, 102, 1137)*

gonocytes migrate to the periphery of the tubules and differentiate into spermatogonia, whereas supporting cells produce Sertoli cells. These changes occur when prepuberal gonadotropin elevation is already evident. The Sertoli cells remain present during the whole sexual life, and their number is the limiting factor in spermatozoon production.

Two types of spermatogonia differentiate from gonocytes, ie, reserve stem cells (A_o) and stem cells (A_s). A_s regularly give rise to a new stem cell and to a line of spermatogonia, spermatocytes, spermatids and spermatozoa. The maintenance of the stem cell is the basis of continuous sperm production. A_o cells contribute to the increase of the A_s population between puberty and adulthood and permit renewal of the A_s population in case of testicular damage (irradiation, radiomimetic drugs, disease) (Fig. 7–5).

During the fetal and neonatal period, the testis grows slowly, mainly through lengthening of the seminiferous cords. Spermatogonial differentiation and formation of the first spermatogenetic lines

mark the beginning of rapid testicular growth. After the attainment of full spermatogenetic activity, a slow growth rate is maintained during some months (ram) or years (bull) in response to the continuous increase of the stem cell population.

Follicular Development. The first antral follicles appear during the prepuberal period (sow, rabbit) or earlier (cow, ewe). However, complete follicular development, resumption of oocyte meiosis and ovulation are observed only when FSH and LH have reached adult profiles.

Age at Puberty

In normal breeding conditions puberty occurs at about three to four months of age in rabbits, six to seven months in sheep, goats and swine, 12 months in cattle and 15 to 18 months in horses.

Age at puberty is breed-dependent. However, attainment of puberty is more closely related to body weight than to age. Dairy cattle reach puberty when the body weight is 30% to 40% that of the adult weight, whereas in beef cattle this per-

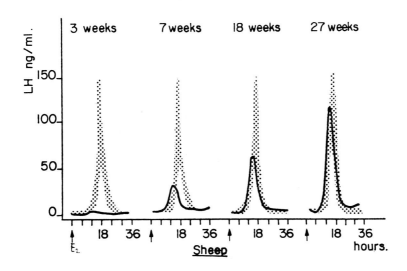

Fig. 7–8. Progressive appearance of positive estradiol feedback (solid line) on LH secretion during puberty in ewe-lamb. Shaded area, comparative LH response in adult ewe. Estradiol injection, ↑ *(Foster and Karsch, 1975. Endocrinology, 97, 1205)*

centage is higher (45 to 55% that of adult body weight) (Roy et al., 1975). The same difference has been shown in sheep (Romney ewes: 40%; Suffolk: 50%; Scottish Blackface: 63% of adult body weight) (Hafez, 1952). Nutritional levels modulate age at puberty. If growth is accelerated by overfeeding, the animal reaches puberty at a younger age. On the other hand, if growth is slowed down by underfeeding, puberty is delayed.

Social and climatic factors, mainly photoperiodism, modify age at puberty. In natural conditions, where reproduction is a seasonal phenomenon, the age at puberty depends on the birth season. Ewes born in January attain puberty eight months later, whereas those born in April become puberal when six months old (during full adult breeding season in both cases). Puberty occurs earlier in gilts bred in a group than alone. The presence of an adult boar hastens puberty in both situations (Mavrogenis and Robinson, 1976).

Full reproductive efficiency is not attained in any species at the first appearance of estrus or ejaculation. There is a period of "adolescent sterility." This period is remarkably short (some weeks) in domestic animals as compared with human beings (1 year or more).

ADULT SEXUALITY

Estrous Cycle

In nonprimate mammals, female mating behavior is limited to the estrous period around ovulation. When the female is not fertilized, estrus occurs at the regular intervals that characterize the estrous cycle.

The estrous cycle lasts from 16 to 24 days in domestic mammals (ie, ewe: 16 to 17 days; cow, sow, goat: 20 to 21 days; mare: 20 to 24 days). The duration of estrus is species-dependent and varies slightly from one female to another within the same species. This is also true in respect to the time of ovulation, which

occurs 24 to 30 hours after the onset of estrus in most ewes and cows, 35 to 45 hours in sows, and 4 to 6 days in mares (Table 7–1).

The length of estrus and the time of ovulation also vary in relation to internal and external factors. In ewes the interval between the onset of estrus and LH ovulatory surge (and therefore the interval between estrus and ovulation) lengthens as the number of ovulations increases (Fig. 7–9). Male stimulation reduces estrus duration (cow, sow, ewe) and decreases the variability of the ovulation timing.

At the ovarian level, the estrous period is characterized by high estrogen secretion from preovulatory graafian follicles. At the end of estrus, ovulation occurs followed by corpus luteum formation resulting in progesterone secretion. This secretion regresses abruptly some days before the next estrus. The period of corpus luteum activity is called the luteal phase; it lasts 14 to 15 days in ewes and 16 to 17 days in cows and sows. The follicular phase, from the regression of the corpus luteum to ovulation, is relatively short, ie two to three days in ewes and

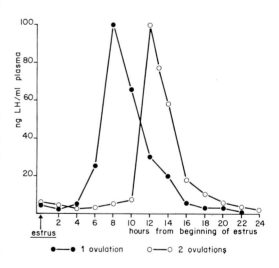

Fig. 7–9. Timing of LH surge in ewe according to number of ovulations (Ile-de-France breed). *(From Thimonier and Pelletier, 1971. Ann. Biol. Anim. Biochim. Biophys. 11, 559)*

Table 7–1.   Estrous Cycle, Estrus and Ovulation in Farm Animals

	Length of Estrous Cycle	Duration of Estrus	Time of Ovulation
Ewe	16–17 days	24–36 hours	24–30 hours from beginning of estrus
Goat	21 days (Also short cycles)	32–40 hours	30–36 hours from beginning of estrus
Sow	19–20 days	48–72 hours	35–45 hours from beginning of estrus
Cow	21–22 days	18–19 hours	10–11 hours after end of estrus
Mare	19–25 days	4–8 days	1–2 days before end of estrus

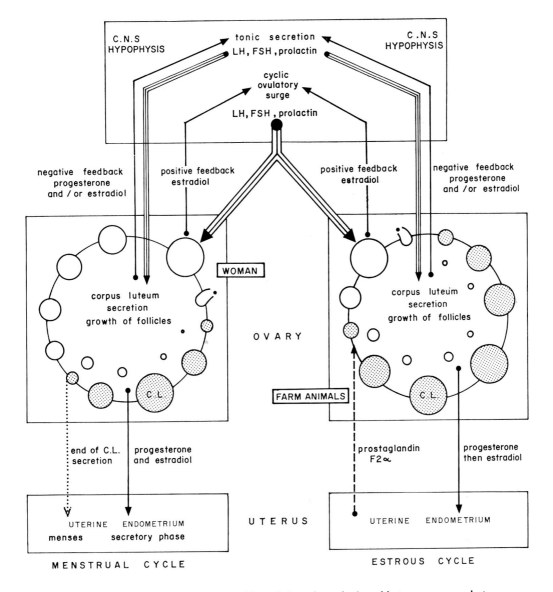

Fig. 7–10.   Interrelationship between ovary and hypothalamo-hypophysis and between ovary and uterus.

goats and three to six days in cows and sows. This short follicular phase does not reflect the true duration of graafian follicle growth (Fig. 7–10). Thus, estrous cycle length is closely related to the duration of the luteal phase. Corpus luteum regression is not caused by a decreased secretion of pituitary luteotropic hormones (LH and prolactin), but by the action of a luteolytic factor, prostaglandin $F_{2\alpha}$ ($PGF_{2\alpha}$).

In domestic mammals the uterus has a basic role in $PGF_{2\alpha}$ production. A high level of $PGF_{2\alpha}$ has been detected during luteal regression in uterine venous blood. As $PGF_{2\alpha}$ can pass selectively through the wall of the uterine vein and the ovarian artery when they are in close contact near the ovary, large amounts of $PGF_{2\alpha}$ are immediately available to the ovary. Experimental increases of estradiol levels during the luteal phase show that estradiol elevation is always followed by $PGF_{2\alpha}$ surge when the endometrium is primed with progesterone.

Resumption of Sexual Activity after Parturition

A relatively high level of progesterone is absolutely necessary throughout gestation. Progesterone is secreted by the corpus luteum and in some species (cow, ewe) mainly by the placenta. This continuous progesterone secretion suppresses estrus and, in most mammals, ovulation. However, silent ovulations occur in the mare during the second month of pregnancy and the resulting corpora lutea have some part in the maintenance of the necessary level of progesterone during pregnancy.

Following parturition, progesterone drops to undetectable levels and estrus and ovulation can resume. The sow exhibits estrus within 48 hours after parturition, but there is no ovulation. The high plasma estrogen rise after farrowing (Shearer et al, 1972) may explain the estrous behavior. In mares there is a fertile estrus one to three

weeks after parturition. In cows, ewes and goats silent ovulations can occur two to three weeks following parturition; however, fertile estrous cycles return later (Casida et al., 1968; Hunter, 1968). This infertile period (postpartum anestrus) lasts longer when the female suckles (lactation anestrus).

The importance of suckling on the duration of postpartum anestrus is demonstrated in sheep by experimentally induced pregnancy during seasonal anestrus so that lambing occurs during the breeding season. Dried-off ewes usually return to estrus after about one month, while suckling ewes present the first estrus some weeks later. In sheep and cattle the duration of postpartum anestrus varies with the breed and seems to be constant for the same female during successive pregnancies.

Fertility is low during the first estrus, particularly when the female suckles. Maximal fertility in the cow occurs 60 to 90 days after calving (Fig. 7–11). In sows fertility is nil during weaning; a highly

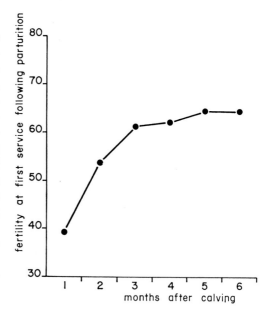

Fig. 7–11. Fertility of dairy cattle at first service following parturition. *(From Casida et al., 1968. Wisc. Expt. Sta. Research Bull. No. 270)*

fertile estrus occurs a few days after weaning.

Seasonality of Sexual Activity

Sexual activity in all wild mammals varies with the season. In domestic mammals such as cattle and swine, seasonality is inconspicuous, while in horses, sheep and goats it is distinct. Numerous experimental studies have shown that the most efficient climatic factor is the variation of the daylight ratio. The action of the light/dark period mechanism is not completely understood; however, photoperiodic manipulation appears to be an efficient tool for increasing fertility.

Annual Breeding Season. In sheep and goats there are important breed differences in the duration of the sexual season. Préalpes and Mérino sheep are long-season breeders, whereas Blackface and Southdown are short-season breeders. The length of the sexual season in these breeds is 260, 200, 139 and 120 days, respectively. A long sexual season is a dominant genetic character. All Mérino crosses exhibit a long sexual season like the Mérino. A cross of Dorset Horn and Persian ewe has produced a breed—the Dorper—which has only a one-month anestrus (Fig. 7–12).

Silent ovulatory cycles always occur at the beginning and end of the sexual sea-

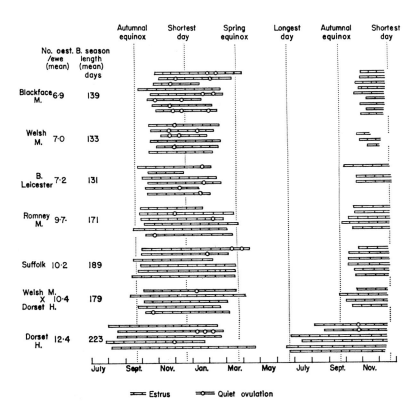

Fig. 7–12. Breed differences in the duration of the sexual season and nonsexual season in adult ewes in Great Britain. Some breeds such as the Dorset Horn had a prolonged sexual season, whereas those such as the Welsh Mountain had a restricted sexual season. In nearly all cases, the sexual season was within the period from the autumnal equinox and the spring equinox and the middle of the season corresponded rather closely to the shortest day of the year, or December 21. This illustrates the close relationship between the sexual season and length of day. Note that some estrous cycles double or triple the usual length occurred due to quiet ovulations or to the failure to detect heat in the nonpregnant females observed. *(After E.S.E. Hafez, 1952. J. Agric. Sci. 42, 305)*

son. These ovarian cycles continue during the anestrous period in a variable number of ewes (Fig. 7–13). In Ile-de-France and Préalpes ewes the frequency of silent cycles rises temporarily in spring. If a ram is present, behavioral estrus appears, thus permitting a second annual sexual season in these breeds. The wild breed, Barbary, may naturally exhibit two sexual seasons, one in October through January and the other in April through June.

Sexual behavior and gonadotropin ovulatory surge are steroid-dependent. Since ovulation can occur without estrus, it seems that the steroid sensitivity of nervous centers regulating sexual behavior is lower than that of the centers involved in gonadotropic discharge.

In goats the sexual season is well defined in temperate climates. The ovaries in the Alpine goat are slightly active from February to March and quiescent from April to July; activity is abruptly resumed in all goats in September (Fig. 7–14). Quiet ovulations are less frequent than in ewes. As in sheep in tropical climates,

Creole goats exhibit continuous sexual activity.

Although rams can mate throughout the year, testis weight, testosterone and gonadotropin levels are minimal from January to May during female anestrus (Fig. 7–15). Similarly, in the billy-goat the plasma testosterone level remains low from January to August, when it rises suddenly at the beginning of the breeding season.

An annual reproductive cycle in horses is also well documented from both hemispheres. In northern temperate countries, ovarian silence in mares and low plasma testosterone and LH levels in stallions are observed from October to February.

In cattle and pigs estrus occurs regularly throughout the year and seasonality is discrete. Local breeding conditions often mask its expression. In cattle a seasonal variation of fertility in temperate climates only becomes evident after studying a large number of herds over a period of several years (Fig. 7–16). Mini-

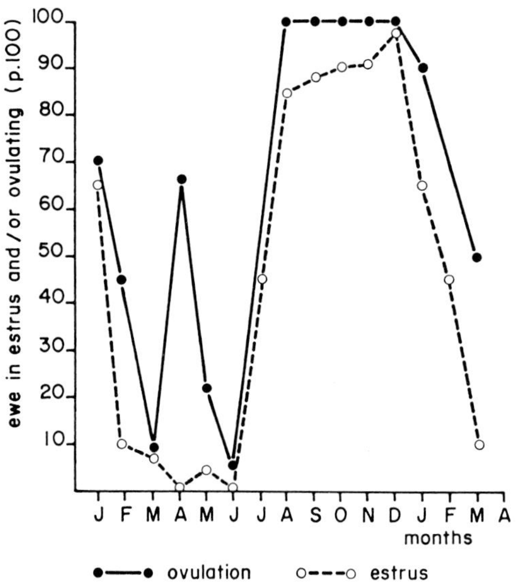

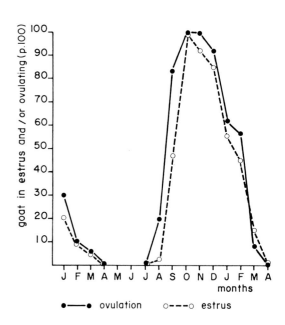

Fig. 7–13. Seasonal variation of estrous behavior and ovulation in the Ile-de-France ewe (47° N). *(From Thimonier and Mauleon, 1969. Ann. Biol. Anim. Biochim. Biophys. 9, 233)*

Fig. 7–14. Seasonal variation of estrous behavior and ovulation in goat of Alpine breed (47° N). *(From Corteel, 1973. World Rev. Anim. Prod., 9, 73)*

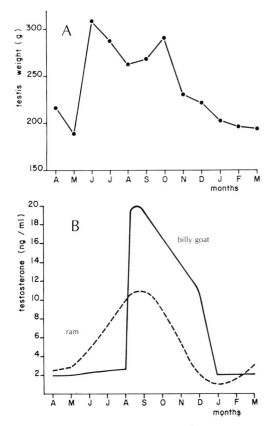

mal fertility occurs in June and maximal fertility in November. This variation may be related to photoperiodism rather than to temperature and feeding, which can fluctuate from year to year. Cows are mainly responsible for the seasonal varia-

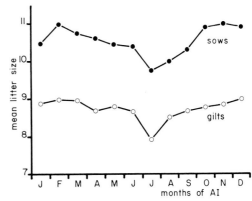

Fig. 7–15. *A*, Seasonal variation of testis weight in Ile-de-France ram (testis weight is adjusted to body weight) (47° N). *(From Pelletier, 1971. Ph.D. Thesis, University of Paris)*
B, Seasonal variations of testosterone plasma levels in Ile-de-France ram and Alpine billy goat. *(Ram: from Attal, 1970. Ph.D. Thesis, University of Paris; billy goat: from Saumande and Rouger, 1972, C.R. Acad. Sci. Paris D, 274, 89)*

Fig. 7–17. Seasonal variation of fertility in cattle: relative influence of male and female. (Sperm has been collected either in spring or in fall, frozen and used for insemination during fall and spring.) *(From Courot et al, 1968. Ann. Biol. Anim. Biochim. Biophys. 8, 209)*

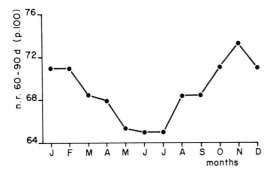

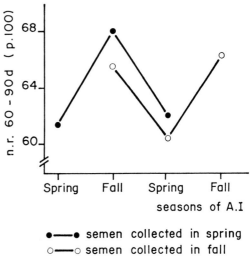

Fig. 7–16. Seasonal variation of fertility in cattle from 320,000 artificial inseminations over seven years (Montbéliard breed in French Jura, 47° N). The non-return rate is lower in spring and higher in autumn. *(From Courot et al., 1968. Ann. Biol. Anim. Biochim. Biophys. 8, 209)*

Fig. 7–18. Evolution of litter size in artificially inseminated sows and gilts in the west of France (46° N). This study includes 4,510 gilts and 13,324 sows and covers three consecutive years. *(From Courot and Bariteau, 1978, personal communication)*
Pig fertility is minimal under conditions of long days and high temperatures, as evidenced in sows and gilts by the smaller litter size in July. Note that adult sows have 1.5 piglets more per litter than gilts all the year round.

tion of fertility, as shown by nonreturn rates after insemination in spring and fall with frozen semen collected in fall or spring (Fig. 7–17). Fertility in the sow is lower in the summer than in other seasons, and the litter size is also smaller then (Fig. 7–18).

Role of Photoperiodism and Temperature. Daylight ratio and temperature are the two main climatic factors influencing the annual sexual cycles; the former is the most efficient factor.

The most conclusive experience showing the control of reproduction by photo-

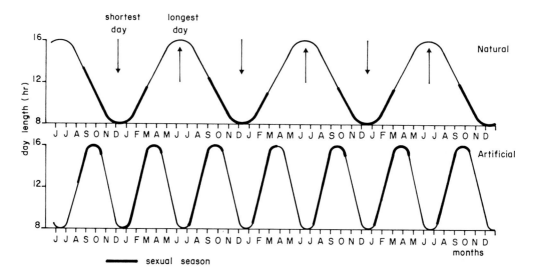

Fig. 7–19.   Periods of sexual activity in Limousine ewes. *Top:* Under natural photoperiodicity, estrus normally occurs during decreasing daylight period. *Bottom:* Under six-month photoperiodic cycles, estrus occurs during the increasing daylight period. *(From Mauleon and Rougeot, 1962, Ann. Biol. Anim. Biochim. Biophys. 2, 209)*

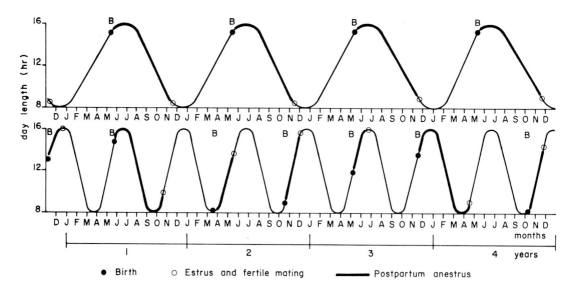

Fig. 7–20.   Lambing and estrus in wild sheep *(Ovis ammon musimon)*. *Top:* Under normal annual photoperiodicity. *Bottom:* Under six-month photoperiodic cycles. *(From Rougeot, 1969. Ann. Biol. Anim. Biochim. Biophys. 9, 441)*

periodism has been conducted in sheep. Using two photoperiodic cycles per year, ewes experience two annual breeding seasons (Fig. 7–19). When the ewe is allowed to mate regularly, lambing occurs every six and one-half months, since the five-month gestation is followed by a lactational anestrus of one and one-half months (Fig. 7–20). In the northern hemisphere, increasing the daylight ratio up to 16 hours in November and December advances the beginning of the sexual season in the mare; first ovulation occurs up to 3 months earlier than under natural photoperiod. Induced cycles are endocrinologically normal and fertile (Oxender et al., 1977).

Similar manipulations are also effective with rams and stallions. Using two photoperiodic cycles per year, the ram exhibits two periods of decreasing spermatogenetic activity coinciding with the two periods of increasing daylength (Ortavant and Thibault, 1956). However, experiments show that photoperiodism is basically a synchronizer of sexual activity. When ewes are placed under 12 hours of daylight every day or under constant illumination for many years, a breeding season is maintained for one or two years, then estrus becomes more random throughout the year.

Seasonal variation of temperature plays a major role in the regulation of sexual function in lower vertebrates, particularly in reptiles. In mammals, when environmental temperatures remain within the limit compatible with thermoregulatory mechanisms, seasonal temperature variation effect on fertility is rarely reported (Hafez, 1968). Nevertheless, the postfertilization period appears to be a critical one in domestic females. Cow, ewe and sow embryos are susceptible to damage during the first 10 days of development (Ortavant and Loir, 1978).

In swine photoperiodism and temperature interact unfavorably on fertility. Sperm output, sperm motility and farrowing rate are severely lowered when boars are submitted to summer temperatures (35°C) under long days (16 hours) (Mazzari et al, 1968).

Mechanisms of Action of Photoperiodism. Changes in daylight ratio always trigger variations in plasma gonadotropin and prolactin levels. In the ram, a reduction in the 16-hour daylight ratio to 8 hours is followed by a decrease of prolactin and a parallel increment of FSH-LH, and then by a rise in testosterone and testis weight.

Photoperiodism effect involves at least two separate mechanisms. First, there is a direct action on the hypothalamic pituitary axis. In the castrated ram and spayed mare, in which negative feedback from sexual steroids does not occur, gonadotropin levels reach a maximum during the normal breeding season and decrease during the nonbreeding season (Fig. 7–21). Secondly, there is a simultaneous change in the sensitivity of the central nervous system to negative feedback from steroids. A quantity of testosterone or estradiol injected into castrated sheep that is able to reduce gonadotropins to undetectable levels during the anestrous season also depresses gonadotropins only slightly

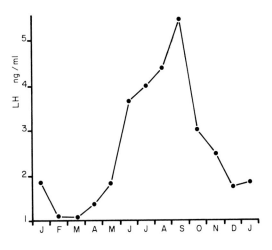

Fig. 7–21. **Seasonal variation of plasma LH levels in spayed mare.** *(From Garcia and Ginther, 1976. Endocrinology, 98, 958)*

during the normal sexual season (Fig. 7–22) or under short days. Similarly, estrus in spayed ewes is more regularly induced by progesterone/estrogen treatment during the normal breeding season than during anestrus (Reardon and Robinson, 1961).

Experiments done on hamsters suggest a third site of light action by pineal gland regulation of the secretion of antigonadotropic substances. However, no proof of such an effect is actually available in domestic mammals.

How is change in daylight ratio perceived by the central nervous system as a stimulus to gonadotropic function? There is no evidence in mammals, as in birds, of direct light perception by hypothalamic photosensitive neurons; the retinal pathway seems the more probable route. However, experiments in all animal species show that the length of the daylight period necessary for sexual stimulation does not mean that light must stimulate the nervous system during the whole light period. Light is efficient only during a precise period of the 24-hour dark/light cycle. According to species, this sensitive period occurs 10 to 20 hours after light on (dawn). The photosensitive period is short

in birds (about one hour) and remarkably broad in the ram (seven hours). In rams submitted to eight hours of light per day given in two parts (seven hours plus one hour given at various intervals in the night), LH level was higher when the one-hour flash was given 11 to 20 hours after the beginning of the seven-hour light period (Fig. 7–23). Under natural conditions, LH levels are higher in the summer than in the winter months.

These results are apparently in opposition with those that show an increase of FSH, LH, testosterone and spermatogenesis in rams when they are put under short days (8L/16D) after being under long days (16L/8D) (Lincoln et al, 1977). However, a correct understanding of the mechanism involved in natural seasonality must include varying levels of other hormones, such as prolactin and thyroxine, which interfere with the gonadotropic function or with gonad responsiveness. Prolactin and thyroxine secretion are also modulated photoperiodically. Under the same photoperiod of eight hours of light given in a seven-hour period plus a one-hour period, maximal prolactin secretion was obtained in the ram when the one-hour flash was given 17

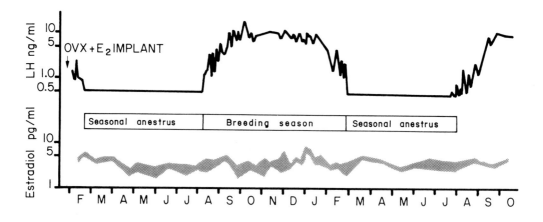

Fig. 7–22. Seasonal changes in sensitivity of negative estradiol feedback on LH secretion. In ovariectomized ewes the same level of estradiol throughout the year repressed LH secretion strongly during anestrus and weakly during the breeding season. Estradiol was given by subcutaneous implant. *(From Legan, Karsch and Foster, 1977. Endocrinology, 101, 818)*

hours after light on (Fig. 7–23). Moreover, a photoperiod enhancing either gonadotropin and prolactin secretion or maximal testis enlargement is never followed by sustained pituitary and gonadal responses. A fatigue or refractory period generally ensues. Such a decrease in response is explained at present as resulting from a skidding of the moment of the diurnal maximal secretion of some of the hormones involved so that the same average levels of secretions may result from different physiologic effects. Moreover, the profiles of gonadotropin and prolactin secretion must be taken into account as well as the average levels. The frequency of LH and FSH pulsatile discharges changes at critical periods of sexual activity. For instance, five peaks of LH per 24 hours have been observed during the summer months in rams as compared to three peaks per 24 hours in the winter. As the physiologic significance of this pulsatile mechanism is unknown at the present time, this knowledge cannot contribute to a better understanding of sexual seasonality.

AGING AND FERTILITY

Herd fertility is determined by the percentage of pregnant females and the litter size. These both increase for a few years after puberty, reach a maximum and then decrease slowly.

The maximal pregnancy rate is reached around three to four years in sows, four to six years in ewes and five to seven years in cows. Maximal litter size occurs in third, fourth and fifth pregnancies in the sow (Legault, 1969). Maximal frequency of twin pregnancies appears from the fifth pregnancy onward in cows (Table 7–2). In ewes the twinning rate increases up to six to seven years and then decreases slowly (Turner and Dolling, 1965) (Fig. 7–24).

As in other mammals, ovulation and fertilization rates decrease only slightly in aged domestic females, but embryonic mortality, stillbirth and postpartum losses increase (ewe: Dolling and Nicolson, 1967). Early embryonic mortality may result from poor egg quality in aging females, as shown in the rabbit by the relatively unsuccessful development of blastocysts transferred from old donors to young foster mothers (Adams, 1970). However, high rates of embryonic and perinatal loss mainly result from the fact that the aging uterus reacts too slowly to the demands of the rapidly growing fetus and to the stimulus initiating parturition (Levasseur and Thibault, 1979). For economic purposes (meat quality, food avail-

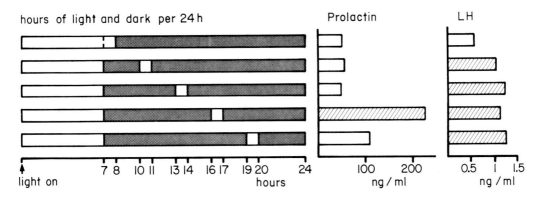

Fig. 7–23. Light régime and LH prolactin levels in rams. Prolactin secretion is only stimulated when the one-hour light flash is given 16 to 17 hours after the morning light on. The light-sensitive period is broader for LH. *(From Ravault and Ortavant, 1977. Ann. Biol. Anim. Biochim. Biophys., 17, 459–472, and Ortavant, 1977. Management of Reproduction in Sheep and Goats Symposium, Univ. Wisconsin)*

Table 7–2.   Twinning Frequencies in the Cow According to Age (Gestation Number)

Gestation No.	Twin Pregnancies (%)	
	Monozygotic	Dizygotic
1	0.15	0.33
2	0.17	1.36
3	0.14	1.96
4	0.24	2.30
5	0.17	2.54

(Swedish breed, from Johansson I., Lindhé B. and Pirchner F. (1974). Hereditas 78, 201.)

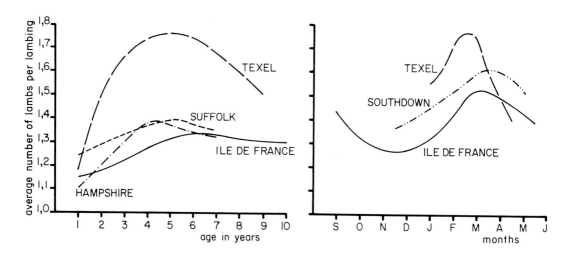

Fig. 7–24.   Effect of age on lambs per lambing and of season on lambing rate. *(Institut technique de l'Elevage ovin et caprin, Paris, France, 1972)*

ability), female domestic animals are slaughtered relatively young (average four years for the cow), long before they have reached maximal fertility, and information on domestic mammals is scarce.

In male domestic animals, semen quality of the first ejaculates is generally low. Spermatozoa are few, motility is poor and the percentage of dead and abnormal spermatozoa can exceed 80%. Even if spermatozoa can be collected early (as in six-month-old lamb, eight-month-old bull calf), the quality of ejaculates is not satisfactory until a few months later (one to four months in ram, three to six months in bull). At that time, 80% to 90% of the spermatozoa are normal and semen can be used for artificial insemination. Increased sperm production continues but at a

slower rate during one or two years in bull, ram and buck (Amann, 1970; Colas and Courot, 1977; Corteel, 1977) and maximal fertility rates are observed.

Apart from pathologic disturbances, male fertility, sperm production and semen quality decrease slowly during aging. In bulls, the daily sperm output falls from 6×10^9 at 3 to 4 years to 4×10^9 between 6 and 13 years (Hahn, Foote and Seidel, 1969). Fertility decreases in the same proportion from 65% of nonreturn at 3 to 4 years to 54% at 12 years (Bishop, 1970).

REFERENCES

Adams, C.E. (1970). Ageing and reproduction in the female mammal with particular reference to the rabbit. J. Reprod. Fertil. Suppl. *12*, 1.

Amann, R.P. (1970). Sperm production rates. *In* The

Testis. Vol. 1. A.D. Johnson, W.R. Gomes and N.L. VanDemark (eds), New York, Academic Press, pp. 433–482.

Bishop, M.W.H. (1970). Ageing and reproduction in the male. J. Reprod. Fertil. Suppl. *12*, 65.

Casida, L.E. (1968). Studies on the postpartum cow. Wisconsin Experimental Station Research Bulletin, No. 270.

Colas, G. and Courot, M. (1977). Production of spermatozoa, storage of semen and artificial insemination in the sheep. In Management of Reproduction in Sheep and Goats Symposium. Madison, University of Wisconsin, pp. 31–40.

Corteel, J.M. (1977). Production, storage and insemination of goat semen. In Management of Reproduction in Sheep and Goats Symposium. Madison, University of Wisconsin, pp. 41–57.

Courot, M. (1971). Etablissement de la spermatogenèse chez l'Agneau *(Ovis aries)*. Etude expérimentale de son contrôle gonadotrope; importance des cellules de la lignée sertolienne. Ph.D. Thesis, Université Paris VI.

Dolling, C.H.S. and Nicolson, A.D. (1967). Vital statistics for an experimental flock of Merino sheep. IV- Failure in conception and embryonic loss as causes of failure to lamb. Aust. J. Agric. Res., *18*, 767.

Foster, D.L., Lemons, J.A., Jaffe, R.B. and Niswender, G.D. (1975). Sequential patterns of circulating luteinizing hormone and follicle-stimulating hormone in female sheep from early postnatal life through the first estrous cycles. Endocrinology *97*, 985.

Hafez, E.S.E. (1952). Studies on the breeding season and reproduction of the ewe. J. Agric. Sci. *42*, 189.

Hafez, E.S.E. (1968). Environmental Effect on Animal Productivity. In Adaptation of Domestic Animals. E.S.E. Hafez (ed), Philadelphia, Lea & Febiger.

Hahn, J., Foote, R.H. and Seidel, G.E. (1969). Testicular growth and related sperm output in dairy bulls. J. Anim. Sci. *29*, 41.

Hunter, G.L. (1968). Increasing frequency of pregnancy in sheep. Anim. Breed. Abstr. *36*, 347, 533.

Land, R.B., Thimonier, J. and Pelletier, J. (1970). Possibilité d'induction d'une décharge de LH par une injection d'oestrogène chez l'agneau femelle en fonction de l'âge. C.R. Acad. Sci., Paris, D, *271*, 1549.

Legault, C. (1969). Etude statistique et génétique des performances d'élevage des truies de la race Large White. I. Effets du troupeau, de la période semestrielle, du numéro de portée et du mois de naissance. Ann. Génét. Sél. Anim. *1*, 281.

Levasseur, M.C. (1977). Thoughts on puberty. Initiation of the gonadotropic function. Ann. Biol. Anim. Biochim. Biophys. *17*, 345.

Levasseur, M.C. and Thibault, C. (1980). La fertilité de la puberté à la sénescence, chez l'Homme et les autres Mammifères. 1 Vol. Masson, Paris (in press).

Lincoln, G.A., Peet, M.J. and Cunningham, R.A. (1977). Seasonal and circadian changes in the episodic release of follicle-stimulating hormone, luteinizing hormone and testosterone in rams exposed to artificial photoperiods. J. Endocrinol. *72*, 337.

Mauleon, P. (1973). Apparition et évolution de la prophase méiotique dans l'ovaire d'embryon de brebis placé dans diverses conditions expérimentales. Ann. Biol. Anim. Biochim. Biophys., Suppl. *12*, 89.

Mauleon, P., Bezard, J., Terqui, M. (1977). Very early and transient 17β-estradiol secretion by fetal sheep ovary. In vitro study. Ann. Biol. Anim. Biochim. Biophys. *17*, 399.

Mavrogenis, A.P. and Robinson, O.W. (1976). Factors affecting puberty in swine. J. Anim. Sci. *42*, 1251.

Mazzari, G., du Mesnil du Buisson, F. and Ortavant, R. (1968). Action de la température et de la lumière sur la spermatogenèse, la production et le pouvoir fécondant du sperme chez le verrat. 6e Congr. Internat. Reprod. Anim. Insem. Artif., Paris, Vol. 1, 305–308. INRA, Paris.

Milewich, L., George, F.W. and Wilson, J.D. (1977). Estrogen formation by the ovary of the rabbit embryo. Endocrinology *100*, 187.

Ortavant, R. and Loir, M. (1978). The environment as a factor in reproduction in farm animals. World Congr. Anim. Prod., Buenos Aires.

Ortavant, R. and Thibault, C. (1956). Influence de la durée d'éclairement sur les productions spermatiques du Bélier. C.R. Soc. Biol. *150*, 358.

Oxender, W.D., Noden, P.A. and Hafs, H.D. (1977). Estrus, ovulation and serum progesterone, estradiol and LH concentrations in mares after an increased photoperiod during winter. Am. J. Vet. Res. *38*, 203.

Reardon, T.F. and Robinson, T.J. (1961). Seasonal variation in the reactivity to oestrogen of the ovariectomized ewe. Aust. J. Agric. Res. *12*, 320.

Roy, J.H.B., Gillies, C.M. and Shotton, S.M. (1975). Factors affecting first oestrus in cattle and their effects on early breeding. In The Early Calving of Heifers and its Impact on Beef Production. J.C. Tayler (ed), Brussels, European Economic Communities, pp. 128–142.

Schlenker, G., Köppe, D. and Siefert, H. (1973). Die präpuberale und präovulatorische Östrogenausscheidung in Beziehung zur Fortpflanzungsregulation weiblicher Schweine. Arch. exp. Vet. Med. *27*, 881.

Shearer, I.J., Purvis, K., Jenkin, G. and Haynes, N.B. (1972). Peripheral plasma progesterone and oestradiol 17β levels before and after puberty in gilts. J. Reprod. Fertil. *30*, 347.

Shemesh, M., Ailenberg, M., Milaguir, F., Ayalon, N., and Hansel, W. (1978). Hormone secretion by cultured bovine pre- and postimplantation gonads. Biol. Reprod., *19*, 761.

Turner, H.N. and Dolling, C.H.S. (1965). Vtial statistics for an experimental flock of Mérino sheep. II. The influence of age on reproductive performance. Aust. J. Agric. Res. *16*, 699.

Van Straaten, H.M.W. and Wensing, C.J.G. (1978). Leydig cell development in the testis of the pig. Biol. Reprod. *18*, 86.

8

Folliculogenesis, Egg Maturation and Ovulation

E.S.E. HAFEZ,
MARIE-CLAIRE LEVASSEUR
and
C. THIBAULT

The ovary performs two major functions: (1) the cyclic production of fertilizable ova and (2) the production of a balanced ratio of steroid hormones that maintain the development of the genital tract, facilitate the migration of the early embryo and secure its successful implantation and development in the uterus. The follicle is the ovarian compartment that enables the ovary to fulfill its dual function of gametogenesis and steroidogenesis.

This chapter discusses the physiologic mechanisms of development and differentiation of the ovarian follicle and the oocyte until ovulation.

FOLLICULOGENESIS

In the primordial follicle reserve, formed during fetal life or soon after birth, some primordial follicles begin to grow continuously throughout life, or at least until the reserve is exhausted. Little is known about the factors that control the mechanisms triggering follicular growth. The role of the oocyte in the initiation of this growth was stressed initially from measurements of oocyte and follicle diameters. However, enlargement of the oocyte follows changes in the shape of the follicular cell from flat to cuboid. When any follicle is released from this reserve it continues to grow until ovulation or until the follicle degenerates, which is the case with the majority of follicles.

The total period required from follicular growth to ovulation is one of the basic parameters in ovarian physiology. Data are available only for small rodents, for which the most acceptable duration is around 20 days, four days being necessary for final growth after antrum formation. Since comparable values for final growth in ewes, cows and sows range between 12 to 34 days, the total duration of follicular growth in domestic mammals is longer than 20 days and presumably about six months.

The growth of the follicle up to the stage of antrum formation is not strictly gonadotropin-dependent. In hypophysectomized females, the formation of preantral follicles continues at a more or less normal rate. On the other hand, antrum formation and final growth are entirely FSH- and LH-dependent (Fig. 8–1).

Endocrine Control of Follicular Growth

In vivo and *in vitro* studies show that FSH plays a major role in the initiation of antrum formation. This gonadotropin stimulates granulosa cell mitosis and follicular fluid formation. Estradiol enhances FSH mitotic effect. FSH stimulates granulosa cells through membrane receptors whose number per cell remains constant during follicular growth. Moreover, FSH induces granulosa cell sensitivity to LH by increasing the number of LH receptors. In sows, LH receptors increase from 300 in small follicles to 10,000 in large preovulatory follicles. This LH-receptor increment prepares the luteinization of granulosa cells in response to LH ovulatory surge.

On the other hand, theca cells are stimulated only by LH, and LH receptors are present from the beginning of theca cell formation. In sows the number of receptors per theca cell only doubles during final follicular growth.

Endocrine Control of Steroidogenesis

Steroidogenic activity of the follicle also depends on FSH and LH acting on granulosa and theca cells, respectively. The primary steroid secreted is usually

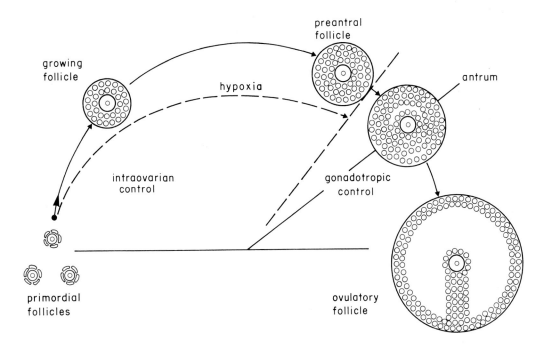

Fig. 8–1. Some follicles begin to grow every day. The growth of the primordial follicle up to the preantral stage occurs in the hypophysectomized female but at a lower rate. The number of primordial follicles beginning growth every day is controlled by an unknown intraovarian factor. Antrum formation and final follicle growth up to the ovulatory size is gonadotropin-dependent. This second phase of follicle growth is shorter than the first one. *(Thibault and Levasseur, 1978. La fonction ovarienne chez les Mammiferes. Masson, Paris)*

Table 8–1.　Steroid Concentration in the Follicular Fluid of Ovine Follicles

Diameter of follicles (nm)	Steroid Hormones in Follicular Fluid (Pmol/ml)		
	Estradiol	Testosterone	Progesterone
3–6 (large, nonatretic follicles)	860	70	150
3–6 (large, atretic follicles)	100	190	120
2–3 (small, nonatretic follicles)	140	280	40

Large healthy follicles are characterized by a high estradiol content and relatively low testosterone value. When large follicles are atretic, estradiol level is lower than that of testosterone. The same low estradiol-testosterone ratio is observed in small healthy follicles; however, their progesterone content is lower than that of large follicles, whether they are atretic or not. Intrafollicular steroid content reflects the steroidogenic potency of these follicles. (From Moor et al, 1978. *J. Endocrinol. 77*, 309.)

estradiol-17β. However, progestins and androgens are also produced (Table 8–1). Follicular androgens are important in the female. In rabbits androgens can naturally overpass estrogen secretion. The androgen-estrogen ratio in the follicular fluid reflects the physiologic integrity and viability of the follicle. In sheep a high androgen content is normal in small, healthy follicles, but it signals atresia in large follicles (Moor et al, 1975) (Table 8–1 and Fig. 8–2). Immunization against androstenedi-

one has revealed the unlikely role of ovarian androgens acting at the hypothalamo-pituitary level, since the ovulation rate doubles in immunized ewes (Van Look et al, 1978).

During follicular growth, estradiol production results from the coordinated steroidogenic activity of the theca interna and the granulosa cells (Table 8–2). *In vitro* experiments have shown that theca cells mainly secrete testosterone, while granulosa cells convert testosterone into estradiol due to high aromatase activity. In the ewe granulosa cells secrete only estradiol when testosterone is present in the culture medium; secretion is higher if FSH is added. On the other hand, theca cells from large follicles of cattle and sheep synthesize testosterone (Moor and Trounson, 1977). Because FSH mainly stimulates granulosa cells and LH stimulates theca cell testosterone production, the FSH-LH ratio is an important endocrine parameter of normal ovarian steroid production.

After fixation on the cell membrane receptors, FSH and LH first stimulate adenylcyclase activity, causing an increase of intracellular cAMP. The next step classically involves cAMP activation of protein kinase which, by unknown mechanisms, enhances specific enzyme activity of the steroidogenic pathways. It is generally accepted that LH stimulates pregnenolone formation resulting from cholesterol side-chain cleavage and that

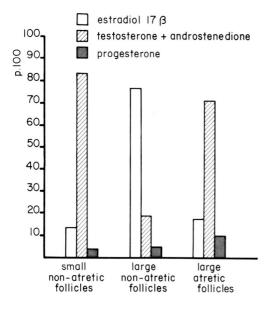

Fig. 8–2.　Ovine follicles. Relative efficiency of steroid synthesis *in vitro*. (Moor et al, 1978. *J. Endocrinol. 77, 309*)

Table 8–2. Relative Role of Theca and Granulosa Cells in Estradiol in the Cow and Ewe

Species	Tissue Culture		Hormone Synthesized	
	Hormone Added	Cells	Estradiol	Androstendione and testosterone
Cow	Pregnenolone as precursor	Granulosa, theca	(%) 0.5–1	(%) 2–3
	Androstendione	Granulosa, theca	30–50 0–5	
Ewe	—	Granulosa	(ng/ml) 3	
	—	Theca	3	
	—	Granulosa and theca	20	
	Testosterone	Granulosa	56	
	FSH	Granulosa	3	
	Testosterone and FSH	Granulosa	82	

(From Lacroix et al, 1974. *Steroids*, 23, 337, and Moor, 1977. *J. Endocrinol.* 73, 143.)

FSH is mainly responsible for aromatase activity (Fig. 8–3).

Follicular Growth During "Follicular" and "Luteal" Phases

Active corpora lutea are present in the ovaries during a large part of the estrous cycle called the "luteal phase." The "follicular phase," the period from corpus luteum regression to the following ovulation, is apparently short (two days in ewes and four to five days in cows and sows). However, the presence of antral follicles throughout the luteal phase suggests that the real duration of the follicular phase is longer than two to five days, if "follicular phase" refers to the period from antral follicle formation to ovulation. Therefore, the luteal phase in domestic mammals would partially overlap the true follicular phase, obscuring the relationship between basal plasma FSH and LH levels and follicular growth.

Figure 8–4 illustrates some species differences regarding these phases: (a) animal species with no luteal phase, such as rodents with four-day estrous cycles; (b) primates with quite distinct follicular and luteal phases; (c) domestic mammals with overlapping "follicular" and "luteal" phases.

In rodents, plasma FSH profiles differ from those of LH. After the preovulatory discharge of both gonadotropins, FSH surges again after ovulation. Preventing the effect of this FSH surge with FSH antibodies shows that the surge is responsible for the formation of the antral follicles destined to ovulate 3.5 days later at the subsequent estrous cycle (Welschen and Dullaart, 1976). In primates, a rise of FSH begins during menstruation at the end of the luteal phase. This elevation presumably initiates antrum formation of the follicle that will ovulate two weeks later and also of others that become atretic at the end of the first week of the follicular phase (Thibault, 1977).

In domestic mammals there is also a second rise of FSH 20 to 30 hours after the preovulatory surge of LH and FSH (ewes: Pant et al, 1977; sows: Rayford et al, 1974). This postovulatory FSH rise triggers antrum formation in the follicle population that includes candidates for ovulation one or two cycles later. In ewes the second peak of FSH is significantly larger in those

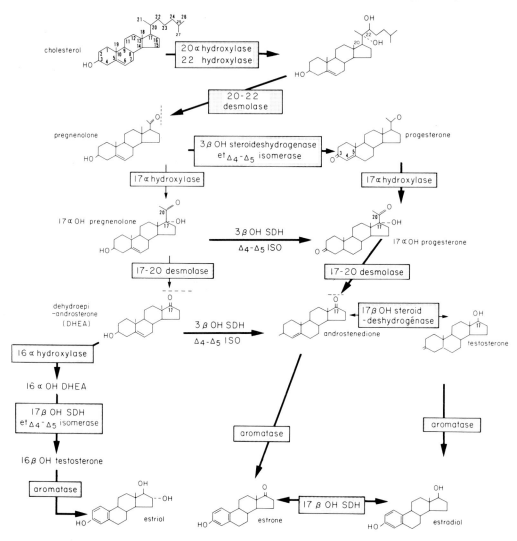

Fig. 8–3. Sex steroid synthesis. Arrows indicate possible pathways with enzymes involved.

animals with a higher ovulation rate, and the magnitude of this peak is highly correlated with the number of antral follicles present in the ovary 17 days later (Cahill et al, 1979). From estimations of the duration of growth of antral follicles, the time lapse from antral formation to ovulation is either 17 or 34 days (Turnbull et al, 1977).

Only a few of these differentiated antral follicles grow to ovulation; the others become atretic and degenerate. Experiments on hamsters show that FSH levels near the

end of the follicular phase are related to the rate of atresia (Grady and Greenwald, 1968). The same physiologic regulation is noted in domestic mammals and superovulation occurs when PMSG (a gonadotropin with high FSH potency) is injected just before atresia begins. Several field trials have shown that this gonadotropin must be administered three to five days before ovulation in order to obtain superovulation in cattle, sheep and swine.

The relatively long duration of the fol-

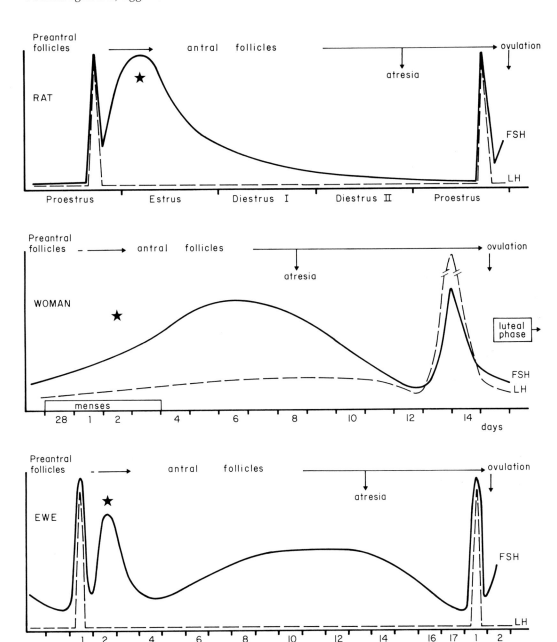

Fig. 8–4. Final growth of follicles destined to ovulate. Whatever the duration of the first follicular growth phase, follicles attain the preantral stage everyday. From this follicle population recruitment of antral follicles destined to ovulate is assumed by FSH postovulatory surge (★) which occurs before the beginning of corpus luteum secretion in the rat and ewe. In primates this surge is postponed and occurs after corpus luteum regression. Some of these follicles become atretic when FSH returns to low basal levels. PMSG injection at this stage rescues these follicles and permits superovulation. FSH appears as the most important gonadotropin to support follicular growth, whereas LH plays a major role in ovulation.

licular phase in domestic mammals as compared to rodents probably results from the slowing down of follicular growth by progesterone from the corpus luteum. When fully functional corpora lutea are induced in rodents by cervical stimulation, the duration of the estrous cycle and of follicular growth increases from 4 days to at least 12 days (Welschen et al, 1975). On the other hand, reduction of the progesterone level during the luteal phase in cows and ewes by corpus luteum enucleation or prostaglandin luteolysis is followed by shortening of the cycle; ovulation occurs within three days, showing an immediate acceleration of follicular growth. This is the physiologic basis of the well-known practice of estrus synchronization in cattle after prostaglandin luteolytic treatment or in sheep after withdrawal of exogenous progestogens (vaginal sponge).

EGG MATURATION

The maturation of mammalian oocytes comprises two stages, (1) a period of growth and (2) a period of final nuclear and cytoplasmic preparation prerequisite to fertilization and normal development.

Oocyte Growth

When a primordial follicle is released from the reserve, the oocyte and its follicle begin to grow. Oocyte growth is almost complete at the time of antrum formation. Through cellular processes, the inner cumulus cells actively cooperate to achieve oocyte growth, as they establish close contact with the oocyte cell membrane. Gap junctions, permeable to small molecules, have been described in cows (Szöllösi, 1975) and rabbits (Gondos, 1970). During the formation of the external membrane of the oocyte (the zona pellucida), the cumulus cell processes are strengthened and the membrane junctions are retained.

RNAs are actively synthesized during oocyte growth and stored for protein synthesis at early stages of zygote development. In sows and cows, ribosomal RNAs are probably synthesized, even after the gonadotropin ovulatory surge; the large nucleolus remains active up to the time of chromosome reappearance (Szöllösi, personal communication).

Oocyte Preparation for Fertilization

From oogenesis onward, the diplotene nucleus of the oocyte remains in a resting stage called the "dictyate nucleus." Meiosis never resumes normally before gonadotropin ovulatory surge. However, in all mammalian species, when the oocyte is removed from the antral follicle and cultured in a gonadotropin-free medium, it spontaneously resumes meiosis up to metaphase I or metaphase II, the stage normally reached at the time of ovulation (Thibault, 1977). Co-culture of oocyte and granulosa or theca cells, as well as culture of the follicle cell-free oocyte in a medium with follicular fluid extracts, has shown that the maintenance of the oocyte nucleus in a dictyate state results from the regressive effect of the granulosa on the oocyte (cattle and swine: Foote and Thibault, 1969; Tsafriri and Channing, 1975) (Fig. 8–5).

The role of the gonadotropin ovulatory discharge is to suppress production of the granulosa cell meiotic-inhibiting factor. Cytologic and ultrastructural changes in granulosa cells after the LH-FSH surge show that the gonadotropin surge is followed by metabolic modification of that follicular layer. When follicles become atretic during the last stages of development, those oocytes that evade granulosa cells resume meiosis to metaphase II and degenerate.

Meiotic resumption (nuclear maturation) is only one aspect of egg maturation; cytoplasmic maturation must also occur. In cows, ewes or rabbits, embryonic de-

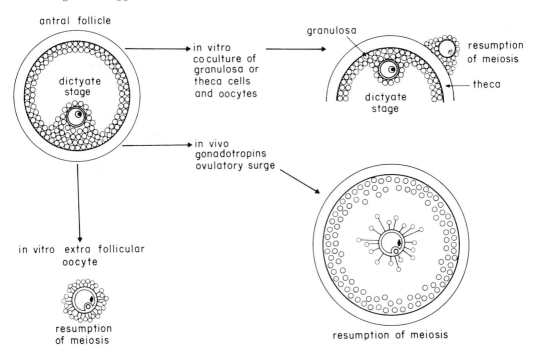

Fig. 8–5. Meiosis only resumes when the oocyte is released from the granulosa cells. *In vivo*, this occurs after the gonadotropin ovulatory surge, which causes loosening of granulosa cell junctions and the release of the oocyte into the follicular fluid. *In vitro*, meiosis resumes when oocytes are cultured, isolated or "grafted" on theca cells. Complete oocyte maturation (resumption of meiosis and cytoplasmic maturation) only occurs in the follicle after gonadotropin ovulatory surge.

velopment never progresses normally when extrafollicular oocytes, which have reached the second meiotic division *in vitro*, are transferred into recipient females. Cytologic examination reveals that the nucleus of the fertilizing sperm transforms into an abnormal pronucleus, although all other aspects of fertilization (second polar body formation, appearance of female pronucleus, timing of the first cleavage) proceed normally in these eggs (Thibault et al, 1975; Motlik and Fulka, 1974). By collecting oocytes at increasing intervals after gonadotropin ovulatory surge, it has been shown that the oocyte becomes able to assume normal male pronucleus formation only a few hours before ovulation. The factors responsible for final cytoplasmic maturation are unknown. This maturation occurs in the follicle when the gonadotropic ovulatory surge has drastically changed granulosa cell metabolism and follicular fluid composition.

The culture of preovulatory follicles in the presence of high FSH and LH levels allows both meiotic resumption and cytoplasmic maturation, as proven by the birth of rabbits (Thibault et al, 1975a) and lambs (Moor and Trounson, 1977) after transfer of *in vitro* matured oocytes into recipient mothers.

OVULATION

Preovulatory follicles undergo three major changes during the ovulatory process: (1) cytoplasmic and nuclear maturation of the oocyte, (2) disruption of cumulus cell cohesiveness among the cells of the granulosa layer and (3) thinning and rupture of the external follicular wall. In the ewe, all these changes result from turning-off the follicular metabolic pathways by the gonadotropin surge.

Ovulatory vs. Atretic Follicles

Little is known about the factors that determine which follicle will reach maturity and ovulate. The distribution of capillary flow to follicles of different sizes was measured; the relative blood flow seemed to be inversely related to the mass of follicular tissue. After the ovulatory surge of gonadotropins, blood flow increases to all classes of follicles. The follicle destined to ovulate, however, receives not only the largest volume of blood in absolute terms (ml/min), but also has capillaries that are more permeable than those in other follicles (Moor et al, 1975). The rapid response of the ovarian microcirculation to LH and the increased metabolic requirements of follicles after gonadotropin stimulation suggest that enhanced vascularity may be an inherent part of the action of LH upon the follicles (Ahren et al, 1969).

Site of Ovulation

The mammalian ovary is normally arranged so that ovulation can occur at any point on its surface, except at the hilus. However, ovulation in mares always occurs in a limited ovarian area called the ovulation fossa. The ovary of horses begins its development in the usual way and germinal epithelium covers the whole ovary, but during the neonatal period this epithelium becomes concentrated in one area, the ovulatory fossa (Fig. 8–6). This observation suggests that the ovarian epithelium plays a basic role in follicular rupture.

In cattle, sheep and horses, ovulation occurs at random with respect to which ovary contains the previous corpus luteum. However, in some mammals ovulation consistently alternates between the ovaries, and in others (eg, whales, mountain viscacha) ovulation may predominate from one ovary (cf Harrison and Weir, 1977). In the ewe the side of ovulation is independent of the location of the corpus luteum of the previous ovarian cycle, and the duration of the estrous cycle is unaffected by the relative locations of the corpus luteum (Wallach et al, 1973). In the rhesus monkey the corpus luteum locally retards subsequent follicular growth, so that alternation of ovulation between the right and left ovaries is associated with a shorter menstrual cycle (29 days) compared with the duration of cycles in which

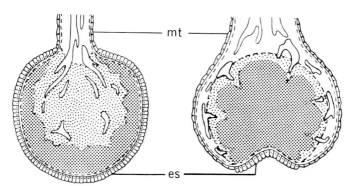

Fig. 8–6. Diagrammatic comparison of the epithelial-medullary arrangement in the mature equine ovary (right) and other mammalian ovaries (left): *es*, epithelium superficiale; *mt*, mesothelium; *close stipple*, cortex; *open stipple*, medulla. Note the reversal of epithelial-medullary areas with only a small area of contact of the cortical area with the ovarian surface (ovulation fossa). *(From Mossman and Duke, 1973: The Ovary. Madison, University of Wisconsin Press)*

the corpus luteum is in the same ovary, as in the previous cycle (35 days). Unlike that of the monkey, the ovine corpus luteum has little or no local inhibitory effect on subsequent ovulation.

Cellular Events

Several tissue layers separate the oocyte from the outside of the follicle. These are the surface epithelium, the collagen-rich tunica albuginea, the theca externa, the thin basement lamina separating the capillary network from the membrana granulosa and the membrana granulosa itself. Before ovulation can occur, all tissue layers must be broken down. Moreover, the necessary increase in follicular elasticity during preovulatory growth is associated with changes in granulosa and theca cell relationships. Such changes are also prerequisite to further corpus luteum organization.

As the enlarging follicle begins to protrude from the surface of the ovary, the vascularity of the follicular surface increases except at its center, which seems devoid of blood vessels. This avascular area is the future point of rupture.

The same processes occur in all mammalian species (Fig. 8–7). The most complete data on the timing of the successive events terminating in follicular rupture have been obtained from studies on rabbits, since ovulation occurs about ten hours after coitus has induced the gonadotropic ovulatory surge.

Oocyte. The cumulus cells of growing graafian follicles are cytologically indistinguishable from the granulosa cells. In rabbits, cavities appear one to two hours after coitus within the cumulus mass and the cells progressively separate from each other. Only the cumulus cells anchored in the zona pellucida remain, surrounding the oocyte and forming the corona radiata. Cumulus cell dissociation frees the oocyte from the granulosa layer and meiosis resumes about three hours after the

gonadotropin surge. This process, called nuclear maturation, ends one hour before ovulation when the first polar body has been extruded.

Cumulus cells actively secrete glycoproteins, which form a viscous mass enclosing the oocyte and its corona. After follicular rupture, the viscous mass spreads at the ovarian surface to facilitate the "pick up" of the oocytes by the fimbriae.

Granulosa Cells. At the same time, a more discrete dissociation takes place between the granulosa cells. The granulosa layer is completely dissociated only at the follicular apex and finally disappears. This process begins six hours after coitus and terminates before ovulation four hours later. About two hours before ovulation, granulosa cell growth processes penetrate through the lamina basalis, preparing the invasion of theca cells and blood vessels into the granulosa after ovulation in the developing corpus luteum (Cherney et al, 1975; Bjersing and Cajander, 1974).

Theca Cells. The follicular volume rapidly increases in the few hours preceding ovulation without any increment of follicular fluid pressure (Rondell, 1970) owing to the increased elasticity of the follicle. This results from a looser cohesion of the theca externa cells owing to the invasive edema of this layer and to collagen fiber dissociation which begins four hours after coitus (Espey and Coons, 1976; Bjersing and Cajander, 1974). Collagenases have been isolated from sow and rabbit follicles due to enzyme release from granulosa cells; follicular fluid enzyme activity increases as ovulation approaches (Espey and Rondell, 1968). Dissociation of bundles of collagen fibers also results from proteolytic enzyme activity on the proteinaceous matrix of the fibers (Espey, 1967). Plasmin activity increases after the gonadotropin surge and this proteolytic enzyme causes an increase in follicular wall elasticity (Beers, 1975).

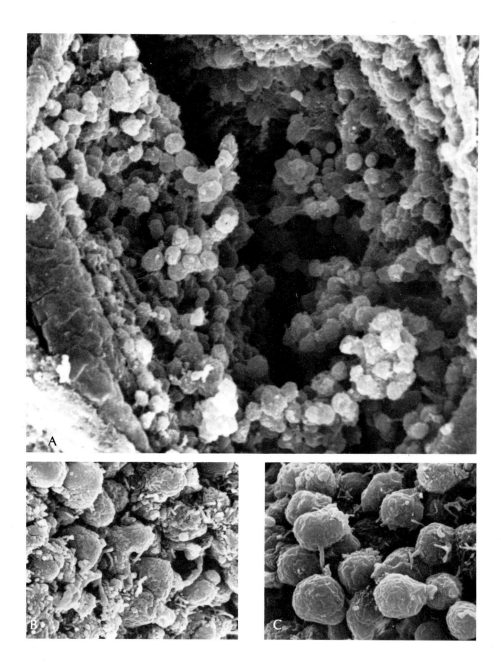

Fig. 8–7. *A,* Scanning electron micrographs of ruptured follicle showing strands of granulosa cells. Note disruption of cells and the remarkable intercellular spaces due to accumulation of follicular fluid. (1000 ×). *B,C,* Granulosa cells from different stages of the estrous cycle (3,000 ×).

Apex Changes. The rupture of the follicle outside the ovary involves interaction between the ovarian epithelium and the underlying follicular wall. Four hours after coitus lysosomes differentiate and enlarge in the epithelial cells covering the follicle. The lysosomes then disrupt and their hydrolases first destroy the underlying albuginea cells and then the theca cells (Cajander, 1976). After lysosome rupture, epithelial cells scale off. The wall of the follicular apex becomes exceedingly thin in a circumscribed area called the "stigma."

The stigma thins out, bulges on the surface of the ovary, and becomes completely avascular. At ovulation the bulging stigma ruptures at the apex, releasing some of the follicular fluid and the viscous glycoprotein mass embedding the oocyte.

Biochemical Mechanisms of Ovulation

The gonadotropin preovulatory surge first induces an immediate and temporary rise in steroid levels due to an increased secretion of progesterone and related progestins. Later, estradiol and prostaglandin F_2 secretion are also augmented. Inhibition of either prostaglandin or steroid secretion prevents ovulation. Thus, the gonadotropin surge induces ovulation by a cascade of biochemical changes.

Changes in Steroid Secretion. The enhancement of steroid secretion and the switch of the estradiol-progesterone ratio that follow the gonadotropin surge are easily detectable in the follicular fluid. In sow follicles, the estradiol-progesterone ratio decreases from 2.0 to 0.15 (Gerard et al, 1979). These changes are slightly noticeable in the ovarian venous blood and are undetectable in the peripheral blood (Eiler and Nalbandov, 1977). Inhibition of progesterone synthesis prevents ovulation (Lipner and Greep, 1971). The role of progesterone is to stimulate collagenase activity in the follicular wall

(sows: Rondell, 1970). Moreover, the thecal edema may result from the transitory but important rise of steroid levels.

Prostaglandins. The increase of $PGF_{2\alpha}$-PGE_2 levels in follicular fluid does not immediately follow the gonadotropin surge as the steroid elevations do. In rabbits prostaglandins begin to increase five hours after coitus and reach a maximum level just before ovulation four hours later (Armstrong and Zamecnik, 1975). In sows an increase of prostaglandins begins only 30 hours after ovulatory discharge and the maximal level occurs about 40 hours later as ovulation approaches (Ainsworth et al, 1975). Prostaglandins play a basic role in follicular rupture; inhibition of their synthesis always prevents ovulation (rabbits, sows: Armstrong, 1975), and their action is exerted at the level of the albuginea and follicular epithelium. When prostaglandin synthesis is inhibited, the oocyte remains inside the luteinizing follicle or may be "ovulated" inside the ovary (Osman and Dullaart, 1976). $PGF_{2\alpha}$ is involved in follicular rupture and PGE_2, in the remodeling of the follicular layers, terminating in corpus luteum formation.

$PGF_{2\alpha}$ contributes to the rupture of epithelial cell lysosomes at the follicular apex, as suggested both by the relation between the elevation of the $PGF_{2\alpha}$ level and lysosomal rupture and by the well-known detrimental effect of $PGF_{2\alpha}$ on lysosomal membrane (Weiner and Kaley, 1972). Increases of $PGF_{2\alpha}$ must also be related to preovulatory enhancement of ovarian contractions. $PGF_{2\alpha}$ stimulates the production of plasminogen activator, thus increasing plasmin activity. Plasmin is generally involved in tissue cell migration and presumably plays a role in the mixing of theca and granulosa cells during corpus luteum formation. These physiologic mechanisms are shown in Figure 8–8.

Gonadotropins. The freeing of the oocyte inside the follicle is the only direct gonadotropic action known; *in vitro* cumulus cell dissociation is exclusively

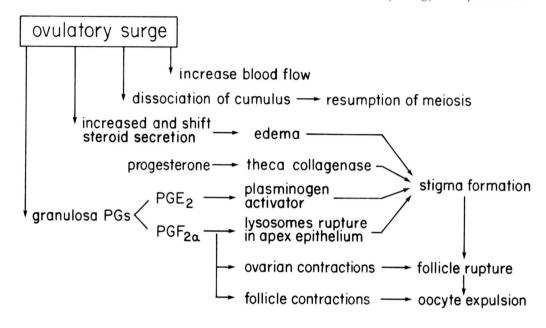

main phenomena having unknown causes
cytoplasmic maturation of the oocyte
lysosomes multiplication in ovarian epithelium
at the apex of the ovulatory follicle

ovulation

Fig. 8–8. Tentative synthesis of biochemical processes in ovulation.

obtained by FSH and LH (Thibault et al, 1975b).

Neuromuscular Mechanisms

The ovarian stroma and the concentric layers of the theca externa of preovulatory follicles contain smooth muscle cells that are richly innervated by autonomic nerve terminals. Drug inhibition of β-adrenergic receptors delays ovulation and reduces ovulation rate, demonstrating a role of neuromuscular systems in follicular rupture (rabbits: Virutamasen et al, 1972). The frequency of spontaneous ovarian contractions begins to increase two to three hours before ovulation, reaching a maximum around ovulation time (rabbits:

Virutamasen et al, 1976). The role of $PGF_{2\alpha}$ is evidenced both by *in vitro* stimulation of ovarian contractions by $PGF_{2\alpha}$ and by the *in vivo* relationship between increasing contraction frequencies and increasing $PGF_{2\alpha}$ levels in follicular fluid (Virutamasen et al, 1972). Thus, ovarian contractions facilitate follicular rupture after the follicular apex has been thinned.

Before rupture the follicle itself does not contract spontaneously, as shown *in vivo* by stable intrafollicular pressure and *in vitro* by the absence of contractions in the isolated follicle. However, strips of follicular wall contract spontaneously and respond to drugs and $PGF_{2\alpha}$ as the entire ovary does (cows: Walles et al, 1975; sheep: O'Shea and Phillips, 1974). Thus, after follicular rupture the thecal neuro-muscular system, stimulated by $PGF_{2\alpha}$, contributes to the extrusion of the oocyte.

Egg "Pick-Up"

The ovary, attached to the back of the broad ligament, lies free in the peritoneal cavity. The oviduct curls over the ovary to facilitate egg "pick-up" by the mucosal folds of the fimbriae. At the time of ovulation the ovum, together with the surrounding cells in a gelatinous mass, protrudes at the ovarian surface and is swept into the ostium of the oviduct by the action of the motile kinocilia of the fimbriae.

Neuroendocrine Control of the Ovulatory Gonadotropic Discharge

A preovulatory gonadotropin surge occurs at the beginning of estrus when progesterone has fallen to its minimal blood levels and when estradiol reaches its highest cyclic values. The usual role of estradiol, to trigger the gonadotropin surge in the absence of progesterone, has been demonstrated by experiments on castrated females (Pelletier and Thimonier, 1975). Estradiol treatment of ovariectomized females always results first in a decreased LH level (estradiol negative feedback effect), and then in a typical LH surge (estradiol positive feedback effect).

A typical preovulatory LH surge is obtained only when the duration of estradiol treatment and blood estradiol levels narrowly mimic the natural estradiol increase before the cyclic LH surge. Simultaneous administration of progesterone always suppresses estradiol positive feedback. Thus, the actively growing graafian follicle is responsible for its own ovulation through estradiol secretion.

Estradiol acts at two levels: the pituitary and the hypothalamus. Estradiol increases the sensitivity of pituitary gonadotropin-producing cells to the competent hypothalamic hormone, GnRH (LH-RH). It has been shown in all mammals that injections of the same quantity of GnRH are followed by increasing releases of LH as the injections are given closer to the time of spontaneous cyclic LH surge. This indicates increasing pituitary sensitivity at the end of the estrous cycle when the estradiol level reaches its maximal value.

On the other hand, indisputable data on rats and macaque monkeys have shown that the LH surge is caused by GnRH discharges by the hypothalamus; such discharges result from estradiol positive feedback on the central nervous system (Fig. 8–9). Although FSH is also released in response to estradiol positive feedback, it is not known whether FSH plays a role in ovulation in domestic mammals (Fig. 8–9).

During postpartum anestrus in lactating females, estradiol positive feedback may be prevented by high prolactin levels related to the suckling stimulus. In fact, estradiol positive feedback is inhibited in castrated ewes in which the prolactin level has been artificially enhanced (Kann et al, 1977).

Anomalies of Ovulation and Reproductive Failure

The absence of ovulation and the subsequent formation of follicular cysts are the main causes of reproductive failure in cows and aged sows. An efficient treatment for preventing cyst formation, which reduces fertility in cystic cows, is the stimulation of gonadotropin release by GnRH or the direct stimulation of the ovary with HCG. The fertility level and changes in steroid secretion indicate that ovulation or luteinization follow these treatments.

Thus, it is probable that the presence of cystic follicles reflects a disorder in gonadotropic function at the hypothalamic level. The main failure is the absence of ovulatory gonadotropic surge, as shown by the absence of detectable levels of progesterone in the blood.

The absence of estradiol positive feedback may be due either to a relative incapacity of the hypothalamus to respond to estradiol by GnRH discharge or to relatively insufficient estradiol secretion by the growing follicle resulting from permanent inadequate gonadotropic secretion.

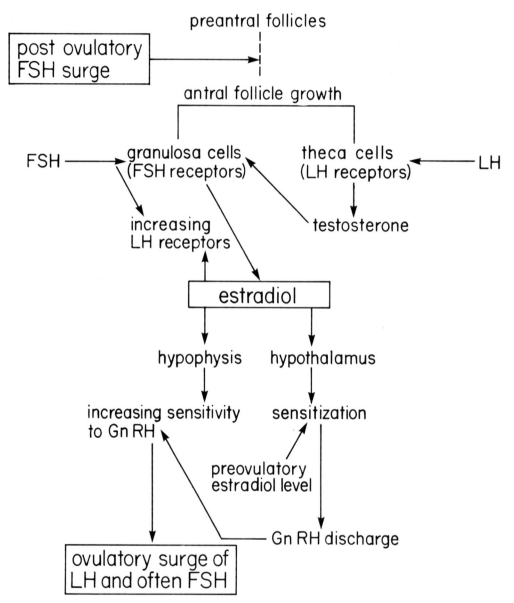

follicle maturation

Fig. 8–9.   Neuroendocrine control of the gonadotropic discharge in relation to ovulatory surge.

REFERENCES

Ahren, K., Hamberger, L. and Rubenstein, L. (1969). Acute in vivo and in vitro effects of gonadotrophins on the metabolism of the rat ovary. *In* The Gonads. K.W. McKerns (ed), New York, Appleton-Century-Crofts, pp. 327–352.

Ainsworth, L., Baker, R.D. and Armstrong, D.T. (1975). Pre-ovulatory changes in follicular fluid prostaglandin F levels in swine. Prostaglandins 9, 915.

Armstrong, D.T. (1975). Role of prostaglandins in follicular responses to luteinizing hormones. Ann. Biol. Anim. Biochim. Biophys. 15, 181.

Armstrong, D.T. and Zamecnik, J. (1975). Preovulatory elevation of rat ovarian prostaglandins F and its blockade by indomethacin. Mol. Cell. Endocrinol. 2, 125.

Beers, W.H. (1975). Follicular plasminogen and plasminogen activator and the effect of plasmin on ovarian follicle wall. Cell 6, 379.

Bjersing, L. and Cajander, S. (1974). Ovulation and the mechanism of follicular rupture. I, II, III. Cell Tiss. Res. 149, 287.

Cahill, L.P., Saumande, J., Ravault, J.P., Blanc, M., Thimonier, J., Mariana, J.C. and Mauleon, P. (1979). Hormonal and follicular parameters in ewes of high and low ovulation rates. J. Reprod. Fertil. (in press)

Cajander, S. (1976). Structural alterations of rabbit ovarian follicles after mating with special reference to the overlying surface epithelium. Cell Tiss. Res. 173, 437.

Cherney, D.D., Didio, L.J.A. and Motta, P. (1975). The development of rabbit ovarian follicles following copulation. Fertil. Steril. 26, 257.

Eiler, H. and Nalbandov, A.V. (1977). Sex steroids in follicular fluid and blood plasma during the estrous cycle of pigs. Endocrinology, 100, 334.

Espey, L.L. (1967). Ultrastructure of the apex of the rabbit graafian follicle during the ovulatory process. Endocrinology 81, 267.

Espey, L.L. and Coons, P.J. (1976). Factors which influence ovulatory degradation of rabbit ovarian follicles. Biol. Reprod. 14, 233.

Espey, L.L. and Rondell, P. (1968). Collagenolytic activity in the rabbit and sow graafian follicle during ovulation. Am. J. Physiol. 214, 326.

Foote, W.D. and Thibault, C. (1969). Recherches expérimentales sur la maturation in vitro des ovocytes de Truie et de Veau. Ann. Biol. Anim. Biochim. Biophys. 9, 329.

Gerard, M., Menezo, Y., Rombauts, P., Szöllösi, D. and Thibault, C. (1979). In vitro studies of oocyte maturation and follicle metabolism in the pig. Ann. Biol. Anim. Biochim. Biophys. 19 (in press).

Gondos, B. (1970). Granulosa cell-germ cell relationship in the developing rabbit ovary. J. Embryol. Exp. Morphol. 23, 419.

Grady, K.L. and Greenwald, G.S. (1968). Studies on interactions between the ovary and pituitary follicle-stimulating hormone in the golden hamster. J. Endocrinol. 40, 85.

Hafez, E.S.E. (1979). Epilogue. In Human Ovulation. E.S.E. Hafez (ed), New York, Elsevier.

Harrison, R.J. and Weir, B.J. (1977). Structure of the mammalian ovary. In The Ovary, 2nd ed., Vol. I. Z. Zuckerman and B.J. Weir (eds), New York, Academic Press, pp. 113–217.

Kann, G., Martinet, J. and Schirar, A. (1977). Modifications of gonadotrophin secretion during natural and artificial hyperprolactinaemia in the ewe. In Prolactin and Human Reproduction. P.G. Crosignani and C. Robyn (eds), New York, Academic Press, pp. 47–59.

Lipner, H. and Greep, R.O. (1971). Inhibition of steroidogenesis at various sites in the biosynthetic pathway in relation to induced ovulation. Endocrinology 88, 602.

Moor, R.M., Hay, M.F. and Seamark, R.F. (1975). The sheep ovary: Regulation of steroidogenic, haemodynamic and structural changes in the largest follicle and adjacent tissues before ovulation. J. Reprod. Fertil. 45, 595.

Moor, R.M. and Trounson, A.O. (1977). Hormonal and follicular factors affecting maturation of sheep oocytes in vitro and their subsequent developmental capacity. J. Reprod. Fertil. 49, 101.

Motlik, J. and Fulka, J. (1974). Fertilization and development in vivo of rabbit oocytes cultivated in vitro. J. Reprod. Fertil. 40, 183.

O'Shea, J.D. and Phillips, R.E. (1974). Contractility in vitro of ovarian follicles from sheep, and the effects of drugs. Biol. Reprod. 10, 370.

Osman, P. and Dullaart, J. (1976). Intraovarian release of eggs in the rat after indomethacin treatment at proestrus. J. Reprod. Fertil. 47, 101.

Pant, H.C., Hopkinson, C.N.R. and Fitzpatrick, R.J. (1977). Concentration of estradiol, luteinizing hormone and follicle-stimulating hormone in the jugular venous plasma of ewes during the estrous cycle. J. Endocrinol. 73, 247.

Pelletier, J. and Thimonier, J. (1975). Interactions between ovarian steroids or progestagens and LH release. Ann. Biol. Anim. Bioch. Biophys. 15, 131.

Rayford, P.L., Brinkley, H.J., Young, E.P. and Reichert, L.E. (1974). Radioimmunoassay of porcine FSH. J. Anim. Sci. 39, 348.

Rondell, P. (1970). Follicular processes in ovulation. Fed. Proc. 29, 1875.

Strickland, S. and Beers, W.H. (1976). Studies on the role of plasminogen activator in ovulation. J. Biol. Chem. 251, 5694.

Szöllösi, D. (1975). Ultrastructural aspects of oocyte maturation and fertilization. In La Fécondation. C. Thibault (ed), Paris, Masson, pp. 13–35.

Thibault, C. (1977). Are follicular maturation and oocyte maturation independent processes? J. Reprod. Fertil. 51, 1.

Thibault, C., Gerard, M. and Menezo, Y. (1975a). Acquisition par l'ovocyte de Lapine et de Veau du facteur de décondensation du noyau du spermatozoïde fécondant (MPGF). Ann. Biol. Anim. Bioch. Biophys. 15, 705.

Thibault, C., Gerard, M. and Menezo, Y. (1975b). Preovulatory and ovulatory mechanism in oocyte maturation. J. Reprod. Fertil. *45*, 605.

Tsafriri, A. and Channing, C. (1975). An inhibitory influence of granulosa cells and follicular fluid upon porcine oocyte meiosis *in vitro*. Endocrinology *96*, 922.

Turnbull, K.E., Braden, A.W.H. and Mattner, P.E. (1977). The pattern of follicular growth and atresia in the ovine ovary. Aust. J. Biol. Sci. *30*, 229.

Van Look, P.F.A., Clarke, J.J., Davidson, W.G. and Scaramuzzi, R.J. (1978). Ovulation and lambing rates in ewes actively immunized against androstenedione. J. Reprod. Fertil. *53*, 129.

Virutamasen, P., Smitasiri, Y. and Fuchs, A.R. (1976). Intraovarian pressure changes during ovulation in rabbits. Fertil. Steril. *27*, 188.

Virutamasen, P., Wright, K.H. and Wallach, E.E. (1972). Effects of prostaglandins E_2 and $F_{2\alpha}$ on ovarian contractility in the rabbit. Fertil. Steril. *23*, 675.

Wallach, E.E., Virutamasen, P. and Wright, K.H. (1973). Menstrual cycle characteristics and side of ovulation in the rhesus monkey. Fertil. Steril. *24*, 715.

Walles, B., Edvinsson, L., Owman, Ch., Sjöberg, N.O. and Svensson, K.G. (1975). Mechanical response in the wall of ovarian follicles mediated by adrenergic receptors. J. Pharmacol. Exp. Ther. *193*, 460.

Weiner, R. and Kaley, G., 1972. Lysosomal fragility induced by prostaglandin $F_{2\alpha}$. Nature New Biol. *236*, 46.

Welschen, R. and Dullaart, J. (1976). Administration of antiserum against ovine stimulating hormone or ovine luteinizing hormone at proestrus in the rat: effects of follicular development during the oncoming cycle. J. Endocrinol. *70*, 301.

Welschen, R., Osman, P., Dullaart, J., De Greef, W.J., Uilenbroek, J.Th.J. and De Jong, F.H. (1975). Levels of follicle-stimulating hormone, luteinizing hormone, estradiol-17β and progesterone, and follicular growth in the pseudopregnant rat. J. Endocrinol. *64*, 37.

9

Spermatozoa

D.L. GARNER
and
E.S.E. HAFEZ

Spermatozoa, the male gametes, are formed within the *seminiferous tubules* of the testes. These tubules contain a complex series of developing germ cells which ultimately form the highly specialized male gametes. The fully-formed spermatozoa are elongated cells consisting of a head which is composed almost entirely of chromatin and a tail which provides the cell with motility (Fig. 9–1). The anterior portion of the sperm head is covered by a thin double-walled structure, the *acrosome* or *acrosomal cap*. A short neck connects the sperm head with its long tail (flagellum), which is subdivided into the middle, principal and tail pieces. These streamlined spermatozoa are mixed at ejaculation with the liquid or semigelatinous secretions of the male accessory sex glands (eg, seminal plasma) to form the cellular suspension known as *semen*.

SEMINIFEROUS EPITHELIUM

The *seminiferous epithelium*, lining the seminiferous tubules, is composed of two basic cell types: the supportive Sertoli cells and the developing germ cells. The germ cells undergo a continuous series of cellular divisions and developmental changes beginning at the periphery of the tubule. The stem cells, which lie adjacent to the basement or outer limit of the tubule, are called *spermatogonia*. The spermatogonia divide several times to produce the specialized cells that form spermatozoa. This cellular division, which reduces the DNA content of the cells to one-half that of somatic cells, is known as *spermatocytogenesis*. The haploid cells resulting from this divisional process are called *spermatids*. The spermatids then undergo a progressive series of structural and developmental changes to form spermatozoa. This metamorphic change of spermatids into spermatozoa is known as *spermiogenesis*.

Spermatogenesis begins at the periphery of the tubule where the spermatogonia divide several times to become *primary spermatocytes*. The primary spermatocytes divide twice to form first *secondary*

167

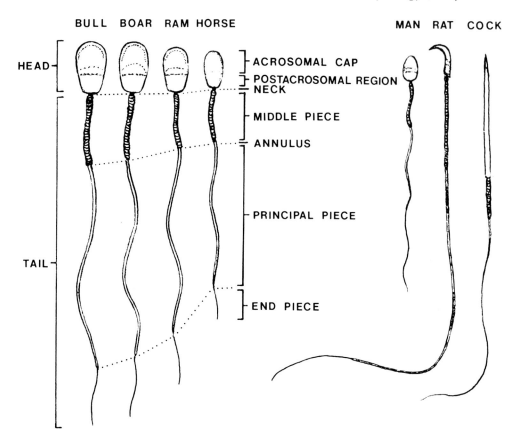

Fig. 9–1. Drawing comparing the spermatozoa of farm animals and other vertebrates. Differences in the relative size and shape as well as the terms used to describe the major structural features are given.

spermatocytes and then spermatids. The most abundant cell type in a cross-section of the seminiferous tubule is that of spermatids. The germinal cells are closely associated with the large sustentacular or *Sertoli cells* that surround them during development (Fig. 9–2).

Spermatocytogenesis

During early embryonic development special cells, called *primordial germ cells*, migrate from the yolk sac region of the embryo into the undifferentiated gonads. After reaching the fetal gonad the primordial cells divide several times before forming special cells called *gonocytes*. In the male, these gonocytes seem to undergo

differentiation just before puberty to form the type A_0 spermatogonia from which the other germ cells originate. The type A_1 spermatogonia divide to form type A_2, type A_3 and type A_4 spermatogonia. The type A_4 divide again to form intermediate spermatogonia (type I_n) and then again to form type B spermatogonia. These various types of spermatogonia can be identified in histologic sections of the seminiferous epithelium.

Some variation exists regarding how spermatogonia are classified and in certain species only three rather than four type A spermatogonia are evident. The type A_2 cells divide not only to produce the many germinal cells that eventually form sperm; but a specific division is thought to be

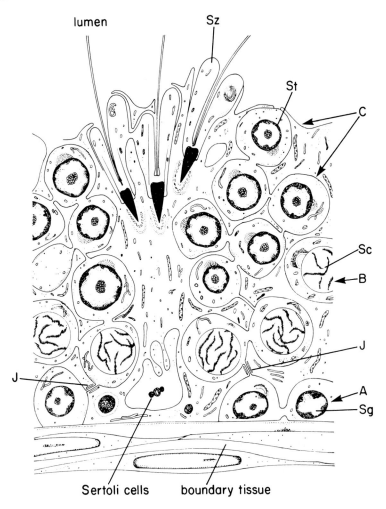

lumen — Sz — St — C — Sc — B — J — A — Sg — J — Sertoli cells — boundary tissue

Fig. 9–2. Diagrammatic illustration showing the relationship of the germinal cells to the Sertoli cells in the seminiferous tubule. The developing germ cells begin at the periphery of the tubule as spermatogonia *(Sg)* and migrate towards the lumen first as spermatocytes *(Sc)*, then as early spermatids *(St)* and finally as elongated spermatids *(Sz)*. The developing germ cells are sandwiched between adjacent Sertoli cells during development. Specialized junctions *(J)* occur between adjacent Sertoli cells. The outer wall of the tubule, boundary tissue, is made up partially of contractile myoid cells. *(From Fawcett, 1974, In: Male Fertility and Sterility, R.E. Mancini and L. Martini [eds], New York, Academic Press)*

utilized to replace the stem cell population of type A_1 spermatogonia. It seems that special reserve stem cells, type A_0 spermatogonia, replace the stem cell population (Clermont and Bustos-Obregon, 1968).

The type B spermatogonia divide at least once and probably twice (Hochereau de Reviers, 1970) to form the primary spermatocytes. The primary spermato-cytes undergo progressive nuclear changes of meiotic prophase known as preleptotene, leptotene, zygotene, pachytene and diplotene before dividing to form secondary spermatocytes. The nuclear changes to meiotic prophase are morphologic evidence of significant DNA synthesis within the cells. In fact, the diplotene primary spermatocytes contain approximately twice the DNA content of

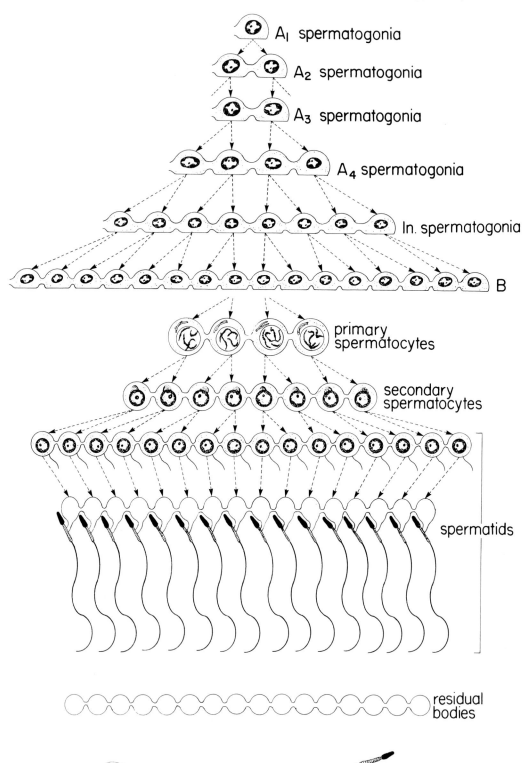

somatic cells. The primary spermatocytes undergo diakinesis and divide to form secondary spermatocytes. The resultant secondary spermatocytes divide again without further DNA synthesis to form the haploid cells known as spermatids. The entire divisional process of spermatocytogenesis, from spermatogonia to spermatid, takes approximately 45 days in the bull (Berndtson and Desjardins, 1974). However, these divisional processes are incomplete in that small cytoplasmic or intercellular bridges are retained between most cells of a series or "clone" of developing germ cells (Fig. 9–3).

Spermiogenesis

Spermatids are transformed into spermatozoa by a series of progressive morphologic changes collectively known as spermiogenesis. These changes include condensation of the nuclear chromatin, formation of the sperm tail or flagellar apparatus and development of the acrosomal cap (Fig. 9–4). The various developmental stages of spermatid transformation are readily classified by using periodic acid-Schiff (PAS) reaction to stain the developing acrosomal components a deep red. Four phases are noted in this developmental process: *Golgi phase, cap phase, acrosomal phase* and *maturation phase*. Early spermatids are spheroidal cells containing fine dispensed nuclear chromatin and strategically located extranuclear organelles. The most notable organelles are the *Golgi apparatus* near the nucleus and the centrioles near the periphery of the cell.

Golgi Phase. The Golgi phase of spermiogenesis is characterized by formation of PAS-positive proacrosomal granules within the Golgi apparatus, the coalescence of the granules into a single acrosomal granule, the adherence of the resultant acrosomal granule to the nuclear envelope and the early stages of tail development at the pole opposite that of the adherence of the acrosomal granule (Fig. 9–4, steps 1–3).

Cap Phase. The cap phase is characterized by a spreading of the adherent acrosomal granule over the surface of the spermatid nucleus. This process continues until nearly two-thirds of the anterior portion of each spermatid nucleus is covered by a thin, double-layered membranous sac that closely adheres to the nuclear envelope. During this cap phase, the continuing development of the tail filaments or *axoneme* is characterized by migration of both centrioles from the periphery of the cell to positions at the pole of the nucleus opposite from the developing acrosome. The proximal centriole migrates closest to the nucleus where it is thought to form a basis for attachment of the tail to the head. The developing tail, which is formed from elements of the distal centriole, elongates well beyond the periphery of the cellular cytoplasm. During the early development, the axoneme closely resembles the structure of a cilium in that it consists of two central tubules surrounded peripherally by nine pairs of tubules (Fig. 9–4, steps 4–7).

Acrosomal Phase. The acrosomal phase of spermiogenesis is characterized by major changes in the nuclei, the acrosomes and the tails of the developing spermatids. The developmental changes

Fig. 9–3. Diagram illustrating the various cell types that occur sequentially during spermatogenesis. The process of spermatocytogenesis includes the cell divisions from type A_1 spermatogonia at the top of the illustration to the production of rounded spermatids. The metamorphosis of round spermatids into elongated cells is spermiogenesis. The clonal nature of the developing germ cells is illustrated by the intercellular bridges that connect large groups of cells during the spermatogenic process. The residual bodies remain interconnected even after the release of the spermatozoa. (From Bloom and Fawcett, 1975)

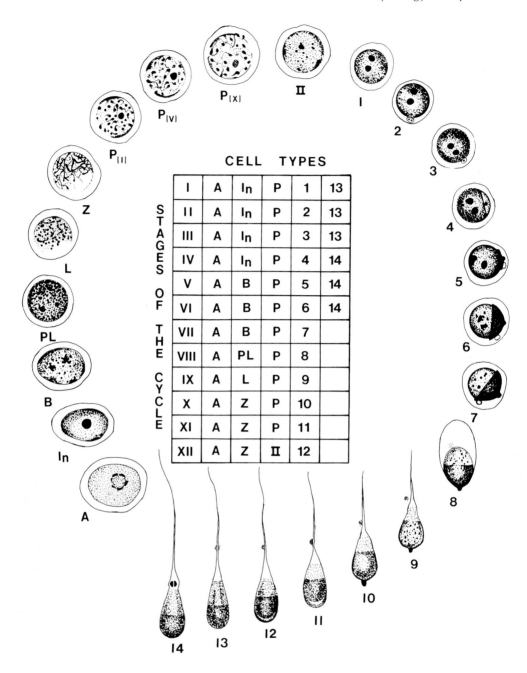

CELL TYPES

I	A	In	P	1	13	
II	A	In	P	2	13	
III	A	In	P	3	13	
IV	A	In	P	4	14	
V	A	B	P	5	14	
VI	A	B	P	6	14	
VII	A	B	P	7		
VIII	A	PL	P	8		
IX	A	L	P	9		
X	A	Z	P	10		
XI	A	Z	P	11		
XII	A	Z	II	12		

(left margin label: STAGES OF THE CYCLE)

Fig. 9–4. Diagrammatic illustration at the various steps of spermatogenesis in the bull beginning with a type A spermatogonium. The table in the center indicates the particular cellular associations of the twelve stages of the cycle of the seminiferous epithelium. The various cell types are: *A, In, B,* successive stages of spermatogonia; *PL,* preleptotene spermatocyte; *L,* leptotene spermatocyte; *Z,* zygotene spermatocyte; *P,* pachytene spermatocytes from stages I, V, and X; *II,* secondary spermatocyte; *I* through *14* are stages of spermiogenesis showing the Golgi phase (steps *1–3*), the cap phase (steps 4–7), the acrosome phase (steps 8–12) and the maturation phase (steps *13–14*). *(Adapted from Berndtson and Desjardins, 1974, Am. J. Anat., 140, 167)*

are facilitated by rotation of each spermatid so that the acrosome is directed toward the basement or outer wall of the seminiferous tubule (Fig. 9–4, steps 8–12). The nuclear changes include migration of the nucleus from the center to a position near the periphery of the cell, condensation of the chromatin into dense granules and reshaping of the spheroidal nucleus into an elongated, flattened structure. The acrosome, which is closely adherent to the nucleus, also condenses and elongates to correspond to the shape of the nucleus. These modifications in nuclear and acrosomal shape appear to be "molded" by the surrounding Sertoli cells. The changes are slightly different for each species and, thus, result in elongated spermatids that are characteristic for each species.

The changes in nuclear morphology are accompanied by displacement of the cytoplasm to the caudal aspect of the nucleus where it surrounds the proximal portion of the developing tail. Within this cytoplasm, microtubules associate to form a temporary cylindrical sheath, called the *manchette*, which projects from the caudal border of the acrosome posteriorly where it loosely surrounds the axoneme. Within the cylindrical manchette a specialized cytoplasmic structure, called the chromatoid body, condenses around the axoneme to form the ring-like structure known as the *annulus*. The annulus forms first near the proximal centriole and then during subsequent development migrates posteriorly. The mitochondria, which were previously distributed throughout the cytoplasm of the spermatid, begin to concentrate close to the axoneme where they eventually form a sheath surrounding the middle piece of the tail.

Maturation Phase. The maturation phase of spermiogenesis involves final transformation of the elongated spermatids into cells to be released into the lumen of the seminiferous tubule. The reshaping of the nucleus and acrosome of each spermatid, initiated during the previous phase, produces spermatozoa characteristic for each species. Within the nucleus the chromatin granules undergo progressive condensation until they form a fine homogeneous material that uniformly fills the entire sperm nucleus (Fig. 9–4, steps 13–14).

During the maturation phase a fibrous sheath containing nine coarse fibers is formed around the axoneme. These coarse fibers appear to be associated individually with the nine pairs of microtubules of the axoneme and are continuous with columns in the neck of the connecting piece of the spermatid. The fibrous sheath covers the axoneme from the annulus to the beginning of the tail piece. The annulus migrates from its position adjacent to the nucleus distally along the tail to a point where it will subsequently separate the middle piece from the principal piece of the tail. The mitochondria, which were previously concentrated around the axoneme, become arranged along the middle piece to form a mitochondrial sheath covering the previously deposited coarse fibers from the neck to the annulus.

During the later stages of spermiogenesis the manchette disappears and the Sertoli cell then forms the cytoplasm remaining after elongation of the spermatid into a spheroidal lobule called the *residual body*. This lobule of cytoplasm, which remains connected to the elongated spermatid by a slender thread of cytoplasm, is also interconnected with other residual bodies by intercellular bridges that resulted from the incomplete division of the germ cells during spermatocytogenesis (Fig. 9–3). Once the residual body has formed the elongated spermatids undergo final maturation and are ready for release as spermatozoa.

Cycle of the Seminiferous Epithelium

The various cell types within any cross section of the seminiferous epithelium form well-defined cellular associations that undergo cyclic changes. As many as 14 distinct cellular associations or stages have been identified in species such as the rat (Leblond and Clermont, 1952), whereas only six stages have been identified in man (Clermont, 1963). In the bull, 12 stages of this cycle have been described (Berndtson and Desjardins, 1974) (Fig. 9–4). A complete, time-dependent cycle of the stages, known as the *cycle of the seminiferous epithelium*, is defined as "a series of changes in a given area of seminiferous epithelium between two appearances of the same developmental stages" (Leblond and Clermont, 1952). The well-defined steps in spermiogenesis are used to classify the various stages of the cycle. The time necessary to complete a cycle of the seminiferous epithelium varies among domestic species. Duration of the cycle is 13.5 days in the bull (Hochereau de Reviers et al, 1964), 8.6 days in the boar (Swierstra, 1968), 10.3 days in the ram (Ortavant, 1959) and 12.2 days in the horse (Swierstra et al, 1974). Depending on the species, four to nearly five epithelial cycles are required before the type A spermatogonia from the first cycle have completed the metamorphosis of spermiogenesis. The rate of spermatogenesis, however, appears to be relatively constant for each species.

Spermatogenic Wave

The stages of the cycle of the seminiferous epithelium change not only with time but also along the length of the tubule. A length of tubule at one stage is usually adjacent to the lengths of tubule in stages just preceding or following it in time (Perey et al, 1961) (Fig. 9–5). This sequential change in stage of cycle along the length of the tubule is known as the *wave of the seminiferous epithelium*. Examination of a loop of seminiferous tubule along its length also reveals that the wave involves a sequence of stages beginning

with the less advanced stages in the middle of the loop to progressively more advanced stages nearer the rete testis. Certain irregularities or breaks in the sequential order are noted. Such breaks in sequence, called *modulations*, occur occasionally but involve relatively short lengths of tubule.

Endocrine Control of Spermatogenesis

The principal endocrine stimulus to spermatogenesis is androgen (Hansson et al, 1975). This steroid-dependency is met by the production of androgens by the interstitial cells of Leydig, which are adjacent to the seminiferous tubules. The Leydig cells are stimulated by the pituitary gonadotropin, LH, to secrete androgens. The androgens produced by the Leydig cells then feedback to both the hypothalamus and pituitary to control LH production. The other principal gonadotropin, FSH, stimulates production of an *androgen-binding protein* (ABP) by the Sertoli cells. The ABP, which is secreted into the lumen of the seminiferous tubule, helps maintain a high level of androgen within the tubule by forming a complex with the steroids produced by the Leydig cells. The two gonadotropins function together to concentrate testosterone and dihydrotestosterone within the seminiferous tubules, where these androgens stimulate development of the germ cells (Hansson et al, 1975) (Fig. 9–6). However, the precise role of individual androgens on the spermatogenic process is not completely understood.

In addition to ABP, the Sertoli cells are thought to produce a hormonal factor that controls the production of FSH by the pituitary gland. This hormonal factor, termed *inhibin*, appears to send feedback to the hypothalamus and/or pituitary to control FSH production (Setchell et al, 1977). Inhibin, which depresses FSH secretion, has been demonstrated in the testis, in rete testis fluid and in semen. It may

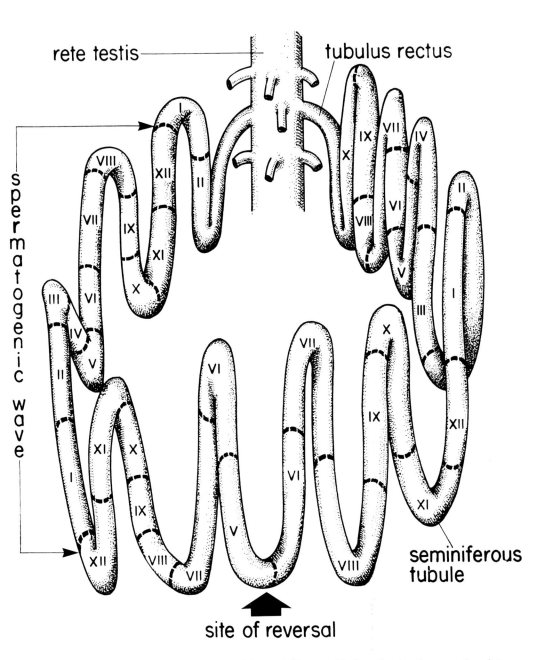

Fig. 9–5. Schematic representation of the wave of the seminiferous epithelium showing the succession of stages, the site of reversal in the middle of the tubules and their relationship to the rete testis. The more advanced of the twelve stages of spermatogenesis, shown in roman numerals, are located nearer the rete testis. (An actual seminiferous tubule may contain 15 or more complete spermatogenic waves [Perey, Clermont and Leblond, 1961].)

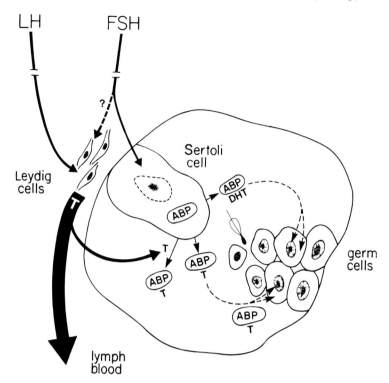

Fig. 9–6. Drawing illustrating the endocrine control of spermatogenesis. The Leydig cells secrete androgens, testosterone (T) and dihydrotestosterone (DHT), in response to LH from the pituitary. The Sertoli cells synthesize androgen-binding protein (ABP) in response to pituitary FSH. Although most of the androgens are secreted into the blood and lymph, the ABP helps maintain a high level of androgen within the seminiferous tubule and duct system by forming a complex with the steroid. *(Hansson et al, 1975, In: Handbook of Physiology. Vol. V. R.O. Greep and E.B. Astwood [eds], Bethesda, American Physiologic Society)*

control FSH production by relaying quantitative information on the rate of sperm production. However, considerable doubt exists regarding the role of inhibin in control of the spermatogenic process because the hormonal milieu appears to influence the number of germ cells produced rather than their rate of production (Setchell et al, 1977).

BLOOD-TESTIS BARRIER

The seminiferous tubules are not penetrated by blood or lymph vessels. In addition, the developing germ cells within the tubules are protected from chemical changes in the blood by a specialized permeability barrier. This *blood-testis* barrier has two principal components: (1) the incomplete or partial barrier of the *myoid cells* that surround the tubule and (2) the unique junctions between adjacent Sertoli cells.

Myoid Layer

The basement membrane or tunica propria that surrounds the seminiferous tubules contains a layer of contractile myoid cells (Fig. 9–2). In some species a majority of the cell junctions of this layer are sealed by tight apposition of the adjacent cell membranes. However, this barrier is not well developed in the bull, ram or boar and may be a relatively unimportant permeability barrier in the testis of farm animals (Fawcett, 1975b).

Sertoli Cell Junctions

The principal permeability barrier between the blood and testis is thought to be the junctional complexes between adjacent Sertoli cells. These Sertoli-Sertoli junctions, which are situated near the cellular base, contain multiple zones of adhesion where the opposing membranes are fused (Fig. 9–7). The occluding junctions divide the seminiferous tubules into two distinct compartments: (1) a *basal compartment* containing spermatogonia

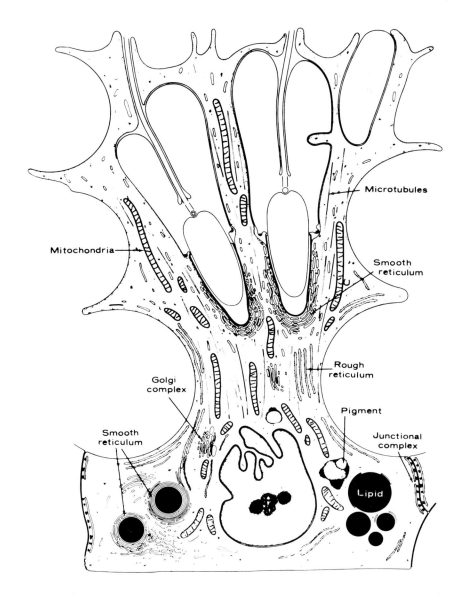

Fig. 9–7. Illustration depicting the specialized junctions between adjacent Sertoli cells. These Sertoli-Sertoli junctions form the principal permeability barrier of the testis of most farm animals. These occluding junctions separate the area of the seminiferous tubule between adjacent Sertoli cells into basal and adluminal compartments. The basal compartment is occupied by spermatogonia and preleptotene spermatocytes and the adluminal compartment contains the more advanced stages of spermatocytes and the spermatids. The adluminal compartment communicates freely with the lumen of the seminiferous tubule. *(Adapted from Fawcett, 1975, In: Handbook of Physiology. Vol. V. R.O. Greep and E.B. Astwood [eds], Bethesda, American Physiologic Society)*

and preleptotene spermatocytes and (2) an *adluminal compartment*, containing the more advanced stages of spermatocytes and spermatids, which freely communicates with the lumen of the tubule (Fawcett, 1975b).

The basal compartment is freely accessible to components that have previously penetrated the myoid layer. However, the second barrier, composed of the occluding junctions between Sertoli cells, demonstrates a wide range of permeability from complete exclusion of some substances to nearly free transfer of others. This differential permeability appears to be important in maintaining an environment suitable for the spermatogenic function of the tubules. The blood-testis barrier not only excludes entry of certain substances but also appears to function in retaining specific levels of certain substances, such as ABP, inhibin and enzyme inhibitors, within the luminal compartments of the tubules.

Another important function of the blood-testis barrier is immunologic isolation of the developing spermatids because these haploid cells are readily recognized as foreign cells by the immune system of the body. This immunologic barrier prevents a male from developing antibodies against his own sperm cells.

TESTICULAR SPERMATOZOA

The mature spermatids produced during the final phase of spermiogenesis are released into the lumen of the seminiferous tubules as spermatozoa. These sperm cells, which are essentially immotile, are transported from the tubules into the epididymis by contractile elements and fluid secretions of the testis and by cilia lining the efferent ducts.

Spermiation

The release of formed germ cells into the lumen of the seminiferous tubules is known as *spermiation*. The elongated spermatids, which are oriented perpendicularly to the tubular wall, are gradually extruded into the lumen of the tubule (Fig. 9–8). The lobules of residual cytoplasm through which large syncytial groups of spermatids are connected by intercellular bridges remain embedded in the epithelium. Extrusion of the spermatozoal components continues until only a slender stalk of cytoplasm connects the neck of the sperm cell to the residual body (Fawcett, 1975) (Fig. 9–8). Breakage of the stalk results in formation of the *cytoplasmic droplet* in the neck region of the spermatozoa and retention of rounded residual bodies. Following release of the spermatozoa the residual bodies are rapidly disposed of by the Sertoli cells. Although the Sertoli cells are actively involved in the process of spermiation, their precise role in the apparent recycling of the protoplasmic components of the residual bodies is not clear (Fawcett, 1975). Not only must the Sertoli cells phagocytize the residual bodies remaining from the spermatogenic process but they also remove considerable numbers of degenerating germ cells. This occurs because the spermatogenic process is relatively inefficient in that large numbers of potential sperm cells degenerate before reaching maturity (Clermont and Bustos-Obregon, 1968).

Transport of Sperm into the Epididymis

Testicular spermatozoa are transported from the testis by means other than their own motility. Sperm transport into the epididymis is aided by active secretion of fluid into the seminiferous tubules and rete testes. This fluid secretion, which can amount to as much as 40 ml per day in the ram, is readily reabsorbed by the epididymis because less than 1 ml is extruded through the vas deferens (Setchell, 1970).

The testicular contractile elements involved in sperm transport include the

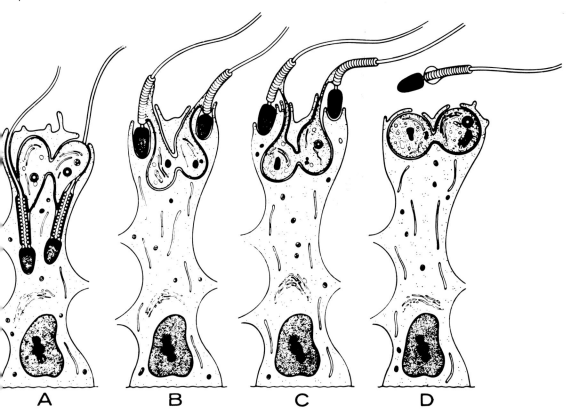

Fig. 9–8. Diagram illustrating sperm release in mammals. The successive stages show a gradual extrusion of the elongated spermatid into the lumen with retention of the interconnecting residual bodies. Release results from attenuation of the slender stalk of cytoplasm connecting the germ cell to the residual body. Once separated from the residual body, the cell becomes a spermatozoon. *(From Fawcett, 1975, In: Handbook of Physiology. Vol. V. R.O. Greep and E.B. Astwood [eds], Bethesda, American Physiologic Society)*

smooth muscle-like cells in the testicular capsule (Davis, et al, 1970) and the cells of the myoid layer in the wall of the seminiferous tubules (Clermont, 1958). However, some variation exists in the relative contribution of the contractile components. In some species, the testicular capsules undergo spontaneous phasic contractions. In other species in which the capsules possess relatively little rhythmic contractility, the seminiferous tubules have regular pronounced contractions (Hargrove et al, 1977). An increase in the flow of spermatozoa from the testis has been demonstrated using either oxytocin (Voglmayr, 1975) or prostaglandin $F_{2\alpha}$ (Hafs et al, 1974) to stimulate the contrac-

tile components. Some variability exists, however, in the physiologic responses generated by administration of these hormonal agents.

The initial segment of the excurrent duct system, called the efferent ducts, is lined with ciliated epithelial cells (Reid and Cleland, 1957). The ciliary action of these cells facilitate sperm transport into the head of the epididymis.

EPIDIDYMAL TRANSPORT, SPERM MATURATION AND STORAGE

Testicular spermatozoa are transported from the testis through a highly convoluted duct known as the epididymis (Fig.

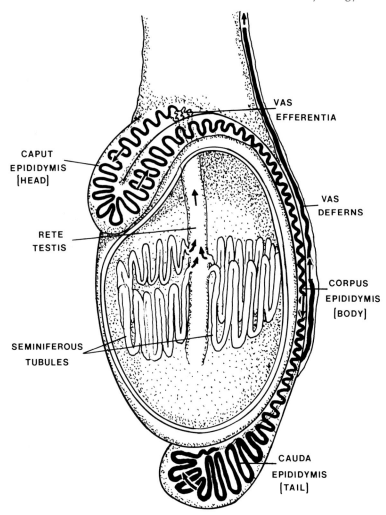

Fig. 9–9. Diagrammatic illustration of the excurrent duct system of the ram. The pathway by which spermatozoa are transported to the exterior is indicated by arrows. *(Modified from Setchell, 1977)*

9–9). The epididymis not only transports spermatozoa distally from the testis into the vas deferens but during this transit the spermatozoa undergo a maturation process in which they gain the potential ability to fertilize ova. This maturation involves several functional changes, including development of the potential for sustained motility, progressive loss of water and distal migration and eventual loss of the cytoplasmic droplet. The functional capabilities of the various epithelial cells lining the epididymis, and hence their influence on the sperm maturation process, are maintained by testicular androgens.

Transport Mechanisms

The passage of spermatozoa through the epididymis depends on localized contractions of the duct wall (Risley, 1963). These contractions, which have been observed *in vitro* in rat epididymides to take place every six to ten seconds (Risley, 1963), move the spermatozoa through the epididymis in about 11 days in the bull (Orgebin-Crist, 1962), 9 to 14 days in the boar (Swierstra, 1968) and 13 days in the ram. The transit time may be reduced by 10% to 20% by an increased frequency of ejaculation. The contractile elements of the epididymal wall show regional differences in that the content of smooth muscle cells increases progressively from the tail of the epididymis to the vas deferens (Bedford, 1975). There is an increase in smooth muscle cells and in the sympathetic innervation of the duct (Sjöstrand, 1965). Bedford (1975) suggested that this innervation may permit more precise control of the emission of spermatozoa during the ejaculatory process.

Maturation and Storage of Spermatozoa

The functional changes occurring during epididymal transit of spermatozoa involve the maturation of several different cell organelles. For instance, the development of the capacity for progressive sperm motility reflects maturational processes associated with both qualitative and quantitative changes in the metabolic patterns of the flagellar apparatus. Although mature epididymal spermatozoa are relatively quiescent within the epididymis, they rapidly demonstrate their motile character upon removal and examination. The maturation process in which epididymal spermatozoa attain the capacity for progressive motility is thought to involve progressive changes in the flexibility and patterns of movement of their flagella. Rapid forward progression occurs only with a reduction of the arc and reduced flexibility of tail movements; this appears first in the middle of the corpus epididymis in a few spermatozoa and becomes the predominating motility pattern in spermatozoa from the cauda and vas deferens (Bedford, 1975). The alteration of structural properties of the tail elements is supported by evidence

indicating that formation of disulfide bonds occurs during epididymal transit. The cyclic nucleotide, specifically the cyclic AMP (adenosine monophosphate) level, increases two-fold in bovine spermatozoa during passage through the epididymis (Hoskins et al, 1974). Testicular spermatozoa differ from ejaculated spermatozoa not only in their capacity for oxygen uptake but also in their pattern of sugar utilization for energy production (Voglmayr et al, 1967).

Transit through the epididymis is associated with significant changes within the chromatin of the sperm nucleus. This DNA-protein complex, which was once thought to be relatively inert following its condensation during the latter phases of spermiogenesis, undergoes a qualitative reduction in reactivity to Feulgen staining during epididymal transit (Gledhill, 1971). As the capacity for progressive motility increases, spermatozoa undergo a progressive loss of water and a corresponding increase in specific gravity. In addition to maturation within the spermatozoa, a significant change also occurs in the surface characteristics of the sperm plasma membrane.

The attainment of motility is not associated with any detectable change in the structure of the sperm tail. However, morphologic changes occur in the position of the cytoplasmic droplet and the shape of the acrosome. During epididymal transit, the droplet migrates from the neck region to a position near the annulus. Although some species variation exists, the migration is usually initiated close to the time when spermatozoa pass from the head to the tail of the epididymis. Presence of the droplet on a significant number of ejaculated spermatozoa is a sign of immaturity because in most species, with the possible exception of the boar, the droplet has separated from the spermatozoa before ejaculation occurs.

Changes associated with maturation of the acrosome have been noted in most species during passage through the epididymis. Although marked changes occur in some species, those occurring in farm animals are limited to a rather subtle reduction in acrosome dimensions (Bedford, 1975).

The major site of sperm storage within the male reproductive tract is the tail of the epididymis. This portion of the tract has a relatively wide lumen in which high concentrations of spermatozoa are stored. The tail of the epididymis contains 70% of the total number of spermatozoa in the excurrent ducts, whereas the vas deferens contains only 2%. Although the environment is favorable to their survival, spermatozoa are not preserved indefinitely in a viable state. They undergo a gradual senescence in which they first lose their fertilizing ability, then their motility and finally disintegrate in the vas deferens. Unless these senescent spermatozoa are eliminated from the male reproductive tract at regular intervals they result in semen of poor quality. Following prolonged sexual rest, there is a high percentage of degenerating or "stale" spermatozoa.

Development of Fertilizing Capacity in the Epididymis

Spermatozoa develop their initial ability to fertilize ova during their transport through the epididymis. Their fertilizing capacity is considered potential since they must undergo capacitation in the female tract before they can penetrate ova. Testicular spermatozoa from the ram are infertile even when inseminated in relatively large numbers (Setchell et al, 1969). Evidence obtained with rabbit spermatozoa indicate that caput epididymal spermatozoa are also infertile (Orgebin-Crist, 1967). The lack of fertility in spermatozoa from the head of the epididymis may be related to type of motility expressed by the immature spermatozoa: spermatozoa from the head of the epididymis possess active circular swimming movement, but yet are incapable of the vigorous unidirectional

movement of spermatozoa possessing the ability to undergo longitudinal rotation (Blandau and Rumery, 1964). Some of the changes that occur during epididymal transport such as droplet movement and loss and the increase in specific gravity are difficult to interpret from a functional standpoint. Changes in motility, chromatin structure, acrosome morphology or plasma membrane charge can be interpreted readily as functional in the fertilization process.

The attainment of potential fertilizing capacity during epididymal transit appears to be a progressive phenomenon. For instance, fertility can be obtained with spermatozoa removed from the distal half of the body of the epididymis (Bedford, 1966) but the fertilization rate is less than that obtained with spermatozoa removed from the tail of the epididymis or vas deferens (Orgebin-Crist, 1967).

The development of fertilizing ability is associated with changes in several aspects of the functional integrity of the spermatozoa: (a) development of the potential for sustained progressive motility, (b) alteration of the metabolic patterns and the structural state of specific tail organelles, (c) changes in nuclear chromatin, (d) changes in nature of the surface of the plasma membrane, (e) movement and loss of the protoplasmic droplet, and (f) modification, at least in some species, of the form of the acrosome (Bedford, 1975).

The attainment of potential fertilizing capacity during epididymal passage of spermatozoa depends directly on androgen support of the epididymal epithelium, since bilateral orchiectomy results in infertile spermatozoa within only a few days (Orgebin-Crist et al, 1972). Fertility is also lost if the epididymis is transferred to the higher ambient temperature of the abdominal cavity.

FATE OF UNEJACULATED SPERMATOZOA

Not all of the spermatozoa formed in the testes are ejaculated in semen. In fact, approximately half of the spermatozoa produced in the testes are apparently reabsorbed in the excurrent duct system of the bull (Amann and Almquist, 1962). Both reabsorption within the male tract and loss into the urine seem to be important mechanisms in sperm disposal.

Resorption in Male Reproductive Tract

Evidence obtained by counting the spermatozoa eliminated in ejaculates and urine suggests that millions of spermatozoa are reabsorbed by the male tract (Koefoed-Johnson and Pederson, 1970). Confirmation of this extensive resorption process by microscopic examination of the reproductive tract has been difficult to obtain because sperm fragments are rarely seen in the epithelium at the epididymis and vas deferens of the bull.

Bovine spermatozoa, however, may be broken down and reabsorbed in the ampullary region of the reproductive tract (Orgebin-Crist, 1926), as judged by high levels of deoxyribonuclease and proteinase in this portion of the tract (Waldschmit et al, 1964). Selective elimination of abnormal spermatozoa by luminal macrophages occurs in the epididymis because fewer abnormal spermatozoa are found in smears obtained from the tail than from the head of the epididymis (Roussel et al, 1967).

Disposal in the Urine

Considerable numbers of spermatozoa are voided in the urine in most species. About 87% of the daily sperm production is detected in the urine of sexually inactive rams (Lino et al, 1967). Apparently, minimal reabsorption of spermatozoa occurs in the ram. Relatively large numbers of spermatozoa are eliminated in the urine or reabsorbed at the ampulla because ligation or blockage of the vas deferens in most species results in considerable distention of the cauda epididymis (Bedford, 1973).

EJACULATED SPERMATOZOA

Spermatozoa, the elongated haploid cells resulting from the spermatogenic and maturation processes of the male, are highly specialized cells with a limited function—to carry genetic information to the oocyte. In fact, spermatozoa are so streamlined that they possess only a limited number of highly modified cellular organelles. Although species differences exist, the spermatozoa of farm animals and nearly all vertebrates have a similar structural plan in that they each contain an acrosome, a highly condensed nucleus and an attached flagellum with its associated mitochondria, annulus, dense fibers and fibrous sheath.

Sperm Structure and Composition

The generalized organization of the spermatozoon includes the head with its nucleus and acrosomal cap and the tail with its four anatomic divisions: the neck, the middle piece, the principal piece and the end piece (Fig. 9–10). These general features are apparently necessary for the basic functions of the spermatozoon.

The Sperm Head. The major feature of the head is the oval, flattened *nucleus* containing the highly compact chromatin (Fig. 9–11). The condensed chromatin, which is protected by an essentially nonporous membrane, is comprised almost entirely of deoxyribonucleic acid (DNA) complexed to specific nuclear proteins known as histones. The chromosome number and hence the DNA content of the sperm nucleus is half that of somatic cells of the same species. This resulted from the reduction division that occurred during spermatocytogenesis. The reduction division also produced two distinct types of spermatozoa as far as sex chromatin is concerned. Gametes carrying the X chromosome produce female embryos, whereas those carrying the Y chromosome produce male embryos. Considerable controversy still exists concerning the possible separation of X-bearing from Y-bearing sperma-

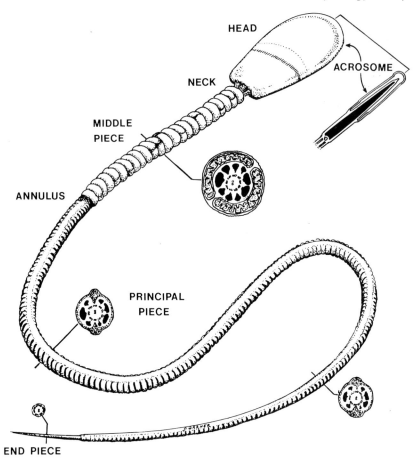

Fig. 9–10. Diagrammatic illustration of the structural features of a bovine spermatozoon showing the head with its acrosomal cap and the tail with its four anatomic divisions. The sagittal section through the midline of the head shows the relationship of the double-walled acrosomal cap to the nucleus. Cross sections of the middle piece, principal piece (2) and tail piece show the axoneme, the mitochondrial sheath, the fibrous sheath, the nine coarse fibers and the dorsal and ventral longitudinal columns. *(Adapted from Saacke, 1970).*

tozoa to provide semen capable of yielding upon insemination offspring of only one sex.

The Acrosome. The anterior end of the sperm nucleus is covered by the *acrosome* —a thin, double-layered membraneous sac that is closely applied to the nucleus during spermiogenesis (Fig. 9–11). This cap-like structure, which contains a number of hydrolytic enzymes such as acrosin, hyaluronidase, corona-penetrating enzyme and several acid hydrolases (McRorie and Williams, 1974; Morton, 1976), is involved in the fertilization process. During fertiliza-

tion the spermatozoon undergoes what is called the acrosome reaction in which the enzymatic contents of the acrosome are released through openings created by fusion of the plasma and outer acrosomal membranes (Fig. 9–11). The released hyaluronidase disperses the cummulus cells that surround the newly ovulated ova, whereas corona-penetrating enzyme is thought to provide a means for the penetrating spermatozoon to pass through the corona radiata of the ovum. Acrosin, a proteinase, seems to digest a pathway through the zona pellucida for the pene-

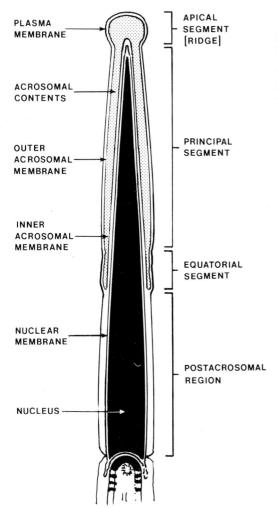

PLASMA MEMBRANE

APICAL SEGMENT [RIDGE]

ACROSOMAL CONTENTS

OUTER ACROSOMAL MEMBRANE

PRINCIPAL SEGMENT

INNER ACROSOMAL MEMBRANE

EQUATORIAL SEGMENT

NUCLEAR MEMBRANE

POSTACROSOMAL REGION

NUCLEUS

Fig. 9–11. Schematic representation of a sagittal section of a bovine sperm head showing the various anatomic subdivisions. The acrosome includes the apical (apical ridge), the principal and the equatorial segments. The outer membranes of the apical and principal segments make up what is called the acrosomal cap. The relationship of the acrosome with its inner and outer membranes, to the nuclear and plasma membranes is also shown. *(Adapted from Saacke, 1970 using the terminology of Fawcett, 1975a).*

trating spermatozoon. However, the specific role of each acrosomal enzyme in fertilization is not completely understood. For instance, both the cummulus oophorous and the corona radiata are absent from the oocyte at the time of fertilization in most farm animals.

The function of the *equatorial segment* is not completely understood because its contents are not released during the acrosome reaction. Its importance to the fertilization process is obvious because it is this part of the spermatozoon, including possibly the anterior portion of the post-acrosomal region, that initially fuses with the oocyte membrane during fertilization.

The Sperm Tail. The tail of the male gamete is composed of the neck and the middle, principal and tail pieces (Fig. 9–10). The neck or connecting piece forms a basal plate that articulates with a depression in the posterior surface of the nucleus. The basal plate of the neck is continuous posteriorly with nine coarse fibers that project posteriorly throughout most of the tail. The slender neck is highly complex and easily fractured.

The region of the tail between the neck and the annulus is the *middle piece.* The central core of the middle piece and, for a matter of fact, the entire length of the tail is the *axoneme.* It is composed of nine pairs of filaments or microtubules that are arranged radially around two central filaments. In the middle piece this 9 + 2 arrangement of filaments is surrounded by nine coarse or dense fibers that appear to be associated with the nine doublets of the axoneme (Fig. 9–10). The outer coarse fibers, which possess both elastic and contractile properties (Phillips, 1973), stain for adenosine triphosphatase activity (Nelson, 1967). The entire middle piece is covered by sausage-shaped mitochondria (Fig. 9–10). This mitochondrial sheath, which is arranged in a helical pattern around the longitudinal fibers of the tail (Fig. 9–10), is believed to generate the energy needed for sperm motility.

The *principal piece,* which is located posteriorly to the annulus and extends to near the end of the tail, is composed centrally of the axoneme and its associated coarse fibers. Surrounding this central core is a tough fibrous sheath. This cylindrical sheath has two prominent longitudinal thickenings or columns that run the length of what is thought to be the dorsal and ventral surfaces of the principal piece (Fig. 9–10). These longitudinal columns, which replace two of the coarse fibers, appear to be connected on each side by numerous ribs. The fibrous sheath is thought to provide stability for the contractile elements of the tail.

The end piece is that portion of the tail posterior to the end of the fibrous sheath. It contains only the central axoneme covered by the plasma membrane.

The protoplasmic droplet, which is usually detached from ejaculated spermatozoa, is composed of residual cytoplasm containing remnants of the Golgi apparatus (Phillips, 1973).

Sperm Metabolism

Although spermatozoa lack many of the organelles associated with metabolic processes, they are active metabolically because they possess the enzymes necessary to carry out the biochemical reactions of glycolysis (Embden-Meyerhof pathway), the tricarboxylic acid cycle, fatty acid oxidation, electron transport and possibly the hexose menophosphate shunt (Mann, 1964). Under anaerobic conditions, that is in the absence of oxygen, spermatozoa break down glucose, fructose or mannose to lactic acid. This glycolytic activity, or more correctly fructolytic activity because fructose is the principal seminal sugar, allows spermatozoa to survive under the anaerobic conditions of storage for artificial insemination. The enzymes that permit this glycolytic metabolism appear to be distributed throughout the tail.

Spermatozoa utilize a variety of substrates in the presence of oxygen. Their respiratory activity provides the means of utilizing the lactate or pyruvate resulting from the fructolysis of sugars to carbon dioxide and water (Mann, 1964). This oxidative pathway, which is apparently located in the mitochondria, is consid-erably more efficient in the production of energy than fructolysis. Using these catabolic processes the spermatozoa convert most of the energy into ATP (adenosine triphosphate). Most of the ATP is utilized for the energy-consuming process of motility. Some of the energy, however, must be utilized to maintain the integrity of the spermatozoal membranes that protect the cell from loss of vital components.

Endogenous respiration of intracellular plasmalogen can provide energy on a short term basis when exogenous substrates are exhausted. Utilization of this reserve source of energy appears to be detrimental to the spermatozoa (Mann, 1964).

Sperm Motility

The motile character of spermatozoa provides an easily discernible means of assessing their physiologic status. This vigorous beating, which provides the spermatozoa with the ability for self-propulsion, is the result of contractile elements located in the longitudinal fibers of the tail. Although specific contractile proteins have been partially characterized at a biochemical level, the mechanism by which spermatozoa convert the chemical energy resulting from degradation of substrates into motility is poorly understood. The energy required for motility is apparently derived from intracellular stores of ATP. The breakdown of ATP seems to be regulated by the endogenous level of cyclic AMP. This cyclic nucleotide not only plays a role in regulation of ATP breakdown, but also has a direct effect on sperm motility (Hoskins and Casillas, 1973). This apparently complex effect of cyclic AMP on sperm motility has been demonstrated in vitro by adding to spermatozoa either dibutyryl cyclic AMP or inhibitors such as methyl xanthines that block the normal intracellular degradation of cyclic AMP (Hoskins and Casillas, 1973).

The motility of spermatozoa provides a

simple means of evaluating the physiologic status of a sample of semen. The motile character of a sperm cell is, by itself, not an accurate predictor of its potential fertilizing capacity. Sperm motility, however, continues to be a useful tool in evaluating the viability of spermatozoa.

REFERENCES

Amann, R.P. and Almquist, J.O. (1962). Reproductive capacity of dairy bulls. VI. Effect of unilateral vasectomy and ejaculation frequency on sperm reserves; aspect of epididymal physiology. J. Reprod. Fertil. 3, 260.

Bedford, J.M. (1966). Development of the fertilizing ability of spermatozoa in the epididymis of the rabbit. J. Exp. Zool. 163, 319.

Bedford, J.M. (1973). The biology of primate spermatozoa. In Advances in Primatology; Reproductive Biology of the Primates. W.P. Luckett (ed), New York, Appleton Century Crofts, Vol. III.

Bedford, J.M. (1975). Maturation, transport, and fate of spermatozoa in the epididymis. In Handbook of Physiology, Section 7, Endocrinology, Vol. V, Male Reproductive System. R.O. Greep and E.B. Astwood (eds), Washington, D.C., American Physiological Society.

Berndtson, W.E. and Desjardins, C. (1974). The cycle of the seminiferous epithelium and spermatogenesis in the bovine testis. Am. J. Anat. 140, 167.

Blandau, R.J. and Rumery, R.E. (1964). The relationship of swimming movements of epididymal spermatozoa to their fertilizing capacity. Fertil. Steril. 15, 571.

Bloom, W. and Fawcett, D.W. (1975). A Textbook of Histology. Philadelphia, W.B. Saunders, p. 824.

Clermont, Y. (1958). Contractile elements in the limiting membrane of the seminiferous tubules of the rat. Exp. Cell. Res. 15, 438.

Clermont, Y. (1963). The cycle of the seminiferous epithelium in man. Am. J. Anat. 112, 35.

Clermont, Y. and Bustos-Obregon, E. (1968). Reexamination of spermatogonial renewal in the rat by means of seminiferous tubules mounted "in toto." Am. J. Anat. 122, 237.

Davis, J.R., Langford, G.A. and Kirby, P.J. (1970). The testicular capsule. In The Testis. Vol. 1. A.D. Johnson, W.R. Gomes and N.L. VanDemark (eds), New York, Academic Press, p. 281, 337.

Fawcett, D.W. (1974). Interactions between Sertoli cells and germ cells. In Male Fertility and Sterility. R.E. Mancini and L. Martini (eds), New York, Academic Press, p. 13.

Fawcett, D.W. (1975a). The mammalian spermatozoon. Devel. Biol. 44, 394.

Fawcett, D.W. (1975b). Ultrastructure and function of the Sertoli cell. In Handbook of Physiology, Sec. 7, Endocrinology, Vol. V. Male Reproductive System. R.O. Greep and E.B. Astwood (eds), Washington, D.C., American Physiological Society, pp. 21–55.

Gledhill, B.L. (1971). Changes in deoxyribonucleoprotein in relation to spermateliosis and epididymal maturation of spermatozoa. J. Reprod. Fertil. Suppl. 13, 77.

Hafs, H.D., Louis, T.M. and Stellflug, J.N. (1974). Increased sperm numbers in the deferent duct after prostaglandin $F_{2\alpha}$ in rabbits. Proc. Soc. Exp. Biol. Med. 145, 1120.

Hansson, V., Ritzén, E.M., French, F.S. and Nayfeh, S.N. (1975). Androgen transport and receptor mechanisms in testis and epididymis. In Handbook of Physiology, Section 7, Endocrinology, Vol. V. Male Reproductive System. R.O. Greep and E.B. Astwood (eds), Washington, D.C., American Physiological Society.

Hargrove, J.L., MacIndoe, J.H. and Ellis, L.C. (1977). Testicular contractile cells and sperm transport. Fertil. Steril. 28, 1146.

Hochereau de Reviers, M.T. (1970). Etudes des divisions spermatogoniales et du renouvellement de la spermatogonic souche chez le taureau. D. Sci. Thesis, University of Paris, Paris.

Hochereau de Reviers, M.T., Courot, M. and Ortavant, R. (1964). Marquage des Cellules Germinales du Bélier et du Taureau par Injection de Thymidine Tritiée dans L'Artère Spermatique. Ann. Biol. Anim. Biochim. Biophys. 4, 157.

Hoskins, D.D. and Casillas, E.R. (1973). Function of cyclic nucleotides in mammalian spermatozoa. In Handbook of Physiology, Section 7, Endocrinology, Vol. V., Male Reproductive System. R.O. Greep and E.B. Astwood (eds), Washington, D.C., American Physiological Society, pp. 453–460.

Hoskins, D.D., Stephens, D.T. and Hall, M.L. (1974). Cyclic adenosine 3′, 5′-monophosphate and protein kinase levels in developing bovine spermatozoa. J. Reprod. Fertil. 37, 131.

Koefoed-Johnson, H.H. and Pedersen, H. (1970). Light and electron microscopic investigation of a case of the so-called Dag defect in the bovine spermatozoa tail. Ann. Rept. Roy. Vet. Agric. Cell. Sterility Inst., Copenhagen, p. 49–62.

Lino, B.F., Braden, A.W.H. and Turnbull, K.E. (1967). Fate of unejaculated spermatozoa. Nature 213, 594.

Leblond, C.P. and Clermont, Y. (1952). Definition of the stages of the cycle of the seminiferous epithelium in the rat. Ann. N.Y. Acad. Sci. 55, 548.

Mann, T. (1964). The Biochemistry of Semen and of the Male Reproductive Tract. London, Methuen.

McRorie, R.A. and Williams, W.L. (1974). Biochemistry of mammalian fertilization. Ann. Rev. Biochem. 43, 777.

Morton, D.B. (1976). Lysosomal enzymes in mammalian spermatozoa. In Lysosomes in Biology and Pathology. Vol. 5. J.T. Dingle and R.T. Dean (eds), New York, American Elsevier Publishing Co., Inc., pp. 203–255.

Nelson, L. (1967). Sperm motility. In Fertilization. C.B. Metz and A. Monroy (eds), New York, Academic Press.

Orgebin-Crist, M.-C. (1962). Recherches experimentales sur la durée de passage des spermatozoides dans l'epididyme du taureau. Ann. Biol. Anim. Biochim. Biophys. 2, 51.

Orgebin-Crist, M.-C. (1967). Sperm maturation in rabbit epididymis. Nature 216, 816.

Orgebin-Crist, M.-C., Davies, J. and Tichenor, P.L. (1972). Maturation of spermatozoa in the rabbit epididymis. In Regulation of Mammalian Reproduction. S.J. Segal, R. Crozier and P.A. Corfman (eds). Springfield, Ill., Charles C Thomas.

Ortavant, R. (1959). Déroulement et Durée Du Cycle Spermatogénétique Chez Le Belier. Ann. Zootech. 8, 183, 271.

Perey, B., Clermont, Y. and Leblond, C.P. (1961). The wave of the seminiferous epithelium in the rat. Am. J. Anat. 108, 47.

Phillips, D.M. (1973). Mammalian sperm structure. In Handbook of Physiology, Section 7, Endocrinology, Vol. 5, Male Reproductive System. R.O. Greep and E.B. Astwood (eds), Washington, D.C., American Physiological Society, pp. 405–419.

Reid, B.L. and Cleland, K.W. (1957). The structure and function of the epididymis. I. The histology of the rat epididymis. Aust. J. Zool. 5, 223.

Risely, R.L. (1963). Physiology of the male accessory organs. In Mechanisms Concerned With Contraception. C.C. Hartman (ed), New York, Macmillan, p. 73.

Roussel, J.D., Stallcup, O.T. and Austin, C.R. (1967). Selective phagocytosis of spermatozoa in the epididymis of bulls, rabbits and monkeys. Fertil. Steril. 18:509.

Saacke, R.G. (1970). Morphology of sperm and its relationship to fertility. Proc. Third Tech. Conf. Artif. Insem. and Reprod., pp. 3–16.

Setchell, B.P. (1970). Testicular blood supply, lymphatic drainage and secretion of fluid. In The Testis. Vol. 1. A.D. Johnson, W.R. Gomes and N.L. VanDemark (eds), New York, Academic Press, pp. 101, 230.

Setchell, B.P. (1977). Male reproductive organs and semen. In Reproduction in Domestic Animals. H.H. Cole, and P.T. Cupps (eds), New York, Academic Press, pp. 229–256.

Setchell, B.P., Scott, T.W., Voglmayr, J.K. and Waites, G.M.H. (1969). Characteristics of testicular spermatozoa and fluid which transport them into the epididymis. Biol. Reprod. Suppl. 1, 40.

Stechell, B.P., Davies, R.V. and Main, S.J. (1977). Inhibin. In The Testis, Vol. IV. A.D. Johnson and W.R. Gomes (eds). New York, Academic Press, pp. 189–238.

Sjöstrand, N.O. (1965). The adrenergic innervation of the vas deferens and the accessory male genital glands. Acad. Physiol. Scand. Suppl. 65, 257.

Swierstra, E.E. (1968). Cytology and duration of the cycle of the seminiferous epithelium of the boar; duration of spermatozoan transit through the epididymis. Anat. Rec. 161, 171.

Swierstra, E.E., Genauer, M.R. and Pickett, B.W. (1974). Reproductive physiology of the stallion. I. Spermatogenesis and testis composition. J. Reprod. Fertil. 40, 113.

Voglmayr, J.K. (1975). Output of spermatozoa and fluid by the testis of the ram and its response to oxytocin. J. Reprod. Fertil. 43, 119.

Voglmayr, J.K., Scott, T.W., Setchell, B.P. and Waites, G.M.H. (1967). Metabolism of testicular spermatozoa and characteristics of testicular fluid collected from conscious rams. J. Reprod. Fertil. 14, 87.

Waldschmit, M., Karg, H. and Kinzler, M. (1964). Vorkommen von Desoxyribonuklease in männlichen Geschlechtssekreten beim Rind. Naturwiss 51, 364.

10

Secretions of the Male Reproductive Tract and Seminal Plasma

I.G. WHITE

Semen as ejaculated consists of spermatozoa coming from the testes suspended in the seminal plasma, which is composed of fluids contributed by the testes and the various accessory organs.

The epididymis or an analogous Wolffian duct is the only male accessory organ common to all mammals, birds and reptiles. The number and size of the other accessory organs varies tremendously from one species to another (Mann, 1964; Price and Williams, 1961; Brandes, 1974). Birds and reptiles lack accessory organs other than the epididymis. All mammals have a prostate of one sort or another, and many have supplementary accessory organs, eg, seminal vesicles, ampullae and Cowper's or bulbourethal glands. This is true for bull, ram, boar and stallion; however, the dog and cat do not have seminal vesicles.

The development and maintenance of the accessory glands and their secretions depend upon androgen. Removal of the androgenic stimulus by castration produces structural alterations and depression of functional activities, most of which can be reversed by the administration of androgens. Structural involution and functional inhibition can also be achieved by administration of antagonistic sex hormones such as estrogens; these changes reverse after stopping estrogen treatment.

TESTICULAR FLUID

Although the fluid produced by the testes constitutes a sort of primordial seminal plasma, it represents only a minor portion of the final ejaculate as it is greatly modified and augmented by the accessory organs during its passage through the male reproductive tract (Fig. 10–1).

Spermatozoa are swept out of the seminiferous tubules through the rete testis and into the epididymis in fluid that is believed to be actively secreted by the Sertoli cells. The mechanism of secretion is not certain, but the Sertoli cells may pump solute from the interstitial spaces around the seminiferous tubules into the intercellular spaces above the narrow junctions between the cells. This would

189

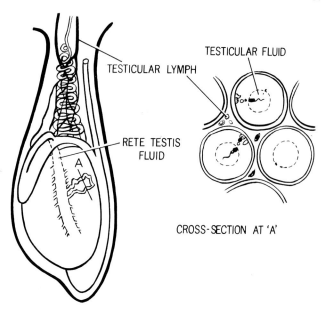

TESTICULAR FLUID

TESTICULAR LYMPH

RETE TESTIS
FLUID

CROSS-SECTION AT 'A'

Fig. 10–1. **Diagram of site of the rete testis and origin of rete testis fluid and testicular lymph.** *(G.M. Waites and B.F. Setchell, 1969. In: The Gonads, McKerns, [ed.], Amsterdam, North Holland)*

make the fluid there hypertonic and water would then be drawn in by osmosis.

The composition of testicular fluid is different from that of the blood and lymph draining the testis owing to a blood-testis barrier (Fig. 10–2). This barrier is formed by occluding junctions between Sertoli cells which divide the cavity of the seminiferous tubule into a "basal" compartment (ie, next to the basement membrane) (Setchell, 1974; Setchell and Waites, 1975; Fawcett, 1975) and an "adluminal" compartment. The selectively permeable barrier restricts the passage of some substances from the blood into the interior of the seminiferous tubules and hence into the rete testis fluid, even though these substances pass into the lymph.

Much of the information on the composition of the rete testis fluid came from analyses of material obtained from slaughtered bulls and rams by severing the testis at the junction of the testis and epididymis. A more elegant technique has since evolved that enables one to cannulate the rete testis and obtain fluid from living animals (Setchell, Scott, Voglmayr and Waites, 1969; White, 1973; Setchell, 1974; Setchell and Waites, 1975; Voglmayr, 1975).

Rete testis fluid has a low spermatozoal concentration; its composition is shown in Table 10–1. The ionic composition of the rete testis fluid differs somewhat from blood plasma. However, perhaps the most striking difference between the rete testis fluid and blood plasma is in the glucose and inositol content. Rete testis fluid normally contains practically no glucose but about 100 times the concentration of inositol that occurs in blood plasma. Therefore, glucose is not available to spermatozoa as an energy source during the two or three hours during which they pass from the seminiferous tubules to the head of the epididymis.

The possibility that inositol might be utilized by ram and bull testicular sperm as a substrate has been investigated. Al-

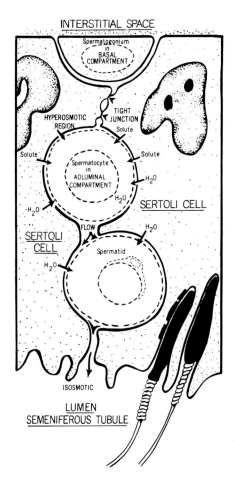

INTERSTITIAL SPACE

Spermatogonium in BASAL COMPARTMENT

TIGHT JUNCTION

HYPEROSMOTIC REGION

Solute

Solute

Solute

Spermatocyte in ADLUMINAL COMPARTMENT

H_2O

SERTOLI CELL

H_2O

H_2O

SERTOLI CELL

FLOW

H_2O

H_2O

Spermatid

ISOSMOTIC

LUMEN SEMENIFEROUS TUBULE

Fig. 10–2. Diagram depicting the localization of the blood-testis barrier and the compartmentalization of the germinal epithelium by tight junctions between adjacent Sertoli cells. The tight junctions delimit a basal compartment in the germinal epithelium, which contains the spermatogonia and early preleptotene spermatocytes, and an adluminal compartment, which contains more advanced spermatocytes and spermatids. Substrates have direct access to cells in the basal compartment, but to reach the cells in the adluminal compartment, substances must pass through the Sertoli cells. The diagram also shows a possible mechanism of secretion of fluid by the Sertoli cells. If solute were pumped into the clefts around the germ cells, a standing osmotic gradient might be created in the adluminal compartment that would move water from base to lumen. *(Fawcett, 1975. In: Handbook of Physiology Vol. V, Section 7, R.O. Greep [ed.], American Physiological Society. Washington D.C., pp. 21–55)*

though testicular sperm oxidize inositol only at a slow rate when it is added to media *in vitro*, *in vivo* inositol may be absorbed into the epididymis and incorporated into phospholipid for subsequent release into the lumen as an energy source for the spermatozoa (Voglmayr, 1975). Of course, lactic acid would be available to spermatozoa from the rete testis fluid if the oxygen tension is sufficiently high for its oxidation, as it is readily oxidized by testicular spermatozoa *in vitro*.

As might be expected from the existence of a blood-testis barrier, the total protein content of the rete testis fluid is less than that of the blood plasma and its nature is different (Waites, 1976). The protein or polypeptide fraction of ram testicular fluid includes (1) androgen-binding protein (ABP) produced by the Sertoli cells, which may be concerned with the transport of androgens to the caput epididymidis (Hansson et al, 1976), (2) the rather controversial "inhibin" which, if absorbed from the caput epididymidis, may regulate spermatogenesis by means of suppression of FSH production by the anterior pituitary and (3) a peptide inhibitor of sperm acrosin, which may inactivate any proteases that leak from the acrosome.

A number of amino acids (eg, glutamic acid, alanine, glycine, lysine, aspartic acid and serine) occur in higher concentrations in the testicular fluid than in the blood plasma of the ram and bull, presumably due to synthesis within the seminiferous tubules. These amino acids are not utilized rapidly by spermatozoa. However, aspartic acid, glycine and glutamine are involved in the synthesis of purine and pyrimidine bases and the high concentrations established in the testicular fluid may reflect especially favorable conditions for nucleic acid synthesis within the seminiferous tubules.

Testosterone, dehydroepiandrosterone, 5α-dihydrotestosterone and also estrogens have been found in the rete testis fluid of

Table 10–1. Chemical Composition of Testicular Fluid*
(After Removing Spermatozoa)

Constituent	Ram	Bull	Boar
Sodium	271	308	347
Potassium	48	35	21
Calcium	4.1	1.9	4.8
Magnesium	0.8	1.0	
Chloride	454	433	476
Bicarbonate	49	42	
Protein	280	540–1300	165
Urea	28	34	
Alanine (mM)	0.33	1.04	0.48
Glycine (mM)	1.87	1.09	1.19
Glutamic acid (mM)	2.03	2.19	1.96
Aspartic acid (mM)	0.33	0.43	0.29
Serine (mM)	0.36	0.73	
Glucose	<2	<2	<2
Inositol	131		126
Lactic acid	5.9		6.6
Ascorbic acid	3.0		2.8
Glycerylphosphorylcholine	80	9–19	52
Testosterone (μg/100 ml)	2.8–8.0	2.3	

* Values are mg/100 ml unless otherwise stated. Data from the literature.

the ram, bull and boar in concentrations at least as high as in the peripheral blood plasma (Waites, 1976). The steroids arise from the interstitial cells and the seminiferous tubules are clearly exposed to testosterone, which plays a part in the maintenance of spermatogenesis and perhaps also directly influences spermatozoa in the epididymis.

THE EPIDIDYMIS AND EPIDIDYMAL FLUID

The epididymis is responsible for the transport, concentration, maturation and storage of spermatozoa, and the cells lining the tubules have both an absorptive and secretory function (Glover and Nicander, 1971).

Transport and Maturation of Spermatozoa

Spermatozoa are swept from the testis and through the efferent ducts by the pressure of rete fluid in the testes, aided by cilia movement. Their passage through

the duct itself probably depends partly on peristaltic movements of the muscle wall. In mammals this passage takes 5 to 25 days, depending on the species (7 to 9 days in the bull); increased frequency of ejaculation probably has only a slight effect on the rate.

During their passage through the epididymis spermatozoa mature or ripen. Although the acrosome may be modified in some species, the most conspicuous change is migration of the remnant of cytoplasm known as the "cytoplasmic droplet." This is attached near the neck of the spermatozoa on leaving the testis, but it moves so that the droplet is located farther down the midpiece when spermatozoa reach the tail of the epididymis. By the time semen is ejaculated most of these droplets have become completely detached from the spermatozoa, but the mechanism by which this occurs is obscure.

Other features of the ripening process in spermatozoa are changes in membrane characteristics and an increase in specific gravity and fertilizing capacity. In some species at least, the spermatozoa also increase their potential for movement. A decrease in DNA, protein and phospholipid content and negative charge have also been reported, along with changes in metabolic pattern and an increased susceptibility to cold shock (see subsequent sections). Studies in which rabbit spermatozoa were retained in the caput or corpus epididymis by ligatures suggest that the morphologic and motility changes may take place without normal passage into the cauda; however, a sojourn in the cauda appears essential for the attainment of full fertility.

The Epididymal Fluid

The fluid in the lumen of the epididymis constitutes the environment not only for the maturation but also for the storage of spermatozoa and it is not unreasonable to suppose that the composition of the fluid may play some part in both processes (Bedford, 1975; Orgebin-Crist et al, 1976).

During the past ten years knowledge has accumulated about the chemical nature of the environment in the epididymis of laboratory animals and domestic species (Table 10–2). Analyses have been made of fluid collected from the epididymis of the ram and bull after slaughter and of fluid collected from fistulae or cannulae in the vas deferens (White, 1973; Setchell, 1974). The volume of fluid flowing through the epididymis of the ram is less than 1 ml per day, which contrasts sharply with a volume of 40 ml per day produced from a cannula inserted into the rete testis. Clearly, much of the fluid produced by the testis is absorbed; this occurs chiefly in the efferent ducts and in the initial segment of the epididymis and leads to a much greater concentration of spermatozoa in the tail (Crabo and Hunter, 1975).

Glycerylphosphorylcholine (GPC). One of the quantitatively most important constituents of epididymal fluid is the androgen-dependent glycerylphosphorylcholine. The epididymis and not the testis

Table 10– 2. Characteristics and Chemical Composition of Epididymal Fluid*
(After Removing Spermatozoa)

| Parameter | Ram | | Bull | | Boar |
	Caput	Cauda	Caput	Cauda	Cauda
pH			6.7	6.8	
Freezing depression(°C)	0.61	0.64			
Sodium	207	131	132	83	88
Potassium	191	152	170	108	188
Calcium	3.3	1.9	1.4	1.4	6
Magnesium	3.5	1.4	7.6	3.2	
Chloride	219	120	327	231	12
Bicarbonate			18.9	0	
Protein		1800		3333	
Urea	21	14	36	23	18
Alanine (mM)		0.84		0.98	
Glycine (mM)		1.36		1.73	
Glutamic acid (mM)		12.21		11.89	
Aspartic acid (mM)		0.89		1.68	
Serine (mM)		1.66		3.59	
Total reducing sugar			13.0	2.0	
Lactic acid	99	60	73	42	
Glycerylphosphorylcholine	497	2091	236	480	3060
Carnitine (mM)				2	

* Values are mg/100 ml unless otherwise stated. Data from the literature.

is responsible for the production and secretion of GPC, which is shown by the progressive increase in its concentration in fluids taken from the rete testis, the head and tail of the epididymis and the vas deferens of the ram and bull (White, 1973). Concurrent analyses indicate that the concentration of sodium and chloride ions decreases while the potassium content remains about the same. At least some sodium chloride may be replaced by GPC as fluid passes from the rete testis through to the vas deferens and the osmotic pressure of the luminal fluid tends to be maintained, despite the loss of sodium chloride, by the secretion of GPC from the epithelial cells lining the lumen.

Proof of the synthesis of GPC by epididymal tissue has been obtained by injecting labeled phosphorus (^{32}P) into laboratory animals and measuring its incorporation into the GPC of the epididymis. Phospholipid has been suggested as a

ably due to selective resorption of fluid in the head of the epididymis.

Carnitine. The epididymis of domestic species (eg, bull and boar) is rich in carnitine (Hamilton, 1975). In the rat, and presumably also in other species, carnitine from the blood is concentrated in the epididymis, and high concentrations of carnitine (63 ± 3.3 mM) occur in epididymal plasma collected from the cauda epididymidis. The carnitine level of the fluid is under androgen control and its high concentration must also contribute to the osmotic pressure of the fluid and at least partly redress the loss of sodium chloride in the proximal region of the duct. Epididymal spermatozoa also contain high levels of carnitine along with the enzyme carnitine acetyltransferase, which probably accounts for the fact that about 5% to 10% of the carnitine in the sperm and epididymal plasma is in the acetyl form.

$$
\begin{array}{ccc}
\text{CH}_2 \cdot \text{CHOH} \cdot \text{CH}_2\text{COOH} & & \text{CH}_2 \cdot \text{CHO} \cdot \text{COCH}_3 \cdot \text{CH}_2\text{COOH} \\
| & \xrightarrow{\text{Acetyl CoA}} & | \\
\text{N}^+(\text{CH}_3)_3 & \overline{\text{Carnitine Acetyltransferase}} & \text{N}^+(\text{CH}_3)_3 \\
\text{Carnitine} & & \text{Acetylcarnitine}
\end{array}
$$

possible precursor of GPC but some experiments suggest a more direct pathway of GPC synthesis in the epididymis that may not proceed by way of lecithin (White, 1972). As fluid passes through the epididymis, large quantities of GPC are added to it, and this raises the question of what is the function, if any, of GPC in that organ. It is possible that GPC might, in part, be responsible for the fact that spermatozoa can survive for long periods in the epididymis; GPC might even play some part in the maturation process that occurs as spermatozoa pass through the epididymis. This might also be true of glutamic acid, which occurs in an even higher concentration in ram epididymal plasma than in testicular fluid, presum-

Carnitine acts as a cofactor in fatty acid oxidation and acetylcarnitine may serve as an energy reservoir and buffer against rapid changes in the concentration of acetyl-CoA in spermatozoa. It has been suggested that the maturation of sperm in the epididymis may be related to accumulation of carnitine in the sperm during epididymal transit.

Sialic Acid and Glycoprotein. Sialic acid forms part of the glycoprotein of epididymal fluid and its concentration falls in the epididymis of the rat after castration or treatment with the antiandrogen, cyproterone acetate.

The glycoprotein in epididymal fluid may act as a lubricant and so reduce the friction between the densely packed sperm in the cauda epididymides (Hamilton, 1975). However, recent reports indicate that a more subtle interaction occurs between the glycoprotein in epididymal fluid and epididymal sperm, suggesting that glycoproteins may be involved in sperm maturation. Motility may be induced in immotile sperm taken from the caput epididymidis of the bull by exposing them to a mixture of a cyclic AMP phosphodiesterase inhibitor (eg, theophylline) and a glycoprotein factor in the seminal plasma. Furthermore, as boar sperm mature in the epididymis they become less permeable to Na+ and K+ due to a factor in the cauda plasma which is probably a glycoprotein. Glycoproteins or similar macromolecules may also contain the active peptide of the decapacitation factor that probably coats sperm as they pass through the epididymis.

Survival and Storage of Sperm in the Epididymis

The cauda epididymidis constitutes a storage reservoir for spermatozoa and experiments on the ligated epididymis in bulls and rams show that conditions in the cauda epididymidis are peculiarly conducive to the survival of spermatozoa. For example, the time that sperm can survive in the ligated epididymis is still greater than that achieved with any diluent at temperatures above the freezing point in routine artificial insemination practice.

The ability of spermatozoa to survive for long periods in the epididymis may be linked to either physical or chemical conditions in that organ; it is generally believed that spermatozoa are immotile, or nearly so, in the epididymis and that their metabolic state is quiescent. The question of what constitutes the substrate for the metabolism of spermatozoa stored in the epididymis is an intriguing one. Conditions in the tubules are probably aerobic, at least immediately next to the epithe-

lium and even if, as is generally supposed, spermatozoa are in a quiescent state, at least some basal metabolism might be expected to maintain the integrity of the cells. The relatively high concentration of lactic acid in the epididymal lumen may provide a substrate for spermatozoal metabolism. Although GPC is present in large quantities in the luminal fluid, it cannot be used directly by spermatozoa; glutamic acid is oxidized only slowly and there is little fructose, glucose or acetate present. Of course, if a constant interchange took place between epididymal fluid and the blood, a substrate such as glucose could still be important in the economy of the spermatozoa, despite a low level at any one time. Recent studies on the ram are of interest in this regard; they show that intravenously infused ^{14}C glucose appears in the fluid of the cauda epididymidis as lactic acid.

It seems probable that the turnover of spermatozoa in the cauda epididymidis is more rapid than previously envisaged. Sufficient spermatozoa are probably lost daily in the urine or by spontaneous discharge, at least in the ram, to account for the continuous production of spermatozoa (Bedford, 1975). There seems no need to postulate massive reabsorption of surplus spermatozoa from the epididymis in these species, although there is some evidence that reabsorption does occur in the bull.

THE PROSTATE AND ITS SECRETIONS

The gross anatomic arrangement of the prostate gland varies among different species. Furthermore, in those species in which the prostate is composed of several lobes, as in the rat, chemical and physiologic differences between the various lobes may be pronounced (Mann, 1964).

Two mechanisms are involved in secretion: merocrine and apocrine (Mann,

1974). In the merocrine type of secretion, the glandular cells remain intact during the formation and discharge of their secretory product. In apocrine secretion, a portion of the cytoplasm is discharged by the secretory cells along with the secretory products, and this is the type of secretion largely responsible for the formation of the prostatic and vesicular fluids. This explains why prostatic and vesicular secretions are rich in enzymes that are generally associated with the interior rather than the exterior milieu of cells (eg, glycolytic enzymes, nucleases, nucleotidases and lysosomal enzymes, including proteinases, phosphatases and glycosidases). Many of the microscopically visible "particles" normally encountered in ejaculated semen or prostatic secretion, such as lipid bodies, prostatic caluli and lipid granules, represent the contents of secretory cells that have undergone desquamation or disruption during their normal secretory activity.

In the dog, the prostatic secretion is usually slightly acid, about pH 6.5, and almost completely lacking in reducing sugar; however, it does contain citric acid and acid phosphatase (Mann, 1964; Smith, 1975). The zinc content of the canine prostate is unusually high and the prostatic secretion is chiefly responsible for the high zinc concentration of the seminal plasma (Byar, 1975).

In rams and goats, the prostate is the primitive "disseminate" type and consists of glands that do not penetrate the muscle surrounding the pelvic part of the urethra. In bulls and boars, the organ possesses the features of both the lobed and the disseminate types (Mann, 1964). However, because of the presence of other well-developed accessory organs, notably the seminal vesicles, the relative contribution of the prostate to the final ejaculate is small.

SEMINAL VESICLES

In several species, including rams and bulls, the seminal vesicles are paired glands situated laterally to the ampullae; the nature of the vesicular secretions has been extensively explored in domestic animals (Mann, 1964, 1975). The vesicular glands of the bull are lobed; those of the stallion form elongated pear-shaped sacs and contribute the gel to the ejaculate. These glands are particularly large and bag-like in the boar and are filled with a milky, highly viscous fluid. There are no vesicular glands in dogs or cats.

Compared with prostatic fluid, the seminal vesicle secretion is usually more alkaline, has a higher dry weight and contains more potassium, bicarbonate, acid-soluble phosphate and protein. A feature of the seminal vesicle secretion is its high content of reducing substances, which include sugars and ascorbic acid (Mann, 1964). The normal seminal vesicle secretion is usually slightly yellowish, but sometimes (as in the bull), it can be deeply pigmented due to riboflavin, which causes the vesicular secretion and seminal plasma to fluoresce strongly in ultraviolet light.

In many species (eg, bull, ram, goat and boar), the bulk of seminal fructose is secreted by the seminal vesicles and in such species the chemical assay of fructose may be used as an indicator of the relative contribution made by the seminal vesicles to the semen (Mann, 1964).

Studies done primarily on ram and bull seminal vesicles show that there are two major pathways of fructose formation from blood glucose (Mann, 1974). One proceeds by way of glucose phosphate or fructose phosphate and the other by way of sorbitol. The two principal enzymes involved in the conversion of glucose via sorbitol are aldohexose reductase (which reduces glucose to sorbitol) and sorbitol dehydrogenase (which oxidizes sorbitol to fructose). In addition to containing large amounts of fructose, semen from several species contains smaller quantities of both glucose (the precursor) and sorbitol (the intermediate).

In man, bull, boar and stallion citric

acid as well as fructose is produced by the vesicular glands, although to different degrees (Mann, 1964).

The vesicular secretion of the boar is characterized by a high inositol content (2% to 3%) that contributes substantially to its osmotic pressure; it also contains ergothioneine, at least part of which appears to originate in feedstuff.

Although the name prostaglandin (PG) would suggest that these compounds originate in the prostate, most of the large amount of prostaglandins that occur in the seminal plasma of the ram originates in the seminal vesicles (Cenderella, 1975).

COWPER'S OR BULBOURETHRAL GLANDS AND AMPULLAE

The paired Cowper's or bulbourethral glands are round, compact bodies in the bull, stallion and ram (Mann, 1964). In the boar, they are much larger, almost cylindrical, and are filled with a viscous, rubberlike, white secretion that is essential for gel-formation in the ejaculated semen. The "dribblings" from the prepuce of the bull before mounting are bulbourethral secretions and their function is to flush the urethra free of urine. The glands are absent in the dog.

In some species, the final segment of the vas deferens becomes thicker owing to the presence of glands in the wall which form the ampullae (Mann, 1964). During courtship and precoital stimulation in some species (eg, the bull), spermatozoa are transported from the tail of the epididymis to the ampulla by peristaltic movements of the vas deferens.

The boar lacks ampullae; however, they are highly developed in the stallion and contribute ergothioneine, a sulfur-containing nitrogenous base, to the ejaculate. The absence of fructose in the ampullary glands of the stallion contrasts with the presence of some of this sugar in the ampullae of other species, for instance the bull.

SEMINAL PLASMA

During ejaculation the various accessory glands discharge their products successively in a well-controlled pattern. Ram and bull semen is normally ejaculated rapidly and there is complete mixing of the seminal components. On the other hand, the stallion (<1 minute), the boar (2 to 10 minutes) and the dog (5 to 20 minutes) emit their semen over a longer period in fractions corresponding to the secretions of different parts of the reproductive tract (Mann, 1964).

As the anatomy and hence the relative contribution of the accessory glands to the seminal plasma varies greatly with the species, it is not surprising to find considerable differences in both the volume and the composition of the semen among species (Mann, 1964, 1975). For example, bull, ram and dog semens have a small volume (1 to 10 ml). Stallion and boar semens are much more voluminous (100 ml or more) with lower sperm densities.

The seminal plasma constitutes the bulk of the ejaculate, particularly in species such as the boar and stallion, and functions as a vehicle for conveying spermatozoa from the male to the female reproductive tract (Rodger, 1975). It has proved of great biochemical interest as it contains many unusual organic compounds (eg, fructose, citric acid, sorbitol, inositol, glycerylphosphorylcholine and ergothioneine—see Fig. 10–3) that are not found elsewhere in the body in such high concentrations. Table 10–3 shows the concentrations of some of these constituents in seminal plasma. However, do not rely greatly on the precision of these figures, as estimates are likely to vary not only with and between animals but also with the method of semen collection. Seminal plasma is an isotonic, neutral medium and in many species it contains a source of energy directly available to spermatozoa (eg, fructose and sorbitol), or one that might be unlocked on mixing with female secretions (ie, glycerylphosphorylcholine [GPC]). The functions,

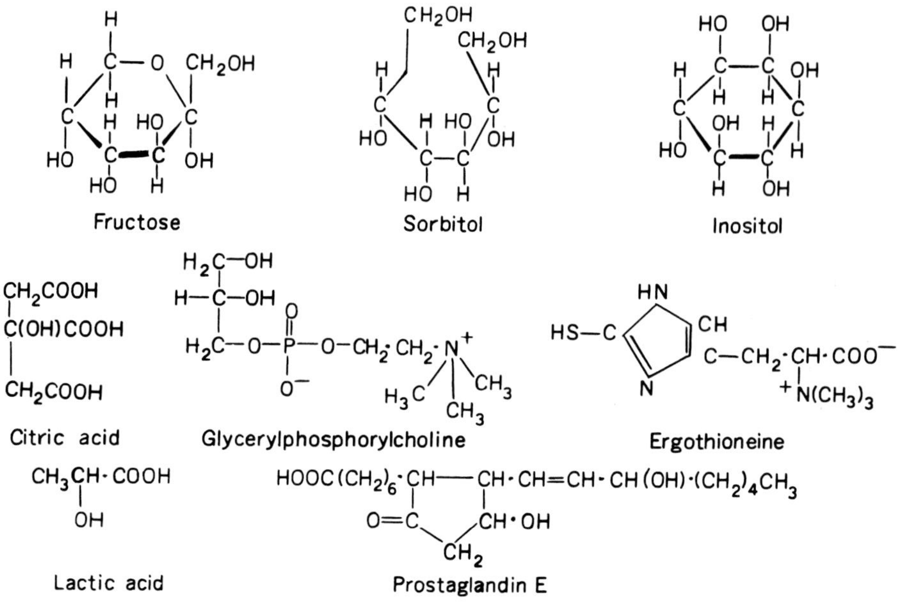

Fig. 10–3. Formulae of some seminal plasma constituents. *(Modified from White, 1976. In: Veterinary Physiology, J.W. Phillis [ed.], Bristol, Wright-Scientechnica)*

if any, of the other unusual constituents of seminal plasma (eg, citric acid, inositol) are not known. These organic substances are produced by various accessory glands in response to testosterone; as already mentioned, their concentrations in ejaculated semen indicate accessory gland function. Thus after castration, they disappear from the seminal plasma but reappear after testosterone has been injected into the animal.

Fructose and Other Sugars

Sugars are present in the semen and accessory sex glands of most mammals that have been examined (Rodgers, 1976; White et al, 1977). The levels of sugars vary widely between species and even individuals, but levels of the order of 100 to 500 mg fructose/100 ml semen or 100 gm of secretory tissue are common. Two Ferungulate orders depart from this normal pattern; Carnivora (eg, dog and cat) and Perissodactyla (eg, horse) not only

lack "normal" fructose levels in semen but have low seminal sugar levels generally (20 to 30 mg/100 ml).

In those species (eg, human, ram and bull) in which the fructose concentration of the semen is high, the sugar is a nutrient for the spermatozoa that break it down. The metabolism of fructose by spermatozoa proceeds by way of the Embden-Meyerhof pathway. Hexose phosphates, triose phosphates and pyruvic acid are involved as transient intermediates leading to the formation of lactic acid, which tends to accumulate although it is further oxidized in the presence of oxygen to carbon dioxide and water over the Krebs tricarboxylic acid cycle (Mann, 1964) (see Chapter 5). Fructose is formed from blood glucose in the accessory organs, and small amounts of glucose and also the intermediate compound sorbitol may accompany fructose in the seminal plasma of some species. Like fructose, glucose can be utilized by sperm over the glycolytic pathway, and

Table 10–3. Characteristics and Approximate Chemical Composition of Seminal Plasma

Constituent or Property	Bull	Ram	Boar	Stallion	Dog	
					SBF	PF
Volume of ejaculate (ml)	5–8	0.8–1.2	150–200	60–100	1–5	5–20
pH	*6.9(6.4–7.8)	*6.9 (5.9–7.3)	*7.5 (7.3–7.8)	*7.4 (7.2–7.8)	6.3	6.8
Sodium	225±13	178±11	587	257	332±11	335±10
Potassium	155± 6	89± 4	197	103	31± 4	34± 3
Calcium	40± 2	6± 2	6	26	5± .5	4± .6
Magnesium	8± .3	6± .8	*11 (5–14)	9	4± 1	2± .5
Chloride	174–320	86	*330 (260–430)	448	444	525
Fructose	460–600 (SV)	*250 (SV)	9 (SV)	*2 (0–6)	1	1
Sorbitol	*(10–140) (SV)	*72 (26–120) (SV)	*12 (6–18)	*40 (20–60)	Trace	
Citric acid	620–806 (SV)	*140 (110–260) (SV)	173 (SV)	*26 (8–53) (SV)	4	<30
Inositol	*35 (25–46)	*12 (7–14)	*530 (380–630) (SV)	*30 (20–47)		
Glycerylphosphorylcholine (GPC)	*350 (100–500) (E)	*1650 (1100–2100) (E)	*(110–240) (E)	*(40–100) (E)	176	20
Ergothioneine	*0	*0	17 (SV)	*(40–110) (A)		
Protein (g/100 ml)	*6.8	*5.0	*3.7	*1.0	3.7	2.8

* Analyses on whole semen. Mean values (mg/100 ml of seminal plasma unless otherwise indicated) are given, S.E. or range.
SBF, Sperm-bearing fraction; PF, prostatic fluid; SV, mainly from seminal vesicles; E, mainly from epididymis; A, from ampullae.

sorbitol can be oxidized to fructose and thus also serve as a nutrient for the sperm.

Recent observations show that N-acetylglucosamine rather than fructose is the sugar in the semen of several Australian marsupials, such as the kangaroo. The N-acetylglucosamine is produced by the well-developed prostate in these species.

Miscellaneous Organic Constituents

As it is produced by the epididymis, glycerylphosphorylcholine (GPC) occurs in the seminal plasma of all animals investigated thus far (Mann, 1964, 1975). Spermatozoa are incapable of attacking GPC as such, but in some species (eg, sheep) an enzyme is present in the secretions of the female genital tract that can break GPC down into phosphoglycerol, which spermatozoa can utilize by way of the glycolytic pathway. Therefore, GPC may act as a source of energy for spermatozoa in the female tract.

Citric acid is present in the seminal plasma of laboratory and domestic animals, often in high concentrations. However, it is not readily utilized by the spermatozoa, presumably owing to the relative impermeability of the spermatozoal and mitochondrial membranes, and it is not important as an energy source. The semen of most species also contains some inositol, which presumably has its origin in the testicular fluid. The high concentration of inositol in boar semen comes from the seminal vesicles and probably replaces sodium chloride in the plasma, at least to some extent. Ergothioneine, which occurs in high concentrations in boar and stallion semen, is formed in the vesicular glands of the boar and the ampullae of the stallion and may have a protective function in preventing the oxidation of an essential sulfhydryl group of the sperm.

Seminal plasma contains appreciable quantities of ascorbic acid and the riboflavin content of some bull ejaculates is sufficiently high to color the semen a deep yellow.

Lipids, Fatty Acids and Prostaglandins

Phospholipid constitutes the major fraction of the lipid in the seminal plasma of the bull, boar and stallion which also contains cholesterol, diglyceride, triglyceride and wax ester. The phospholipid fraction of ram, bull and boar seminal plasma contains all of the fractions found in spermatozoa, that is, phosphatidylcholine and ethanolamine and sphingomyelins, the latter being quantitatively most important in the boar. Since the accessory organs of the ram anterior to the epididymis secrete a phospholipase A, free fatty acids as well as phospholipids might be expected to be found in the seminal plasma after ejaculation. The extent to which the plasma phospholipids and their breakdown are utilized by spermatozoa is not known, but they clearly represent a potential energy source, at least in the ram.

The prostaglandins (PGs) are derivatives of unsaturated fatty acids and most mammalian species have levels of seminal PGs less than 100 ng/ml (eg, bull, horse, pig, rabbit), whereas the ram and goat have at least 40 μg/ml PGE plus some PGF (Cenderella, 1975), produced chiefly in the seminal vesicles. The role of seminal PGs remains elusive and they do not appear to have notable effects on the metabolism or motility of spermatozoa. The known actions of PGs on smooth muscle have led to the suggestion that PGs aid the transport of spermatozoa in the female reproductive tract and some recent experiments indicate that the fertility of diluted ram semen on artificial insemination can be increased by the addition of prostaglandins.

Proteins, Amino Acids and Enzymes

The protein content of mammalian seminal plasma varies from about 3% to 7%

depending on the species. In addition, the plasma may contain polypeptides, probably arising from the breakdown of protein after ejaculation, and free amino acids, which originate chiefly in the testes.

Many enzymes have been found in the seminal plasma (Murdoch and White, 1968). Although most have their origin in the accessory glands, at least some might be the result of leakage from spermatozoa. This could be the case, for instance, with lactic dehydrogenase (LDH), which occurs in much greater concentrations in ram and bull spermatozoa than in the seminal plasma. In general, when enzymes occur in higher concentrations in the spermatozoa than in seminal plasma they are "leached" from the cells upon cold-shock or deep-freezing; loss of such important enzymes as LDH might be a factor in explaining the depressed metabolism of cold-shocked or deep-frozen spermatozoa.

Much of the glutamic-oxalacetic transaminase (GOT) activity in the seminal plasma of the ram comes from the epididymal fluid and seminal vesicles. The epididymis is one source of GOT and GPT (glutamic-pyruvic transaminase) in bull seminal plasma, but there also may be "leakage" from the spermatozoa. In the ram, the cauda epididymidis appears to be the important site for production of both acid and alkaline phosphatase; however, the seminal vesicles also seem to contribute a considerable proportion of acid phosphatase to the seminal plasma.

Inorganic Ions

The advent of the atomic spectrophotometer and the development of a more efficient method for separating spermatozoa from seminal plasma have stimulated renewed interest in the cation composition of semen and the interchange of cations between spermatozoa and their environment. Sodium and potassium are the predominant cations in mammalian seminal plasma, which contains lower concentrations of calcium and magnesium (Quinn et al, 1965). The concentration of sodium is greater in the seminal plasma than in the spermatozoa while the reverse is true for potassium. Some of these cations, particularly potassium, may influence the viability of the spermatozoa. The chief inorganic anion in seminal plasma is chloride.

Significance of Seminal Plasma

Since it has been possible to induce pregnancy in some animals by artificial insemination of epididymal spermatozoa, the functional significance of the accessory secretions, at least those produced beyond the epididymis, is debatable. However, the natural mating process could scarcely function efficiently without the provision of the seminal plasma as a carrier of the spermatozoa from the male to the female reproductive tract or as a diluent for the highly concentrated epididymal spermatozoa. In some species (eg, sow and mare) a considerable amount of seminal plasma passes into the uterus with the spermatozoa and the plasma may constitute an important medium for their transport through the female reproductive tract. In other species (eg, sheep and cow) in which the semen is ejaculated into the vagina, the seminal plasma is likely to be less important in this respect.

REFERENCES

Bedford, J.M. (1975). Maturation, transport, and fate of spermatozoa in the epididymis. *In* Handbook of Physiology, Section 7, Vol V. D.W. Hamilton and R.O. Greep (eds), Washington, D.C., American Physiological Society, pp. 303–317.

Brandes, D. (1974). Fine structure and cytochemistry of male sex accessory organs. *In* Male Accessory Sex Organs. D. Brandes (ed), New York, Academic Press, pp. 18–113.

Byar, D.P. (1975). Zinc in male sex accessory organs: distribution and hormonal response. *In* Male Accessory Sex Organs. D. Brandes (ed), New York, Academic Press, pp. 161–171.

Cenderella, R.J. (1975). Prostaglandins in male reproductive physiology. *In* Molecular Mechanisms of Gonadal Action. J.A. Thomas and R.L. Singhal (eds), Baltimore, University Park Press, pp. 325–358.

Crabo, B.J. and Hunter, A.G. (1975). Sperm maturation and epididymal function. *In* Control of Male Fertility. J.J. Sciorra, C. Marblad and J.J. Speidel (eds), Hagerstown, Md., Harper and Row, pp. 2–23.

Fawcett, D.W. (1975). Ultrastructure and function of the Sertoli cell. *In* Handbook of Physiology, Section 7, Vol. V. D.W. Hamilton and R.O. Greep (eds), Washington, D.C., American Physiological Society, pp. 21–55.

Glover, T.D. and Nicander, T. (1971). Some aspects of structure and function in the mammaliam epididymis. J. Reprod. Suppl. *31*, 39.

Hamilton, D.W. (1975). Structure and function of the epithelium lining the ductuli efferentes, ductus epididymis, and ductus deferens in the rat. *In* Handbook of Physiology, Section 7, Vol. V. D.W. Hamilton and R.O. Greep (eds), Washington, D.C., American Physiological Society, pp. 259–301.

Hansson, F., Weddington, S.C., French, F.S., McLean, W., Smith, A., Nayfeh, Ritzen, E.M. and Hagenas, L. (1976). Secretion and role of androgen binding proteins in the testis and epididymis. J. Reprod. Fertil. Suppl. *24*, 17.

Mann, T. (1964). The Biochemistry of Semen and of the Male Reproductive Tract. London, Methuen.

Mann, T. (1974). Secretory function of the prostate, seminal vesicle and other male accessory organs of reproduction. J. Reprod. Fertil. *37*, 179.

Mann, T. (1975). Biochemistry of semen. *In* Handbook of Physiology, Section 7, Vol. V. D.W. Hamilton and R.O. Greep (eds), Washington, D.C., American Physiological Society, pp. 461–471.

Murdoch, R.N. and White, I.G. (1968). Studies of the distribution and source of enzymes in mammalian semen. Aust. J. Biol. Sci. *21*, 483.

Orgebin-Crist, M.-C., Danzo, B.J. and Cooper, T.G. (1976). Reexamination of the dependence of the epididymal sperm viability on the epididymal environment. J. Reprod. Fertil. Suppl. *24*, 115.

Price, D. and Williams-Ashman, W.G. (1961). The accessory reproductive glands of mammals. *In* Sex and Internal Secretions. W.C. Young (ed), Baltimore, Williams & Wilkins, pp. 366–448.

Quinn, P.J., White, I.G. and Wirrick, B.R. (1965). Studies of the distribution of the major cations in semen and male accessory secretions. J. Reprod. Fertil., *10*, 379.

Rodger, J.C. (1975). Seminal plasma, an unnecessary evil? Theriogenology *3*, 237.

Rodger, J.C. (1976). Comparative aspects of the accessory sex glands and seminal biochemistry of mammals. Comp. Biochem. Physiol. *55B*, 1.

Setchell, B.P. (1974). Secretions of the testis and epididymis. J. Reprod. Fertil. *37*, 165.

Setchell, B.P., Scott, T.W., Voglmayr, J.K. and Waites, G.M.H. (1969). Characteristics of testicular spermatozoa and the fluid which transports them into the epididymis. Biol. Reprod. *1*, 40.

Setchell, B.P. and Waites, G.M.H. (1975). The blood-testis barrier. *In* Handbook of Physiology, Section 7, Vol. V. D.W. Hamilton and R.O. Greep (eds), Washington, D.C., American Physiological Society, pp. 143–172.

Smith, E.R. (1975). The canine prostate and its secretion. *In* Molecular Mechanisms of Gonadal Action. J.A. Thomas and R.L. Singhal (eds), Baltimore, University Park Press, pp. 167–204.

Voglmayr, J.K. (1975). Metabolic changes in spermatozoa during epididymal transit. *In* Handbook of Physiology, Section 7, Vol. V. D.W. Hamilton and R.O. Greep (eds), Washington, D.C., American Physiological Society, pp. 437–460.

Waites, G.M.H. (1976). Permeability of the seminiferous tubules and the rete testis to natural and synthetic compounds. J. Reprod. Fertil., Suppl. *24*, 49.

White, I.G. (1972). Metabolism of sperm with particular reference to the epididymis. Adv. Biosci. *10*, 157.

White, I.G. (1973). Biochemical aspects of spermatozoa and their environment in the male reproductive tract. J. Reprod. Suppl. *18*, 225.

White, I.G., Rodger, J.C., Morris, S.R. and Marley, P.B. (1977). Role of the secretions of the male accessory organs of mammals. *In* Reproduction and Evolution. J.H. Calaby and C.H. Tyndale-Biscoe (eds), Canberra, National Academy of Science, pp. 183–193.

11

Transport and Survival of Gametes

E.S.E. HAFEZ

The mammalian spermatozoon and egg undergo various physiologic maturational changes in preparation for fertilization (Table 11–1). While the female sheds one or two ova (or 10 to 15 ova in the case of swine) each estrous cycle, the male dis-

Table 11–1. Physiologic Phenomena of Sperm and Egg Related to Fertilization

Parameters	Oogenesis and Characteristics of Ova	Spermatogenesis and Characteristics of Sperm
Mitosis in the gonad	Ceases during fetal life; no new eggs are formed after birth	Continues throughout reproductive life of the male
Meiosis in the gonad	Begins during fetal life	Begins at puberty and continues throughout reproductive life
First maturational division in gamete	First maturational division is completed in preovulatory follicle	First meiotic division results in two cells of equal size
Second maturational division	Second maturational division is completed only when egg is penetrated by sperm	Not comparable
Number of gametes produced during reproductive life	Thousands of oogonia are found in neonate ovary	Millions of sperm are produced in each ejaculate from puberty, with reduced numbers during senility
Sex chromosome in gamete	X	X or Y
Amount of cytoplasm in gamete	As oocyte matures, the amount of its cytoplasm increases	As spermatid develops into sperm, the amount of cytoplasm decreases; acrosome and tail develop in late spermatid
Motility of gamete	Oocytes, surrounded by follicular cells, are immotile	Sperm motility develops gradually in various parts of epididymis and increases at ejaculation
Plasma membrane at fertilization	Egg acquires plasma membranes from sperm	Sperm loses its plasma membranes to egg
Survival in female reproductive tract	12 to 24 hours after ovulation	Fertilizability is maintained 24 hours after ejaculation

charges massive numbers of spermatozoa at each copulation. Since the survival time of ova and spermatozoa is relatively short (20 to 48 hours), fertilization depends primarily on the synchronous transport of the gametes in the female reproductive tract. Gamete transport is the result of the inherent contractility of the female tract as modified by central nervous system reflexes and hormonal activity. Pharmacologically active substances in the semen stimulate and modulate the contractility of the female reproductive tract. The oviductal cilia and fluids, cervix, uterotubal junction and ampullary-isthmic junction may play roles that are yet undetermined.

This chapter considers the physiologic mechanisms involved in the transport, survival and loss of spermatozoa, and the transport and development of eggs in the oviduct.

SPERM TRANSPORT
IN THE FEMALE TRACT

Species differences exist in the sites at which the ejaculate is deposited in the female reproductive tract during copulation (Table 11–2). In cattle and sheep, the small volume of semen is ejaculated into the cranial end of the vagina and onto the cervix (Fig. 11–1). In horses and swine the

voluminous ejaculate is deposited through the relaxed cervical canal into the uterus. The spermatozoa are unique because they are transported through various luminal fluids of completely different physiologic and biochemical characteristics (eg, testicular fluid, epididymal fluid, seminal plasma, vaginal fluid, cervical mucus, uterine fluid, oviductal fluid and peritoneal fluid) (Fig. 11–2).

Physicochemical and immunologic factors in the vagina and cervix at the time of insemination play an important role in sperm survival and transport into the uterus and oviduct (Table 11–3), (Figs. 11–3, 11–4). The vaginal secretions immobilize spermatozoa within one to two hours of insemination. The rapid elimination and immobilization of spermatozoa in the vagina make the rapid transport of sperm to a more favorable environment essential.

Seminal plasma plays a major role in the transport and physiology of spermatozoa. However, spermatozoa removed from the vas deferens and epididymis are successfully used in artificial insemination, and the removal of various accessory organs of the male tract rarely decreases fertility so long as ejaculation still results in the release of a few million spermatozoa through the urethra.

Table 11–2. Species Differences in the Site of Ejaculation and Semen Characteristics in Several Mammals

Site of Ejaculation		Semen Characteristics	Species
Vagina	Incipient plug	Slight coagulation of ejaculate	Man / Rabbit
	Incipient plug	Instant coagulation of ejaculate	Monkey
	Little accessory fluid	Semen with high sperm concentration	Cattle / Sheep
Uterus	Voluminous	Distention of cervix	Horse
	Voluminous	Retention of penis during copulation	Dog / Pig
	Vaginal plug	Spasmodic contraction of vagina	Rodents

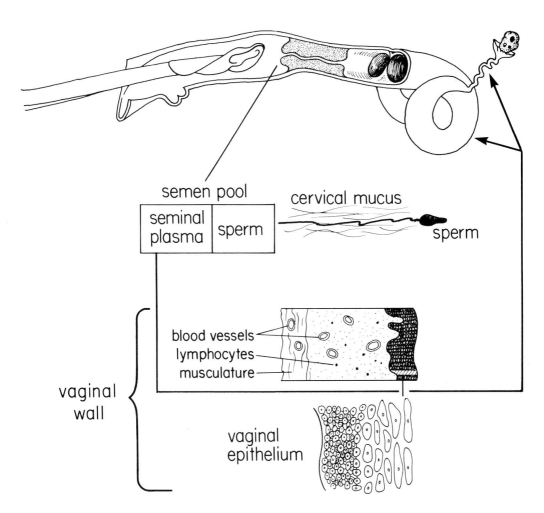

Fig. 11-1. Diagrammatic illustration showing the deposition of the ejaculate in the cow. Spermatozoa are transported through the micelles of cervical mucus into the uterine lumen. The seminal plasma is partly or completely absorbed through the vaginal wall. Some biochemical components of seminal plasma are absorbed to exert an effect on the contraction of musculature of oviduct and uterus, which may facilitate sperm transport to the site of fertilization.

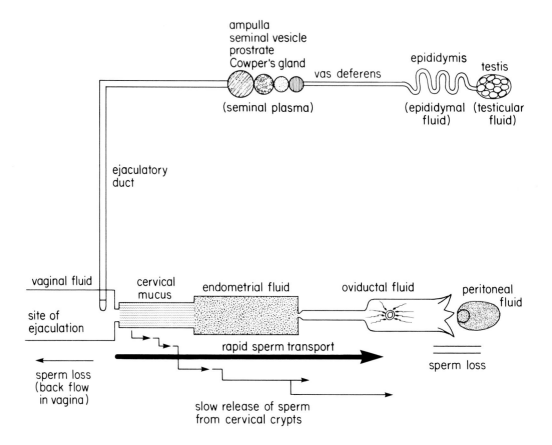

Fig. 11–2.   Diagrammatic illustration of the various luminal fluids in which spermatozoa are suspended from the time of sperm production in the seminiferous tubules to the time of fertilization in the oviduct.

Table 11–3. **Summary of Sequence of Major Physiologic Phenomena Associated with Sperm Transport in the Male and Female Reproductive Tract**

Site	Physiologic Phenomena	Mechanisms Involved
Male reproductive tract	Sperm stored in cauda epididymidis undergo maturation	Neuromuscular
	At ejaculation, sperm released from epididymis are mixed with male accessory secretions	Metabolic
Vagina	Semen deposited in several ejaculatory pulsations ⎱ Semen mixed with vaginal and cervical secretions ⎰	Copulatory motor activities
Cervix	Sperm migrate through micelles of cervical mucus	Biophysical
	Abnormal sperm filtered (gross selection of sperm) through cervical canal	Biochemical
	Cervical crypts establish "sperm reservoir" or rid excessive sperm causing massive reduction in sperm number	Mechanical (kinocilia of epithelium)
Uterus	Sperm separated from seminal plasma and transported to oviduct	Myometrial contraction
	Surface plasma of sperm removed	Agglutination of sperm
	Metabolic changes and capacitation of sperm	Phagocytosis of sperm by leukocytes
	Acrosomal proteinase (trypsin-enzyme) inactivated by trypsin inhibitors from seminal plasma	Enzymatic
Uterotubal junction	Quantitative selection of sperm	Mechanical
Isthmus	Sperm numbers reduced	
Ampullary-isthmic junction	Control of egg transport in oviduct	Neural
	Sperm plasma membrane changes (acrosome reaction), sperm capacitation	Biochemical
Ampulla	Sperm motility increases in oviductal fluid to be able to penetrate corona radiata and zona pellucida	Mechanical
	Reduction division of gametes completed	Metabolic
	Acrosomal proteinases released	Enzymatic
	Selection at egg surface (receptors?) by sperm	Biophysical
Fimbriae	Excessive sperm lost into peritoneal cavity	Sperm motility

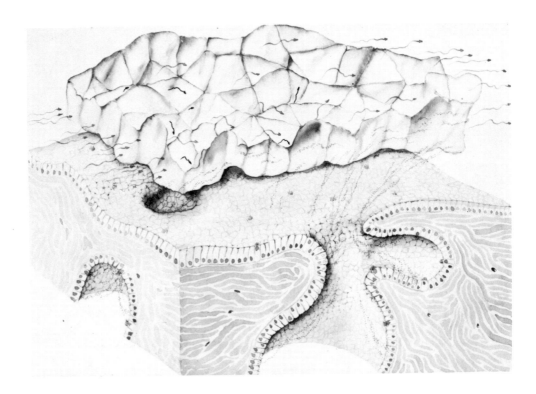

Fig. 11–3. Diagrammatic illustration showing the production of cervical mucus. Macromolecules of glycoprotein (shown in dots) are elaborated by the secretory cells in the cervical crypts to form the mesh of cervical mucus (honeycomb structure shown on the top). A few ciliated cells are shown within the crypts and at the openings of the crypts. During estrus, the mesh of the cervical mucus is quite wide to allow the transport of sperm from the site of ejaculation to the uterine lumen. It is possible that the micelles of the cervical mucus filter the ejaculate when nonviable sperm (shown in dark color) adhere to the mucus.

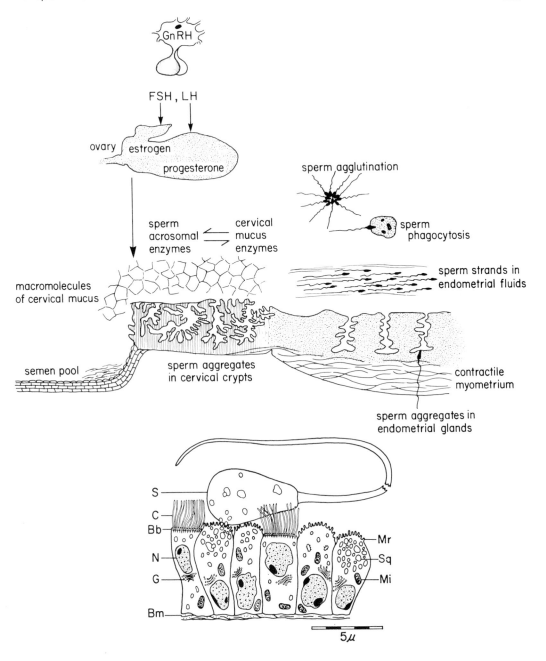

Fig. 11-4. Sperm transport through the cervix to the uterus.
Top, Sperm transport through the cervix to the uterine lumen involves biochemical mechanisms as well as biophysical and physiologic changes in cervical mucus, which in turn are controlled by endocrine factors.
Bottom, The size of spermatozoa *(S)* in relation to the nonciliated secretory cells and ciliated cells of the cervical epithelium. *Bb,* Basal bodies of cilia; *Bm,* basal membrane of epithelium; *C,* cilia; *G,* Golgi apparatus; *Mi,* mitochondria; *Mr,* Microvilli of nonciliated cells; *N,* nucleus; *Sg,* Secretory granules.

Unlike the vagina, the epithelial lining of the cervix, uterus, and oviduct is composed of nonciliated secretory cells and kinociliated cells. In general, secretory cells have a dome-shaped surface covered with numerous microvilli, and their cytoplasm contains numerous secretory granules. The percentage of kinociliated cells in the epithelium, which varies in different parts of the reproductive tract, is maximal in the fimbriae and oviductal ampulla and minimal in the uterus and uterine cervix. The ciliated cells are covered with kinocilia that beat rhythmically toward the vagina.

Sperm Distribution in the Female Reproductive Tract

Three stages are recognized in sperm transport in the female reproductive tract: short, rapid sperm transport; colonization of reservoirs; and slow, prolonged release.

Rapid Transport. Immediately after insemination, spermatozoa penetrate the micelles of the cervical mucus where some are quickly transported through the cervical canal. This phase takes two to ten minutes and may be facilitated by increased contractile activity of the myometrium and mesosalpinx during courtship and coitus. Some spermatozoa reach the internal os of the cervix within 1.5 to 3 minutes after insemination. Thus, some sperm can reach the site of fertilization rapidly. Whether the first spermatozoa entering the oviduct participate in fertilization of the ovum is not known; it has been proposed that fertilization occurs only when a critical number of spermatozoa reach the site of fertilization.

Colonization of Sperm Reservoirs. Massive numbers of spermatozoa are trapped in the complex mucosal folds of the cervical crypts. This process is facilitated by the fact that the micelles of the cervical mucus help to direct spermatozoa to the cervical crypts where the reservoir is formed. Fewer leukocytes are found in the cervical secretions compared with those of the vagina or uterus; this suggests that less phagocytosis of spermatozoa takes place in the cervix. Concentration gradients of spermatozoa that occur in different segments of the reproductive tract are important for fertility. The more spermatozoa that enter the cervical reservoir, the more that will reach the oviduct, thus increasing the chance of fertilization. In addition, the larger the reservoir, the longer an adequate population of spermatozoa will be maintained in the oviduct. Spermatozoa may leave the cervix by means of their own motility or be passively transported by cervical and uterine contractions.

In species in which ejaculation occurs in the uterine horns, sperm reservoirs are localized in the uterotubal junction, as in the pig, or in the endometrial glands, as in the dog. There is no evidence to show that spermatozoa are released after their entry into the endometrial glands of any species.

Slow Release and Transport. After adequate sperm reservoirs have been established within the reproductive tract, the spermatozoa are released sequentially for a prolonged period. This slow release, which involves the innate motility of spermatozoa and the contractile activity of the myometrium and mesosalpinx, ensures the continued availability of spermatozoa for entry to the oviduct to effect fertilization of the egg. However, various anatomic and physiologic barriers prevent massive numbers of spermatozoa in the ejaculate from reaching the site of fertilization (Fig. 11–5), presumably to avoid polyspermy, a condition that is lethal to the fertilized egg.

Sperm Transport in the Cervix

The endocervical mucosa is an intricate system of clefts, grooves and crypts grouped together (Figs. 11–6, 11–7). Several functions have been ascribed to the

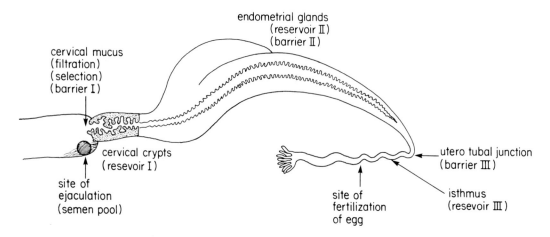

Fig. 11–5. Diagram showing the various anatomic and physiologic barriers that prevent massive numbers of spermatozoa in the ejaculate from reaching the site of fertilization, presumably to avoid polyspermy.

cervix and its secretion: (a) it is receptive to sperm penetration at or near ovulation and inhibits migration at other phases of the cycle, (b) it acts as a sperm reservoir, (c) it protects spermatozoa from the hostile environment of the vagina and from phagocytosis, (d) it provides spermatozoa with energy requirements, (e) it filters defective and immotile spermatozoa, and (f) it possibly participates in the capacitation of spermatozoa.

Cervical Mucus. Cervical mucus that accumulates in the vaginal pool may contain endometrial, oviductal, follicular and peritoneal fluids as well as leukocytes and cellular debris from uterine, cervical and vaginal epithelia. The cervical mucus is a hydrogel, which consists of water and a solid component composed of three or more units forming a three-dimensional network. The secretions are heterogeneous in composition, due to the presence of two types of low- and high-viscosity components. The low-viscosity fraction, a yellowish fluid readily aspirated into a capillary tube, is composed of nonmucin proteins (characteristic of serum proteins), salts, lipids and carbohydrates. The high-viscosity fraction, a clear, white substance, contains mucins that are the gly-

coproteins or glycopeptides responsible for the gel formation.

The cervical mucus at the time of ovulation is composed of macromolecules arranged into micellar units that contain 100 to 1,000 chains of micromolecules. The micromolecules of mucins of cervical mucus have a polypeptide backbone with oligosaccharide and sialic acid side-chains. Proteolytic hydrolysis of the mucus or its mucins by proteolytic enzymes causes physical and biochemical changes. Cervical mucus has rheologic properties such as viscosity, flow elasticity, spinnbarkeit, thixotropy and tack (stickiness). Low-molecular-weight organic components include free simple sugars (glucose, maltose and mannose) and amino acids. The mucus also contains proteins, trace elements and enzymes.

A physiologic balance of ovarian steroids is necessary for the initiation and maintenance of a cervical population of spermatozoa following artificial insemination, particularly after estrus synchronization (Hawk and Conley, 1975). An imbalance of progesterone and estrogen can affect the transport of spermatozoa by altering either the quantitative and qualitative characteristics of the cer-

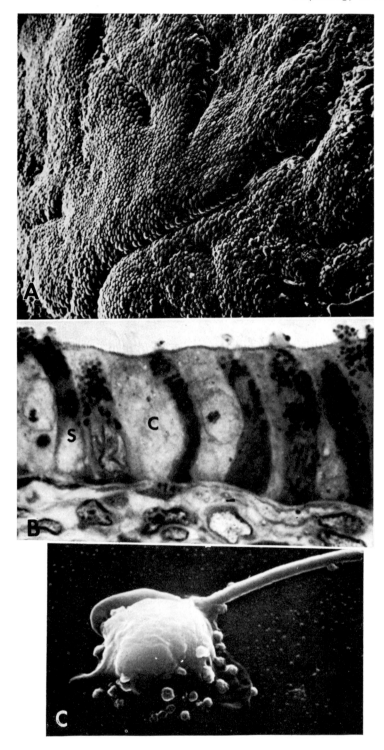

Fig. 11–6. Scanning electron micrograph of cervical epithelium showing cobblestonelike cells. *A*, Cervical epithelium showing secretory cells *(S)* with secretory granules, and ciliated cells *(C)*. *B*, Scanning electron micrograph of phagocytosis of the mammalian spermatozoa by a leukocyte. *(Hafez and Kanagawa, 1973. Fertil. Steril. 25, 776.)*

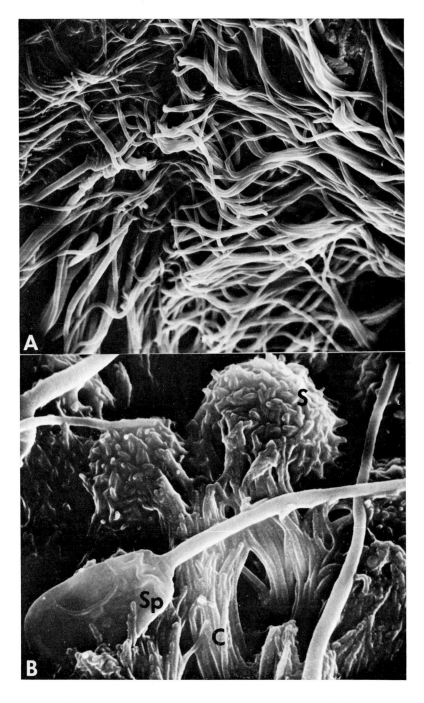

Fig. 11–7.   Scanning electron micrographs. *A*, Spermatozoa on the cervical epithelium one hour postcoitum. Note the arrangement of sperm tail in a parallel formation. *(Hafez, 1973. J. Reprod. Med.) B,* Cervical epithelium showing nonciliated secretory cells *(S)* with microvilli, ciliated cells *(C)* with kinocilia. Note the relative size of the spermatozoa *(Sp)* and kinocilia.

vical secretions or the motility of the reproductive tract.

Sperm Penetration in Cervical Mucus. Ejaculated spermatozoa rapidly penetrate watery cervical mucus during midcycle, aided principally by sperm motility as well as the micro- and macrorheologic properties of mucus. The rate of sperm penetration in the mucus varies throughout the estrous cycle.

Spermatozoa are mechanically oriented toward the internal os. As the flagellum beats and vibrates, the sperm head is propelled forward in the channels of least resistance. The tail frequency sets up a mechanical resonance between itself and the oscillation frequency of the molecular lattice. Hydrodynamic principles seem to apply to sperm motility; thus, motile spermatozoa are in dynamic equilibrium with the viscous force of the medium rather than being affected by the inertial force that influences large moving objects.

Dead pig spermatozoa inseminated in pigs were transported to the oviduct less efficiently than live spermatozoa; it appears that sperm mobility facilitates penetrability but is not absolutely necessary.

The aqueous spaces between the micelles allow the passage of sperm as well as the diffusion of soluble substances. Proteolytic enzymes (proteinases in seminal plasma) may hydrolyze the backbone protein or some of the crosslinkages of the mucin and thus reduce the network to a less resistant mesh with more open channels for the enhancement of sperm migration. It is also possible that the release of spermatozoa from the cervical crypts opposed by filaments of glycoprotein depends on hydrolysis of these chains by proteases in the sperm head.

Although spermatozoa appear to move at random in the cervical secretion, they probably follow the path of least resistance along strands of cervical mucus. When migration of a spermatozoon in mucus is impeded, the sperm usually resumes its forward course with a sudden deflection into an adjacent parallel path. When semen is mixed with cervical mucus *in vitro*, a sharp boundary occurs between the two fluids and the cervical mucus is penetrated by finger-like phalanges. Phalanx formation may function to increase the surface area between semen and cervical mucus, provide pockets of semen within the mucus to protect spermatozoa from the hostile vaginal environment or facilitate sperm migration into the uterine cavity. Sperm phalanges develop significant degrees of arborization, the terminal aspects of which consist of canals through which only one or two spermatozoa can pass.

Sperm Transport in the Uterus

The contractile activity of the vagina and myometrium plays a major role in the transport of spermatozoa into and through the uterus. Massive numbers of spermatozoa invade the endometrial glands. It is believed that the presence of spermatozoa in the uterus induces endometrial leukocytic response, which enhances phagocytosis of excessive numbers of living and probably dead spermatozoa.

Transport in the Oviduct

The oviduct has the unique function of conveying spermatozoa and eggs in opposite directions almost simultaneously. The pattern and rate of sperm transport through the oviduct are controlled by several mechanisms, such as peristalsis and antiperistalsis of oviductal musculature, complex contractions of the oviductal mucosal folds and the mesosalpinx, fluid currents and countercurrents created by ciliary action, and possibly the opening and closing of the intramural portion. The relative importance of these mechanisms in sperm transport through the oviduct is unknown. Oviductal contractions alter the configuration of the oviductal compartments momentarily, so that fluids and suspending spermatozoa may be transported toward the fimbriae from one compartment to the next. In the oviducts of the pigeon and the tortoise, there are two systems of kinocilia: one beats toward the ovary and the other toward the cloaca. These two ciliary systems are capable of moving particles in opposite directions.

The rate and pattern of sperm transport through the oviduct are attributed to peristalsis and antiperistalsis of musculature and contractions of the mucosal folds and mesosalpinx. The frequency and amplitude of contractions of oviductal circular and longitudinal musculature, mesosalpinx and mesotubarium are controlled by ovarian hormones, adrenergic and nonadrenergic activity, and such components of seminal plasma as prostaglandin.

The pattern and amplitude of contractions vary in different segments of the oviduct. In the isthmus, peristaltic and antiperistaltic contractions are segmental, vigorous, and almost continuous. In the ampulla strong peristaltic waves move in a segmental fashion toward the midportion of the oviduct. Whether sperm are transported to the ampulla by momentary relaxation of the ampullary-isthmic junction or by their own motility is not known.

Massive numbers of sperm disappear within the female reproductive tract, and only a few viable sperm are found in the ampulla at the time of fertilization.

Endocrine Control of Sperm Transport

More is known about the endocrine control of ovulation and spermatogenesis than about the endocrinology of sperm transport in the female tract. Ovarian hormones affect (a) the structure, ultrastructure and secretory activity of the cervical, uterine and oviductal epithelia; (b) the contractile activity of the uterotubal musculature; (c) the quantitative and qualitative characteristics of cervical mucus and uterine and oviductal secretions. Changes are noted in the protein content, enzyme activity, electrolyte composition, surface tension and conductivity of these fluids. Increasing the amount of endogenous estrogen during the preovulatory phase of the cycle or administering synthetic estrogens produces copious amounts of thin, watery, cervical secretions. Endogenous progesterone during the luteal phase of the cycle or in pregnancy produces scanty, viscous, cellular cervical mucus with low spinnbarkeit and ferning properties. The penetrability of spermatozoa is inhibited greatly in progestational cervical mucus. It is possible that the cyclic changes that occur in cervical mucus are mechanisms to protect the female from unnecessary exposure to the foreign proteins of semen.

Sperm transport in the female genital tract is also controlled by oxytocin and the sympathetic and parasympathetic nervous systems. Epinephrine, acetylcholine, histamine and various vasoconstrictors can alter uterine contraction, but the effects are transient.

Sperm Transport and Fertility

The continuous flow of spermatozoa from the cervix is associated with phagocytosis of spermatozoa within the uterus and sperm loss into the peritoneal cavity. Thus, a population of fertile spermatozoa is maintained at the site of fertilization near the ampullary-isthmic junction of the oviduct. The percentage of morphologically normal spermatozoa is higher in the oviducts and uterus than in the ejaculate. Some morphologically abnormal spermatozoa may reach the oviduct, although to a lesser extent than normal spermatozoa. The filtering of dead, abnormal and incompetent sperm during their passage through the reproductive tract ensures the greatest viability of the zygote. Ejaculates that contain a high concentration of abnormal spermatozoa might be associated with a high rate of abortion.

Head-to-head and tail-to-tail sperm agglutination may occur, causing inhibition in sperm transport (Fig. 11–8). The immunologic significance of sperm agglutination in relation to infertility is not known.

Effect of Estrus Synchronization. The survival and transport of spermatozoa in the female reproductive tract are generally decreased after alteration of the estrous cycle. When exogenous progestogen has been used to prolong the estrous cycle in the ewe, the number of spermatozoa present in the oviducts 24 hours after insemination during the regulated estrus is drastically reduced (Hawk and Cooper, 1977). When prostaglandin has been administered to ewes in the luteal phase to cause regression of corpora lutea and shorten the estrous cycle, the number of spermatozoa in the oviducts 24 hours after mating at the induced estrus is much lower than normal (Hawk, 1973). After the use of either prostaglandin or progesterone to regulate estrus, the number of spermatozoa present in the anterior cervix two hours after mating is only a small fraction of that found in the anterior cervix of untreated ewes.

Survival of Spermatozoa

Once ejaculation has occurred, spermatozoa have a finite life span. Certain components of the seminal plasma stimulate sperm motility whereas others inhibit motility. Much information is known about the duration of spermatozoal motility, but little is known about the duration of fertilizing capacity, which is lost long before motility. A relationship exists between the pH of the intravaginal seminal pool and the motility of the spermatozoa. Contamination with mucus at times alters the pH of the posterior fornix of the vagina and prolongs the survival time of ejaculate and sperm.

When migrating in the genital tract, sperm are separated rapidly from the seminal plasma and resuspended in the female genital fluid. In the oviduct, spermatozoa are greatly diluted. Since only a few spermatozoa appear in the oviduct, their survival time is difficult to estimate, and if they remain motile, they migrate into the peritoneal cavity. The oxygen uptake of rabbit spermatozoa in oviductal fluid was higher than that of spermatozoa in saline diluent plus glucose.

During transport to the site of fertilization, spermatozoa are significantly diluted with luminal secretions from the female reproductive tract and are susceptible to changes in the pH of luminal fluids. Acidity or excessive alkalinity of the mucus immobilizes spermatozoa, whereas moderately alkaline mucus enhances their motility.

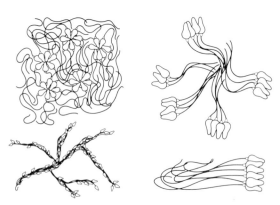

Fig. 11–8. Different patterns of head-to-head and tail-to-tail sperm agglutination, a phenomenon that interferes with sperm transport. Little is known about the immunologic significance of sperm agglutination in relation to infertility in farm animals.

Glycolytic and metabolic enzymes in the sperm tail and respiratory enzymes in the mitochondria are required for the biochemical reactions of the Embden-Meyerhof pathway, the tricarboxylic acid cycle, fatty acid oxidation, and the electron transport system.

The cervical mucus secreted at the time of ovulation provides an environment suited to the maintenance of metabolic activity of spermatozoa. This mucus undergoes biochemical changes, such as a decrease in albumin, alkaline phosphatase, peptidase, antitrypsin, esterase and sialic acid, as well as an increase in mucins and sodium chloride. The main components rendering mucus suitable as a "culture medium" for sperm have not yet been identified, although hexosamines and carbohydrates existing in either a free or a polysaccharide form may contribute to sperm longevity in the cervix.

Transport of spermatozoa into the uterus may influence capacitation because the sperm are separated from an excess of "decapacitation factor" and from other enzyme inhibitors in the seminal plasma.

Loss of Spermatozoa

Although millions of spermatozoa are deposited into the reproductive tract of the female, few ever reach the egg at the site of fertilization. Most spermatozoa perish at the selective barriers: uterine cervix, uterotubal junction and oviductal isthmus. Little is known about the fate of those that invade the endometrial glands in large numbers (Fig. 11–5). In the uterine cavity, spermatozoa undergo phagocytosis by leukocytes. A continual loss of sperm also occurs in the vaginal and peritoneal cavities.

The introduction of semen into the uterine cavity initiates the leukocytic response: the appearance of polymorphonuclear leukocytes. The biologic relationship between leukocytes and spermatozoa with respect to capacitation and/or sperm survival is not known. In the bovine cervix, the majority of leukocytes occurs in the central mass of the mucin, a fact indicating that most of them have invaded the cervix from the uterus. Most viable spermatozoa, lodging in the cervical crypts, escape the leukocytes, so that an adequate population survives.

Damaged spermatozoa are carried passively back through the ectocervix with the help of ciliated cells beating toward the vagina. Such spermatozoa, advancing only a short distance into the cervical mucus core, do not reach the cervical crypts and greatly decrease in number within a few hours after coitus. Since spermatozoa that become immotile elsewhere are not rapidly eliminated, the ratio of immotile spermatozoa that are being eliminated is higher in the cervix than in other segments of the female reproductive tract. Large amounts of cervical mucus are produced and numerous spermatozoa are expelled with the mucus through the vulva in cattle. Spermatozoa that reach the fimbriae are released into the peritoneal cavity.

RECEPTION OF EGGS (OVA PICKUP)

The viscid mass of cumulus oophorus that contains oocyte and corona cells adheres to the stigma and remains attached to it unless it is removed by the action of the kinocilia of the fimbriae. Ovum transport through the ostium itself and the first few millimeters of the ampulla is effected by the action of the cilia.

The physiologic mechanism by which freshly ovulated eggs are picked up into the oviducts depends on four main factors: (a) the structural characteristics of the fimbriae of the infundibulum and its relationship to the surface of the ovary at the time of ovulation, (b) the pattern of release of the cumulus oophorus and its contained egg from the follicle at the time of ovulation, (c) the biophysical properties of the follicular fluids and the fluids that comprise the matrix of the cumulus oophorus and (d) the coordinated contraction of the fimbriae and the utero-ovarian ligaments.

At the time of ovulation, the fimbriae are engorged with blood and are brought into close contact with the surface of the ovary by the muscular activity of the meso-tubarium. The ovary is moved slowly to and fro and around its longitudinal axis by contractions of the ligamentum ovarii proprium. This chain of reactions is controlled by anatomic and hormonal mechanisms.

The ovary is located inside the ovarian bursa to which the ampulla of the oviduct and part of the fimbriae are attached. The ovary can move readily from this location to the surface of the fimbriae, which is positioned at the open portion of the ovarian bursa. This movement is controlled by both the *ligamentum ovarii proprium* and the mesovarium which hold the ovary and oviduct in position.

The fimbriae and the infundibulum are basically composed of an erectile structure that is rich in vascular and muscular tissues. During estrus the fimbriae distend from the increased blood flow; furthermore, the margins of the fimbriae become edematous and translucent.

The contractile activities of the fimbriae, oviduct and ligaments are partly coordinated by hormonal mechanisms involving the estrogen-progesterone ratio. Egg reception is most efficient about the time of estrus, but it occurs to some degree throughout the cycle. In some species neurohormonal mechanisms also stimulate the contractile activity of the fimbriae at the time of copulation; however, these are not yet understood.

EGG TRANSPORT IN THE OVIDUCT

The transport time of ova in the oviduct varies with the species (Table 11–4). In cattle, sheep and swine, the transport time ranges from 72 to 90 hours (Fig. 11–9). Unfertilized ova are retained in the oviduct of the mare for several months

Table 11–4. Transport Time of Ova in the Oviduct of Farm Animals Compared with Some Other Mammals

Species	Time in Oviduct (Hours)
Cattle	90
Sheep	72
Horse	98
Pig	50
Cat	148
Dog	168
Monkey, rhesus	96
Opossum	24
Woman	48–72

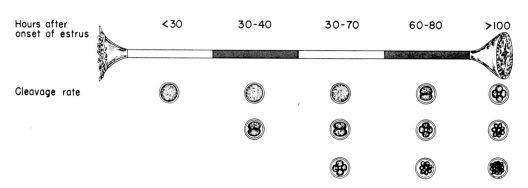

Fig. 11–9.   Rate of transport and cleavage of eggs in swine. Eggs pass through the first half of the oviduct rapidly and they remain in the third quarter which contains the ampullary-isthmic junction until 60–75 hours after onset of estrus. The eggs enter the uterus between 66 and 90 hours after onset of estrus. *(Data from Oxenreider and Day, 1965. J. Anim. Sci. 24, 413.)*

(see Chapter 19). It is critical that fertilized eggs reach the uterus at an appropriate progestational stage of the estrous cycle.

The rate of egg transport is faster from the infundibulum to the ampullary-isthmic junction than through the isthmic portion. This delay in ovum transport appears to be required for subsequent implantation of the embryo. The time of entry of the ovum into the uterus is relatively precise compared with the movement of the ovum past the ampullary-isthmic junction into the isthmus.

The transport of the egg through the oviduct is regulated by four primary forces: (1) the frequency, force and programming of contraction of oviductal musculature and related ligaments, as influenced by endocrine, pharmacologic and neural mechanisms, (2) the direction and rate of currents and countercurrents of luminal fluids as affected by the rate and direction of the beat of kinocilia lining the mucosal folds, (3) the secretory activity of nonciliated cells in the oviductal epithelium as influenced by the estrogen-progesterone ratio, and (4) the hydrodynamics and rheologic properties of luminal fluids at the critical times that ova are being transported (Figs. 11–10, 11–11). It is possible that propulsive forces such as muscular and ciliary activity are generally adequate for ovum transport but more precise transport depends on the size of the isthmic lumen.

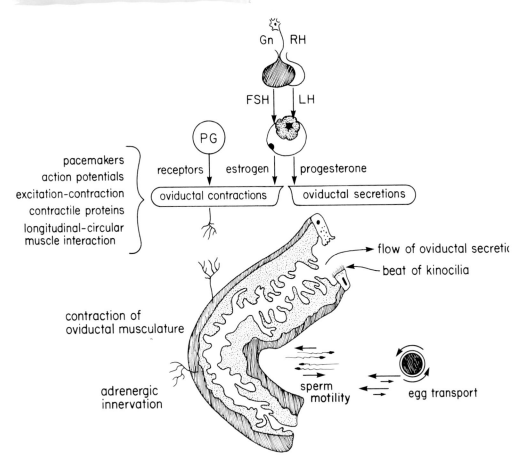

Fig. 11–10. Anatomic and physiologic mechanisms that regulate egg and sperm transport in the oviduct.

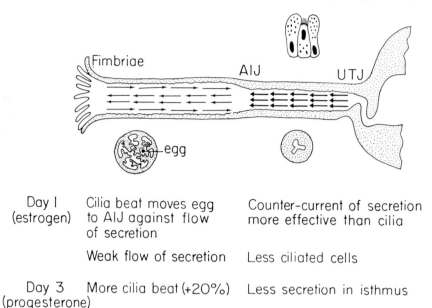

Day 1 | Cilia beat moves egg | Counter-current of secretion
(estrogen) | to AIJ against flow | more effective than cilia
| of secretion |

| Weak flow of secretion | Less ciliated cells

Day 3 | More cilia beat (+20%) | Less secretion in isthmus
(progesterone) |

Fig. 11–11. Diagram to show the rate of estrogen and progesterone on the flow of oviductal secretions in the ampulla and isthmus (shown by arrows toward the fimbriae), the flow of oviductal cilia (shown by arrows toward the uterotubal junction [UTJ]), and egg transport in the oviduct. AIJ = ampullary-isthmic junction.

Oviductal Contraction

The various patterns of oviductal contraction regulate to some extent the rate of egg transport (Table 11–5). However, there is little evidence that ova are transported primarily by muscle contractions. The mesovarium, mesosalpinx, mesotubarium, and utero-ovarian ligaments also undergo rigorous contractile activity. These contractions continuously change the position of the ovaries relative to the fimbriae as well as the relation of the various subdivisions of the oviduct to one another. The patterns of contraction vary significantly at various times in the estrous cycle.

Neural Mechanisms and Pharmacologic Response

The abundant adrenergic innervation of the oviductal musculature in conjunction with its response to adrenergic drugs has suggested that the sympathetic nervous

Table 11–5. Patterns of Egg Transport in the Oviduct as Affected by Type of Oviductal Contractility

Type of Oviductal Contractility	Pattern of Egg Transport
Peristaltic contractions from ampulla to ampullary-isthmic junction	Fast progress of egg
Peristaltic contraction from uterotubal junction to ampullary-isthmic junction	Obstruction of egg transport
Segmental contraction	Forward and backward shuttling of egg
Outbursts of spastic contraction of circular musculature	Complete obstruction of egg transport at a sphincter
Contraction of related ligaments causing bending of the oviduct	Regulation of rate of egg transport

(Data from Aref and Hafez, 1971; Takeda and Doteuchi, 1976; and Hafez, unpublished observations)

system is involved in egg transport. There is more norepinephrine in the isthmus than in the ampulla (Owman et al, 1967). Myosalpingeal contraction mediated by adrenergic innervation to the circular muscle of the oviductal isthmus is believed to be involved in delaying ovum transport at the isthmus (Chatkoff and Pauerstein, 1975). The mechanisms underlying this phenomenon, including modulation of adrenergic factors by ovarian steroids, are unknown. In the rabbit, such outbursts occur 36 hours post-coitum, when the majority of ova are in the oviduct. A drop in oviductal motility occurs at 72 hours post coitum, when most ova are in the uterus (Aref and Hafez, 1973). This decrease in oviductal motility is presumably due to a postovulatory drop in norepinephrine levels and relaxation of oviductal musculature (Bodkhe and Harper, 1972).

The α-adrenoceptors (stimulatory) mediate oviductal contraction, whereas the β-adrenoceptors (inhibitory) mediate oviductal relaxation. Nerve stimulation and various autonomic drugs cause variable responses of the oviduct (Table 11–5). There is no pharmacologic evidence of cholinergic innervation; however, acetylcholine produces a contractile response of the oviduct.

The adrenergic nerves of the oviduct can be destroyed surgically or chemically. The administration of 6-hydroxy-dopamine (2,4,5,-tri-hydroxyphenylethylamine) causes the degeneration of adrenergic nerve endings within two to three days after administration. This chemical denervation is associated with depletion of local norepinephrine and reduction in tyrosine hydroxylase. Depleting the adrenergic nerves of norepinephrine with reserpine causes functional adrenergic denervation. The extent of the depletion induced by reserpine depends on the species, organ and time. One intravenous injection of reserpine causes catecholamines to disappear more slowly

and recover more quickly from organs innervated by short adrenergic neurons such as the oviduct, than from organs innervated exclusively by long adrenergic neurons as in the ovary.

When the muscular activity of the oviduct is pharmacologically blocked with isoproterenol, ampullary egg transport changes its typical discrete "to and fro" character and becomes continuous and unidirectional. This implies that the driving forces provided by the ciliary action are indeed asymmetric.

Prostaglandins have a remarkable effect on the smooth musculature of the oviducts and the uterus (Horton, 1972). The formation and release of prostaglandins is brought about by nerve activity. PGE_1 and PGE_2, released by nerve stimulation, inhibit the responses of the effector tissue, whereas PGE_1 blocks the response to hypogastric nerve stimulation.

"Locking" and "Unlocking" of Ova in Oviduct

Coutinho (1976) summarized the various physiologic and pharmacologic mechanisms that might influence oviductal locking and unlocking of ova:

1. mechanical blocking (eg, edema)
2. myogenic blocking from
   a. sustained contraction of circular musculature of the isthmus
   b. contractions regulated by myogenic pacemakers which are somehow coupled in time and space so that they do not propel ova predominantly to the uterus
   c. local or general relaxation of muscle so that no motive force exists to propel the ova toward the uterus
3. neurogenic blocking controls one of the foregoing myogenic mechanisms which affect:
   a. sustained release of epinephrine onto circular muscle cells which respond by contraction
   b. increased intracellular Ca^{++}

4. hormonal blocking, which acts on a myogenic or neurogenic mechanism:
   a. the amount and function of intracellular organelles that control Ca^{++} binding and release
   b. the predominance of alpha or beta adrenergic receptors in oviductal muscle.

Possible Role of Ampullary-Isthmic Junction. Several mechanisms have been suggested to account for the closure of the isthmus or ampullary-isthmic junction that is responsible for the retention of ova within the oviduct before they are rapidly transported into the uterus. However, no structural basis for an isthmic block has been demonstrated by histologic techniques.

In rabbits the ampullary-isthmic junction is biochemically characterized by high acid phosphatase activity as compared to the isthmus or ampulla, but during passage of the ova a "surge" of activity of this enzyme is noted uniformly throughout the oviduct (Gupta et al, 1970). Acid phosphatase may play a role in ova denudation, and removal of denuded cumulus and corona cell debris.

Ovarian hormones control the activity of an electric pacemaker that is localized in a discrete area around the ampullary-isthmic junction (Talo and Brundin, 1971). In goats a pronounced activity and groups of positive pulses are recorded during estrus. In diestrus, the activity is abolished and groups of negative pulses are dominant. The electric pulses and the changes in their directions are probably involved in the underlying control of ovum transport in the isthmus.

Possible Role of Uterotubal Junction. The uterotubal junction may exert a regulatory influence on ova entry into the uterus. In rabbits this junction opens only on days 3, 4 and 30 of pregnancy, when the ova are transported into the uterus, and at the time of parturition. In sheep, estrogen-induced edema and flexure of the uterotubal junction causes a valve-like action that prevents the movement of fluid and presumably ova from the oviduct into the uterus (Edgar and Adsell, 1960).

GAMETE TRANSPORT AND CONCEPTION RATE

The rapid transport of live or dead spermatozoa to the upper oviduct within a matter of minutes infers the importance of smooth muscle contractions in this situation. In natural breeding, spermatozoa are deposited during estrus at least 10 to 12 hours before ovulation, and the rate of sperm transport is unlikely to become a critical factor determining conception. In the case of artificial insemination, particularly when this forms part of an estrous synchronization and/or gonadotropin treatment program, the rate and efficiency of sperm transport to the site of fertilization are of fundamental importance.

In cattle, the semen is ejaculated in the anterior vagina near the external cervical os. Penetration of spermatozoa into the cervical canal may depend upon muscular activity in the female reproductive tract which is enhanced by oxytocin released at coitus, but entry into the canal also depends on sperm motility. The condition of the cervical mucus is critical for successful colonization of the cervix, the latter region providing the principal post-coital reservoir of spermatozoa in ruminants, at least during the first 24 hours (Thibault et al, 1973). Spermatozoa are stored for two to three days in the endometrial glands and in the folds of the lower isthmus and region of the uterotubal junction. Thus spermatozoa may be protected from phagocytosis.

Significant numbers of spermatozoa enter the oviducts of estrous cows within two hours of mating, although they are found in the ampulla by eight hours (Thibault et al, 1973).

The distribution of spermatozoa in the

isthmus of the bovine fallopian tube is regular at eight hours after mating, but muscular activity has clustered the sperm by 18 hours (Thibault et al, 1973). Sperm counts on sectioned material reveal that only a small fraction of the ejaculate reaches the upper tube (Thibault et al, 1973). This phenomenon is important to avoid the pathologic condition of polyspermic fertilization. Microscopic examination of the eggs indicates that few spermatozoa are present in the ampulla at the time of fertilization, and counts of spermatozoa at a series of levels in the isthmus reveal that sperm numbers decrease abruptly a few centimeters below the ampulla (Thibault et al, 1973).

Thus the probability of postovulatory aging of the egg before sperm penetration is extremely low under conditions of natural mating. However, if artificial insemination is employed or under systems of controlled breeding, the eggs may deteriorate before spermatozoa reach the ampulla, even though the fertilizable life of the egg is 20 to 24 hours.

An accelerated descent of eggs through the oviduct occurs after superovulation treatment in cattle, although experiments with PMSG have not produced this particular response. The disturbance of egg transport is well known in laboratory animals under steroid hormone treatment, so excessive progesterone production or progestogenic treatments shortly after ovulation may produce a similar result in farm animals, leading to temporary infertility.

FERTILIZABLE LIFE AND AGING OF EGGS

The fertilizable life of the egg is the maximal period during which it remains capable of fertilization and normal development. In most species the egg is capable of being fertilized for some 12 to 24 hours (Table 11–1). It rapidly loses its fertilizability upon reaching the isthmus and is completely nonfertilizable after reaching the uterus.

The egg may be fertilized near the end of its fertilizable life as a result of delayed breeding. Such eggs may or may not implant and, if so, they produce mostly nonviable embryos. Guinea pigs show a high percentage of abnormal pregnancies and decrease in litter size as the age of the egg increases prior to fertilization (Fig. 11–12). Fertilization of aged eggs in swine is associated with polyspermy and hence abnormal embryonic development. In single-bearing animals, aging of the egg may cause abortion, embryonic resorption or abnormal development of the embryo. Similar abnormalities may result from aged sperm.

In general, fertilization of aged gametes involves one of the following possibilities: aged egg plus freshly ejaculated sperm; aged egg plus aged sperm; freshly ovulated egg plus aged sperm.

Nonviable embryos resulting from any of the foregoing combinations may cause low conception rates in certain herds and flocks. There is good evidence that fertilization with aged sperm increases sub-

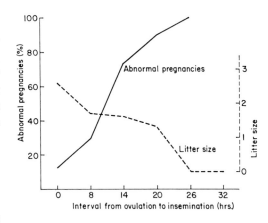

Fig. 11–12. **The effect of aging of ova (delayed insemination) on percentage of abnormal pregnancies and litter size in guinea pigs. Note that the ova were fertilized and implanted when the animals were inseminated at 26 hours after ovulation, yet the embryos did not continue development. *(Data from Blandau and Young, 1939. Am. J. Anat. 64, 303.)***

sequent embryonic mortality in swine and poultry. At present there is insufficient evidence in farm mammals concerning the relative deleterious results of gamete aging on fertilization, implantation and prenatal development. It is possible that some of the congenital abnormalities in postnatal life are a consequence of aged gametes.

If the egg is not fertilized, it fragments into several cytoplasmic segments of unequal size, and in some cases it may even resemble a fertilized egg. All unfertilized eggs eventually disappear through complete disintegration or phagocytosis in the uterus.

The egg may never reach the infundibulum due to many causes. For example, eggs entrapped in the ruptured follicles can be found in the developing corpora lutea. The egg may also be lost into the peritoneal cavity; such an egg usually degenerates but in rare cases may result in ectopic pregnancy (pregnancy located outside the uterus). Egg loss in the peritoneal cavity may be caused by the immobilization of the oviduct as a result of faulty rectal palpation of the ovaries, postpartum or postabortum infections, endometritis or nonspecific abdominal infections.

TRANSUTERINE MIGRATION AND LOSS OF EGGS

Transuterine migration of the egg through the common body of the uterus is common in ungulates. For example, when one of the ovaries is removed from a sow, approximately half of the embryos develop in each uterine horn, irrespective of which ovary was removed. There is also a tendency in the normal sow for the number of embryos to be equalized between the two horns. Transuterine migration is more common in swine and horses than in cattle and sheep. Nonetheless, cattle and sheep that have double ovulations from one ovary usually have one embryo in each uterine horn. The physiologic mechanisms that govern the movement of eggs, both within individual horns and between horns, is unknown.

Transperitoneal migration of eggs can be accomplished by suitable experimental conditions; for example, removal of one ovary leaving the fimbriae and oviduct intact and ligation of the other oviduct. In this case, the remaining oviduct has the ability to pick up the ovum released by the contralateral ovary and a normal pregnancy may follow. Transperitoneal migration may be avoided by currents and surface tension of the peritoneal fluid.

EMBRYONIC DEVELOPMENT IN OVIDUCT

The oviduct seems to take an active part in maintaining eggs and preparing them for fertilization and segmentation. The oviductal fluid is rich in substrates and cofactors involved in ovum development, such as pyruvate and bicarbonate, free amino acids, oxygen, CO_2, carbohydrates, perhaps lipids, nucleosides, steroids and other compounds. Apparently these substances are contributed by the cells of the oviductal mucosa to the luminal fluid milieu.

Oviductal epithelium is most active near the developing embryo. In the rabbit, the epithelial uptake of radioactive sulfate increases in the region containing the egg. Endocrine factors are important in the early development of embryos in the oviduct. The biochemical composition of the uterine fluid is very different from that of the oviductal fluid. This seems to indicate that eggs at early cleavage stages require specific substances provided by the oviduct for their development. Therefore, premature entry of morulae into the uterus will cause their degeneration. After a certain time, the blastocysts need to enter the uterus for final development and implantation.

REFERENCES

Aref, I. and Hafez, E.S.E. (1973). Oviductal contractility in relation to egg transport in the rabbit. Obstet. Gynecol. *42*, 165.

Blandau, R.J. and Verdugo, P. (1976). An overview of gamete transport—comparative aspects. *In* Ovum Transport and Fertility Regulation. M.J.K. Harper, C.J. Pauerstein, C.E. Adams, E.M. Coutinho, H.B. Croxatto and D.M. Paton (eds), Copenhagen, Scriptor, p. 138.

Bodkhe, R.R. and Harper, M.J.K. (1972). Mechanisms of egg transport: Changes in the amount of adrenergic transmitter in the genital tract of normal and hormone treated rabbits. *In* The Regulation of Mammalian Reproduction. S.J. Segal, et al (eds), Chap. 26, Springfield, IL, Charles C Thomas, pp. 364–374.

Chatkoff, M.L. and Pauerstein, C.J. (1975). Biochemistry of adrenergic receptors in ovum transport. Gynecol. Invest. *6*, 43.

Coutinho, E.M. (1976). Summary of discussion on oviductal contractility. *In* Ovum Transport and Fertility Regulation. M.J.K. Harper, C.J. Pauerstein, C.E. Adams, E.M. Coutinho, H.B. Croxatto, and D.M. Paton (eds), Copenhagen, Scriptor, p. 237.

Edgar, D.G. and Adsell, S.A. (1960). Spermatozoa in the female genital tract. J. Endocrinol. *21*, 321.

Gupta, D.N., Karkun, J.N. and Kar, A.B. (1970). Biochemical changes in different parts of the rabbit fallopian tube during passage of ova. Am. J. Obstet. Gynecol. *106*, 833.

Hawk, H.W. (1973). Uterine motility and sperm transport in the estrous ewe after prostaglandin induced regression of corpora lutea. J. Anim. Sci. *37*, 1380.

Hawk, H.W. and Conley, H.H. (1975). Involvement of the cervix in sperm transport failures in the reproductive tract of the ewe. Biol. Reprod. *13*, 322.

Hawk, H.W. and Cooper, B.S. (1977). Sperm transport into the cervix of the ewe after regulation of estrus with prostaglandin or progestogen. J. Anim. Sci. *44*, 638.

Horton, E.W. (1972). Female reproductive tract smooth muscle. *In* Prostaglandins. E.W. Horton (ed), Monographs on Endocrinology, 7, New York, Springer Verlag, pp. 87–104.

Owman, C., Rosengren, E. and Sjoberg, N.O. (1967). Adrenergic innervation of the human female reproductive organs: A histochemical and chemical investigation. Obstet. Gynecol. *30*, 763.

Talo, A. and Brundin, J. (1971). Muscular activity in the rabbit oviduct; a combination of electric and mechanic recordings. Biol. Reprod. *5*, 67.

Thibault, C. (1973). Sperm transport and storage in vertebrates. J. Reprod. Fertil. Suppl. *18*, 39.

Thibault, C., Gerard, M. and Heyman, Y. (1973). Transport and survival of spermatozoa in cattle. *In* Transport, Survival and Fertilizing Ability of Spermatozoa. E.S.E. Hafez and C. Thibault (eds), Basel, S. Karger.

12

Fertilization, Cleavage and Implantation

ANNE McLAREN

FERTILIZATION

The entire process of sexual reproduction is centered around the act of fertilization, yet fertilization is not itself a reproductive process. On the contrary, it consists essentially of the fusion of two cells, the male and female gametes, to form one single cell, the zygote. Fertilization is a dual process:

(a) In its *embryologic* aspect, it involves activation of the ovum by the sperm. Without the stimulus of fertilization, the ovum does not normally begin to cleave, and no embryologic development occurs. In some animals experimental treatments are known that mimic this aspect of fertilization, inducing development of the ovum without fertilization.

(b) In its *genetic* aspect, fertilization involves the introduction of hereditary material from the sire into the ovum. By this means it is possible for beneficial characters arising far apart in time and space eventually to become combined in a single individual. The importance of this process for natural and artificial selection can hardly be overestimated. According to current genetic belief, the essential hereditary material is the chromosomal DNA in the sperm nucleus: fusion of male and female nuclei in the process of syngamy is therefore often thought of as the central process of fertilization. Although attempts have been made to inject foreign DNA into the ovum experimentally, this aspect of fertilization has not yet been mimicked in the laboratory.

In fertilization, two cells combine to form one, the first cell of the new individual, yet the number of chromosomes remains constant in every generation. This is because the two gametes each contain only half the number of chromosomes characteristic of the species.

In this chapter, use is made of information derived from the study of mice, rats and rabbits, since more direct work has so far been done on fertilization and early development in these laboratory species than in the larger and more expensive farm animals.

Description of Fertilization Process

The Ovum: Its Position and State. In most mammals, fertilization begins after the first polar body has been extruded, so that the sperm penetrates the ovum while the second reduction division is in progress (see Chapter 4 for a description of

226

meiosis). In the horse and dog, however, the sperm may enter the ovum before the second reduction division has begun.

The site of fertilization in all farm and most other mammals is the ampullary-isthmic junction. Transport of ova to this region of the oviduct is facilitated by peristaltic contraction, and takes no more than 30 to 45 minutes in swine. When it enters the ampulla, the ovum in its mucoprotein coat (the *zona pellucida*) is still surrounded by a cluster of granulosa cells that were shed with it from the ovarian follicle, and which at this stage are often called *cumulus cells*. In swine, the cumulus cells surrounding the several ova join together in the fimbriae of the oviduct to form a single cluster, the "egg plug." This normally disintegrates soon after ovulation, but may persist until after fertilization when ovulation is induced with hormones. In farm animals other than swine, cumulus cells are absent from those ova that have been examined a few hours after ovulation (eg, within 9 to 14 hours of ovulation in the cow).

In most species, ova that are not fertilized degenerate within a few days, but in the horse they remain in the fallopian tube for several months.

The Sperm: The Encounter With the Ovum. The entry of sperm into the female genital tract and their transport to the site of fertilization have been dealt with previously (see Chapter 11). Here I wish to emphasize three points only: (1) Although the total number of sperm in an ejaculate is measured in hundreds or thousands of millions, the number travelling as far as the ampulla is relatively small, probably not much more than 1000 in any mammal. A major role in regulating sperm numbers is played by the isthmus of the oviduct in the pig, and by the uterotubal junction and cervix in the cow and sheep. (2) Some sperm reach the site of fertilization quickly, within 15 to 30 minutes of mating. (3) In the cow, sheep, pig, ferret, cat, rabbit and rhesus monkey, as well as in laboratory rodents, the sperm have to undergo some change or set of changes (called capacitation) before they can undergo the acrosome reaction necessary for penetration through the zona pellucida. For rapid capacitation of boar sperm, a brief exposure to the uterine environment is required, followed by a more prolonged sojourn in the oviduct.

The fertile life of ova is short (Table 12–1), usually less than 24 hours. In all

Table 12–1. Estimates of the Fertile Life of Sperm and Ova, and the Tempo of Embryonic Development

Species	Fertile Life* (hours of)		Days After Ovulation†				
	Sperm	Ovum	2-Cell	8-Cell	Into Uterus	Blastocyst	Birth
Cattle	30–48	20–24	1	3	3–3½	7–8	278–290
Horse	72–120	6–8	1	3	4–5	6	335–345
Man	28–48	6–24	1½	2½	2–3	4	252–274
Rabbit	30–36	6–8	1	2½	3	4	30–32
Sheep	30–48	16–24	1	2½	3	6–7	145–155
Swine	24–72	8–10	16–20hr	2½	2	5–6	112–115

* "Fertile life" is a relative concept, since fertility declines progressively over a period of hours. For sperm, only the period in the female genital tract is included. The life of ova is timed from ovulation. For both, longevity probably depends on a variety of factors, including the hormonal state of the female. It is therefore impossible to give precise figures.

† These estimates are only approximate, since developmental rate is subject to considerable variation both among individuals and among breeds. In addition, accurate information on the time of ovulation is lacking in several species.

mammals except bats, the fertile life of sperm is also fairly short, though successful fertilization five days after insemination has been reported in mares, and survival may be longer in certain specialized anatomic sites, such as the uterotubal junction of swine, than in the lumen of the uterus or oviduct. Possibly sperm may lose their ability to induce viable embryos before they lose their ability to fertilize. The relatively brief fertile life of both sperm and ova renders timing a matter of the utmost importance in mating and artificial insemination. For instance in the cow, which normally ovulates about 14 hours after the end of heat, the conception rate from inseminations made at the time of ovulation is low, and the best time for insemination is between 6 and 24 hours before ovulation (Fig. 12–1).

Fertilization rate in most mammals is remarkably high. In one strain of rabbits, only 1.4% of ova failed to be fertilized, though the overall prenatal loss amounted to 30%. Data on fertility in farm animals relate mainly to overall pregnancy rates, and little is known about fertilization

losses, but fertilization rates of 94% to 98% have been reported for swine after artificial insemination.

It has long been assumed that fertilization is an entirely random process, ie, that there is an equal chance of any sperm fertilizing any ovum. However, experiments with mixed inseminations demonstrate that the sperm from different males may differ in their fertilizing capacity. In cattle, heterospermic insemination offers an efficient method of comparing the fertility of different bulls, and has been used by Beatty and his colleagues (1976) to show that freezing depresses the fertility of the semen of different bulls unequally. Bateman has reported a case of true selective fertilization in the mouse, in which sperm of a particular type, presented with a choice of ova, united more frequently with one type than with another.

Entry of Sperm into Ovum. To enter the ovum, the sperm has first to penetrate (a) the cumulus mass, if this is still present; (b) the zona pellucida; and (c) the vitelline membrane.

Disintegration of the cumulus mass

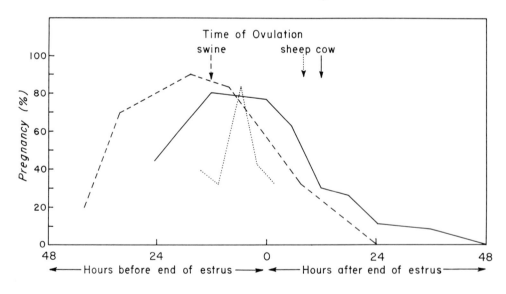

Fig. 12–1. Relationship between time of insemination and fertility in the cow, sheep and swine. _____ Cow; ------ Swine; • • • • • • Sheep. *(After Trimberger and Davis, 1943, Neb. Agric. Expt. Sta. Res. Bull. No 129, for cow. Dziuk, 1970, J. Reprod. Fertil. 22, 277, for sheep. Willemse and Boender, 1967, Tijdschr. Diergeneesk, 92, 18, for swine.)*

usually precedes fertilization in farm animals. In swine, for example, sperm are rarely found in the cumulus and attachment of sperm to the zona pellucida occurs after the ovum is nearly completely denuded. The enzyme *hyaluronidase*, released from the sperm, helps to dissolve the cumulus mass and the sperm then makes its way toward the ovum by virtue of its own motility.

The next obstacle to sperm entry is the zona pellucida (Fig. 12–2, A). At this stage the acrosome, loosened during capacitation, is finally lost, exposing the *perforatorium*. Probably the action of a proteolytic enzyme associated with the perforatorium facilitates passage through the zona. An extract of acrosomes isolated from ram, bull or rabbit sperm has been

reported effective in dissolving the zona, as well as dispersing the corona radiata of rabbit ova. Sperm penetrate the zona pellucida obliquely (Fig. 12–2, B).

The last stage in the penetration of the ovum involves the attachment of the sperm head to the surface of the vitellus (Fig. 12–2, B). This period, which lasts for up to 30 minutes, is a vital one because it is at this time that *activation* occurs. Stimulated by the proximity of the sperm, the ovum awakes from its dormancy and development begins. The sperm head, and in some species the tail as well, enters the vitellus. A projection on the surface of the vitellus marks for some hours the point of sperm entry. In the mouse, most of the surface of the vitellus is covered with microvilli, but in the region where the

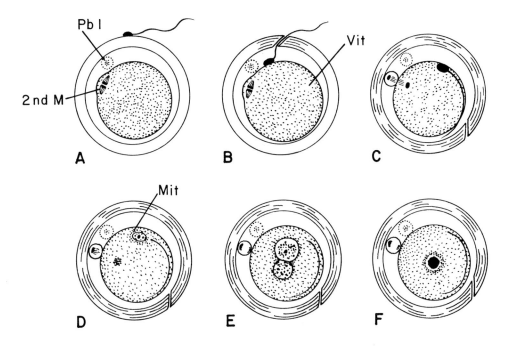

Fig. 12–2.   Diagram illustrating the processes occurring during fertilization in swine. *A,* The sperm in contact with the zona pellucida *(Z.p.).* The first polar body *(Pb. 1)* has been extruded; the nucleus of the ovum is undergoing the second meiotic division *(2nd M). B,* The sperm has penetrated the zona pellucida, and is now attached to the vitellus *(Vit).* This evokes the zona reaction, which is indicated by shading as it passes round the zona pellucida. *C,* The sperm has now been taken almost entirely within the vitellus. The head is swollen (see Fig. 12–3, *A).* The vitellus has decreased in volume, and the second polar body has been extruded. The zona has rotated relative to the vitellus. *D,* Male and female pronuclei develop. Mitochondria *(Mit.)* gather around the pronuclei. *E,* The pronuclei are fully developed and contain numerous nucleoli. The male pronucleus is larger than the female. *F,* Fertilization is complete. The pronuclei have disappeared and been replaced by chromosome groups, which have united in the prophase of the first cleavage division.

first polar body was extruded the surface is smooth. Sperm entry rarely takes place in the smooth region, which amounts to about one third of the total egg surface.

A combination of transmission and scanning electron microscopy has been used to study the actual process of penetration of the sperm into the vitellus. In the hamster, studies by Yanagimachi and Noda (Fig. 12–3) suggest that the microvilli on the surface of the ovum actively participate in sperm-ovum association. The microvilli grasp the sperm head; the plasma membranes of sperm and ovum then rupture and fuse with one another to form a continuous cell membrane over the ovum and outer surface of the sperm. As a result, the sperm comes to lie inside the vitellus, leaving its own plasma membrane incorporated into the vitelline membrane.

In the pig, the initial fusion between the sperm and the vitelline surface is accompanied by a constriction in the sperm head. The inner acrosomal membrane remains in the perivitelline space, and the nuclear material of the sperm thus makes direct contact with the cytoplasm of the ovum, and starts immediately to swell.

Experiments suggest that the major block to the penetration of ova by foreign sperm resides in the zona pellucida. Capacitated sperm of one species are unable to get through the zona of another species but may penetrate the vitellus if the zona has first been removed.

Pronucleus Formation. One striking result of activation in some species (eg, pig, cow) is that the vitellus shrinks in volume, expelling fluid into the perivitelline space. At the same time the sperm head in the vitellus swells and acquires the consistency of a gel, losing its characteristic shape (Fig. 12–2, D; Fig. 12–4, A,B). The perforatorium and the tail drop off. Within the sperm nucleus, a number of nucleoli appear and subsequently coalesce, and a nuclear membrane develops around its periphery. The final structure, which resembles the nucleus of a somatic cell much more closely than it does a sperm nucleus, is termed the male pronucleus (Fig. 12–2, E; Fig. 12–4, C).

Little is known of the subsequent fate of sperm constituents other than the nucleus. In some species the numerous mitochondria of the sperm midpiece all go to one of the two daughter cells at the first cleavage division; in others they are released into the cytoplasm of the ovum

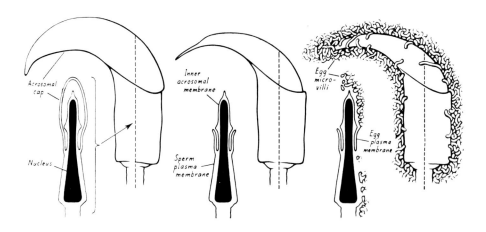

Fig. 12–3. Changes in hamster sperm before and during fertilization. *A,* intact epididymal sperm; *B,* sperm after capacitation; *C,* initial stage of sperm entry into the vitellus. *(After Yanagimachi and Noda, 1972. Scanning electron microscopy of golden hamster spermatozoa before and during fertilization. Experientia 28, 69)*

and are distributed equally. In the pig, the sperm midpiece and tail can often be detected, until the stage of late syngamy, in close association with the male pronucleus.

In most species the second polar body is extruded from the ovum soon after sperm entry, and formation of the *female pronucleus* then begins (Fig. 12–4, C). This resembles the male pronucleus in the appearance of nucleoli and the formation of a nuclear membrane. The two pronuclei develop synchronously, increasing in volume during the course of several hours to an extent that was estimated at 20-fold for the rat. In many species, including swine and cattle, the male chromatin begins its decondensation and swelling before the female chromatin, and hence the male pronucleus is larger than the female (Fig. 12–2, E; Fig. 12–4, C). In the pig, the formation of pronuclei occurs 6 to 18 hours after ovulation, and the female pronucleus is characterized by an asymmetric

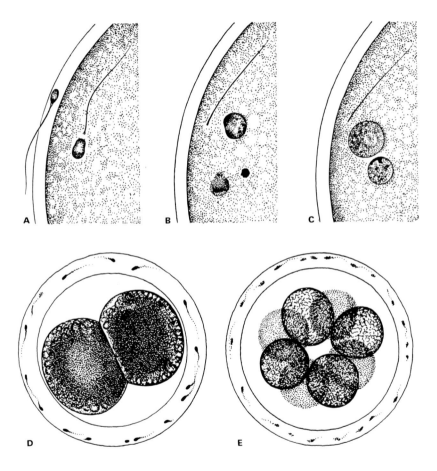

Fig. 12–4. Sections of pig ova during fertilization. *A,* Four hours postcoitum, showing a nonfertilizing sperm in the zona pellucida, and the penetrating sperm within the vitellus (note its swollen head, and the detachment of the head from the flagellum). *B,* Five hours postcoitum, showing an early male pronucleus and a portion of the sperm tail, and (below) the nuclear apparatus of the ovum, with the recently formed nucleus of the 2nd polar body to the left, and the female pronucleus to the right. *C,* Six hours postcoitum, showing the male and female pronuclei, and a portion of the sperm tail. Condensed chromatin can be seen on the membrane of the smaller, female pronucleus. *D,* Two-celled pig ovum from the fallopian tube about 18 hours after ovulation, with sperm embedded in the zona pellucida. *E,* Eight-celled pig ovum from the uterine horn about 55 hours after ovulation. The blastomeres are all similar in size and appearance.

distribution of its chromatin (Fig. 12–4, C).

Syngamy. At some stage during their maximal development, the male and female pronuclei come into contact. After a time they begin to shrink, and at the same time to coalesce. The nucleoli and nuclear membrane disappear, and the pronuclei can no longer be seen. In those mammals in which it has been measured, the total lifespan of the pronuclei extends over a period of 10 to 15 hours. As the time of the first cleavage draws near, two chromosome groups become visible. These are the maternal and paternal chromosomes respectively. They unite to form a single group, which represents the prophase of the first cleavage mitosis (Fig. 12–2, F). Fertilization is now complete.

As in any other mitotic division, each chromosome splits longitudinally, and the halves separate to opposite ends of the cleavage spindle. The fertilized ovum undergoes its first cleavage to produce a two-celled embryo. Each daughter cell now contains the normal diploid number of chromosomes characteristic of the species, half having been derived from the ovum and half from the sperm.

The duration of fertilization, ie, the total time interval from the penetration of sperm to the metaphase of the first cleavage division, has been estimated as 12 hours in the rabbit, 12 to 14 hours in the pig, 16 to 21 hours in the sheep, 20 to 24 hours in the cow and approximately 36 hours in man.

Zona Reaction and Vitelline Block. If ova are classified according to the number of sperm they contain, those penetrated by one sperm only are found to be more common than would be expected by chance. Often ova are observed with several sperm clustering around the outside of the zona pellucida, but only a single one within. It is therefore inferred that the zona pellucida can undergo some change after the passage of the first sperm that renders it less easy for sperm to penetrate subsequently. This change is termed the *zona reaction*. The reaction consists of a propagated change in the zona, set off when the first sperm makes contact with the surface of the vitellus, and mediated by some substance liberated by the *cortical granules*. These particles, 0.1 to 0.5μ in diameter, come to lie in the cortical region of the ovum, and release their contents as soon as the first sperm penetrates the ovum.

Extra sperm that succeed in passing through the zona pellucida into the perivitelline space are called *supplementary sperm*. In some species (sheep, dog, hamster) the zona reaction is relatively quick and effective, and supplementary sperm are found rarely if at all. In other species (mouse, rat) they are more common. In the pig, the zona reaction is restricted to the inner region of the zona pellucida, so that extra sperm enter the zona (Fig. 12–4, D,E) but do not normally succeed in passing right through it (eg, Hunter, 1977). The rabbit shows no zona reaction, and up to 200 supplementary sperm have been observed in the perivitelline space of the fertilized ovum.

The other defense mechanism against entry of more than one sperm is shown by the vitellus itself, and is termed the *vitelline block* or the *block to polyspermy*. The fertilizing sperm is actively engulfed by the vitellus, but subsequently the vitelline surface becomes unresponsive to contact and no further sperm are engulfed. Sperm that have been damaged by x-irradiation may make contact with the vitelline surface without activating the ovum. In this case the contact also fails to induce the vitelline block. In the golden hamster ovum, the vitelline block is completed two to three hours after sperm penetration.

Extra sperm that succeed in entering the vitellus, in spite of both the zona reaction and the vitelline block, are called *super-numerary sperm*, and the ovum is said to show *polyspermy*. The effectiveness of the vitelline block varies from species to species. When polyspermic ova are found, but supplementary sperm are seldom or never seen (eg, sheep, dog, hamster), the vitelline block either must be absent or must be delayed until after the zona reaction is in operation. On the other hand, in a species such as the rabbit, with many supplementary sperm in the perivitelline space but a low incidence of polyspermy, there must exist a rapid and efficient vitelline block.

In mice the incidence of polyspermy depends upon the strain of the female; on the other hand the proportion of ova containing supplementary sperm is related to the strain of the male and does not depend on the female.

Polyspermy

The existence of the zona reaction and the vitelline block would lead one to expect that polyspermy in mammals was a disadvantageous state; this is indeed so. All the processes of fertilization, including the factors that regulate the number of sperm reaching the ampulla, are coordinated to ensure that, while every ovum is fertilized, polyspermy is kept to a minimum. The incidence of polyspermic ova in most mammalian species is normally only 1% to 2%. In birds, on the other hand, polyspermy is usual.

The incidence of polyspermy may be increased experimentally, either by increasing the number of sperm in the ampulla, or by lowering the barriers that prevent extra sperm from entering the ovum. Conditions that reduce the zona reaction also tend to reduce the vitelline block. These include aging of the ovum, heating of the ovum or heating of the whole animal. Delayed copulation, since it leads to fertilization of aged ova, is an effective method of increasing the incidence of polyspermy in pigs, rats and rabbits. In swine, resection of most of the isthmus followed by end-to-end anastomosis of the remaining portions of the oviduct interferes with the normal mechanisms for restricting the number of sperm at the site of fertilization, and hence leads to polyspermy. A high incidence of polyspermy has also been observed in hamster ova fertilized *in vitro*.

When polyspermy does occur, the one or more supernumerary sperm often form pronuclei in the normal way, though in such a case all the pronuclei are reduced in size. Ova with three or occasionally four pronuclei have been observed in many species, including the cow, sheep and pig. Sometimes it is possible to prove that two of the three pronuclei are male in origin and hence that polyspermy has occurred; however, sometimes the extra pronucleus is female in origin (digyny), having arisen through failure of polar body formation at one or another reduction division. Digyny is rare in cows and ewes, even after delayed fertilization, but in sows, Thibault (1959) has found that if copulation takes place more than 36 hours after the beginning of estrus, more than 20% of the ova are digynic.

At syngamy the three pronuclei, whatever their origins, give rise to three chromosome groups which then unite. In this way a triploid embryo is formed, with three sets of chromosomes in every cell instead of the normal two sets. Triploid rat and rabbit embryos can survive to midgestation, and three children and one cat with three sets of chromosomes in most of their cells have been reported, but the great majority of triploid embryos die at an early stage of development. McFeely (1967) reported that 10% of pig blastocysts recovered ten days after mating had chromosome abnormalities including a high incidence of both triploidy (35%) and tetraploidy (18%). In swine, neither triploid nor tetraploid embryos survive implantation. The disadvantage of polyspermy to the organism is thus that it leads to triploidy, which is a lethal condition.

Fertilization *in Vitro*

Experiments on the mechanism of fertilization in mammals require a reliable procedure for achieving fertilization of mammalian ova outside the body of the female. Numerous attempts to obtain fertilization *in vitro* before the phenomenon of sperm capacitation had been discovered were largely vitiated by the use of uncapacitated sperm. Cleavage of the supposedly fertilized ovum in culture may have been due to parthenogenetic activation, while ova treated with sperm *in vitro* and shortly afterward transferred to the fallopian tubes of a recipient female may have had sperm attached to their surfaces which only achieved fertilization after transfer.

These difficulties were first overcome in the work done on rabbits by Dauzier and his associates (1954). Chang, using the same technique, returned the embryos to recipient female rabbits and obtained live young (see Chang, 1974). Fertilization *in vitro* has since been achieved in the mouse, rat, Syrian hamster, Chinese hamster, guinea pig, cat, squirrel, monkey and man. Only in the rabbit, mouse, rat and man has *in vitro* fertilization been followed by cleavage and birth of live young.

Earlier work on *in vitro* fertilization in farm animals was not encouraging, but recently Iritani and Niwa (1977) succeeded in achieving fertilization of cow ova *in vitro*. Oocytes were removed from follicles in the ovary at the germinal vesicle stage and matured in culture; they were then fertilized with bull sperm that had been capacitated in uteri isolated from estrous cows or in live estrous rabbit uteri. Sperm penetration was followed by normal pronuclear formation.

Sex Determination

Every cell in the mammalian body except the gametes contains a pair of *sex chromosomes*. In females the two members of the pair resemble one another and are known as X *chromosomes*; in males the sex chromosomes differ, one being an X chromosome, the other smaller, known as a Y *chromosome*. The sex-chromosome constitutions of females and males are therefore referred to as XX and XY respectively. The gametes, being haploid, contain only a single sex chromosome: an X chromosome in the female (*homogametic sex*) and either an X or a Y chromosome in the male (*heterogametic sex*). Faulty sharing-out (*nondisjunction*) of sex chromosomes, either to the gametes (Fig. 12–5) or, after fertilization, to the products of early cleavage, occasionally gives rise to individuals whose cells contain only a single X chromosome (XO), or an extra X or Y chromosome (XXX, XXY or XYY). XXY individuals are male and XO individuals female, proving that maleness must be determined in the first instance by factors on the Y chromosome. Cases of intersexuality in farm animals, such as freemartinism in cattle, arise during the course of development rather than at fertilization (see Chapter 24).

		OVA			
		Normal X		Nondisjunctive XX O	
SPERM Normal	X	XX (= normal female)		XXX	XO
	Y	XY (= normal male)		XXY	YO
SPERM Nondisjunctive	XY	XXY			
	XX	XXX			
	YY	XYY			
	O	XO			

Fig. 12–5. Diagram showing how normal and abnormal sex-chromosome constitutions can arise at fertilization. An O sperm or ovum is one that carries neither an X nor a Y chromosome. Nondisjunctive gametes arise through faulty sharing-out (nondisjunction) of the sex chromosomes. YO individuals probably are not viable; XXX individuals, in humans, are abnormal females.

In normal fertilization, the embryo develops as a female or as a male according to whether the ovum (carrying an X chromosome) is fertilized by a sperm carrying an X or Y chromosome. If the two types of sperm are present in equal numbers, the ratio of males to females at conception (the *primary sex ratio*) should be equal to one. Attempts to control the sex ratio in farm animals usually depend on treating the semen in such a way as might be expected to alter the proportions of X-bearing and Y-bearing sperm (Chapter 3). An alternative approach involving embryo transfer has been used in cattle.

CLEAVAGE

After syngamy is completed, there ensues a period of several days during which the fertilized ovum, zygote or *embryo* (as it should be called once development has started) leads a free-living existence in the oviduct and uterus of the mother. Metabolism of the developing embryo may be influenced by the nature and volume of the oviductal secretions. The ampullary portion of the oviduct shows greater secretory activity than the isthmus, and is a more favorable environment for embryos artificially retained in the oviduct. Once in the uterus, the embryo may be nourished by uterine secretions, but not until implantation has taken place does the embryo derive any nourishment from the maternal circulation.

At the beginning of the free-living period the ovum is a single cell, of relatively enormous volume compared with other cells of the body and therefore with a large ratio of cytoplasm to nuclear material. Reserve nutrients are stored in the cytoplasm in the form of yolk (*deutoplasm*). This single cell divides and redivides many times without any accompanying increase in volume of cytoplasm, though some increase in volume occurs through the uptake of water. The total amount of cellular material in the embryo actually decreases by about 20% in the cow, and by as much as 40% in the sheep; the total protein content of the embryo decreases by 25% in the mouse. The process of cellular division without growth is called *cleavage*. It continues until blastocyst formation, by which time cell size has been reduced more or less to the size characteristic of the species. During the early stages of cleavage, up to the appearance of the blastocoele, the embryonic cells are often known as *blastomeres*.

Normal Course of Cleavage

The unfertilized ovum possesses some polarity. The female chromatin lies to one side of the ovum, near to the point where the first polar body was extruded, and is often surrounded by a region of cytoplasm dense and rich in ribonucleoproteins and mitochondria. In the opposite half the cytoplasm is more vacuolated and contains fewer mitochondria. In species (eg, guinea pig) in which the ovum contains fat globules, they chiefly accumulate in the more vacuolated region.

In swine and horses, in which the ovum is rich in yolk, the surplus is eliminated (*deutoplasmolysis*) into the perivitelline space during cleavage and later is found in the blastocoele. In the horse, this yolk is extruded asymmetrically on the side of the ovum that is farthest from the nucleus.

The plane of the first cleavage division is not related to the plane of symmetry of the ovum, but usually passes through the area where the male and female pronuclei were situated at the beginning of syngamy. The second cleavage divisions occur at right angles to the first, the third more or less at right angles to the second (Fig. 12–4, *D,E*). The divisions are not perfectly synchronized, so that three-celled and five-, six-, seven-celled stages may be found. All the divisions are mitotic, and consequently each cell of the embryo, from the fertilized ovum onward, contains the diploid number of chromosomes (2n).

At some stage during early cleavage (late eight-cell in the mouse), the process of *compaction* occurs. The cells change shape, becoming wedge-shaped rather than spherical, so that the amount of cell-to-cell contact increases. By the 16- to 32-celled stage, the cells are crowded together into a compact group within the zona pellucida. The embryo is now known as a *morula.*

Fluid begins to collect in the intercellular spaces, and an inner cavity or *blastocoele* appears. Once this has begun to expand, the embryo is known as a *blastocyst.* A single peripheral layer of large flattened cells, the *trophectoderm* layer, surrounds a knob of smaller cells that lie to one side of the central cavity. This knob, the *inner cell mass*, will give rise to the adult organism, while the cells of the trophectoderm form the *trophoblast*, the placenta and the embryonic membranes. The elegant cell reassociation experiments of Graham and others on mouse embryos (see Gardner and Rossant, 1976) have established that the differentiation of cells to form trophectoderm or inner cell mass is not due to any cytoplasmic heterogeneity preexisting in the cytoplasm of the fertilized ovum, but depends on differences in microenvironment within the embryo. Any cell on the periphery about the time of the 16-celled stage differentiates as trophectoderm, while interior cells give rise to the inner cell mass.

Electron microscopic investigations have shown that the membranes of trophectoderm cells are closely apposed and interdigitated, and are linked by the tight junctions characteristic of epithelial tissues. In sheep, Calarco and McLaren (1976) have observed these tight junctions as early as the 8- to 16-celled stage around the periphery of the embryo. Their formation may constitute part of the compaction process.

Until about the eight-celled stage, all the cells of the embryo are thought to have equal developmental potential. In sheep and swine, normal embryos can develop from single blastomeres isolated at the four- to six-celled stage, while in mice and rabbits, development of normal adult animals has been achieved from blastomeres isolated at the two- and eight-celled stages respectively.

Cleavage Rates. Approximate estimates of the time taken by the embryos of various mammals to reach certain stages of development are given in Table 12-1. The rate of cleavage up to the blastocyst stage tends to be faster in species in which the total gestation period is short. In the pig, the two-celled stage lasts a mere six to eight hours, but the four-celled stage lasts 24 to 48 hours. Subsequent cell divisions occur about every 12 hours. Little is known about the influence of the maternal environment, although Wintenberger-Torres (1974) has shown that in sheep the rate of cleavage of blastocysts after seven days is significantly increased either by superovulation, in which there is an abnormally high number of corpora lutea, or during treatment with exogenous progesterone.

Steroid hormones, whether administered experimentally or ingested in the diet, also affect the rate of transport of embryos along the oviduct in cows, sheep and swine.

Activation of the Embryonic Genome. In mammals, we still have little information on the extent to which early development depends on maternal products transmitted through the cytoplasm of the ovum, but we know that the embryonic genome is active soon after first cleavage. The first embryonic genes to be activated are those concerned with the protein-synthesizing apparatus. In the mouse ribosomal RNA is synthesized at the two-celled stage and transfer RNA, at the four-celled stage. Messenger RNA is probably produced from the two-celled stage onward; the action of mutant genes has been detected as early as the two-celled stage, and paternal variants of enzymes and antigens—positive evidence of embryonic gene action—have been detected by the six- to eight-celled stage. In the rabbit, on the other hand, although transfer and messenger RNA are again synthesized early in development, no ribosomal RNA is synthesized until the morula stage, by which time the embryo contains several hundred cells. Presumably the larger rabbit ovum contains more storage products than the mouse ovum, and hence can support development for a longer period. The three major classes of DNA-dependent RNA polymerases have all been identified in preimplantation mouse embryos. No information is available yet on the stage at which the embryonic genome is activated in any farm animal.

The Blastocyst. Trophectoderm cells are characterized by their ability to pump fluid inward, and since the tight junc-

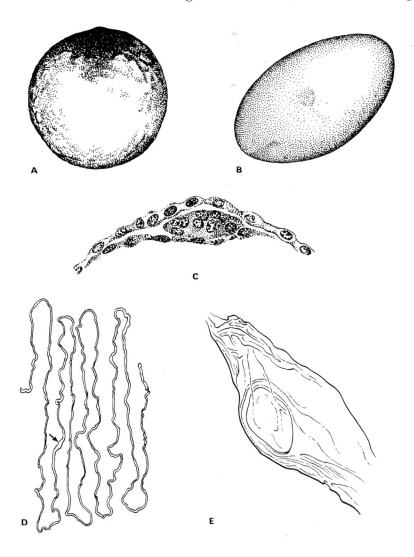

Fig. 12-6. *A–C.* Cattle blastocysts. *A,* Hatched blastocyst, still spherical, with distinct inner cell mass. *B,* Elongating blastocyst. The trophoblast has enlarged greatly relative to the inner cell mass. *C,* Section of hatched blastocyst, showing trophoblast above inner cell mass and primary endoderm below. *D,* Pig blastocyst 13 days after mating at the beginning of estrus. The total length of this blastocyst was 157 cm. The position of the embryonic disc is indicated by an arrow. *E,* Enlarged view of a pig embryonic disc 14 days after mating.

tions joining them form an effective permeability seal, the blastocoele swells up with fluid. The extent of this enlargement varies from species to species. In farm mammals it is considerable; within a few days the blastocyst becomes a thin-walled fluid-filled sac, more or less filling the uterine lumen. Pig blastocysts are classified into spherical, ovoid, tubular and filamentous. Up to ten days after mating, all blastocysts are spherical; after 16 days, only filamentous blastocysts are found. Blastocyst elongation involves cell division in the trophectoderm rather than in the inner cell mass, so that in cattle embryos, for example, a negative correlation exists between the size of the inner cell mass and the total cell number.

At the blastocyst stage, the metabolic rate increases in all species that have been studied. Bovine blastocysts possess Krebs cycle dehydrogenases by day 10, when they are beginning to elongate, and thus they are capable at this time of switching to the more efficient aerobic pathways of energy metabolism. Measurement of glucose consumption by the bovine blastocyst *in vitro* has been used as an effective method of predicting embryo quality in transfer experiments.

The loss of the zona pellucida has been studied extensively in rats and mice. In normal pregnancy, the zona is dissolved at the onset of implantation, probably by a proteolytic enzyme emanating from the estrogen-sensitized uterus. During lactation, or when females are ovariectomized early in pregnancy so that the uterus is deficient in estrogen, or when blastocysts are cultured *in vitro*, loss of the zona is delayed. Under these conditions no lysis of the zona occurs, but instead the zona ruptures and the blastocyst "hatches," probably aided by the rhythmic contractions and expansions that have been observed in culture. Bovine blastocysts will also "hatch" from the zona in culture, but only if glucose is present in the culture medium. The zona pellucida of the guinea

pig becomes penetrated at the time of implantation by slender pseudopodial processes from the abembryonic trophoblast cells of the blastocyst, but it is not lost until after attachment has been effected. In farm mammals, the zona is lost before attachment, about six days after ovulation in swine, eight days in cattle. The mechanism of loss is not known, but the occasional finding of empty ruptured zonae in the cow uterus supports the view that, in this species at least, hatching takes place *in vivo* as *in vitro*.

Twinning

Two distinct types of twins are known: *monozygotic (identical)* twins and *dizygotic (fraternal)* twins.

Dizygotic twins originate from a double ovulation in a normally monotocous species. Two ova may be shed in the same estrous period, and fertilized by two different sperm. The resulting young do not resemble one another genetically any more closely than do ordinary brothers and sisters. In fact, a pair of dizygotic twin lambs of opposite sex differ on average more from one another in birth weight than do male and female lambs born in twin pairs of like sex, suggesting that competition occurs between embryos in the uterus. Multiple ovulation may be induced, and hence the frequency of dizygotic twins increased, by the injection of pituitary or chorionic gonadotropins. This technique is already finding practical application in sheep breeding. A high rate of twinning has also been achieved in cattle, by transfer of one embryo to each uterine horn.

Monozygotic twins, on the other hand, originate from a single fertilized ovum. In theory they could occur in all species, including species such as the pig which normally bear several young at a time, but so far they have been recognized in only a few species, notably man and cattle. Even here they are relatively rare: up to 5%

(depending on breed, age and environment) of all births in cattle will be twin births, but only about one in every thousand births will be monozygotic twins. Monozygotic twins resemble one another closely, indeed in all genetically determined characteristics. For instance, they are always of the same sex. Because of their identical hereditary equipment, monozygotic twin pairs of cattle are particularly valuable experimental material for studying the effects of various environmental conditions on such characters as milk yield in dairy breeds and weight gain and conformation in beef breeds.

Many cases of monozygotic twinning probably originate fairly late in development, after implantation. A single blastocyst implants, and the single inner cell mass then differentiates two primitive streaks, giving rise to two separate individuals. Such twins have a common chorion, and also sometimes a common amnion. Alternatively, the inner cell mass in a single blastocyst may duplicate before implantation; this condition has been reported in both sheep and pigs. The resulting embryos have separate amnions and placentas. It is not impossible that some monozygotic twins may originate still earlier, as a result of blastomeres separating inside the zona pellucida. Monozygotic twin lambs have been produced experimentally, by separating the two blastomeres at the two-celled stage and transferring them independently to foster mothers (Willadsen, 1978).

In swine, the pregnant uterus normally contains several embryos. There is rarely fusion between either the allantois or the chorion of neighboring embryos. In sheep, in which more than one embryo occurs frequently but not always, there is fusion of the chorion but not of the allantois; in cattle, in which multiple pregnancies occur relatively seldom, there is usually fusion of both chorion and allantois, with consequent anastomosis of blood vessels between neighboring embryos. This means that in cattle, even most dizygotic twins have a common blood circulation, while in other farm mammals the proportion of twins sharing a common circulation is much lower. Where the twin partners are of opposite sex, the common blood circulation leads to the development of freemartins.

The fusion of the embryonic blood vessels is responsible for an unexpected difficulty in distinguishing monozygotic from dizygotic twins in cattle. In general, monozygotic twins are recognized by their striking similarity in all traits that depend mainly or only on heredity. Sex, coat color and pattern, nose prints, serum globulin type and presence or absence of horns are particularly useful traits for distinguishing types of cattle twins. However, the genetically determined blood groups have to be used with caution in classifying cattle twins, because the common embryonic circulation results in a mixture of the blood-forming cells, so that each dizygotic twin shows not only his own blood group but also that of the other twin. If detected, this mosaicism is itself evidence of dizygosity. It has been estimated that, if all diagnostic tests are used, the error in diagnosing monozygosity in cattle twins can be reduced to about 1%.

Interspecific Hybrids

Crosses between related species have been reported from many different mammalian groups (Gray, 1972). Some are economically important, such as crosses of cattle with zebu or buffalo. Species crosses, even when viable, tend to be partly or totally sterile. Fertility is most affected in the heterogametic (XY) sex (ie, in mammals, the male).

Differences between the results of reciprocal crosses have often been reported. Thus, in ewes inseminated with goat semen, the ova are either not fertilized or they arrest during cleavage, but in goats inseminated with ram semen, the hybrid embryo survives until about halfway through pregnancy. The death of the hybrid embryo is thought to be due at least partly to immunologic rejection by the mother. The successful gestation of sheep-goat hybrids by both ewes and nanny goats has been claimed by Bratanov and Dikov (1962) after experimental treatment of the females.

IMPLANTATION

The embryo is said to be implanted or attached when it becomes fixed in position and physical contact with the maternal organism is established.

The term implantation seems most appropriate for those species in which the embryo becomes buried in the wall of the uterus. In many rodents, the blastocyst comes to lie in a pocket (crypt) of the uterine wall, in intimate association with the maternal tissues; in other species, including man, the blastocyst passes through the uterine epithelium and is thus entirely cut off from the uterine cavity. In farm mammals, on the other hand, the embryo remains in the uterine cavity, and whatever attachment it forms with the wall of the uterus prior to the formation of the placenta is extremely loose. Movement of the blastocyst within the uterus becomes increasingly restricted as it expands; in the sheep, a mucous substance sticking the blastocysts to the uterine wall has been reported.

The loose and gradual nature of the attachment process in farm mammals has led to considerable controversy about when implantation actually begins. Estimates have ranged from the 10th to the 22nd day postcoitum for sheep, and from the 11th to the 40th day for cattle.

The Embryo

Spacing. In polytocous species, the blastocysts become distributed down the length of the uterine horn as a result of the muscular churning movements of the uterine wall. In sheep with a single corpus luteum, blastocysts rarely pass from one uterine horn to the other, but if two ova are released from the same ovary, migration of one to the contralateral horn almost always occurs. In swine, blastocysts pass freely between the two horns. Dziuk and his associates (see Dziuk, 1977) timed the distribution of pig embryos entering the uterus from one side only. The proportion of the total uterus that was occupied increased from 13% on the sixth day of gestation to 86% on the 12th day.

The presence of large blastocysts such as those of the rabbit may modify the muscular movements of the uterus to result in some regularity of spacing of embryos during implantation. In the sow also, the distribution of embryos between the two horns is much more even than would be expected by chance, and once elongation has occurred there is little or no overlap between adjacent blastocysts.

An implanting blastocyst appears to have no inhibitory effect upon its neighbors. Implantation rate is therefore unlikely to be a limiting factor in attempts to increase litter size artificially in swine (eg, by inducing superovulation with gonadotropic hormones). Such attempts are more likely to founder at a later stage of pregnancy, owing to the inadequacy of the vascular supply or to the mechanical effects of crowding on the embryos; also, the young that are born may be undersized and difficult to rear.

Orientation. In any given species, the position in which the blastocyst implants is usually fixed. In cattle, sheep and swine, the embryonic disc is always situated on the antimesometrial side of the uterine horn.

If the mesometrial-antimesometrial axis of the rat uterus is reversed surgically, blastocysts still implant on the antimesometrial edge. The position of implantation is therefore determined by the relationship between blastocyst and uterine wall, rather than by the action of any external factor, such as gravity, on the blastocyst. The uterus itself plays the chief determining role, because glass beads or pieces of muscle or tumor put into the uterus at the appropriate time will be positioned on the antimesometrial side, as are blastocysts.

Gastrulation. Gastrulation is a stage of embryonic development that occurs, though in different guises, in all vertebrates. It succeeds formation of the blastocyst and precedes organ formation. Essentially, gastrulation consists of movements of cells or groups of cells in such a way that (a) the embryo converts from a two-layered to a three-layered structure composed of *ectoderm, mesoderm* and *endoderm*; and (b) the future organ-forming regions are brought into their definitive positions in the embryo.

In swine and sheep, the trophectoderm overlying the inner cell mass ruptures soon after the loss of the zona pellucida, and the inner cell mass pushes through to form the *embryonic disc* (Fig. 12–7, A–C). Gastrulation involves the cells of the em-

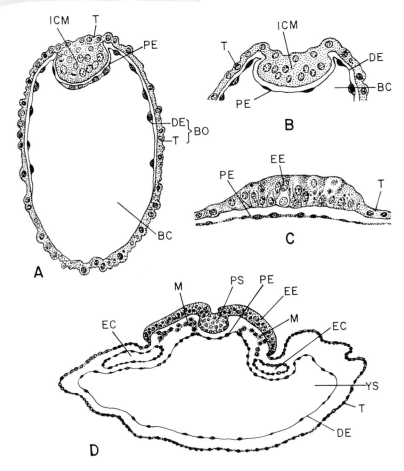

Fig. 12–7. Gastrulation in the sheep. *A,* Section of sheep blastocyst 10 days postcoitum. The endoderm can be seen lining the upper half of the blastocyst cavity. *B,* A slightly older blastocyst than A. The inner cell mass is becoming intercalated into the trophoblast. *C,* Section of embryonic disc of a sheep blastocyst 12 days postcoitum. The embryonic ectoderm is now bulging above the level of the adjacent trophoblast. *D,* Section through Hensen's node, 14 days postcoitum. The yolk sac is now competely formed, mesoderm cells are interposed between the embryonic ectoderm and endoderm, and the extra-embryonic coelom can be seen at each side.

BC, blastocyst cavity; *BO,* bilaminar omphalopleure; *DE,* distal endoderm; *EC,* extra-embryonic coelom; *EE,* embryonic ectoderm; *ICM,* inner cell mass; *M,* mesoderm; *PE,* primary endoderm; *PS,* primitive streak; *T,* trophoblast; *YS,* yolk sac cavity.

bryonic disc only. From the embryonic disc three types of tissue differentiate (see prenatal development in Chapter 13). The endoderm, delaminated from the surface of the inner cell mass facing the blastocyst cavity, spreads around to line the inner surface of the trophectoderm and thus form the *yolk sac* (Fig. 12–7; Fig. 12–6, C). The outer part of the inner cell mass or embryonic disc differentiates as *embryonic ectoderm*. The notochord and mesoderm are formed by invagination of cells in the *primitive streak* region of the embryonic disc. The primitive streak defines the anteroposterior axis of the embryo. As the mesoderm spreads outwards from the primitive streak between the endoderm and ectoderm, it splits into two layers, separated by the coelom (Fig. 12–7, D). The notochord develops from the anterior end of the primitive streak (Hensen's node).

Dorsal to the notochord, the ectoderm thickens to form a *neural plate*. After a few days, *neural folds* grow up and fuse to form a *neural tube*, the forerunner of the brain and spinal cord. Meanwhile, somites, paired condensations of dorsal mesoderm, have appeared on either side of the notochord.

The Uterus

Preimplantational Changes. While the embryo is undergoing cleavage and blastocyst formation, the uterus too is undergoing changes, preparing the way for implantation. During this *progestational* period, the muscular activity and tonicity of the uterus decrease, which may help to retain the blastocysts in the uterine lumen. At the same time, an increased blood supply to the uterine epithelium develops. In some species the increased vascularity is greatest along the side of the uterus at which implantation takes place. The amino acid and protein content of the uterine fluid changes at the time of implantation. Certain protein fractions, de-

tectable by electrophoresis, appear in uterine fluid only at this time. The first such fraction to be observed, in the rabbit, was termed *blastokinin* or *uteroglobin*; its biologic function has not yet been established. The "purple protein" produced by the pig uterus under the influence of progesterone is thought to have a nutritional role for the embryo (Bazer, 1975).

Associated with the increased blood supply to the uterus, changes occur in the secretory activity of glandular and surface epithelia of the endometrium. High-molecular-weight compounds (proteins, carbohydrates, mucopolysaccharides) are broken down, and low-molecular-weight derivatives, along with glycogen and fats, accumulate. This material, along with cellular debris and extravasated leukocytes in the uterine lumen, forms the *histotrophe* (uterine milk) which provides nourishment for the embryo in the early period of uterine life, before the chorioallantoic placenta is established. There is evidence that histotrophe plays an important role in embryonic nutrition from about 80 hours postcoitum in the rabbit, and from the ninth-day blastocyst stage onward in the sheep. In farm animals, in which the placenta is epitheliochorial (swine, horse) or syndesmochorial (sheep, cow), the association between fetal and maternal blood is not close. Histotrophic nutrition is important, therefore, not only in the early stages of uterine life, but throughout gestation.

The hormonal basis of implantation has been studied extensively. Progesterone plays the leading part in determining the preimplantational changes in the uterus, but the estrogen released at estrus also seems to be important for priming the endometrium. A further release of estrogen from the ovaries just prior to implantation is required in some rodents, but not in farm animals. In almost all mammals, ovariectomy immediately after ovulation prevents both implantation and the uterine developments associated with it.

Relation Between Embryo and Maternal Environment. The importance of the tubal environment in embryonic development has been underlined by the findings of Winterberger-Torres. If the passage of sheep embryos through the fallopian tubes to the uterus is hastened by hormonal treatment, development is slowed down and cleavage rate does not return to normal for nine days. On the other hand, if the embryos are retained in the tubes by a ligature, they continue to develop normally for a week, and for the next two days they develop more slowly but will recover if transferred to the uterus; if they remain for more than ten days in the tubes, they undergo no further development even if transferred to the uterus.

In species other than the sheep, the role of the tubal environment has been little studied. Data on the role of the uterine environment, on the other hand, have been obtained by experiments on embryo transfer (see Chapter 29). The critical importance of *synchronization* between the stages of development of the embryo and of the uterus has been established in several species. Sheep embryos transferred to the uterus between the 2nd and 11th day after estrus (Rowson and Moor, 1966) gave a high success rate when donors and recipients were synchronized exactly, but few pregnancies ensued when they were three days or more out of phase. This is the more surprising because, from

the seventh day onward, the level of progesterone in the ovarian venous blood of the ewe remains relatively constant. In cattle, synchronization requirements for successful egg transfer seem even more acute than in sheep. Synchronization is also required in swine.

In farm mammals, the period of implantation accounts for a significant proportion of all reproductive losses. Delayed development of the embryos, their delayed passage into the uterus or precocious development of the endometrium will all lead to failure of synchronization between blastocyst and endometrium at the critical time, and hence may be responsible for failures of implantation. The more precise the synchronization requirements (eg, cow), the more likely it is that reproductive losses will be due to this cause.

Maternal Recognition of Pregnancy. In farm animals, the presence of embryos must be signaled before implantation begins to prevent the regression of the corpora lutea. Times after ovulation at which the presence of an embryo must be recognized if luteal regression is to be prevented are given in Table 12–2. The requirement for precise synchronization suggests that a complex interaction occurs between embryo and mother, with each partner providing stimuli essential for the further development of the other. It seems likely that the striking expansion of the

Table 12–2. Implantation in Farm Animals

Species	Maternal recognition of pregnancy begins (days)	Attachment begins (days after ovulation)	Attachment complete (days after ovulation)	Degree of elongation	Location of first embryo-maternal association	Type of placenta
Cattle	16–17	28–32	40–45		Uterine caruncles (convex)	Cotyledonary
Sheep	12–13	14–16	28–35	10–20cm	Uterine caruncles (concave)	Cotyledonary
Swine	10–12	12–13	25–26	Up to 1m	Deep folds of uterine wall	Diffuse
Horse	14–16	35–40	95–105	6–7cm	Chorionic girdle	Diffuse

trophoblast serves a dual role, not only by ingesting "uterine milk" for the nutrition of the embryo, but also by synthesizing some hormone-like substance that either stimulates the corpus luteum or inhibits the uterus from producing a luteolysin.

Culture of pig blastocysts together with fragments of endometrium have shown

that the uterus can produce substances that activate specific protein synthesis by the embryo; similarly, the presence of a blastocyst alters the pattern of protein synthesis by the uterus. Heap and his colleagues (see Perry et al, 1976) have shown that the unimplanted blastocysts of sheep, cow, pig and horse are all capable

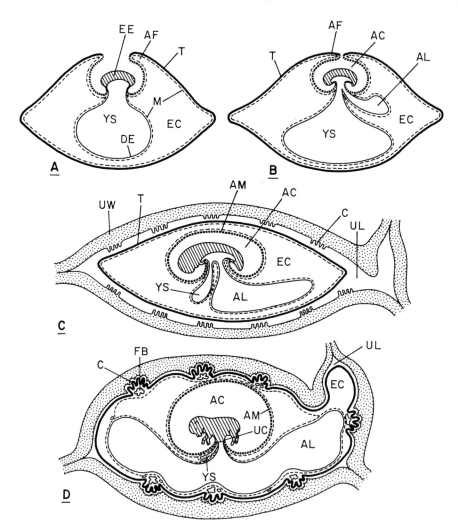

Fig. 12–8. Implantation in the sheep. *A*, Elongated blastocyst, with amniotic folds beginning to close over the embryo. *B*, Formation of yolk sac placenta, with vascularized yolk sac wall (distal endoderm and mesoderm) closely apposed to chorion (trophoblast and medoserm). *C*, Relationship of fetal membranes to uterine cavity. Amniotic folds have closed over the embryo, to enclose the amniotic cavity, which is expanding and pushing the enlarging allantois into line with the regressing yolk sac. *D*, Formation of chorioallantoic placenta. The expanding allantois presses the chorion up against the uterine caruncles.

AC, Amniotic cavity; *AF*, amniotic folds; *AL*, allantois; *AM*, amniotic membrane; *C*, caruncle; *DE*, distal endoderm; *EC*, extra-embryonic coelom; *EE*, embryonic ectoderm; *FB*, fetal blood vessels; *M*, mesoderm; *T*, trophoblast; *UC*, umbilical cord; *UL*, uterine lumen; *UW*, wall of uterus; *YS*, yolk sac.

of synthesizing estrogen. At least in the pig, it seems likely that this estrogen, produced by the blastocyst on day 12, inhibits the uterine prostaglandin production that would cause regression of the corpus luteum in the absence of an embryo. In sheep, on the other hand, the antiluteolytic activity is thought to be exerted by a glycoprotein (Findlay et al, 1979).

Course of Implantation

Some of the temporal parameters and morphologic processes concerned with implantation in farm mammals are summarized in Table 12–2. Because implantation is such a gradual process in this group, the times given are approximate. The differentiation of the embryonic membranes, by which the embryo is connected to the mother, and the subsequent formation of the chorioallantoic placenta, are described in Chapter 13.

The trophoblast with its underlying layer of mesoderm is termed the *chorion* (Fig. 12–8, *A, B*). Once the zona pellucida is lost, the chorion, its surface densely covered with microvilli, is in direct contact with the uterine wall, and is responsible for ingesting histotrophe. In sheep and swine, junctional complexes are formed between trophoblast and uterine epithelial cells.

The next stage of association involves smoothing out the chorionic microvilli, and adhesion of the chorion to the uterine wall. Cattle and sheep have specialized and permanent "attachment organs" in the uteri, the caruncles, 80 or more in number (Fig. 12–8, *C, D*). Growth of the caruncles is stimulated by progesterone. In the pig and horse, the association is less intimate, and chorionic folds or *villi* are distributed diffusely over the entire surface, rather than being restricted to certain areas (*cotyledons*). The breakdown of the uterine epithelium is thought to be due to cytolytic action in the epithelial cells

themselves, rather than to phagocytosis by trophoblast. In the pig, the absorptive surface is increased by the great length of the trophoblastic tube (up to 1 m), which is closely apposed to the deeply folded uterine wall; the chorion becomes extensively vascularized by allantoic blood vessels, and attachment is then complete.

In the horse, by 35 days the yolk sac is ringed by a band of elongated trophoblast cells, the *chorionic girdle;* these cells invade the endometrium to form the *endometrial cups,* which by 38 to 40 days are actively secreting gonadotropin. By 150 days of gestation, after attachment has been completed, the endometrial cups regress, probably owing to immunologic rejection by the mare.

REFERENCES

Anderson, L.L. (1978). Growth, protein content and distribution of early pig embryos. Anat. Rec. *190,* 153.

Bateman, N. (1960). Selective fertilization at the T-locus of the mouse. Genet. Res. Camb. *1,* 226.

Bazer, F.W. (1975). Uterine protein secretions: relationship to development of the conceptus. J. Anim. Sci. *41,* 1376.

Beatty, R.A., Stewart, D.L., Spooner, R.L. and Hancock, J.L. (1976). Evaluation by the heterospermic insemination technique of the differential effect of freezing at −196°C on fertility of individual bull semen. J. Reprod. Fertil. *47,* 377.

Bratanov, C. and Dikov, V. (1962). Fécondation entre les espèces brebis et chèvres et obtention d'hybrides interespèces. *In* Proc 4th Int. Congr. Anim. Reprod., *4,* The Hague, 744.

Calarco, P.G. and McLaren, A. (1976). Ultrastructural observations of preimplantation stages of the sheep. J. Embryol. Exp. Morphol. *36,* 609.

Chang, M.C. and Rowson, L.E.A. (1965). Fertilization and early development of Dorset Horn sheep in the spring and summer. Anat. Rec. *152,* 303.

Chang, M.C. (1974). Recent advances in fertilization of mammalian eggs *in vitro. In* The Functional Anatomy of the Spermatozoon. B.A. Afzelius (ed), New York, Pergamon Press.

Cook, B. and Hunter, R.H.F. (1978). Systemic and local hormonal requirements for implantation in domestic animals. J. Reprod. Fertil. *54,* 471.

Dauzier, L., Thibault, C. and Winterberger, S. (1954). La fécondation *in vitro* de l'oeuf de la lapine. C.R. Acad. Sci. (Paris), *238,* 844.

Dziuk, P.J. (1977). Reproduction in pigs. *In* Reproduction in Domestic Animals. 3rd ed. New York, Academic Press.

Findlay, J.K., Cerini, M., Sheers, M., Staples, L.D. and Cumming, I.A. (1979). The nature and role of pregnancy-associated antigens and the endocrinology of early pregnancy in the ewe. In Maternal Recognition of Pregnancy, Ciba Foundation Symposium 64. M. O'Connor and J. Whelan (eds), Amsterdam, Elsevier/Excerpta Medica/North-Holland.

Gardner, R.L. and Rossant, J. (1976). Determination during embryogenesis. In Embryogenesis in Mammals. Ciba Foundation Symposium 40. K. Elliott and M. O'Connor (eds), Amsterdam, Elsevier/Excerpta Medica/North-Holland.

Gray, A.P. (1972). Mammalian Hybrids. 2nd ed. CAB Tech. Commun. No. 10 (rev). England, Commonwealth Agricultural Bureaux.

Hunter, R.H.F. (1977). Physiological factors influencing ovulation, fertilization, early embryonic development and establishment of pregnancy in pigs. Br. Vet. J. 133, 461.

Iritani, A. and Niwa, K. (1977). Capacitation of bull spermatozoa and fertilization in vitro of cattle follicular oocytes matured in culture. J. Reprod. Fertil. 50, 119.

McFeely, R.A. (1967). Chromosome abnormalities in early embryos of the pig. J. Reprod. Fertil. 13, 579.

McLaren, A. (1973). Blastocyst activation. In Regulation of Mammalian Reproduction. S.J. Segal, R. Crozier, P. Corfman and P. Condliffe (eds), Bethesda, National Institutes of Health. pp. 321–328.

Norberg, H.S. (1973). Ultrastructural aspects of the preattached pig embryo: cleavage and early blastocyst stages. Arch. Anat. Entwickl. Gesch. 143, 95.

Perry, J.S., Heap, R.B., Burton, R.D. and Gadsby, J.E. (1976). Endocrinology of the blastocyst and its role in the establishment of pregnancy. J. Reprod. Fertil. Suppl. 25, 85.

Polge, C. and Baker, R.D. (1976). Fertilization in pigs and cattle. Canad. J. Anim. Sci. 56, 105.

Rowson, L.E.A. and Moor, R.M. (1966). Embryo transfer in the sheep: The significance of synchronizing oestrus in the donor and recipient animals. J. Reprod. Fertil. 11, 207.

Szollosi, D. and Hunter, R.H.F. (1973). Ultrastructural aspects of fertilization in the domestic pig: sperm penetration and pronucleus formation. J. Anat. 116, 181.

Thibault, C. (1959). Analyse de la fécondation de l'oeuf de la truie après accouplement ou insemination artificielle. Ann. Zootechn. 8, (Suppl.). 165.

Wales, R.G. (1975). Maturation of mammalian embryo: biochemical aspects. Biol. Reprod. 12, 66.

Willadsen, S.M. (1978). A method for culture of micromanipulated sheep eggs and its use to produce monozygotic twins. Nature 277, 298.

Wintenberger-Torres, S. and Fléchon, J.E. (1974). Ultrastructural evolution of the trophoblast cells of the pre-implantation sheep blastocyst from day 8 to day 18. J. Anat. 118, 143.

13

Gestation, Prenatal Physiology and Parturition

M.R. JAINUDEEN
and
E.S.E. HAFEZ

GESTATION

Length of gestation is calculated as the interval from fertile service to parturition (Table 13–1). The duration of gestation is genetically determined although it can be modified by maternal, fetal and environmental factors (Fig. 13–1).

Maternal Factors. The age of the dam influences the duration of pregnancy in different species. A two-day extension from the normal occurs in the eight-year-old ewe. Heifers that conceive at a relatively young age carry their calves for a slightly shorter period than those that conceive at an older age.

Fetal Factors. An inverse relation between the duration of gestation and litter size is well-documented in several polytocous species except the pig. Multiple fetuses in monotocous species also have shorter gestation periods. Twin calves are carried three to six days less than single calves. The sex of the fetus may also determine gestation length; male calves and foals are carried one to two days longer than females. The size presumably affects gestation length by hastening the

Table 13–1. Differences in Gestation Periods of Farm Mammals

Animal	Average (Range)
Cattle (dairy breeds)	
Ayrshire	278
Brown Swiss	290 (270–306)
Dairy Shorthorn	282
Friesian	276 (240–333)
Guernsey	284
Holstein-Friesian	279 (262–359)
Jersey	279 (270–285)
Swedish-Friesian	282 (260–300)
Zebu (Brahman)	292 (271–310)
Cattle (beef breeds)	
Aberdeen-Angus	279
Hereford	285 (243–316)
Beef Shorthorn	283 (273–294)
Sheep	148 (140–159)
Swine	
Domestic	114 (102–128)
Wild Pig	(124–140)
Horse	
Arabian	337 (301–371)
Belgian	335 (304–354)
Clydesdale	334
Morgan	344 (316–363)
Percheron	(321–345)
Shire	340
Thoroughbred	338 (301–349)

(From the literature.)

time of parturition initiation. The duration of pregnancy may be influenced by the endocrine functions of the fetus.

Genetic Factors. The small variations in pregnancy duration among breeds (Table 13–1) may be due to genetic, seasonal or local effects. The extreme expression of genetically prolonged gestation is known among dairy cows that carry a fetus homozygous for an autosomal recessive gene.

The influence of equine fetal genotype on gestation length can be observed in hybrids between the horse and the donkey. For example, the duration of pregnancy of a mare carrying a foal to a stallion is 340 days, whereas that to a jack is 355 days. This influence may be mediated either through a hormonal mechanism or may merely reflect the influence of fetal size. A sex-linked gene in Arabian mares or fetuses affects the duration of their pregnancy. Differences in gestation between mutton and wool breeds of sheep are due to genetic factors.

MATERNAL PHYSIOLOGY IN PREGNANCY

Early pregnancy is marked by processes that prolong the life span of the cyclic corpus luteum. These processes suggest a maternal recognition of pregnancy.

Maternal Recognition of Pregnancy

In many species hysterectomy prolongs the life of the corpus luteum. This suggests that the uterus produces a substance, probably prostaglandin $F_{2\alpha}$, which normally acts through a direct or local pathway between the uterine horn and the adjacent ovary (cattle, sheep, swine) or through systemic channels (horse) to terminate the life of the corpus luteum. When administered to the female, prostaglandin $F_{2\alpha}$ causes a reduction in ovarian progesterone secretion and a regression of the corpus luteum.

Apparently during early pregnancy, the developing conceptus exerts an antiluteolytic effect and either prevents the release or neutralizes the uterine luteolytic effect. The cyclic corpus luteum is converted into a corpus luteum of pregnancy, which continues to function and secrete progesterone during the early stages if not throughout pregnancy.

The time of maternal recognition varies among species. Recognition is apparently early in the mare, where only fertilized ova enter the uterus. This indicates that

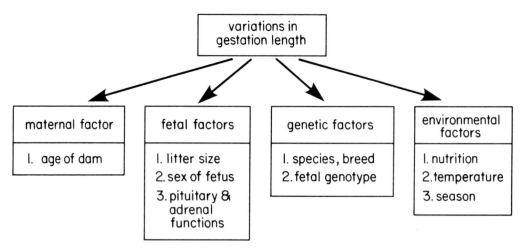

Fig. 13–1. Schematic representation of variations in length of gestation due to maternal, fetal, genetic and environmental factors. Whereas many of these factors within a species cause minor variations, hypofunction of the pituitary-adrenal axis of the fetus is associated with prolonged gestation in the ewe and cow.

maternal recognition of pregnancy may occur even before the conceptus could exert a luteotropic effect. However, in most species this recognition is postponed until the blastocyst has been transported to the uterine lumen. If no blastocyst is present in the uterus of the ewe by day 12 of the cycle, the corpus luteum begins to regress. In the sow, the embryos suppress the luteolytic effect of the uterus, and a minimum of about four embryos is needed to maintain pregnancy (Polge et al, 1966).

Reproductive Organ Changes

Vulvar and Vaginal Changes. Edema and vascularity are the major reactions of the vulva to the progress of pregnancy. These vulval changes, noted more in cattle than in horses, occur around the fifth and seventh months of gestation in heifers and cows respectively. The vaginal mucosa is pale and dry during most of gestation but becomes edematous and pliable toward the end of pregnancy.

The Cervix. During pregnancy the external os of the cervix remains tightly closed. The endocervical crypts increase in number and produce a highly viscid mucus that seals the cervical canal. This so-called mucous plug of pregnancy liquifies prior to parturition and is discharged in strings.

Uterine Changes. As pregnancy progresses, the uterus undergoes gradual enlargement to permit expansion of the fetus, but the myometrium remains quiescent to prevent premature expulsion. Three phases can be identified in the adaptation of the uterus to accommodate the products of conception—proliferation, growth and stretching; the duration of each varies with the species. The mechanisms that permit the enormous increase in size are unknown but are probably hormonal.

Endometrial proliferation occurs before blastocyst attachment and is characterized by a preparatory progestational sensitization of the endometrium. Characteristic changes of the endometrium that are initiated by hormones, mainly progesterone, include increased vascularity, growth and coiling of the uterine glands, and leukocyte infiltration of the uterine lumen.

Uterine growth commences after implantation and comprises muscular hypertrophy, an extensive increase in connective tissue ground substance and an increase in fibrillar elements and collagen content. Modification of the ground substance is significant both in uterine adaptation to the conceptus and in the process leading to involution. The structural changes that take place in the gravid uterus are reversible but are differentially restored after parturition.

During the period of uterine stretching, uterine growth diminishes while its contents are growing at an accelerating rate.

Ovarian Changes. Ovarian changes commence with the transformation of the graafian follicle to a corpus luteum. Whereas the corpus luteum regresses in a nonfertile estrous cycle it persists as the corpus luteum of pregnancy (corpus luteum verum) and as a result estrous cycles are suspended. However, some cows may show estrus during early pregnancy due to follicular activity in the ovaries. In the mare, as many as 10 to 15 follicles develop between the 40th and 160th day of pregnancy. These follicles luteinize to form accessory corpora lutea (see Chapter 19).

The corpus luteum of pregnancy in the cow persists at a maximal size throughout pregnancy, but in the mare both the primary as well as the accessory corpora lutea regress by the seventh month of pregnancy.

Pelvic Ligaments and Pubic Symphysis. Relaxation of the pelvic ligaments, which occurs gradually during the course of pregnancy, becomes more rapid with approaching parturition. This relaxation is more notable in the cow and ewe than in the mare and is related to high levels of estrogens in late pregnancy and to the action of relaxin. The caudal part of the sacrosciatic ligament, which is cordlike in the nonpregnant cow, becomes more relaxed and flaccid with approaching parturition.

Hormones of Pregnancy

Maintenance of Pregnancy. A balance of certain hormones is necessary for pregnancy. Surgical procedures such as ovariectomy, hypophysectomy and fetectomy, have given valuable information on the nature of the fetal luteotropic complex and the role of the corpus luteum for maintenance of pregnancy (Table 13–2). The corpus luteum persists throughout pregnancy in all farm animals except the horse. Ovariectomy at any stage terminates pregnancy in the cow, goat and sow. Bilateral ovariectomy during the latter half of pregnancy does not cause abortion in the mare and the ewe because the placenta produces progesterone in these species. Hypophysectomy before implantation terminates pregnancy and after the 44th day of gestation causes abortion in the goat.

The sheep conceptus must also neutralize or inhibit the production and/or release of uterine luteolysin. Until 50 days of pregnancy the conceptus is unable to produce sufficient luteotropins to maintain pregnancy; hypophysectomy before that time causes abortion (Fig. 13–2). Women tolerate ovariectomy and hypophysectomy from a very early stage of pregnancy owing to the production of progesterone and human chorionic gonadotropin (hCG) by the placenta.

Blood and Urinary Concentrations of Hormones. Species differences occur in the urinary excretion of estrogens (Table 13–3). In the mare, plasma estrogen concentrations (Fig. 13–3) remain low during the first three months of pregnancy, then rise steadily to reach a peak between the ninth and eleventh months, thereafter declining rapidly to term (Nett et al, 1973). The rapid decrease in urinary estrogen concentrations following bilateral fetal gonadectomy suggests that fetal gonads participate in the production of estrogens by the pregnant mare (Raeside et al, 1973). In the sow, total urinary estrogen (estrone) rate shows an increase between the sec-

Table 13–2. Effect of Ovariectomy and Hypophysectomy on the Maintenance of Gestation in Farm Animals and Other Mammals

Animal Species	Length of Gestation (Days)	Approximate Stage of Pregnancy When Operation Performed			
		Ovariectomy		Hypophysectomy	
		First Half	Second Half	First Half	Second Half
Cow	282	−	±	n.d.	n.d.
Ewe	148	−	+	−	+
She-goat	148	−	−	−	−
Sow	113	−	−	−	−
Mare	350	−	+	n.d.	n.d.
Woman	280	+	+	+	+
Monkey	165	+	+	+	+
Rat	22	−	±	−	+
Rabbit	29	−	−	−	−
Bitch	61	−	n.d.	−	±

+ = Fetuses survive; − = abortion; ± = some fetuses survive; n.d. = not determined.

(Adapted from the literature and Heap, 1972. In: Reproduction in Mammals. C. R. Austin and R. V. Short (eds), Cambridge, Cambridge University Press.)

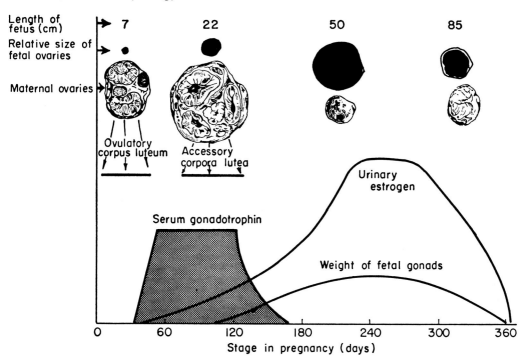

Fig. 13–2. Sequence of events during gestation in the mare: the rise and fall of serum gonadotropin and urinary estrogen in pregnant mares, and the effects of these hormones on the fetal gonads. Note that the corpus luteum verum degenerates and that accessory corpora lutea are formed when the serum gonadotropin level rises. Fetal ovaries *(upper row)* grow in response first to gonadotropin and later to estrogen and become larger than the atrophic maternal ovaries. *(Redrawn from Nalbandov's Reproductive Physiology, 1964. Cole et al., 1933, Anat. Rec. 56, 275, 1955, Brit. Med. Bull. 11.)*

ond and fifth weeks of gestation, a decline between the fifth and eighth weeks and a rapid increase to a peak at the time of parturition, which declines rapidly thereafter (Edgerton and Erb, 1972). In the cow, maximal excretion of 17 α-estradiol and to

Table 13–3. Estrogen and Related Compounds in the Urine During Pregnancy

Female	Estrone	17β-Estradiol	17α-Estradiol
Cow	+	−	+
Ewe	−	−	+
Goat	−	−	+
Mare*	+	+	+
Sow	+	−	−

* Mare's urine also contains equilin, equilenin, 17α- and 17β-dihydroequilenin.

(From the literature)

a lesser degree estrone occurs at nine months of gestation (Hunter et al, 1970).

The blood progesterone level remains constant throughout pregnancy in the ewe and cow and attains a high level early in pregnancy in the sow. Pregnanediol, the urinary metabolite of progesterone in the mare, has not been detected in other farm species.

In the mare, progesterone concentration (Fig. 13–3) up to day 35 reflects secretion by the primary corpus luteum. A rise in the level then occurs with the development of the secondary corpora lutea and this concentration is maintained until the secondary corpora lutea begin to regress at 150 to 180 days. Subsequently the plasma progestogen level remains low, but during the last two months of gestation, contrary to previous views, it rises steadily to reach a second peak that is significantly higher than previous concentrations (Barnes et al, 1975).

Physical Changes

Pregnant animals gain weight due to growth of the conceptus as well as to increases in maternal body weight. In young females, nutrient retention due to growth may mask actual weight increase due to pregnancy and would necessarily be continued in weight gain during pregnancy. Litter size and maternal weight gain in swine appear to be independent of each other.

Alteration in the distribution of body water occurs in pregnancy. Some of this is mechanical and related to the increase in venous pressure of the enlarging uterus. Edema extending from the udder to the umbilicus is observed frequently during late gestation in cows and mares.

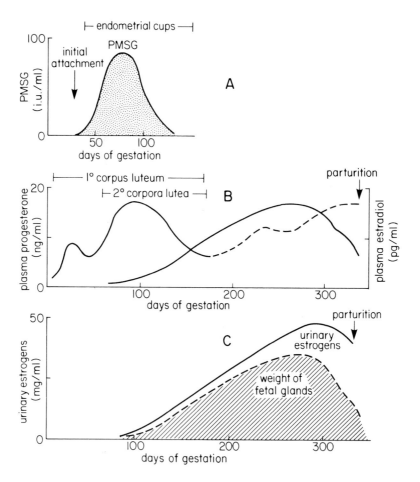

Fig. 13–3.   The relationship of the changes occurring in the uterus and ovaries of the mare and the fetal gonads to endocrine events during pregnancy. *A,* The level of PMSG reflects the functional activity of the endometrial cups. *B,* With the regression of the primary (1°) and secondary (2°) corpora lutea during pregnancy, the plasma progesterone level drops but pregnancy is maintained by placental progestagens (------). *C,* The fetal gonads respond to the increased secretion of estrogens by the fetomaternal unit by increasing in size and weight. Note that at parturition, the progestogen level continues to remain high while estrogen levels drop. *(Adapted from Hay & Allen, 1975. J. Reprod. Fertil., Suppl. 23, 557; Squires et al, 1974. J. Anim. Sci. 38, 759; Nett et al, (1973). J. Anim. Sci. 37, 962.)*

Maternal Adjustments

Blood Volume and Composition. As pregnancy advances blood volume increases in ewes and cows due to an increase in plasma volume. Unlike human pregnancy, this increase in plasma volume is not associated with a decrease in hemoglobin concentrations in the blood. The phenomenon of "physiologic anemia of pregnancy" observed in man does not occur in farm animals.

Cardiovascular Dynamics. The pregnant uterus depends on its circulation to perform various functions. In sheep, cardiac output increases during pregnancy and provides the gravid uterus with additional blood supply. The uterine blood flow, which amounts to 2% of the cardiac output in nonpregnant ewes, increases to 20% at term in pregnancy but does not keep pace with the growth of the fetus, which extracts increasing amounts of oxygen from the maternal blood allotted to it. The quantity of blood in the uterus tends to increase in proportion to its contents and is related more to fetal than placental weight (Barron, 1970).

The blood pressure of the ewe tends to fall during late gestation. An increasing cardiac output with a decrease in blood pressure indicates a decrease in peripheral resistance. Probably the uterus contains an area of low vascular resistance similar to an arteriovenous fistula.

PLACENTA

A unique feature of early mammalian development is the provision of nutrients from the maternal organism by way of the placenta. The placenta is an apposition, or fusion, of the fetal membranes to the endometrium to permit physiologic exchange between fetus and mother. The placenta differs from other organs in many respects. It originates as a result of various degrees of fetal-maternal interactions and is connected to the embryo by a cord of blood vessels. The size and function of the placenta change continually during the course of pregnancy and the organ is eventually expelled. For the fetus, the placenta combines in one organ many functional activities that are separate in the adult.

Placental Development

Fetal Membranes. The morphogenesis of the placenta during early gestation is closely related to those extraembryonic or fetal membranes that are differentiated into amnion, allantois, chorion and yolk sac (Table 13–4). The fetal membranes

Table 13–4. The Fetal Membranes of Farm Animals

Membrane	Origin	Functions
Yolk sac	Early entodermal layer	Vestigial
Amnion	Cavitation from inner cell mass	Encloses fetus in a fluid-filled cavity
Allantois	Diverticulum of hindgut	Blood vessels connect fetal with placental circulation
		Fuses with chorion to form the chorioallantoic placenta
Chorion	Trophoblastic capsule of blastocyst	Encloses embryo and other fetal membranes
		Intimately associated with lining of uterus to form placenta
Umbilical cord	Amnion wraps about the yolk stalk	Encloses allantoic vessels and acts as the vascular link between mother and fetus

participate in the formation of the placenta either separately or in certain combinations and give rise to three basic types of placentation, which differ in regard to the identity of the fetal membranes involved: *chorionic, chorioallantoic* and *yolk sac* placentation. Among these types, the chorioallantoic placentation is characteristic of all farm animals. By fusion of the outer layer of the allantois to the chorion in the chorioallantoic placenta, the fetal vessels in the allantois come into close apposition to the umbilical arteries and veins located in the connective tissue between the allantois and chorion (Fig. 13–4).

Chorionic Villi. A feature of the chorioallantoic placenta is the highly in-

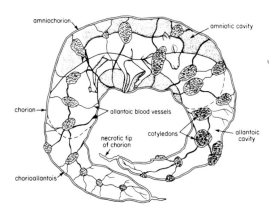

Fig. 13–4. Diagram of the fetal membranes of a 105-day fetal calf to show the allantoic and amniotic cavities. The cotyledons are distributed over the chorioallantoic membrane and the amniochorion.

creased area at the feto-maternal junction, either by the formation of chorionic villi protruding into uterine crypts or by the formation of chorionic labyrinths. The chorionic villi consist of vascular mesenchymal cones surrounded by cuboidal trophoblastic and giant binucleate cells. These either penetrate directly into the endometrium or simply interdigitate with vascular foldings of the endometrial surface (eg, in farm animals). The function of the villi is to bring the fetal (allantoic) vessels into proximity with the maternal blood vessels.

Classification of Placenta

The placenta may be classified according to morphology, microscopic characteristics of the maternal-fetal barrier, and loss of maternal tissue at birth (Table 13–5).

Gross Shape. The definitive shape of the placenta is determined by the distribution of villi over the chorionic surface (Fig. 13–5). In ruminants the fetal *cotyledons* fuse with *caruncles* or specialized projections of the uterine mucosa to form *placentomes* or functional units. The caruncles are convex in the cow and are concave in the ewe and goat (Fig. 13–6). In early pregnancy in the mare, the placenta consists of a simple apposition of fetal and maternal epithelia, but between 75 and 110 days of gestation it becomes more complex with the formation of microcotyledons.

Table 13–5.   Classification of Chorioallantoic Placentas

Species	Chorionic Villous Pattern	Maternal-fetal Barrier	Loss of Maternal Tissue at Birth
Pig	Diffuse	Epitheliochorial	None (nondeciduate)
Mare	Diffuse and Microcotyledonary	Epitheliochorial	None (nondeciduate)
Sheep, goat, cow	Cotyledonary	Epitheliochorial	None (nondeciduate)
Dog, cat	Zonary	Endotheliochorial	Moderate (deciduate)
Man, monkey	Discoid	Hemochorial	Extensive (deciduate)

(From the literature)

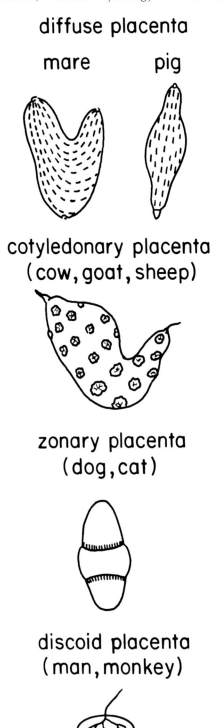

diffuse placenta

mare   pig

cotyledonary placenta
(cow, goat, sheep)

zonary placenta
(dog, cat)

discoid placenta
(man, monkey)

Fig. 13–5. The distribution of chorionic villi as the basis of classifying placental shape.

In sheep, the number of placentomes ranges between 90 and 100, which are evenly distributed between the pregnant and nonpregnant horns. In cattle, 70 to 120 placentomes develop around the fetus and progress toward the distal limit of the chorioallantois in the nongravid horn (Fig. 13–4). During pregnancy these placentomes enlarge to several times their original diameter. Those situated about the middle of the gravid horn develop to a larger size than those at the extremities. During this growth, they change from flat, plaquelike bodies to rounded, pedunculated, mushroomlike structures which, except for an area around the pedicle, are engulfed completely by the chorioallantois. Normally the chorioallantois extends into the nongravid horn; its caruncles hypertrophy, but the degree of development is usually less than that in the gravid horn.

The chorionic sacs of adjacent pig fetuses are in apposition and chorionic attachment between one or more fetuses is encountered frequently. Though there is a high incidence of chorionic fusion during multiple pregnancy, vascular anastomosis between allantoic circulations occurs rarely in sheep. In contrast, a high incidence of vascular anastomosis is encountered between twin bovine fetuses, giving rise to the well-known intersexual condition of "freemartinism."

Placental Barrier

The membranes separating the fetal and maternal circulations are collectively known as the placental barrier. This barrier is named according to the maternal and fetal tissues actually in contact, in the order maternal ⟶ fetal (Table 13–6).

Using electron microscopy the junctional zone between fetal and maternal tissues of the epitheliochorial placenta shows an interdigitation of fetal and maternal microvilli with little direct contact between fetal and maternal cell mem-

branes. Wide structural variations occur
in the epitheliochorial placenta of farm
species (Fig. 13–7). For example, the
uterine epithelium forms a partial syncytium in the ewe and large binucleate
cells are a feature of the chorionic
epithelium of the cow and ewe (Steven,
1975).

Placental Circulation

In the placenta, two circulations lie in
parallel to the fetal and maternal circulations, but the fetal and the maternal blood
do not intermingle in the epitheliochorial
placentas of farm animals.

Placental Blood Supply. The maternal
blood supply to the placenta is derived

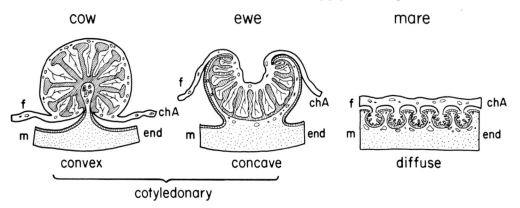

Fig. 13–6. Epitheliochorial placenta of mare, ewe and cow. The placenta of mare is diffuse and microcotyledonary; those of the ewe and cow are cotyledonary. *chA,* Chorioallantois; *end,* endometrium; *F,* fetal; *M,*
maternal. *(Redrawn from Silver et al, 1973. In: Proceedings of Sir Joseph Barcroft Centenary Symposium, K.S.
Comline et al [eds], Cambridge, Cambridge University Press.)*

Table 13–6.   Tissues Forming the Placental Barrier in Chorioallantoic Placentas

Tissues	Placental Barrier		
	Epitheliochorial	Endotheliochorial	Hemochorial
Maternal			
Endothelium	+	+	−
Connective tissue	+	−	−
Epithelium	+	−	−
Fetal			
Trophoblast	+	+	+
Connective tissue	+	+	+
Epithelium	+	+	+

Fig. 13–7. *Left.* The ultrastructural features of the epitheliochorial placenta of mare, sow, ewe and cow. Note
that the fetomaternal junction *(j)* is characterized by interdigitating microvilli between fetal *(f)* and maternal *(m)*
epithelia *(ep).* Note the presence of binucleate giant cells *(bc)* on the fetal epithelium of cow and sheep, and the
syncytial *(syn)* nature of the maternal epithelium in sheep. *bm,* basement membrane. *(Adapted and redrawn from
Steven, 1975. In: Comparative Placentation, Steven [ed]. Copyright by Academic Press, Inc. London Ltd) Right.*
Diagrammatic representation of the placental barrier. The dashed lines (cow) represent variations in arrangement
that are occasionally found. Note that the maternal capillaries *(bvm)* lie at some distance from the incomplete
syncytium that forms the maternal epithelial boundary. Note also that both maternal *(bvm)* and fetal capillaries
(bvf) are closer to the microvillous junctional zone in the mare than in the ewe or cow. *(Adapted from Silver et al,
1973. In: Proceedings of Sir Joseph Barcroft Centenary Symposium, Comline et al [eds], Cambridge, Cambridge
University Press.)*

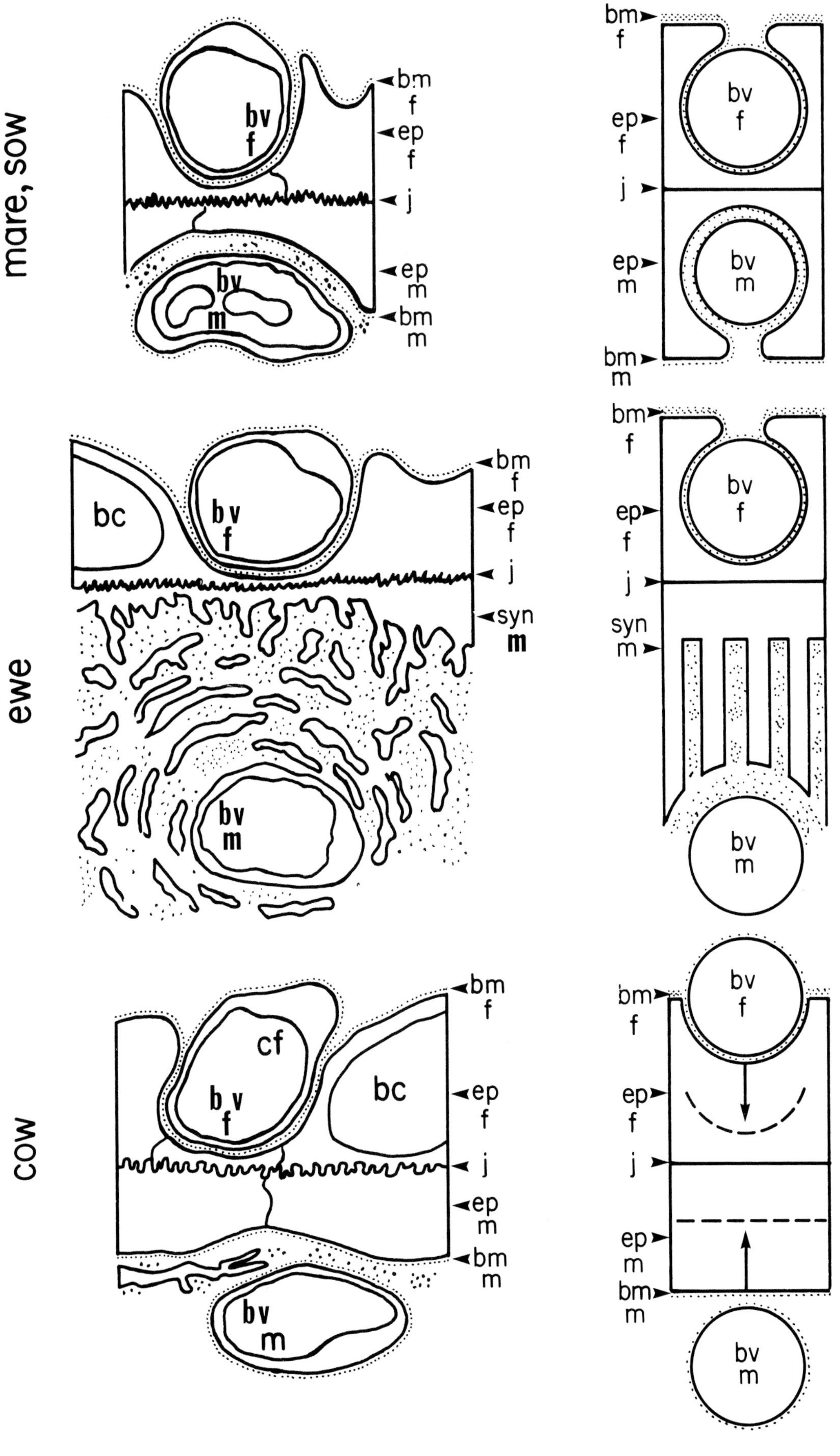

mare, sow

ewe

cow

bm
f
ep
f
j
ep
m
bm
m

bm
f
ep
f
j
syn
m

bm
f
ep
f
j
ep
m
bm
m

bv
f

bv
m

bc

bv
f

bv
m

cf
bv
f

bc

bv
m

bm
f
ep
f
j
ep
m
bm
m

bv
f

bv
m

bm
f
ep
f
j
syn
m

bv
f

bv
m

bm
f
ep
f
j
ep
m
bm
m

bv
f

bv
m

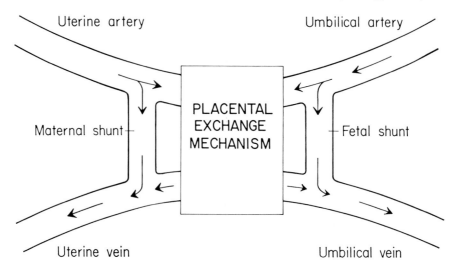

Fig. 13–8. A scheme illustrating the presence of shunts in maternal and fetal circulation through which blood flows without participating in gas exchange.

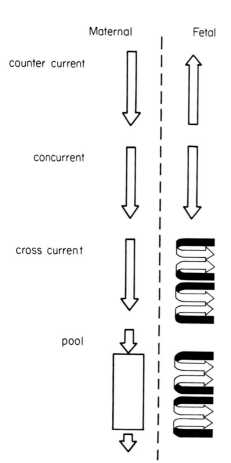

Fig. 13–9. Schematic representation of theoretic models for direction of maternal and fetal blood flow in the placenta *(Adapted from Dawes, 1968. In: Fetal and Neonatal Physiology. Chicago, Year Book Medical Publishers.)*

from the uterine arteries and veins, but because it is difficult to partition the placental supply from that to the uterine wall it is usually referred to as *uteroplacental* circulation. On the fetal side, the placental circulation is represented by the umbilical arteries and veins (Fig. 13–8).

Placental Microcirculation. Before discussing placental microcirculation, it is necessary to understand the various theoretic models that have been proposed for the direction of maternal and fetal blood flow in the placenta. Blood flow in adjacent maternal and fetal vascular channels could be *countercurrent, concurrent, crosscurrent (multivillous)* or *pool* (Fig. 13–9). In the pool flow, maternal blood enters a large space in which it is exposed to fetal capillaries.

In the ovine placenta, the uterine arteries and veins show a characteristic coiling pattern, running through the maternal septa of the caruncles and crypts. As the maternal arterioles course through the septa they give off a small number of branches to supply a network of maternal capillaries surrounding the fetal villi (Fig. 13–10, B). The mean direction of blood flow in the maternal capillaries is parallel to the long axis of the fetal villi. On the other hand, branches of the umbilical arteries and veins penetrate the villi without coiling. Each villus is supplied by a single centrally situated artery which penetrates to its tip (Fig. 13–10, A). The mean direction of flow in the fetal capillaries is at right angles to the long axes of the villi. Thus the pattern of blood flow in the ovine placenta is either crosscurrent or a mixture of crosscurrent and countercurrent.

In the mare, the microcotyledons, like the cotyledons of the sheep and cow, are highly vascularized. Long, straight arteries pass between the uterine glands to the endometrium where they branch (Fig.

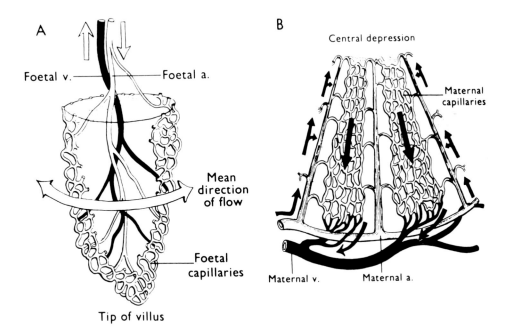

Fig. 13–10. Plan of the placental circulation in sheep. *A*, The arrangement of fetal vessels at the tip of a villous branch. Note that the collecting vein runs centrally in the villus and the mean direction of flow in the fetal capillaries is at right angles to the long axes of the villi. *B*, The arrangement of maternal vessels surrounding the fetal villi. *(From Steven, 1966. J. Physiol. Lond., 187, 18–19)*

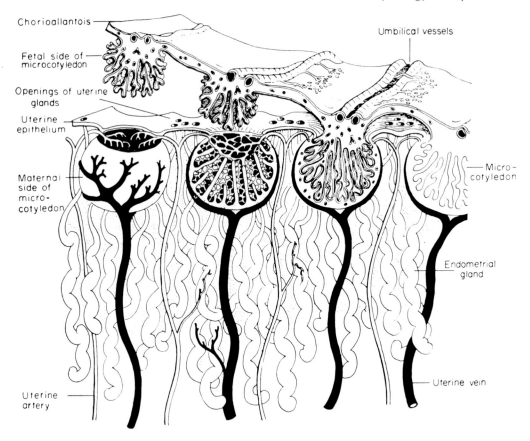

Chorioallantois

Fetal side of microcotyledon

Openings of uterine glands

Uterine epithelium

Maternal side of micro-cotyledon

Uterine artery

Umbilical vessels

Micro-cotyledon

Endometrial gland

Uterine vein

Fig. 13–11. Diagram of the mature equine placenta illustrating the structure of the microcotyledons, which are formed between 75 and 100 days of gestation *(Steven and Samuel, 1975. J. Reprod. Fertil. Suppl. 23, 580)*

13–11). As each branch passes over the rim of the microcotyledon, it gives rise to a dense vascular network in the walls of the maternal crypts. This network drains to a single vein which passes to the sub-endometrial complex. On the fetal side, the chorionic villi are supplied by branches of the umbilical arteries and veins. The direction of the blood flow in the fetal capillaries is from the tip to the base of each villus and is opposite to the flow in the maternal capillaries (counter-current).

Uterine Blood Flow and Distribution. In sheep the rate of uterine blood flow increases and is related to fetal weight as gestation advances (Barron, 1970). About 84% of the total uterine flow near term passes to the placentomes; the remainder supplies the endometrial and myometrial layers of the placenta (Fig. 13–12).

Umbilical Blood Flow. The mean umbilical blood flow in goat, sheep and cow is higher than that in the mare (Table 13–7). The reasons for the low umbilical blood flow in the fetal foal *in utero* are not clear, but may be related to the absence of a ductus venosus, to an increased resistance to umbilical venous return and to a low fetal heart rate. The relative distribution of umbilical blood flow in a pregnant sheep at term (Fig. 13–12) shows that 94% of the total umbilical flow is distributed to the cotyledons, while only 6% supplies the chorioallantois.

The rate of umbilical flow increases

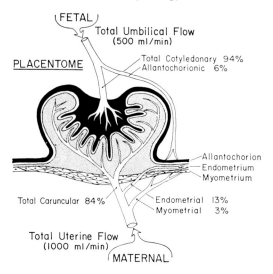

FETAL

Total Umbilical Flow
(500 ml/min)

PLACENTOME

Total Cotyledonary 94%
Allantochorionic 6%

Allantochorion
Endometrium
Myometrium

Total Caruncular 84%

Endometrial 13%
Myometrial 3%

Total Uterine Flow
(1000 ml/min)

MATERNAL

Fig. 13–12. Distribution of uterine and umbilical blood flow in a pregnant sheep at term. Percentage of blood flow is given in parentheses. Note that in late gestation only a small part of blood is shunted away from placenta. *(Adapted from Makowski et al., 1968. Circulation Res. 23, 623.)*

Table 13–7.　Umbilical Blood Flow in Conscious Farm Animals During Late Gestation

Species	Umbilical Blood Flow (ml/kg/min)
*Sheep and goat	302
Cow	232
Mare	171

Based on Crenshaw et al (1968). Quart. J. Exp. Physiol. 53, 66; Silver and Comline (1975). J. Reprod. Fertil. Suppl., 23, 589–594.

with advancing pregnancy to meet the growing demands of the fetus. The overall increase in umbilical flow is achieved by a decreased umbilical vascular resistance earlier in pregnancy and by an increased arterial blood pressure later.

Placental Functions

The placenta performs many functions and substitutes for the fetal gastrointestinal tract, lung, kidney, liver and endocrine glands. In addition, the placenta separates the maternal and fetal organisms, thus ensuring the separate development of the fetus.

Placental Exchange

The blood of the fetus and dam never come into direct contact. Yet, the two circulations are close enough at the junction of the chorion and endometrium for oxygen and nutrients to pass from maternal to fetal blood, and waste products in the opposite direction.

The placental membrane controls the transfer of a wide range of substances by several processes, such as simple diffusion, active transport, *phagocytosis* (engulfing food) and *pinocytosis* (engulfing water). The mechanism of transfer between mother and fetus can be based on the physiologic significance of the material to be transferred (Table 13–8). To

Table 13–8.　Classification of the Transfer of Substances Across the Placenta

Group	Physiologic Role	Substances	Exchange Mechanism
I	Maintenance of biochemical homeostasis or protection against sudden fetal death	Electrolytes, water and respiratory gases	Rapid diffusion
II	Fetal nutrition	Amino acids, sugars and most water-soluble vitamins	Predominantly by active transport systems
III	Modification of fetal growth or the maintenance of pregnancy	Hormones	Slow diffusion
IV	Immunologic or toxic importance	Drugs and anesthetics	Rapid diffusion
		Plasma proteins, antibodies and whole cells	Pinocytosis or leakage through pores in placental membrane

(Adapted from Page, 1957. Am. J. Obstet. Gynecol. 74, 705)

drive these processes, energy is derived from energy-yielding metabolic reactions. The placenta contains enzymes for glycolysis, the citric acid cycle and the pentose phosphate pathway. Placental oxygen consumption is comparable to that of such organs as the liver or kidney.

Respiration. Many similarities exist between the gas exchange across the placenta and that across the lungs. The major difference, however, is that in the placenta it is a fluid-to-fluid system whereas in the lung it is a gas-to-fluid system. The umbilical arteries carry unoxygenated blood from the fetus to the placenta, while the umbilical veins carry oxygenated blood in the reverse direction.

Several factors affect the placental transfer of oxygen: maternal and fetal oxyhemoglobin dissociation curves, diffusion distance and total surface area for diffusion, O_2 use by placental tissue, O_2 affinity and capacity, the spatial relation of the maternal and fetal exchange vessels, and the maternal and fetal blood flow rates (Longo, 1972; Comline and Silver, 1974). The fetal dissociation curve for oxygen lies to the left of the maternal curve in sheep, goat, pig and cow but not in horse. This means that at any given oxygen tension the fetal blood contains more oxygen than the maternal blood. In sheep, goat and cow, the minimum P_{O_2} gradient between fetus and mother is about 20 mm Hg, whereas in the mare this difference is only 2 to 4 mm Hg. This high affinity of fetal blood to O_2 may be related to a specific hemoglobin type that has a high O_2 affinity (ruminants) or to the presence of a high concentration of 2, 3-diphosphoglycerate (DPG), which shifts the maternal O_2 dissociation curve to the right (pig). On the contrary, fetal and maternal hemoglobin types are identical but the smaller difference in DPG concentration between maternal and fetal erythrocytes accounts for the small P_{O_2} difference between maternal and fetal blood.

Four basic systems have been suggested for gas exchange in the placenta (Fig. 13–9). The efficiency of oxygen exchange varies with the particular system. It is maximal in the countercurrent system and minimal in the concurrent system. The efficiency of the multivillous system is intermediate between the previously mentioned systems. Gas exchange in the pool system is less than that in the multivillous system but comparable to that of a concurrent system. It is difficult to ascertain which of these systems is primarily involved in a particular species and some species probably contain more than one system. For example, gas exchange in the sheep placenta is either crosscurrent or a mixture of countercurrent and crosscurrent systems.

Carbon dioxide diffuses freely from the fetal to the maternal circulation as in the blood of nonpregnant adults. The transfer of CO_2 from the fetus to the mother is facilitated by certain physiologic mechanisms. For example, the maternal and fetal dissociation curves for CO_2 show that fetal blood has a lower affinity for CO_2 than maternal blood during placental oxygen transfer. This favors the diffusion of CO_2 from fetal to maternal blood. P_{CO_2} gradients in all species are much smaller than P_{O_2} gradients, but differences between the ruminant and equine placental P_{CO_2} gradients still exist.

In sheep, cow and mare, no change in fetal blood gases, pH, packed cell volume, O_2 saturation or O_2 capacity occurs during the last third of pregnancy.

Nutrition. The placenta permits the transport of sugars, amino acids, vitamins and minerals to the fetus as substrates for fetal growth; it serves as a storage organ for glycogen and certain other substances such as iron.

Water and Electrolytes. The placental membranes are freely permeable to water and electrolytes. Circulation of water and exchange of solutes also occurs between fetus and the amniotic and allantoic sacs in addition to the exchange across the placenta.

Large quantities of sodium and other electrolytes cross the placenta by simple diffusion and by "solvent drag." Although the osmolarity of the fetal plasma is lower than that of the maternal plasma, the [Na$^+$] shows no consistent difference.

Minerals. Iron tends to be more abundant in the fetus than in the mother and is stored in the fetal liver, spleen and bone marrow. The pig fetus obtains its iron from the uterine lumen by way of the areolae. Copper readily traverses the bovine placenta. Both copper and iron accumulate in the fetal liver but apparently manganese does not accumulate. Calcium and phosphorus enter fetal blood against a concentration gradient.

Carbohydrates, Fat and Proteins. Glucose is the major metabolic fuel of the fetus and is transferred across the placenta by means of an active transport system. The fetal blood glucose level is lower than that of the mother. For example, the fetal plasma glucose level is only 25% of that in the ewe; the percentage is somewhat higher in the cow, while fetal levels as high as 60% of those in maternal plasma occur in the mare (Comline and Silver, 1974).

In man, glucose is the principal fetal carbohydrate and the placenta contains large amounts of glycogen synthesized mainly from maternal glucose. In ruminants, fructose is the main carbohydrate in fetal blood and the placenta contains only small amounts of glycogen. Fetal fructose is produced by the placenta from glucose and its function in the fetus is obscure. Fructose comprises about 70% to 80% of the sugar in fetal blood, while glucose is predominant in maternal blood.

Fetal fat is probably derived from free fatty acids (FFA) transferred across the placenta and by fetal synthesis from carbohydrates and acetate. FFA are transported across the placenta by simple diffusion. The levels of FFA in maternal and fetal blood are closely correlated in the mare, but FFA transfer across the ruminant placenta is minimal.

Proteins as such are not transferred. The fetus probably synthesizes most of its proteins from amino acids. Amino acids cross readily against a concentration gradient and are found in higher concentrations in fetal than in maternal plasma. Immunoglobulins are transmitted in man and some animals but not in farm animals. This can be explained by structural differences in the various placental types.

Vitamins and Hormones. The lipid soluble vitamins (A, D and E) are impeded by the placenta; thus at birth the concentrations of these are lower in the fetus than in the mother. Water soluble vitamins (B and C) cross the placental barrier more readily than those that are lipid soluble.

Polypeptides cross the placenta slowly. In sheep, although iodine crosses the placenta readily, there is little or no transfer of thyroid hormones or thyroid stimulating hormone. Insulin also probably crosses only slowly and in insignificant amounts.

Cortisol is not transferred from mother to fetus in goats and sheep. The unconjugated steroids, progesterone and estrogens cross the placental barrier readily. Many steroids undergo enzymic alteration in moving across the placenta and such alterations play a major role in their transport.

Hormone Production

The placenta is a transient endocrine organ like the corpus luteum. It secretes both tropic and steroid hormones that are released into the fetal as well as the maternal circulations. Some of these hormones enter the amniotic fluid or are reabsorbed by the fetus or mother. Pregnant mare serum gonadotropin (PMSG), estrogens and progesterone are produced by the placenta. (Concentrations of these hormones in maternal blood and urine have been presented in a previous section.)

The concept of a fetoplacental unit was proposed to explain the various mechanisms by which large amounts of progesterone and estrogens are produced during pregnancy. Both the placenta and the fetus lack certain enzymatic functions that are essential for steroidogenesis. But enzymes absent from the placenta are present in the fetus and *vice versa*. Thus by sequential integration of the fetal and placental steroidogenetic functions, the fetoplacental unit can elaborate most, if not all, hormonally active steroids.

The ewe and mare, unlike the cow, the goat and the sow, are capable of synthesizing sufficient amounts of progesterone to maintain pregnancy by utilizing acetate and cholesterol derived from the maternal circulation.

During the latter half of gestation a high rate of estrogen production occurs in the placentas of the mare, cow, sow and sheep. The placenta relies on the fetus to provide precursors for the synthesis of estrogens, whereas fetal assistance is not required for the synthesis of progesterone.

Placental Immunology

As a result of the paternal contribution, the mammalian fetus differs genetically from the host, and the placenta may be considered as an *allograft* or tissue from a different individual of the same species. The fetus does provide antigenic stimulation to the maternal immune system, but unlike the immunologic destruction of allografts in other parts of the organism, the placenta is not rejected until parturition, a period far in excess of the time taken to elicit an allograft reaction (usually in two to three weeks). The failure of maternal tissue to reject the placenta has puzzled immunologists and has led to a

Table 13–9. Hypotheses Proposed to Account for Protection of the Fetus from Allograft Reaction During Pregnancy

Site	Proposed Mechanism
Fetus	Antigenically immature
Mother	Immunologically inert
Uterus	Immunologically privileged site
Placenta	Immunologic barrier

number of theories (Table 13–9) to explain the unique relationship between mother and fetus.

The hypothesis that the fetus is antigenically immature can be discarded because transplantation antigens have been demonstrated in the embryos of mouse and other species. Despite this antigenic stimulation, the maternal immune system apparently remains ineffective.

The mother's immunologic activities may be reduced or suspended during pregnancy since hCG, estrogens, and corticosteroids interfere with local immunologic reactions during pregnancy, at least in man and rats. However, a fetus that is removed from a pregnant rat or rabbit and grafted onto the flank muscle of its mother is subjected to an allograft reaction (Woodruff, 1958). Thus, a reduction or suspension in the activity of the maternal immune system is not sufficient to explain the tolerance of the fetus.

The uterus is also not an immunologically privileged site. Protection of the fetus is not a property conferred by the uterus because normal fetal development occurs in completely abnormal sites such as the oviduct, anterior chamber of the eye and peritoneum. Ectopic (extrauterine) pregnancies are not uncommon in man. Also normal embryos develop from mouse eggs transplanted to kidney, brain, spleen and abdominal cavity.

More important is the antigenic status of the placenta. At this site, maternal and fetal tissues come into physical contact with one another, and hence they are the primary focus of any immunologic interaction. Thus the fetus might owe its privileged position as an allograft to the presence of a maternal-fetal physical barrier.

The exact location of this immunologic barrier is controversial but one possibility might be the sialo-mucin (fibrinoid) layer that exists between the trophoblast cells and maternal tissue in most chorioallantoic placentas. Other sites rich in sialomucin (eg, anterior chamber of the rabbit eye) have unique immunologic properties. This suggests that the sialo-mucin coat plays a role in protecting the trophoblast from a maternal lymphocyte attack. However both fetal and maternal cells can cross the fibrinoid barrier and survive in a hostile environment. Also, the fibrinoid layer is not continuous between maternal and fetal tissues; it is absent around the villi of the human placenta. Thus if the fibrinoid layer is the sole immunologic barrier, the trophoblastic cells would be exposed to attack in these unprotected sites.

An alternative possibility is that the trophoblastic cells themselves are the site of the immunologic barrier. Unlike fibrinoid, the trophoblastic cells form a continuous barrier around each fetus and are present in all types of chorionic placentas.

The exact nature of the immunologic mechanism remains to be established. Apparently two factors operate to protect the fetus (Fig. 13–13). Graft rejection is mediated by sensitized lymphocytes (cell-bound immunity). The destructive ability of these cells can be blocked, primarily by the presence of circulating humoral antibody by a process known as "immunologic enhancement." In this process, a humoral antibody is produced that protects rather than destroys the target cells (graft). The target cells are probably trophoblastic in origin. Presumably the humoral antibodies combine with all the available antigenic sites on the target cells, which are then no longer

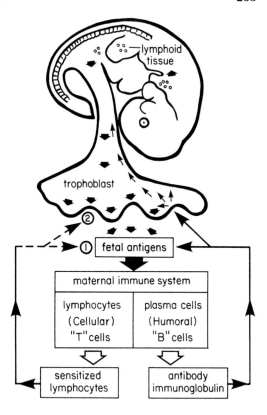

Fig. 13–13. Diagram of the processes occurring to protect the fetus from an immunologic attack by the sensitized lymphocytes of the mother. (1) "Immunologic enhancement" involves the combination of humoral antibodies with sites on the trophoblast cells, thereby blocking the sensitized lymphocytes from reaching these cells. (2) The sialo-mucin coat and/or the trophoblast forming a physical barrier between mother and fetus act to mask transplantation antigens and to repel the lymphocytes.

available for attack by more powerful antibodies (sensitized lymphocytes). Thus in the mammalian embryo, the cellular immune reaction at the trophoblast is inhibited.

PRENATAL PHYSIOLOGY

Prenatal Periods

The prenatal development of farm animals may be divided into three main periods. The *period of the ovum* culminates with the initial attachment of the blastocyst but is prior to the establishment of an intraembryonic circulation. The *embryonic period* extends from day 15 to day 45 of gestation in the cow; day 11 to about day 34 in the sheep, and day 12 to day 60 in the mare. In this period, rapid growth and differentiation occur during which the major tissues, organs and systems are established and the major features of external body form are recognizable. The *fetal period* extends from about day 34 of gestation in sheep, day 45 in cattle, and day 60 in horses until birth. This period is characterized by growth and changes in the form of the fetus.

Fate of Germ Layers. During early differentiation the cells at one pole of the blastocyst, the germ disc, differentiate into an outer *ectodermal* and inner *endodermal* layer separated by an intermediate layer of *mesoderm*. These three primary germ layers make specific contributions to the formation of different tissues and organs (Fig. 13–14).

External Form of the Embryo. The somites or body segments develop from the outer layer (somatic layer) of the mesoderm. They are the bases from which the greater part of the axial skeleton and

musculature are developed. The somites differentiate into three regions. The first region develops into the vertebrae, which encase the neural tube. The second region, the upper part near the neural tube, forms the skeletal muscles. The third region, which is the lower part of the somites, forms the connective tissues of the skin. Differentiation of the somites commences 19 days after ovulation in cattle, and the number of somites increases rapidly to 25 on day 23, 40 on day 26 and 55 on day 32 of gestation.

Organogenesis. In cattle most organs and body parts are formed between the second and sixth week of gestation. During this period the digestive tract, lungs, liver, and pancreas all develop from the primitive gut. The beginnings of muscular, skeletal, nervous and urogenital systems are established. On day 21 of gestation, the heart begins to function and the circulation is initiated.

Fetal Nutrition and Metabolism

Whereas the blastocyst and the early embryo are nourished by endometrial fluid, the fetus receives its supply of nu-

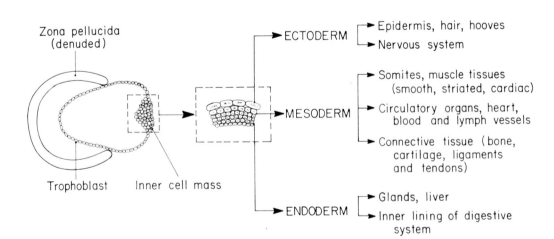

Fig. 13–14. The derivation of various body organs by progressive differentiation and divergent specialization. The origin of all fetal organs can be traced back to the primary germ layers which originate from the inner cell mass.

trients from the maternal circulation across the placenta. The fetus may be regarded as a parasite living within the mother and it has priority in the event of insufficient maternal nutrition so that its development can proceed unimpaired. It needs carbohydrates, proteins, vitamins and minerals for maintenance, differentiation and subsequent development and growth.

The fetus receives a continuous supply of glucose from its mother through the placenta. Glucose is the major metabolic fuel of the fetus. The level of glucose in fetal blood is lower than that in maternal blood. Placental metabolism of glucose probably accounts for part of the maternal-to-fetal glucose concentration difference. Toward the end of gestation, the normal fetus accumulates glycogen in its liver and skeletal muscles to assist it to overcome the transitional period after birth until efficient suckling is established. Although the blood of fetal ungulates (cattle, sheep, goats) contains large concentrations of fructose, its utilization is negligible, but more is utilized when blood glucose levels are low.

The fetus synthesizes all its proteins from the amino acids derived from the mother; proteins are used mainly for synthesis rather than oxidation or gluconeogenesis. However, as much as 40% of the total amount of nitrogen that passes from the mother to the fetus in sheep is returned to the maternal circulation, and this strongly suggests that amino acid catabolism plays an important role in the metabolic needs of the fetus; it could account for at least 25% of the total fetal oxygen consumption.

Throughout gestation the retention of calcium, phosphorus and iron increases relative to fetal body weight. The fetus has the unique ability to deplete maternal skeletal stores of calcium if feeds are low in calcium. Iron is used for hemoglobin synthesis but little is known about its distribution and metabolism.

Fetal Growth

Growth Rate. As it grows from the spherical fertilized ovum to the full-term fetus the embryo not only increases in size and weight but also undergoes many changes in form. The rate of growth, ie, the percentage increase in weight and dimensions per unit of time (*relative growth*), is most rapid in the earlier stages and declines as gestation advances, whereas the absolute increment per unit of time (*absolute growth*) increases exponentially, reaching a maximum during late gestation. In cattle, over one half of the increase in fetal weight occurs during the last two months of gestation. At term the weight of the fetus contributes to approximately 60% of the total weight of the conceptus.

The growth rates of the fetus and its component organs and tissues vary during different stages of intrauterine life. For example, during early fetal development the cephalic region grows rapidly and, consequently, the fetal head is disproportionately large. Later in gestation, cephalic growth slows. At birth, the head and limbs are relatively more developed than the muscles.

Whereas certain fetal organs grow rapidly early in prenatal life, others begin to grow later. For each organ and tissue, growth rate increases to a maximum, then declines. These maximal rates of growth occur in a definite sequence: first, the central nervous system, then bones and lastly muscle and adipose tissue. If the fetus is deprived of nutritional supplies, some fetal organs are affected more than others according to the stage of pregnancy when nutritional deprivation occurs. Some organs, however, seem to have nutritional preference regardless of stage of deprivation. In growth retarded fetuses, the heart and brain are less affected, and the liver more so, than body weight.

Estimation of Age. Unless precise information of coitus and ovulation is available, the main criteria used in determining age of embryos and fetuses are based on size and general features of development. However, these measurements and features are variable and only approximate the fetal age.

The measurements commonly employed are: (1) the crown-rump length (CR), which is the measurement from the vertex of the skull to the root of the tail; (2) the length from the tip of the nostril to the tip of the tail over the back in a sagittal plane (this is considered a reliable estimate of fetal age in sheep); and (3) the length of the radius and tibia in cattle and sheep. All these measurements vary from one fetus to another owing to such factors as breed, strain, maternal age, litter size and season of birth.

An ideal method of estimating age would be to use the differentiation and development of embryonic and fetal structures, or the so called "*developmental horizons*," as a guide, but this information is incomplete for farm mammals (Table 13–10).

Factors Affecting Fetal Growth. The rate of fetal growth depends primarily upon the feed supply and the ability of the fetus to utilize the feed. Species, breed and strain differences in fetal size are due to differences in the rate of cell division which is determined genetically. Thus there is close integration between the feed supply to the fetus (*environmental factors*), the rate of cell division (*genetic factors*) and hence the rate of growth.

Genetic Factors. Holstein fetuses at birth weigh about 35% more than Jersey calves and about 15% more than the average dairy calf. Similarly, Romney sheep

Table 13–10. Some Outstanding Horizons in Development of Embryos and Fetuses of Farm Mammals

Developmental Horizons	Mare (Days)	Cow (Days)	Ewe (Days)	Sow (Days)
Morula		4–7	3–4	3.5
Blastula	14	7–12	4–10	4.75
Differentiation of germ layers	—	14	10–14	7–8
Elongation of chorionic vesicle	—	16	13–14	9
Primitive streak formation	14	18	14	9–12
Open neural tube	—	20	15–21	13
Somite differentiation (first)	18	20	17	14
Fusion of chorioamniotic folds	—	18	17	16
Chorion elongates in nonpregnant horn	—	20	14	—
Heart beat apparent	24	21–22	20	16
Closed neural tube	—	22–23	21–28	16
Allantois prominent (anchor-shaped)	—	23	21–28	16–17
Forelimb bud visible	24	25	28–35	17–18
Hindlimb bud visible	24	27–28	28–35	17–19
Differentiation of digits	50	30–45	35–42	28+
Nostril and eyes differentiated	40	30–45	42–49	21–28
Cotyledons first appear on chorion	—	30	—	—
Allantois replaces exocoelom of pregnant horn	—	32	21–28	—
First attachment (implantation)	40	33	21–30(?)	24
Allantois replaces all of exocoelom	—	36–37	—	25–28
Eyelids close	60	60	49–56	—
Hair follicles first appear	—	90	42–49	28
Horn pits apparent	—	100	77–84	—
Tooth eruption	—	110	98–105	—
Hair around eyes and muzzle	160	150	98–105	—
Hair covering body	220	230	119–126	—
Birth	340	280	147–155	112

(For mare: *Adapted from Douglas and Ginther, 1975, J. Reprod. Fertil. Suppl. 23, 503;* for cow: *Adapted from Salisbury and VanDemark, 1961. Physiology of Reproduction and Artificial Insemination of Cattle. San Francisco, Freeman & Co.;* for ewe: *Cloette, 1939. Onderstep. J. Vet. Sci. Anim. Ind. 13, 417;* for sow: *Patten, 1948. Embryology of the Pig. Philadelphia, The Blakiston Co.*)

fetuses grow faster than Merino fetuses. The maternal contribution to variability in fetal size is greater than the paternal contribution.

Environmental Factors. The major factors that affect prenatal growth include size, parity and nutrition of the mother, litter size, placental size and climatic stress. Of these factors, maternal size is important. The size of the young at birth from reciprocal crosses between the large Shire horses and the small Shetland ponies depends mainly upon the size of the mother. The size of the sire only begins to exert much influence on growth after birth. Similar observations have been made in cattle and sheep.

Fetal size is also influenced by the age of the dam. Young dams that have not reached adult size continue to grow during their first pregnancy and thus compete with the fetus for the available nutrients.

Maternal nutrition exerts an important influence on fetal growth, notably in sheep. Undernutrition of the ewe during the latter part of gestation leads to the production of stunted lambs, even though a normal level of nutrition had been present earlier. Conversely, a reversed type of feeding program results in normal-sized lambs.

In polytocous species such as the pig, during early gestation, feed and uterine accommodation are adequate, but in the later stages, the number of fetuses sharing the uterine blood supply can have a profound influence upon their size at birth. The length of gestation in the pig is not reduced by increases in litter size, suggesting that the small birth weight with large litters must be related to the availability of nutrition to individual fetuses. In monotocous species, notably cattle, twin fetuses are generally smaller than single fetuses, probably because the length of gestation is reduced.

In sheep, fetal weight is related to placental weight. However, in several species including man, the difference in fetal size between singletons and twins in late pregnancy is not wholly accounted for by the difference in placental weight.

High ambient temperature during pregnancy affects fetal size. Exposure of pregnant ewes to heat stress reduces fetal growth, the degree of reduction being proportional to the length of exposure. This dwarfing is a specific effect of temperature and not due to reduced feed intake during pregnancy.

Fetal Circulation

Course of the Circulation. The fetal circulation (Fig. 13–15) is essentially similar to that of the adult except that oxygena-

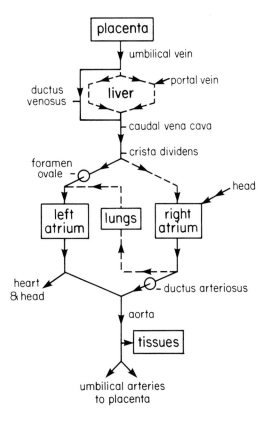

Fig. 13–15. Schematic representation of fetal circulation. The ductus venosus, crista dividens, foramen ovale and ductus arteriosus act as shunts directing oxygenated blood away from the liver, right ventricle and functionless lungs respectively.

tion of blood occurs in the placenta rather than in the lungs. In addition, the circulation has several shunts or bypasses which direct oxygenated blood to the tissue. To avoid metabolism by the liver, a major portion of the blood in the umbilical vein is shunted through the *ductus venosus* in the liver into the caudal vena cava. In the pig and the horse, a ductus venosus never develops and the umbilical venous blood passes through the liver sinusoids. The *crista dividens*, a structure projecting from the border of the foramen ovale, separates the caudal vena cava flow into two separate streams before the atria are reached; the stream from the ductus venosus is guided through the foramen ovale largely into the left atrium, thereby directing oxygenated blood to the head and developing the left ventricle in the neonatal period. The ductus arteriosus shunts most of the pulmonary arterial blood flow into the aorta away from the functionless lungs. Blood leaving the aorta above the ductus is distributed to the descending aorta. The two umbilical arteries are long, highly contractile with thick muscular layers and originate from the caudal end of the descending aorta. They carry blood to the placenta.

Cardiovascular Changes. The blood volume of fetal lambs increases during gestation with a linear relationship between blood volume and fetal body weight.

Regional Distribution of Blood. The proportion of blood (34%–91%) passing through the ductus venosus increases with increasing umbilical blood flow. The shunting of blood through the foramen ovale depends on whether the fetus is hypoxic or acidemic. Total cardiac output and output per kilogram body weight are higher in the latter stages of gestation than at near term. Shunting the blood through the foramen ovale and ductus arteriosus returns approximately 55% of the total cardiac output of the fetus directly to the placenta, 10% to perfuse the lungs and

35% to perfuse the body tissues (Rudolph and Heymann, 1967).

Blood Pressure. Striking differences exist between fetal and adult blood pressures. The higher blood pressure in the right side of the fetal heart than in the left side keeps the foramen ovale patent. Likewise the pressure in the right ventricle is greater than in the left ventricle. The higher pressure in the pulmonary artery than in the aorta, probably due to the high vascular resistance existing in the fetal lungs, causes blood to flow from the pulmonary artery into the aorta by way of the ductus arteriosus.

The mean aortic pressure of the fetal lamb rises gradually during the latter stages of pregnancy while the biggest drop in pressure in the fetal circulation occurs in the fetal placenta (Fig. 13–16).

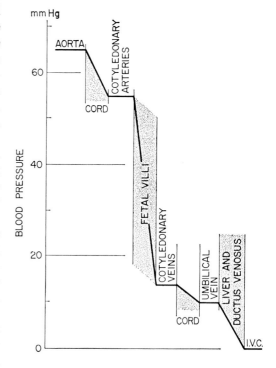

Fig. 13–16. Diagram illustrating the fall in blood pressure from the aorta through the umbilical circulation and liver to the inferior vena cava *(IVC)* in mature lamb. The biggest drop in pressure occurs in the placenta. *(Redrawn from Dawes, 1962. Am. J. Obstet. Gynecol. 84, 1634.)*

Fetal Heart Rate. In general heart rates are higher in the fetus than in the adult. Fetal heart rates differ in various species as well as at different stages of gestation within each species. The fetal heart rate ranges from 170 to 220 per minute in sheep and from 120 to 140 per minute in cattle.

Fetal Fluids

Origin. The origin of fetal fluids (amniotic and allantoic) and the secretions that contribute to them are complex (Table 13–11). In the fetal lamb, urine formed by the mesonephros passes into the allantoic cavity through the urachus up to about 90 days of gestation. Thereafter urine passes in increasing quantities into the amniotic sac due to occlusion of the urachus and patency of the urethra. Thus fetal urine forms a major source of amniotic fluid in the latter part of pregnancy in sheep. In other species sources other than the fetal kidney may influence the amount and composition of amniotic fluid: (a) secretions from fetal salivary glands, buccal mucosa, lungs and trachea; and (b) dynamic interchange between maternal, fetal and amniotic fluid compartments. In the pig, the initial accumulation of allantoic fluid is the result of the secretory activity of the allantoic membrane; later in gestation, however, the fetal urine provides most of the allantoic fluid (Table 13–11).

In cattle, the physical characteristics of the fetal fluids gives some indication regarding their sources. During early pregnancy the fetal fluids resemble urine. Probably the fetal urine enters the allantoic and amniotic cavities by way of the urachus and the urethra respectively. At midpregnancy, the amniotic fluid changes from a watery to a mucoid fluid. This might be due to an increase in the secretions from the respiratory tract and buccal cavity, and/or to a decrease in the flow of urine into the amniotic cavity as the sphincter of the fetal bladder begins to function.

A rapid exchange of water occurs between the maternal circulation, the fetal circulation and the amniotic fluid with a net water circulation: mother $\longrightarrow$ fetus $\longrightarrow$ amniotic fluid $\longrightarrow$ mother. The fetus also removes fluid by swallowing or by drawing amniotic fluid into the fetal lungs during respiratory movements.

Volume. The relative volumes of fluid in the amniotic and allantoic cavities show much fluctuation during pregnancy

Table 13–11. Origin, Composition and Functions of Fetal Fluids in Farm Animals

Fluid	Origin	Composition	Functions
Amniotic	Fetal urine Secretions from respiratory tract and buccal cavity Maternal circulation	A solution with suspended particulate material Low levels of K^+, Mg^{++}, glucose creatinine, uric acid and urea High levels of Na^+, Cl^-, P^{+++} and fructose Enzymes, iron, amniotic plaques, cells	Protects fetus from external shock Prevents adhesion between fetal skin and amniotic membrane Assists in dilating cervix and lubricating birth passages during birth
Allantoic	Fetal urine Secretory activity of allantoic membrane	Ultrafiltrate Low levels of Na^+, Cl^-, P^{+++}, glucose High levels of K^+, Mg^{++}, Ca^{++}, fructose, creatinine, uric acid and urea	Brings allantochorion into close apposition with endometrium during initial steps of attachment Stores fetal excretory products not readily transferred back to the mother Helps to maintain osmotic pressure of fetal plasma

(Fig. 13–17). These variations probably reflect the contributions of the fetal and maternal compartments. Fetal fluids increase throughout gestation in all species, but in the pig they tend to decline at term. The amniotic fluid in the ewe reaches a maximum during midpregnancy, falling thereafter; in the mare it equals the volume of the allantoic fluid during the latter stages of pregnancy. Similarly the allantoic fluid increases during the course of gestation and, especially in the cow, a considerable increase occurs a few weeks prior to calving. The volume of allantoic fluid is relatively higher than amniotic fluid during pregnancy, the exception being the ewe at midgestation.

Functions. Amniotic fluid is not a stagnant pool but a vital fluid bathing the fetus and performing several functions (Table 13–11). Allantoic fluid, composed of hypotonic urine, maintains the osmotic pressure of the fetal plasma and prevents fluid loss to maternal circulation. In the pig, the chorioallantoic membrane possesses secretory properties and it is capable of actively removing sodium from the allantoic cavity, thereby maintaining the allantoic fluid hypotonic relative to bladder or serum.

Composition. Amniotic and allantoic fluids contain metabolic constituents, electrolytes, enzymes, hormones, cells and other structures. Numerous biochem-

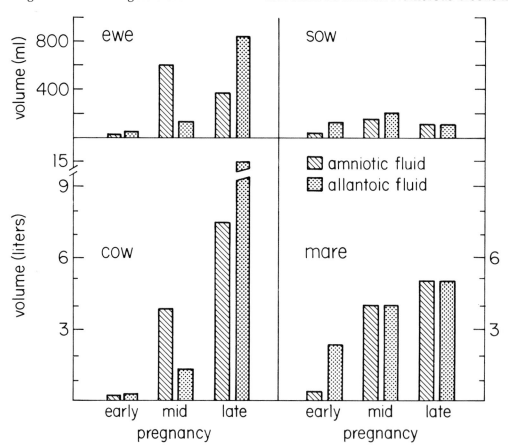

Fig. 13–17. Volume of amniotic and allantoic fluids at three stages of gestation. Note that the volume of fetal fluids increases throughout gestation in ewe, cow and mare, whereas in the sow it decreases. In the cow allantoic fluid increases considerably during the last few weeks of gestation. *(Data from Arthur, 1969. J. Reprod. Fertil. Suppl. 9, 45)*

ical constituents have been investigated in cattle (Baetz et al, 1976). Allantoic fluid has lower concentrations of sodium, chloride and phosphorus, and higher levels of potassium, calcium, magnesium, creatinine and uric acid, than amniotic fluid (Table 13–11). High concentrations of fructose are also present in both fluid compartments. The urea concentrations are similar in both fluids except that the amounts in amniotic fluid increase with fetal maturity.

In ruminants, the inner lining of the amnion, particularly near the umbilicus, contains numerous raised, discrete, round foci called *amnionic pustules,* which are rich in glycogen and disappear late in gestation. Amniotic fluid also contains cells that may be utilized for a prenatal diagnosis of sex. *Hippomanes* are smooth, discoid, rubber-like, dark brown masses floating in the allantoic fluid, and are probably aggregations of fetal hair and meconium.

PARTURITION

Parturition or labor is defined as the physiologic process by which the pregnant uterus delivers the fetus and placenta from the maternal organism.

The Pregnant Uterus at Term

Before normal labor begins, the fetus usually assumes a position in the uterus characteristic of the species. It is presented at the time of parturition in a position that poses the least amount of difficulty for passing through the pelvic girdle. In monotocous species, the fetus lies on its back during intrauterine life. Prior to labor it rotates to an upright position with its nose and forelegs directed toward the posterior end of the dam. The anterior presentation is the most common in ruminants; the front feet of the fetus emerge first with the nose between them; the head is extended and the dorsum of the fetus is in contact with the sacrum of the dam. This presentation coordinates the natural curvature of the birth canal and the curvature of the fetus. Posterior presentation or entry of the fetus with hind feet first, hocks up, occurs less frequently in cattle.

The foal is presented essentially the same as a calf. In swine, delivery of the individual fetus from the two uterine horns proceeds in an orderly fashion beginning at the cervical end. The fetus is presented either anteriorly or posteriorly with equal facility.

Initiation of Parturition

The onset of parturition is regulated by a complex interaction of endocrine, neural and mechanical factors (Table 13–12) but their precise roles and interrelationships are not fully understood. For each major factor identified in a specific

Table 13–12. Some Theories on the Initiation of Parturition

Theory	Possible Mechanism(s)
Fall in progesterone concentration	Blocks myometrial contractions during pregnancy; near term the blocking action of progesterone decreases
A rise in estrogen concentration	Overcomes the progesterone block of myometrial contractility and/or increases spontaneous myometrial contractility
Increase of uterine volume	Overcomes the effects of progesterone block of myometrial contractility
Release of oxytocin	Leads to contractions in an estrogen-sensitized myometrium
Release of prostaglandins ($PGF_{2\alpha}$)	Stimulates myometrial contractions; induces luteolysis leading to a fall in progesterone concentration (corpus luteum-dependent species)
Activation of fetal hypothalamic-pituitary-adrenal axis	Fetal corticosteroids cause a fall in progesterone, a rise in estrogen and a release of $PGF_{2\alpha}$. These events lead to myometrial contractility

species for the initiation of parturition, there are known exceptions among other mammalian species that precludes drawing generalizations. The physiologic mechanisms for the initiation of parturition are summarized (Liggins et al, 1973; Cox, 1975; Thorburn et al, 1977), and possible mechanisms consistent with available evidence are considered for farm species.

Fetal Pituitary-Adrenal System. The physiologic character of the fetus can affect the duration of gestation. Prolonged gestation is associated with a pituitary-adrenal defect in dairy cows and congenital malformations and fetal adrenal hypoplasia in sheep. It appears that the fetal pituitary-adrenal axis plays a primary role in determining the length of gestation and timing of parturition (Liggins et al, 1973). Fetal hypophysectomy or adrenalectomy *in utero* prolongs gestation, whereas infusion of ACTH or glucocorticoids into the fetal lamb terminates pregnancy prematurely.

A significant increase in the fetal plasma concentration of cortisol occurs during the final stages of gestation in sheep. Whether the increase in fetal cortisol production is due to a progressive maturation (sensitivity) of the fetal adrenal to "basal" levels of ACTH or is a response to an increase in the tropic stimulus (ACTH) remains to be determined.

A similar mechanism involving the activity of the fetal pituitary-adrenal axis and changes in cortisol secretion is intimately involved in triggering parturition in goats, cattle and swine (Thorburn et al, 1977). In the horse, some increase in adrenal activity probably occurs in the fetus near term, but there is no spectacular rise in fetal plasma cortisol concentration before birth as in other species (Nathanielsz et al, 1975).

Maternal Hormonal Changes. Changes in the maternal plasma concentrations of progesterone and estrogen occur as a major part of the physiologic changes associated with onset of parturition. In farm animals there is a fall in the progesterone level and a rise in the levels of estrogen and prostaglandin $F_{2\alpha}$ ($PGF_{2\alpha}$), except in the mare, after the initial rise in fetal plasma cortisol. However, there are species differences in the sequence of these changes depending upon whether the source of progesterone is from the placenta (sheep, mare) or the corpus luteum (goat, cow, sow).

Sheep. Plasma progesterone levels decline during the last seven to ten days of pregnancy, reaching very low levels at parturition (Fig. 13–18), primarily due to a decline in placental production rate and/or an increase in metabolic clearance rate. Plasma estrogen levels, which are low throughout gestation, rise sharply during the 48 hours prior to parturition. Apparently the decline in progesterone levels is not due to the increase in estrogen levels. The increase in fetal cortisol may be responsible for both the decrease in progesterone and the increase in estrogen production rates (Table 13–13).

These changes in steroid levels, particularly estrogens, stimulate $PGF_{2\alpha}$ synthesis and release from the maternal cotyledons (Fig. 13–19) and account for the rapid increase of $PGF_{2\alpha}$ in uterine venous blood during the last 24 hours of gestation. The decline in progesterone levels at term is essential before $PGF_{2\alpha}$ output from the placenta can occur.

Goat, Cow, Sow. The corpus luteum remains the major source of progesterone throughout pregnancy in the goat and sow, and up to about 200 days in the cow. Plasma progesterone levels fall rapidly in these species just prior to the onset of parturition (Fig. 13–18); this decrease is believed to be due to regression of the corpus luteum (Table 13–13). In the goat and the cow significant levels of $PGF_{2\alpha}$ occur in the uterine venous blood at the time or immediately before the decrease in plasma progesterone levels. Thus $PGF_{2\alpha}$

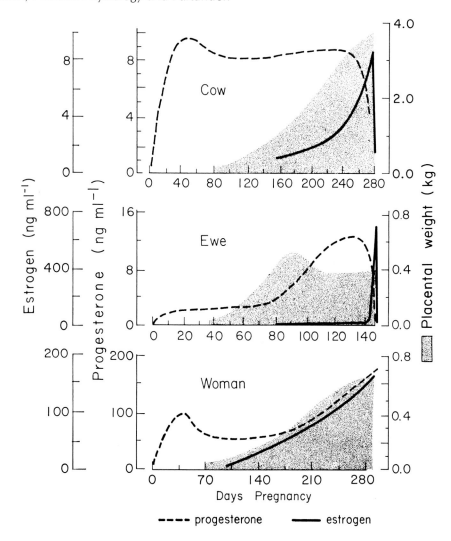

Fig. 13–18. Circulating hormonal levels and placental weights during pregnancy. In the cow and ewe the rise in estrogen coincides with a decline in progesterone. In women, estrogen and progesterone continue to rise at parturition. *(Redrawn from Bedford, 1972. J. Reprod. Fertil. Suppl. 16, 1; placental weight of cow from Swett et al., USDA Tech. Bull. 964.)*

released from the placenta or uterus may be the luteolytic factor in the goat and the cow. Placental estrogens stimulate $PGF_{2\alpha}$ secretion, since estrogen levels rise abruptly during the same period of $PGF_{2\alpha}$ output and progesterone decline. The ability of a synthetic analogue of $PGF_{2\alpha}$ to induce parturition suggests that $PGF_{2\alpha}$ may be responsible for luteal regression preceding parturition in the pig.

Mare. Changes in maternal hormonal levels do not seem to play a major role in parturition in the mare. The relatively high levels of progestagens and low levels of estrogens (Barnes et al, 1975) are in contrast to those in other farm species (Figs. 13–3, 13–18). $PGF_{2\alpha}$ levels increase in the peripheral blood during foaling.

Activation of the Myometrium. The roles of estrogen and especially proges-

Table 13–13. The Sequence of Hormonal Changes and Mechanism of Parturition in Farm Animals

Species	Fetus		Placenta		Probable Mechanism(s)
			Fetal	Maternal	
Sheep	Cortisol	$+$ → Estrogen $-$ → Progesterone	$+$ → PGF$_{2\alpha}$	$+$ →	Myometrial contractility
Goat, cow	Cortisol	$+$ → Estrogen	$+$ → PGF$_{2\alpha}$	$+$ → $+$ →	Luteolysis —— progesterone Myometrial contractility
Pig	(?)	→ Estrogen	$+$ → PGF$_{2\alpha}$	$+$ → $+$ →	Luteolysis —— progesterone Myometrial contractility
Mare	?	$-$ Estrogen $+$ Progesterone	? → PGF$_{2\alpha}$? →	$+$ →	Myometrial contractility

Key: + increase; − decrease; (?) probably increase but not proven.

(Adapted from Thorburn et al, 1977; Biol. Reprod. 16, 18; Nathanielsz et al, J. Reprod. Fertil. Suppl. 23, 625.)

terone in controlling uterine activity during gestation and parturition remain controversial (Table 13–12). The major determinant of uterine contractility at parturition in pregnant sheep is $PGF_{2\alpha}$ synthesized in the maternal component of the placenta and in the myometrium (Liggins, 1974). Synthesis in the former site may depend on an increase in the ratio of estrogen to progesterone, whereas synthesis in the latter site may be a response to stretch during uterine contractions and may augment rather than initiate parturition. Probably $PGF_{2\alpha}$ interacts with the smooth muscle adenyl cyclase system to lower cyclic AMP levels and elevate cyclic GMP levels, leading to myometrial contractions (Fig. 13–20).

Oxytocin causes strong uterine contractions during the later stages of birth but its precise role in parturition is not well-defined. For example, parturition occurs after maternal hypophysectomy or pituitary stalk section in sheep (Chard, 1972). Oxytocin levels show little change during the initial stages of parturition but rise to a peak during expulsion of the fetus and decrease thereafter. This oxytocin release may account for the massive release of $PGF_{2\alpha}$ which occurs at this stage and the potentiation of uterine activity. Therefore, oxytocin plays an augmenting role in the second stage of labor and not in the initiation of parturition (Thorburn et al, 1977).

Labor

Labor commences with the onset of regular, peristaltic uterine contractions accompanied by progressive dilation of the cervix.

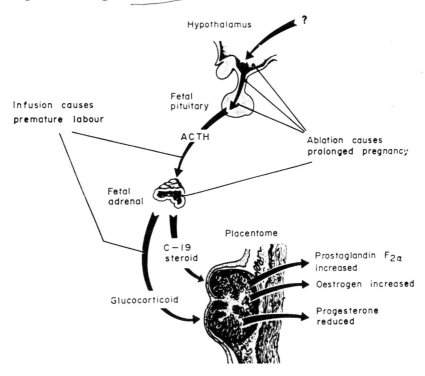

Fig. 13–19. Diagrammatic illustration of fetal mechanisms regulating parturition in sheep. Fetal hypothalamic-pituitary-adrenal axis initiates parturition by stimulating the production of prostaglandin $F_{2\alpha}$ and unconjugated estrogens in the placenta. Following a fall in progesterone prior to parturition induced by prostaglandin, the unconjugated estrogens initiate spontaneous uterine contractility. Oxytocin secreted by the maternal posterior pituitary alone or in combination with prostaglandin causes a rapid delivery of the fetus. *(From Liggins et al. 1972. J. Reprod. Fertil. Suppl. 16, 85.)*

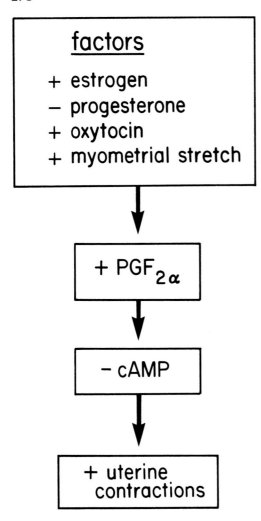

factors

+ estrogen
− progesterone
+ oxytocin
+ myometrial stretch

↓

+ PGF$_{2\alpha}$

↓

− cAMP

↓

+ uterine contractions

Fig. 13–20. Sequence of events leading to uterine contractions of parturition in sheep. Hormonal changes and myometrial stretch stimulate PGF$_{2\alpha}$ synthesis and release from the maternal caruncles and myometrium. The elevated levels of PGF$_{2\alpha}$ lower the threshold to oxytocin and also interact with the smooth muscle adenylcyclase system to lower cAMP levels, resulting in uterine contractions.

Stages of Labor. For descriptive purposes three stages of labor may be recognized: (1) dilation of the cervix; (2) expulsion of the fetus; and (3) expulsion of the placenta (Table 13–14). In the pig, the placentas of adjacent piglets are usually fused. Therefore the placentas are expelled as one or more masses interspersed with the birth of piglets. However, the largest mass of placentas is usually expelled three to four hours after the last piglet has been born. The time required for the expulsion of the fetus is the shortest of the three stages in monotocous species (Table 13–15).

Forces of Delivery. Labor requires a large amount of mechanical energy, which is provided mainly by the myometrium, and leads to cervical dilatation, vaginal stretch and distention, and delivery of the fetus. In monotocous species, myometrial contractions start at the apex of the cornua, while the caudal part remains quiescent. In polytocous species, contractions begin just cranial to the fetus nearest the cervix; the rest of the uterus remains quiescent.

The distention of the cervix and vagina by conceptus initiates a reflex—*Ferguson's reflex*—which produces the expulsive force of abdominal muscular contractions (straining) and the release of oxytocin, which in turn accentuates myometrial contractions. The combined forces of intraabdominal and intrauterine pressure propel the fetus through the birth canal (second stage).

Rhythmic uterine contractions originating at the apex of the uterine horn continue after birth (third stage) and cause the inversion of the chorioallantois in ruminants. The presence of the detached placenta within the birth canal then initiates further straining and the expulsion of the placenta.

Neonatal Physiology

A complex series of structural and physiologic changes adjusts the fetus for extrauterine life.

Cardiovascular Changes. During fetal life, the cardiovascular system is modified to bypass the unexpanded lungs. Thus when the placenta functions as the respiratory organ the lungs are in parallel with the systemic circulation; the foramen ovale allows blood to pass from the right

Table 13–14. Stages of Labor and Related Events in Farm Animals

Stage of Labor	Mechanical Forces	Period	Related Events
I Dilation of Cervix	Regular peristaltic uterine contractions	Beginning of uterine contractions until cervix is fully dilated and continuous with vagina	Maternal restlessness Changes in fetal position and posture
II Expulsion of Fetus	Strong uterine and abdominal contractions	From complete cervical dilation to end of delivery of fetus	Maternal recumbency and straining Appearance of amnion (water bag) at vulva Rupture of amnion and delivery of fetus
III Expulsion of Placenta	Uterine contractions	Following delivery of fetus to expulsion of placenta	Loosening of chorionic villi from maternal crypts Inversion of chorioallantois Straining and expulsion of fetal membranes

Table 13–15. Average Duration of the Three Stages of Labor in Farm Animals (Hours)

	Stage of Labor		
	I	II	III
Animal	Dilation of Cervix	Expulsion of Fetus(es)	Expulsion of Placenta(s)
Mare	1–4	0.2–0.5	1
Cow	2–6	0.5–1.0	4–5
Ewe	2–6	0.5–2.0	0.5–8
Sow	2–12	2.5–3.0	1–4

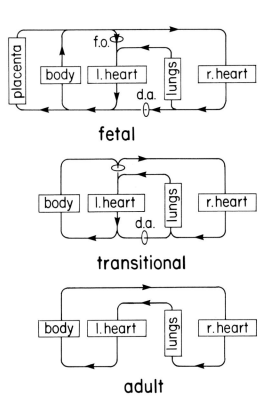

fetal

transitional

adult

Fig. 13–21. Diagrams illustrating the changes in fetal circulation at birth. In the fetus both ventricles work in parallel. At birth the placental circulation ceases and the foramen ovale closes *(f.o.)* to give rise to the neonatal (transitional) circulation. During the next few days the lungs become fully functional and the ductus arteriosus *(d.a.)* closes to produce the adult circulation. *(After Born, Dawes, Mott, and Widdicombe, 1954. Cold Spring Harb. Symp. Quant. Biol. 19, 102.)*

auricle to the left, and the ductus arteriosus shunts blood from the pulmonary artery to the aorta (Fig. 13–21).

Closure of Ductus Arteriosus. With the cessation of the umbilical circulation and the commencement of lung ventilation at birth, the flow in the ductus arteriosus is reversed, mainly owing to a drop in pulmonary arterial pressure, and blood then flows from left to right. This results in a distinctive murmur over the left thoracic wall in the neonate. This flow ceases promptly after the first few breaths owing to an increase in the blood oxygen tension. Once this vessel is totally obliterated, the ventricles and the lungs begin to function in series (Fig. 13–21). The closure of the ductus arteriosus is one of the most important adjustments for extrauterine life in the fetal circulation. The whole volume of blood now traverses the lungs during each circuit of the body.

Closure of Foramen Ovale. The rapid decline in blood pressure in the right auricle due to the interruption of the umbilical flow and the increasing left auricular pressure causes closure of the foramen ovale within a few hours in the foal and toward the end of the first week of life in the lamb.

Lung Maturation. After it is cut off from the respiratory tissue of the placenta, the neonate's survival depends on the rapid establishment of efficient gaseous exchange in the lungs. The modifications that make this possible are anticipated by the fetal lung. The membrane lining the fetal alveoli elaborates a surface active material (surfactant) during the later stages of gestation which reduces the surface tension at the air-fluid system created with the establishment of postnatal breathing. A relationship may exist between the time of first appearance of surfactant and fetal adrenal activity. Fetal lambs delivered by cesarean section before days 125 to 130 of gestation are unable to establish satisfactory respiration. However, when premature parturition was induced at that age with intrafetal glucocorticoids, breathing was initiated, suggesting that the glucocorticoids had accelerated lung maturation and pulmonary surfactant production.

Thermoregulatory Adjustments. Upon birth the neonate must make thermoregulatory adjustments to fluctuating environmental conditions, in contrast to the relatively constant temperature and nutrient supply present *in utero* during pregnancy. The efficiency of such adjustments depends primarily on the degree of physiologic immaturity of the species at birth, glycogen reserves and the presence of brown adipose tissue. Swine and sheep are particularly susceptible to low am-

bient temperatures; the rectal temperature of lambs falls 2° to 3°C, while that of piglets declines 2° to 5°C in the first hour after birth.

The neonate is not well adapted to withstand high temperatures early in life, lambs and calves being especially susceptible. For example, lambs between two and seven days of age cannot survive longer than two hours at 38°C, or more than three hours of solar radiation exposure.

Energy Metabolism After Birth. During the period between birth and suckling, the neonate depends for energy metabolism on its own resources of glycogen stored in the liver and skeletal and cardiac muscles. The rapid fall in liver glycogen concentration after birth suggests that it is mobilized rapidly to maintain the blood glucose levels.

Immune Status. The offspring is born without a supply of maternal antibodies or immunoglobulins in its blood. This is probably due to the impermeability of the epitheliochorial placenta of farm animals to maternal antibodies. However, immediately after birth, immunoglobulins are transferred to the newborn by way of the colostrum, the small intestine being permeable to protein for a period of 24 to 36 hours after birth.

PUERPERIUM

The puerperium extends from the time of expulsion of the placenta until the maternal organism returns to its normal nonpregnant stage. Among the most important changes that occur during this period are regeneration of the endometrium, uterine involution and resumption of estrous cycles.

Regeneration of the Endometrium

Cow. Lochia, or the uterine discharge that normally occurs during the puerperium, is composed of mucus, blood, shreds of fetal membranes and caruncular tissue. For the first two or three days, lochia is blood stained and then becomes paler in color; between the seventh and fourteenth days it is mixed with an increased quantity of blood due to hemorrhage from sloughing caruncular tissue. The involution of the maternal caruncle involves degenerative vascular changes, peripheral ischemia, necrosis and sloughing (Fig. 13–22). The surface of the bovine caruncle, which is devoid of epithelium immediately after parturition, begins to regenerate 12 to 14 days postpartum by proliferation from the surrounding tissue; it is completely reestablished in most normal cows within 30 days postpartum.

Mare and Sow. During the first week postpartum the mare may discharge small quantities of lochia. At "foal heat" (11 days postpartum) the endometrium is highly disorganized and contains large numbers of leukocytes. By 13 to 25 days it is fully regenerated in normal foaling mares. Regeneration of the uterine epithelium in the sow begins at one week and is completed by the third week postpartum.

Involution of Uterus

Following expulsion of the fetus and placenta, the return of the uterus to its normal nonpregnant size is termed uterine involution. This process is associated with enzymatic breakdown of mucopolysaccharides, a rapid shrinkage of cells and at the end of the involution period, a grouping of the nuclei of the muscle cells.

Cow. Uterine contractions gradually diminish during the first few days postpartum. These contractions cause shortening of the elongated uterine muscle cells. Involution of the uterus and cervix can be detected by rectal palpation and is completed by 30 to 45 days postpartum. The nongravid horn regresses almost completely, whereas the pregnant horn and cervix remain larger than before, even after involution is completed. Such a uterus in a heifer that has not calved indicates that she has conceived and subsequently aborted. Involution of the uterus is faster in cows suckling calves and in primiparous cows, and it is delayed after dystocia, twin births and retained placenta.

Mare, Ewe, Sow. Involution of the

Fig. 13–22. Diagram showing postpartum regression of maternal caruncles in cow. Numbers indicate days after calving. Note the dissolution and sloughing of the caruncle between days 10 and 13, and also regression of a caruncle to normal size by day 20. *(Redrawn from Rasbech, 1950. Nord Vet. Med. 2, 265.)*

uterus in the mare is rapid, although the uterus may not be completely involuted by the onset of the foal heat. In the ewe, at least 24 days are needed for complete involution and another 10 to 12 days before ewes could conceive; as in the cow, retention of the placenta delays uterine involution. In the sow involution is completed within 28 days of farrowing.

Resumption of Estrous Cycles

Cow. The corpus luteum of the previous pregnancy regresses rapidly. The interval from parturition to first estrus ranges from 30 to 72 days for dairy cows and 46 to 104 days for beef cows. The interval is prolonged when a calf is suckled and by increasing the frequency of milking (four milkings vs. two milkings per day). The removal of the calf shortens this interval. During the postpartum period the first ovulation occurs earlier than the first observed estrus. Following the reestablishment of ovulatory cycles, a short first cycle occurs in the early postpartum period, especially in high-producing dairy cows. This short estrous cycle is related to a deficiency of progesterone production by the corpus luteum. Ovarian activity after calving occurs more often on the ovary on the side of the previously nongravid horn. This tendency decreases as the interval from parturition to ovulation increases.

Mare. Most mares exhibit a *foal heat* within 6 to 13 days postpartum. It is a routine practice to breed mares at the foal heat despite the lower conception rates and higher incidence of nonviable foals and abortions.

Sow. Corpora lutea of pregnancy regress rapidly following parturition. Anovulatory estrus occurs three to five days after farrowing. However, in the majority of animals, estrus and ovulation are generally inhibited throughout lactation, although some nursing sows may exhibit estrus 40 days after farrowing.

Sows that do not nurse their litters during the first week of farrowing show estrus and ovulation within two weeks. Removing the piglets or weaning them at any time induces estrus and ovulation in three to five days.

REFERENCES

Baetz, A.L., Hubert, W.T. and Graham, C.K. (1976). Changes of biochemical constituents in bovine fetal fluids with gestational age. Am. J. Vet. Res. *37*, 1047.

Barnes, R.J., Nathanielsz, P.W., Rossdale, P.D., Comline, R.S. and Silver, M. (1975). Plasma progestagens and oestrogens in fetus and mother in late pregnancy. J. Reprod. Fertil. Suppl. *23*, 617.

Barron, D.H. (1970). The environment in which the fetus lives: lesson learned since Barcraft. *In* Prenatal Life. H.C. Mack (ed), Detroit, Wayne University Press.

Borland, R. (1975). Placenta as an allograft. *In* Comparative Placentation. D.H. Steven (ed), New York, Academic Press.

Chard, T. (1972). The posterior pituitary in human and animal parturition. J. Reprod. Fertil. Suppl. *16*, 121.

Comline, R.S. and Silver, M. (1974). Recent observations on the undisturbed foetus in utero and its delivery. *In* Recent Advances in Physiology. R.J. Linden (ed), London, Churchill Livingstone.

Cox, R.I. (1975). The endocrinologic changes of gestation and parturition in the sheep. *In* Advances in Veterinary Science and Comparative Medicine. C.A. Brandly and C.E. Cornelius (eds), New York, Academic Press.

Edgerton, L.A. and Erb, R.E. (1972). Metabolite of progesterone and estrogen in domestic sow urine. I. Effect of pregnancy. J. Anim. Sci. *32*, 515.

Hunter, D.L., Erb, R.E., Randel, R.D., Garverick, H.A., Callahan, C.J. and Harrington, R.B. (1970). Reproductive steroids in the bovine. I. Relationships during late gestation. J. Anim. Sci. *30*, 47.

Liggins, G.C. (1974). Parturition in the sheep and human. *In* Physiology and Genetics of Reproduction. Part B. E.S. Coutinho and F. Fuchs (eds), New York, Plenum Press.

Liggins, G.C., Fairclough, R.J., Grieves, S.A., Kendall, J.Z. and Knox, B.S. (1973). The mechanism of initiation of parturition in the ewe. Recent Prog. Horm. Res. *29*, 111.

Longo, L.D. (1972). Disorders of placental transfer. *In* Pathophysiology of Gestation, Vol II. N.S. Assali (ed), New York, Academic Press.

Nett, T.M., Holtan, D.W. and Estergreen, V.L. (1973). Plasma estrogens in pregnant and postpartum mare. J. Anim. Sci. *37*, 962.

Polge, C., Rowson, L.E.A. and Chang, M.C. (1966). The effect of reducing the number of embryos during early stages of gestation on the maintenance of pregnancy in the pig. J. Reprod. Fertil. *12*, 395.

Raeside, J.I., Liptrap, R.M. and Milne, F.J. (1973). Relationship of fetal gonads to urinary estrogen excretion by the pregnant mare. Am. J. Vet. Res. 34, 843.

Rudolf, A.M. and Heyman, M.A. (1967). The circulation of the fetus in utero: Methods for studying distribution of blood flow, cardiac output and organ blood flow. Circ. Res. 27, 163.

Steven, D. (1975). Anatomy of the placental barrier. *In* Comparative Placentation. D.H. Steven (ed), New York, Academic Press.

Thorburn, G.D., Challis, J.R.C. and Currie, W.B. (1977). Control of parturition in domestic animals. Biol. Reprod. 16, 18.

Woodruff, M.F.A. (1958). Transplantation immunity and the immunological problem of pregnancy. Proc. Roy. Soc. Lond. [Biol] 148, 68.

14

Lactation

A. T. COWIE
and
H. L. BUTTLE

Lactation is the final phase of the reproductive cycle of mammals and the physiologic state of the mammary gland is linked to the reproductive state of the animal. The duration of lactation varies considerably in different species but in nearly all, the milk provides the only nourishment available to the young in the postnatal period. Lactation is thus an essential phase of reproduction and, with few exceptions, failure to lactate means failure to reproduce. The milk, and in particular the colostrum, is in ruminants, horse and pig the chief route whereby antibodies are transmitted from mother to offspring.

The cow, the goat, the sheep and the water buffalo have been domesticated and selectively bred by man so that they produce milk in quantities far in excess of the needs of their young; this is harvested by man for his own and his children's nurture.

In this chapter the cow and small ruminants will be used as the main models in discussing the physiology of lactation,

but brief reference will be made to other species when such becomes pertinent. The basic components of the functional mammary gland, ie, the alveoli and ducts embedded in a stromal or supporting tissue, are common to the mammary glands of all mammals, from the lowly monotremes which have retained the reptilian practice of laying eggs to the higher placental mammals. There are, however, species differences in the numbers and positions of the mammary glands and in their shape and detailed architecture.

ANATOMY OF THE UDDER

Gross Structure

The bovine udder consists of four mammary glands or "quarters." Each gland is a separate entity drained by its own duct system and with its own storage cistern and teat. Under no normal circumstances can milk secreted in one gland pass into an adjacent gland. Within the udder the four glands are in close apposition; on the ventral surface the right and left "halves" are demarcated by a distinct groove on the skin; internally they are separated by the double-layered medial suspensory ligament. The boundary between the glands of the same side cannot be distinguished but the glandular systems are separate, as can be demonstrated by the injection of suitable dyes through the teat into the duct system.

The opening through the tip of the teat (the *teat* or *streak canal*) leads into the teat cistern or cavity within the teat (Fig. 14–1). Where the canal opens into the teat cistern there occurs a series of four to eight radiating folds in the mucosal lining

of the sinus known as the *Fürstenberg rosette*. Within the teat cistern there are numerous irregular annular and longitudinal folds in the mucosal lining. At its upper end the teat cistern communicates through a circular opening with the gland cistern. The size and shape of the gland cistern can vary considerably; it has a multilocular appearance because of the pockets formed by the openings of the large ducts. The cistern region merges into the more solid glandular substance, the presence of numerous smaller ducts giving sections of the glandular substance a sponge-like appearance; more dorsally the glandular substance becomes dense and fleshy.

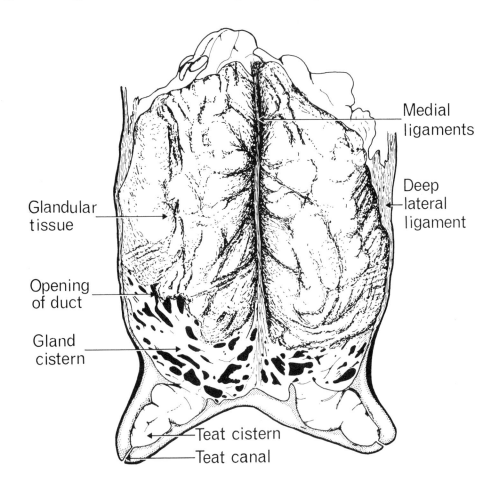

Fig. 14–1. Section of bovine udder showing teat canal, teat cistern, gland cistern and the openings of large ducts.

Microscopic Structure

In the mammary gland are two main types of tissue: first the parenchyma or glandular tissue and, secondly, the stroma or supporting tissue. In the lactating animal the parenchyma comprises the alveoli in which the milk is secreted and the duct system through which the milk flows to reach the cisterns.

The alveoli are minute, saclike or pear-shaped structures, the walls of which consist of a single layer of epithelial cells (Fig. 14–2). The alveoli occur in clusters or lobules in which there may be up to 200 alveoli, bounded by a thin fibrous septum. The majority of the alveoli within the lobule open individually into their respective terminal intralobular ducts, but groups of two or three may open together into a common duct and occasionally even large clusters may form a common opening into a terminal duct. The lobules

of alveoli themselves form larger clusters or lobes bounded by thicker septa.

In the cow the mammary parenchyma is arranged in the form of a series of leaves lying more or less parallel to the surface of the gland. The connective tissue separating these lobes is connected to the ligaments forming the suspensory mechanism. Immediately overlying the alveoli are the myoepithelial cells. These are stellate cells with long contractile processes (Fig. 14–3) that contract in response to oxytocin in the blood and compress the alveoli, thereby ejecting the milk into the duct system. Each alveolus is surrounded by a delicate stroma in which lies a fine capillary network. As the ducts become confluent and increase in diameter their epithelial lining changes from a single to a double layer of epithelial cells. On the outer surface of the ducts the myoepithelial cells are arranged in a longitudinal

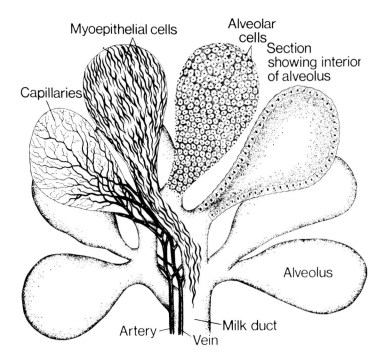

Fig. 14–2. Diagram of clusters of alveoli. *(From Cowie, 1972. In: Reproduction in Mammals. Austin and Short (eds), Cambridge, Cambridge University Press.)*

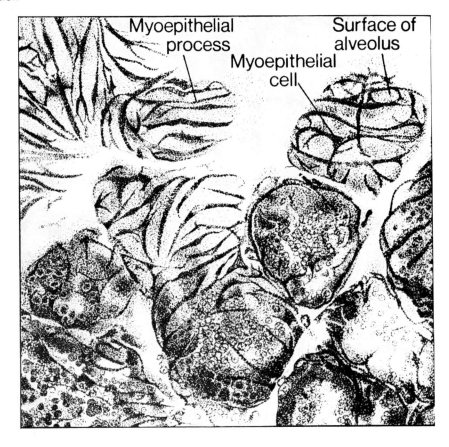

Fig. 14–3. Myoepithelial cells on the outer surface of alveoli. *(From Cowie, 1972. In: Reproduction in Mammals. Austin and Short (eds), Cambridge, Cambridge University Press.)*

manner so that, on contraction, they cause the ducts to shorten and thereby increase in diameter, so facilitating the flow of milk. A double-layered epithelium lines the gland and teat cisterns and is continuous with that of the ducts.

Accessory glandular tissue in the form of small periductal lobules may occur in the wall of the teat cistern and even in the wall of the teat canal. At the junction of the teat cistern with the teat canal the two-layered epithelium of the cistern changes abruptly into a thick squamous epithelium similar to and continuous with that of the skin of the teat. There is no sharply defined sphincter muscle surrounding the teat canal as often alleged, but the canal is kept closed by an ill-

defined circular network of smooth muscle and elastic fibers (Fig. 14–4).

Vascular System

In high-yielding animals one volume of milk is produced from about 500 volumes of blood passing through the mammary gland. This ratio decreases to 1000:1 in low-yielding animals. Most of the blood is supplied through the large paired external pudendal (or external pudic) arteries, which leave the abdomen by way of the inguinal canal.

The main venous drainage of the udder is through paired external pudendal veins that pass through the inguinal canal. At the base of the udder these veins have

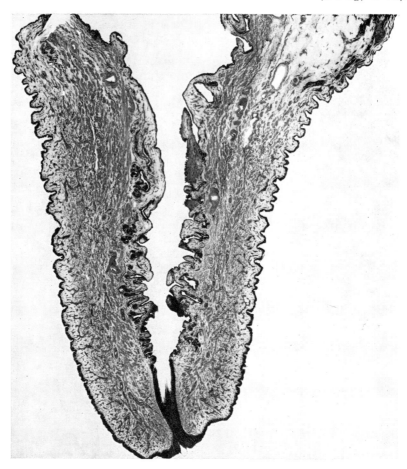

Fig. 14–4. Section through the teat of a cow. Note the thick stratified epithelium lining the teat canal and the absence of any well-defined sphincter muscle around the teat canal. *(From Cowie and Tindal, 1971. In: The Physiology of Lactation. London, Monographs of the Physiological Society, Arnold)*

anastomotic connections with the caudal superficial epigastric (subcutaneous abdominal or milk) veins and with the perineal vein; these connections form the so-called venous circle at the base of the udder. In the heifer the caudal superficial epigastric vein drains into the venous circle as does the perineal vein, but in the pregnant or lactating animal the valves in the caudal superficial epigastric vein become incompetent and these paired veins then drain blood away from the venous circle (Linzell, 1974).

The lymphatic system carries tissue fluid and lymph from the connective tissue spaces in the interlobular and interalveolar areas to the lymph nodes and thence by way of the thoracic duct into the venous system. In the normal lactating animal the composition of the mammary lymph indicates that it is formed by diffusion of plasma from the capillaries with no back diffusion from the mammary alveoli or ducts. However, if the udder becomes acutely distended with milk there may be a movement of protein, lactose and probably other milk constituents into the lymph. The rate of lymph flow greatly increases soon after parturition when copious milk secretion is being established.

Nervous System

The mammary nerves contain somatic sensory and sympathetic motor fibers; there is no evidence of a parasympathetic innervation. The sympathetic components supply the smooth muscle elements in the walls of the larger ducts and cisterns. Sensory nerve endings are present in the teat but the degree of sensory innervation of mammary tissue proper is still uncertain (Linzell, 1971). The main nerves are the first and second lumbar, the inguinal and the perineal nerves.

Suspensory Mechanisms

The udder of a high-yielding cow is a relatively large organ which, inclusive of blood and milk, may weigh over 40 kg; it therefore requires adequate support. The skin plays only a minor role in supporting and stabilizing the udder, and the suspensory mechanism proper consists of a series of strong ligaments and tendons by which the udder is attached and suspended both directly and indirectly to the bony pelvis (Fig. 14–5). Superficial and deep lateral ligaments arise from the subpelvic tendon; the superficial ligaments sweep downward and forward over the sides of the udder and then reflect off the udder to the inner aspect of the thigh; the deep lateral ligaments extend down over the sides of the udder and virtually envelop it. These deep ligaments join with the medial ligaments, which pass down between the two udder-halves. The elastic medial ligaments arise from the strong tendons of the abdominal wall at a point located over the center of gravity of the udder. The deep lateral and medial ligaments thus form slings for the suspension of each udder-half. The connective tissue septa of the udder tissues are also connected to these ligaments, so that the

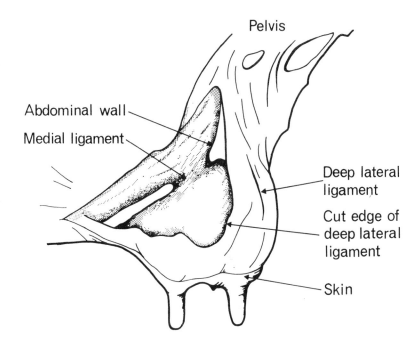

Fig. 14–5. Ligaments of the udder.

glandular tissue is supported in a series of layers, thereby preventing the lower part of the glands from being compressed by the weight of the upper part.

Species Other Than Cow

The udder of the sheep and goat is made up of two mammary glands separated by the medial suspensory ligament. The mammary glands of the sow extend over the entire abdominal wall, the number ranging from four to nine pairs. Each nipple is traversed by two canals; each of these canals leads into a small dilatation of the duct or sinus, beyond which the duct ramifies into its own sector of lobulo-alveolar tissue, ie, there are two separate sectors to each gland, each of which is drained by its own duct system and each opening separately at the tip of the nipple. There is a pair of inguinal mammary glands in the mare; each nipple has two canals, each leading to a sinus beyond which the duct ramifies in its own sector of lobulo-alveolar tissue.

MAMMARY GROWTH

Embryonic and Fetal Period

The glandular tissues of the mammary gland are derived in the embryo from ectoderm, the various stages of their morphogenesis being similar in all mammals. At an early age two parallel ridges of ectoderm appear on either side of the ventral midline of the fetus—the *milk lines*; these lines then diminish in length and become interrupted to form a series of nodules of ectodermal cells, the number and positions depending on the species. These nodules sink into the dermis to form the *mammary buds*. Initially these buds are lenticular in shape but become spherical and then conical. Generally, a distinct pause of about a month occurs in

cattle then in their development, after which the deep end of the bud (ie, the apex of the cone) elongates to form a cord-like sprout—the *primary mammary cord or sprout*.

The primary cord becomes canalized and the lumen so formed at the growing tip dilates to form a miniature gland cistern, which becomes well-outlined when the fetus is four to five months old; also about this time canalization of the base of the primary cord has formed the rudiment of the teat cistern. Secondary cords grow out from the gland cistern, representing the future ducts; later tertiary cords may appear. At birth the ducts are still restricted to a relatively small zone around the gland cistern. The stroma of the udder is now, however, well-developed; even as early as 13 weeks the stromal tissue has assumed the characteristic udder form.

In the goat and sheep, apart from the retention of only one pair of mammary buds, the pattern of fetal mammary development is similar to that described for the cow. In the pig embryo the number of pairs of mammary buds corresponds to the number of paired mammary glands of the adult; two primary sprouts grow from each bud, these become canalized and secondary sprouts appear.

Factors Regulating the Growth of the Mammary Rudiments in the Fetus. Little information is available so far about the mechanisms regulating fetal mammary growth in ruminants and other farm animals. However, in rats and mice certain sex differences in mammary development (eg, the absence of nipples in male rats and mice) are brought about by modifications in the growth pattern induced by androgens from the fetal testes in the male. Whether sex differences in growth of the mammary glands in fetal ruminants, such as the failure of the stroma to form an udder in the male fetus, are induced by sex hormones is not known (see Sonstegard, 1972; Anderson, 1978).

Postnatal Mammary Growth

Before Pregnancy. At birth the mammary gland of the female calf has gland and teat cisterns that are essentially mature in form, further changes being largely increases in size. The mammary ducts are still short and confined to the region of the gland cistern. The gland stroma is well-organized, forming a large pad of fatty and connective tissues. For some time after birth there is minimal mammary growth. Slight extension of the ducts occurs, although in some animals there may be a considerable increase in size of the stromal pad. However, about two months in advance of the first estrus there begins a period of rapid parenchymal growth that lasts for four months, declining in rate when the heifer is about one year old.

Cyclic changes have been described in the duct system during the estrous cycle. At estrus, a secretion is present in the lumina of the smaller ducts and their epithelium is cuboidal, whereas in the progestational phase of the cycle the ducts are empty and shrunken and their epithelium is columnar. These changes suggest that some cellular proliferation and fluid exudation into the ducts occurs at estrus, but that later in the cycle some regression occurs.

During Pregnancy. In the early months of pregnancy further extension of the duct system occurs but the intensity of this depends on the age of the heifer; in older animals, considerable duct growth will have occurred before conception. There is further branching of the duct system, the small interlobular ducts are formed and the alveoli begin to appear. The stroma now contains numerous islands of parenchyma made up of collections of small ducts and alveoli. By the fourth to the fifth months the glandular lobules are well-formed. These lobules increase in size, both through the formation of new alveoli (ie, true growth or hyperplasia) and

through hypertrophy or increase in volume of the existing alveolar cells and distention of the alveoli with the onset of secretory activity. Secretion containing fat globules is present within the alveoli during the fifth month. By the sixth month much of the stroma has been occupied by lobules that have so increased in size that they are now separated from each other only by thick bands of stromal tissue. During the last two months of pregnancy the alveoli become further distended with secretion rich in fat globules, and the stroma is represented only by the thin sheets of connective tissue that divide the parenchyma into lobes and lobules.

Mammary Growth in Species Other Than Cow. The general pattern of postnatal mammary growth in the goat and sheep is similar to that in the cow. In virgin goatlings there can be great individual variations in the degree of mammary development. Generally the udder pad is compact, containing well-defined gland cisterns and limited duct systems, but in some individuals the cisterns and large ducts become greatly distended with secretion thereby increasing the size of the udder. Small areas of true alveoli adjacent to the cistern may be found in the mammary stroma of such animals. In the pregnant goat a rapid extension of the lobulo-alveolar tissue begins early in the second half of pregnancy, i.e. 80 to 100 days; there is also secretory activity at this period, the alveoli becoming distended with secretion rich in fat globules (Fig. 14–6).

In the pig at birth the mammary glands contain relatively few secondary ducts; by five months of age the duct system is well-developed though still mostly confined to the region of the gland cistern. Lobulo-alveolar development is present by day 45 of pregnancy and is well-developed by day 60. Traces of secretion appear in the alveoli but they do not become distended as in ruminants; even

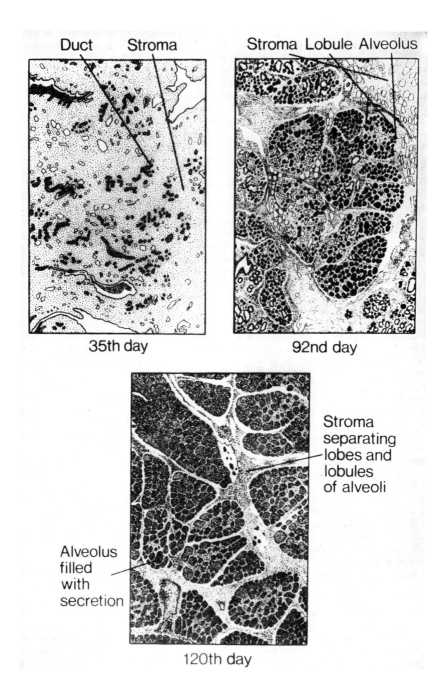

Fig. 14–6. Sections of the mammary gland of the goat during pregnancy (gestation length 150 days). *35th day,* Small collections of ducts scattered throughout the stroma. *92nd day,* Lobules of alveoli forming into groups or lobes; secretion is present in many of the alveoli. *120th day,* Lobules and lobes are now well-developed; the alveoli are full of secretion; stroma is reduced to thin bands of tissue. *(From Cowie, 1972. In: Reproduction in Mammals. Austin and Short (eds), Cambridge, Cambridge University Press.)*

nine days before parturition, the alveoli are small, their secretion is still scanty and hyaline in appearance with no evidence of fat globules; the intralobular stroma is broad. Four days before parturition the alveoli have become wider and the intralobular stroma much thinner; fat globules first appear in the alveolar cells and in the secretion about two days before parturition (Cross et al, 1958).

Hormonal Control of Mammary Growth. Analysis of the hormones required for full lobulo-alveolar growth has been carried out in hypophysectomized-ovariectomized-adrenalectomized rodents; the removal of the endocrine glands avoids the difficulties of interpretation that occur when endogenous hormones are also present. In such triply-operated rodents, the atrophied mammary duct system can be stimulated to full lobulo-alveolar growth equivalent to that of late pregnancy by suitable injections of estrogen + progesterone + growth hormone + prolactin + adrenocortical hormones. The anterior pituitary hormones by themselves will induce some lobulo-alveolar growth, whereas the steroids in the absence of the anterior pituitary hormones have little effect. The anterior pituitary hormones thus appear to be of major importance. In pregnancy in rodents it is most probable that a protein hormone secreted by the placenta also participates in the stimulation of mammary growth. There is evidence that the steroids act in part by sensitizing the mammary tissue to the growth-stimulating effects of the pituitary and placental hormones. Estrogens, moreover, can in some species act indirectly by stimulating the release of anterior pituitary hormones. Although an extensive analysis of the hormones involved in mammogenesis has not been made in ruminants, a preliminary study in hypophysectomized-ovariectomized goatlings suggests that the general pattern may be similar to that described for rodents. The anterior pituitary and placental hormones appear to be of major importance, and if these are absent the steroid hormones are ineffective.

The ruminant placenta synthesizes and secretes a placental lactogen during pregnancy; this hormone is chemically and biologically similar to the pituitary gland hormones prolactin and growth hormone. In sheep and goats the increase in plasma concentration of placental lactogen is coincident with the phase of rapid lobulo-alveolar proliferation during pregnancy, and the mammogenic properties of the placental hormones have recently been demonstrated by the observation of normal lobulo-alveolar development in hypophysectomized but pregnant goats (Buttle et al, in press). These two observations strongly suggest that it is the placental lactogen, which is produced by the fetal side of the placenta, that is primarily responsible for the stimulation of mammary development during pregnancy.

Artificial Induction of Mammary Growth. In the intact nonpregnant ruminant the ovarian steroids can be used to induce mammary growth and milk secretion (Cowie and Tindal, 1971; Erb, 1977). Estrogen alone will induce extensive lobulo-alveolar growth, although the alveoli tend to be abnormally large, resulting in a reduction in the normal area of secretory epithelium. Treatment with estrogen-progesterone combinations induces a more normal alveolar structure. This effect may be associated with the ability of progesterone to inhibit the onset of secretory activity by the alveolar cells. Goats treated with estrogen alone tend to produce milk more quickly, whereas those receiving estrogen + progesterone, although they come into milk production more slowly, eventually give higher milk yields. In the hormonal induction of lactation in ruminants it is important to recognize the role played by the milking stimulus itself. Indeed, mammary growth and milk secretion may be induced in virgin ovariectomized goatlings merely by the regular application of the milking stimulus. When attempting to induce udder growth and lactation in ruminants by ovarian hormone therapy it would seem desirable to start regular milking at an early stage in the treatment.

Although considerable milk yields have been obtained from ruminants in which udder growth and milk secretion have been induced by ovarian hormone therapy, in few animals has the degree of udder growth or subsequent milk yield been comparable to those occurring with normal pregnancy and lactation.

LACTATION

Characteristics

The onset of copious lactation occurs around parturition; with some high-yielding cows, however, it may be necessary to begin the regular withdrawal of milk before parturition to relieve the pressure that develops in the udder. Initially the secretion is of colostrum which is high in fat, protein and immunoglobulins and low in lactose (Table 14–1). Over the first four days or so of lactation (in cattle), the composition of the secretion changes from that of colostrum to that of normal milk.

The yield in cattle, under present husbandry methods, increases steadily to reach a peak at eight or nine weeks and then steadily declines over the remainder of the lactation period. Lactation can be maintained for long periods at a reduced level as long as the animal remains non-pregnant, receives enough food, stays healthy and is milked regularly. However, cows are usually mated at the first or second estrous period after parturition (8 to 12 weeks), and if pregnancy ensues the animal will be dried off around the 28th week of pregnancy (some 40 weeks after the previous parturition) when the milk yield will be low and she will remain dry until the next parturition. This dry period is essential if high yields are to be obtained in the next lactation.

The composition of the milk changes slightly during lactation. Once the changes from colostrum to milk have taken place there is a small but steady decline in protein, fat and lactose (ie, total solids) until the peak of lactation occurs; thereafter there is a slight increase in the total solids until cessation of lactation. In other words, within an individual there is an inverse relationship between yield and composition. Factors that limit the increase in daily yield to a peak at eight to nine weeks, such as genetic capabilities, disease, accidents or poor husbandry, will adversely affect the daily yield for the remainder of that lactation.

In other domestic species the lactation characteristics vary according to the reproductive cycles of the animal. In sheep and goats lactation is seasonal because of the nature of the breeding cycle, but otherwise similar in characteristics to that of cattle. In pigs, litters obtain milk from their mothers once every hour throughout the day, and the daily yield increases to a peak at four weeks after parturition and then declines. The sow remains anestrous for the duration of lactation except for the sterile estrous period immediately postpartum; cyclic ovarian activity is resumed after the removal of the litter.

Histology and Cytology of Milk Secretion

The secretory changes occurring in the mammary gland both at the level of the

Table 14–1. Components of Colostrum

Components	Cow	Ewe	Goat	Sow	Mare
			(grams/liter)		
Water	733	588	812	698	851
Lipid	51	177	82	72	24
Lactose	22	22	34	24	47
Protein	176	201	57	188	72
Ash	10	10	9	6	6

Data taken from Long, C. (ed), 1961. *Biochemists' Handbook*, London, E. & F. N. Spon, Ltd.

light and electron microscope are essentially the same in all the farm animals and indeed in most species studied so far. In the recently milked gland the alveolar cells are tall, the alveolar walls are wrinkled, the alveolar lumina are cleft-like and the interlobular septa are conspicuous and wide. In the gland fixed when full of milk, the alveoli have stretched walls, their cells are flattened, the lumina are wide and the interlobular septa are inconspicuous and thin (Fig. 14–7).

The milk fat is believed to be synthesized by the rough endoplasmic reticulum. The fat droplets then move toward the apex of the cell, possibly aided by the microtubules and microfilaments within the cell, and eventually cause the apical cell membrane to protrude. As the process of protrusion continues the cell membrane envelops the droplet and constricts behind it so that a narrow neck is formed, the walls of the neck fuse and the fat droplet enveloped in membrane drops free into the lumen (Fig. 14–8). The extru-

sion of the droplet may be facilitated by the numerous vesicles of Golgi origin which come to underlie the globule and which progressively open to the surface, their membranes fusing with the apical cell membrane (Wooding, 1977). Not infrequently a portion of cytoplasm becomes entrapped within the enveloped fat droplet.

The milk protein appears as fine granules within the vesicles of the Golgi apparatus. These vesicles move to the apex of the cell and fuse with the cell membrane, which then ruptures and allows the granules to escape into the lumen.

Formation of Milk Constituents

The concentrations of components in the milk of the cow, sheep, goat, pig and horse are given in Table 14–2. The major component of milk is water, in which the solid constituents are dissolved or suspended. The solid constituents—the pro-

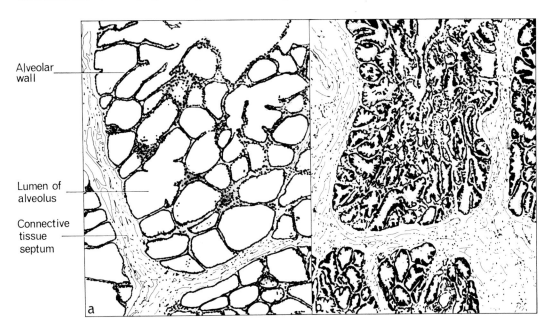

Alveolar wall

Lumen of alveolus

Connective tissue septum

Fig. 14–7. (a) Part of a lobule from left half of a goat's udder fixed when distended with milk. (b) Part of a lobule from the right half of udder of the same goat which was milked out before fixing. Note the contracted lobules with collapsed alveoli.

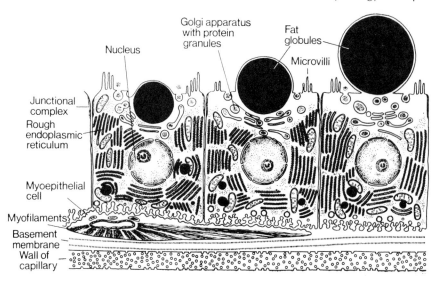

Fig. 14–8. Diagram of the ultrastructure of alveolar epithelium. *(From Cowie, 1972. In: Reproduction in Mammals. Austin and Short (eds), Cambridge, Cambridge University Press.)*

Table 14–2. Components of Milk

	Unit	Cow	Ewe	Goat	Sow	Mare
		(Values given per liter)				
Water	gm	873	837	866	788	890
Lipid	gm	37	53	41	96	16
Lactose	gm	48	46	47	46	61
Total proteins	gm	33	55	33	61	27
Caseins	gm	27.3				
Albumin	gm	3.0				
Globulin	gm	1.3				
Proteose	gm	1.4				
Calcium	mg	1250	1930	1300	2100	1020
Chloride	mg	1030	540	1590	—	—
Magnesium	mg	120	—	160	—	90
Phosphorus (total)	mg	960	995	1060	1500	630
Potassium	mg	1380	1900	1810	—	640
Sodium	mg	580	—	410	—	—
Sulfur	mg	300	310	160	800	323
Vitamin A	I.U.	1460	1460	1340	1760	—
Ascorbic acid	mg	16	40	14	110	118
Biotin	µg	35	—	63	14	—
Choline	mg	130	43	130	122	30
Folic acid	µg	2.3	2.2	2.7	3.9	1.3
Inositol	mg	130	—	210	—	—
Nicotinic acid	µg	850	3930	2730	8350	580
Pantothenic acid	mg	3.5	3.7	2.9	4.3	3.3
Pyridoxine	µg	480	—	70	200	—
Riboflavin	µg	1570	4360	1140	1450	400
Thiamine	µg	420	600	480	980	160
B_{12}	µg	5.6	1.4	0.2	1.05	0.02

Data taken from Long, C. (ed), 1961. *Biochemists' Handbook,* London, E. & F. N. Spon, Ltd.

teins, fats and lactose—are all formed within the one type of alveolar cell.

All components of milk are derived from the blood, either in the same chemical form or in different chemical forms as precursors. In general, the major components, protein, fat and lactose, are formed within the epithelial cell from precursors, and the minor components such as vitamins, salts and immunoglobulins, are secreted by selective passage of materials, without change of chemical form, across the epithelial cells into the lumen along with the water. During active secretion the mammary gland utilizes some energy sources (glucose and acetate) for its own tissue metabolism.

The cell content of milk is used as an index of the health of the gland. The cells are polymorphonuclear leukocytes and sloughed epithelial cells. The cells and their contents do not contribute much to the solid constituents of milk; in diseased states the cell content increases sharply above the normal level of 10^4 to 10^5 cells/ml milk for the majority of individual quarters.

Proteins. Casein forms the major part of the milk proteins and this protein has been classified into four subtypes: α-casein, κ-casein, β-casein and γ-casein. The caseins are aggregated to form insoluble granules called micelles. The other major proteins in milk, β-lactoglobulin and α-lactalbumin, are present in soluble form. All the major milk proteins are formed from free amino acid precursors in the blood, with little or no milk protein being formed from the breakdown of the plasma proteins. However, the minor protein constituents of milk (γ-casein, immunoglobulins and serum albumin) are apparently absorbed as such from the blood, but these only constitute 5% to 10% of the total milk proteins. The essential amino acids are absorbed as such and used in the formation of milk proteins, whereas a certain amount of interconversion of the nonessential amino acids does occur, both from the essential amino acids and between the nonessential amino acids.

The amino acids absorbed by the epithelial cells are assembled into short chain peptides by both the free and the bound ribosomes of the cells, from whence the peptides migrate in soluble form to the Golgi apparatus of the cell. Further condensation of the peptides occurs within the Golgi apparatus to form the various insoluble granules of the caseins and also the soluble β-lactoglobulin. The Golgi vesicles containing the casein granules then migrate to the luminal surface of the cell.

Lactose. Lactose is the main carbohydrate occurring in milk and is also peculiar to milk, not occurring elsewhere in mammalian tissues. Lactose is formed by the condensation of one glucose molecule with one galactose molecule and the reaction is catalyzed by the two-protein enzyme, lactose synthetase. Glucose is absorbed as such from the blood supply and the majority of galactose is formed from glucose, although there is also a minor conversion from acetate into galactose within the mammary gland cells. Hence the lactose in milk is formed almost entirely from glucose in the blood.

The enzyme lactose synthetase is a complex of two proteins, known as the A protein, which is a galactosyltransferase, and the B protein which is α-lactalbumin. The A protein is present in the epithelial cells within the Golgi apparatus, but α-lactalbumin is formed within the endoplasmic reticulum, after which it migrates to the Golgi apparatus and complexes with the A protein (which is presumably bound to or is part of the Golgi membranes), so that the complex is able to act as lactose synthetase.

The lactose content is largely responsible for the osmotic pressure exerted by milk; hence the lactose concentration will determine to a great extent the amount of water secreted or resorbed by the epithelial cells.

Fats. Almost all of the fat content of milk occurs in the form of triglycerides, which are elaborated within the granular endoplasmic reticulum of the epithelial cell into the form of fat droplets. It has been suggested that fat droplets form at sites in the cell where the granular endoplasmic reticulum is in close association with mitochondria.

The fatty acids composing the triglycerides are C_4-C_{18} fully saturated acids, with oleic acid being the only significant unsaturated fatty acid. The proportions of fatty acids in milk can be altered substantially by the diet. The medium-chain fatty acids (C_8-C_{12}) are peculiar to milk; they do not occur in other mammalian tissue. The mammary gland develops the ability to produce these medium-chain fatty acids during the middle of pregnancy, with the appearance of an enzyme in mammary tissue that halts fatty acid synthesis at medium chain length (Dils, Clark and Knudsen, 1977). The precursors of the C_4-C_{18} fatty acids in cow and goat milk appear to be obtained from the breakdown of triglycerides contained in the blood chylomicra and of lipoproteins. Acetate and hydroxybutyrate are also utilized to a considerable extent in the formation and interconversion of these C_4-C_{16} acids by the mammary gland cells. The saturated C_{18} acid (stearic acid) is obtained as such entirely from this residue in the blood triglycerides: the oleic acid (unsaturated C_{18} acid) is obtained both from the same source and by conversion from stearic acid. In ruminants synthesis of fatty acids by the mammary gland cells from glucose does not occur to any extent. The glycerol component of the milk triglycerides is obtained mainly by conversion from glucose and also from the glycerol of the blood triglycerides. Further information on the biochemistry of milk secretion will be found in the review by Schmidt (1971).

Milk Removal

Milk-Ejection Reflex. Milk removal, the second phase of lactation, is the process by which the milk stored in the mammary gland is made available to the suckling young or, in the dairy cow, to the milking machine. In the cow during the interval between milkings a considerable volume of milk, up to half that secreted, passes into the larger ducts and cisterns whence it is readily available to the milker. The remainder of the milk, being the portion stored in the fine ducts and alveoli, cannot be obtained until it is ejected or expelled from these regions into the larger ducts and cisterns. The process by which this transfer of milk occurs is known as *milk ejection* or in farm parlance—the "letdown" of milk. Milk ejection is brought about by the operation of a reflex; stimulation of the teat triggers the nerve receptors in the skin and nerve impulses ascend the spinal cord to reach the hypothalamus where they cause the release of the hormone *oxytocin* into the circulation from the posterior lobe of the pituitary gland. Oxytocin is carried in the blood to the mammary gland, where it causes the myoepithelial cells to contract, thereby expelling the milk from the alveoli, forcing it along the duct system toward the gland and teat cisterns and causing the internal pressure in the cisterns to rise. Note that, unlike the classic reflex in which both afferent and efferent arcs are nervous, the efferent arc in the milk-ejection reflex is hormonal and the reflex is thus known as a neuroendocrine reflex.

In the dairy cow the usual stimulus for triggering the reflex is the application of the teat cups. Like other reflexes the milk-ejection reflex can become conditioned so that the reflex occurs in response to visual or sound stimuli that the cow has come to associate with the act of milking, for example the appearance of the milker or sight or sound of the milking apparatus. For efficient milking the cow should be milked as soon as possible after milk ejection has occurred and it is thus important to maintain a regular routine in preparation for milking. Studies on the levels of oxytocin in the jugular blood reveal that further releases of oxytocin may occur during milking.

Dairy cattle have been selectively bred for high milk production and this selection has therefore favored cows in which milk ejection is readily induced. In the more primitive breeds of cattle the induction of milk ejection to ensure efficient milking can be troublesome and difficult, and for over 4000 years primitive peoples have practiced various stratagems to facilitate milk ejection; for example, the cow may be milked in the presence of its calf or the calf may be allowed to suckle at one teat while the other teats are hand milked. An odder technique, referred to by Herodotus and one that is still used by primitive peoples in both Africa and Asia, is the induction of milk ejection by vaginal stimulation—usually blowing air into the vagina. It is now known that vaginal stimulation can cause the release of oxytocin, so that this ancient practice has a sound physiologic basis.

As is true for other reflexes, the milk-ejection reflex can be inhibited under conditions of stress, hence the importance of a disturbance-free environment in the milking parlor. Inhibition is believed to be effected through a block of the release of oxytocin.

The importance of the reflex, however, is not the same in all species. In some breeds of goats and sheep complete milking can be achieved in the absence of the reflex, although recent studies on the blood levels of oxytocin indicate that the reflex does commonly operate particularly in early lactation. Why the reflex should be less essential in the small ruminants than in the cow is not known, but it has been suggested that slight differences in the duct architecture permit the milk to drain more readily from the alveoli and fine ducts. It is also possible that the myoepithelial cells may be more sensitive in these species and that they may contract in response to direct physical or mechanical stimuli. This occurs in the rabbit, in which a sharp tap on the skin overlying the lactating mammary gland will cause some contraction of the myoepithelial cells; this has been termed the "tap reflex." In the lactating sow the proper functioning of the milk-ejection reflex is vital to the maintenance of lactation and to the survival of the litter, for unless milk ejection occurs virtually no milk can be removed from the mammary gland.

Ascending Paths of the Suckling (Milking) Stimulus. The pathways in the spinal cord carrying the impulses from the mammary gland have yet to be properly delineated (Tindal, 1978). In the brainstem the paths concerned in oxytocin release have been traced in some detail for the goat, guinea pig and rabbit. These have been found to pass through the midbrain and diencephalon to the lateral hypothalamus and thence to the paraventricular nuclei, from which the neurosecretory axons sweep down to the posterior pituitary gland. Studies on the pathways to the anterior pituitary are still in progress; in the rabbit a prolactin release path has been traced to the medial forebrain bundle in the lateral hypothalamus. There is, moreover, a prolactin release pathway from the orbitofrontal region of the neocortex. This latter path from a "higher" region of the brain may be involved in modulating anterior pituitary function and hence lactation in response to environmental changes.

The Mechanism of Suckling and Milking. It is a common belief that the young obtain milk by sucking the teat or nipple of the mother but this is not so; only the milking machine obtains milk in this manner. Cineradiographic studies have shown that the young ruminant strips the milk from the teat with its tongue in an action similar to that of the fingers in hand milking. The base of the teat is compressed between the tongue and hard palate, the milk trapped in the teat is stripped out by the tongue compressing the teat from the base toward its tip against the hard palate; the pressure on the base of the teat is then released to allow the teat cistern to refill with milk and the whole action is repeated. A negative pressure is certainly created within the mouth which undoubtedly aids in the removal of the milk, but suction is not an essential component of the act of suckling.

Cineradiographic studies of the action of the milking machine have shown that once the teat is placed in the teat cup all the surface of the teat except the tip is in contact with the liner, and that the tip of the teat is exposed to the constant vacuum in the milk liner throughout the whole period of milking. When the liner is in the expanded position the milk is sucked out of the teat; when the liner collapses below the teat the milk flow ceases because the pressure of the liner walls closes the teat canal. There is thus no squeezing of the milk from the teat in machine milking; the rhythmic collapse of the liner around and below the teat helps to maintain a reasonable normal blood circulation in the teat.

The Maintenance of Lactation

Once lactation has been established it can be maintained for long periods of time, especially in ruminants, as long as conditions within the animal do not preclude lactation. The need for adequate food, maintenance of health and regular milking are self-evident factors concerned with adequate husbandry, but the hormonal factors necessary for the maintenance of lactation require some further discussion.

The evidence that implicates hormones in the maintenance of lactation is based upon the cessation of lactation after the removal of endocrine glands and the subsequent maintenance or restoration of lactation by administering the hormone in question. The conditions required by mammary gland cells for their maintenance and continued secretion when in culture *in vitro* also give an indication of the requirements for maintenance of the cells *in vivo*.

The removal of the pituitary gland results in the immediate and complete cessation of lactation and hence it is considered to be of primary importance in the maintenance of lactation. The removal of other endocrine glands such as the adrenals or the destruction of the islets of Langerhans will also result in a depression and eventual cessation of lactation, but as the specific effects of ablation of these glands on the mammary gland cannot be readily distinguished from the general metabolic effects (ie, death of the animal), the hormones secreted by these glands may or may not have definite effects on the maintenance of lactation.

Removal of the thyroids or parathyroids diminishes milk secretion. In the hypophysectomized rabbit, prolactin alone is capable of restoring milk secretion. In the goat and sheep—the only ruminants in which the restoration of lactation after hypophysectomy has been studied— prolactin, growth hormone, thyroid hormone and adrenal steroids are all necessary for full milk restoration. Once lactation has been restored in the hypophysectomized goat, it may be maintained at least temporarily in the absence of prolactin, provided that growth hormone, thyroid hormone and adrenal steroids are given. Recent studies in normal lactating ruminants with bromocriptine—an ergot alkaloid that inhibits the release of prolactin—have also shown that blood prolactin levels can be depressed without seriously affecting milk yields, and it is thus possible that the main role of prolactin in ruminants is in the initiation of lactation rather than in the maintenance of milk secretion.

Placental lactogens are not normally present during the first part of lactation and so cannot be considered of importance in maintaining lactation.

Various attempts to determine the hormones concerned in the control and maintenance of lactation have been made by injecting intact lactating ruminants with hormone preparations and measuring any change in milk yields. Thyroid hormone or growth hormone can boost declining milk yields, whereas prolactin is usually ineffective; adrenal steroids tend to depress milk yields. However, such studies in intact animals are difficult to interpret, since the injection of hormones may depress the secretion of the animal's own hormones.

Additional evidence concerning the role of the various hormones in lactation comes from the comparison of the concentrations of the hormones in blood during milking and at various stages of lactation with the concentrations in nonlactating animals; such studies have become possible in recent years with the advent of radioimmunoassay techniques. In all species so far investigated the suckling or milking stimulus causes an immediate increase in the levels of circulating prolactin, but the physiologic significance of this in ruminants is uncertain. The levels of prolactin in the blood of ruminants are seasonally controlled, being dependent on day length and temperature; the lowest values occur in winter and the highest, in summer.

Recent investigations have revealed an important role for hormones in the utilization of dietary energy. Nutritional studies indicate that, at any level of feeding, the high-yielding dairy cow will preferentially direct energy toward milk production and away from the deposition of body tissue, whereas the converse applies to the low-yielding cow. This partitioning of dietary energy is probably the result of inherited patterns of endocrine function. A recent study on groups of high- and low-yielding cows subjected to intensive investigation of hormone and metabolite levels in the blood during lactation indicates that concentrations of growth hormone are higher throughout lactation in high-yielding cows than in low-yielders, while the converse is true for the levels of insulin. When the animals were dried-off, these differences in hormone concentrations were no longer significant, whereas the levels of prolactin did not differ between the two groups. The high-yielding cows had higher levels of nonesterified fatty acids and β-hydroxybutyric acid during lactation (see Hart et al, 1978). To date, experimental evidence strongly suggests that in ruminants, levels of growth hormone are of major importance in the maintenance of high milk yields.

Initiation of Lactation

At or around the time of parturition the mammary gland changes from actively growing tissue which, according to the species, is nonsecretory or is secreting only small amounts of colostrum, to one that has almost ceased to grow but is secreting large volumes of milk. The likely stimuli for these changes are the changes in blood hormone concentrations associated with parturition. During pregnancy high blood levels of progesterone, estradiol, adrenal steroids and placental lactogen occur, whereas prolactin levels are variable but on the whole low. After parturition the levels of all these hormones change. Estradiol and progesterone concentrations are low, the levels of adrenal steroids decrease somewhat, and placental lactogen is absent, but prolactin is present in high concentrations. Also during pregnancy, a corticoid-binding globulin is present in plasma in large quantities and this may be responsible for "inactivating" the high levels of adrenal steroids; after parturition this corticoid-binding protein disappears from the circulation, thus "liberating" adrenal steroids for use by the mammary gland and other tissues.

The precise hormonal mechanisms involved in the initiation of copious milk secretion in ruminants have yet to be determined. In the rat the high levels of progesterone in the blood inhibit the onset of secretory changes in the mammary gland during pregnancy by directly preventing the alveolar epithelium from responding to the lactogenic effects of anterior pituitary and placental hormones. A change in the steroid metabolism of the ovary shortly before parturition results in a fall in blood progesterone and a removal of this block to the action of the lactogenic hormones. In ruminants other mechanisms are involved, since secretory activity first occurs in the mammary alveoli about mid-pregnancy when progesterone levels in the blood are still high. In the ruminant, moreover, prepartum milking is possible although increases in yield will still occur soon after parturition.

Regression of the Mammary Gland

Regression of parenchymal tissue of the udder may be a rapid or slow process depending on the pertinent circumstances. It will occur rapidly if milk removal is suddenly stopped in the fully lactating animal, but usually it is a gradual process, associated with the natural decline in milk yield, becoming somewhat accelerated when weaning finally occurs or milking ceases. In animals that are allowed to live out their natural lives there occurs another type of regression—senile involution of the gland, when reproductive activities have waned.

The histologic and cytologic changes in the acute or rapid type of mammary regression have been studied extensively in laboratory animals and to a lesser extent in ruminants. The alveoli soon become distended, their walls stretched and the alveolar cells flattened, the capillaries become compressed and the blood supply is greatly reduced. Within three to four days the alveolar cells break up, lysosomal hydrolytic enzymes are released and there is digestion of the cellular components and resorption of the secretory products, which diffuse into the interstitial spaces and are carried away in the lymph. By day 5 the alveoli collapse and disappear and the gland is infiltrated by phagocytic cells. In ruminants macrophages play a major role in the removal of fat from the regressing gland. Eventually the lobular structure disappears and the stroma predominates, the parenchyma being reduced to a duct system.

The tissue and cellular changes associated with the "drying off" of the udder at the end of lactation have been little studied. In the course of normal lactation both the amount of parenchyma and its secretory activity decline despite continued regular milking. The decline in yield is more marked if a new pregnancy supervenes, since the resulting hormonal patterns of raised estrogen and progesterone levels tend to depress secretory activity while maintaining or even stimulating parenchymal growth. In the lactating cow that is pregnant and "dried off" two months before parturition, there is little evidence that any marked regression occurs in the lobulo-alveolar tissue. Although little is known about the histologic and cellular changes, the customary dry period of two months is of considerable physiologic importance, for if it be unduly curtailed there will be a serious depression of milk yield in the subsequent lactation. This deleterious response is due to some local effect, as yet undetermined, which apparently interferes with the normal renewal or regeneration of the alveolar cells.

The histologic picture of the mammary

tissue in the nonpregnant cow that is being "dried off" at the end of lactation is somewhat different, since the hormone levels are not sufficient to stimulate or even maintain mammary growth. The lobules decrease in size, the alveoli become collapsed and folded and there is an increase in stromal tissue. The changes progress until ultimately the lobules are reduced to a few branching ducts.

Further information on the regression of the mammary gland will be found in the review by Lascelles & Lee (1978).

REFERENCES

Anderson, R.R. (1978). Embryonic and fetal development of the mammary apparatus. *In* Lactation, vol. IV. B.L. Larson (ed). New York, Academic Press, pp. 3–40.

Buttle, H.L., Cowie, A.T., Jones, E.A. and Turvey, A. (1979). Mammary growth during pregnancy in hypophysectomized or bromocriptine-treated goats. J. Endocrinol. *80*, 343.

Cowie, A.T. and Tindal, J.S. (1971). The Physiology of Lactation. London, Monographs of the Physiological Society, Arnold.

Cross, B.A., Goodwin, R.F.W. and Silver, I.A. (1958). A histological and functional study of the mammary gland in normal and agalactic sows. J. Endocrinol. *17*, 63.

Dils, R., Clark, S. and Knudsen, J. (1977). Comparative aspects of milk fat synthesis. *In* Comparative Aspects of Lactation, Symposia of the Zoological Society of London, no. 41. M. Peaker (ed). New York, Academic Press, pp. 43–55.

Erb, R.E. (1977). Hormonal control of mammogenesis and onset of lactation in cows—a review. J. Dairy Sci. *60*, 155.

Hart, I.C., Bines, J.A., Morant, S.V. and Ridley, J.L. (1978). Endocrine control of energy metabolism in the cow: comparison of the levels of hormones (prolactin, growth hormone, insulin and thyroxine) and metabolites in the plasma of high- and low-yielding cattle at various stages of lactation. J. Endocrinol. *77*, 333.

Lascelles, A.K. and Lee, C.S. (1978). Involution of the mammary gland. *In* Lactation, vol. IV. B.L. Larson (ed). New York, Academic Press, pp. 115–177.

Linzell, J.L. (1971). Mammary blood vessels, lymphatics and nerves. *In* Lactation. I.R. Falconer (ed), London, Butterworths, pp. 41–50.

Linzell, J.L. (1974). Mammary blood flow and methods of identifying and measuring precursors of milk. *In* Lactation, vol. I. B.L. Larson and V.R. Smith (eds). New York, Academic Press, pp. 143–225.

Schmidt, G.H. (1971). Biology of Lactation. San Francisco, Freeman & Co.

Sonstegard, K.S. (1972). The fetal mammary gland in cattle: normal development and response to hormones *in vitro*. Ph.D. Thesis, University of Guelph.

Tindal, J.S. (1978). Neuroendocrine control of lactation. *In* Lactation, vol. IV. B.L. Larson (ed). New York, Academic Press, pp. 67–114.

Wooding, F.B.P. (1977). Comparative mammary fine structure. *In* Comparative Aspects of Lactation, Symposia of the Zoological Society of London, no. 41. M. Peaker (ed). New York, Academic Press, pp. 1–41.

Recommended Further Reading

Larson, B.L. and Smith, V.R. (eds) (1974). Lactation, A Comprehensive Treatise, vols. I, II, III. New York, Academic Press.

Larson, B.L. (ed) (1978) Lactation, A Comprehensive Treatise, vol. IV. New York, Academic Press.

15

Sexual, Maternal and Neonatal Behavior

G. ALEXANDER
J.P. SIGNORET
and
E.S.E. HAFEZ

The behavior of animals plays an important role in reproduction, affecting both the success of mating and survival of the young. Behavioral patterns, associated with courtship and copulation, with birth and with maternal care and suckling attempts of the newborn have a dramatic quality that has attracted students of mammalian behavior and has led to the development of an extensive literature that covers wild and domestic animals.

This chapter deals with the patterns in the domestic ungulates, patterns that have been muted by domestication and restricted or modified by conditions imposed in accordance with husbandry requirements. These requirements include confinement in paddocks, yard or indoor pens, segregation of sexes, controlled mating, cesarean delivery, enforced weaning, imposed proximity with other individuals, and the inescapable presence of man, his dogs and machinery.

By comparison with other aspects of physiology, animal behavior is still in the observational stage; there are comparatively few aspects of behavior that can yet be adequately explained in physiologic terms. Much of the documented knowledge is derived from observations made under restricted environmental conditions, with animals of a single breed or strain, or a narrow age range and a unique background of early experience. Conclusions about species differences should therefore be made with great caution.

ETHOLOGY OF SEXUAL BEHAVIOR

Various patterns of courtship, display, motor activities and postures are directed to bring the male and female gametes together to ensure fertilization, pregnancy and propagation of the species. The coordination of motor patterns leading to insemination of the female has been achieved by the evolution of an orderly series of responses to specific stimuli. Each response becomes a stimulus in turn, and thus leads to other responses and stimuli, a phenomenon known as a behavioral chain or sequence.

Social Structure, Home Range and Reproduction

The encounter of sexual partners is the first step of reproductive behavior. In free-living animals, this occurs largely under the influence of preexisting social structure and the territorial or home range behavior of males and females, and leads to an organized pattern of reproduction that varies with the sociospatial or territorial characteristics of the species. In the roe deer and muntjak antelope, males and females live in a limited area, the boundaries of which are defended against any intruder of the same sex. The territories of males and females are overlapping, with permanent association between potential sexual partners. In other species, as in the wild rabbit and beaver, the territory is occupied by a permanent couple or harem, and the male avoids any encounter outside his territory. This pattern persists under artificial environments. For example, male rabbits breeding in cages display sexual behavior toward receptive females only after the male occupies the cage for a sufficiently long time to consider it as his territory. Territorial behavior is intensified during the season of reproduction and in fact in a number of species such as the seal it exists only at that time.

Under feral conditions, farm animals do not defend defined territories against intruders, but herds and flocks tend to occupy a "home range." The basic unit is matriarchal consisting of a female, her adult female offspring and their immature young. Such a matriarchal herd is remarkably stable. It persists after a temporary dispersion of its members, or mixing in large groups of several hundreds of individuals (African antelopes, bisons). This stability is the consequence of strong interindividual bonds resulting from contacts occurring during infancy. Experimentally, cows reared together from birth form such a stable group even in the middle of a large herd. Such a bond, limited to the dams and their female offspring, could be the basis of the social organization in ungulates.

In horses, each matriarchal herd is the permanent harem of a dominant stallion, whereas younger males form a permanent "bachelor" herd (Klingel, 1967). In other species, the males either aggregate in groups (feral sheep and goats) or even stay solitary, with occasional contacts with animals of the same species (bisons, wild pigs) (Fig. 15–1). In such cases, the herds of males are only temporary associations in which the interindividual bonds are loose.

In those different cases the evolution of the social groups minimizes the risk of mating between too closely related individuals, which could have an adaptive value in eliminating the close consanguinity. In horses, as in the packs of wolves or other canids, the young females tend to migrate in another harem, preventing them from being mated by their father.

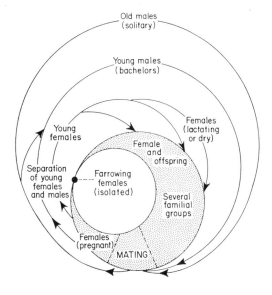

Fig. 15–1. Diagram of the evolution of the social organization in pigs.

Sequence of Sexual Behavior

The motor patterns of courtship behavior are stereotyped and are not altered by experience, which acts mainly on the latency and efficiency of mating. The components of copulatory patterns are sexual arousal, courtship (sexual display), erection, penile protrusion, mounting, intromission, ejaculation, dismounting and refractoriness. The duration of courtship and copulation varies with the species; both events are shorter in cattle and sheep than in swine and horses.

Courtship. The sequence of courtship tends to be simpler and shorter in mammals than in birds, fish or arthropods. The management of farm animals facilitates the contact of males and receptive females. Nevertheless, even in the wild, the patterns of courtship in mammals are simple, involving only a limited number of ritualized displays.

Male. In the male, sniffing and licking the female are the most frequent patterns, suggesting an important function of chemical communication through olfaction (Table 15–1). Except in swine, the male of domestic ungulates smells the female's urine and then raises his head, with lips curled, in the ritualized "Flehmen" reaction. In sheep, goats and cattle, tactile stimulation of the female is made by nuzzling and licking the perineal region, whereas with the horse, the stallion often bites the female's neck and with swine the boar noses her flanks.

The characteristic odor of the male does not appear to be emitted through specific postures or displays. However, urination is used by the stallion for marking the place where an estrous female has urinated, and boars urinate in a rhythmic pattern during sexual activity.

Female. In most species, the estrous

Table 15–1. Species-specific Patterns of Courtship in the Male

Behavior	Cattle	Sheep	Goat	Swine	Horse
Sniffing		Sniffing to female's genitalia and urine		Sniffing to female's head	
Ritualized reaction to female's urine	"Flehmen": male stands rigidly and holds his head in horizontal position which he may move slowly from side to side with his neck extended and upper lip raised (Fig. 15–2); "Flehmen" lasts 10–30 seconds			Absent	Same as in cattle, sheep and goat
Urination	No	No	Frequent miction on foreleg during excitement	Rhythmic emission of urine during sexual excitement	Marks with urine the place where a mare has urinated
Vocalization	No	Courting grunts during sexual approach		Courting grunts	Neighing during sexual excitement
Tactile stimulation of female		Licks female's genitalia		Noses female's flanks	Licks female's body; bites back and neck
Postures during courtship	Head on female's back	"Nudging": ritualized approach of female, head turned on side with a motion of foreleg (Fig. 15–2)			
Postures during copulation	Presses head on female's back, "leap" at ejaculation	Rapid movement backwards of head at ejaculation		Motionless during ejaculation, scrotal contraction	Bites female's neck
Postcoital reactions	No	Stretches head and neck	Licks penis	No	No

female shows increased motor activity, becoming restless and moving at the slightest disturbance. Receptive cows and goats exhibit increased frequency of nonspecific bellows or bleats, whereas the sow utters a typical estrous grunt. Cows, goats and sows tend to mount and to be mounted by other females but this is exceptional in ewes and mares. In the presence of a male, the female sniffs at his perineum or scrotal region. Mutual sniffing leads both animals to circling motions in a reverse parallel position. Receptive sows also display interest in the boar's head. Frontal contact between the estrous cow or sow and the boar may be associated with "mock fighting." Estrous mares tend to urinate frequently in the presence of a stallion.

When approached and stimulated by the male, the female domestic ungulate assumes a mating posture. This entails immobilization, often accompanied by tail deviation, and some minor species-specific features such as turning the head back in the goat and ewe, cocking the ears in the sow, and exposing the clitoris in the mare (Table 15–2).

Mating. The posture of the sexually receptive female terminates courtship behavior by allowing mating to take place. The female stands immobile, and the male mounts and ejaculates (Table 15–3).

Mounting. In the presence of a proestrous female, the male attempts several mounts; the penis becomes partially erect and protrudes from the prepuce. These mounts are usually unsuccessful. During this activity, the male, especially the bull, excretes "dribblings" of accessory fluid, derived from the Cowper's gland and differing from the seminal plasma emitted from the vesicular glands during ejaculation. However, if the female is receptive, copulation may occur rapidly. The male rests his chin on the female and she in turn responds by "standing." The male then mounts, "fixes" his forelegs around the female, grasps her firmly and performs rhythmic pelvic thrusts (Fig. 15–2). Some

Table 15–2. Species-specific Patterns of Courtship Behavior in the Female

Behavior	Cattle	Sheep	Goat	Swine	Horse
Sniffing	Sniffs male's body and genitals				
Urination	More frequent when the female is teased by the male; increased urination is not characteristic of estrus in sheep and pigs but is specific of sexual receptivity in horses				
Vocalization	Increased frequency of nonspecific bellows	Increased frequency of nonspecific bleats		Estrus grunts	?
Motor activity	Increased general motor activity. Attempts to sniff at the male's genitals lead to a reverse parallel position and circling movements				
	Frontal head-to-head contact; mounts other females	Tail wagging No mounts	Mounts other females	Head-to-head, sniffs; "mock fighting"	No mounts
Postures	Immobility when approached and teased by the male				
	Turning head back Tail deviation			Typical position of ears	Clitoris exposed by contraction of labia; extends hind legs, lifts tail sidewise; lowers croup
Postcoital reactions	Arching of back; elevation of tail	No	No	No	No

Table 15–3.   Patterns of Mating in Farm Mammals

	Cattle	Sheep	Goat	Swine	Horse
Duration of estrus	15 hrs (5–30)	24 hrs (12–50)	32 hrs (24–96)	50 hrs (24–72)	7 days (2–10 days)
Time of ovulation	4–15 hrs after onset of estrus	30 hrs after onset of estrus	30–36 hrs after onset of estrus	40 hrs after onset of estrus	24–48 hrs before end of estrus
Male anatomy: Penis	Fibroelastic	Fibroelastic with filiform process		Fibroelastic spiral tip	Vascular-muscular
Scrotum		Pendulous		Close to body	
Accessory glands	Vesicular gland large, prostate and and Cowper's glands proportionally small			Vesicular and Cowper's gland large, important urethral gland, prostate relatively small	Vesicular and prostate gland large, Cowper's gland small
Mating: Duration		Brief (a second or less)		5 minutes	40 seconds
Site of semen ejaculation		Near os cervix		Cervix and uterus	Uterus
Volume of ejaculate	5–8 ml	1 ml	1 ml	150–200 ml	60–100 ml
Number of ejaculations to exhaustion, average	20	10	7	3	3
Maximum	60–80	30–40	14	8	20

boars and stallions mount and dismount the female repeatedly before copulation, whereas others mount once and copulate.

Intromission. At mounting, the abdominal muscles of the male, particularly the rectus abdominis muscles, contract suddenly. As a result, the pelvic region of the male is quickly brought into direct apposition to the external genitalia of the female. The boar, with the penis partially out of the prepuce, thrusts his pelvis until the tip of the penis penetrates the vulva; only then is the penis fully unsheathed and intromission accomplished. One intromission takes place per copulation in farm animals. The stallion oscillates the pelvis several times resulting in engorgement of the penis with blood and making it rigid for maximal intromission. In contrast, ejaculation in several species of rodents is preceded by a series of mounts and intromissions.

The duration of intromission varies widely between species. Bulls and rams represent one extreme in that they typically ejaculate instantaneously on the first intromission. Boars, however, may maintain intromission for as long as 20 minutes with a single ejaculation.

Ejaculation. Semen is ejaculated near the os cervix in the case of cattle and sheep, into the uterus in swine and partially into the uterus in horses. Abortive ejaculations may occur if the female refuses intromission or the penis fails to penetrate the vulva. In the ram, the goat and the bull, an intense generalized muscular contraction takes place at ejaculation. Often the force is so strong in the bull that the hind legs of the male leave the ground, giving the appearance of an active leap. In the ram and the goat the male's head is suddenly moved backward, whereas in the bull it is pressed down on

Sniffing

"Flehmen"

Nudging
and kicking

Mounting

Copulation

Fig. 15–2. Sexual behavior patterns in cattle, sheep and goats. *(Y. Rouger, unpublished data)*

the female's neck (Fig. 15–2). During ejaculation itself, the boar is quiet, presenting only slight rhythmic contractions of the scrotum; such periods of immobility are followed by some thrusts at irregular intervals.

After ejaculation, the male dismounts, and the penis is soon retracted into the prepuce. Postcoital displays are scarce in the domestic mammals: the cow arches her back and elevates the tail after copulation and keeps this posture for several minutes. The male goat usually licks the penis after ejaculation.

Refractoriness. Most males show no sexual activity immediately following copulation. The duration of the refractory period is highly variable and increases gradually when several copulations are allowed successively with the same female. This period is modified by environmental stimuli.

Frequency of Copulation. The frequency of copulation varies with the species, the breed, the ratio of males and females present, available space, period of sexual rest, climate and nature of sexual stimuli. The maximal number of ejaculations is higher in bulls and rams than in stallions and boars: some bulls have been observed to copulate over 80 times within 24 hours or 60 times within six hours; an average of 21 copulations before exhaustion was observed (Wierzbowski, 1966).

After a long sexual rest, a ram may copulate up to 50 times on the first day after joining with the ewes, but this frequency is greatly reduced on subsequent days. The goat, the stallion and the boar reach exhaustion after a smaller number of ejaculations than in the ram and bull (Table 15–3). The number of copulations with the same estrous female is lower in natural mating than when the semen is collected in an artificial vagina. The bull will copulate only five to ten times with a free estrous female, the ram three to six times, the stallion and the boar, two to four times daily.

Duration of Estrus. The duration of estrus is influenced by species, breed, climate and management. Estrus is limited to about a day in sheep and cattle but to longer periods in the sow and the mare (Table 15–3). In species in which the period of sexual receptivity is short, ovulation takes place after its end, but in species that remain receptive for long periods, ovulation occurs during estrus.

Measurement of Sexual Behavior

Measurement of mating behavioral responses involves both copulation and the whole sequence of courtship. However, the intensity of a response expressed at any specific time may not reflect the potential, since intensity depends greatly upon environmental factors. Hence conditions of testing should be highly standardized, with due regard to management of animals, to the individuality of sexual partners, to time of day and to other environmental factors.

The nature and time of any interaction between animals must be recorded during the test, and sufficient time must be allowed for complete expression of the whole sequence of activities. Accessory tests can be developed for the study of details of the sequence, such as interattraction, postural responses to isolated stimuli and the control of specific and environmental stimuli.

Male. The interest in farm animals has been generally limited to copulation; assessment of the intensity of their sexual behavior is based on the latency to ejaculate, the interval between successive ejaculations, the number of copulations required for satiation, environmental stimuli remaining unchanged and, finally, on the period required to recover after sexual satiation, when the same or different female partners are made available.

The use of several teaser females minimizes the influence of individual preferences, and has been used in experimental studies. Recording the main patterns of the sexual sequence appears necessary (sniffing, nudging, mounting, mating). The nudging/mounting and mounting/mating ratios are useful criteria for measuring the efficiency of the motivation and the mating dexterity, respectively.

Female. The intensity of estrus is subjectively classified according to the degree of sexual receptivity to the male. However, no objective measure of female sexual behavior is yet available for research, though some aspects could be developed. The attraction of female to male can be measured easily in the T maze with swine, but not with cattle or sheep. The frequency with which the female displays particular behavior patterns during courtship could also be measured; such patterns include sniffing and head turning, but they are few and not highly specific. The willingness to adopt the mating posture as a response to the sexual approach of the male could be measured in terms of latency or by the percentage of accepted mountings.

MECHANISMS OF SEXUAL BEHAVIOR

The physiologic signal that originates sexual motivation is the gonadal steroid balance; it is clearly identified and well-known. Transmitted by the blood flow, the hormones activate the central nervous system. The humoral signal is transformed into sexual motivation or sex drive. The motor patterns of copulatory activity are programmed according to preexisting species-specific neuronal circuits. The sensory information allows the initial searching for sexual partners and the identification of their physiologic stages, and releases the appropriate motor reactions.

The following discussion deals with the sensory information that organizes the sequence of courtship and copulation, and with the endocrine, neural and genetic control of the reproductive behavior.

Sensory Organization of Sexual Behavior

The behavioral interactions leading to copulation can be divided into four major phases: mutual searching for the sexual partner; identification of the physiologic state of the partner; the sequence of behavioral interactions resulting in the adoption of the mating posture by the female; and the mounting reaction of the male leading to copulation.

Finding a Sexual Partner. Despite the existence of variations in the intensity of the sex drive, the male's sexual reactivity is grossly permanent or undergoes slow and limited seasonal variations, whereas the receptivity of the female is restricted to a few hours or a few days. Thus, the behavioral sequence appears to be initiated by the female. Her role during estrus can be passive in emitting specific signals for the male or changing her motor or postural reactions to attract him.

The Role of the Female. During estrus, general activity of the female increases and is oriented toward exploration. In all domestic species, such exploratory activity of the estrous female is definitively oriented to the male. In an attempt to study the role of attraction, Lindsay and Robinson (1961) compared the mating efficiency of rams either free with the ewes in a pasture or tethered with a five-meter chain. One third of estrous ewes were not mated by the tethered rams; the other two thirds actively sought out the male.

The sow placed in a T maze is strongly attracted to the boar during estrus. The stimuli eliciting the approach of the estrous female appear to be essentially olfactory in nature. When the sow is unable to see the boar, the attraction is not modified, but the ability of the sow to discriminate between male and female animals is drastically impaired by the removal of her olfactory bulbs. The production of the attracting male pheromone is under the control of androgens; after daily injection of testosterone, both castrated boars and ovariectomized sows attract estrous females as efficiently as intact males. In the sheep, the impairment of any sensory information in estrous females reduces the percentage of animals mated, indicating that the senses of sight and hearing, as well as olfaction, help the estrous female to be oriented to the male.

The Role of the Male. A male placed with a group of females is initially active, testing them intensively. However, the male may actively pursue an anestrous female, whereas an estrous one is temporarily neglected. Thus the male is hardly able to identify the estrous female with certainty in the absence of close physical interactions. In addition, the discrimination between potential sexual partners is more efficient in the sow than in the boar placed in a similar testing situation in a T maze. However, odor does play a role in attracting males to females. In sheep, for instance, attracting pheromones appear to be emitted by estrous ewes; receptive and nonreceptive females are approached at random by anosmic rams but not by intact rams (Lindsay, 1965). Hearing and sight are not effective in the identification of estrous ewes.

Both males and females search for social contact with homospecific animals and the search is selectively oriented toward the potential sexual partner. Both participate in this orientation, but the female's role seems to be the most important. During estrus, the orientation to the male is strong, highly selective, and results from the cumulative action of several long-range clues, among which the pheromones appear to be of utmost importance. Conversely, the male is less sensitive to the stimuli from the estrous female. His approaches, although not random, tend not to be selectively oriented.

Identification of Female Receptivity by

the Male. The identification of the receptive female is not made from a distance but takes place during the course of the courtship sequence. The patterns of interaction are the same when a male is testing an estrous or an anestrous female. It is possible that the male engages in courtship behavior upon contact with any female, and the essential clue appears to be the female's postural and motor reaction to the testing male. For example, the various interactions between the boar and the sow take place rapidly with a succession of nasonasal and nasogenital contacts, and tactile stimulation through nosing the flanks and mounting attempts (Signoret, 1970). The sequence is resumed after any of the patterns, provided that the sow has not "stood" when tested. Immobility or standing is the final clue that the female is receptive.

Stimulation of Reactions Allowing Copulation. *The Female's Mating Posture.* The mating posture is often adopted spontaneously when a female is approached by a male or by another female. In sows it may appear under nonspecific tactile stimulation by the herdsman or even without external stimulation; however, an intense amount of stimulation is necessary to elicit the response, especially during the beginning and the end of estrus. Stimuli from the male are much more effective in facilitating and releasing the postural response. The "mating stance" of the sow—clear, long lasting and easy to release by an experimenter—is especially suitable for a study of releasing mechanism. During the "standing reaction" the receptive sow is absolutely immobile, arches her back and cocks the ears, and this reaction may be exhibited when an estrous female is touched on the back. However, only 48% of estrous gilts will "stand" in the absence of the male. The frequency of the response increases from less than 40% at the beginning of estrus to 60% between the 24th and 36th hours. Stimuli from the boar, especially olfactory and auditory, are effective in increasing the rate of response to 100% in estrous females; the standing reaction is observed in 90% of estrous gilts when the boar cannot be seen or touched. The addition of visual and tactile stimuli increased the number of gilts responding by 7% and 3% respectively.

Over 60% of estrous gilts that were previously negative to the experimenter exhibit the standing reaction when placed in the boar's pen, or presented with the odor of the boar's preputial secretion (Signoret, 1970) or that of the compound, 5 α-androst-16-ene-3-one, which is responsible for the boar odor (Melrose et al, 1971). Broadcasting tape-recorded "courting grunts" is similarly effective in 50% of previously negative females. Thus the stimuli emitted during precopulatory interactions facilitate the release of the female's postural response. This appears to be true for ungulates in general.

Mounting by the Male. In general, the female in the mating stance is mounted immediately, and this reaction seems to be released mainly by visual and tactile clues. A restrained female, although not in estrus, is immediately mounted even by a sexually experienced bull or ram. Similarly, a ram does not copulate selectively with an estrous ewe when presented with two restrained anestrous females. Sexual reactions of the male toward stimuli other than those emanating from the female are common. For example, the bull or the boar reacts rapidly to a restrained male or to a dummy. Some 90% of bulls mount a "dummy cow," and 75% of sexually naive young boars react to a simplified dummy when presented to it for the first time (Signoret, 1970).

It is possible that the sexual releaser for mounting is the overall shape of the female and her immobility. The other visual, olfactory or acoustic information from the estrous female may be of minor but complementary importance. This scheme for the bull, termed a "Torbogen"

or archway, seems to apply to naive as well as to experienced males. Other sensory information from the female enhances sexual responsiveness of the male. The need for different types and amount of stimulation according to the species, the breed and the individual may explain various reactions to dummies or anestrous females.

Endocrine Mechanisms of Sexual Behavior

The steroid hormones in the male and the female have close biochemical similarities, but the rhythm of their release into the bloodstream is totally different. The secretion of androgens is not permanent. In the male, it takes place in the form of several peaks within 24 hours, reflecting the pulsatile release of pituitary gonadotropins. However, the total amount secreted is practically constant from day to day. Any seasonal fluctuations are progressive and slow. In the female, however, estrogens are only present during a few days of the estrous cycle. The effects of the deprivation of the gonadal hormones and therapy in males are completely different from those in females.

Hormonal Mechanisms of Male Sexual Behavior. Castration of males is routinely used in animal husbandry but little information is available on its effects on sexual behavior. The depressing action of castration varies with the species, the individual and the physiologic and behavioral status of the animal at the time of operation.

Generally, some mounting activity is retained after prepubertal castration in bulls and rams, but the under-development of the genital tract, due to the lack of androgen during ontogeny, drastically inhibits mating. There seems to be wide species variation in the effect of castration. Banks (1964) reported variable activity in rams castrated prior to puberty. In contrast, male rodents and cats rarely mount if castrated prepubertally. Following postpubertal castration, erection, intromission and even ejaculation may persist for a long time, but with a decreased frequency. Among eight tropical male goats, only one had lost the ejaculatory response one year after castration.

The loss of ejaculation potency may be due to modifications of androgen-dependent structures of the copulatory organs. The persistence of the other behavioral patterns cannot be attributed to the presence of androgen from adrenal origin; for example, adrenalectomy does not change the sexual behavior of castrated dogs.

Precastration sexual experience influences the persistence of libido. The patterns of sexual behavior gradually reappear under daily androgen treatment in an order inverse to their disappearance. Their frequency reaches a plateau similar to the precastration level of activity. The only effect of increasing the daily dose of hormone is to accelerate the recovery of the precastration activity. The hormonal therapy allows the male to recover the preoperative level of copulatory activity, but the preexisting differences cannot be overcome by an extra dosage of hormone. The endocrine balance appears only to reveal the potential intensity of reaction without being able to modify it.

Hormonal Mechanisms of Female Sexual Behavior. Female sexual behavior depends on an appropriate endocrine balance resulting in the development of the ovarian follicles. Ovariectomy inhibits sexual behavior, but in the cow and sow sexual behavior is restored in ovariectomized females after the injection of a minimal dose of estrogen following 8 to 12 days of progesterone pretreatment. In the sow and ewe there is a linear dose-response relationship between the duration of estrus and the logarithm of the dose of estrogen (Fig. 15–3). There is also a relationship between duration of natural

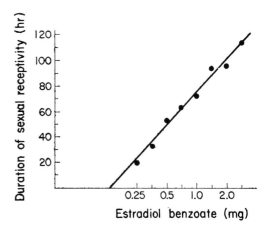

Fig. 15–3. Effect of the dose of estrogen on the duration of sexual receptivity in the sow. *(Signoret, 1967. Ann. Biol. Anim. Bioch. Biophys. 7, 1)*

estrus and the number of the ovulations in the sow and the ewe.

Progesterone inhibits the female's sexual reaction when it is injected after the appropriate hormonal sequence during the long-lasting period of latency.

The action of exogenous hormones in the intact female depends on her physiologic state at injection. When estrogens are injected during proestrus, sexual receptivity is hastened. During the luteal phase of the cycle, the inhibitory action of progesterone prevents an estrous response. The effects of hormonal treatment during anestrus are similar to those observed in spayed females.

Sex-specificity of the Endocrine Control of Behavior. The existence of sex-specific steroid hormones—androgens for males, estrogens and progestogens for females—raises the question of a direct influence of the hormone on the behavioral response. However, the rhythm of secretion is different between males and females, and a possible sexualization of the brain may influence the reaction. The treatment of gonadectomized animals with the hormone of the other sex shows that the sex-specificity of the hormone is limited as far as the behavioral responses are concerned. Daily injections of estrogen allow a complete recovery of male activity in castrated rams, whereas a single treatment with testosterone induces a normal female receptivity in the ovariectomized female. The biochemical transformation of androgens to estrogens in the brain could account for such a similarity of action, the estrogen being the active form at the brain level, as hypothesized to occur in rats.

Neural Mechanisms of Sexual Behavior

The physiologic signal that initiates sexual motivation is the secretion of steroid hormones. Once released in the bloodstream, hormones are rapidly bound to receptor sites in the CNS. Maximal estrogen levels in the blood of the ewe and sow occur about 24 hours before onset of estrus. When the animal is sexually motivated, behavioral events are initiated. Specific or unspecific sensory stimuli acting on the sense organs, through innate or acquired mechanisms, are integrated in the brain to elicit appropriate motor reactions.

The experimental investigation of the role of nervous structures implies the classic techniques of neurophysiology: lesions, stimulation and recording of electrical activity. Additional and specific methods are simultaneously used: the participation of a given nervous structure can be analyzed by local implantation of minute amounts of steroids. Furthermore, the use of radioactive labeled hormones allows one to determine the sites of action both at the gross anatomic and the cellular level.

The nervous organization of the sexual behavior, integrating humoral and physiologic signals as well as elaborated sensory information, involves the participation of the various levels.

Hypothalamus. The role of the hypothalamus in reproduction involves both the triggering effect of steroid hormones on sexual behavior and simultaneously the control of the secretion of pituitary gonadotropins. Such a duality makes it difficult to analyze separately each mechanism, as any experimental modification of pituitary function will interfere with sexual behavior through the control of sexual steroid secretion. However, in several experiments, a disruption of sexual behavior has been observed, without modification of the endocrine or gametogenic function of the gonad, in the male rat (Soulairac, 1963) and the female sheep (Domanski, 1970). Conversely, in the ovariectomized female, sections among hypothalamic structures can eliminate the feedback response to exogenous steroids without interfering with the development of sexual receptivity in estrogen-treated females. The local action of steroids administered by intrahypothalamic implant or microinjection induces in the gonadectomized animal the development of the sexual behavior. The selective fixation of radioactive labeled steroids on a neuronal population supports the assessment that the hypothalamus plays a critical role in initiating the mechanisms of sexual behavior as a response to hormones.

The neurons involved in such a "sex center" do not form an anatomically individualized nucleus, but appear to be scattered in an area extending from slightly posterior to the optic chiasm to the anterior preoptic area (Fig. 15–4). The nervous pathways that control pituitary function extend from the retrochiasmatic area posteriorly near the floor of the third ventricle to the median eminence, whereas those involved in sexual behavior seem to be located in more anterior and superior parts of the hypothalamus.

No experimental evidence has shown whether a single population of steroid-fixating neurons controls both behavioral and pituitary responses, or whether the

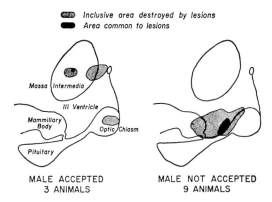

Inclusive area destroyed by lesions
Area common to lesions

Massa Intermedia

III Ventricle

Mammillary Body

Optic Chiasm

Pituitary

MALE ACCEPTED
3 ANIMALS

MALE NOT ACCEPTED
9 ANIMALS

Fig. 15–4. Effect of hypothalamic lesions on estrous behavior in the ewe. The male was accepted when the lesions were in the massa intermedia or in the optic chiasm of the hypothalamus. *(Clegg et al., 1958. Endocrinology 62, 790)*

receptor cells are specialized in only one function.

Another type of interference between both mechanisms can result from a possible direct role of the neurohormones or releasing factors acting as neurotransmitters. LRH, which stimulates the release of gonadotropins, reportedly induced female sexual receptivity (lordosis) in the rat, just as other neuropeptides interfere with learning and other activities.

The activation of copulatory patterns seems to be under varying degrees of inhibitory control by the CNS. Lesions in the junction of diencephalon and mesencephalon cause an increase in the copulatory performance of male rats. The hypothalamus may control sexual behavior in several ways: (a) fixation of sexual steroids and slow-acting elaboration of sexual motivation; (b) direct control of sexual activity; and (c) sexual satisfaction.

Rhinencephalon. The phylogenically older parts of the telencephalon (such as olfactory bulbs and olfactory tract, septum, amygdala, hippocampus, cingulum) are derived from primary olfactory structures. They are involved in the specialized behaviors (eg, feeding and sexual and emotional reactions such as arousal, aggression) (Karli, 1968). Specific EEG pat-

terns are recorded in the rhinencephalon after coitus in the female rabbit. Estrogen-fixating neurons exist in amygdala and hippocampus. Lesions in the amygdaloid and piriform cortex cause chronic hypersexuality in male cats; this response is androgen-dependent. MacLean and Ploog (1962) obtained erection in a squirrel monkey following electrical stimulation of various structures of the limbic system (hippocampus, septum, cingulate gyrus).

Cortex and Sensory Capacities. Deprivation of sensory capacity can limit sexual behavior, reduce the ability to detect the partner and/or impair orientation. Inexperienced males are impaired to a greater degree than experienced ones. If one sense is inhibited, another sense that is ordinarily used to a lesser degree may be augmented. Thus elimination of the stimuli to visual receptors in males results in the use of tactile and olfactory receptors. Copulation in domestic mammals is not suppressed with the elimination of vision, smell or hearing provided contact with the partner has been established. Tactile stimuli are involved in the organization of postural responses of copulation (eg, immobilization of the estrous sow and lordosis in the estrous rat in response to flank palpation).

Medulla. The postural reflexes of mating behavior are organized in the spinal cord of adult female mammals. In the female cat, elevation of the pelvis, treading of back legs and lateral deviation of the tail are observed after total spinal cord transection. Erection and ejaculation can be elicited in paraplegic male dogs (transection of spinal cord) as a response to genital stimulation. Sexual refractory period in the male following ejaculation seems to be partially due to refractoriness of spinal cord. The intensity of certain reactions of males and females with transection of spinal cord appears to be under the influence of sexual steroids. Furthermore, the recovery of normal reactions in castrated male rats with spinal transection can be obtained by testosterone implants in the medulla.

Neural Mechanisms of Erection and Ejaculation. Erection is predominantly under the influence of the parasympathetic system. The parasympathetic nerves in the bull, which supply the external genitalia, arise from the sacral segments of the spinal cord and are connected to the ventral roots of the spinal segments. These fibers are distributed by way of the pelvic nerves and pelvic plexus to all the reproductive organs, except perhaps the testes. Thus, drugs that affect the autonomic nervous system can be used to alter the ejaculatory process. Atropine reduces the volume of the ejaculate by blocking the secretion of the bulbourethral glands of the boar and the bull. Electrical stimulation of the sacral nerves causes erection and/or ejaculation, a phenomenon that has been put to practical use in collecting semen.

The copulatory patterns of the male are primarily governed by the neuromuscular anatomy and blood supply of the penis. The bull, ram and boar have a fibroelastic penis that is relatively small in diameter and rigid when nonerect. Although the penis becomes more rigid upon rapid erection, it enlarges little and the amount of contractile tissue is limited. Protrusion is effected mainly by straightening the S-shaped flexure and relaxation of the retractor muscle.

On the other hand, the stallion has a typical vascular penis with no sigmoid flexure. The function of the penis as an organ of intromission depends on the power of erection as a result of sexual excitement. The size, shape and length of the penis vary greatly between the flaccid and the erect state (Fig. 15–5).

Intromission and ejaculation are elicited by tactile stimuli (warmth of vagina and slipperiness of mucus) acting on the

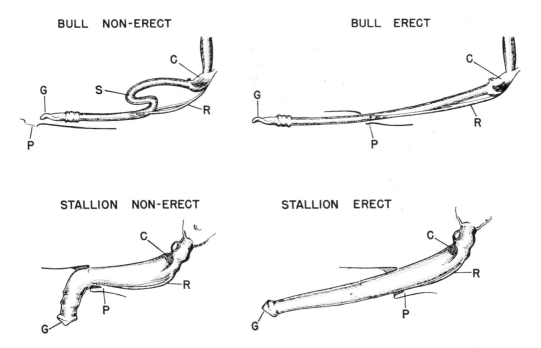

Fig. 15–5. Diagram of the anatomy of fibroelastic type penis (bull) and a vascular-muscular type penis (stallion) in the nonerect and erect positions. The anatomy of the penis determines, to a great extent, the ejaculatory responses of the species. *C,* cavernous muscle; *G,* glans; *P,* prepuce; *R,* retractor penile muscle; *S,* sigmoid flexure. *(Redrawn from Hafez, 1960. Cornell Vet. 50, 384)*

penile receptors. The penis of the bull and the ram is sensitive to temperature, whereas that of the stallion is more sensitive to pressure exerted by the vaginal walls. In the boar, the corkscrew-shaped tip of the penis is engaged in the cervix during mating. The pressure exerted by this is sufficient to elicit ejaculation even without any thermal stimulation.

Biochemical Mechanism: Role of Biogenic Amines in Sexual Behavior. Experimental studies have stressed the importance of a specialized neuronal population in some nervous structures, but the sensory and motor organization of sexual behavior implies the existence of nonspecialized circuits. The existence of some kind of "sex center" from which sexual motivation could be initiated in response to steroids is supported by analysis of the role of some hypothalamic neurons. However, in every structure involved in sexual behavior—hypothalamus, rhinencephalon, medulla—a specific fixation of sexual steroids by specialized neurons occurs, which suggests that the action of the hormones is highly diversified. Similarly, the endocrine balance interferes with general biochemical mechanisms of the central nervous system and vice versa. Biogenic amines are involved in behavioral activities as neurotransmitters. Recent advances of neuropharmacology allow the experimenter to interfere through specific inhibitors or precursors with different metabolic reactions of any neurotransmitter. For example, Kobayashi et al (1946) observed a decrease in brain monoamine levels at estrus in the rat. Monoamine-oxidase inhibitors prevent such reduction and impair sexual receptivity.

FACTORS AFFECTING SEXUAL BEHAVIOR

The patterns and intensity of sexual behavior are affected by genetic, physi-

ologic and environmental factors as well as previous experience.

Genetic Factors

Breed and strain differences in libido are frequently observed. Males of dairy breeds are more active than beef males, whereas Brahman bulls are sluggish. Yorkshire boars are easier to train for semen collections than Durocs. More differences in the pattern of sexual behavior occur between pairs of identical twin bulls than between members of the pair (Fig. 15–6). Breed differences in the duration of estrus in sheep and pigs may be partly due to differences in ovulation rates but similar differences occur in estrogen-

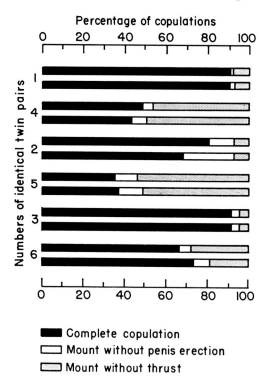

Fig. 15–6.　Ejaculatory behavior in identical twin dairy bulls showing percentage distribution of complete copulations, mounts without penis erection and mounts without thrust. Note the great similarities in the ejaculatory pattern of the twin brothers and the great variability between the twin pairs. (*Adapted from Bane, 1954. Acta Agric. Scand. 4, 95*)

induced estrus in ovariectomized females (Signoret, 1967b).

Individual differences in the amount of sexual stimulation required to elicit "immobilization reaction" in the sow are independent of sexual experience, the female or the amount of estrogen injected (Signoret, 1967a). A genetic influence may account for such differences.

Sexual Differentiation of the Nervous System

Hormones secreted during fetal and neonatal life cause sexual differentiation of the nervous system. A female treated neonatally with androgen does not show the cyclic patterns of gonadotropin secretion that are characteristic of the female. The secretory activity of the hypophysis is more or less constant as in the male, but males castrated in their neonatal life exhibit a cyclic pattern of pituitary secretion, as in the female. The neonatal androgen appears to differentiate the hypothalamic structures controlling gonadotropin secretion from a neutral cyclic pattern of the female into a tonic type of the male.

Anatomic evidence for sexual dimorphism in the hypothalamus has been reported. Untreated male and neonatally testosterone-treated female rats showed reduced numbers of dendritic spine synapses in the strial part of the preoptic area. In most mammalian species, the female also exhibits male patterns of sexual behavior, although with reduced frequency, which suggests the absence of sexualization of the nervous structures involved in the realization of such patterns. In fact, a female receiving a male-like hormonal treatment (daily injections of androgen) exhibits the complete repertoire of male sexual behavior, including the copulatory patterns and the subsequent refractory period. The frequency of the patterns, their sequential organiza-

tion, and the dose-reaction relations are similar.

The possibility of female responses occurring in the male has not been studied in detail until recently. In the rat, the lordosis reaction occurs only after estrogen-progesterone has been administered in neonatally castrated subjects. The conclusion was that the neonatal secretion of androgen had sexualized the potentially bisexual brain of the male through "defeminization." However, recent studies have shown that after long-term treatment with estrogen, lordosis can be observed in adult castrated male rats. Furthermore, domestic pigs castrated at four months of age, exhibit a complete female immobilization to the boar when treated as adults with a single dose of estradiol.

Environmental Factors

The effect of external stimulation on sexual behavior is more pronounced in the male than in the female.

Effect of Novelty of Stimulus Females (Coolidge Effect). Sexual activity of the male increases when new females in the herd become receptive. If four receptive ewes are available, the ram mates three times as much as he does when only one ewe is in estrus. The enhancing effect of a new stimulus animal should be kept in mind under modern husbandry conditions. For instance, changing the teaser cow is an effective way to increase sexual behavior of a sluggish male.

Nonspecific Stimuli. Nonspecific external stimuli may lead to sexual activity during the refractory period that follows ejaculation in males of low libido. In the rat, both painful stimuli such as an electric shock and gentle handling increase the frequency of ejaculations and reduce the postejaculatory interval. Changing the place of semen collection by moving the teaser animal, or "encouraging" the bull, are all effective in sluggish bulls.

Presence of Other Animals. The presence of other males while teasing a female or copulating improves sexual libido of the male. However, social hierarchy may interfere with sexual activity when several males compete for one receptive female. The dominant male performs most of the copulations and restricts the sexual performance of his subordinates. When females are in excess, however, dominant males cannot effectively control the activity of their inferiors. Adult rams usually dominate or "boss" yearling rams, and the degree of their dominance is greater than the dominance of yearling rams over other yearlings (Hulet et al, 1962). Unfortunately social dominance is not correlated with fertility of males. Thus a "bossy" male who is infertile or diseased may depress the conception rate of the entire herd. The size of the pasture also affects the competition among males and the number of copulations per female. The proportion of ewes bred by three rams compared to those bred to one ram was higher in a 1/5-acre pasture than in a 17-acre one (Lindsay and Robinson, 1961).

Season and Climate. Seasonal variations in sexual behavior of sheep, goats and horses are mostly due to seasonality of pituitary function controlling the secretion of gonadal hormones. Seasonal changes are also reported in the responsiveness of ovariectomized ewes and sows to exogenous hormones, showing a direct effect of the season on the sensibility of the nervous system. The intensity of sexual behavior is reduced in hot climates. The plane of nutrition per se does not seem to affect sexual behavior. However, any physical trouble may seriously affect sexual expression (eg, inflammation of the hooves or joints, change of teeth, eczema, pains from accidents or certain diseases).

Effect of Experience

The efficiency of copulation of males and females is improved by experience. Individual contacts before puberty can have an organizing effect on subsequent sexual performance.

Social Contacts Before Puberty. Social deprivation during infancy drastically impairs adult sexual behavior in primates (Harlow et al, 1966). Sexually motivated male and female rhesus monkeys reared in isolation cannot perform postural adjustments for copulation. In other mammals, the female's sexual behavior does not appear modified by deprivation of social contacts. However, social deprivation during ontogeny may account for some cases of sexual inhibition of domestic males. In boars and rams, rearing in isolation or in unisexual groups has been reported as having a detrimental effect on subsequent libido. Such results remain controversial, and in many cases rehabilitation of the inhibited male appears possible.

Sexual Experience. Young inexperienced males are usually awkward during their first contact with a receptive female: they approach hesitantly, spend a long time exploring the genitalia, mount with erection, descend and try to mount again. Erection and ejaculation are weak and the volume of semen is small. After the first ejaculation, the motor patterns are rapidly organized and normal mating efficiency is reached. The search and identification of an appropriate sexual partner improves gradually by experience. On the other hand, sexual reactions in the female are immediately adapted with little improvement due to sexual experience.

EFFECTS OF SEXUAL INTERACTIONS ON REPRODUCTIVE PHYSIOLOGY

Female

The presence of the male may affect the physiologic events at the time of estrus and interfere with the mechanisms regulating the sexual season and puberty.

Effects of Male on the Duration of Es-

trus and Time of Ovulation. The presence of the male reduces the duration of the sexual receptivity of the ewe by 50%. This effect is not related to the possible changes in ovulation or steroid secretion, since similar effects occur in ovariectomized ewes and sows treated with progesterone and estrogens. With the exception of the alpaca and possibly some members of the Camelidae family, farm animals are known to be spontaneous ovulators. However, permanent association with the male hastens LH release and ovulation in the ewe and the sow.

Uterine Motility and Sperm Transport. Sterile matings with a vasectomized male stimulate sperm transport in the rabbit (Dandekar et al, 1972) but not in sheep (Hawk, 1972). However, stimulation of the genitalia or precoital stimuli cause contraction of the cervix and uterus of the ewe (Lightfoot, 1970) and the cow, as a result of the release of oxytocin. Vaginal distention and precoital stimulation causes maximal oxytocin release in sheep and goats (MacNeilly and Folley, 1970). Oxytocin release often occurs before actual coitus has taken place. The exteroceptive factors that stimulate sexual behavior, in order of decreasing effectiveness, are presence of the male, smell of the male, sounds emitted by the male and sight of the male.

Effect of Male on Anestrous Females. Many domestic females undergo periods of anestrus. It is well-documented that by the end of seasonal anestrus in sheep and goats, the introduction of the male results in an earlier and synchronized appearance of estrous cycles. Even the Merino sheep, which is reputed to breed throughout the year, ovulates and breeds spontaneously for only a restricted period during autumn. The influence of the ram is rapid; the preovulatory peak of LH in Merino sheep has been demonstrated at a mean of 27 hours after joining. However, the ovulations occurring under such conditions are of poor quality, often not producing a functional corpus luteum and without sexual behavior. The peak of estrus observed 17 to 18 days following the introduction of the male represents in fact the second cycle, and this allows normal fertilization.

An androgen-dependent pheromone from the male is responsible for similar synchronization of estrus in mice (Bronson and Whitten, 1968) and possibly in sheep. Neither sight nor contact is necessary for the synchronization of the first estrus of the breeding season in sheep.

Nursing the young delays estrus compared with milking in cows and ewes. The presence of the ram results in an earlier postpartum estrus in nursing ewes, making the postpartum estrus similar in dry and milked females.

The introduction of the boar shortly before spontaneous puberty in a group of previously isolated gilts results in earlier onset of estrus.

Male

Precoital stimulation affects both composition of the ejaculate and androgen secretion. A period of restraint for 2 to 20 minutes causes an increase in semen volume and concentration and number of sperm in bulls, with sperm motility being unaffected. False mounts cause further increases in semen characteristics. The presence of another bull, changing the teaser, or using the bull as a teaser prior to collection has no such augmenting effect on semen characteristics despite a great increase in sexual excitement. Thus the stimuli that influence semen composition differ from those that cause sexual excitement.

In the rabbit, copulation or the presence of a doe causes a remarkable increase in blood testosterone within 30 minutes (Haltmeyer and Eik-Nes, 1969). Similar observations reported in bulls and rams remain controversial, because they have not been repeated by others.

ABNORMAL SEXUAL BEHAVIOR

Homosexuality, hypersexuality, hypo-sexuality and autoerotic behavior are not uncommon. These syndromes may be due to genetic factors, disturbance in the endocrine or nervous systems or faulty management. Unadapted sexual reactions are more frequent among domestic animals and under conditions of captivity in the zoo than in the wild. Homosexuality refers to sexual behavior among males, particularly at puberty and when young males are housed together. The stimuli eliciting the male's sexual response are essentially visual. A releaser of an appropriate shape and size presented to a highly motivated male may elicit mounting. An immobile anestrous female, another male or an inanimate object may release sexual reactions. Most homosexual males in sex-segregated groups become heterosexual when placed with females and again homosexual when segregated.

Hypersexuality in males consists of increased sexual excitement, increased frequency of copulation and attempted copulations with young males and females of the same or different species. Hyposexuality is characterized by abnormalities in the ejaculatory pattern. Certain males may fail to ejaculate in spite of protrusion of erection, whereas others cannot mount or exhibit no sexual desire for varying periods of time.

Autoerotic behavior refers to self-arousal of sexual responses, which is called masturbation in males. The motor patterns vary with the species. The stallion rubs his rigid erected penis against the hypogastrium (anterior median of the abdomen) and lowers the loin region rapidly. This is followed by several forward movements of the pelvis, resulting in abortive ejaculation. Masturbation is less common in rams and most common among bulls on high protein ration (eg, bulls prepared for shows). As a result of such diets, the peripheral mucosa of the penis becomes more sensitive to tactile stimulation.

The most common abnormal female behavior is nymphomania in cattle.

MATERNAL AND NEONATAL BEHAVIOR

The behavioral events at birth and shortly afterward (Table 15–4) have an important influence on the survival of the newborn, and hence on the successful outcome of reproductive processes. This is especially true for sheep, cattle, deer, horses and pigs in which the initial suckling and development of a bond between mother and young frequently occur in the outdoor environment, often under adverse conditions such as inclement weather and the presence of predators, or in artificial conditions of close confinement indoors.

Maternal Behavior in Sheep

Prepartum Behavior. Ewes tend to cease grazing within an hour or so before lambing and wander about as if searching for a lamb. Parturition, at least in Merinos, usually begins while the ewe is with the flock, and the lambing ewe is left behind as the flock grazes on. There is no satisfactory documentation that ewes normally leave the flock to lamb at a sheltered or isolated spot and so facilitate survival of the lamb. However, ewes can be induced to lamb in shelter by shearing shortly before lambing (Fig. 15–7).

Role of Fetal Fluids. The birth site appears to be determined fortuitously by where the placental fluids are first spilled. These fluids are attractive to ewes near the time of lambing and the ewe usually remains at the site of spillage, licking and pawing the ground. The fluids appear to play a critical role in attracting the ewe to her newborn lamb. Ewes that have not yet lambed are attracted to the fluids and to newborn lambs of other ewes, leading to "lamb stealing." This adoption sometimes

Table 15–4. Maternal and Neonatal Behavior in Domestic Ungulates—Semiquantitative Comparison of Species*

	Sheep	Goats	Cattle	Deer	Horses	Swine
Prepartum behavior						
Restless	+	Probably the same as sheep	+	+	+	+
Seeks isolation	±		±	±	(?)	+
Milk ejection	−		±	−	±	−
Builds nest	−		−	−	−	+
Time of most births	Variable	(?)	(?)	(?)	Night	Early hours of darkness
Duration of birth (approximate hours)	1	1 (?)	1 (?)	1	1 (?)	3 (from first to last piglet)
Delivery of placenta						
Time from delivery of young to delivery of placenta (approximate hours)	Several	Several	Several	Several	1 (?)	Several
Eating of placenta by dam	±	± (?)	± (?)	+	−	±
Time for dam to stand up after birth (if not already standing) (approximate minutes)	<1	<1 (?)	<1 (?)	<1 (?)	10	3
Grooming						
Occurrence	+	+	+	+	+	Nil
Duration (approximate hours)	1	1 (?)	1 (?)	1	Several	
Maternal solicitude for alien young	−	(?)	±	−	(?)	+
Abnormal maternal behavior						
Desertion of young	+	Probably the same as sheep	+	+	(?)	−
Moves from suckling	+		+	(?)	+	NA
Attacks young	+		+	+	+	NA
Cannibalism	−		−	−	−	+
Susceptibility of young to chilling	+	+	− (?)	+ (?)	− (?)	+
Progress of young						
Time to stand (approximate minutes)	15	Probably the same as sheep	40	30	40	5
Time to first suckle (approximate hours)	1		2	1	2	0.3
Mother consumes feces and urine	−	(?)	± (?)	+	(?)	(?)
Hiding phase	−	±	+	+	−	NA
Frequency of suckling (times per day approximately)						
First four days	30	(?)	4 (?)	3	(?)	10
Midlactation	15	(?)	3 (?)	3	(?)	25

Key: Behavior usually present, +; behavior sometimes present, ±; behavior usually absent, −; behavior not documented, ?; not directly applicable, N.A.
*Behavior, even with a group of homogeneous individuals of the one species, is extremely variable, so comparison between species must be made with great caution.

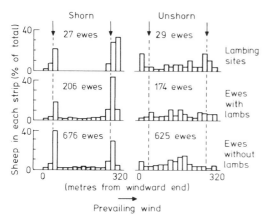

Fig. 15–7. Distribution of lambing sites at night and the simultaneous distribution of unlambed ewes and recently lambed ewes between 20-m strips along the length of a paddock 320 m × 25 m, containing grass hedges (↓) across the 25-m width, 40 m from each end. A much higher proportion of shorn ewes than unshorn ewes lambed in the lee of the shelters. *(Adapted from Alexander, Lynch and Mottershead, 1979. Appl. Anim. Ethol., 5, 51)*

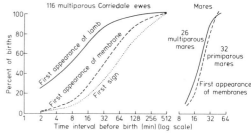

Fig. 15–8. Cumulative distribution curves (log time scale) of the duration of birth in ewes and mares measured from the first sign of impending birth, from the time of first appearance of the membranes or from the time the fetus first appeared at the vulva. Although the modal time for the interval from first appearance of the membranes to birth is similar in ewes and mares, the variability is much wider in ewes and includes some very long intervals. *(Drawn from data of Alexander [unpublished] and Rossdale, 1967. Brit. Vet. J. 123, 470)*

results in lambs being left without maternal care, and the rearing of lambs by ewes that are not their natural mothers can have serious effects on genetic experiments in which lambs are identified with ewes many hours or even days after their birth.

The chemical basis of the attraction of placental fluids is unknown.

Lambing. There is no consistent peak of lambing at any particular time of the day. Behavior during parturition largely depends on the ease of the process, but generally, the initial restlessness is broken by periods of lying with abdominal straining. Most lambs are born with head and forefeet foremost. Lambing usually occurs while the ewe is recumbent, but it can also occur while the ewe is standing; most ewes are on their feet within a minute after birth. The umbilical cord is broken simply by stretching. The fetal placenta, or "afterbirth," is delivered two to five hours later and is frequently eaten by some breeds, but rarely by others.

The duration of birth varies widely (Fig. 15–8) within the one flock. Sometimes an extended birth is associated with a single lamb being too large for the birth canal, with twins being impacted in the canal, or with an abnormal presentation resulting in an increase in the effective diameter of the lamb (George, 1976). Birth also tends to be protracted in ewes lambing for the first time or in ewes debilitated by ill-health or undernutrition. Protracted labor can exhaust the ewe and have adverse effects on maternal behavior and the viability of the lamb.

Twin births are usually more rapid than single births because twins are usually smaller than singles, but the interval between delivery of twins varies widely from a few minutes to an hour or more.

Postpartum Behavior. *Grooming and Maternal Attachment to Lamb.* Vigorous licking (grooming) and eating of any amniotic and allantoic membranes adhering to the lamb usually commence immediately after birth. During this phase of intense olfactory and gustatory contact, which persists for little more than an hour (Bareham, 1976), the ewe learns to distinguish her own lamb from aliens, which are soon rejected by vigorous butting. Experiments with lambing ewes in which the olfactory bulbs have been destroyed confirms that this attraction is largely olfactory (Baldwin and Shillito, 1974). However, maternal behavior is not abolished by destroying the sense of smell. Other factors, such as warmth and movement, may be important in maintaining the attraction of the mother to the newborn lamb; ewes rapidly lose interest in an immobile, chilled lamb.

This "critical period" of attachment of the ewe to the lamb is short; if the lamb is removed at birth it will be rejected by the ewe if presented to her six to twelve hours later.

Grooming of the lamb by the ewe may remove some of the 0.5 kg of fluid present in the coat at birth, so reducing heat loss. Grooming may also play an important role in stimulating the lamb to stand and suckle for the first time.

Maternal behavior appears to be under endocrine control; it can be induced in sheep under the influences of a rapidly declining progesterone level and rapidly increasing estrogen level, conditions that normally apply around birth.

Finding the Teats for the First Time. Most lambs are standing within 15 minutes of birth, and with increasing coordination make exploratory approaches to the ewe, moving from the head along her side and nosing into body angles. During the initial approaches the ewe tends to keep the lamb in front of her head to facilitate grooming, but within an hour or two most ewes will allow the lamb to move towards the udder. This brings the lamb's anal region near to the ewe's head, and this region is frequently nuzzled and sniffed by the ewe.

Both maternal orientation and grooming appear to facilitate the lamb's finding the teats for the first time, but neither is essential (Alexander and Williams, 1964). The teats seem to be found by trial and error. For example, in goats the teats are found with equal success whether the udder is in the normal position or has been grafted onto the neck (Stephens and Linzell, 1974). Visual guidance is not essential, as lambs will suckle for the first time in a light-proof room; however, visual cues appear to facilitate suckling subsequently (Bareham, 1975). There is no information on any role of odor.

In the absence of the mother during the initial critical hours, the lamb will follow and become attached to a substitute maternal figure, even if the figure is not associated with feeding. During this initial phase, the lamb also learns to follow the ewe as it moves from the birth site. For one or two days after birth this following response can be elicited by any large moving object, but it is then replaced by a fear response. The sheep is a classic example of a "follower" species as distinct from a "hider" species: that is, the young tend to follow their mothers from birth rather than lie hidden for several days while the dams are absent. The tendency of lambs to follow their mothers long distances to water in hot climates can lead to death from heat stroke.

Aberrant Behavior. Ewes that are exhausted by a difficult parturition may remain prone for some hours after birth and the lamb may stray. Some ewes lambing for the first time display little interest and abandon their newborn lambs. Other forms of misbehavior in inexperienced ewes include butting the newborn lamb as it moves, and a tendency to move to maintain the initial head-to-head orientation. When these behavioral patterns persist, the lamb's chances of finding the teats and suckling can be reduced, since the teat-seeking activity declines rapidly from about two hours after birth in the absence of successful suckling (Fig. 15–9). Some lambs, particularly after poor fetal nutrition, a long birth process, birth injury or chilling, are slow to stand and suckle.

Aberrant patterns are more common with twin births than with single births, especially among Merinos. For example,

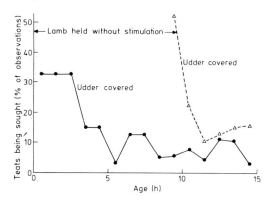

Fig. 15–9. Mean teat-seeking activity in lambs prevented from sucking by an udder cover. Ten lambs were left with their ewes from birth, and ten were first held in a dark box under conditions of minimal stimulation for nine hours after birth. The teat-seeking activity declined rapidly after birth but was preserved by depriving the lamb of stimulation. *(Adapted from Alexander and Williams, 1966. Animal Behav. 14, 166)*

if twins are born several meters apart, one lamb may be neglected and become lost or may be repelled as an alien if contact is remade after the critical receptive period of the ewe has passed; imprinting with one lamb does not guarantee acceptance of others in the same litter. However, twin losses are more likely to occur when the ewe moves from one site to another or is frightened away by the presence of man or other animals; one lamb is likely to be left behind, since some ewes, especially Merino ewes, appear to take several days to recognize that they have more than one lamb. Aberrant behavior in the postpartum period is a significant cause of lamb mortality (Arnold and Morgan, 1975).

Behavior and Thermoregulation. The temperature regulatory mechanisms of lambs are remarkably well developed at birth (Alexander, 1975), but some lambs tend to chill at birth either because of a physiologic deficiency or through excessive heat loss. For example, many wet newborn lambs in windy outdoor conditions would chill even at 10°C air temperature. Even when body temperature is normal, cold conditions suppress teat-

seeking activity, and severe chilling can lead to inactivity and ultimately to maternal desertion and lamb death.

The Suckling Phase. Ewes with lambs tend to form distinct groups away from the main flock, and for some days after birth the ewes remain within earshot of their lambs. Accidental separation results in considerable agitation of both ewe and lamb, so it is important that there are effective mechanisms for reunion of the two.

Mutual Recognition by Ewes and Lambs. The means whereby ewes and their lambs recognize each other has been the subject of much research (see volumes 1–4 of *Applied Animal Ethology*). The experiments were largely based on manipulation of the cues to recognition provided by the partner (Alexander, 1977). Voices were blocked by local anaesthesia and visual cues were removed by screening or altered by dusting pigments into the coat, or by shearing. The effects were observed in a small enclosure when a ewe and a lamb were simultaneously released from opposite ends.

The experiments showed that ewes are attracted by a lamb's bleat and the attraction is particularly strong for the lamb's own mother, but it is not essential for a ewe to hear her lamb's voice for her to accept it. The appearance of the lamb is of paramount importance; for example, if lambs are disguised by blackening or coloring, many ewes actively avoid the approaches of the disguised lamb when initially presented with it, even if the lamb is the ewe's own. Similar experiments with blackening various parts of the lamb show that critical visual cues emanate largely from the head (Fig. 15–10). The scent of the lamb is also important, but only at close quarters (Fig. 15–11); scent appears to be the final criterion by which a ewe identifies a lamb as her own.

Initially, lambs are insensitive to identity cues provided by the mother, and they will attempt to suckle from any ewe.

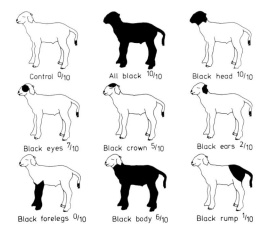

Fig. 15–10. The effect of blackening various regions of lambs' bodies on the incidence of hesitation or dodging away by their dams when confronted with their partially blackened lamb for the first time. The number of ewes reacting out of the ten tested with each treatment is indicated. (Adapted from Alexander and Shillito, 1977. Appl. Anim. Ethol. 3, 137)

There is a nonspecific attraction of ewe's voices initially, but the specificity increases with age. Visual cues become more important than auditory cues by the time the lamb is about three weeks old. Whether an odor from the ewe plays any role in recognition by the lamb is unknown.

These experiments on maternal behavior using disguised lambs provided the basis of tests that showed that sheep have some color vision. Ewes that had become accustomed to their own lambs colored red, yellow or orange, were presented with disguised alien lambs, from which they were able to select their own color from a variety of grey shades (Alexander and Stevens, 1979).

Suckling Behavior. After the newborn lamb has achieved satiety, the frequency of suckling stabilizes at once or twice per hour (Fig. 15–12). During the first week,

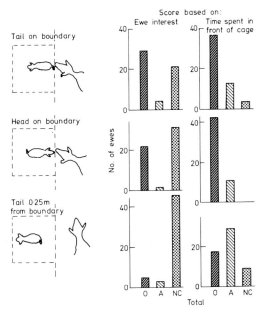

Fig. 15–11. Numbers of ewes choosing own lamb (O), an alien (A) or making no choice at all (NC) when presented with three anaesthetized lambs (own and two aliens) beyond a wire-netting barrier. Each ewe was tested three times, each time with the three lambs all in the same position (tail against wire, head against wire or tail 0.25 m from the wire). Ewes were unable to identify their own lamb unless the head or the tail was against the wire. (Adapted from Alexander, 1978. Appl. Anim. Ethol. 4, 153)

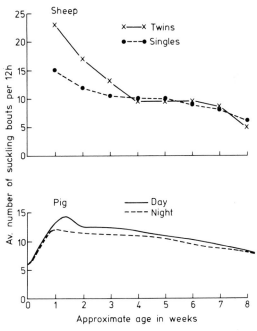

Fig. 15–12. Frequency of suckling bouts in sheep and pigs. Peak frequency is reached later in swine than in sheep. (From Ewbank, 1967. Anim. Behav. 15, 251, and data of Niwa et al., 1951. Bull. Nat. Inst. Agric. Sci. Tokyo (Ser. G) 1, 135)

suckling is usually initiated and termi-nated by the lamb. Twin lambs may be suckled individually and more frequently than singles.

As the lamb becomes older and mutual recognition improves, ewes and lambs stray further apart (Morgan and Arnold, 1974) and the ewe now tends to terminate suckling, although she may initiate suck-ling by approaching and calling the lamb. After a week or so twins are not usually permitted to suckle unless both are pres-ent, so the ewe now recognizes that she has two lambs. With advancing age, the frequency and duration of suckling de-cline (Fig. 15–12). Under natural condi-tions these changes culminate in wean-ing, with maternal solicitude being sud-denly replaced by antagonistic behavior; weaning of domestic ungulates is usually enforced prematurely by man.

There is no fierce maternal protective behavior in sheep; the mere presence of the ewe appears to act as a deterrent to predators such as foxes and crows. Mater-nal care consists primarily of supplying milk exclusively for her own lamb. A small minority of ewes will allow an alien lamb to suckle, and some lambs become adept at stealing milk from an alien ewe; these abnormal patterns may be due to disruption of the normal sequence of events immediately after birth.

During artificial rearing, some lambs spend considerable time sucking the scrotum, navel or ears of other lambs. This interferes with feeding of the suckling lamb, and sometimes results in ingestion of feces and in diarrhea. The social in-teractions with other animals may also be affected.

Maternal Behavior in Goats

Maternal and neonatal behavior of goats is similar to that of sheep. However, goats may be more efficient than sheep in rec-ognizing that they have twins. Also the doe spends more time vigorously groom-

ing and orienting to the first-born, so that the second-born, which is usually the weaker of the two, has the greater oppor-tunity to suckle (Klopfer and Klopfer, 1977).

Goats appear to be "hiders" for the first few days after birth, as judged by the behavior of feral goats in mountainous country, and there is a clear preference for maternal isolation at the time of birth (Rudge, 1970).

Kids' voices are initially similar, when analyzed by a sonagraph, and the does cannot distinguish between kids by audi-tory means until the sounds begin to di-verge, at about four days of age.

Maternal Behavior in Cattle

Both normal and abnormal behavioral patterns of the cow and calf are remark-ably similar to those of sheep (Selman et al, 1970) (Table 15–4), although there are some points of difference. A day or two before calving the cow may become rest-less and keep to a small isolated area, which she defends against other cows. Confinement and interference at this time can result in prolonged birth and poor calf survival (Dufty, 1972). Cows tend to eat the afterbirth more frequently than do sheep. Inexperienced heifers can be par-ticularly aggressive toward their newborn calf when it moves, although this behav-ior is usually transient. Newly calved cows are also much more aggressive than sheep in defending the territory around the calf. The progress of newborn calves is somewhat slower than that of lambs. Most new calves take at least 45 minutes to stand, and may take upward of four hours to suckle for the first time.

Unconfined cattle exhibit a short "hid-ing" phase during the first few days after birth; the calf seeks seclusion and remains in isolation while the cow grazes well out of earshot. In tropical environments this can lead to exposure of calves to intense solar radiation and the risk of heat stroke.

However, most calves move to shade with the cow within four or five days of birth. "Hiding" avoids the necessity for calves to travel long distances to water with their dams, thus running the risk of heat stress through exercise. After the initial hiding phase, it is common in large paddocks to see a group of calves apparently being cared for by one or two cows, the other mothers being absent grazing or watering. Also in contrast to sheep, unconfined cows suckle single calves about four times daily, usually at dawn, dusk and at the start of grazing periods; the frequency is greater in closely confined animals and with twins.

Cows seem to be less aggressive toward alien calves than ewes, and appear to rely less on olfaction and more on vision than do ewes in identifying their offspring. These tendencies may have facilitated the now common practice of rearing calves by fostering several onto a single dairy cow. Blind-folding the cow for several days is a recommended aid in this procedure (Crowley and Darby, 1969), but a high level of success is achieved if the cow's own calf is exchanged at birth for the foster calves and these are smeared with the fetal fluids (Hudson, 1977). With fostering several days after birth, the cow-calf bond is frequently weak, resulting in suboptimal calf growth rates.

The artificial rearing of calves leads to behavioral abnormalities, such as inability to find the teats on a cow, cross suckling as with lambs, and subsequent poor maternal behavior if the calf has been reared in isolation.

Maternal Behavior in Deer

Red deer and some other species are now being managed as farm animals. The behavior of mother and young is similar to that of conventional domestic ruminants (Kelly and Whateley, 1975; Clutton-Brock and Guinness, 1975). The dam may seek isolation from birth, but births appear to be equally distributed between night and day. Some dams are particularly aggressive toward alien fawns and may beat them to death. Like calves, fawns have a marked preference for concealment. Suckling frequency is similar to that of cattle, and the hiding phase may last for more than a week. Ingestion of the feces and urine of the newborn is common and may persist for weeks. Deer are the only ungulates in which imprinting of the young on a species-specific odor has been demonstrated (Müller-Schwarze and Müller-Schwarze, 1971).

Maternal Behavior in Horses

Studies on horses have been confined to thoroughbreds in small paddocks or in loose-boxes and often with man nearby (Rossdale, 1968). Approaching birth may be indicated by milk ejection and restlessness, but these signs are not confined to the immediate prepartum period.

Foaling. In mares, most births are recorded at night, but whether this is due to photoperiodic effects or routine husbandry procedures is not clear. Delays in parturition have been attributed to human presence and interference, as well as to the presence of spectator groups of mares. In contrast to the situation in sheep and cattle, the duration of birth is little affected by whether or not the mare has foaled before (Fig. 15–8).

Abnormally long parturition is usually associated with malpresentation of the fetus, though aged mares may not sustain their efforts to foal. Mares tend to remain recumbent longer than sheep, often not standing until more than ten minutes after birth.

Postpartum Behavior. Most foals take an hour or more to maintain a stable stance and most have suckled within three hours of birth. Grooming continues for several hours. Premature young are more common in horses than in ruminants and the premature foal tends to be weak and slow to progress, although the suckling reflex is well developed.

The use of visual, olfactory and auditory cues in mutual recognition between mares and foals is similar to that in sheep.

Maternal Behavior in Pigs

Studies with pigs have also been made mostly with the animals in close confinement, conditions under which the behavior observed is muted by comparison with that of pigs in the wild (Frädrich, 1974). Behavior in the pig with its large litters of small, almost naked young contrasts with that in other domestic ungulates, and appropriate behavior patterns are of particular importance in swine because of the susceptibility of the young to starvation, chilling and accidental injury by the sow.

Prepartum Behavior. In the field the approach of parturition is indicated by characteristic nest-building activity, but in practice, sows are so restricted by the lack of nest-building material and by modern farrowing pens, designed to protect the piglets, that this activity is inhibited and the sows may become disturbed. As parturition approaches, characteristic vocalizations become evident and the "nest" area is defended as if piglets are present.

Farrowing. Parturition in the sow is most frequent during the hours after sunset. Though most domestic sows tolerate the presence of an observer at farrowing, some become highly disturbed. Sows normally farrow lying on one side, and delivery is accomplished with much less apparent effort than in other farm animals.

The rupture of the membranes and the voiding of the fetal fluids is not well-defined as in other ungulates. Most piglets are born partly covered with fetal membranes, and in contrast with the larger young of other ungulates, the piglet must escape from these without maternal aid or perish. The umbilical cord is broken by the piglet moving away from the vulva. At the end of delivery the sow usually stands to urinate, and in the process of lying down again, and also during earlier bouts of restlessness, the piglets are prone to be overlain and injured.

The average duration of farrowing is about three hours, but can range from a half to eight hours or more. The interval between birth of individual piglets ranges from less than a minute to three or more hours, and piglets in the last half of the litter are prone to be stillborn, perhaps because of premature rupture of the cord and prolonged hypoxia (Fig. 15–13). More than 50% of piglets are born head foremost, but posterior birth does not appear to be unphysiologic in this species (Randall, 1972), in contrast with other ungulates.

Postpartum Behavior. Newborn piglets are almost immediately mobile. They rapidly find their way to the udder and may be suckling within five minutes; most have obtained milk within half an hour of birth. At this early stage, milk is available on demand, possibly because of continuous milk letdown due to circulating oxytocin associated with the birth process. Behavior of litters immediately

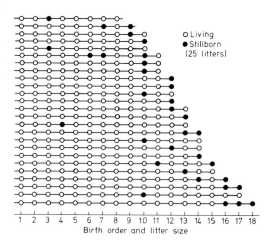

Fig. 15–13. Mortality in piglets in relation to birth order in 25 randomly selected litters. Later born piglets are likely to be stillborn. *(Drawn from unpublished data of P. English.)*

after birth is variable, and piglets may be attracted to infrared heaters before moving to the udder. Attraction to heaters is probably undesirable at this early stage when a teat-order is being established, but may be desirable subsequently in keeping piglets from being overlain by the sow. The mechanism of attraction to the udder region is unknown. In the search for the teats, the piglets tend to concentrate on the pectoral region of the udder and explore vertical surfaces with their noses until a teat is contacted, grasped and suckled (Scheel et al, 1977); visual cues appear to be unimportant in the initial search for the teats. The piglets show no aggression toward litter mates during their teat-seeking phase. Piglets that do not find a functional teat soon after birth rapidly deplete their energy reserves in cold weather and die from hypothermia.

The Suckling Phase. *Establishment of Teat Preference.* Having suckled initially, the piglets tend to sample several teats in the same row as the teat first sucked; they locate these teats readily, having rapidly learned the appropriate orientation and teat height. Contact with another piglet may now result in fighting, in the form of pushing or biting with the sharp canine teeth. Injuries result, but appear to be of little significance (Fraser, 1975). In most litters, the frequency of these encounters rapidly declines during the first 12 hours (Fig. 15–14) and individual piglets tend to confine their interest to specific areas of the udder near the teat initially suckled, and finally to specific teats. Conflicts at this stage are largely confined to neighboring litter-mates. The time for teat-order to stabilize varies widely from a day or two to one or two weeks (Hemsworth et al, 1976). It is shortest with sows that remain lying on the one side for suckling during the first day or so; changes in position unsettle the order and the fighting phase is prolonged. The teat-order tends to be unstable with large litters or with a poor milk supply. Identification of teats by pig-

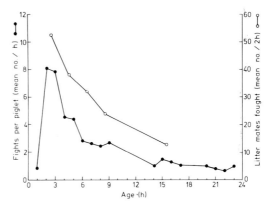

Fig. 15–14. Fighting by piglets during the first day of life. *(From Hartstock and Graves, 1976. J. Anim. Sci. 42, 235.)*

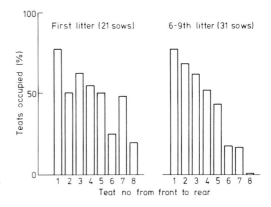

Fig. 15–15. Occupation of teats by piglets during established lactation. *(Drawn from unpublished data of P. English)*

lets does not seem to be based on taste or smell, since thorough washing of the teats on a dummy sow has no effect on the teat-order; recognition of neighbors and the position relative to the sow may be important, as a litter of piglets attempting to suck from an alien sow tend to maintain the same teat-order. In small litters, some piglets have the regular use of adjacent teats, though one is usually preferred. There is clear preference for the anterior teats (Fig. 15–15), and although their control is usually gained by the larger, dominant piglets, there appears to be no major advantage in suckling from them (Fraser and Jones, 1975).

Suckling Behavior. Suckling bouts are initiated by the sow, either spontaneously or by the piglets squealing or attempting to suckle. There is a distinctive food call, a series of soft grunts to which the litter rapidly becomes conditioned, and which initiates suckling. The bouts can be stimulated by a disturbance in the farrowing house or by suckling by a neighboring sow. Suckling positions tend to be characteristic of the individual sow; the standing position is more common in feral than domestic pigs. Sows about to move use a "warning" call to which piglets respond by moving away to avoid the risk of being crushed; sows that suckle lying down usually give the food call after they have lain down.

The food call initiates a period lasting several minutes, of intense udder massage by the noses of the piglets; at the same time, a recumbent sow rotates her body to expose the teats. The grunt frequency increases to a peak and the phase of active movement by the piglets is suddenly replaced by a quieter phase of milk letdown and suckling which lasts for a mere minute or two. The frequency of suckling increases during the first week to about once per hour, but thereafter declines (Fig. 15–12). Stimulation of the anterior teats plays a critical role in promoting normal suckling behavior (Fraser, 1973).

Maternal Recognition. Experiments in which sows' olfactory bulbs were destroyed show that sows recognize their litters largely through smell (Meese and Baldwin, 1975), but they are more tolerant of alien young than other domestic ungulates, and will permit alien piglets to suckle, particularly during the first days of lactation. Piglets do not appear to discriminate between sows. Fostering is not difficult, and litters can be reared together, with piglets suckling more than one sow. Lactating sows may become aggressive if piglets are disturbed or threatened.

Behavior and Thermoregulation. Newborn piglets, with their small body size, sparse pelage and skin wet with fluids, are prone to chill in air temperatures as high as 20°C with a 5-km per hour wind, despite a vigorous thermogenic response (Mount, 1968). Young piglets huddle together or against the sow to minimize heat loss; feral piglets appear to be more cold-resistant than domestic piglets. Deaths from direct or indirect effects of chilling are common in unheated farrowing houses, but the resistance to cold increases during the two or three days that piglets would remain in the nest.

Behavioral Abnormalities Contributing to Piglet Mortality. Sows occasionally kill and eat their young during parturition. If aggressive behavior is detected as farrowing approaches, the piglets can be removed at birth and returned in relative safety after farrowing is complete, but cannibalism can also occur later in lactation.

Starvation accounts for nearly half of the 10% mortality in liveborn domestic piglets. Young piglets will suckle anything that contacts their noses, including the maternal vulva and other piglets, and excessive time may be spent in these activities while littermates are laying claim to the teats. Sometimes there are more piglets in the litter than teats on the udder. The number of teats available is effectively reduced in some sows, particularly in older animals in which the teats can be hidden beneath the udder. Piglets that are displaced from their established position at the udder after the first day or two do not readily accept a vacant teat and may die. These deaths are concentrated among the smallest piglets in the litter.

REFERENCES

Alexander, G. (1975). Body temperature control in mammalian young. Br. Med. Bull. *31*, 62.

Alexander, G. (1977). Role of auditory and visual cues in mutual recognition between ewes and lambs in Merino sheep. Appl. Anim. Ethol. *3*, 65.

Alexander, G. and Stevens, D. (1979). Discrimination of colors and grey shades by Merino ewes: tests using colored lambs. Appl. Anim. Ethol., *5*, 215.

Alexander, G. and Williams, D. (1964). Maternal facilitation of sucking drive in newborn lambs. Science NY, *146*, 665.

Arnold, G.W. and Morgan, P.D. (1975). Behavior of the ewe and lamb at lambing and its relationship to lamb mortality. Appl. Anim. Ethol. *2*, 25.

Baldwin, B.A. and Shillito, E.E. (1974). The effects of ablation of the olfactory bulbs on parturition and maternal behavior in Soay sheep. Anim. Behav. *22*, 220.

Banks, E.M. (1964). Some aspects of sexual behavior in domestic sheep (vis aries.) 469–1. Behavior Int. J. Comp. Ethol. *23*, (Part 3–4), 249.

Bareham, J.R. (1975). The effect of lack of vision on suckling behavior of lambs. Appl. Anim. Ethol. *1*, 245.

Bareham, J.R. (1976). The behavior of lambs on the first day after birth. Br. Vet. J. *132*, 152.

Bronson, F.H. and Whitten, W.K. (1968). Estrus-accelerating pheromone of mice: Assay, androgen-dependency and presence in bladder urine. J. Reprod. Fertil. *15*(1), 131.

Cutton-Brock, T.H. and Guinness, F.E. (1975). Behavior of red deer *(Coruus elaphus L.)* at calving time. Behaviour *55*, 287.

Crowley, J.P. and Darby, T.E. (1969). Observations on the fostering of calves for multiple suckling systems. Vet. J. *126*, 658.

Dandekar, P., Vaidya, R. and Morris, J. (1972). The effects of coitus on transport of sperm in the rabbit. Fertil. Steril. *23*, 759.

Domanski, E. (1970). Hypothalamic areas involved in the control of ovulation and lactation in the sheep. Final Report, FAO 202.

Dufty, J.H. (1972). Maternal causes of dystocia and stillbirth in an experimental herd of Hereford cattle. Aust. Vet. J. *48*, 1.

Fernandez-Baca, S., Madden, D.H.L. and Novoa, D. (1970). Effect of different mating stimuli on induction of ovulation in the Alpaca. J. Reprod. Fertil. *22*, 261.

Frädrich, H. (1974). A comparison of behaviour in the Suidae. In The Behavior of Ungulates and Its Relation to Management. Proceedings of an International Symposium, Calgary, November 1971. V. Geist and F. Walther (eds), Morges, Switzerland, International Union for Conservation of Nature and Natural Resources.

Fraser, D. (1973). The nursing and suckling behavior of pigs. I. The importance of stimulation of the anterior teats. Br. Vet. J. *129*, 324.

Fraser, D. (1975). The "teat order" of suckling pigs. II. Fighting during suckling and the effects of clipping the eye teeth. J. Agric. Sci. (Cambridge) *84*, 393.

Fraser, D. and Jones, R.M. (1975). The "teat order" of suckling pigs. I. Relation to birth weight and subsequent growth. J. Agric. Sci. (Cambridge) *84*, 387.

George, J.M. (1976). The incidence of dystocia in Dorset Horn ewes. Aust. Vet. J. *52*, 519.

Haltmeyer, G.C. and Eik-Nes, K.B. (1969). Plasma levels of testosterone in male rabbits following copulation. J. Reprod. Fertil. *19*, 273.

Harlow, H.F., Joslyn, W.D., Senko, M.G. and Dopp, A. (1966). Behavioral aspects of reproduction in primates. J. Anim. Sci. Suppl. *25*, 49.

Hawk, H.W. (1972). Failure of vasectomized rams to improve sperm transport in inseminated ewes. J. Anim. Sci. *35*, 69.

Hemsworth, P.H., Winfield, C.G. and Mullaney, P.D. (1976). A study of the development of the teat order in piglets. Appl. Anim. Ethol. *2*, 225.

Hudson, S.J. (1977). Multiple fostering of calves onto nurse cows at birth. Appl. Anim. Ethol. *3*, 57.

Hulet, C.V., Ercanbrack, S.K., Blackwell, R.L., Price, D.A. and Wilson, L.O. (1962). Mating behavior of the ram in the multi-sire pen. J. Anim. Sci. *21*, 865.

Karli, P. (1968). Systeme limbique et processus de motivation. J. Physiol. *60*(1), 3.

Kelly, R.W. and Whateley, J.A. (1975). Observations on the calving of red deer *(Cervus elaphus)* run in confined areas. Appl. Anim. Ethol. *1*, 293.

Klingel, H. (1967). Soziale Organization und Verhalten freilebender Steppenzebras. Z. Tierpsychol. *34*, 580.

Klopfer, P. and Klopfer, M. (1977). Compensatory responses of goat mothers to their impaired young. Anim. Behav. *25*, 286.

Kobayashi, T., Kato, J. and Minaguchi, H. (1964). Fluctuations in monoamineoxydase activity in the hypothalamus of the rat during estrus cycle and after castration. Endocrinol. Jpn. *2*, 283.

Lightfoot, R.J. (1970). The contractile activity of the genital tract of the ewe in response to oxytocin and mating. J. Reprod. Fertil. *21*, 376.

Lindsay, D.R. (1965). The importance of olfactory stimuli in the mating behavior of the ram. Anim. Behav. *13*(1), 75.

Lindsay, D.R. and Robinson, T.J. (1961). Studies on the efficiency of mating in sheep. J. Agric. Sci. *57*, 137.

MacLean, P.D. and Ploog, D.W. (1962). Cerebral representation of penile erection. J. Neurophysiol. *25*(1), 29.

MacNeilly, A.S. and Folley, S.J. (1970). Blood levels of milk ejection activity (oxytocin) in the female goat during mating. J. Endocrinol. *48*(1), IX-X.

Meese, G.B. and Baldwin, B.A. (1975). Effects of olfactory bulb ablation on maternal behavior in sows. Appl. Anim. Ethol. *1*, 379.

Melrose, D.R., Reed, H.C.B. and Patterson, R.L.S. (1971). Androgen steroids associated with boar odor as an aid to the detection of estrus in pig artificial insemination. Br. Vet. J. *137*, 497.

Morgan, P.D. and Arnold, G.W. (1974). Behavioral relationships between Merino ewes and lambs during the first four weeks after birth. Anim. Prod. *19*, 169.

Mount, L.E. (1968). The Climatic Physiology of the Pig. London, Arnold.

Müller-Schwarze, D. and Müller-Schwarze, C. (1971). Olfactory imprinting in a precocial mammal. Nature (London) *229*, 55.

Randall, G.C.B. (1972). Observations on parturition in the sow. I. Factors associated with the delivery of the piglets and their subsequent behavior. Vet. Rec. *90*, 178.

Rossdale, P.D. (1968). Abnormal perinatal behavior in the thoroughbred horse. Br. Vet. J. *124*, 540.

Rudge, M.R. (1970). Mother and kid behavior in feral goats (Capra hircus L.). Z Tierphychol. *27*, 687.

Scheel, D.E., Graves, H.B. and Sherritt, G.W. (1977). Nursing order, social dominance and growth in swine. J. Anim. Sci. *45*, 219.

Selman, I.E., McEwan, A.D. and Fisher, E.W. (1970). Studies on natural suckling in cattle during the first eight hours post-partum. Anim. Behav. *18*, 276.

Signoret, J.P. (1967a). Attraction de la femelle en estrus par le male chez les porcins. Rev. Comp. Anim. *4*, 10.

Signoret, J.P. (1967b). Duree du cycle oestrien et de l'oestrus chez la Truie; action du benzoate d'oestradiol chez la femelle ovariectomisee. Ann. Biol. Anim. Biochim. Biophys. *7*, 1.

Signoret, J.P. (1970). Reproductive behavior of pigs. J. Reprod. Fertil., Suppl. *11*, 105.

Soulairac, M.L. (1963). Etude experimentale des regulations hormono nerveuses du comportement sexual du rat male. Ann. Endocrinol. *24(3)* Suppl., 5.

Stephens, D.B. and Linzell, J.L. (1974). The development of sucking behavior in the new born goat. Anim. Behav. *22*, 628.

Wierzbowski, S. (1966). The scheme of sexual behavior in bulls, rams and stallions. World Rev. Anim. Prod. *2*, 66.

III. reproductive cycles

16

Beef and Dairy Cattle

H.W. HAWK
and
R.A. BELLOWS

Most of the present-day beef and dairy breeds were developed from common ancestry in Europe and the British Isles during past centuries. Consequently, beef and dairy cattle possess similar reproductive characteristics. Differences between these two types of cattle can be attributed to physiologic variations arising from several hundred years of selection or to differences in management practices.

PUBERTY

Female

Puberty is the stage of development characterized by the production and release of functional gametes and by the desire and ability to mate. Puberty covers the period during which the functional hypothalamo-pituitary-gonadal relationships and interactions are being established. In the heifer, puberty is considered to be the first estrus accompanied by spontaneous ovulation.

As puberty nears in beef heifers, a cyclic pattern of LH release is established that causes some luteinization of follicles, from which low levels of progesterone are secreted. Progesterone appears to be critical for establishing the pubertal release of luteinizing hormone that causes ovulation (Gonzalez-Padilla et al, 1975). One or more "quiet" ovulations may occur before heifers show overt signs of estrus in conjunction with ovulation, but the frequency of "quiet" ovulations can depend largely on the efficiency of estrus detection. Dairy heifers that were watched closely exhibited signs of behavioral estrus several weeks before the first ovulation (Morrow et al, 1976).

The age of first estrus in heifers varies considerably, mostly owing to breed and differences in growth rates. A low level of nutrient intake and slow growth delays puberty in heifers for weeks, and a high level of nutrition and rapid growth hastens puberty. The average age of puberty for groups of heifers on recommended

337

Table 16–1.   Age in Weeks at Attainment of Several Characteristics of Puberty in Beef and Dairy Bulls

Characteristic	Angus and Hereford Bulls	Charolais Bulls	Holstein Bulls
Protrusion of penis	34	33	31
Separation of penis from sheath	38	37	35
First collection of sperm	41	38	37
Puberty*	45	41	39

* Defined as the first ejaculate containing 50 million sperm with 10% progressive motility. *(Adapted from Almquist and Amann, 1976. J. Dairy Sci. 59, 986)*

levels of nutrition falls between 300 and 360 days for dairy breeds and between 320 and 460 days for beef breeds. Heifers with Zebu breeding attain puberty at 500 to 800 days of age. If heifers are provided with adequate nutrition, estrus normally recurs regularly after the pubertal estrus.

Male

Cellular differentiation occurs gradually in the seminiferous epithelium of the testes during calfhood, with mature spermatozoa present in the seminiferous tubules by about five months of age. The testes produce increasing numbers of spermatozoa as puberty nears (Hafs and McCarthy, 1979).

Bull calves will mount or attempt to mount females within a few months after birth. During calfhood, the penis is firmly adhered within the sheath by the preputial frenulum and cannot be extended; two to four months before puberty, partial protrusion of the penis occurs during mounting, followed by separation of the penis from the sheath and complete erection, and eventually by mating and ejaculation of spermatozoa. In the data summarized in Table 16–1, Holstein bulls reached puberty earlier than Angus or Hereford bulls, with Charolais bulls falling between the others. Crossbred beef bulls generally reach puberty earlier than straightbred bulls. This difference is presumably asso-

ciated with the more rapid growth rate of crossbred bulls. After bulls reach puberty, the testes continue to grow and the number of sperm per ejaculate increases until 18 to 24 months of age in both beef and dairy bulls.

Bulls vary in sexual aggressiveness, but in general, dairy bulls respond more quickly than beef bulls to sexual stimuli. In one comparison, the reaction time of bulls being used for semen collection was considerably shorter for dairy than beef bulls (Table 16–2). Sperm production also varies widely among individual bulls, with some variation among breeds. Charolais and dairy bulls tended to produce more sperm and to hold more sperm in extragonadal reservoirs (epididymis and ductus deferens) than Hereford and Angus bulls (Table 16–3).

Table 16–2.   Reaction by Beef and Dairy Bulls to a Stimulus Animal

Reaction Measured	Reaction Time (minutes)	
	Angus and Hereford Bulls	Holstein Bulls
Time to:		
First mount	13	1
First ejaculation	20	5
Second ejaculation	33	13

(Adapted from Almquist, 1973. J. Anim. Sci. 36, 331)

Table 16–3. Sperm Production and Sperm Reserves in Sexually Rested Dairy and Beef Bulls

Characteristic	Dairy Bulls	Beef Bulls		
		Angus	Hereford	Charolais
Testes weight (g)	725	726	646	773
Daily sperm production (billions)	7.5	6.9	5.9	8.9
Extragonadal sperm reserves (billions)	69	55	40	64

(Adapted from Weisgold and Almquist, 1979. J. Anim. Sci. 48, 351)

ESTRUS, OVULATION AND ESTROUS CYCLES

Estrus and Estrous Cycles

Estrus is often classified as standing or nonstanding. A female in standing estrus stands still to be mounted by other cows or by a bull. Standing to be mounted is the most certain indicator of estrus. A female in standing estrus will often mount or attempt to mount other females, her vulva is swollen, and clear mucus often flows from the vulva and becomes smeared on the tail or flanks. Physical activity increases on the day of estrus, and milk production in dairy cows decreases. Standing estrus lasts for approximately 20 hours, with considerable normal variation in length (Table 16–4).

A cow in nonstanding estrus sometimes has mucus discharges and mounts other cows but moves away when cows attempt to mount her. Signs of nonstanding estrus must be considered with caution. Mounting other females is not a particularly good indicator of estrus because many females that are not in estrus will mount estrous females. Mucus is sometimes discharged on the day before estrus as well as on the day of estrus. Cows may show nonstanding estrous activity for periods up to 24 hours but fail to stand at any time. Nonstanding estrus occurs most frequently within the first few weeks after calving.

The modal length of estrous cycles is 21 days for cows and 20 days for heifers, although cycles of 17 to 24 days in length are normal. The average estrous cycle length is always greater than the mode of 20 or 21 days because cows often ovulate without estrus being detected, and two or more ovulation cycles may be counted as one estrous cycle.

Table 16–4. Characteristics of Estrous Cycles and Postpartum Intervals in Cows

Characteristic	Beef Cows		Dairy Cows	
	Mean	Range	Mean	Range
Duration of estrus (hr)	20	(12–30)	15	(13–17)
Ovulation, after beginning of estrus (hr)	31	(18–48)	29	(25–32)
Length of estrous cycle (days)	21	(17–24)	21	(17–24)
Parturition to first ovulation (days)	62	(35–105)	20	(10–40)
Parturition to first estrus (days)	63	(40–110)	34	(20–70)
Parturition to uterine involution (days)	45	(32–50)	45	(32–50)

(Postpartum data from Bellows and Thomas, 1976. J. Range Management 29, 192; and Menge, Mares, Tyler and Casida, 1962. J. Dairy Sci. 45, 233)

Ovulation

Cattle ovulate on the average at 28 to 32 hours after the beginning of estrus, or 10 to 12 hours after the end of estrus (Table 16–4). The time from the end of estrus until ovulation has been decreased experimentally by about two hours by service with a vasectomized bull. Similarly, manual massage of the clitoris for ten seconds following artificial insemination of beef cows shortened the interval from beginning of estrus until the ovulatory surge of LH by more than four hours, shortened the time from beginning of estrus until ovulation by four hours, and increased the conception rate by 6% (Randel et al, 1975).

Follicles ovulate on the right ovary about 60% of the time and on the left ovary about 40% of the time. The first ovulation after parturition occurs more frequently on the ovary opposite to the uterine horn that previously carried the fetus. Occasionally, a follicle fails to ovulate and becomes cystic. Such follicles sometimes grow to several centimeters in diameter and may cause nymphomania. At the opposite extreme, the follicle walls may luteinize and cause prolonged periods of anestrus.

Normally, one follicle ovulates per estrous cycle in cattle. Two follicles ovulate approximately 10% of the time, and three follicles ovulate infrequently.

Metestrous bleeding occurs in the uterus of most cattle about one day after ovulation (50 to 70 hours after the beginning of estrus). Some of the blood reaches the exterior, and small amounts of blood can be seen in mucus hanging from the vulva, on the tailhead or around the rear quarters of more than 50% of females. Intrauterine bleeding results from the rupture of small blood vessels in the endometrium, probably caused by increased fragility of the endometrial vasculature as the stimulatory effects of estrogen secreted during proestrus and estrus subside. Metestrous bleeding apparently has no relationship to the chances of conception.

PARTURITION AND DYSTOCIA

Several clinical changes in the pregnant female indicate approaching parturition. The muscles and ligaments of the rump and tailhead soften and relax, the tailhead is elevated 24 to 48 hours before calving, and the vulva swells. As calving nears, the vulva discharges thick, stringy mucus, the udder enlarges, and the teats appear to be distended with milk.

Dystocia

Beef cattle experience dystocia in 30% to 60% of births in primiparous dams, in 8% to 25% of second-calf births, and in 2% to 8% of births in mature dams. Dystocia is caused by abnormal presentation of the calf in 2% to 6% of all parturitions, with the frequency tending to be higher in young dams. The remaining cases of dystocia are due to prolonged or difficult labor, with the most important causes being high birth weight of the calf and small pelvic area of the dam (Fig. 16–1). Of calves lost at or shortly after birth, approximately 60% are males, which reflects the fact that male calves are larger than females at birth. As many as 70% of the calves lost at parturition have functional lungs, indicating that they were not stillborn. Rather, these calves died as a result of injuries and physiologic trauma during prolonged and difficult delivery.

Calf deaths at parturition reduce the calf crop considerably, with dystocia causing up to 80% of these perinatal calf losses. A further consequence of dystocia is a reduced chance of conception during the next breeding period. The major causes of dystocia—small pelvic area of the dam, large size of the calf and male sex of the calf—are either known or determined at the time of conception (Price and Wiltbank, 1978). Matings should be planned

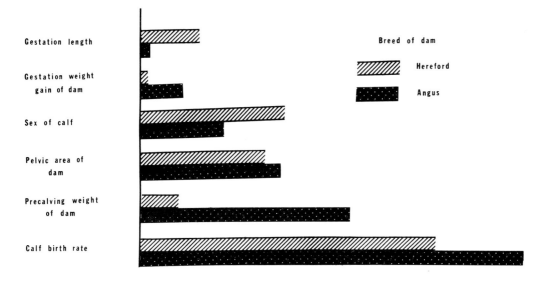

Fig. 16–1. Relative importance of factors affecting dystocia in primiparous beef heifers. *(From data of Bellows et al, 1971. J. Anim. Sci. 33, 407.)*

to avoid disproportionately large calves in dams with small pelvic areas.

Twins are born in about 1.9% of births in dairy breeds and about 0.4% of births in most beef breeds. Twinning frequencies as high as 10% have been reported in some strains of the Simmental breed in Europe. Most twins result from double ovulations and are called fraternal twins. Identical twins, resulting from the division of a single zygote, are rare.

THE POSTPARTUM PERIOD

The postpartum period is the time following parturition during which lactation is initiated and reproductive cycles are established.

Involution of the Uterus

Immediately after calving, the uterine horn that carried the fetus (gravid horn) is considerably larger than the opposite nongravid horn. Both horns lack firm muscle tone. The weight and size of the uterus soon decrease, the number and size of myometrial cells decrease, and the muscle tone of the uterus gradually improves. The uterus is considered to be involuted when it decreases in size to that characteristic for the age and parity of the animal and both horns are approximately equal in size. Involution is judged by palpation per rectum. The uterus involutes most quickly in primiparous cows and cows that are suckling calves.

Ovarian Function

Ovarian follicular growth begins soon after calving. In experimental dairy herds, the number of days from parturition to first ovulation has averaged about 20 days and the number of days to first estrus about 34 days. These intervals are longer in beef cattle, largely owing to the fact that most beef cows nurse their calves. Nursing delays postpartum follicular development and ovulation and prolongs the interval from parturition to first estrus. In one study, the interval to first estrus for Angus cows was 29 days for those not suckling calves and 73 days for those suckling calves; for Holstein cows, the intervals were 30 and 45 days respectively (Casida, 1968). In another study of postpartum ovarian activity in beef cows, weaning the calf at birth reduced the interval to first estrus from 67 to 25 days. Surgical removal of the udder during the last trimester of pregnancy, thus removing all inhibitory effects of lactation or nursing on pituitary or ovarian function, reduced the postpartum interval to only 12 days (Short et al, 1972). The physiologic mechanisms controlling this effect have not been established.

The postpartum interval from parturition to first estrus and ovulation in beef cattle depends not only on the intensity of the suckling stimulus but also on the plane of nutrition. Low nutrient intake that causes loss of weight in beef cattle both before and after calving results in infrequent estrus and low fertility during the subsequent breeding season (Wiltbank, 1978).

12-month intervals are seldom achieved. Fertility is often measured as the net calf crop for naturally bred beef cattle and as the calving rate to first service for artificially inseminated dairy cattle. The net calf crop for beef cattle, or the percentage of cows that wean calves each year, is estimated at 70% to 75% (Bellows, 1979). For dairy cattle the percentage of cows calving to first service is only about 50% (Spalding et al, 1975; Pelissier, 1976).

Approximately 25% of all milk cows are culled annually from dairy herds; in most sections of the U.S. more than one-fourth of these cows are culled for reproductive failures. High conception rates and efficient reproduction are so important to the success of dairy operations that semen from genetically average bulls with high conception rates in artificial insemination centers is as valuable as semen from genetically superior bulls with average conception rates (McGilliard, 1978).

REPRODUCTIVE EFFICIENCY

General Fertility

The optimal calving interval for both beef and dairy cattle is 12 months, but

Reproductive Failures

The major factors that reduce the net calf crop for beef cattle and cause infertile services to dairy cattle are listed in Table 16–5.

Table 16–5.   Causes of Loss of Potential Calves for Beef and Dairy Cattle

Beef Cattle—Bred Naturally[a]		Dairy Cattle—Inseminated Artificially[b]	
Status After Breeding Season	Percentage of Females	Outcome of First Service	Percentage of Females
Cow weaned calf	71	Cow calved	50
Potential calves lost	29	Reproductive failure	50
Females not pregnant	17	Nonestrous cow inseminated	10
Females lost pregnancy	2	Anatomic abnormalities	2
Perinatal calf death	7	Ovulation failure	2
Calf deaths, birth to weaning	3	Lost or ruptured ova	5
Total	100	Fertilization failure	13
		Embryonic death	15
		Fetal death	3
		Total	100

[a] A 14-year summary of data from the Livestock and Range Research Station, Miles City, Montana, including females 14 months to 10 years of age, with breeding seasons of 45 or 60 days.
[b] Adapted from Hawk (1979).

Failure of the beef female to be pregnant at the end of the breeding season, because of lack of estrus and infertile services, reduced the potential net calf crop by 17% (Table 16–5) and accounted for about 60% of the loss of potential calves. The calf crop was reduced further by loss of diagnosed pregnancies, perinatal calf deaths, and calf deaths between birth and weaning. The calving rate to first service is generally not known for naturally bred beef cattle; however, infertile services to beef cattle are caused primarily by ovum fertilization failure and embryonic death, as discussed below and listed in Table 16–5 for dairy cattle (Bellows, 1979; Hawk, 1979).

Three major and several minor factors cause infertility in dairy cattle (Table 16–5). Major reasons include insemination of cows at time periods outside the lifespan of sperm or ova in the cow (insemination of nonestrous cows), failure of ova to be fertilized in cows inseminated at or near the proper time, and death of embryos developing in the uterus.

Insemination of nonestrous cows is a problem only in artificially inseminated animals. The frequency of insemination of nonestrous cattle varies greatly, depending upon the accuracy of estrus detection. Measurements of progesterone concentrations in blood or milk have indicated that as many as 20% of cows may be inseminated when progesterone concentrations are high, indicating that the animal was not actually in estrus (Appleyard and Cook, 1976). In one study, the calving rate to first service was 47% in dairy cows inseminated artificially and 63% in cows mated naturally (Pelissier, 1976). The difference was probably partially due to artificial insemination of nonestrous cows. In Table 16–5, the frequency of insemination of nonestrous cows is rather arbitrarily listed as 10%.

About 13% of the ova recovered from the oviducts three days after insemination of estrous cattle have not cleaved, indicating fertilization failure. The figure is similar for beef and dairy cattle. The fertilization failure rate is much higher than 13% during the first few weeks after parturition, presumably due to failure of some phase of sperm transport in the female reproductive tract.

The embryonic death rate is the difference between the proportion of cows with fertilized ova at three days after insemination or breeding and the proportion of cows with live embryos at 35 to 40 days after insemination. The rate of embryonic death has been estimated in several studies to be about 15% in both beef and dairy cattle. Most embryonic deaths occur within three weeks after insemination.

Less frequent causes of infertile service include ovulation failure, anatomic defects in the female, such as occluded oviducts or adhesions around the ovaries, defective or lost ova, and resorption or abortion of the fetus.

The conception rate usually decreases by about 5% with each successive service. In cows with four or more successive infertile services, the rate of both ovum fertilization failure and embryonic death is much higher than the rate in first-service cattle (Hawk, 1979).

Reproductive efficiency of cattle varies with season of the year. In temperate climates of the northern United States, fertility tends to be highest in the spring and somewhat lower during summer. In the southernmost states, fertility is often low during the summer months in both beef and dairy cattle. Low fertility correlates with high temperature and humidity within one or two days before or after breeding (Thatcher, 1974). Fertility rate has been improved somewhat by artificially cooling the environment. In beef cattle in Louisiana, the body temperature of the cow increased when the air temperature rose above 27°C. The first-service conception rate for cows with rectal temperatures below 39.7°C at the time of insemination was 55%, compared to 24% for cows with rectal temperatures above 39.7°C (Vincent, 1972).

Improving Reproductive Performance

Because 95% of the beef females in the U.S. are bred by natural service, the bull must produce semen of high potential fertility and possess sufficient libido and physical stamina to detect estrus and mate repeatedly. Reproductive performance of beef cattle can often be improved considerably by thorough evaluation of bulls for breeding soundness.

Failure of the beef female to express either a pubertal or postpartum estrus early in the breeding season is a serious problem. Conception early in the breeding season for both heifers and cows permits early calving and the weaning of older, heavier calves. Beef cattle must receive adequate nutrition so that heifers attain puberty and maximal pelvic growth at an early age and postpartum females express estrus and have acceptable fertility early in the breeding season (Wiltbank, 1978).

Reproductive problems in dairy herds are caused mainly by inefficient detection of estrus and by infertile services. Consequently, reproductive efficiency can be improved in most dairy herds by accurate estrus detection, insemination at the proper time in relation to estrus and ovulation, and use of semen from highly fertile bulls. Insufficient attention to estrus detection results in detection of only about half of the expected estrous periods in commercial dairy herds and causes missed opportunities to inseminate cows, insemination of cows that are not in estrus, reduced ovum fertilization rates because of poorly timed insemination, and increased embryonic mortality because of fertilization of aged ova. Inseminating cows with semen from bulls of high fertility improves the ovum fertilization rate and perhaps reduces the embryonic death rate.

Most estrous periods can be detected by careful observation of cattle at least twice daily. During checks for estrus, any distractions to cattle, such as feeding, should be avoided. Detection of estrus is improved by the use of bulls; they may be vasectomized or their penises may be surgically deflected or locked mechanically in the sheath (Foote, 1975). Other aids to detecting estrus include pressure-sensitive indicators placed on the rump of females and chin-ball markers or marking harnesses on bulls. When bulls equipped with chin-ball markers or harnesses mount an estrous female, an easily visible mark of dye or pigmented grease is left on the rump and tailhead of the female. Although nonestrous cows will occasionally be marked, the marking aids identify cows that should be observed closely for confirmation of estrus.

REFERENCES

Appleyard, W.T. and Cook, B. (1976). The detection of oestrus in dairy cattle. Vet. Rec. 99, 253.

Bellows, R.A. (1979). Research areas in beef cattle reproduction. In Beltsville Symposia in Agricultural Research III. Animal Reproduction. H. W. Hawk (ed), Montclair, NJ, Allenheld, Osmun.

Casida, L.E. (1968). Studies on the postpartum cow. Wisc. Res. Sta. Res. Bull. No. 270, p. 15.

Foote, R.H. (1975). Estrus detection and estrus detection aids. J. Anim. Sci. 58, 248.

Gonzalez-Padilla, E., Wiltbank, J.N. and Niswender, G.D. (1975). Puberty in beef heifers. 1. The interrelationship between pituitary, hypothalamic and ovarian hormones. J. Anim. Sci. 40, 1091.

Hafs, H.D. and McCarthy, M.S. (1979). Endocrine control of testicular function. In Beltsville Symposia in Agricultural Research III. Animal Reproduction. H. W. Hawk (ed), Montclair, NJ, Allenheld, Osmun.

Hawk, H.W. (1979). Infertility in dairy cattle. In Beltsville Symposia in Agricultural Research III. Animal Reproduction. H. W. Hawk (ed), Montclair, NJ, Allenheld, Osmun.

McGilliard, M.L. (1978). Net returns from using genetically superior sires. J. Dairy Sci. 61, 250.

Morrow, D.A., Swanson, L.V. and Hafs, H.D. (1976). Estrous behavior and ovarian activity in prepuberal heifers. Theriogenology 6, 427.

Pelissier, C.L. (1976). Dairy cattle breeding problems and their consequences. Theriogenology 6, 575.

Price, T.D. and Wiltbank, J.N. (1978). Dystocia in cattle. A review and implications. Theriogenology 9, 195.

Randel, R.D., Short, R.E., Christensen, D.S. and Bellows, R.A. (1975). Effect of clitoral massage after artificial insemination on conception in the bovine. J. Anim. Sci. *40*, 1119.

Short, R.E., Bellows, R.A., Moody, E.L. and Howland, B.E. (1972). Effects of suckling and mastectomy on bovine postpartum reproduction. J. Anim. Sci. *34*, 70.

Spalding, R. W., Everett, R.W. and Foote, R.H. (1975). Fertility in New York artificially inseminated Holstein herds in Dairy Herd Improvement. J. Dairy Sci. *58*, 718.

Thatcher, W.W. (1974). Effects of season, climate, and temperature on reproduction and lactation. J. Dairy Sci. *57*, 360.

Vincent, C.K. (1972). Effects of season and high environmental temperature on fertility in cattle: A review. Am. J. Vet. Med. Assoc. *161*, 1333.

Wiltbank, J.N. (1978). Management of heifer replacements and the brood cow herd through the calving and breeding periods. In Commercial Beef Cattle Production. 2nd ed. C.C. O'Mary and I. A. Dyer (eds), Philadelphia, Lea & Febiger.

17

Sheep and Goats

C.V. HULET
and
M. SHELTON

Most sheep and goat breeds originated at the higher elevations in the mountains of Eurasia. These species were among the first domesticated. Although matings between the two species can lead to conception, the embryos usually die at 30 to 50 days (Moore and Eppleston, 1977; Bunch et al, 1976). Consequently, important differences in physical and reproductive characteristics exist. Comparative distinguishing characteristics of these two species are shown in Table 17–1. Important breed differences also exist, which can generally be attributed to physiologic differences resulting from hundreds of years of selection for specific adaptations, production characteristics or differences in management.

Table 17–1.   Comparative Distinguishing Characteristics of Sheep and Goats

Characteristic	Sheep	Goats
Chromosome number	54	60
Fertility of matings		
Buck goat × female	Sterile	Fertile
Ram × female	Fertile	Fertile (embryo mortality general)
Domestication	Among first	Among first
Gregarious nature	Strong to moderate	Weak to moderate
Most adapted area	Arid temperature or tropical regions	Arid tropics
World population	1 billion	400 million
Physical characteristic		
Tail length	Generally long, some short	Short
Tail carriage	Drooping	Erect
Interdigital glands	Present	Absent
Lacrimal pits	Present	Absent
Beard	Absent	Present
Protective coat	Wool or hair	Hair

BREEDING SEASON

Both sheep and goats can be generally characterized as seasonally polyestrous with recurring estrous periods in the fall of the year. However, there are important exceptions to this basic pattern. In general, those genotypes that have evolved in equatorial regions, having been subjected to less variation in photoperiod and temperature, are less seasonally restricted than those from temperate and polar regions.

Apparently both present location and point of origin are involved, as animals that are relocated tend to gradually adapt or drift to breeding habits characteristic of the area where they are located (Hulet et al, 1974). Animals that have benefited from long adaptation to the tropics often respond more to other factors of the environment, such as temperature or feed supply, than to daylength. When sheep from a common breed and origin are relocated either near the equator or near the poles, those sheep near the equator experience a much longer breeding season than those located in polar regions.

Thus there is adequate and conclusive evidence for the existence of an inherent rhythm that controls estrous activity in sheep in the absence of adequate changes in daylength. Changes in daylength modify the inherent rhythm. Increases in daylength cause a cessation of estrous activity, and decreases stimulate the onset of estrous activity. The reaction interval varies with breed and ranges from 9 to 25 weeks. Ewes subjected to a six-month light rhythm exhibited two periods of sexual activity each year. Temperature and the introduction of the ram at the approach of the breeding season or intermittent exposure to the ram at the end of the breeding season are factors affecting the time of onset and the end of the breeding season in sheep. A strain of Australian Merinos selected for high fecundity had a much lower decline in ovarian activity in midsummer than did

an unselected strain (Binden and Piper, 1976).

Seasonal breeding in sheep is associated with the frequency and size of episodic LH secretions (Yuthasastrakosal et al, 1977). The mean LH level and the frequency of release is high at the onset of the breeding season. Also, with the approach of the breeding season, the negative feedback sensitivity to estrogen is less and the positive feedback effect on the release of LH is greater.

In temperate regions both sexes of goats tend to be seasonal in sexual activity with a reasonably high degree of synchrony between the sexes. By contrast, the ovine male generally remains sexually active throughout the year, but with some variation in the level of breeding activity. Daylength variations clearly control spermatogenic activity in the ram. Seasonal variations in semen production and quality are evident, with total production and quality highest in the fall and lowest in spring and summer. A change from long to short daylengths induced testicular growth after three weeks (Lincoln, 1976). Plasma LH levels increased following exposure to short daylengths and decreased following long daylengths. Both temperature and length of day affect total semen production and quality. High ambient temperatures above 27°C tend to depress semen quality and may result in summer sterility. Shearing rams before the start of breeding and providing shade or an artificially cool environment can be helpful in preventing or correcting summer sterility.

The Alpine dairy goat breeds tend to be seasonal breeders but have extended their breeding season under intense management. Early literature indicated that most kids were dropped from September to December matings. Currently kids may be born at any month of the year. However, April to June tends to be an anestrous season. The Angora, known for fiber production, is highly seasonal, with matings generally occurring from September to January; however, most kids are dropped from October and November matings. Goats raised primarily for meat do not fall into well-defined breeds, and length of breeding season appears to depend on location. The Dorset, Merino and Rambouillet sheep breeds are generally recognized as having extended breeding seasons, with some ewes showing estrus every month of the year, whereas the Southdown, Shropshire and Hampshire are examples of breeds having shorter breeding seasons. Some wild breeds of sheep have only one or two estrous cycles per year. Figure 17–1 shows the seasonal breeding pattern of meat-type goats in the southern part of the United States (near 30° N latitude) as compared to two types of sheep (Shelton, 1978).

Finewool or Merino type sheep and the Angora goat often fail to reach puberty during the first breeding season. Consequently they may be 18 to 20 months of age at first estrus. Early maturing breeds such as the pigmy goat or the Finnsheep may reach puberty as early as three or four months of age. As in other species a cyclic pattern of LH release with progesterone secretion from luteinized follicles occurs before the first estrus and the first estrus is not always accompanied by ovulation (Edey et al, 1977).

Testis size increases when ram lambs reach eight to ten weeks of age at body weights of 16 to 20 kg. This coincides with the appearance of primary spermatocytes and an enlargement of seminiferous tubules. Copulation with ejaculation of live sperm and resulting fertility can occur as early as 112 to 185 days of age with live weight of 40% to 60% of mature weight. In the buck goat, copulation with ejaculation of live sperm commences about four to six months of age under good conditions.

PUBERTY

As a rule female goats show estrus as early as five to seven months, whereas six to nine months is more typical for sheep.

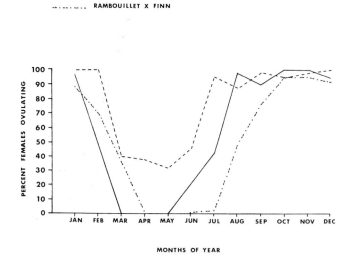

Fig. 17–1.   Seasonal cycle of reproductive activity in goats and two types of sheep.

Table 17–2. Comparative Estrus and Estrous Cycle Data in Sheep and Goats

Species	Estrous Cycle Length (days)		Length of Estrus (hrs)	Time of Ovulation
	Mean	(Normal Range)		
Sheep	16.7	(14–19)	24–36	Near end of estrus
Goats	20.6	(18–22)	26–42	Shortly after end of estrus

ESTRUS, OVULATION AND ESTROUS CYCLES

Length of Estrus

The length of the estrous period in both species is about 35 hours but varies from 24 to 48 hours. Angora goats appear to be an exception to the above rule; they are reported to have estrous periods averaging only 22 hours. Also, wool breeds of sheep may tend to have longer estrous periods than meat breeds. Hanrahan and Quirke (1975) observed that Finnsheep had significantly longer estrous periods than Galways and Finnsheep × Galway crosses. Estrus is generally shortest for ewe lambs and young does in their first season; estrus in yearlings is intermediate in length. It is often shortest near the beginning and the end of the breeding season. Estrus appears to be shorter when rams are with ewes continually rather than intermittently.

Length of Estrous Cycles

Comparative estrous cycle data are shown in Table 17–2. Many investigators have observed estrous cycles in both sheep and goats that vary widely from those just reported. However, these data are generally considered to be abnormal and are often explained on the basis of multiples or fractions of the normal cycle length. Conception and early embryo mortality, retained corpora lutea and cystic ovaries can explain other deviations from the normal cycle length.

Ovulation

Ewes normally ovulate near the end of estrus. The time of ovulation varies from as long as 11 hours before the end of estrus to seven hours after the end of estrus, but generally it occurs before the end of estrus and is more closely related to the end than the beginning of estrus. Ovulation occurs in the goat a few hours after the end of standing estrus.

Factors known to affect ovulation rate are the same as in other species. The effects of season and level of nutrition are clearly illustrated in Figure 17–2. Both season and level of nutrition affect the ovulation rate in goats also, especially those of the Angora breed.

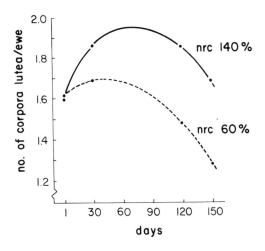

Fig. 17–2. Effect of two levels of feed and advance of the breeding season on average number of corpora lutea per ewe. (Feed levels were 60% and 140% of National Research Council recommendations for maintenance; Day 1 = Sept. 9.)

Ovulation rate increases in ewes to about four or five years of age depending on the early maturing nature of the breed. It then declines with advancing age. The same or a similar ovulation curve is thought to apply to goats also.

Large genetic variations in ovulation rate exist among breeds of sheep. For example the Australian Merino, even at mature ages, produces primarily single lambs, whereas the mature Finnish Landrace or Finnsheep breeds average about three lambs per birth, with quadruplets and quintuplets common. Both ovulation and lambing rate are under genetic control, as evidenced by large breed differences and also by the fact that highly fertile strains within breeds have been developed. The heritability of ovulation or lambing rate at a single or specific time is considered to be low (approximately 10%), whereas the heritability of lifetime records is about 20%.

Data are inadequate to attempt to characterize the ovulation rate in goats by breed. The Angora typically has one ovulation under most production conditions, but under good conditions it may have two. Most other types of goats tend to have relatively high ovulation rates. Under good conditions twin ovulations may be the rule. Some breeds, especially the Nubian, frequently produce triplets.

BREEDING AND CONCEPTION

Natural

Rams may copulate two or three times in a few minutes when first turned with ewes in heat. They generally mate more frequently when with more than one ewe in heat. The number of matings per day is also greatly affected by individual variations, the climate and the time at which the rams are introduced into breeding. Certain breeds tend to copulate more frequently than other breeds. Infertile or sexually inhibited rams are often the cause of poor lambing rates. Such factors as age, physical condition, heredity, temperature, disease and sex drive may affect the number of ewes that can be successfully mated to one ram. The physical condition of the ram, including malnutrition, is important in conditioning his desire to mate. Parasitism or disease that may cause the male to be in poor condition may result in decreased fertility and a depressed breeding activity. Diseases that affect the feet, prepuce, penis or testicles may render the ram incapable of serving a ewe or cause complete infertility.

Natural synchronization of goat matings may occur in which large numbers of does are in estrus in one day. This can result in some does not being settled. Heavy breeding activity due to synchronized estrus in does may result in a period of impotency in bucks which are not in strong condition.

At ovulation the ovum is picked up by the infundibulum and starts its course through the oviduct to the uterus. Copulation usually occurs before ovulation, and therefore many of the sperm that have been traveling through the reproductive tract of the female for some time at an average of 4 cm per minute may be present in the oviduct by this time. Other sperm remain in the cervix where conditions are best for preservation (up to three days) and are continually released into the uterus, where survival is limited to about 30 hours. The primary mode of transport is by motility of the female reproductive tract, although sperm motility may be important in the cervix (Lightfoot and Restall, 1971). Estrogen (Hawk and Cooper, 1975) facilitates the transport of spermatozoa, whereas progestogens inhibit transport in the reproductive tract. However, progesterone preceding estrogen is required for optimal transport. Estrogenic pastures and hormonal or artificial synchronization of estrus may interfere with sperm transport. Eggs may remain viable for 10 to 25 hours, but abnormal development and lowered viability appear to increase with age of either the sperm or egg. Although many sperm cells may enter the zona pellucida of the egg, only one enters into the nucleus and effects fertilization. The mechanism of penetration and rejection is not clearly understood but enzyme systems are known to be involved.

Pronuclei are formed after three to nine hours and the fertilized egg undergoes its first cleavage about 19 to 20 hours after ovulation. The fertilized egg enters the uterus on the average three to four days after the ewe first shows estrus. Holst (1974), using improved techniques, has concluded that the majority of ova are present in the uterus 66 hours after ovulation. Sheep eggs have been fertilized *in vitro* but with a low incidence of success. Cultured ova have been transferred to recipient ewes and have developed normally to term (Peters et al, 1977).

The fertilizability of sheep ova is lower in the early part of the breeding season than at the peak of the season. This is due partially to a higher percentage of abnormal ova. Embryo mortality is also high during the early breeding season, resulting in a higher incidence of repeat breeding and lower conception rates. Hot weather, high body condition and the insulating properties of wool, which increase body temperature, appear to increase the incidence of abnormal ova and embryo mortality, leading to the lower conception rates. Midseason conception rates for healthy sheep in temperate zones should be close to 85% for a single service.

Synchronization

Exogenous progestogens such as progesterone or flurogestone acetate (Cronolone), administered in physiologic doses by injection, vaginal pessary, subcutaneous implant or feeding over a 12- to 14-day period during the normal breeding season, permit time for the corpora lutea (CL) at various stages of development to complete their natural lifespan and inhibit follicular growth and estrogen production. Thus, when the progestogen is withdrawn the follicles grow rapidly, producing an estrogen peak. This estrogen, in the presence of declining levels of blood progesterone, causes behavioral estrus at an average of 48 hours and triggers LH release, which in turn causes ovulation.

This could be an excellent technique for facilitating artificial insemination except that fertility is lower at the first post-treatment estrus (60% vs. 85%). The use of prostaglandins as a luteolytic agent has not given the hoped-for improvement in synchronization and fertility. However, a combination of a progestogen followed 4 to 16 days later with prostaglandin may give the hoped-for combination of good synchrony and high fertility (Serna et al, 1978). In addition to facilitating artificial insemination, estrus synchronization in combination with induced parturition could provide a management tool for timing lambing to more nearly fit the availability of labor, such as weekend lambing for the part-time farmer.

One unique characteristic of the goat is that, under certain conditions, the male can be made to synchronize estrus in females. The sudden introduction and teasing activity of the male goat during the transition period between the breeding and nonbreeding seasons can initiate LH release, followed by synchronized estrus and ovulation in the exposed group of does about eight days after introduction (Fig. 17–3). In contrast, ewes may ovulate synchronously following the sudden introduction of the ram, but the ovulatory response is less uniform than in the goat and is not accompanied by estrus. However, approximately 23 days after exposure to the ram many ewes show synchronized estrus. This is a product of six days to the first ovulation not accompanied by estrus plus an additional

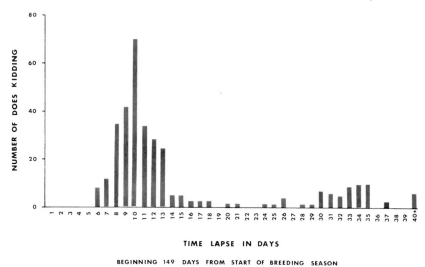

Fig. 17–3. Natural synchronization of kidding in Angora goats.

17-day estrous cycle culminating in estrus and ovulation.

Artificial Insemination

Artificial insemination in sheep has been generally limited owing to the high cost of labor, difficulty of accurately identifying superior sires, and low conception rates, especially with frozen semen. The use of frozen semen has been much more successful with goats than sheep. This is probably due to an improved ability to practice deep cervical or intrauterine insemination with multiparous goats. Successful uterine inseminations of sheep have been achieved using long hypodermic needles with a dulled or rounded tip (Anderson et al, 1973).

Goat semen freezes better than sheep semen. In contrast to the limited use of AI in sheep, it has been used relatively widely in dairy goats. Goat semen has been successfully frozen using sterile skim milk as a diluent, a glycerol level of 6% to 9%, a 1½- to 2-hour equilibration and a slow rate of freezing. An enzyme present in seminal plasma reacts with constituents of egg yolk to produce a

product that is toxic to sperm when this diluent is used (Watson and Martin, 1973). Also, seminal plasma itself has been identified as a limiting factor in freezability of goat semen. This can be overcome by washing the sperm cells to remove the plasma before freezing. Advances in techniques for deep-freezing ram semen have been documented by Colas (1975). Tests by Salamon and Visser (1974) have shown that deep frozen semen can retain its fertilizing capacity for at least five years.

Comparative semen characteristics and artificial insemination specifications for sheep and goats are shown in Table 17–3. The decline of fertility with storage of the semen at 0° to 12° C is rapid, decreasing from 60% with fresh semen to 50% at 24 hours, 30% at 48 hours and 20% at 72 hours. The rate of cooling to reach the storage temperature may be a major factor affecting fertility.

AI should be timed to provide a maximal number of fertile sperm at the site of fertilization when the ova reach this area shortly after ovulation, which usually occurs near the end of heat in ewes and shortly after the end of heat in does. Females exhibiting estrus should be removed and inseminated at least once and preferably twice daily. Fertility is usually reduced in females synchronized with progesterone. Conception rate is normally lower in yearling than in mature animals. Fertilization with AI has been more effective in the fall than in the spring.

GESTATION AND PARTURITION

Gestation

One distinct difference between the two species is in the endocrine profile during pregnancy. During the first trimester of pregnancy, both species depend on the presence of a functional corpus luteum. Subsequently the placenta becomes a primary source of progesterone in the sheep, whereas placental production of progesterone either does not occur in the goat or is inadequate to maintain pregnancy (Van Rensburg, 1970). Another difference between the two species is in the endocrine picture at parturition. Parturition in the goat is preceded by a precipitous decline of progesterone 12 to 24 hours prior to initiation of labor. Even a low level of exogenous progesterone at this time will cause prolonged gestation and dystocia. There is no such obvious and consistent change in the progesterone pattern with sheep, and a high level of exogenous progesterone is required to prolong gestation (Thorburn et al, 1977).

The normal gestation length for the two species is almost exactly the same, about 149 days. Variation among breeds has been reported from 143 to 151 days; individual normal pregnancies vary from 138 to 159 days. Extremely short and long periods are of questionable normalcy. The early maturing meat breeds of sheep (eg, Southdown or Hampshire) and the highly fertile or prolific breeds (eg, Finnsheep or Romanov) have short gestation periods averaging about 144 to 145 days. Slow-maturing finewool breeds (eg, Merino or Rambouillet) have long periods of 150 to 151 days. Crossbred types (eg, Columbia or Targhee) have intermediate periods of 148 to 149 days. Individual gestation periods within a breed vary up to 13 days, with a standard deviation of about 2.2 days. Although most breeds of goats vary within two days of the species average for gestation length, the Black Bengal breed has an average gestation of only 143 days.

Heredity ($h^2=0.34-0.65$) plays an important role in determining the length of gestation (Prud'Hon et al, 1970). The genotype of the fetus is the major determinant of the duration of gestation (Bradford et al, 1972). The fetus accounts for at least two-thirds of the genetic variation in the gestation period in the sheep. Male lambs are carried longer than ewe lambs, spring-born lambs longer than fall-born lambs, and singles longer than twins. Length of gestation appears to increase with age of dam.

Table 17–3. Normal Range of Semen Characteristics and Artificial Insemination Specifications for Sheep and Goats

Trait	Sheep	Goat
Semen volume (ml)	0.3–2.0	0.1–1.5
Sperm concentration (billions/ml)	1–5	2–6
Motility (%)	60–90	60–80
Abnormal spermatozoa (%)	10–30	11
Volume/insemination (ml)	0.1	0.02–0.2
Spermatozoa/insemination (millions)	120–150	100–150
Insemination site	Cervix	Cervix or uterus

Development of Embryo and Fetus

Differentiation of sexual organs commences about the 35th day following conception and the scrotum is apparent in the 50- to 60-day fetus. The testes are generally descended at birth, but retained testicles (cryptorchidism) may occur with both species (goats, 4.6%; sheep, 0.1% to 0.2%; Ercanbrack and Knight, 1978). This defect generally is much more common with Merino-type sheep that have been selected for the absence of horns. However, in an extensive study in Idaho, some polled lines of sheep were completely free of the condition whereas some horned lines produced cryptorchid lambs. One polled line had a high incidence (2.6%) of cryptorchidism. These data indicate that the polled gene is not necessarily linked to cryptorchidism as some have suggested. In all strains of goats that have been studied, the absence of horns is associated with a high incidence of intersex offspring (Hancock and Louca, 1975). These individuals have been identified as masculinized females.

Parturition

Uterine contractions begin about 12 hours before parturition in the ewe; these contractions are responsible for dilation of the cervix. Actual parturition after dilation of the cervix requires only about 15 minutes per lamb. The placenta, which is the cotyledonary type, is expelled soon after the fetus. The placenta of the first lamb of a pair of twins is expelled before the second lamb is born.

Dystocia is not a serious problem in sheep or goats, but difficult births requiring assistance do occur. These are usually associated with abnormal presentations, such as one or both front legs turned backward, front legs presented but head back, and the breech presentation. Big heads and shoulders are a problem in some breeds and crosses, especially with single births. Hampshires and ewes giving birth to Hampshire crossbred lambs seem to have a greater frequency of dystocia due to size of head and shoulders. Occasionally the cervix does not dilate, requiring cesarean section.

The maturing fetus plays the key role in determining the time of parturition. Studies by Liggins et al (1973) have established that the fetal pituitary-adrenal axis signals the initiation of labor in sheep and goats. In both species, the fetal adrenal cortices grow rapidly during the last 10 to 15 days of gestation. This growth is accompanied by an increase in functional activity, producing rapidly rising levels of corticosteroids in the fetal circulation. Treatment of younger fetuses, up to a week before parturition is expected (Rawlings and Ward, 1978), with corticotrophin or glucocorticoids such as dexamethasone or flumethazone simulates the same course of events, but the changes occur more rapidly than those occurring naturally. The treatment of both sheep and goats with appropriate dosages of glucocorticoids during the last week of gestation constitutes an effective method of shortening and synchronizing parturition. Unlike cattle, sheep seldom have retained placentas following treatments with glucocorticoids and parturition is relatively devoid of complications, with good survival of the lambs.

POSTPARTUM PERIOD

Call et al (1976) have shown that the sheep uterus has involuted (ie, returned to normal size and tone) by day 27 postpartum. It is common to find rather large amounts of cellular debris in the uterus before involution. This may partially account for the relatively low fertility in the early postpartum ewe. The relatively slow recovery of the estrogen (E_2) positive feedback mechanism may also be an important factor in low early postpartum fertility (Wright and Finlay, 1977). There appears to be little direct information available on the time of the first ovulation following parturition. The first ovulation postpartum, with few exceptions, is probably unaccompanied by estrus. Therefore, the best estimate of the time of this ovulation is 16.7 days before the first postpartum estrus. The average number of days from lambing to first estrus in a group of lactating, fall-lambing ewes was 35.4 days. This suggests that the first ovulation in these lactating, October-lambing ewes occurred on the average about 18 days postpartum. The average postpartum interval to conception in these same ewes was 44.3 days. The postpartum interval is much longer in ewes lambing in mid- or late winter. Shevah et al (1975) have shown that lactation inhibits the occurrence of estrus in the early postpartum period.

REPRODUCTIVE EFFICIENCY

Conception Rates of Sheep and Goats

Conception rates in mature sheep and goats in temperate zones during mid-breeding season are about 85%. However, fertility is depressed near the equator,

near the beginning and end of the breeding season, during hot weather, in undernourished or overly fat females, in young and old animals, when estrogenic content of the forage is high, and when the females are parasitized or suffering from disease or other stress.

Ewes in normal body condition seem to respond to flushing (increased level of nutrition) during the early part of the breeding season and during late season breeding but not during midseason.

The heritability of fertility and fecundity appears to be low. However, several workers (Piper et al, 1976) have been able to increase the prolificacy of sheep through careful selection, taking into account the lifetime reproductive performance of the dam. There also exists an apparent relationship between age at puberty and fecundity. Therefore, selecting for early puberty, which appears to be moderately heritable, might permit acceleration of progress in improving fecundity if added to the selection criteria suggested previously. Reducing temperature stress by providing shade, adequate fresh water, breeding at the optimum season of the year (Fig. 17−4) and keeping the ewes in good breeding condition by

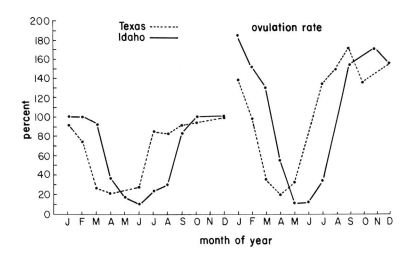

Fig. 17−4. The effects of location on incidence of estrus and ovulation rate throughout the year in Rambouillet ewes on identical feed treatments.

proper nutrition can all contribute to maximizing reproductive efficiency.

Under intensive management it is possible to reduce the lambing interval from once per year to three times in two years (8-month interval) or to twice a year (6-month lambing interval). Although a few Dorset and Rambouillet flock owners are having reasonable success in breeding for three lamb crops in two years without hormones, many have not been successful. In many cases when natural methods are unsatisfactory, hormone therapy has proved successful. Figure 17–5 illustrates the successful implementation of twice-a-year lambing using hormone therapy at the desired time following each parturition. Some success in twice-a-year lambing has also been achieved by using hormones only during the off-season breeding (McNeal et al, 1978). Using controlled lighting to induce two lambings per year is also being investigated. Although the potential efficiency of twice-a-year lambing looks good, the problems of low fertil-

ity in the early postpartum period and reduced fertility near the beginning and end of the breeding season will delay the successful implementation of this accelerated system for some time. Currently, efforts are being made through selection to achieve twice-a-year lambing without hormone therapy or light modification. This will be achieved primarily by lengthening the natural breeding season. Improved early postpartum fertility and shorter gestation length could also contribute to successful twice-a-year lambing.

REFERENCES

Alberio, R. (1976). Rôle de la photoperiode dans le developpement de la fonction de reproduction chez l'agneau Ile-de-France, de la naissance à 21 mois. Thèse Doc. 3e, Univ. Paris VI, 57 pp.

Anderson, V.K., Aamdal, J. and Fougner, J.A. (1973). Intrauterine and deep cervical insemination with frozen semen in sheep. Zuchthyg 8, 115.

Binden, B.M. and Piper, L.R. (1976). Seasonality of ovulation rate in Merino ewes differing in fecundity. Theriogenology 6, 621.

Bradford, G.E., Hart, R., Quirke, J.F. and Land, R.B. (1972). Genetic control of the duration of gestation in sheep. J. Reprod. Fertil. 30, 459.

Bunch, T.D., Foote, W.C. and Spillett, J.J. (1976). Sheep-goat hybrid karyotypes. Theriogenology 6, 379.

Call, J.W., Foote, W.C., Ecker, C.D. and Hulet, C.V. (1976). Postpartum uterine and ovarian changes, and estrous behavior from lactation effects in normal and hormone treated ewes. Theriogenology 6, 495.

Colas, G. (1975). Effect of initial freezing temperature, addition of glycerol and dilution on the survival and fertilizing ability of deep-frozen ram semen. J. Reprod. Fertil. 42, 277.

Edey, T.N., Chu, T.T., Kilgour, R., Smith, J.F. and Tervit, H.R. (1977). Estrus without ovulation in puberal ewes. Theriogenology 7, 11.

Ercanbrack, S.K., and Knight, A.D. (1978). Frequency of various birth defects of Targhee and Columbia sheep. J. Hered. 69, 237.

Hancock, J. and Louca, A. (1975). Polledness and intersexuality in the Damascus breed of goat. Anim. Prod. 21, 227.

Hanrahan, J.P. and Quirke, J.F. (1975). Repeatability of the duration of oestrus and breed differences in the relationship between duration of oestrus and ovulation rate of sheep. J. Reprod. Fertil. 45, 29.

Hawk, H.W. and Cooper, B.S. (1975). Improvement of sperm transport by the administration of estradiol to estrous ewes. J. Anim. Sci. 41, 1400.

Holst, P.J. (1974). The time of entry of ova into the uterus of the ewe. J. Reprod. Fertil. 36, 427.

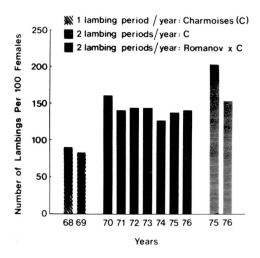

Fig. 17–5. Effects on number of lambings per year by changing from one breeding period/year (68–69) to two breeding periods/year using hormone therapy to induce estrus at the desired time postpartum. (Data from Charmoises (C) and C × Romanov females include lambs 7 months of age.) (From Management of Reproduction in the Sheep and Goat, Symposium, Madison, Wisc., July 24–25, 1977, p. 113.)

Hulet, C.V., Shelton, M., Gallagher, J.R. and Price, D.A. (1974). Effects of origin and environment on reproductive phenomena in Rambouillet ewes. I. Breeding Season and Ovulation. J. Anim. Sci. *38*, 1210.

Liggins, G.C., Fairclough, R.J., Grieves, S.A., Kendall, J.Z. and Knox, B.X. (1973). The mechanism of initiation of parturition in the ewe. Prog. Horm. Res. *28*, 111.

Lightfoot, R.J. and Restall, B.J. (1971). Effects of site of insemination, sperm motility and genital tract contractions on transport of spermatozoa in the ewe. J. Reprod. Fertil. *26*, 1.

Lincoln, G.A. (1976). Secretion of LH in rams exposed to two different photoperiods. J. Reprod. Fertil. *47*, 351.

McNeal, L.G., Hulet, C.V., Walters, J.L. and Foote, W.C. (1978). Range ewe production under twice-a-year lambing: I. Reproductive performance. Proc. West. Sect. Am. Soc. Anim. Sci. *29*, 278.

Moore, N.W. and Eppleston, J. (1977). Inter-species fertilization between sheep and goats. *In* Reproduction and Evolution. Fourth International Symposium on Comparative Biology of Reproduction. Canberra, Australian Academy of Science, pp. 45–46.

Peters, D.F., Anderson, G.B. and Cupps, P.T. (1977). Culture and transfer of sheep embryos. J. Anim. Sci. *45*, 350.

Piper, L.R., Allison, A.J., Bindon, B.M.,Gheradi, P., Kelly, R.W., Killeen, I.D., Lindsey, D.R., Oldham, C., Robertson, D. and Stevenson, J.R. (1976). Ovulation rate in high fecundity Merino crosses. Theriogenology *6*, 622.

Prud'Hon, M., Desvignes, A. and Devoy, I. (1970). IV. Duration of pregnancy and birth weight of lambs. Ann. Zootech. *19*, 439.

Rawlings, N.C. and Ward, W.R. (1978). Fetal and maternal endocrine changes associated with parturition in the goat. Theriogenology *9*, 109.

Salamon, S. and Visser, D. (1974). Fertility of ram spermatozoa frozen-stored for 5 years. J. Reprod. Fertil. *37*, 433.

Schanbacher, B.D. and Ford, J.J. (1976). Seasonal profiles of plasma luteinizing hormone, testosterone and estradiol in the ram. Endocrinology *99*, 752.

Serna, J.A., Bosu, W.T.K. and Barker, C.A.V. (1978). Sequential administration of cronolone and prostaglandin F2α for estrus synchronization in goats. Theriogenology *9*, 177.

Shelton, M. (1978). Reproduction and breeding of goats. J. Dairy Sci. *61*, 994.

Shevah, V., Black, W.J.M. and Land, R.B. (1975). The effects of nutrition on the reproductive performance of Finn × Dorset ewes. II. Post-partum ovarian activity, conception and plasma concentrations of progesterone and LH. J. Reprod. Fertil. *45*, 289.

Snyder, D.A. and Dukelow, W.R. (1974). Laparoscopic studies of ovulation, pregnancy diagnosis and follicle aspiration in sheep. Theriogenology *2*, 143.

Thorburn, G.D., Challis, J.R.C. and Currie, W.B. (1977). Control of parturition in domestic animals. Biol. Reprod. *16*, 18.

Van Rensburg, S.J. (1970). Reproductive physiology and endocrinology of normal and habitually aborting Angora goats. Thesis, Dept. of Physiology, Faculty of Veterinary Science, University of Pretoria, South Africa.

Watson, P.F. and Martin, I.C.A. (1973). The response of ram spermatozoa to preparations of egg yolk in semen diluents during storage at 5° or −196°C. Aust. J. Biol. Sci. *26*, 927.

Wright, P.J. and Finlay, J.K. (1977). LH release due to LH-RH or oestradiol-17β (E2) in post-partum ewes. Theriogenology *8*, 191.

Yuthasastrakosal, P., Palmer, W.M. and Howland, B.E. (1977). Release of LH in anoestrous and cyclic ewes. J. Reprod. Fertil. *50*, 319.

18

Pigs

L.L. ANDERSON

SEXUAL DEVELOPMENT AND MATURATION

Genetic sex determines the development of gonadal sex, which in turn determines phenotypic and reproductive capacities. Thus, in mammalian species, the presence of the Y chromosome is critical to normal development of gonads eventually capable of spermatogenesis, and it is necessary indirectly for the normal development of the male reproductive system, body sex and typical male behavioral characteristics.

Both testes and ovaries originate from the primordium developing on the inner aspect of the mesonephros, which consists of coelomic epithelium, underlying mesenchyme and primordial germ cells. The germ cells have an extra-regional origin. Sex differentiation of mammalian gonads from the undifferentiated primordium begins when gonads differentiate into genetic males.

In testicular organogenesis, the germ cells are attracted and encapsulated with somatic cells initially in arrangements as seminiferous cords, which are delineated by connective tissue. The cords of cells eventually hollow into seminiferous tubules with connections to mesonephric tubules. Somatic cells within the seminiferous tubules differentiate into sustentacular cells (supporting or Sertoli cells). Between seminiferous tubules, mesenchymal cells differentiate into interstitial cells (Leydig cells).

In the genetic female, oogenesis occurs later with the obvious feature being the absence of testicular-inducing patterns of organogenesis. The surface epithelium of the presumptive ovary becomes separated from the central cellular mass. The ovarian cortex proliferates and germ cells (oogonia) within this region transform into oocytes and enter early phases of premeiosis before they become separated from mesenchyme by a single layer of differentiating follicle cells. Proliferation of rete cords is arrested in the medullary region of the ovary. The follicles presumably prevent oocytes from entering meiotic processes beyond the diplotene phase. The primary follicles remain in this resting stage of development until puberty.

Mesonephric tubules stabilize in the normal male embryo as the wolffian ducts (eg, epididymides, vas deferens, seminal vesicles, prostate and bulbourethral glands), whereas in the normal female embryo the paramesonephric tubules survive as the müllerian ducts (eg, oviducts, uterus, cervix and upper part of vagina). Both duct systems are present in early embryonic development and are capable of developing as male or female internal and external genitalia as well as secondary sex characteristics. It seems that testes are body sex differentiators because they impose masculinity on the developing genital tract and on secondary sexual features and behavior (Jost, 1970; Jost et al, 1973). In the absence of the testes, or in the presence or absence of the ovaries, the genital tract and secondary sex characteristics develop as a normal female. Although the production of androgen from embryonic gonads influences development of internal and external genitalia, the secretion of antimüllerian hormone (AMH) by the gonads seems essential for inducing the regression of the müllerian ducts in genetic male embryos (Josso et al, 1977). The AMH is a glycoprotein synthesized by Sertoli cells as soon as they are recognizable in seminiferous tubules and before the appearance of fetal Leydig cells.

The gonadal primordium (ridge) in pig embryos at 21 and 22 days is composed of the surface epithelium, proliferating tissue of the primitive gonadal cords and mesenchyme (Pelliniemi, 1975a). The surface epithelium consists of varying columnar cells in a single layer and smaller cuboidal cells arranged in two or three layers. The increase in cell height is often the first histologic sign of epithelial differentiation. Primitive cord cells are derivatives of the epithelial cells and resemble them morphologically. The cords are first found in the posterior part of the gonadal ridge and by 22 days their numbers have increased and they extend deeper into the mesenchyme. These cords are continuous with the surface epithelium. Mesenchymal cells in the gonadal ridge are more densely arranged than in neighboring areas. Primordial germ cells are round or elongated with diameters of 10 to 20 μm (Fig. 18–1, c and e). These germ cells undergo mitotic divisions in the gonadal ridge at this time. The nucleus of these germ cells is round (ie, 10 μm) and contains one or two nucleoli about 3 μm in diameter. The pair of centrioles is located near the nuclear envelope. The cytoplasm contains polysomes, mitochondria and granular endoplasmic reticulum. Furthermore, the Golgi complex is extensive but conventional secretory vesicles are not found in the cytoplasm; the granular endoplasmic reticulum occurs as solitary cisternae. The free polysomes are evenly distributed in the entire cytoplasm. Fine filaments, coated vesicles and occasional dense lipid droplets are characteristic features of the cytoplasm. The fine morphology of these germ cells is similar in both sexes at 21 and 22 days of embryonic life (Fig. 18–1, e and f). The basic components required for development of a mature gonad are already present in the gonadal ridge of both sexes at this time.

By 24 days, gonads of both sexes show further development of primitive cords in continuity with the surface epithelium (Fig. 18–1, c and d) (Pelliniemi, 1976). At this time the gonadal blastema occupies space between the surface epithelium and the mesenchyme in the basal part of the gonad. In the central region, the gonadal blastema consists primarily of irregularly organized cells and is called the blastema proper. The shape of blastema cells varies greatly in irregularly arranged areas, and they are usually columnar in cord-like portions.

At 26 and 27 days, porcine gonads appear as longitudinal protrusions along the medial mesonephric surface of both sexes (Fig. 18–1, a and b) (Pelliniemi, 1975b).

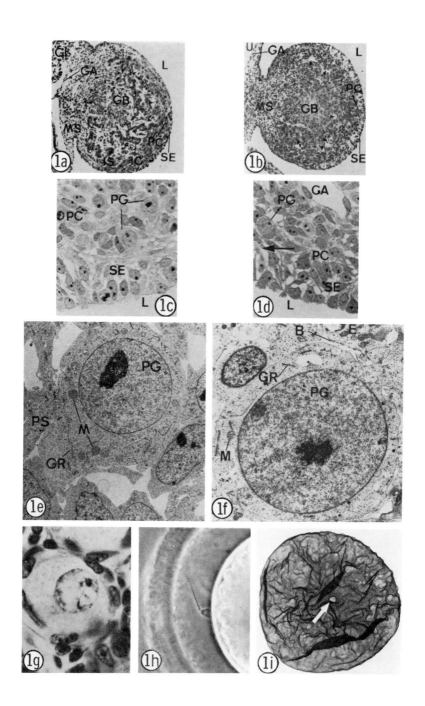

The female gonad consists of three tissues: surface epithelium, gonadal blastema and mesenchyme, and these structures are similar to those found in earlier stages of development. In male gonads, four tissues are differentiated at this time and these consist of surface epithelium, mesenchyme, testicular cords and interstitium. The testicular cords and interstitium are derived from gonadal blastema. Sustentacular cells of the testicular cords resemble primitive cord cells and spermatogonia are similar to primordial germ cells. Interstitial cells have not yet differentiated into Leydig cells. Cells of the surface epithelium, primitive cords, mesenchyme and primordial germ cells retain ultrastructural features that are similar in both sexes (Pelliniemi, 1975a, b; 1976).

Gametogenesis and ovarian development in pig embryos from day 13 post coitum to birth and during the early neonatal period indicate patterns of mitotic and meiotic activities throughout these periods, unlike those found in several other mammalian species (Black and Erickson, 1968; Fulka et al, 1972). The numbers of germ cells increase dramatically from approximately 5,000 at day 20

to a peak of 5,000,000 by day 50 (Fig. 18–2). Thereafter, germinal mitotic activity decreases and necrosis of germ cells increases. At birth the population of germ cells is approximately 500,000. Throughout these same stages of embryonic development, somatic mitotic activity follows a higher rate but similar pattern to germinal mitotic activity.

On the basis of ^{3}H-thymidine incorporation into premeiotic DNA synthesis, transformation of oogonia to oocytes in the pig continues at least to 35 days of postnatal life (Fulka et al, 1972). Meiosis begins as early as *day 40* of embryonic development

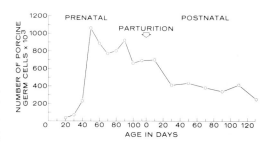

Fig. 18–2. Quantification of female porcine germ cells during embryonic and neonatal growth. Only mean values are presented, representing 5 to 20 animals. *(Adapted from Black and Erickson, 1968. Anat. Rec. 161, 45)*

Fig. 18–1. *a.* Light micrograph of porcine fetal testis in transverse section at *day 26.* SE, Surface epithelium; PC, primitive cords; IC, interstitial cell; IS, interstitium; L, coelomic cavity; GB, gonadal blastema; MS, mesenchyme. *b.* Light micrograph of porcine fetal ovary at *day 27.* (Refer to *a* for explanation of symbols.) *c.* Light micrograph of male gonad in pig fetus at *day 24.* CA, Capillary; GB, gonadal blastema; L, coelomic cavity; PC, primitive cords; PG, primordial germ cells; SE, surface epithelium. (480×) *d.* Light micrograph of female gonad in pig embryo at *day 24.* A radiate pattern of surface epithelial cells is seen on border (arrow). Along upper side of capillary (CA), columnar cells have a cord-like arrangement. GB, gonadal blastema; L, coelomic cavity; PC, primitive cords; PG, primordial germ cells; SE, surface epithelium. (480×) *e.* An electron micrograph of a cortical portion of a male gonad in pig embryo at *day 24.* PG, primordial germ cell; M, mitochondria; GR, granular endoplasmic reticulum; PS, pseudopod of cytoplasm. (2,640×) *f.* An electron micrograph of a primordial germ cell *(PG)* of female gonad in pig embryo at *day 21* with a prominent nucleolus. E, surface epithelial cell; B, surface epithelial basal lamina; GR, granular endoplasmic reticulum; M, mitochondria. (3,040×) *g.* Porcine oocyte in diplotene. Note large eccentric nucleus containing the single acidophilic nucleolus and diffuse chromosomes. A single layer of flattened follicular cells surrounds the oocyte, thus forming a primordial follicle. (960×) *h.* Porcine spermatozoan in perivitelline space with head attached to vitelline membrane of ovum. Note the tangential path of sperm tail through the zona pellucida. (432×) *i.* Spherical porcine blastocyst (8.5 mm diameter with fluid-filled cavity) at *day 12* shows irregular protrusions of surface-membrane. Embryonic disc is indicated by arrow. (7.4×) *(a and b from Pelliniemi, 1975. Am. J. Anat. 144, 89; c, d and e from Pelliniemi, 1976. Cell & Tissue 8, 163; f from Pelliniemi, 1975. Anat. Embryol. 147, 19; g from Black and Erickson, 1968. Anat. Rec. 161, 45; h from Hunter and Dziuk, 1968. J. Reprod. Fertil. 15, 199; and i from Anderson, 1978. Anat. Rec. 190, 143)*

and is evident in most ovaries by day 50. The premeiotic resting stage (diplotene) of porcine oocytes first appears by day 50 of embryonic life and almost all germ cells in the ovaries are diplotene by 20 days after birth. The paucity of oogonia and absence of oogonial mitoses indicate completion of the process of oogenesis by day 100 of embryonic development. Cellular and nuclear growth of porcine germ cells increases greatly from the oogonial stage to the oocyte within a primordial follicle (Table 18–1). Cell sizes during early stages of meiotic prophase (leptotene, zygotene and pachytene) are larger than those in oogonia. During transition from pachytene to diplotene, cellular and nuclear diameters increase two- to threefold. The oocyte remains in the diplotene stage up to the time of ovulation, but it continues to increase in diameter and volume during follicular maturation. Diameters of growing follicles increase approximately three-fold, with little further growth evident during the final stages of follicular maturation.

Porcine oocytes obtained from mature graafian follicles near the time of ovulation are about 120 μm diameter with a zona pellucida consisting of a homogeneous matrix approximately 8.6 μm in thickness (Fig. 18–1, h). Fine structural features of the oocyte cytoplasm include fine filaments and granules embedded in its matrix. Surface area of the cell membrane (vitelline membrane) is increased by irregularly spaced microvilli in contact with cell processes arising from corona radiata banding the zona pellucida (Norberg, 1972). Cortical granules about 0.20 μm diameter are numerous immediately beneath the oocyte membrane wall and near the Golgi complex. Other prominent features of the oocyte cytoplasm include homogeneous yolk globules, mitochondria often located near these globules, both granular and agranular endoplasmic reticulum in the form of dilated-continuous channels, free ribosomes distributed singly or in aggregates, and membrane-bound granules scattered in the cytoplasmic matrix.

The nucleus is about 35 μm in diameter and consists of an inner and outer nuclear membrane with numerous nuclear membrane pores. The nucleoplasm is rather clear with evenly distributed fine fibrils and small dense granules. Within the nucleus an eccentrically located nucleolus (about 7 μm in diameter) contains fibrils, granules and spherical vacuoles.

Corona radiata cells are porcine granulosa cells on the outer aspect of the zona pellucida, which are arranged in a radial pattern. These granulosa cells usually adhere to each other and they project long cell processes (microvilli) through the zona pellucida and perivitelline space, and these microvilli terminate as end bulbs in contact with the oocyte membrane wall. Granulosa cells immediately surrounding oocytes are regarded as nurse cells for the growing oocyte during oogenesis; they convey their nutritive material by means of these extensive cytoplasmic processes. Corona radiata cells disappear soon after ovulation. Mitochondria formed in oocytes may include contributions of precursor mate-

Table 18–1. Growth of Porcine Germ Cells

Cell Stage	Diameter, μm	
	Nuclear	Cellular
Oogonia	9	13
Leptotene	10	18
Zygotene	11	16
Pachytene	8	16
Diplotene in follicle*		
Primordial	16	27
Growing	39	84
Vesicular	34	88

*Primordial follicle consists of oocytes with none or one layer of follicle cells; growing follicle consists of an oocyte with two or more layers of follicle cells; and vesicular follicle is an oocyte with several layers of follicle cells and antrum formation. (From Black and Erickson, 1968. Anat. Rec. 161, 45)

rials from corona radiata cells, but these mechanisms are not clearly defined. The dense, membrane-bound granules synthesized near the vitelline membrane often appear fused with mitochondria. These cortical granules are extruded through the vitelline membrane on contact with the fertilizing sperm.

Suitable *in vitro* conditions allow pig oocytes to resume meiosis, but there is a graded increase in the competence of the oocyte in this process that seems related to follicular growth (Tsafriri and Channing, 1975). The ability of granulosa cells to luteinize and bind gonadotropins (Channing, 1970) in culture also increases as the follicles mature. Porcine follicular fluid has an inhibitory action upon maturation of isolated porcine oocytes; the inhibitor is a peptide of about 2,000 daltons (Channing and Tsafriri, 1977).

Normal testicular development in the pig from the early fetal period to sexual maturity indicates that testicular growth lags behind body growth from 7 to 14 weeks postcoitum (Fig. 18–3). From 14 weeks postcoitum to three weeks postpartum, testicular growth exceeds body growth, primarily because of Leydig cell development (van Straaten and Wensing, 1977a). Although testicular weight increases in the prenatal period, a significant decrease in weight occurs three to seven weeks postpartum as a result of Leydig cell regression. After seven weeks postpartum, testicular growth again exceeds body growth largely as a result of increases in the length and diameter of the seminiferous tubules.

Aberrant development of the gubernaculum is the primary cause of cryptorchidism in pigs. In unilaterally cryptorchid pigs, testicular development in both the abdominal and the scrotal testes progresses normally until birth (van Straaten and Wensing, 1977b). During the first month postpartum, aberrations in the abdominal testis become evident, as indicated by impaired Leydig cell develop-

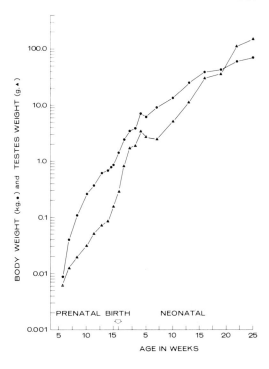

Fig. 18–3. Body weight and testis weight during fetal development and during the first 25 weeks after birth in the pig. Mean values are presented, representing three to six animals. *(Adapted from van Straaten and Wensing, 1977. Biol. Reprod. 17, 467)*

ment, a decrease in doubling rate of germ cells, and a reduction in the diameter of the seminiferous tubules. The pubertal increase in seminiferous tubule diameter is delayed while increase in the tubular length almost ceases. The number of germ cells increases in the abdominal testis, but only spermatocytes are occasionally present while all stages of spermatogenesis are found in the remaining scrotal testis during the prepubertal period.

In normal pigs testicular development of Leydig cells and seminiferous tubules, expressed relative to body weight, compensates for marked changes in growth rate (Fig. 18–4). A decrease in relative weight results from a slowing of absolute growth rate of the structure. Relative testis weight shows two periods of increase: in the prenatal period the increase results primarily from Leydig cell development,

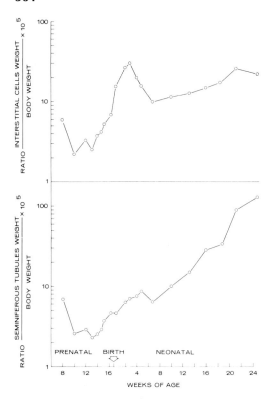

Fig. 18–4. Development of relative weights of interstitial cells and seminiferous tubules during fetal development and neonatal growth in the pig. Relative weights are expressed by the ratio of cell or tubular weight and body weight, logarithmic scale. *(Adapted from van Straaten and Wensing, 1977. Biol. Reprod. 17, 467)*

and in the neonatal period, mainly from seminiferous tubule development. Leydig cell weight approximates its prenatal level again at 22 weeks after birth (Fig. 18–4). The increased growth of the seminiferous tubules is a result of increased length and diameter of the tubules. From 14 weeks postcoitum to three weeks after birth the numbers of germ cells per testis show a constant doubling rate, but their numbers per tubule cross section decrease as a result of increasing tubular length. Morphogenesis of the normal pig testis is nearly complete by 25 weeks after birth.

HORMONE REGULATION IN THE BOAR

The interstitial cells of Leydig are the primary source of testicular steroids, and in the fetal pig their histologic development occurs in a biphasic pattern with the first peak of activity near the time of gonadal sex differentiation (crown rump length, 4.5 cm) and the second increase in cell size and numbers in late prenatal life (crown rump length, 20 cm) (Moon and Hardy, 1973). Mature Leydig cells arise from mesenchymal cells, which differentiate initially into immature Leydig cells, and the process is repeated late in fetal life. Porcine blastocysts are capable of synthesizing estrogens by day 12 (Perry et al, 1973; (Fig. 18–1, *i*), and fetal pig testes contain hydroxysteroid dehydrogenase activity and thus steroidogenic capabilities before differentiation of Leydig cells (Moon and Raeside, 1972; Moon et al, 1973). Fetal pig testes in culture secrete testosterone, with peak activity occurring at the stage of sexual differentiation of external genitalia and coincident to the initial peak in mature Leydig cell development (Stewart and Raeside, 1976). Testosterone content and concentration in pig testis during fetal development (crown rump length 2 to 12 cm) indicate a pattern of higher steroid production at sexual differentiation (approximately days 34 to 39), followed by a brief decline and then a subsequent rise during late stages of fetal life (Fig. 18–5). A sharp peak in plasma testosterone levels in male pig fetuses at about day 56 corresponds with the period of stabilization and development of the wolffian ducts (Meusy-Dessolle, 1975). Blood serum testosterone and luteinizing hormone (LH) concentrations in fetal pigs beginning 49 days post coitum and continuing to 25 weeks into the prepubertal period are shown in Figure 18–6. From 49 to 80 days post coitum, serum LH concentrations are below detectable levels, but detectable LH levels are found after day 85 and increase

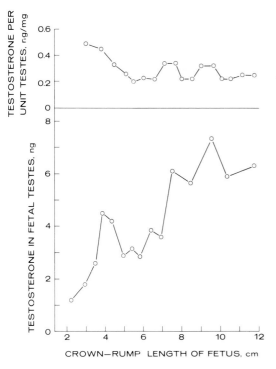

Fig. 18–5. **Testosterone content in fetal testes and testosterone concentrations per unit testis during fetal development in the pig.** *(Adapted from Raeside and Sigman, 1975. Biol. Reprod. 13, 318)*

(Elsaesser et al, 1976). These high plasma levels of progesterone during the late fetal period may indicate either reduced catabolism of this steroid or a reduction in biosynthesis of testosterone, which is low during this time.

During prepubertal hypertrophy, the Leydig cells reach maximal diameters of 30 μm and show typical fine structural features that include an increase in numbers of mitochondria, development of agranular endoplasmic reticulum and numerous cytoplasmic filaments (Dierichs et al, 1973). These and other cytoplasmic organelles resemble those found in adult boars (Belt and Cavazos, 1967). Hypophysectomy in immature boars causes regression of the testes, epididymides, prostate, seminal vesicles and bulbourethral glands (Anderson et al, 1976). In sexually mature boars hypophysectomy results in testicular regression, decreased diameter of seminiferous tubules with ablation of

during the remainder of fetal development (Colenbrander et al, 1977, 1978). During the first week after birth, LH remains elevated and then gradually decreases by the seventh week. Testosterone concentrations in peripheral serum are elevated between 40 and 60 days post coitum, decline between 60 and 100 days, and increase again during late fetal and neonatal periods as well as 13 weeks after birth. Patterns of LH and testosterone concentrations parallel testicular development during these prenatal and postnatal periods. Although high LH levels in plasma of male fetuses were found beginning somewhat earlier (day 61), plasma concentrations of progesterone by day 110 average >20 ng/ml, decrease (3 ng/ml) within the first week after birth and decline to <0.5 ng/ml by nine weeks of age

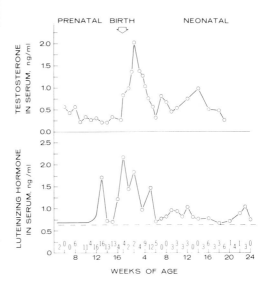

Fig. 18–6. **Testosterone and luteinizing hormone concentrations in blood serum of the male pigs during fetal development and during neonatal growth. Only mean values are presented. Dashed line in lower figure indicates minimum detectable level and the numbers indicate determinations within this nondetectable range.** *(Adapted from Colenbrander et al, 1977. Biol. Reprod. 17, 506; and Colenbrander et al, 1978. J. Reprod. Fertil. 53, 377)*

spermatogenesis, and a reduction in numbers of Leydig cells (Dufaure et al, 1974). Although daily injections of human chorionic gonadotropin re-establish the volumes of Leydig cells in these hypophysectomized animals, the effect is transitory and essentially lost after one month. Hourly blood sampling during a period of 24 hours indicates no significant changes in plasma LH concentrations, but testosterone levels are lower during the dark period as compared with those in the morning hours (Ellendorff et al, 1975).

On the basis of blood sampling at ten-minute intervals, it was found that copulation causes a significant increase in plasma LH but not in testosterone concentrations as compared with precopulatory hormone levels. Liptrap and Raeside (1978) found that either exposure of an aggressive boar to an estrous sow for several hours or copulation results in a transient increase in plasma levels of corticosteroids followed by an increase of testosterone. Estrogens also are secreted by the testes, as indicated by high urinary levels of the steroid. Relaxin is found in interstitial cells of sexually mature boars by immunocytochemical techniques, but it is absent in prepubertal and cryptorchid animals (Dubois and Dacheux, 1978). Physiologic actions of relaxin in the male are unknown.

In both pubertal and sexually mature boars, the intravenous injection of synthetic luteinizing hormone-releasing hormone (LHRH) causes an immediate release of pituitary LH to peak values within ten minutes (Pomerantz et al, 1974). The pituitary response to LHRH is dose-dependent, and the subsequent LH decline follows a parallel pattern in both immature and mature animals. Thus, pubertal and adult male pituitaries have similar sensitivities to LHRH stimulation, and the circulating levels of testosterone in adult boars do not appreciably modify this response. In castrated males, plasma

LH is sustained at higher basal levels but the immediate release of LH in response to LHRH follows a pattern similar to that in pubertal and mature boars. Testosterone and its metabolite, 5α-dihydrotestosterone (5α-DHT), provide both stimulatory and inhibitory effects on LH secretion in the boar. Direct intracerebral implantation of large amounts of 5α-DHT or testosterone into the mediobasal hypothalamus or amygdala inhibit release of LH from the pituitary gland, as reflected in lowered peripheral blood plasma levels of the gonadotropin, whereas implantation of lower amounts of these steroids in the brain elevates plasma LH levels (Parvizi et al, 1977). Thus, androgens modulate the secretion of hypothalamic hormones, which in turn control the release of LH from the adenohypophysis in the boar.

SPERM PRODUCTION

Boars reach puberty about 125 days of age. Spermatozoa are found in the testes but there may be further delay before they are capable of fertilizing ova. Mounting activity occurs early (erection by four months), but sequential patterns of sexual behavior culminate after five months. First ejaculates occur at five to eight months of age. The number of spermatozoa and semen volume continue to increase during the first 18 months of life. Within the seminiferous tubules of sexually mature boars, the germ cells are organized in well-defined cellular associations or stages that succeed one another over time.

The cycle of the seminiferous epithelium describes a series of changes in a given area of the tubule between two successive appearances of the same cellular association. In the boar, the duration of one cycle of the seminiferous epithelium requires 8.6 days (Swierstra, 1968). On the basis of histologic features of the cells, eight stages can be described within this period, with the duration of stages one through eight being 0.9, 1.2, 0.3, 1.0, 0.8, 1.7, 1.6, and 1.0 days, respectively. The duration of the spermatogenic cycle (spermatogonia to mature spermatozoa) is approximately 34.4 days in the boar; thus, four cycles of the seminiferous epithelium occur during one spermatogenic cycle. The life span of primary spermatocytes is 12.3 days; secondary spermatocytes, 0.4 days; spermatids with round nuclei, 6.3 days; spermatids with elongated nuclei, 1.5 days; and spermatozoa, 6.2 days. The duration for transit of the spermatozoa through the epididymis is about 10.2 days in this species. This period within the epididymides is necessary for spermatozoa to mature and acquire the ability to fertilize ova (Holtz and Smidt, 1976; Hunter et al, 1976).

Spermatozoa obtained from the caput epididymidis in the boar have poor fertilization rates, whereas those from the corpus epididymidis have improved fertilization abilities. Highest fertilization rates result with spermatozoa obtained from the proximal and distal cauda epididymidis. Testosterone sustains secretory activities of the accessory glands (e.g., seminal vesicles, prostate, bulbourethral glands) and these seminal fluids constitute a large proportion of the total ejaculate in the boar. Surgical removal of the seminal vesicles, however, does not impair the fertilizing ability of the spermatozoa (Davies et al, 1975).

Maternal influences, as evaluated by crossfostering techniques, have no significant effect on testicular growth and development (Eden et al, 1978). Evaluation of male reproductive efficiency based on the influence of the breed of sire on conception rate and number of living pigs at birth is not clearly defined for pure breeds of boars. Comparisons of reproductive characteristics of purebred and crossbred boars indicate that testes from crossbreds are significantly heavier and contain more spermatozoa than those from purebred boars (Table 18–2). Reproductive efficiency of Yorkshire gilts mated to crossbred boars did not significantly increase conception rate, mean number of living embryos or embryonic survival rates as compared with those parameters in gilts mated to purebred boars. A greater proportion of the crossbred boars mated at each exposure to estrous females.

HORMONES AND PUBERTY IN GILTS

During the prepubertal period the ovaries contain numerous small follicles (2 to 4 mm diameter) and several (8 to 15) medium-sized follicles (6 to 8 mm). The uterus responds to increasing ovarian steroidogenic activity during late stages of the prepubertal period; the uterus weighs 30 to 60 gm during infantile stages as compared with 150 to 250 gm in prepubertal gilts. As ovarian follicles develop in prepubertal gilts, there are corresponding increases in ovarian weights. Puberty is

Table 18–2. Reproductive Characteristics of Boars

	Testes		Reproductive Efficiency		
	Weight (gm)	No. of Spermatozoa	Gilts Pregnant (%)	No. Embryos per Gilt	Embryonic Survival at Day 30 (%)
Purebred boars*	590	57×10^9	56	10.7	70
Crossbred boars†	685	72×10^9	64	11.3	77

*Duroc and Hampshire breeds.
†Duroc and Hampshire crosses.
(Wilson et al, 1977. J. Anim. Sci. 44, 939)

characterized by first estrus, ovulation of graafian follicles and release of ova capable of fertilization.

Age at puberty may be influenced by level of nutrition, social environment, body weight, season of year, breed, disease or parasite infestation and management practices. Limiting energy intake to half that of full-fed controls delays puberty more than 40 days. Restricting energy intake to 60% to 70% that of ad libitum feed for a few days to several weeks has resulted in hastening the onset of puberty (eg, −11 days) in some trials and delaying puberty (eg, +16 days) in other trials. Protein supplementation seems to have little influence on the onset of puberty; age at puberty averaged 168 days in Yorkshire gilts (Friend, 1977). Augmentation of protein and energy in the diet usually increases body weight gain, resulting in heavier gilts, but does not reduce age at puberty. The presence of boars reduces age (191 vs 232 days) and body weight (105 vs 116 kg) to puberty (Mavrogenis and Robison, 1976) and agrees with the findings of Brooks and Cole (1970) and Zimmerman et al (1974). Puberty also is delayed in gilts penned individually as compared with those maintained in groups of 30 animals. Gilts born in the fall season reach puberty at a younger age and lower body weight than those born in the spring (Mavrogenis and Robison, 1976). Pubertal age in eight linebred breeds averages 209 days. Inbreeding increases pubertal age (eg, 243 days) whereas crossing inbred lines usually lowers pubertal age (eg, 228 days). There is a considerable range in age at puberty, from as early as 116 days (Holness, 1972) to more than 250 days.

Progesterone concentrations in peripheral plasma during a prepubertal period of >20 days are low (eg, average 2.4 ng/ml), and remain low throughout the proestrous, estrous and early metestrous periods of the first heat (Shearer et al,

1972). Since prepubertal gilts have no corpora lutea, the origin of the progesterone is unknown and may be at least partly adrenal in origin. Profiles of progesterone during the luteal phase of the first estrous cycle follow a pattern typical of normal cycling gilts. FSH and LH concentrations in the anterior pituitaries of 17-day-old gilts are significantly higher than during the estrous cycle (Parlow et al, 1964). Furthermore, pituitary FSH concentrations in gilts 110 to 119 days old exceed those of any stage of the estrous cycle. Basal levels of peripheral serum LH are low (eg, 1.3 ng/ml) in gilts 77 days old (Chakraborty et al, 1973) and are similar to those in prepubertal gilts at 160 days of age (Rayford et al, 1971). Infusion of synthetic LHRH induces a significant increase in serum LH concentrations to peak values (6.4 ng/ml) 15 or 30 minutes later, and is followed by a decline of LH to basal levels within 240 minutes. Sequential infusions of the LHRH elicit LH release, but do not induce significant ovarian changes in the 77-day-old animals. Exogenous estradiol benzoate seems to provide a positive feedback by inducing preovulatory surges of LH in prepubertal gilts at 60 days of age, but these surges become less pronounced at 160 days of age (Foxcroft and Elsaesser, 1977). Other evidence indicates that estrogen treatment suppresses pituitary response to exogenous LHRH as reflected in patterns of LH release (Foxcroft et al, 1975; Pomerantz et al, 1975). Although the prepubertal gilt synthesizes gonadotropins and can release them, further maturation of the hypothalamo-hypophyseal system is required for autonomous regulation of repetitive cycles.

Maturation of ovarian follicles and ovulations can be induced by exogenous gonadotropins in prepubertal gilts after 60 days of age. For example, a single injection of pregnant mare serum gonadotropin (PMSG) followed by human chorionic gonadotropin (HCG) induces ovulation in

90% of gilts 90 to 130 days of age, but few of them exhibit estrus or remain pregnant (Dziuk and Gehlbach, 1966). However, in gilts 9 to 12 months old who have not exhibited a previous heat, a similar gonadotropin regimen induces estrus, ovulation, recurrent estrous cycles and normal fertility in a high percentage (Dziuk and Dhindsa, 1969). In young prepubertal gilts, the administration of progestins or gonadotropins following induced ovulation increases the proportion of animals remaining pregnant to at least day 25 (Shaw et al, 1971; Rampacek et al, 1976b). Injection of LHRH also induces ovulations in prepubertal gilts and the corpora lutea from those ovulations sustain pregnancies in a similar proportion of animals (Rampacek et al, 1976b; Guthrie, 1977). In prepubertal gilts hysterectomized on day 10, the previously induced corpora lutea sustain secretion of progesterone to day 30 (Rampacek et al, 1976a). After partial hysterectomy the remaining nongravid uterine segment induces luteal failure in prepubertal gilts (Puglisi et al, 1978) similar to that found in sexually mature gilts (du Mesnil du Buisson, 1961; Anderson et al, 1961). Exogenous estrogen in cycling gilts sustains functional corpora lutea for periods exceeding 120 days (Anderson et al, 1973), but similar estrogen treatment is not luteotropic in prepubertal gilts (Rampacek and Kraeling, 1978).

In mature gilts, uterine-specific proteins are secreted during the luteal phase of the estrous cycle, as well as in progesterone injected-ovariectomized gilts (Squires et al, 1972; Knight et al, 1973). Prepubertal gilts do not respond to exogenous progesterone by secreting uterine-specific proteins until about 140 days of age (Murray and Grifo, Jr., 1976). After induction of ovulation and luteal formation in prepubertal gilts, uterine-specific proteins are produced in amounts considerably less than those found in mature animals (Segerson, Jr. and Murray,

1977). Quantitative and qualitative deficiencies in uterine-specific proteins in prepubertal gilts may relate to the inability to sustain pregnancy.

ESTRUS AND ESTROUS CYCLE

Estrus

Onset of estrus is characterized by gradual changes in behavioral patterns (eg, restlessness, mounting other animals, lordosis response), vulva responses (eg, swelling, pink-red coloring) and occasionally a mucous discharge. Sexual receptivity lasts an average of 40 to 60 hours. The pubertal estrous period usually is shorter (47 hours) than later ones (56 hours), and gilts usually have a shorter period of estrus than sows. Breed, seasonal variation (eg, longest in summer, shortest in winter) and endocrine abnormalities affect the duration of heat.

Ovulation

Ova are released 38 to 42 hours after onset of estrus, and the duration of this ovulatory process requires 3.8 hours (du Mesnil du Buisson et al, 1970). Ovulations occur about four hours earlier in mated than in unmated animals (Signoret et al, 1973).

Estrous Cycle

The length of the cycle is about 21 days (range: 19 to 23 days). The pig is polyestrous throughout the year; only pregnancy or endocrine dysfunction interrupts this cyclicity.

Cyclic Changes

Throughout the estrous cycle the interdependence of the ovary, hypothalamus, pituitary gland and uterus is reflected in their secretory functions as indicated by morphologic and hormonal changes. The estrous cycle may be categorized into a follicular phase (proestrus and estrus) and a luteal phase (metestrus and diestrus).

Ovarian Morphology and Hormone Secretion

There are about 50 small follicles (eg, 2 to 5 mm in diameter) per animal during the luteal and early follicular phases of the cycle. During the proestrous and estrous phases about 10 to 20 follicles approach preovulatory size (8 to 11 mm), while the number of smaller follicles declines (those <5 mm). Between days 5 and 16, the luteal phase of the cycle, numbers of follicles 2 to 5 mm in diameter (with a few up to 7 mm) increase, whereas after day 18 (proestrous phase) an increase occurs primarily in the growth of preovulatory follicles (those ≥8 mm in diameter).

Soon after ovulation there is rapid proliferation of primarily granulosa and a few theca cells lining the follicle wall. These cells become luteinized to form luteal tissue, thus the corpus luteum. Initially the corpus is considered a corpus hemorrhagicum because of the blood-filled central cavity, but within six to eight days the corpus luteum is a solid mass of luteal cells with an overall diameter of 8 to 11 mm. The relatively long luteal phase (about 16 days) is characterized by rapid development of the corpus luteum to its maximal weight (eg, 350 to 450 mg) by days 6 to 8, maintenance of cellular integrity and secretory function to day 16, and then rapid regression to a nonsecreting corpus albicans.

Characteristic cytologic features of a steroid-secreting cell include a large Golgi complex, few cisternal profiles of granular (rough) endoplasmic reticulum and extensive agranular (smooth) endoplasmic reticulum, whereas a protein-secreting cell contains prominent granular endoplasmic reticulum with well-developed

cisternae (Fawcett et al, 1969). Fine structural changes in the lutein cell indicate a close correlation between its morphology and steroid secretion during the estrous cycle (Cavazos et al, 1969).

During luteinization (day 1) granulosa cells at the periphery of the ruptured follicle are cuboidal to columnar and separated by irregular extracellular spaces that contain precipitated liquor folliculi. The cytoplasm in these peripheral cells contains granular endoplasmic reticulum and free polysomes. Deeper cells within the corpus are hypertrophied with an eccentrically located nucleus; the cytoplasm contains abundant granular endoplasmic reticulum. By day 4, luteinization is essentially complete; the cells are hypertrophied with masses of agranular endoplasmic reticulum. These cells typify the secretory phase (days 4 to 12) by their protein and steroid production. Small, coated vesicles are found near the Golgi complex and larger ones are found in peripheral locations near the membrane wall; these vesicles may be related to cellular transport. During cell regression (days 14 to 18) there is an increase in cytoplasmic lipid droplets, cytoplasmic disorganization and vacuolation of the agranular endoplasmic reticulum. At the terminal phase of the cycle there is an increase in the number of lysosomes, vacuolation of agranular endoplasmic reticulum and invasion of connective tissue; these events result in formation of the corpus albicans.

The vaginal epithelium proliferates in response to steroids from the ovaries. Histometric values for cell layers of vaginal epithelium are highest at estrus (ie, 10 to 12 cell layers), decline to lowest levels during the luteal phase (ie, 3 to 5 cell layers) and then increase during proestrus. Vaginal smears also indicate a cyclic pattern in the distribution of leukocytes (luteal phase), and epithelial and cornified cells (proestrous and estrous phases).

Steroid-secreting activity of the corpora lutea is indicated by concentrations of progesterone and estrogen throughout the cycle (Fig. 18–7). Progesterone levels are low at estrus (day 0), begin to increase abruptly after day 2 to peak values by days 8 to 12, and then decline precipitously thereafter to day 18. These profiles of progesterone secretion in peripheral blood are similar to those reported by others (Tillson et al, 1970; Henricks et al, 1972; Shearer et al, 1972) and correspond with patterns of progesterone secretion in ovarian venous blood throughout the cycle (Masuda et al, 1967). These steroid levels follow a pattern similar to the morphologic development and decline of the corpus luteum as well as ultrastructural changes in luteal cells. In porcine luteal tissue a decline in adenylate cyclase activity coincides with the onset of luteal regression (Andersen et al, 1974). LH stimulates adenylate cyclase activity and progesterone production in luteal tissue during midcycle, but enzyme activity fails to increase after day 16. The life span and secretory function of this ephemeral structure in the pig can be prolonged by pregnancy or hysterectomy. Estrogen concentrations in peripheral plasma begin to increase coincident with the decline and disappearance of progesterone (Fig. 18–7). Peak values occur two days preceding estrus and reflect rapid growth and maturation of graafian follicles during the late proestrous phase of the cycle. Soon after estrus estrogen declines and remains low during the luteal phase of the cycle; estrone in 24-hour urine collections follows a similar pattern (Lunaas, 1962).

Ovarian follicles depend upon secretion of adenohypophyseal gonadotropins for their growth and maturation; hypophysectomy (du Mesnil du Buisson and Léglise, 1963) or hypophyseal stalk transection (Anderson et al, 1967) results in abrupt regression of these follicles. The adenohypophysis synthesizes but secretes little luteinizing hormone during the luteal phase of the cycle. FSH levels increase in the pituitary gland during the luteal phase, a time when follicles are growing. Adenohypophyseal LH and FSH concentrations peak at proestrus and estrus and are low again during the early luteal phase (Parlow et al, 1964). FSH in peripheral serum reaches peak levels on days 2 and 3 following onset of behavioral estrus and may reflect decreased utilization of FSH by the remaining, small ovarian follicles at this time (Fig. 18–7). Peripheral plasma levels of LH show one sharp peak at estrus and drop to low levels during the remainder of the cycle (Fig.

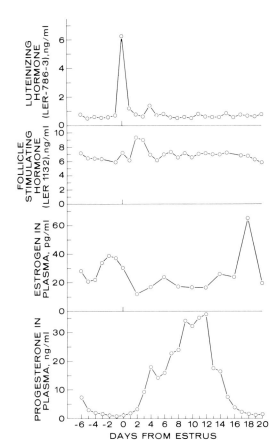

Fig. 18–7. Peripheral blood concentrations of progesterone, estrogen, follicle stimulating hormone and luteinizing hormone during the estrous cycle in the pig. *(Adapted from Parvizi et al, 1976. J. Endocrinol. 69, 193; Guthrie et al, 1972. Endocrinology 91, 675; and Rayford et al, 1974, J. Anim. Sci. 39, 348)*

18–7). Concentrations of FSH and LH in peripheral blood and adenohypophyseal tissue suggest a pattern of synthesis, storage and release. According to Brinkley et al (1973), prolactin in peripheral plasma peaks during the estrous cycle when plasma levels of estrogen are highest, and during estrus, prolactin profiles coincide better with the FSH peak than with the preovulatory LH peak.

Relaxin activity remains low in the luteal phase throughout the estrous cycle and shows no relationship to the high levels of progesterone secreted by these same cells during this brief period (Anderson et al, 1973).

Prostaglandin F (PGF) concentrations in utero-ovarian venous plasma increase during the estrous cycle to peak values (eg, ≥ 3 ng/ml) between days 12 and 16, a period coinciding with onset of luteal regression (Gleeson et al., 1974; Moeljono et al, 1977; Frank et al, 1977). Porcine endometrium also synthesizes *in vitro* highest levels of PGF during the mid- to late-luteal phase of the cycle (Patek and Watson, 1976).

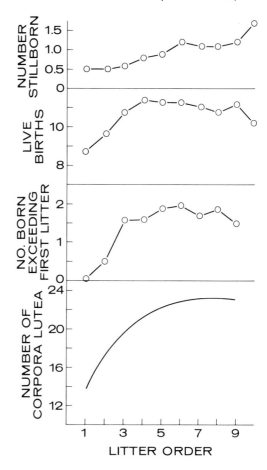

Fig. 18–8. **Reproductive performance in relation to parity in the pig.** *(Adapted from Perry, 1954. J. Embryol. Exp. Morph. 2, 308; Lush and Molln, 1942. Tech. Bull. U.S. Dept. Agric., 836; Rasbech, 1969, Br. Vet. J. 125, 599)*

OVULATION RATE

Ovulation rate is associated with breed (lines or crosses), amount of inbreeding, age at breeding and weight at breeding. In inbred lines there is an average increase of 0.8 ova from the first to second estrous periods; ovulation continues to increase (1.1 more ova) at the third estrus, but little if any additional increase occurs beyond the fourth postpubertal estrus. Reproductive experience correlates with ovulation rate; ovulations increase with parity to seven or more litters (Fig. 18–8). Inbreeding reduces ovulation rate, whereas crossing inbred lines increases the number of ovulations. For example, ovulation rates in crossbred pigs increase 0.55 ova for each 10% of inbreeding of their parent strains. The age at breeding in young gilts is positively correlated with ovulation rate. Weight at breeding is positively associated with ovulation rate when compared with weaning weight or weight at 154 days. Selection experiments based on a controlled gene pool over seven generations indicate that heritability of ovulation rate is 0.40 on weighted cumulative selection differential (Zimmerman and Cunningham, 1975; Newton et al, 1977). Ovulation rate in the select line increased from 14.4 to 16.1 by the seventh generation. Regression of line difference on generation number was not significantly correlated with age at puberty, weight at puberty, daily gain, backfat thickness, as well as four carcass traits (Newton et al, 1977; England et al, 1977).

Methods to Increase Ovulation Rates

Ovulations can be induced by injection of PMSG or PMSG followed by an injection of HCG. The ovulatory response depends primarily upon the dosage of PMSG. For example, 750 to 1,000 IU PMSG yield 12 to 25 ovulations. HCG induces ovulation in cycling gilts but causes little if any increase in ovulation rate. After an intramuscular injection of HCG (eg, 500 IU) during the proestrous period, ovulations occur in most of the animals 44 to 46 hours later (Hunter, 1972). The injection of PMSG induces superovulatory responses when given on days 15 or 16 of the cycle. The gonadotropins usually reduce the length of the cycle, increase the duration of estrus and may increase the incidence of cystic follicles, but the ova shed are capable of acceptable fertilization rates.

Nutrition and Ovulation Rate

High-energy diets induce a higher ovulation rate in the pig when the diets are fed for a restricted duration (Anderson and Melampy, 1972). Although the number of ovulations is predominantly affected by genetic background, ovulation rate is usually affected in a positive way with increasing levels of energy intake. The levels of energy restriction before feeding the pigs a high-energy diet is an important factor influencing ovulation rate. A low level of energy intake (eg, 3,000 to 5,000 kcal) is usually given before high-energy (eg, 8,000 to 10,000 kcal) diets. Durations for feeding the high-energy diets have ranged from 1 to >21 days. The optimal duration of a high-energy regimen seems to be 11–14 days before expected estrus or mating. Results from 14 trials indicated an additional 2.2 ova shed

as compared with ovulation rates in pigs on restricted diets. A single-feed flush seems to have little beneficial effect on ovulation rate (Staigmiller and First, 1973). There is little evidence that increased protein intake during brief periods increases ovulation rate. In contrast, subjecting gilts to starvation beginning day 10 of the estrous cycle results in 37% of them failing to return to estrus, mate and ovulate at the next expected estrus (Anderson, 1975). In those gilts that conceive under these conditions, the corpora lutea remain functional but ovarian follicles become atretic as the period of starvation increases.

CONCEPTION RATE

Fertilization rate in pigs is usually high (>90%). Low or high ovulation rates have little or no effect on fertilization rate. Loss of the whole litter may result from fertilization failure or death of all the embryos. Estimates indicate that approximately 5% of the litters are lost during the remainder of gestation. Early embryonic death results in resorption of the conceptus, whereas losses occurring after day 50 may result in abortion, fetal mummification or delivery of stillborns at term.

EMBRYO SURVIVAL

Embryo loss is an important factor in the pig. At least 40% of the embryos are lost before parturition and a major part of this loss occurs during the first half of gestation. Within the first 18 days embryonic survival is reduced by 17% (Anderson, 1978). By day 25 approximately 33% of the embryos die and this increases to 40% by day 50. Although sows have greater fecundity than gilts, they also lose a greater proportion of their embryos during the first 40 days. With each 10% inbreeding in the dam there results 0.55 to 0.76 fewer ova, loss of 0.53 more fertilized ova, and 0.8 fewer embryos by day 25.

Crossing these inbred lines results in 0.55 more ova, 0.33 increase in number of fertilized ova and 0.8 more embryos at day 25.

Uterine Capacity and Embryo Survival

The relationship between uterine capacity and embryonic development has been examined experimentally by several methods, which include superovulation, embryo transfer, superinduction and compensatory ovulation in unilaterally ovariectomized-hysterectomized animals.

Overcrowding during early pregnancy is not a major factor limiting litter size in the pig, as indicated by transferring either 12 or 24 embryos to recipient gilts and determining embryo survival before day 30; the number and percentage survival were 6.8 (57%) and 16.3 (68%) (Pope et al, 1972). By utilizing the technique of unilateral ovariectomy-hysterectomy to accomplish a compensatory ovulation rate in the remaining ovary and a limited amount of uterus, Knight et al (1977)

found that placental insufficiency was the primary cause of increased fetal death and decreased fetal growth after day 35 in the unilaterally hysterectomized-ovariectomized gilts (Fig. 18–9). All placental measurements were significantly greater in the control gilts at all stages of gestation. They conclude that between days 20 and 30 the extent of placental development seems to influence subsequent fetal growth and survival.

LITTER SIZE

Reproductive performance is measured primarily by the number of living pigs at birth or by the total farrowing or weaning weight of pigs produced by the dam within one year. Ovulation rates continue to increase with subsequent gestations but litter size reaches maximal levels by the fourth or fifth parity (Fig. 18–8). The number of pigs farrowed increases between the first and fourth litters, but by the eighth litter the number of live births declines while the number of stillborn increases. When litter size is related to age of dam, reproductive performance begins to decline after 4.5 years. The genetic contribution (heritability) to litter size is estimated as 0.17; most variation is attributed to environmental factors. Evaluation of several breed combinations provides estimates of heterosis and the average direct and maternal effects of the breeds. For example, pure lines of Duroc and Yorkshire animals average 13.8 corpora lutea, and when Duroc dams are bred to a boar of another breed, litter size increases by 1.44 pigs at farrowing and Yorkshire dams with crossbred litters have 0.37 more pigs at farrowing than their purebred litters (Johnson and Omtvedt, 1973). Comparisons of reproductive performance of purebred and two-breed cross Duroc, Hampshire and Yorkshire females indicate that the number of embryos increases 0.71 in crossbred gilts at day 30, and that these

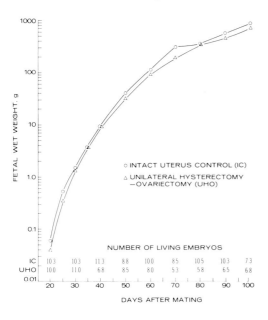

Fig. 18–9. Fetal development in intact and unilaterally hysterectomized-ovariectomized gilts. *(Adapted from Knight et al, 1977. J. Anim. Sci. 44, 620)*

crossbred females farrow 0.93 and wean 1.24 more pigs per litter than purebred females (Johnson et al, 1978). Survival rates at day 30, farrowing and weaning of progeny from three-breed crosses exceed those progeny from two-breed crosses.

Growth of Dam

When food intake is optimal, pregnancy is an anabolic process benefiting development of the conceptuses as well as growth of the dam. The increase in body weight of the dam through several gestation cycles results from anabolic effects of pregnancy (Fig. 18–10). Maternal and conceptus weight increases during pregnancy and maternal weight decreases during lactation, but the overall weight change is positive during each reproductive cycle.

Nutrition and Litter Size

Although embryonic losses of 40% *normally* occur during a rather brief gestation of approximately 115 days in pigs, the roles of maternal nutrition and ovarian hormones on fetal survival and development are undefined. Numerous trials in-

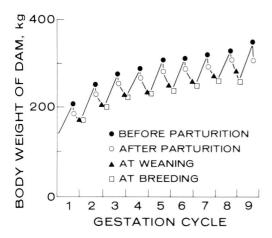

Fig. 18–10. Body weight change in sows in relation to gestation. *(Adapted from Salmon-Legagneur et al, 1966. Ann. Zootech. 15, 215)*

dicate that restricting metabolizable energy intake (eg, <5,000 vs >9,000 kcal/day) throughout gestation has no effect on embryo survival rates (10.4 vs 10.5 live embryos) and only a slight decrease results in piglet birth weight (Anderson and Melampy, 1972). There is an increase in litter size of about two pigs between the first and third gestation, but this is unrelated to the nutritional status of the dam. Dams given restricted energy levels deliver more pigs during the second and third gestations than do full-fed controls. Thus high energy intake during pregnancy has no beneficial effect on fetal survival. Long-term effects of low (3,000 kcal) compared to high (7,500 kcal) energy intake on reproductive performance through three consecutive gestation periods indicate that the total number of pigs farrowed declines (10.6 vs 8.7) but birth weight of living pigs (1.2 vs 1.5 kg) increases with increasing levels of metabolizable energy for the dam (Frobish et al., 1973).

Protein intake in nutritionally adequate gestation rations is 200 to 300 gm/day. Protein-free diets fed during gestation reduce birth weight, and postnatal growth remains depressed even when the offspring are provided a full diet (Pond, 1973; McCance and Widdowson, 1974). Severe protein restriction during gestation and lactation impairs subsequent reproductive efficiency in the increased number of days from weaning to estrus and reduced ovulation rate.

Progesterone secretion is essential throughout pregnancy in the pig; ovariectomy even as late as day 110 results in abrupt abortion (Belt et al, 1971), but in well-nourished dams there is no evident relationship between peripheral blood levels of progesterone and fetal survival rates (Webel et al, 1975). Inadequate maternal nutrition during brief periods of pregnancy has little effect on fetal survival. For example, pregnancies and embryonic survival rates are maintained in a

Table 18–3. Effect of Inanition on Embryonic Survival in the Pig

| Days After Mating | Days of Inanition | No. of Pigs | | Embryonic Survival Rate (%) |
		Mated	Remaining Pregnant	
		Inanition		
14	25	6	6	81
18	29	6	5	92
22	33	6	4	96
26	37	6	4	89
30	41	6	2	94
34	45	6	1	83
		Inanition plus progesterone and estradiol benzoate		
34	45	6	6	87

(Anderson, 1975. Am. J. Physiol. 229, 1687)

high proportion of gilts subjected to 37 days of inanition (0 kcal/day; water only), beginning ten days before mating (Table 18–3). Only a few dams remain pregnant when starvation continues to 45 days. All gilts remain pregnant, however, when progesterone and estradiol benzoate are given during 45 days of starvation in early gestation (Anderson, 1975). Embryonic and placental development in starved and full-fed dams remain similar in both groups to day 26 (equivalent to 37 days of inanition; Fig. 18–11). It is only after a

prolonged period of >40 days inanition that embryonic and placental growth are adversely affected; however, exogenous progesterone and estradiol benzoate significantly increase both embryonic and placental development during 45 days starvation. When dams are deprived nutritionally (eg, 2,208 kcal/day) during the middle third (days 30 to 70) or last two-thirds (days 30 to 110) of pregnancy, most (93%) remain pregnant, but exogenous progesterone and estrogen do not increase fetoplacental growth (Anderson and Dun-

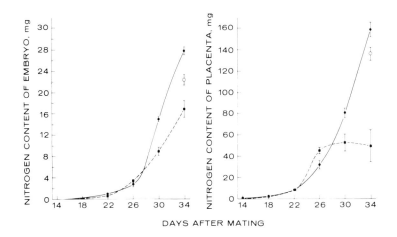

Fig. 18–11. Effect of prolonged starvation on embryonic and placental development in gilts beginning 11 days before mating and continuing to *days 14, 18, 22, 26, 30,* and *34* after mating. The periods during pregnancy correspond to 25, 29, 33, 37, 41 and 45 days' starvation. Symbols for full-fed controls are (●), those for gilts subjected to prolonged starvation are (■), and those for gilts starved 45 days, ovariectomized on *day 22* of pregnancy and given progesterone and estradiol benzoate are (○). *(Adapted from Anderson, 1975. Am. J. Physiol. 229, 1687)*

seth, 1978). Pregnancies are maintained in 81% of Yorkshire gilts during 40 days of starvation in either the middle third (days 30 to 70) or last third (days 70 to 110) of pregnancy (Anderson et al, 1979). Fetal survival rates average 65% in starved dams and 63% in full-fed controls; mean number of living fetuses is 9.9 in starved and 9.6 in control dams. Progesterone in peripheral serum of dams starved either during middle or late pregnancy is maintained at levels similar to those in controls (eg, 18 to 22 ng/ml). Abortion occurs in starved dams only when serum progesterone concentrations drop to <10 ng/ml. Concentrations of maternal serum proteins, electrolytes and iron are sustained at adequate levels to maintain normal development of the conceptuses during prolonged starvation in middle or late pregnancy (Kertiles et al, 1979). When gilts are starved 40 days either in middle third or last third of pregnancy and then gradually realimentated to a full diet and allowed to advance to term, 74% of them complete pregnancies with an average of 9.4 living pigs at parturition (Hard and Anderson, 1979). Progesterone concentrations in peripheral serum are maintained at similar levels in starved dams and full-fed controls. These results clearly indicate that ovarian progesterone secretion is maintained at relatively normal levels in pigs subjected to the extreme conditions of prolonged starvation during pregnancy.

Furthermore, inadequate maternal nutrition during relatively brief segments of gestation is not a major limitation to fetal survival in this litter-bearing species.

PREGNANCY AND LACTATION

Growth of Conceptuses

Ova are fertilized in the ampulla of the oviduct and their arrival to this region may be aided by the rapid beat, in a downward direction, of cilia on the mucosal surface; in the isthmus region of the oviduct there is an extensive upward ciliary current that may aid sperm ascent (Gaddum-Rosse and Blandau, 1973). Embryos are usually in the four-celled stage when they enter the uterus. Cleavage advances to morula stage by day 5 and then blastocyst formation by days 6 to 8 (Hunter, 1974). Hatching describes at least partial escape of the embryo from the zona pellucida and occurs on the sixth day; pig blastocysts may reach ≥150 cells before hatching. Blastocysts are unevenly distributed throughout both uterine horns. Rapid development of conceptuses is indicated by intrauterine migration of the embryos, spacing of the embryos and transition of blastocysts from spherical to extremely elongated forms and subsequent embryogenesis (Table 18–4). By day 11 half the blastocysts rapidly elongate to filamentous forms, often exceeding

Table 18–4. Embryonic Development as Related to Day in the Pig

Days After Mating	Spherical Blastocyst	Ovoid Blastocyst	Tubular Blastocyst	Filamentous Blastocyst	Embryogenesis
9	+				
10	+				
11	+	+	+	+	
12	+	+	+	+	
13		+	+	+	+
14		+	+	+	+
15			+	+	+
16			+	+	+
18				+	+

(Anderson, 1978. Anat. Rec. 190, 143)

60 cm, and by day 13 most embryos have completed this process (Vincent et al, 1976; Anderson, 1978). These conceptuses become regularly spaced with no overlap of tubular membranes from other embryos in that horn. Protein content in individual conceptuses denotes exponential growth between days 9 and 18, and is independent of the developmental stage or potential loss of those neighbors nearest that conceptus. Embryo recovery for immediate transfer and the requirements of synchrony between donor and recipient pigs (Hunter et al, 1967; Webel et al, 1970), as well as development of media for *in vitro* culture of embryos during cleavage stages (Wright, Jr., 1977; Pope and Day, 1977) and incubation in rabbit oviduct (Polge et al, 1972) have been described. Methods for culture of four-cell eggs to blastocyst stages and successful transfer to recipient gilts have been detailed by Davis and Day (1978).

Patterns of normal embryonic and fetal growth from days 20 to 100 of gestation are shown in the control group in Figure 18–9. Fetal wet weight is highly correlated with placental length (r = 0.64), placental surface area (r = 0.72) and total areolae surface area per placenta (r = 0.65).

Hormones During Pregnancy

Corpora lutea are essential for maintenance of pregnancy to term in the pig. The corpus luteum develops to maximal weight by day 8 and is sustained to late pregnancy (Fig. 18–12). By utilizing techniques of cell dissociation and sedimentation, Lemon and Loir (1977) have described two porcine luteal cell populations of 30 to 50 μm in diameter and 15 to 20 μm in diameter during pregnancy. Production of progesterone seems associated with cell size and stage of pregnancy. After day 114, soon after delivery, there is a precipitous decline in luteal weight. Progesterone concentrations in peripheral

blood increase to peak values by day 12, gradually decreasing to levels of 20 to 25 ng/ml by day 104 (Fig. 18–12). Relaxin gradually accumulates in luteal tissue to peak values during late pregnancy (days 105 to 110). Plasma levels of unconjugated estrone and estradiol-17β in peripheral blood are measurable by day 80 and rapidly increase to peak values just before parturition. Two peaks of estrone sulfate occur at day 30 and day 112, respectively, and then drop at onset of parturition (Fig. 18–12). Estrone in uterine vein blood remains low from days 20 to 60 (eg, 8 to 27 pg/ml) and then increases to >3,000 pg/ml by day 100 (Knight et al, 1977). The fetoplacental unit is the major source of estrogen production, as indicated by finding similar urinary excretory patterns in intact controls as in sows after ovariectomy, hypophysectomy (Fèvre et al, 1968) or adrenalectomy (Fèvre et al, 1972).

Adenohypophyseal concentrations and contents of FSH remain similar throughout pregnancy, whereas LH increases from day 13 to a maximum at day 18 and then decreases during the remainder of pregnancy (Parlow et al, 1964; Melampy et al, 1966). Immunocytochemical and ultrastructural evidence in the porcine pituitary indicate that most gonadotropic cells secrete both FSH and LH (Dacheux, 1978). Pituitary prolactin as well as growth hormone also remains at similar levels and total contents during three stages of pregnancy (Anderson et al, 1972). In gilts hypophysectomized at day 4, pregnancy and luteal function are maintained to day 12 (du Mesnil du Buisson et al, 1964), but pregnancy fails soon after hypophysectomy at day 70 (du Mesnil du Buisson and Denamur, 1969) and days 80 and 90 (Kraeling and Davis, 1974). Pregnancy is maintained at least 20 days, however, after hypophyseal stalk transection at day 70 or 90 (du Mesnil du Buisson and Denamur, 1969). The pig requires adenohypophyseal luteotropic support during a major part of gestation.

During early pregnancy (days 12 to 21), endogenous concentrations of prostaglandin F in utero-ovarian plasma remain lower in pregnant than nonpregnant gilts (0.52 vs 1.37 ng/ml; Moeljono et al, 1977). There also is a greater frequency of PGF peaks in nonpregnant than pregnant pigs during this time. Furthermore, estradiol concentrations in utero-ovarian venous plasma are greater on days 12 to 17 of pregnancy as compared with nonpregnant animals. The blastocysts maintain luteal function during critical phases of early pregnancy by overcoming uterine luteoly-

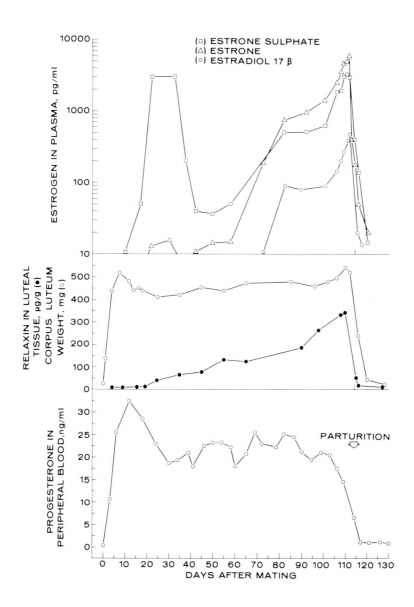

Fig. 18–12. Peripheral blood concentrations of progesterone and estrogens, relaxin concentrations in luteal tissue, and corpus luteum weight throughout pregnancy in the pig. *(Adapted from Anderson et al, 1979. Am. J. Physiol. 236, E335; Guthrie et al, 1972. Endocrinology 91, 675; Anderson et al, 1973. Am. J. Physiol. 225, 1215; Belt et al, 1971. Endocrinology 89, 1; and Robertson and King, 1974. J. Reprod. Fertil. 40, 133)*

tic action, and may contribute to the luteotropic effect by their production of estrogen (Perry et al, 1976). In nonpregnant gilts, exogenous estrogen given on days 11 through 15 of the cycle maintains corpora lutea for prolonged periods (Gardner et al, 1963), presumably by its action on the endometrium. The increase in PGF concentrations in uterine flushings from estrogen-treated gilts may prevent uterine luteolytic action by suppressing PGF secretion into the uterine venous system (Frank et al, 1978). When the uterus is removed (hysterectomy) the corpora lutea are maintained for a period exceeding that of pregnancy, and they produce progesterone and relaxin in a manner similar to that found in pregnant gilts (Anderson et al, 1969; 1973). Porcine corpora lutea are sensitive to the luteolytic effects of pharmacologic dosages of synthetic analogues of $PGF_{2\alpha}$ at day 10 or later in the estrous cycle (Guthrie and Polge, 1976), and $PGF_{2\alpha}$ induces luteolysis in gilts with corpora lutea maintained by estrogen (Kraeling et al, 1975). Exogenous $PGF_{2\alpha}$ is luteolytic and induces abortion from day 23 of pregnancy onward (Diehl and Day, 1974).

During late pregnancy corticosteroid concentrations increase within 24 hours of parturition and decrease during early lactation (Fig. 18–13; Ash and Heap, 1975; Baldwin and Stabenfeldt, 1975). Progesterone levels decline during the last days of pregnancy and drop abruptly to 0.5 ng/ml by day 1 postpartum (Fig. 18–13; Killian et al, 1973). Estrone increases to peak concentrations until day 2 prepartum and then falls to basal levels after delivery of conceptuses. The rise in estrone as well as estradiol is associated with fetal maturity and is primarily of placental origin.

Relaxin is produced and accumulated in porcine corpora lutea throughout pregnancy, and then released before parturition (Belt et al, 1971; Anderson et al, 1973). The accumulation of cytoplasmic

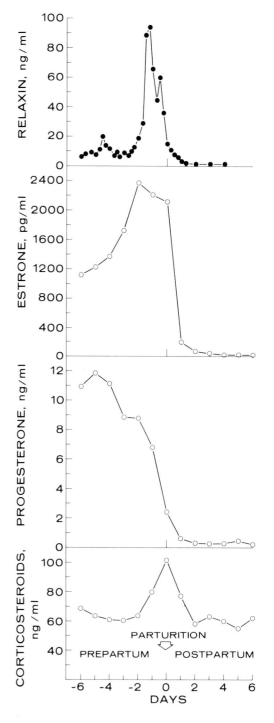

Fig. 18–13. Peripheral blood concentrations of corticosteroids, progesterone, estrone and relaxin during late pregnancy, parturition and early lactation in the pig. *(Adapted from Molokwu and Wagner, 1973. J. Anim. Sci. 36, 1158; and Sherwood et al, 1978. Endocrinology 102, 471)*

granules and relaxin activity in corpora lutea, beginning approximately on day 28, and the disappearance of these granules correlate with the rise and fall of relaxin activity in ovarian venous blood just preceding delivery. Immunocytochemical localization of relaxin in granulosa lutein cells at the ultrastructural level also is correlated with changes in the numbers of granules and relaxin levels (Belt et al, 1971; Kendall et al, 1978). Relaxin concentrations in peripheral plasma remain consistently low (<2 ng/ml) during the first 100 days of pregnancy, then increase to peak values two days before parturition and signal the discharge of accumulated relaxin from the corpora lutea (Fig. 18–13; Sherwood et al, 1975). There is an abrupt decline in relaxin just before delivery. When corpora lutea are induced 15 days before expected parturition, they have inadequate time to accumulate relaxin (Sherwood et al, 1977). The surge in relaxin levels that occurs in pigs experiencing progesterone-delayed parturition is not a sufficient stimulus to initiate parturition within 24 hours (Sherwood et al, 1978). Although the luteal cells produce both progesterone and relaxin, their secretion profiles are different during the last days of pregnancy. Porcine relaxin given daily, beginning either 105 or 107 days after mating and before surgical enucleation of corpora lutea (luteectomy) on day 110, significantly (a) induces premature cervical dilatation, (b) decreases the interval from surgery to delivery of the first neonate, and (c) reduces duration of delivery of all neonates in the litter, compared with those parameters in luteectomized controls (Kertiles and Anderson, 1979).

Autotransplantation of ovaries or uterus indicates that the usual connections of the ovaries to the uterus seem unessential for the maintenance of pregnancy and normal parturition in this species (Martin et al, 1978). Premature parturition in intact gilts or sows can be induced by the administration of exogenous PGF_{2_α} (Diehl et al, 1974;

Coggins et al, 1977; Wettemann et al, 1977). Dexamethasone injections induce premature delivery, whereas feeding the dams methallibure delays onset of parturition (Coggins and First, 1977; Coggins et al, 1977). The role of the fetal pituitary and adrenal glands in initiating processes of parturition is implicated by the effects of fetal hypophysectomy (Bosc et al, 1974) or fetal decapitation (Stryker and Dziuk, 1975; Coggins et al, 1977) on prolongation of gestation beyond term.

Postpartum

Immediately after delivery of the conceptuses peripheral blood levels of progesterone, estrone, estradiol and relaxin decline to basal levels during early lactation (Fig. 18–13). Estrus frequently occurs within one to three days after parturition. If mated, the sow fails to conceive at this estrus because the ovarian follicles are immature and ovulation usually does not occur. The postpartum estrus is observed in sows with low concentrations of estrogen in peripheral plasma, and even after ovariectomy (Ash and Heap, 1975; Holness and Hunter, 1975). The early postpartum estrus may result from peak levels of fetoplacental estrogens just preceding parturition.

Lactation

With the exception of the postpartum estrus, sows rarely exhibit estrus during lactation. Ovarian morphology during the anestrous period indicates an absence of gonadotropic stimulation. Average diameters of the ovarian follicles decrease (eg, from 4.6 to 2.7 mm) during the first week after parturition and then gradually increase (eg, >5 mm) by the fifth week of lactation (Palmer et al, 1965). Uterine weight and length decline rapidly for 21 to 28 days following parturition; thereafter, both remain constant. The endometrium is thinner and the uterine glands are less numerous, particularly in the basal region near the myometrium.

Pituitary FSH activity is high during early and late lactation while LH activity remains low throughout lactation. The lactational anestrus may be a period of depressed FSH release and reduced LH synthesis. Prolactin concentrations in peripheral plasma are high at parturition, increase in response to suckling by the piglets, and decline soon after weaning (van Landeghem and van de Wiel, 1978; Bevers et al, 1978). By the third week of lactation, estrus and ovulation can be induced by separating the sow from the litter for periods of 12 hours for three consecutive days and injecting PMSG to induce maturation of follicles and ovulation (Crighton, 1970).

Journal Paper No. J-9413 of the Iowa Agriculture and Home Economics Experiment Station, Ames Iowa. Projects 2092 and 2093.

REFERENCES

Andersen, R.N., Schwartz, F.L. and Ulberg, L.C. (1974). Adenylate cyclase activity of porcine corpora lutea. Biol. Reprod. *10*, 321.

Anderson, L.L. (1973). Effects of hysterectomy and other factors on luteal function. In Handbook of Physiology. R.O. Greep, (ed), Vol. II, Part 2, pp. 69–86. Washington, American Physiological Society.

Anderson, L.L. (1975). Embryonic and placental development during prolonged inanition in the pig. Am. J. Physiol. *229*, 1687.

Anderson, L.L. (1978). Growth, protein content and distribution of early pig embryos. Anat. Rec. *190*, 143.

Anderson, L.L., Bland, K.P. and Melampy, R.M. (1969). Comparative aspects of uterine-luteal relationships. Recent Prog. Horm. Res. *25*, 57.

Anderson, L.L., Butcher, R.L. and Melampy, R.M. (1961). Subtotal hysterectomy and ovarian function in gilts. Endocrinology *69*, 571.

Anderson, L.L. and Dunseth, D.W. (1978). Dietary restriction and ovarian steroids on fetal development in the pig. Am. J. Physiol. *234*, E190.

Anderson, L.L., Dyck, G.W., Mori, H., Henricks, D.M. and Melampy, R.M. (1967). Ovarian function in pigs following hypophysial stalk transection or hypophysectomy. Am. J. Physiol. *212*, 1188.

Anderson, L.L., Feder, J. and Bohnker, C.R. (1976). Effect of growth hormone on growth in immature hypophysectomized pigs. J. Endocrinol. *68*, 345.

Anderson, L.L., Ford, J.J., Melampy, R.M. and Cox, D.F. (1973). Relaxin in porcine corpora lutea during pregnancy and after hysterectomy. Am. J. Physiol. *225*, 1215.

Anderson, L.L., Hard, D.L. and Kertiles, L.P. (1979). Progesterone secretion and fetal development during prolonged starvation in the pig. Am. J. Physiol. *236*, E335.

Anderson, L.L. and Melampy, R.M. (1972). Factors affecting ovulation rate in the pig. In Pig Production. D.J.A. Cole (ed). London, Butterworths.

Anderson, L.L., Peters, J.B., Melampy, R.M. and Cox, D.F. (1972). Changes in adenohypophysial cells and levels of somatotrophin and prolactin at different reproductive stages in the pig. J. Reprod. Fertil. *28*, 55.

Ash, R.W. and Heap, R.B. (1975). Oestrogen, progesterone and corticosteroid concentrations in peripheral plasma of sows during pregnancy, parturition, lactation and after weaning. J. Endocrinol. *64*, 141.

Baldwin, D.M. and Stabenfeldt, G.H. (1975). Endocrine changes in the pig during late pregnancy, parturition and lactation. Biol. Reprod. *12*, 508.

Belt, W.D., Anderson, L.L., Cavazos, L.F. and Melampy, R.M. (1971). Cytoplasmic granules and relaxin levels in porcine corpora lutea. Endocrinology *89*, 1.

Belt, W.D. and Cavazos, L.F. (1967). Fine structure of the interstitial cells of Leydig in the boar. Anat. Rec. *158*, 333.

Bevers, M.M., Willemse, A.H. and Kruip, T.A.M. (1978). Plasma prolactin levels in the sow during lactation and the postweaning period as measured by radioimmunoassay. Biol. Reprod. *19*, 628.

Black, J.L. and Erickson, B.H. (1968). Oogenesis and ovarian development in the prenatal pig. Anat. Rec. *161*, 45.

Bosc, M., du Mesnil du Buisson, F. and Locatelli, A. (1974). Mise en évidence d'un contrôle foetal de la parturition chez la truie. Interactions avec la fonction lutéale. C.R. Acad. Sci., Paris *278D*, 1507.

Brinkley, H.J., Wilfinger, W.W. and Young, E.P. (1973). Plasma PRL in the estrous cycle of the pig. J. Anim. Sci. *37*, 303.

Brooks, P.H. and Cole, D.J.A. (1970). The effect of the presence of a boar on the attainment of puberty in gilts. J. Reprod. Fertil. *23*, 435.

Cavazos, L.F., Anderson, L.L., Belt, W.D., Henricks, D.M., Kraeling, R.R. and Melampy, R.M. (1969). Fine structure and progesterone levels in the corpus luteum of the pig during the estrous cycle. Biol. Reprod. *1*, 83.

Chakraborty, P.K., Reeves, J.J., Arimura, A. and Schally, A.V. (1973). Serum LH levels in prepubertal female pigs chronically treated with synthetic luteinizing hormone-releasing hormone/follicle-stimulating hormone-releasing hormone (LH-RH/FSH-RH). Endocrinology *92*, 55.

Channing, C.P. (1970). Effects of stage of the estrous cycle and gonadotropins upon luteinization of porcine granulosa cells in culture. Endocrinology *87*, 156.

Channing, C.P. and Tsafriri, A. (1977). Mechanism of action of luteinizing hormone and follicle-stimulating hormone on the ovary in vitro. Metabolism 26, 413.

Coggins, E.G. and First, N.L. (1977). Effect of dexamethasone, methallibure and fetal decapitation on porcine gestation. J. Anim. Sci. 44, 1041.

Coggins, E.G., Van Horn, D. and First, N.L. (1977). Influence of prostaglandin $F_{2\alpha}$, dexamethasone, progesterone and induced corpora lutea on porcine parturition. J. Anim. Sci. 46, 754.

Colenbrander, B., de Jong, F.H. and Wensing, C.J.G. (1978). Changes in serum testosterone concentrations in the male pig during development. J. Reprod. Fertil. 53, 377.

Colenbrander, B., Kruip, T.A.M., Dieleman, S.J. and Wensing, C.J.G. (1977). Changes in serum LH concentrations during normal and abnormal sexual development in the pig. Biol. Reprod. 17, 506.

Crighton, D.B. (1970). The induction of pregnancy during lactation in the sow: The effects of a treatment imposed at 21 days of lactation. Anim. Prod. 12, 611.

Dacheux, F. (1978). Ultrastructural localization of gonadotrophic hormones in porcine pituitary using the immunoperoxidase technique. Cell Tissue Res. 191, 219.

Davies, D.C., Hall, G., Hibbit, K.G. and Moore, H.D.M. (1975). The removal of the seminal vesicles from the boar and the effects on the semen characteristics. J. Reprod. Fertil. 43, 305.

Davis, D.L. and Day, B.N. (1978). Cleavage and blastocyst formation by pig eggs in vitro. J. Anim. Sci. 46, 1043.

Diehl, J.R. and Day, B.N. (1974). Effect of prostaglandin F_{2_a} on luteal function in swine. J. Anim. Sci. 39, 392.

Diehl, J.R., Godke, R.A., Killian, D.B. and Day, B.N. (1974). Induction of parturition in swine with prostaglandin F_{2_a}. J. Anim. Sci. 38, 1229.

Dierichs, R., Wrobel, K.-H. and Schilling, E. (1973). Licht- und elektronenmikroskopische Untersuchungen an den Leydigzellen des Schweines während der postnatalen Entwicklung. Z. Zellforsch. 143, 207.

Dubois, M.P. and Dacheux, J.L. (1978). Relaxin, a male hormone? Immunocytochemical location of a related antigen in the boar testis. Cell Tissue Res. 187, 201.

Dufaure, J.-P., du Mesnil du Buisson, F., Morat, M., Chevalier M. and Locatelli, A. (1974). Effects de l'hypophysectomie et de l'administration d'hormone gonadotrope (HCG) sur les cellules de Leydig du testicule de verrat. C.R. Acad. Sci., Paris 279D, 1907.

du Mesnil du Buisson, F. (1961). Régression unilatérale des corps jaunes après hysterectomie partielle chez la truie. Ann. Biol. Anim. Biochim. Biophys. 1, 105.

du Mesnil du Buisson, F., and Denamur, R. (1969). Mécanismes de contrôle de la fonction lutéale chez la truie, la brébis et la vache. IIIrd Intern. Congr. Endocr., Mexico, Excerpta Medica Intern. Congr. Ser. 184, 927.

du Mesnil du Buisson, F. and Léglise, P.C. (1963). Effet de l'hypophysectomie sur les corps jaunes de la truie. Résultats préliminaires. C.R. Acad. Sci., Paris 257, 261.

du Mesnil du Buisson, F., Léglise, P.C., Anderson, L.L. and Rombauts, P. (1964). Maintien des corps jaunes et de la géstation de la truie au course de la phase preimplantatoire après hypophysectomie. Intern. Congr. Anim. Reprod. Artif. Insem., Trento 3, 571.

du Mesnil du Buisson, F., Mauleon, P., Locatelli, A. and Mariana, J.C. (1970). Modification du moment et de l'étalement des ovulations après maîtrise du cycle sexuel de la truie. Colloque Sté Nat. Etude Steril-Fertil "L'inhibition de l'ovulation," Paris, Masson, 225.

Dzuik, P.J. and Dhindsa, D.S. (1969). Induction of heat, ovulation and fertility in gilts with delayed puberty. J. Anim. Sci. 29, 39.

Dzuik, P.J. and Gehlbach, G.D. (1966). Induction of ovulation and fertilization in the immature gilt. J. Anim. Sci. 25, 410.

Eden, C.W., Johnson, B.H. and Robison, O.W. (1978). Prenatal and postnatal influences on testicular growth and development in boars. J. Anim. Sci. 47, 375.

Ellendorff, F., Parvizi, N., Pomerantz, D.K., Hartjen, A., König, A., Smidt, D. and Elsaesser, F. (1975). Plasma luteinizing hormone and testosterone in the adult male pig: 24 hour fluctuations and the effect of copulation. J. Endocrinol. 67, 403.

Elsaesser, F., Ellendorff, F., Pomerantz, D.K., Parvizi, N. and Smidt, D. (1976). Plasma levels of luteinizing hormone, progesterone, testosterone and 5α-dihydro-testosterone in male and female pigs during sexual maturation. J. Endocrinol. 68, 347.

England, M.E., Cunningham, P.J., Mandigo, R.W. and Zimmerman, D.R. (1977). Selection for ovulation rate in swine: Correlated response in carcass traits. J. Anim. Sci. 45, 983.

Fawcett, D.W., Long, J.A. and Jones, A.L. (1969). The ultrastructure of endocrine glands. Rec. Prog. Horm. Res. 25, 315.

Fêvre, J., Léglise, P.C. and Reynaud, O. (1972). Rôle de surrénales maternelles dans la production d'oestrogènes par la truie gravide. Ann. Biol. Anim. Biochim. Biophys. 12, 559.

Fêvre, J., Léglise, P.C. and Rombauts, P. (1968). Du rôle de l'hypophyse et des ovaires dans la biosynthèse des oestrogènes au cours de la gestation chez la truie. Ann. Biol. Anim. Biochim. Biophys. 8, 225.

Foxcroft, G.R. and Elsaesser, F. (1977). The ontogeny and characteristics of positive oestrogen feedback on luteinizing hormone secretion in the domestic female pig. J. Endocrinol. 75, 45.

Foxcroft, G.R., Pomerantz, D.K. and Nalbandov, A.V. (1975). Effects of estradiol-17β on LH-RH/FSH-RH-induced, and spontaneous, LH release in prepubertal female pigs. Endocrinology 96, 551.

Frank, M., Bazer, F.W., Thatcher, W.W. and Wilcox, C.J. (1977). A study of prostaglandin F_2 as the luteolysin in swine: III Effects of estradiol valerate on prostaglandin F, progestins, estrone and estradiol concentrations in the utero-ovarian vein of nonpregnant gilts. Prostaglandins *14*, 1183.

Frank, M., Bazer, F.W., Thatcher, W.W. and Wilcox, C.J. (1978). A study of prostaglandin $F_{2\alpha}$ as the luteolysin in swine: IV an explanation for the luteotrophic effect of estradiol. Prostaglandins *15*, 151.

Friend, D.W. (1977). Effect of dietary energy and protein on age and weight at puberty of gilts. J. Anim. Sci. *44*, 601.

Frobish, L.T., Steele, N.C. and Davey, R.J. (1973). Long term effect of energy intake on reproductive performance in swine. J. Anim. Sci. *36*, 293.

Fulka, J., Kopecný, V. and Trebichavský, I. (1972). Studies on oogenesis in the early postnatal pig ovary. Biol. Reprod. *6*, 46.

Gaddum-Rosse, P. and Blandau, R.J. (1973). In vitro studies on ciliary activity within the oviducts of the rabbit and pig. Am. J. Anat. *136*, 91.

Gardner, M.L., First, N.L. and Casida, L.E. (1963). Effect of exogenous estrogens on corpus luteum maintenance in gilts. J. Anim. Sci. *22*, 132.

Gleeson, A.R., Thorburn, G.D. and Cox, R.I. (1974). Prostaglandin F concentrations in the utero-ovarian venous plasma of the sow during the late luteal phase of the oestrous cycle. Prostaglandins *5*, 521.

Guthrie, H.D. (1977). Induction of ovulation and fertility in prepuberal gilts. J. Anim. Sci. *45*, 1360.

Guthrie, H.D. and Polge, C. (1976). Luteal function and oestrus in gilts treated with a synthetic analogue of prostaglandin F-2α (ICI 79, 939) at various times during the oestrous cycle. J. Reprod. Fertil. *48*, 423.

Hard, D.L. and Anderson, L.L. (1979). Maternal starvation and progesterone secretion, litter size, and growth in the pig. Am. J. Physiol. *237*, E273.

Henricks, D.M., Guthrie, H.D. and Hanlin, D.L. (1972). Plasma estrogen, progesterone and luteinizing hormone levels during the estrous cycle in pigs. Biol. Reprod. *6*, 210.

Holness, D.H. (1972). Aspects of puberty in the indigenous gilt. S. Afr. J. Anim. Sci. *5*, 85.

Holness, D.H. and Hunter, R.H.F. (1975). Postpartum oestrus in the sow in relation to the concentration of plasma oestrogens. J. Reprod. Fertil. *45*, 15.

Holtz, W., and Smidt, D. (1976). The fertilizing capacity of epididymal spermatozoa in the pig. J. Reprod. Fertil. *46*, 227.

Hunter, R.H.F. (1972). Ovulation in the pig: Timing of the response to injection human chorionic gonadotrophin. Res. Vet. Sci. *13*, 356.

Hunter, R.H.F. (1974). Chronological and cytological details of fertilization and early embryonic development in the domestic pig, Sus scrofa. Anat. Rec. *178*, 169.

Hunter, R.H.F., Holtz, W. and Henfrey, P.J. (1976). Epididymal function in the boar in relation to the fertilizing ability of spermatozoa. J. Reprod. Fertil. *46*, 463.

Hunter, R.H.F., Polge, C. and Rowson, L.E.A. (1967).

The recovery, transfer and survival of blastocysts in pigs. J. Reprod. Fertil. *14*, 501.

Johnson, R.K. and Omtvedt, I.T. (1973). Evaluation of purebreds and two-breed crosses in swine: reproductive performance. J. Anim. Sci. *37*, 1279.

Johnson, R.K., Omtvedt, I.T. and Walters, L.E. (1978). Comparison of productivity and performance for two-breed and three-breed crosses in swine. J. Anim. Sci. *46*, 69.

Josso, N., Picard, J.-Y. and Tran, D. (1977). The antimullerian hormone. Rec. Prog. Horm. Res. *33*, 117.

Jost, A. (1970). Hormonal factors in the sex differentiation of the mammalian foetus. Phil. Trans. Roy. Soc. London B *259*, 119.

Jost, A., Vigier, B., Prépin, J. and Perchellet, J.P. (1973). Studies on sex differentiation in mammals. Rec. Prog. Horm. Res. *29*, 1.

Kendall, J.Z., Plopper, C.G. and Bryant-Greenwood, G.D. (1978). Ultrastructural immunoperoxidase demonstration of relaxin in corpora lutea from a pregnant sow. Biol. Reprod. *18*, 94.

Kertiles, L.P. and Anderson, L.L. (1979). Effect of relaxin on cervical dilatation, parturition and lactation in the pig. Biol. Reprod. *21*, 57.

Kertiles, L.P., Anderson, L.L., Parker, R.O. and Hard, D.L. (1979). Maternal serum metabolites during prolonged starvation in pregnant pigs. Metabolism *28*, 100.

Killian, D.B., Garverick, H.A. and Day, B.N. (1973). Peripheral plasma progesterone and corticoid levels at parturition in the sow. J. Anim. Sci. *37*, 1371.

Knight, J.W., Bazer, F.W., Thatcher, W.W., Franke, D.E. and Wallace, H.D. (1977). Conceptus development in intact and unilaterally hysterectomized-ovariectomized gilts: interrelations among hormone status, placental development, fetal fluids and fetal growth. J. Anim. Sci. *44*, 620.

Knight, J.W., Bazer, F.W. and Wallace, H.D. (1973). Hormonal regulation of porcine uterine protein secretion. J. Anim. Sci. *36*, 546.

Kraeling, R.R., Barb, C.R. and Davis, B.J. (1975). Prostaglandin-induced regression of porcine corpora lutea maintained by estrogen. Prostaglandins *9*, 459.

Kraeling, R.R. and Davis, B.J. (1974). Termination of pregnancy by hypophysectomy in the pig. J. Reprod. Fertil. *36*, 215.

Lemon, M. and Loir, M. (1977). Steroid release in vitro by two luteal cell types in the corpus luteum of the pregnant sow. J. Endocrinol. *72*, 351.

Liptrap, R.M. and Raeside, J.I. (1978). A relationship between plasma concentrations of testosterone and corticosteroids during sexual and aggressive behaviour in the boar. J. Endocrinol. *76*, 75.

Lunass, T. (1962). Urinary oestrogen levels in the sow during oestrous cycle and early pregnancy. J. Reprod. Fertil. *4*, 13.

Martin, P.A., Bevier, G.W. and Dziuk, P.J. (1978). The effect of disconnecting the uterus and ovary on the length of gestation in the pig. Biol. Reprod. *18*, 428.

Masuda, H., Anderson, L.L., Henricks, D.M. and Melampy, R.M. (1967). Progesterone in ovarian venous plasma and corpora lutea of the pig. Endocrinology *80*, 240.

Mavrogenis, A.P. and Robison, O.W. (1976). Factors affecting puberty in swine. J. Anim. Sci. *42*, 1251.

McCance, R.A. and Widdowson, E.M. (1974). The determinants of growth and form. Proc. Roy. Soc. London Ser. B *185*, 1.

Melampy, R.M., Henricks, D.M., Anderson, L.L., Chen, C.L. and Schultz, J.R. (1966). Pituitary follicle-stimulating hormone and luteinizing hormone concentrations in pregnant and lactating pigs. Endocrinology *78*, 801.

Meusy-Dessolle, N. (1975). Variations quantitatives de la testosterone plasmatique chez le porc mâle de la naissance à l'age adulte. C.R. Acad. Sci., Paris *D281*, 1875.

Moeljono, M.P.E., Thatcher, W.W., Bazer, F.W., Frank, M., Owens, L.J. and Wilcox, C.J. (1977). A study of prostaglandin F$_{2_a}$ as the luteolysin in swine: II Characterization and comparison of prostaglandin F, estrogens and progestin concentrations in utero-ovarian vein plasma of non-pregnant and pregnant gilts. Prostaglandins *14*, 543.

Moon, Y.S. and Hardy, M.H. (1973). The early differentiation of the testis and interstitial cells in the fetal pig, and its duplication in organ culture. Am. J. Anat. *138*, 253.

Moon, Y.S., Hardy, M.H. and Raeside, J.I. (1973). Biological evidence for androgen secretion by the early fetal pig testes in organ culture. Biol. Reprod. *9*, 330.

Moon, Y.S. and Raeside, J.I. (1972). Histochemical studies on hydroxysteroid dehydrogenase activity of fetal pig testes. Biol. Reprod. *7*, 278.

Murray, F.A. and Grifo, A.P., Jr. (1976). Development of capacity to secrete progesterone-induced protein by the porcine uterus. Biol. Reprod. *15*, 620.

Newton, J.R., Cunningham, P.J. and Zimmerman, D.R. (1977). Selection for ovulation rate in swine: correlated response in age and weight at puberty, daily gain and probe backfat. J. Anim. Sci. *44*, 30.

Norberg, H.S. (1972). The follicular oocyte and its granulosa cells in domestic pig. Z. Zellforsch. *131*, 497.

Palmer, W.M., Teague, H.S. and Venzke, W.G. (1965). Histological changes in the reproductive tract of the sow during lactation and early post-weaning. J. Anim. Sci. *24*, 1117.

Parlow, A.F., Anderson, L.L. and Melampy, R.M. (1964). Pituitary follicle-stimulating hormone and luteinizing hormone concentrations in relation to reproductive stages of the pig. Endocrinology *75*, 365.

Parvizi, N., Elsaesser, F., Smidt, D. and Ellendorff, F. (1977). Effects of intracerebral implantation, microinjection, and peripheral application of sexual steroids on plasma luteinizing hormone levels in the male miniature pig. Endocrinology *101*, 1078.

Patek, C.E. and Watson, J. (1976). Prostaglandin and progesterone secretion by porcine endometrium and corpus luteum in vitro. Prostaglandins *12*, 97.

Pelliniemi, L.J. (1975a). Ultrastructure of gonadal ridge in male and female pig embryos. Anat. Embryol. *147*, 19.

Pelliniemi, L.J. (1975b). Ultrastructure of the early ovary and testis in pig embryos. Am. J. Anat. *144*, 89.

Pelliniemi, L.J. (1976). Ultrastructure of the indifferent gonad in male and female pig embryos. Tissue Cell *8*, 163.

Perry, J.S., Heap, R.B. and Amoroso, E.C. (1973). Steroid hormone production by pig blastocysts. Nature (London) *245*, 45.

Perry, J.S., Heap, R.B., Burton, R.D., and Gadsby, J.E. (1976). Endocrinology of the blastocyst and its role in the establishment of pregnancy. J. Reprod. Fertil. Suppl. *25*, 85.

Polge, C., Adams, C.E. and Baker, R.D. (1972). Development and survival of pig embryos in the rabbit oviduct. VII Congr. Intern. Reprod. Anim. Insem. Artif. Munich *4*, 60.

Pomerantz, D.K., Ellendorff, F., Elsaesser, F., König, A. and Smidt, D. (1974). Plasma LH changes in intact adult, castrated adult and pubertal male pigs following various doses of synthetic luteinizing hormone-releasing hormone (LH-RH). Endocrinology *94*, 330.

Pomerantz, D.K., Foxcroft, G.R. and Nalbandov, A.V. (1975). Acute and chronic estradiol-17β inhibition of LH and release in prepubertal female pigs; time course and site of action. Endocrinology *96*, 558.

Pond, W.G. (1973). Influence of maternal protein and energy nutrition during gestation on progeny performance in swine. J. Anim. Sci. *36*, 175.

Pope, C.E., Christenson, R.K., Zimmerman-Pope, V.A. and Day, B.N. (1972). Effect of number of embryos on embryonic survival in recipient gilts. J. Anim. Sci. *35*, 805.

Pope, C.E. and Day, B.N. (1977). Transfer of preimplantation pig embryos following in vitro culture for 24 or 48 hours. J. Anim. Sci. *44*, 1036.

Puglisi, T.A., Rampacek, G.B. and Kraeling, R.R. (1978). Corpus luteum function following subtotal hysterectomy on the prepuberal gilt. J. Anim. Sci. *46*, 707.

Rampacek, G.B. and Kraeling, R.R. (1978). Effect of estrogen on luteal function in prepuberal gilts. J. Anim. Sci. *46*, 453.

Rampacek, G.B., Kraeling, R.R. and Ball, G.D. (1976a). Luteal function in the hysterectomized prepuberal gilt. J. Anim. Sci. *43*, 792.

Rampacek, G.B., Schwartz, F.L., Fellows, R.E., Robison, O.W. and Ulberg, L.C. (1976b). Initiation of reproductive function and subsequent activity of the corpora lutea in prepuberal gilts. J. Anim. Sci. *42*, 881.

Rayford, P.L., Brinkley, H.J. and Young, E.P. (1971). Radioimmunoassay determination of LH concentration in the serum of female pigs. Endocrinology *88*, 707.

Segerson, E.C., Jr. and Murray, F.A. (1977). Appearance of the uterine specific proteins following induction of ovulation in prepubertal gilts. J. Anim. Sci. *45*, 355.

Shaw, G.A., McDonald, B.E. and Baker, R.D. (1971). Fetal mortality in the prepubertal gilt. Can. J. Anim. Sci. *51*, 233.

Shearer, I.J., Purvis, K., Jenkin, G. and Haynes, N.B. (1972). Peripheral plasma progesterone and oestradiol-17β levels before and after puberty in gilts. J. Reprod. Fertil. *30*, 347.

Sherwood, O.D., Chang, C.C., Bevier, G.W. and Dziuk, P.J. (1975). Radioimmunoassay of plasma relaxin levels throughout pregnancy and at parturition in the pig. Endocrinology *97*, 834.

Sherwood, O.D., Martin, P.A., Chang, C.C. and Dziuk, P.J. (1977). Plasma relaxin levels in pigs with corpora lutea induced during late pregnancy. Biol. Reprod. *17*, 97.

Sherwood, O.D., Wilson, M.E., Edgerton, L.A. and Chang, C.C. (1978). Serum relaxin concentrations in pigs with parturition delayed by progesterone administration. Endocrinology *102*, 471.

Signoret, J.P., du Mesnil du Buisson, F. and Mauleon, P. (1973). Effect of mating on the onset and duration of ovulation in the sow. J. Reprod. Fertil. *31*, 327.

Squires, G.D., Bazer, F.W. and Murray, F.A. (1972). Electrophoretic patterns of porcine uterine protein secretions during the estrous cycle. Biol. Reprod. *7*, 321.

Staigmiller, R.B. and First, N.L. (1973). The effect of a single-flush on ovulation rate in gilts. J. Reprod. Fertil. *35*, 573.

Stewart, D.W. and Raeside, J.I. (1976). Testosterone secretion by the early fetal pig testes in organ culture. Biol. Reprod. *15*, 25.

Stryker, J. and Dziuk, P.J. (1975). Effects of fetal decapitation on fetal development, parturition and lactation in pigs. J. Anim. Sci. *40*, 282.

Swierstra, E.E. (1968). Cytology and duration of the cycle of the seminiferous epithelium of the boar; duration of spermatozoan transit through the epididymis. Anat. Rec. *161*, 171.

Tsafriri, A. and Channing, C.P. (1975). Influence of follicular maturation and culture conditions on meiosis of pig oocytes *in vitro*. J. Reprod. Fertil. *43*, 149.

Tillson, S.A., Erb, R.E. and Niswender, G. D. (1970). Comparison of luteinizing hormone and progesterone in blood and metabolites of progesterone in urine of domestic sows during the estrous cycle and early pregnancy. J. Anim. Sci. *30*, 795.

van Landeghem, A.A.J. and van de Wiel, D.F.M. (1978). Radioimmunoassay for porcine prolactin: plasma levels during lactation, suckling and weaning and after TRH administration. Acta Endocrinol. *88*, 653.

van Straaten, H.W.M. and Wensing, C.J.G. (1977a). Histomorphometric aspects of testicular morphogenesis in the pig. Biol. Reprod. *17*, 467.

van Straaten, H.W.M. and Wensing, C.J.G. (1977b). Histomorphometric aspects of testicular morphogenesis in the naturally unilateral cryptorchid pig. Biol. Reprod. *17*, 473.

Vincent, F., Wintenberger-Torres, S., Paquignon, M. and du Mesnil du Buisson, F. (1976). Développement embryonnaire chez la truie au 17éme jour de la gestation. Relation avec la taille des cornes uterines. Rech. Porcine France *1*, 185.

Webel, S.K., Peters, J.B. and Anderson, L.L. (1970). Synchronous and asynchronous transfer of embryos in the pig. J. Anim. Sci. *30*, 565.

Webel, S.K., Reimers, T.J. and Dziuk, P.J. (1975). The lack of relationship between plasma progesterone levels and number of embryos and their survival in the pig. Biol. Reprod. *13*, 177.

Wettemann, R.P., Hallford, D. M., Kreider, D.L. and Turman, E.J. (1977). Influence of prostaglandin F_2 on endocrine changes at parturition in gilts. J. Anim. Sci. *44*, 106.

Wright, R.W., Jr. (1977). Successful culture *in vitro* of swine embryos to the blastocyst stage. J. Anim. Sci. *44*, 854.

Zimmerman, D.R., Carlson, R. and Lantz, B. (1974). The influence of exposure to the boar and movement on pubertal development in the gilts. J. Anim. Sci. *39*, 353.

Zimmerman, D. R. and Cunningham, P.J. (1975). Selection for ovulation rate in swine: population, procedures and ovulation response. J. Anim. Sci. *40*, 61.

19

Horses

E.S.E. HAFEZ

The various breeds of the domestic horse *(Equus caballus)* are members of the family *Equidae*, which belongs to the order *Perissodactyla*. The horse has several unique aspects of reproductive endocrinology and pregnancy. Whereas other large farm species such as the cattle, swine and sheep have been highly selected for reproductive efficiency, as well as other productive traits, the only selection practiced with horses has been their ability to walk or run.

Race horses are usually aged from January 1 in the year that they are born, and it has been the practice to breed them as early in the year as possible so that, as two-year-olds, the offspring have maximal physical advantage.

BREEDING SEASON

Seasonal variations in sexual behavior affect both stallions and mares.

Stallion

The breeding season of stallions is not well marked and semen can be collected throughout the year. However, remarkable seasonal variations are noted in reaction time, number of mounts per ejaculate, volume of gel-free semen, total number of spermatozoa per ejaculate, sperm agglutination and motility in fresh and in diluted semen.

The effects of season on seminal plasma is greater than those on spermatozoa. Spermatozoa in first ejaculates are less affected by season than those in second ejaculates. This differential effect on first and second ejaculates is noted for most semen characteristics (Pickett et al, 1975).

Mare

The reproductive cycle of the mare is subject to the greatest variability of all the domestic animals. Some mares appear to be truly polyestrous; they can produce offspring at any time of the year. However, the great majority of the mare population are seasonally polyestrous. Although many mares in the northern hemisphere show behavioral estrus in February, March and April, estrus during this time is often unaccompanied by ovulation, and conception rates in mares bred during the period are low. In the northern hemisphere the best conception rates usually occur in mares bred from May to July. The same trends occur in mares in the southern hemisphere for the corresponding seasons. Although mares who feed primarily on grass normally breed only during summer and go into anestrus in winter, those that are well fed and stabled tend to cycle throughout the year. The onset of the fertile breeding season is closely associated with management.

Mares can be classified into three categories according to their breeding season: (1) defined breeding season: the wild breeds of horses manifest several estrous cycles during a restricted breeding season that coincides with the longest days of the year; the foals are born during a restricted foaling season; (2) transitory breeding season: some domestic breeds and some individual mares manifest estrous cycles throughout the year, but ovulation accompanies estrus only during the breeding season, and the foals are born during a limited foaling season; (3) year round breeding: some domestic breeds and some individual mares exhibit estrous cycles accompanied by ovulation throughout the year. Thus it is evident that although some mares, at certain latitudes, may show estrous cycles throughout the year, they do not necessarily conceive during all estrous periods.

In localities where there is a breeding season, the two transitory periods preceding and following the breeding season are characterized by variability of ovarian activity and sexual behavior. At this time the ovarian follicles develop only to limited degrees and then undergo atresia. Also there is a high frequency of prolonged estrus or estrus of short duration as well as irregular estrous cycles during these periods.

Near the equator, there is little seasonal variation in the length of the estrous cycle. In temperate regions, mares are seasonally polyestrus, with the breeding and nonbreeding seasons occurring during the summer and winter months respectively. Photoperiod is perhaps the most important environmental signal that entrains the pituitary-gonadal axis, since artificial photoperiod treatments hasten follicular development and onset of the breeding season. The mare exhibits a photoperiodically entrained seasonal pattern of LH secretion. The cyclic reproductive behavior during the breeding season is mediated by the stimulatory and inhibitory actions of estradiol and progesterone (Garcia and Ginther, 1978). The exposure of mares to additional hours of light during winter will induce estrus and advance the onset of the breeding season. The ovaries of the anestrous mare cannot be activated even by the injection of fairly large doses of pregnant mare serum gonadotropin (PMSG) or human chorionic gonadotropin (HCG), in contrast to anestrous ewes, which respond to either hormone. However, estrogen therapy may be useful.

REPRODUCTIVE PARAMETERS IN STALLIONS

Reproductive parameters include sexual maturity, semen production, and ejaculation.

Sexual Maturity

The testes of the stallion descends into the scrotum at one to three weeks of age. In a few cases the testes are already in the scrotum at birth. Postnatal growth of the testes begins during the 11th month, and the left testis develops earlier and grows faster than the right. At this time, there is also a gradual outward development of the seminiferous tubules around the right testis (Fig. 19–1). The age at which stallions are first used for natural or artificial breeding is determined primarily by managerial conditions.

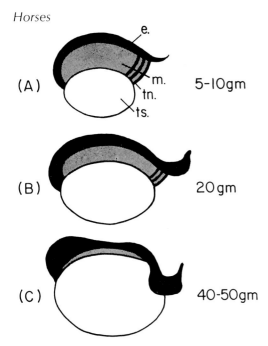

Fig. 19–1. Developmental changes in the attachment of epididymis with the testis in relation to sexual maturity. Figures on right indicate testis weight. e., epididymis; *m.,* thin membrane; *tn.,* tendon; *ts.,* testis. *A,* Loose attachment between testis and epididymis; note tendon attachment at head of epididymis; *B,* elongation of epididymis; *C,* epididymis fully developed and completely attached to testis. *(Adapted from Nishikawa, 1959. Studies on Reproduction in Horses. Tokyo, courtesy of Japan Racing Association)*

Semen Production

The cycle of the seminiferous epithelium can be divided into eight stages on the basis of meiotic divisions, shape of the spermatid nuclei and location of spermatids with elongated nuclei. The characteristics of semen and spermatogenesis are summarized in Tables 19–1, 19–2 and Figure 19–2.

The ejaculate is composed of six to nine jets resulting from the contractions of the urethra. The volume of each succeeding jet in the ejaculate reduces to about 50% of its initial value; 70% or more of the spermatozoa and the basic biochemical constituents are contained in the first three jets.

The gelatinous material in semen, secreted by the seminal vesicles, has no effect on the motility or the fertilizing ability of spermatozoa. The volume of gel, composing about one-third of the ejaculate, varies considerably and is not characteristic of the individual stallion. This is in contrast to gel obtained from boar ejaculates, which is a constant feature of the ejaculates. Ejaculates containing gel

Table 19–1. Some Reproductive Parameters in the Stallion

	Reproductive Parameters or Characteristics	Values (Average)
Sexual Maturity	Postnatal growth of testis	1 year
	Sperm appear in testis	1 year
	Sperm appear in ejaculate	13 months
	Sexual maturity	2 years
Testicular and Epididymal Morphology	Testicular weight (gm)	150–170 gm
	Epididymal weight (gm)	20–30 gm
	Volume of tubules/testis	55–70%
	Length of tubules/testis	2300–2600 m
	Weight (without tunica albuginea)	14–20 m
Spermatogenesis & Sperm Transport in Male Tract	Duration of seminiferous tubule cycle (judged by ^{3}H-thymidine injection)	13 days
	Life span of primary spermatocytes	19 days
	Life span of secondary spermatocytes	0.7 days
	Life span of spermatids with round nuclei	8.7 days
	Life span of spermatids with elongated nuclei	10 days
	Interval for labeled sperm to enter input of epididymis	35 days
	Interval from isotope injection until it appears in ejaculate	40 days
	Time of transport of sperm in excurrent ducts	8–11 days

(Data collected from Swierstra, E.E., Gebauer, M.R., and Pickett, B.W., 1974. Reproductive Physiology of the Stallion, I. Spermatogenesis and Testis Composition. J. Reprod. Fertil. 40, 113.)

Table 19–2. Biochemical Characteristics of Semen Fractions of the Stallion

Semen Fraction	Origin	Physical Characteristics	Biochemical Characteristics
Presperm	Urethral glands	Watery	High NaCl content, no ergothioniene, citric acid or GPC
Sperm-rich	Epididymal and ampullary glands	Milky, nonviscous	High sperm concentration, ergothioneine and GPC; little NaCl and citric acid
Postsperm	Seminal vesicle	Highly viscous	Low sperm concentration High content of citric acid Low ergothioneine and GPC
Penile drip	Tail-end sample	Watery	No spermatozoa, lacks any secretory products of epididymis, ampulla and seminal vesicles (eg, GPC, citric acid)

(Adapted from Mann, T., Leone, E. and Polge, C., 1965. The composition of the stallion's semen. J. Endocrinol. 13, 279; Mann, T., Short, R. V. and Walton, A., 1957. The tail-end sample of stallion semen. J. Agric. Sci. Camb, 49, 301.)

seem to require fewer mounts and a shorter reaction time and possess a slightly larger volume of gel-free semen than ejaculates without gel.

Fructose is present in negligible quantities in stallion semen, whereas relatively large quantities of ergothioneine and citric acid are present. Although the majority of the lipids of stallion spermatozoa are phospholipids, the percentage of phospholipid is much lower than that for the bull or the boar. The lower percentage of phospholipid may be related to the great sensitivity of stallion spermatozoa to stress as compared with spermatozoa of other species.

Semen characteristics are influenced by the degree of sexual stimulation, frequency of ejaculation, age, testicular size and method of semen collection. Season of the year influences the physical and biochemical characteristics of semen as well as blood hormone levels, sexual behavior and fertility of both sexes. Spermatozoal output and libido of stallions are greatest during spring and summer and least during fall and winter. These changes in reproductive capacity coincide with the natural breeding season of mares.

The concentration of plasma testosterone of stallions is also influenced by season and may mediate the patterns of seminal and behavioral characteristics.

The minimal number of motile spermatozoa necessary for maximal conception rate has not been established, although 500×10^6 motile spermatozoa per insemination have been recommended.

Ejaculation

During ejaculation the urethral diverticulum of the penis is in close apposition to the external cervical orifice of the mare. Semen is ejaculated under high pressure directly into the uterus. The final seminal jets ejaculated when erection is ceasing and the penis is being withdrawn are probably deposited in the vagina. The patterns of ejaculation and emission have been carefully studied by Tischner et al (1974).

The process of emission is variable since the number of jets per ejaculate varies from five to ten with an average of eight. The early jets occur under high pressure in a stream with characteristic spatter. The later jets, accompanied by declining erection and withdrawal of the penis from the vagina, are associated with low pressure. Of the total time of ejaculation, 24% involves actual emission of semen; the rest comprises intervals between successive seminal jets. The first three jets contain 80% of the ejaculated spermatozoa. The total number of spermatozoa and the ergothioneine content gradually decrease in successive jets (Tischner et al, 1974). The terminal jets (from four to ten, with low concentrations of sperm cells and ergothioneine, consist mostly of the so-called mucous fraction and correspond to fraction three described by Mann et al (1957).

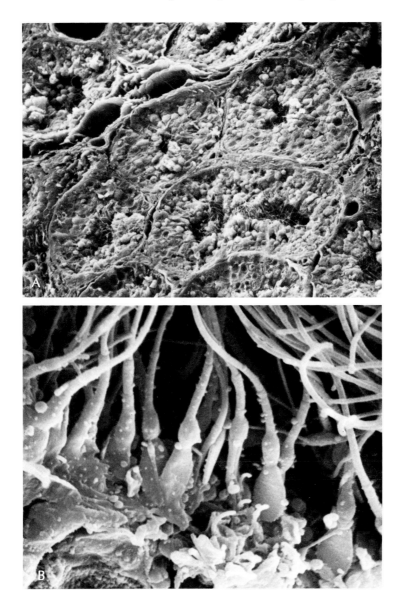

Fig. 19–2. Scanning electron micrograph of equine testis. *A*, Note irregular shape of seminiferous tubules and tails of spermatozoa protruding in the lumen. (440 ×) *B*, Spermiation: spermatozoa with cytoplasmic droplets are released from the Sertoli cells (5,280 ×). *(Courtesy Dr. Larry Johnson, 1978. Fertil. Steril. 17, 21)*

Table 19-3.   Main Characteristics of the Reproductive Cycle in the Mare

	Reproductive Parameters	Value Range (Average)
Estrous Cycle	Duration of estrus	2–11 days (6)
	Length of estrous cycle—normal	16–24 days (22)
	short	10–16 days
	long	29 days
	Number of estrous cycles/year	3–16
	Length of diestrus	15 days
Ovulation	Time from onset of estrus to ovulation	4.3 ± 1 day
	Time from end of estrus to ovulation	1–2 days
	Size of preovulatory follicles (cm)	3–6 (4.5)
	Frequency of twin ovulations	3–30%
	Incidence of ovulations without estrus	50%
Corpus Luteum	Corpus luteum palpable after ovulation	1–18 days (9)
	During nonpregnancy	2–3 months
Pregnancy	Stage of pregnancy when corpus luteum regresses	5–7 months
	Stage when ovulation and new corpora lutea formed in gestation under influences of PMSG	40–60 days
	Interval from foaling to foal heat	5–18 days

ESTROUS CYCLES

The main characteristics of the reproductive and ovulation cycles in the mare are summarized in Table 19–3. The unique reproductive parameters of the mare are compared to those of the cow, ewe and sow (Table 19–4).

Estrus

During estrus the vulva becomes large and swollen; the labial folds are loose and readily open on examination. The vulva becomes scarlet or orange, wet, glossy and covered with a film of transparent mucus. The vaginal mucosa is highly vascular, and thin watery mucus may accumulate in the vagina. The cellular components of the vaginal smear have no value for detecting estrus. During estrus, the cervix dilates enough to admit two to four fingers; during diestrus only one finger can be inserted (Fig. 19–3).

During estrus, the mare assumes a stance characteristic of urination. The tail is raised, urine is expelled in small amounts and the clitoris is exposed by prolonged rhythmic contractions.

The duration of estrus varies among individuals and also among estrous cycles of the same mare. Long duration of estrus in the mare may be due to the following factors: (1) the ovary is surrounded mostly by a serous coat and some follicles have to migrate to reach the ovulation fossa to rupture; (2) the ovary is less sensitive to exogenous FSH than other species (eg, cattle, sheep), so that the preovulatory follicle requires a longer time to reach maximal size; and (3) the level of LH is low compared with FSH and this delays ovulation. The intensity of behavioral estrus varies both throughout the estrous period and among individual mares at comparable stages of the period.

The duration of estrus is prolonged in old mares, in mares underfed during the early part of the breeding season and during twin ovulations.

Follicular Growth and Ovulation

In cattle and sheep, there is a well-defined pattern of follicular development involving one or occasionally two follicles during the last part of the luteal phase

Table 19–4. **Unique Characteristics of the Reproductive Cycle of the Mare as Compared With the Cow, Ewe and Sow**

Reproductive Parameters	Mare	Cow-Ewe-Sow
Development of ovary	During fetal life: cortical tissue entirely surrounds the medullary portion During neonatal life: cortical tissue becomes confined to one area and is nearly surrounded by medullary tissue	Cortex of fetal ovary develops so that ovulation can occur at any point on the surface except at the hilus
Site of ovulation on ovarian surface	Follicles can only ovulate through the surface of the ovary adjacent to cortical tissue ("ovulation fossa")	
Accessory ovulation during luteal phase of estrous cycle	Ovulation unaccompanied by estrus occurs	
Pattern of follicular growth	A small group of follicles develops during late diestrus, some of which enlarge differentially to ovulate during subsequent estrus; remaining follicles may ovulate during early luteal phase and others regress without ovulation	One or two follicles develop during last part of luteal phase of estrous cycle
Gonadotropins associated with ovulation	Prolonged rise of LH surge during estrus which continues after ovulation	Brief surge of LH occurs 12 to 24 hours before ovulation
Life span of corpus luteum during nonpregnancy	Spontaneous prolongation of CL lifespan is common; corpora lutea that fail to regress at normal time persist for two months	Uncommon
FSH content of pituitary	High	Low
Ovarian response to FSH	Low	High
Egg transport in oviduct	Mare can discriminate between fertilized and unfertilized egg. Nonfertilized eggs may be retained in oviducts during pregnancy and nonpregnancy up to seven months	Fertilized and unfertilized eggs are transported through the oviduct within three days
Presence of placental hormone (PMSG)	PMSG found in large quantities; secreted by endometrial cups during early pregnancy	Not recorded
Secondary corpora lutea	Remain functional until midpregnancy	Nil

Adapted from Stabenfeldt, G. H., Hughes, J. P., Evans, J. W. and Geschwind, I. I., 1975. Unique aspects of the reproductive cycle of the mare. J. Reprod. Fertil. Suppl. 23, 155.

of the ovarian cycle, culminating in ovulation. In the mare the pattern is less well defined, since a small group of follicles develops during late diestrus, one of which enlarges differentially and which matures and ovulates during the ensuing estrus. Meanwhile, the remaining follicles may continue to develop; one or more may ovulate during the early luteal phase, and the others regress without ovulating (Hughes et al, 1972). This pattern of follicular growth causes a high incidence of multiple ovulations.

Ovulation in the mare occurs, at the ovulation fossa, equally in the right and left ovaries (Fig. 19–4). The frequency of twin ovulations may range from 3% to 30% depending on the breed and season of year, being rare in pony breeds and during the spring. The incidence of twin

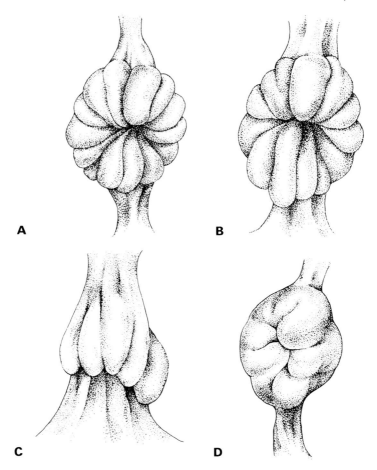

Fig. 19–3. Diagrammatic illustration of the os cervix in the mare as seen by clinical examination during estrous cycle and pregnancy. *A,* Diestrus ten days after ovulation: the cervix is hard and the folds are well defined. *B,* Onset of estrus (first day); the cervix is somewhat swollen, the folds are shallow and less defined, and the orifice of the cervix is opened. *C,* End of estrus (sixth day): the cervix is markedly swollen and relaxed and the upper folds hang down in a "membranous" appearance and cover the orifice. *D,* Pregnancy (four to six months): the cervix is hard and budlike in appearance and is covered with pasty mucus and the orifice is tightly closed.

pregnancy varies from 1% to 5%. Usually one member of the twins dies before birth and twin live births occur only in a few cases.

Preovulatory follicles increase in size from 3 to 6 cm between days 6 and 1 before ovulation, but just before ovulation they measure 3 to 6 cm in diameter. Rarely can ovulation occur from follicles less than 2 cm in diameter. A few anovulatory follicles develop to 10 cm in diameter, persist up to two months and then regress. They do not affect the normal length of

estrus or inhibit ovulation (Hughes et al, 1975). Some follicles soften and then become turgid again before ovulation, whereas others remain turgid throughout. Palpation of the ovary just before and after ovulation occasionally may provoke pain. In half of estrous periods, two or more follicles other than the one about to ovulate are over 25 mm in diameter and are palpable. Half of these regress before ovulation and the rest regress after ovulation (Hughes et al, 1975).

Ovulation occurring during the luteal

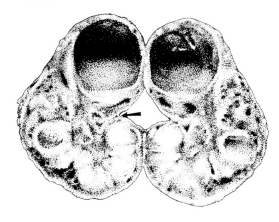

Fig. 19–4. Ovary of the mare cut in halves. Note ovulation fossa (arrow).

phase of the estrous cycle is not accompanied by estrus and may cause confusion during routine rectal examinations of mares. These ovulations do not affect plasma progestogen concentration or any other characteristic of the estrous cycle. It is not known whether diestrous ovulations are fertile.

Owing to the lengthy estrus in mares, copulation must be synchronized with ovulation in order to ensure fertilization. The time of ovulation can be detected by repeated rectal palpations. Prior to ovulation, one of the ovaries enlarges and the developing follicles occupy most of the ovarian stroma. The injection of 2,000 IU of HCG hastens the time of ovulation and shortens the duration of estrus. Repeated injection of HCG does not appear to cause refractoriness.

Corpus Luteum

Most ovulations occur on days 3, 4 or 5 of estrus, 24 to 48 hours prior to the end of behavioral estrus (ie, the time of ovulation is more closely related to the end than to the onset of estrus). Follicular size and the day of ovulation are consistent in mares from one cycle to another. Ovulation takes place after the first meiotic division. The fertility of mating gradually rises to a peak

about two days before the end of estrus and then falls sharply on the last day.

The ruptured follicle can be palpated up to 24 hours after ovulation as a soft fluctuant area. The developing corpus luteum, however, cannot be detected by rectal examination 48 hours after ovulation because it develops within the ovarian stroma.

The corpus luteum reaches only one-half to three-fourths the size of the follicle at the time of ovulation. The maximal size is attained at 14 days, when luteal cells enlarge and have a peripheral vacuolation (Fig. 19–5). Spontaneous prolongation of the corpus luteum, accompanied by follicular activity and without any signs of estrus for periods of two to three months, is common (Hughes et al, 1975). Corpora lutea that fail to regress at the normal time persist for about two months and are characterized by having a white connective tissue core. Most of the granulosa cells that undergo luteinization 24 hours after ovulation become secretory.

Endocrine Control of Estrous Cycles

FSH. In normally cycling mares, two FSH surges occur at approximately 20 and 11 days prior to ovulation. Both FSH and LH concentrations surge around the time of ovulation. FSH and LH surges cause development of follicles from less than 2 mm in diameter through to ovulation. The late estrus/early diestrus surge of FSH appears to initiate development of up to 20 follicles. The mid-diestrus surge may be important for the subsequent development of follicles destined to ovulate 10 to 13 days later.

The ovary of the mare is less sensitive to FSH than is that of the cow, ewe and goat. Injection of massive doses of PMSG during the nonbreeding season is ineffective for inducing ovulation of follicles. Injection of PMSG toward the end of the estrous cycle is also ineffective in promoting follicular development. The administration of HCG shortens estrus and hastens ovulation. Injection of HCG is frequently used in an attempt to decrease the number of inseminations or matings/cycle and to synchronize ovulation time more accurately with mating.

LH. The regulation of LH secretion involves two mechanisms: (1) a central nervous system pituitary component responsible for a basal circannual rhythm of LH release, entrained to an environmental parameter (most probably photoperiod) and independent of ovarian influences; and (2) an ovarian (steroidal) component that modifies the primary LH rhythm during the breeding season (Garcia and Ginther,

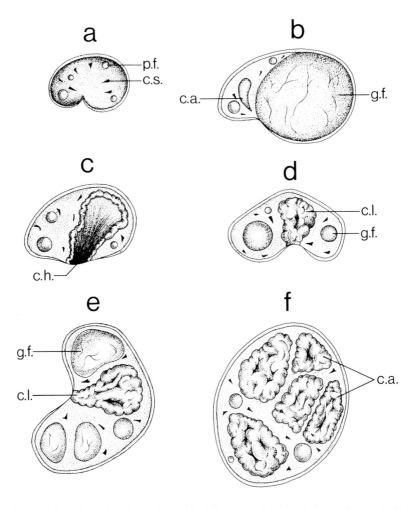

Fig. 19–5. Cross section of ovaries (drawn to scale) of the mare at different stages of reproduction.
A, Nonbreeding season. The ovary is small and contains primary follicles *(p.f.)* and scars *(c.s.)* from degenerating corpora lutea.
B, Breeding season. During estrus, the ovary contains a mature graafian follicle *(g.f.)* and corpus albicans *(c.a.)* from previous ovulation.
C, Three days after ovulation. The corpus haemorrhagica *(c.h.)* develops from the walls of ruptured graafian follicle.
D, Ten days after ovulation. Fully developed corpus luteum *(c.l.)*; graafian follicles *(g.f.)* start to develop for subsequent cycle.
E, Pregnancy—60 days. The corpus luteum of pregnancy *(c.l.)* is maintained and graafian follicles *(g.f.)* develop as a result of circulating PMS.
F, Pregnancy—80 days. Accessory corpora lutea *(c.a.)* develop from the unruptured follicles.

1978). In ovariectomized mares, concentrations of LH follow a seasonal pattern (low basal concentrations in winter; high basal concentrations in summer). The cyclic changes in LH concentrations during the breeding season seem to result from the modulation of high basal LH concentrations by ovarian steroids.

Levels of LH increase by the onset of estrus, reaching a maximum one to two days after ovulation. This rise in LH may stimulate maturation of the follicle throughout estrus. An inverse relationship between progesterone and LH levels occurs, suggesting a negative feedback, but there is no such relationship between FSH and progesterone concentrations. LH levels show a prolonged rise during estrus that continues after ovulation. A significant decrease from peak levels is not observed until the third day after ovulation, from which time levels continue to decline toward diestrus values.

The pattern of plasma LH in the mare differs from that in other species and it is possible that persistence of high concentrations of LH results from a long half-life of the endogenous LH. This in turn may be responsible for the relatively large number of second ovulations detected in many estrous cycles (Geschwind et al, 1975). It seems that LH may be sensitive to progesterone feedback control in that levels do not start to rise until after the corpus luteum has regressed completely. On the other hand, high circulating progesterone levels apparently do not constitute an effective negative-feedback control mechanism once LH release has begun.

This administration of 2000 IU of HCG shortens estrus and hastens ovulation. Injection of HCG is used to accurately synchronize ovulation time with mating in an effort to decrease the number of inseminations or matings per cycle. Repeated HCG injection does not cause refractoriness.

In the mare, the uterus exerts its luteolytic effect on the corpus luteum primarily through a systemic utero-ovarian pathway. This conclusion is supported by the following findings in mares: (1) hysterectomy prolonged the life of the corpus luteum but unilateral hysterectomy failed to indicate that a local utero-ovarian relationship was involved; (2) local administration of prostaglandin $F_{2\alpha}$ ($PGF_{2\alpha}$), a postulated uterine luteolysin, into the uterus did not improve its luteolytic efficacy over systemic administration given intramuscularly; (3) the vascular anatomy of uterus and ovaries provides limited potential for the local transfer of a luteolysin between uterus and ovary through a veno-arterial pathway (Douglas et al, 1976).

Progesterone. In the absence of ovulation, with or without follicular growth, there are prolonged periods of low progesterone levels. The daily intramuscular injection of 100 mg or more of progesterone during the midcycle prevents estrus and ovulation, but a dose of 50 mg per day inhibits only estrus. The interval between termination of treatment and estrus appears to depend upon dosage. To counteract habitual abortion in the mare, it is usual to administer 250 to 500 mg every 10 to 30 days.

Induction of Estrus and Ovulation

Saline Infusion. The technique of intrauterine saline infusion has been used routinely to induce estrus in anestrous mares. Anestrous mares are affected only near the beginning and end of the breeding season when anovulatory heats are induced. Diestrous mares infused between days 5 and 9 return to heat four days earlier than expected, and induced estrus is accompanied by ovulation. Mares in prolonged diestrus may show ovulatory heats within three to nine days of infusion (Arthur, 1975). Repeated infusions are clinically harmless, but postinfusion bacteriologic swabs from the uterus are positive.

Synchronization of Estrus. Mares are

usually mated or artificially inseminated after estrus has been detected with a teaser stallion and follicular development has been assessed by rectal palpation of the ovaries. Synchronization of estrus and ovulation would allow mares to be inseminated at predetermined times without the need to detect estrus or palpate the ovaries.

Two treatments effectively influence ovarian activity in the mare: (1) induction of luteolysis with prostaglandin $F_{2\alpha}$ ($PGF_{2\alpha}$) or analogues of $PGF_{2\alpha}$ and (2) induction of ovulation during the follicular phase by an injection of HCG.

Prostaglandins. Of the farm species studied, the mare is most sensitive, on a body-weight basis, to the luteolytic effects of systemically (intramuscular or subcutaneous) administered $PGF_{2\alpha}$. Systemically administered $PGF_{2\alpha}$ is as effective in causing luteolysis in hysterectomized as in intact mares, indicating that the principal site of action of exogenous $PGF_{2\alpha}$ is not at the uterine level.

Prostaglandin $F_{2\alpha}$ ($PGF_{2\alpha}$) and its analogues have been used to control the estrous cycle of the mare. The treatment causes a prompt cessation of secretion by the corpus luteum as indicated by a rapid fall in plasma progesterone levels. The infusion of 10 mg of $PGF_{2\alpha}$ on days 7 to 9 after ovulation causes a sharp fall in plasma progesterone levels and induces estrus and ovulation. This induced estrus is longer than the natural cycle but the time of ovulation in relation to the end of estrus is normal. The time of return to estrus following luteolysis does not depend on the amount of $PGF_{2\alpha}$. Luteolysis can be induced as early as day 5 following natural ovulation (Oxender et al, 1975). Prostaglandin $F_{2\alpha}$ ($PGF_{2\alpha}$) and its synthetic analogues are luteolytic in mares and cause abortion.

Intramuscular administration is as effective as subcutaneous administration and 1.25 mg $PGF_{2\alpha}$ is the minimal effective systemic dose for inducing luteolysis.

Administration of $PGF_{2\alpha}$ into the uterus or directly into the corpus luteum does not improve the luteolytic efficacy of the intramuscular injection of $PGF_{2\alpha}$.

Ova and Ova Transport. At ovulation the ovum is without corona radiata but is enclosed in a large irregular gelatinous mass of ovarian origin, which separates from the egg within two days. Fertilized ova are transported in the uterus, whereas unfertilized ova are trapped for several months in the isthmus of the oviduct. The ovum undergoes degeneration and fragmentation during the ensuing months. If a mare has a succession of sterile estrous cycles followed by a fertile mating, the developing embryo may outrun the unfer-

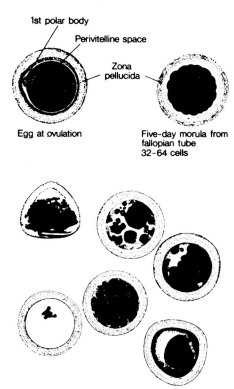

Fig. 19–6. A normal horse egg before and after fertilization, and degenerating oviductal eggs ranging in age from 1 day to 7.5 months. The mechanisms by which the equine unfertilized ova are trapped in the oviduct are unknown. *(From van Niekerk and Gerneke, 1966. Onderstep. J. Vet. Res., 33, 195; redrawn by Short, 1972. In: Reproduction in Mammals. Austin and Short [eds.], Cambridge, Cambridge University Press)*

tilized eggs trapped in the oviduct and enter the uterus.

Nonfertilized eggs may be retained in the oviducts of pregnant and nonpregnant mares for up to seven months (Fig. 19–6). This indicates that the mare can discriminate between fertilized and nonfertilized eggs, allowing the fertilized egg to pass into the uterus while retaining nonfertilized eggs. Retained eggs are more common in heavy than in light breeds and are found more frequently in early than in late pregnancy.

The time at which fertilized ova arrive in the uterus in the mare is much later (>144 hours) than in the cow, and the cleavage stage of equine ova at arrival is more advanced than that in cattle. Transuterine migration of ova occurs in 50% of cases.

GESTATION

Ovarian Function

Four distinct stages of ovarian function are recognized.

1. During early pregnancy, a single corpus luteum verum is present; it was believed that this regressed at approximately the 40th day of gestation, but according to recent observations (Squires et al, 1974), the primary corpus luteum persists beyond day 40.

2. Between the 40th and 150th day of pregnancy ovarian activity occurs. As many as 10 to 15 follicles (over 1 cm in diameter) undergo luteinization to form the accessory corpora lutea. The recovery of recently ovulated ova from the oviducts suggests that some of these follicles ovulate even though unfertilized ova are retained in the oviducts for several months. Usually each ovary contains three to five accessory corpora lutea (Fig. 19–5).

3. From the fifth to the seventh month, though both primary and secondary corpora lutea and the large follicles regress completely, the mare does not show signs of estrus due to placental secretion of progesterone until the end of gestation.

4. From the seventh month onward only vestiges of the corpora lutea and small follicles are present, but during the last two weeks of gestation follicular activity commences in preparation for the postpartum estrus ("foaling heat").

The primary and secondary corpora lutea and the placenta all contribute to the total progesterone pool during pregnancy. The corpora lutea progesterone concentration in the peripheral plasma is closely correlated with morphologic changes in the corpora lutea.

Similarities and differences in ovarian function observed between pregnant and hysterectomized mares suggest that, while PMSG does not appear to stimulate follicular development, it does prolong the life span and stimulate the secretory activity of the primary corpus luteum and induced ovulation and/or luteinization of secondary follicles in pregnant mares (Squires and Ginther, 1975). Ovariectomy does not terminate pregnancy if performed after day 70 of pregnancy.

Placenta

The placenta of the mare is classified as diffuse, microcotyledonary and epitheliochorial (Fig. 19–7). The outer surface of the chorion is closely studded with tufts of branching villi that enter into corresponding invaginations in the endometrium to form small globular structures known as microcotyledons. Microcotyledons, which are a distinctive feature of the mature equine placenta, are fully formed by the fifth month of gestation. The primary folds of trophoblast become elaborately subdivided as gestation proceeds. These changes are reflected in the structure of the maternal crypts, which receive the fetal villi.

Within each microcotyledon, chorionic and uterine epithelia are in intimate con-

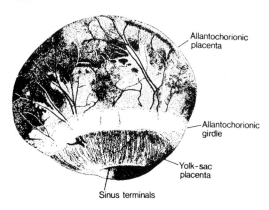

Allantochorionic
placenta

Allantochorionic
girdle

Yolk-sac
placenta

Sinus terminals

**Fig. 19–7.   Equine conceptus at seven weeks of gesta-
tion.** *Top,* A drawing of an equine embryo and its
membranes in section and in surface view. The conceptus is about the size of an orange, and cells become
detached around the allantochorionic girdle at this
stage to invade the endometrium and form the endometrial cups. *(Short, 1972. In: Reproduction in
Mammals. Austin and Short [eds.], Cambridge, Cambridge University Press)*
Bottom, The chorion is oval rather than cyclindrical
as in the cow, and the chorion is more distended with
fluid, which facilitates early pregnancy diagnosis. Note
the amnion in relation to yolk sac and chorioallantois.
*(Arthur, 1964. Wright's Veterinary Obstetrics, London,
Bailliere, Tindall & Cox)*

tact and a microvillous junction is formed
at the fetal-maternal boundary.

Placental Barrier. The following mechanisms control the passage of gases, nutrients, ions or hormones across the placental barrier: (1) those related to diffusion
across the placenal tissues, ie, permeability, diffusion distances, total surface area
and concentration gradients between
maternal and fetal blood vessels; (2) rates
of supply and removal on either side of
the placenta, the presence or absence of
specialized exchange areas, the direction
of blood flow in these areas and the existence of shunts or unequal flows; and (3)
specialized mechanisms that assist the
passage of substances and, in the case of
oxygen, differences in hemoglobin O_2 affinity and O_2 capacity between fetus and
mother (Silver and Comline, 1975).

The primary corpus luteum that develops at the site of the fertile ovulation
regresses about day 160 of gestation and is
replaced by secondary corpora lutea. The
primary and secondary corpora lutea and
the placenta all contribute to the total
progesterone pool during pregnancy. The
luteal progesterone concentration in the
peripheral plasma is closely correlated
with morphologic changes in the corpus
luteum.

Similarities and differences in ovarian
function observed between pregnant and
hysterectomized mares suggest that,
while PMSG does not appear to stimulate
follicular development, it does prolong
the life span and stimulate the secretory
activity of the primary corpus luteum and
induce ovulation and/or luteinization of
secondary follicles in pregnant mares
(Squires and Ginther, 1975).

Endocrine Profile

Steroids. Plasma progestogen levels decline in midpregnancy and remain low
until a few days before parturition, when
they increase again. During pregnancy
there are two peaks of plasma progestogens. The first, which occurs during the
third month, coincides with high levels of
PMSG and is probably produced by the
endometrial cups or the secondary corpus
luteum. The second peak occurs in the
11th month and probably represents the
secretion of placental progestogens. Blood
estrogen levels are highest at midgestation and decrease significantly before parturition. The pregnant mare, moreover,
excretes some other estrogens that are
peculiar to the mare and other equids.

During the fourth to eighth months,
estrone levels exceed 100 ng/ml and then
decline toward parturition. The high concentrations of estrone in midpregnancy
are associated with gradually increasing
concentrations of equilin, which tend to
plateau after the sixth month at just under
100 ng/ml and decline only in the last
month of pregnancy.

Gonadotropins (PMSG). The endometrial cups, which are present from the

second to the fourth months of pregnancy, are the source of the gonadotropin PMSG, which circulates in the maternal blood between 40 and 130 days of gestation. The maternal ovaries are thus stimulated, forming ovarian follicles, many of which ovulate and form accessory corpora lutea, while others luteinize without ovulating. The accessory corpora lutea persist until about 180 days of gestation and then degenerate. During the remainder of gestation, no lutein tissue is formed. If a mare aborts about the 35th day of gestation, when the fetal cells have already invaded the endometrium, the endometrial cups continue to develop normally and secrete PMSG, and the corpora lutea that are present may be maintained.

Several factors influence PMSG levels: season, maternal size, parity and fetal genotype. A high level of PMSG occurs in the serum of a mare carrying twin fetuses if a set of endometrial cups develops in both uterine horns. The sudden fall in the blood level of this hormone in a mare which aborts compared to the maintenance of a high blood level in another in which resorption has taken place suggests that the conceptus may be necessary for the continued function of this hormone.

The fetal genotype has a pronounced effect on PMSG concentration in the mother. Blood serum of mares bred to a jack donkey contains only about 10% of the PMSG concentration in mares bred to a stallion. It appears that the allantochorion may provide a stimulus, possibly chemical in nature, that regulates the secretory activity of the endometrial cups.

PMSG disappears from the blood of mares carrying mule fetuses by the third month of gestation, which is much earlier than in mares carrying horse fetuses. PMSG concentration is high in a donkey mare carrying a hinny fetus. In surgical removal of the conceptus from the uterus on or after day 38 of gestation or fetal death occurring after day 40, PMSG production continues normally as though the

mare is still pregnant (Hay and Allen, 1975). The endometrial cups, once functional, have a life span that is independent of the conceptus.

PMSG may play a major role in stimulating secondary follicular development and accessory CL development during the first half of gestation, but similarities in follicular development in hysterectomized and pregnant mares indicate that other factors may be responsible for follicular development in pregnant mares (Squires et al, 1974).

FOALING

The time of foaling, which usually takes place between nightfall and daybreak, seems to be influenced by photoperiods and quietness in the stable. In England, 86% of foalings occur between 1900 and 0700 hours, with a maximal incidence between 2200 and 2300 hours.

The imminence of foaling is suggested by the degree of mammary hypertrophy, waxing of the teats and possibly the discharge of milk from the udder. The best indication that the first stage has begun is the onset of patchy sweating behind the elbows and about the flanks. This sweating commences about four hours before foaling and increases as the stage progresses. The tail is frequently raised or held to one side.

Induced Parturition

Parturition can be induced by various doses of estrogen, prostaglandins (PGF$_{2\alpha}$) and oxytocin. The time of appearance and degree of expression of the major clinical signs of parturition, and the time for completion of delivery and the passage of the placenta are influenced by increasing doses of oxytocin. Estrogen is useful in softening and relaxing the cervix when it is tight, but is not essential to induction when the cervix is already soft and dilating.

Progesterone (500 mg/day) administered daily from day 318 of pregnancy shortened gestation while estrogen (50 mg/day) administered on the same schedule as progesterone had no such effect. Parturition can be induced in the mares of large-sized breeds by daily treatment with 100 mg of dexamethasone for four days. However, parturition has not been successfully induced in pony mares. Single or low doses of dexamethasone do not induce parturition. The mode of action of dexamethasone in causing premature birth in the equine is unknown. Dexamethasone seems to induce a normal parturition since all prepartum signs, including vulval dilatation, mammary development and sinking in the flank area, are normal.

Indications for elective induction of parturition in the mare include delayed parturition due to uterine atony, prolonged gestation, prevention of injury at foaling, preparturient colic, injury to the mare and impending rupture of the prepubic tendon, and obtaining colostrum-deprived foals for research purposes. Fertility is not adversely affected by induced parturition.

Postpartum Estrus (Foal Heat)

Postpartum estrus usually occurs 5 to 15 days after foaling. Some mares, however, may show estrus as late as 45 days after parturition; such estrus may have been preceded by a quiet ovulation. The interval between postpartum estrus and the following estrus may be affected by the milk yield.

Breeding at the postpartum estrus may cause an increased percentage of abortion, dystocia, stillbirth and retained placenta. This may be due to the introduction of bacteria into the uterus before it is completely involuted and while it still lacks contractility.

Involution of the uterus after normal foaling is rapid. Regression in size is almost complete by the first day of "foal heat." The relatively low conception rate from copulation during this period appears to indicate that involution of the endometrium is not complete in all mares.

Twinning

Twinning is rare. In most dizygotic twin pregnancies there is invagination of the adjacent allantochorions. Identical blood groups are found in twin foals, indicating chimerism and macroscopic or microscopic anastomosis between both chorions. Stillbirth is frequent in twin pregnancies and only one-half of the born foals survive. Twin pregnancies are not desirable in view of the high peri- and postnatal losses and the poor viability and racing performance of twins. Subsequent conception rate is not affected by twin birth if the mare foals at full term but it decreases following abortion.

REPRODUCTIVE FAILURE IN MARES

Conception and foaling percentages differ widely. In a few pony studs conception rate and foaling rates are 100%. In most studs the foaling rate is much lower than conception rate. In general, conception rate is influenced by the breed, nutrition and age of dam and management practices. For example, conception rate from first service is lower in thoroughbred mares than in most other breeds. The conception rate of foaling mares served during foal heat is lower than the first-service pregnancy rate of foaling mares served after the foal heat.

Reproductive efficiency in horses is low compared to that in other farm animals. Reproductive failure in mares may be due to hormonal dysfunction, genital infection, inadequate management, and improper detection of estrus.

Estrous Irregularities

Irregularities in the estrous cycle are associated with seasonal changes in the photoperiod, nutrition and climate. Variations in patterns of cyclic behavior include cycle length and estrous behavior, failure of ovulation and follicular development, and spontaneous prolongation of the corpora lutea. "Quiet ovulation," anovulatory estrus, "split estrus" and prolonged estrus are not uncommon.

Split estrus is commonly observed in healthy mares during the early spring. In such cases, behavioral estrus is interrupted by an interval of sexual nonreceptivity. Fairly frequently the follicle that is present at the beginning of the first part of the split estrus continues its growth normally throughout the period and will ovulate.

Prolonged estrus, lasting 10 to 40 days, may also occur in early spring and in mares used for heavy draft.

Intrauterine infusion of saline may induce estrus in some mares with irregular or no ovarian activity. The distention of the uterus by the infused saline acts as a stimulus to release pituitary gonadotropins. Intrauterine saline infusion in diestrus shortens the ovulatory interval by inducing premature luteolysis but during estrus it has little effect on cycle length.

Anestrus

There are three forms of anestrus: (1) a winter anestrus that lasts 11 weeks in which follicles rarely reach 35 mm before regressing without ovulation and plasma levels of progestogen are 1 ng/ml; (2) a corpus luteum that persists for 5 to 13 weeks with elevated plasma progestogen levels, usually in November and December; (3) cyclic ovulatory patterns unaccompanied by estrus (Hughes et al, 1975).

Evans and Irvine (1977) designed a regimen of exogenous gonadotropin releasing hormone (GnRH) to reproduce in acyclic (anestrous) mares the sequence of changes in plasma FSH and LH that occur in the normal cycle, in an attempt to induce follicular development culminating in ovulation. A single injection of GnRH in anestrous mares will cause an increase in serum FSH to 3.7 times baseline, which is comparable to the peak increases occurring during the estrous cycle; however, the induced increase in LH is much less than that of the cyclic peak.

GnRH in combination with appropriate progesterone treatment in the acyclic mare treated toward the end of the non-breeding season can consistently induce normal cyclic pituitary and ovarian activity culminating in ovulation (Evans and Irvine, 1977).

Prenatal Mortality

Prenatal mortality is frequent in lactating mares mated early in the season or mated after foal heat. Prenatal death is also common in certain horse families. Higher prenatal mortality in yearlings compared to adult mares may be due to inadequate progestogen levels to maintain pregnancy, immaturity of the yearlings, their greater nutritional requirements for growth and maintenance and the physical stresses imposed upon them by the husbandry procedures. Infection of the fetus *in utero* is a significant cause of prenatal mortality in the later months of pregnancy.

Ovarian Neoplasms

Neoplasms of the equine reproductive organs are uncommon whereas teratoma and granulosa-cell tumors are not uncommon. Ovarian tumors have been classified according to their histogenetic origin: (1) stromal and derived from the mesenchymal core of the germinal ridge, with possible contributions from connective and vascular tissues of the hilus; (2)

differentiated epithelial cells, such as the granulosa, theca and lutein cells, a syndrome associated with hormonal imbalance; (3) dysgerminoma, a rare tumor similar to the seminoma of stallions, and (4) cystadenoma derived from the surface coelomic epithelium and of tubular ingrowths in the gonadal cortex.

Granulosa-cell tumor is a common neoplasm that affects only one ovary. Its incidence increases with age, and it is usually associated with overproduction of steroid hormones leading to clinical signs of excessive estrogenization. The surface of the granulosa-cell tumor is smooth, the cut surface may be either solid or cystic, and its white or yellow color depends on the amount of ovarian lipids. The teratoma, a true tumor composed of multiple tissues foreign to the part in which they arise, are noted in mares one to five years of age.

Endometrial Cysts and Related Pathologic Conditions

Focal enlargements of the uterus palpable through the rectal wall have various causes. Lymphatic lacunae seem to be common in older mares and may occasionally give rise to a large endometrial cyst or cause widespread change throughout the uterine horns. In cases of myometrial atony and endometrial atrophy, repeated uterine infusions of hot (40° to 45°C) saline with or without antibiotics may enhance myometrial tone. The response is rapid but transient.

Genital Infections

Uterine infections cause shortening of the estrous cycles. Experimental intrauterine inoculation of *Streptococcus zooepidemicus* into mares during diestrus may shorten the length of the estrous cycle. Chronic endometritis includes latent, purulent, hypertrophic and atrophic types and pyometra. Several factors may cause endometritis in the mare: breeding

at the wrong stage of the estrous cycle; too frequent matings during estrus; poor hygiene at mating or foaling; excessive work during the breeding season, especially at estrus; atony of the uterus as a result of dystocia, general debility or old age; and retention of part of the placenta at abortion or foaling. Prognosis and treatment of infertile mares are based on the breeding history, clinical data and histopathologic findings of endometrial biopsy and cervical mucus.

The internal reproductive organs can be examined by endoscopy. A human rhinolaryngoscope can be used. The induction of artificial pneumoperitoneum and the installation of an endoscopic peritoneal fistula device are used to prolong the observation period. New concepts of endoscopic instrumentation and surgical techniques have enlarged the scope of medical endoscopy in clinical and basic research.

Excessive or prolonged intrauterine infusion of antibiotics in the treatment of chronic endometritis in mares is usually followed by the establishment of fungi and yeasts in the genital tract.

Intrauterine Growth Retardation. Virus infection of the fetus may lead to stunted development in some species, although viruses have not yet been implicated in the pathogenesis of fetal growth retardation in the horse. However, chronic placental infection may interfere with placental exchange and thereby impair fetal growth. For example, birth weight is greatly reduced in cases of rhinopneumonitis and fungal placentitis.

Excessive Length of Umbilical Cord

The length of the umbilical cord is not correlated with gestational age, foals' body weight, sex or viability, dams' age or parity, or surface area, width or length of the allantochorion. However, cord length is correlated with weight of the allantochorion and allantoamnion and the length of the nonpregnant horn. Several pathologic conditions of unknown origin cause excessive elongation of the umbilical cord. Reproductive failures associated with these conditions include: (1) strangulation by the cord around the fetus, (2) excessive twisting around the amniotic or allantoic portion of the fetus, causing vascular occlusion and/or urinary retention, or (3) necrosis of the chorioallantois at the cervix. Strangulation of the fetus may cause abortion, with deep grooves present around the head, neck, thorax and back with apparent local edema.

Mild urachal obstruction is compatible with normal pregnancy. The umbilical cord may become twisted with multiple urachal dilatations that prevent normal sealing of the bladder apex at birth. Excessive twisting of the umbilical cord threatens the integrity of the urachus and umbilical vessels; the effects depend on the completeness, duration and site of the compression.

Abortion

The average abortion rate in mares is high (10%). This may be due to peculiarities in the hormonal balance of the mare during pregnancy. Abortion is lowest between three and six years of age and most abortions due to infection occur during the later stages of pregnancy. During the fifth and tenth months of pregnancy, the mares are endocrinologically susceptible to abortion owing to hormonal deficiencies. It is recommended to avoid sudden changes in the diet or the amount of physical exercise at these times. Postabortum conception rate is low, especially in older mares.

Abortion in older mares may be due to uterine inadequacies. The normal endometrium is thrown into more or less longitudinal folds that vary in size with the estrous cycle. Implantation occurs at the junction of the body and horns of the uterus and the folds in this region can undergo atrophy after a number of pregnancies. The atrophic areas appear to have a reduced number of endometrial glands and, in some cases, contain collagen and/or inflammatory cells.

Neonatal Abnormalities and Neonatal Mortality

Microphthalmos (button eye) and entropion are regular congenital abnormalities in Thoroughbreds. The degree of microphthalmos is variable and may be so severe that the globe is obscured behind a well-developed third eyelid. The cornea is often distorted and pigmented and the palpebral aperture is usually smaller than normal; the condition can be bilateral or unilateral. Entropion is both congenital and hereditary, and may lead to corneal opacity and ulceration.

Barker Syndrome (Convulsive Foal Syndrome, or Neonatal Maladjustment Syndrome). This syndrome affects Thoroughbred foals, often those that have experienced an easy birth, and occurs within minutes after birth to 24 hours later. Clinical symptoms include jerking movements of the head, limb and body musculature, hyperexcitability, inability to stand, convulsions, opisthotonus and erection of the tail. There may also be collapse of the external nares, deep inspiratory movement and a barking sound. If able to stand, the foal may walk around aimlessly. Recovery occurs in about 50% of cases and is usually complete.

The syndrome is associated with necrosis of the cerebral cortex, diencephalon and brain stem, and with severe hemorrhage in the white and gray matter of the cerebral cortex and in the cerebellum.

Ocular Changes. The eye of the foal, which is open at birth, has a clear cornea and ocular media. The fundus is differentiated into tapetum lucidum and tapetum nigrum and is similar to the fundus of the adult horse. Ocular changes in the convulsive foal syndrome include asymmetry of pupils, apparent blindness, variable pupil size, scleral splashing and retinal petechiae. These clinical signs are not always present, even in severe cases. Small round retinal hemorrhages may occur, which are clearly visible as red dots against the background of the tapetal fundus. These hemorrhages occur at one and two days of age in convulsive foals and persist for only a few days.

Neonatal Mortality. Neonatal mortality may be a result of weakness of the mother or the foal or bacterial infection through the umbilical cord of the young. Proper management, clean stables for foaling and sanitary precautions at foaling are the common preventive methods of neonatal mortality.

Recommendations for Breeding Techniques

Careful testing for estrus with the stallion, routine examination of the vagina and rectal palpation of the reproductive organs may help to improve conception rate. Whenever possible, the time of ovulation should be predicted since the duration of estrus and the time of ovulation from the onset of estrus may differ between individuals. Conception rate depends primarily on the time and number of inseminations.

On occasion, the mare may strain (as in micturition and defecation) after mating and evacuate most of the semen from the uterus. This may be prevented by having the mare walk for a while after mating. The mare is susceptible to endometritis, especially after foaling, since the cervix of the mare is not a strong barrier to the introduction of bacteria. The mare is more prone to a deficiency in LH than other farm animals; such deficiency may be alleviated by the use of exogenous LH. The intravenous injection of 1500 to 3000 IU of HCG may cause ovulation during anovulatory estrus, provided the ovarian follicle is at least 3 cm in diameter.

REPRODUCTIVE FAILURE IN STALLIONS

Reproductive failure in stallions includes abnormal sexual behavior, ejaculatory disturbances and poor semen characteristics.

Abnormal Sexual Behavior and Ejaculatory Disturbances

Abnormal sexual behavior of stallions includes: (1) failure to attain or maintain an erection with poor or excellent libido, (2) incomplete intromission or lack of pelvic thrusts after intromission, poor libido or pain from injuries incurred during breeding, (3) dismounting at onset of ejaculation due to injury or pain, (4) failure to ejaculate in spite of a complete prolonged erection and repeated intromissions, (5) good ejaculation for a short time, but no further ejaculation without sexual rest, although libido remains high, (6) masturbation (Pickett and Voss, 1975). Impotent stallions respond well to retraining and recovery can be achieved without pharmacologic treatment. Masturbation in breeding stallions, an abnormal sexual behavior, may be treated by the use of a stallion ring, but this may cause hemospermia.

Ejaculatory disturbances are manifested differently in individual stallions, from normal copulation with or without occasional ejaculation. In most cases, penile erection is associated with several copulatory movements that terminate in complete or incomplete failure of ejaculation. Ejaculatory disturbances—transitory, intermittent or permanent—may occur during the first two or three breeding seasons or after several seasons of normal activity. To ensure that ejaculation has taken place after intromission, one holds the hand under the base of the penis. In the absence of ejaculation a few weak urethral waves may be felt, but when ejaculation takes place it feels like the contents of a 10-ml syringe being transported along the urethra. In a stallion of good fertility it is usual to feel about five of these waves; in the stallion with lower fertility, one sometimes feels only about one and a half waves.

Ejaculatory disturbance may arise from direct blocking of nerve impulses or from fatness, poor condition and exhaustion due to frequent services. Stallions usually also react strongly to unfamiliar surroundings, and psychic factors of this nature may inhibit the normal stimulation from the supraspinal centers. Ejaculatory disturbance may be due to failure of contraction of smooth muscles in the reproductive tract as a result of refractoriness of these cells to norepinephrine, exhaustion of the norepinephrine depots, or failure to release norepinephrine from the sympathetic nerve endings.

Poor Semen Characteristics

This may include one of the following: azoospermia, absence of sperm in the ejaculate; oligozoospermia, decreased sperm concentration per milliliter of semen; teratospermia, increased percentage of morphologically abnormal spermatozoa; asthenospermia, decreased sperm motility; hemospermia, hemorrhage in semen.

Hemospermia results from urethritis in the ejaculatory ducts. Hemospermia may occur occasionally, in isolated instances or in each ejaculate, irrespective of frequency of ejaculation. Affected stallions frequently require several mounts to ejaculate and often exhibit pain upon ejaculation. Semen quality as determined by motility, sperm numbers and morphology is usually unaffected and the cause of the infertility is unknown. Urethroscopic examination, urethrography, bacterial and viral cultures, biopsy, surgery of the urethra and histocytologic examination are used for diagnosis (Voss and Pickett, 1975). The exact cause and location of the hemorrhage should be known before treatment is initiated.

Several species of microflora have been found in the semen: *Pseudomonas* spp., *E. coli*, *Klebsiella*, *Aerobacter* (*Enterobacter*), *Proteus* spp., *Staphylococcus* spp., *Streptococcus* spp. and other gram-positive and gram-negative rods. These microorganisms, however, do not seem to affect the fertility. Most *Streptococcus* spp. in the semen are contaminants from the prepuce.

Recommendations for Breeding Techniques

The detrimental effect of frequent ejaculations on the number of sperm per ejaculate is pronounced in the stallion. In natural breeding, where several mares may exhibit estrus simultaneously, a stallion may copulate several times on one day; this causes a decline in fertility. The use of artificial insemination during such periods of mating will improve the conception rate. The stallion ejaculates directly into the uterus and in most cases there is little semen left in the vagina. The transfer of semen from the vagina to the uterus following insemination is seldom necessary, but this is recommended with wriggling mares or if a stallion is apt to dismount the mare with the penis still erect, as this pulls semen back into the vagina.

REFERENCES

Arthur, G.H. (1975). Influence of intrauterine saline infusion upon the estrous cycle of the mare. J. Reprod. Fertil. Suppl. *23*, 231.

Douglas, R.H., Del Campo, M.R. and Ginther, O.J. (1976). Luteolysis following carotid or ovarian arterial injection of prostaglandin F$_{2\alpha}$ in mares. Biol. Reprod. *14*, 473.

Evans, M.J. and Irvine, C.H.G. (1977). Induction of follicular development, maturation and ovulation by gonadotropin in releasing hormone administration to acyclic mares. Biol. Reprod. *16*, 452.

Garcia, M.C. and Ginther, O.J. (1978). Regulation of plasma LH by estradiol and progesterone in ovariectomized mares. Biol. Reprod. *19*, 447.

Geschwind, I.I., Dewey, R., Hughes, J.P., Evans, J.W. and Stabenfeldt, G.H. (1975). Plasma LH levels in the mare during the estrus cycle. J. Reprod. Fertil. Suppl. *23*, 207.

Hay, M. and Allen, W.R. (1975). An ultrastructural and histochemical study of the interstitial cells in the gonads of the fetal horse. J. Reprod. Fertil. Suppl. *23*, 557.

Hughes, J.P., Stabenfeldt, G.H. and Evans, J.W. (1972). Clinical and endocrine aspects of the estrous cycle of the mare. Proc. A. Meeting, Am. Ass. Equine Pract. pp. 119.

Hughes, J.P., Stabenfeldt, G.H. and Evans, J.W. (1975). The estrous cycle of the mare. J. Reprod. Fertil. Suppl. *23*, 161.

Mann, T., Short, R.V. and Walton, A. (1957). The "tail-end sample" of stallion semen. F. Agric. Sci. Camb., *49*, 301.

Oxender, W.D., Noden, P.A. and Hafs, H.D. (1975). Estrus, ovulation and plasma hormones after prostaglandin $F_{2\alpha}$ in mares. J. Reprod. Fertil. Suppl. *23*, 251.

Pickett, B.W. and Voss, J.L. (1975) Abnormalities of mating behavior in domestic stallions. In Equine Reproduction. I.W. Rowlands, W.R. Allen and P.D. Rossdale (eds), Oxford, Blackwell Scientific Publications.

Pickett, B.W., Faulkner, L.C. and Voss, J.L. (1975). Effect of season on some characteristics of stallion semen. In Equine Reproduction. I.W. Rowlands, W.R. Allen and P.D. Rossdale (eds), Oxford, Blackwell Scientific Publications.

Silver, M. and Comline, R.S. (1975). Transfer of gases and metabolites in the equine placenta: a comparison with other species. J. Reprod. Fertil. Suppl. *23*, 589.

Squires, E.L., Garcia, M.C. and Ginther, O.J. (1974). Effects of pregnancy and hysterectomy on the ovaries of pony mares. J. Anim. Sci. *38*, 823.

Squires, E.L. and Ginther, O.J. (1975). Follicular and luteal development in pregnant mares. J. Reprod. Fertil. Suppl. *23*, 249.

Tischner, M., Kosiniak, K. and Bielanski, W. (1974). Analysis of the pattern of ejaculation in stallions. J. Reprod. Fertil. *41*, 329.

Voss, J.L. and Pickett, B.W. (1975). Diagnosis and treatment of haemospermia in the stallion. In Equine Reproduction. I.W. Rowlands, W.R. Allen and P.D. Rossdale (eds), Oxford, Blackwell Scientific Publications.

20

Laboratory Animals

E.S.E. HAFEZ

The domestic rabbit is derived from the European rabbit (*Oryctolagus cuniculus*) of the order Lagomorpha, which includes hares, rabbits, and pikas. Lagomorphs, like rodents, lack canine teeth and possess rootless, chisel-shaped incisors separated from the premolars by a space (diastema). Most rabbit breeds are smaller than hares and have less well-developed hind limbs.

Rats (*Rattus norwegicus*) and mice (*Mus musculus*) are represented in many genera. Mice have been the subject of much selective inbreeding to exaggerate certain characteristics for research in reproductive physiology and genetics. Consequently, the number of characterized strains is enormous. Well-authenticated strains include Hooded, Wistar and Sprague-Dawley. Hamsters, lemmings, voles, deer mice and gerbils are included in the family, *Cricetidae*, or the order *Rodentia*.

The domestic cat (*Felis catus*) originated in Europe or Western Asia from a cross between *F. sylvestris* and *F. lybica*. The dog (*Canis familiaris*) is characterized by remarkable variation in size, life span, age of maturity, behavioral char-

acteristics and breeding norms among the breeds.

Several reports have been published on the care and management of laboratory animals (Merck's, 1961; Gay, 1965; Growth, 1962; Hafez, 1970; Harvey et al., 1961; Laboratory Animals, 1964; Kent, 1968; Lane-Petter, 1970; UFAW, 1967.)

FEMALE REPRODUCTIVE ORGANS

There are remarkable species differences in the anatomy of the reproductive organs (Tables 20–1, 20–2). Laboratory mammals fall into three anatomic categories based largely on differences in the prominence of their two parts: the body and the two uterine horns. Rats, mice, rabbits and guinea pigs have a duplex uterus. There are two separate uterine horns and two cervices (Figs. 20–1, 20–2).

Uterine weight increase in rodents has been used as a bioindicator of the presence of estrogens, whether administered exogenously, as in estrogen bioassays, or secreted endogenously in response to gonadotropic stimulation of the ovaries (Fig. 20–3). Increased edema, hyper-

409

Table 20–1.   Female Reproductive Anatomy of Some Laboratory Mammals

Organ	Main Anatomic Types	Species
Ovarian bursa	Infundibulum funnel-shaped; lies close to ovary, forming an open ovarian bursa.	Rabbit
	Infundibulum and ovary enclosed by a fold of mesosalpinx, which forms a periovarial sac.	Rat, mouse
Fimbriae	Extensively developed fimbriae enclose the ovarian surface.	Rabbit, cat
	Poorly developed fimbriae make limited contact with ovarian surface.	Rat, mouse
Uterotubal junction	Very complex; glandular mucosal lip; projecting papilla.	Guinea pig
	Rosette type projections; no intramural portion.	Rabbit
	Single papilla; short intramural portion of oviduct.	Rat
	Single mound; short intramural portion of oviduct.	Dog
	Simple fold that projects into a pocket in uterine wall; no mound.	Mink
Uterus, cervix and vagina	Two uteri; two cervices; two vaginas.	Opossum, kangaroo
	Two uteri; two cervices; one vagina.	Rabbit, rat (Norwegian)
	Two well-developed horns, small uterine body; one cervix; one vagina.	Cat, dog

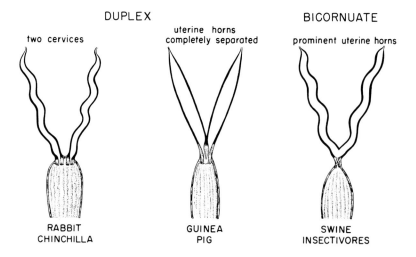

Fig. 20–1.   The relative length of the uterine horns and the uterine body varies among laboratory mammals. In litter-bearing species, the uterine horns are elongated and the uterine body is small or absent. In ruminants, a septum separates the uterine horns and is especially prominent in the uterine body.

Table 20–2. Some Anatomic Characteristics of the Female Reproductive Organs in Laboratory Mammals

Species	Some Anatomic Characteristics	Number of Mammary Glands
Chinchilla	Mesosalpinx tends to enclose ovary; accessory corpora lutea during pregnancy; vaginal closure membrane	6 (2 inguinal and 4 lateral thoracic)
Dog	Ovary is flattened and completely enclosed in a roomy peritoneal pouch. Slender uterine horns are long and straight. Cervix is a short, thick-walled segment. Vagina is wider above (cranially) than below.	10 (arranged in two ventrolateral series)
Guinea pig	Two internal cervical openings, but only one common external os. Intestinal and urinary tracts open into a groove (the "fossa anovaginourethralis"). Lower end of the vagina is closed by an epithelial membrane, but opens periodically at estrus and during parturition.	2 (inguinal)
Hamster	Ovary is compact and encapsulated; oviducts and uterus similar to those of the mouse. Two cervical canals remain separate for about two-thirds of the length of the cervix, but then fuse. Vagina has a mucified type of epithelium; its wall contains urethral glands similar to those in the male prostate.	12 or 14 (thoracic and abdominal)
Mink	Ovary has abundant interstitial tissue; fimbriae only slightly developed. Uterine glands are sparse. External os of the cervix is a transverse uterine slit. Vagina is long and has a transverse fold across its dorsal wall.	6 or 8 (30% nonfunctional)
Mouse	Ovaries lie ventrally just below the kidneys within transparent ovarian capsules. A narrow, tunnel-like passage connects the periovarial space with the peritoneal cavity.	10 (6 thoracic and 4 abdomino-inguinal)
Rabbit	Complete duplication of the uterine segments; two long uterine horns and two entirely separate cervical canals, each of which has an internal and external os. Endometrium arranged in numerous transverse and longitudinal folds, which are particularly prominent along the mesometrial borders. Cervical canals have a narrower lumen and a more extensively folded mucous membrane than the uterine horns. Vaginal portions of the cervical segments are surrounded by a complete ring of fornices.	8 (arranged in ventrolateral series)
Rat	Ovary lies within ovarian bursa. Periovarial space opens into the peritoneal cavity through a slit on the antimesometrial side of the bursa at the tip of each uterine horn.	12 (two ventrolateral series along thoracic and inguinal regions)

3 mm

V
M
E

I

U

attachment to other cervix

Fig. 20–2. Longitudinal section in the uterine cervix of the rabbit. Note the complexity of the mucosa folds. *E*, external os; *I*, internal os; *M*, mucosa; *U*, uterus; *V*, vagina.

trophy and hyplasia of the endometrium (both stroma and mucosa) and hypertrophy of the myometrium contribute to the uterotropic effects of estrogens. Androgens also are potent stimulators of uterine growth.

There are remarkable anatomic and histologic differences in the uterotubal junctions of different species (Figs. 20–4, 20–5): (1) the degree of flexure and the angle at which the oviduct and the uterine cornu meet; (2) the narrowing of the

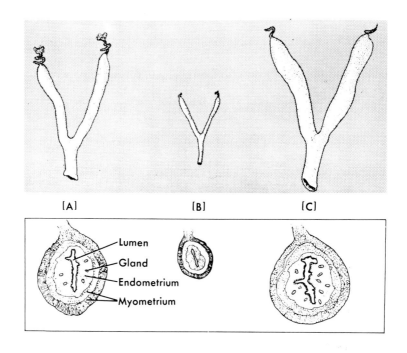

Fig. 20–3. Effects of estrogen on the growth of the uterus and vagina of the rat. *A*, Normal female rat; *B*, rat that has been castrated for four weeks; *C*, rat that has been castrated for four weeks and given estrogen during the last ten days. The lower drawings are cross sections of the corresponding uteri. *(From Frye, 1967. Hormonal Control in Invertebrates. New York, Macmillan)*

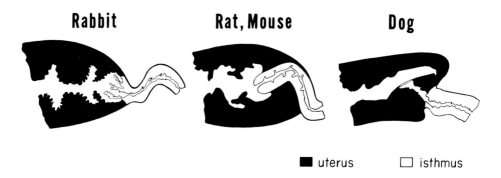

Fig. 20–4. Species differences in the morphology of the uterotubal junction (not drawn to scale). Various anatomic relationships exist between the isthmus and the uterus: the conspicuous folds of the rabbit, the moundlike papilla of the dog and the archiform junction of the rat. *(From Hafez and Black, 1969. In: The Mammalian Oviduct. Hafez and Blandau [eds], Chicago, University of Chicago Press)*

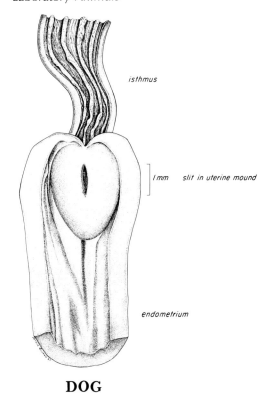

isthmus

1 mm   slit in uterine mound

endometrium

DOG

Fig. 20–5. The uterotubal junction of the dog is a slit in an expansion of the uterus.

lumen in the caudal isthmus; (3) blood and lymph supplies; (4) the number and morphology of ciliated cells; (5) the cytology of mucosal secretory cells and uterine glands; (6) ciliary activity during stages of the reproductive cycle; (7) the relative thickness of the circular and longitudinal bands that make up the tunica muscularis; and (8) the blending of the oviductal musculature with the myometrium (Hafez and Black, 1969). The uterotubal junction is a papilla (colliculus tubaris) in the rat, an elevated slit in the dog, and a rosette-like structure in the rabbit.

The mammary glands are distributed within loose fatty tissue along the lateral aspects of the thoracic and inguinal regions. Their number and position vary with the species (Fig. 20–6). In the rat, for example, the first thoracic gland is anterior to the forelimb; the second and third are thoracic and extend into the axillae; the abdominal and the first and second inguinal glands are separated from the thoracic glands and generally lie between the iliac crest and the ischial tuberosity.

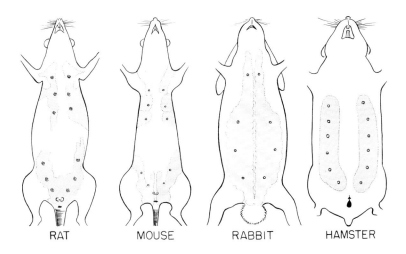

RAT   MOUSE   RABBIT   HAMSTER

Fig. 20–6. Location of the nipples and size of the mammary glands in common laboratory mammals during lactation. *(Drawings of rat and mouse, partly adapted from Eckstein and Zuckerman, 1956. In: Marshall's Physiology of Reproduction, A.S. Parkes [ed.], London, Longman's)*

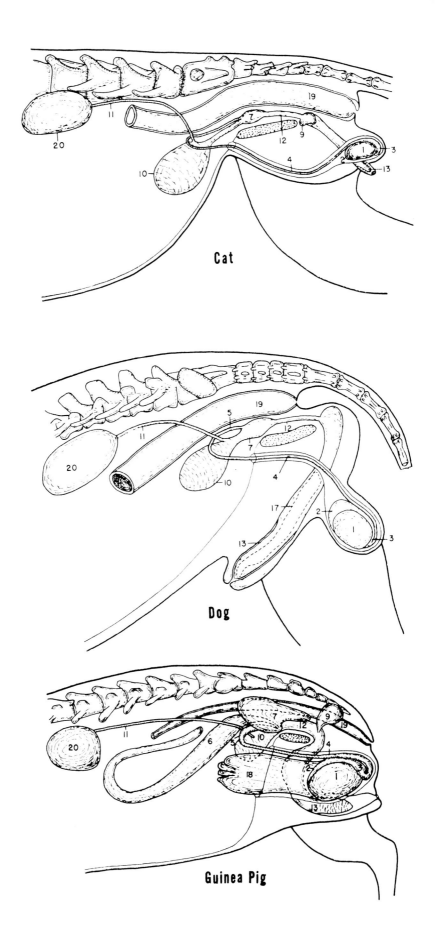

Cat

Dog

Guinea Pig

The abdominal glands are usually the largest and produce the most milk.

MALE REPRODUCTIVE ORGANS

Species differences in the male reproductive organs are particularly noted in the relative size of accessory glands and copulatory organs (Fig. 20–7). These anatomic differences are reflected in semen characteristics (Table 20–3) and copulatory behavior (Table 20–4).

There are also species differences in the morphology of spermatozoa, particularly that of the acrosome (Fig. 20–8). As in other mammalian species, ejaculate volume and sperm concentration increase after puberty (Fig. 20–9).

Table 20–3. Age and Weight of First Breeding and Semen Characteristics

Species	Beginning of Breeding Life		Volume of Ejaculate		Sperm Concentration 10^8 per ml	
	Age (Months)	Body Weight	Range	Mean	Range	Mean
Cat	9	3.5 kg	0.01–0.3	0.04	1.5 –28	14
Dog	10–12	varies	2–25	9.0	0.6 –5.4	1.3
Guinea pig	3–5	450 gm	0.4 –0.8	0.6	0.05–0.2	0.1
Rabbit	4–12	varies	0.4 –6	1.0	0.5 –3.5	1.5

(From Hamner, C. E., 1970. The semen. In: Reproduction and Breeding Techniques for Laboratory Animals. E. S. E. Hafez [ed], Philadelphia, Lea & Febiger)

Table 20–4. Species Differences in Copulatory Behavior

Species	Copulatory Pattern
DOG	Estrous female courts male by jumping, biting his ears, presenting her rear to male, raising her tail and turning it to one side, and lifting her vulva in a rhythmic fashion. Male clasps her hindquarters at rear flank with his forelegs; rapid pelvic copulatory movements in attempting intromission. Within two to three minutes after entrance, pelvic motions cease, and the pair are locked together in a "tie," caused by vascular engorgement of bulbus glandis of penis and active contraction of constrictor vestibuli muscles of bitch. Pair remain locked for 5 to 20 minutes, then separate spontaneously.
RABBIT	Receptive doe lies in mating position, raising her hindquarters to allow copulation. Buck moves quite suddenly, mounts and rests his head on her flank or nuzzles her hindquarters, performs 8 to 12 rapid copulatory movements to effect intromission and ejaculates; copulatory thrust is so vigorous that the buck falls backwards or sideways and may emit a characteristic cry. Gelatinous substance in the ejaculate often forms a vaginal plug, which is expelled a few minutes after copulation.
RAT	Precopulatory behavior: nibbling of female's head or body or examination of her anogenital region; Male mounts female from the rear and clasps his forelegs about her laterolumbar region. During the clasping, male palpates her sides with rapid movements of his forelimbs; his pelvic region is simultaneously moved in rapid piston-like thrusts. Male slips off the female's back rather weakly.

Fig. 20–7. Reproductive structures, *in situ*, of the cat, the dog and the guinea pig. *1*, Testis; *2*, caput epididymis; *3*, cauda epididymis; *4*, ductus deferens; *5*, ampullary gland; *7*, prostate gland; *10*, bladder; *11*, ureter; *12*, urethra; *13*, penis; *17*, baculum; *18*, fat body; *19*, rectum; *20*, kidney. *(From S. McKeever, 1970. Male reproductive organs, In: Reproduction and Breeding Techniques for Laboratory Animals. Hafez, E.S.E. [ed], Philadelphia, Lea & Febiger)*

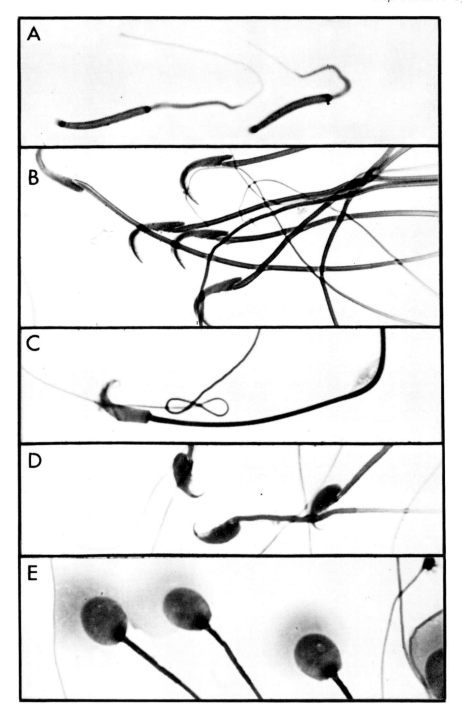

Fig. 20–8. Photomicrographs of sperm of different species. Sperm suspended in 0.15M sodium chloride and stained with rose bengal in 5% phenol solution. *A*, Cock sperm; note elongated cylindrical head (1625×). *B*, Rat sperm; note elongated hooked head with hood extending over the neck (1125 ×). *C*, Hamster sperm; note elongated hooked head (1125 ×). *D*, Mouse sperm; note spicule-like appearance of hook on the head (1750 ×). *E*, Guinea pig sperm; note large wrap-around acrosome (1500 ×). *(From C.E. Hamner, 1970. The Semen, In: Reproduction and Breeding Techniques for Laboratory Animals, E.S.E. Hafez [ed], Philadelphia, Lea & Febiger)*

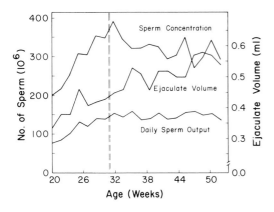

Fig. 20–9. Age changes in the mean gel-free ejaculate volume (ml), sperm concentration (value × 10⁶ sperm/day) for New Zealand white rabbits which ejaculated twice every 48 hours. The daily sperm output reaches adult values when the rabbit is 32 weeks of age. *(Data from Amann and Lambiase, 1967. J. Reprod. Fertil. 14, 329)*

BREEDING SEASON

Wild rabbits show a definite period of anestrus and experience seasonal variations in reproductive capacity. Does show variations in other than the spring and summer breeding seasons. Among domestic rabbits, the duration of anestrus varies with colonies and individual rabbits. Some does and bucks are fertile throughout the year, but most exhibit anestrus for one to two months. Nonetheless, domestic rabbits also manifest seasonal patterns of reproductive activity.

Unlike rodents and farm mammals, rabbits do not show regular estrous cycles, although a certain rhythm exists in their sexual receptivity. Under favorable conditions, the doe shows signs of estrus for long periods during which ovarian follicles are continuously developing and regressing, so that approximately constant numbers are available for ovulation. If the doe is not bred, the follicles in the ovary remain large and active for 12 to 16 days. They then begin to regress and new ones grow to replace them. Active follicles are consequently present at all times during the breeding season except perhaps in

transitional periods, when the new set of follicles is growing and the old set is retrogressing. A doe may show no sexual receptivity when she is molting, lactating, or poorly nourished because follicle development is suspended.

Rats and mice have been kept in captivity for generations and can breed all the year round. Dogs and cats breed only during spring and fall (Table 20–5).

ESTROUS CYCLE

The estrous cycle classically consists of four stages: proestrus, estrus, metestrus and anestrus.

Estrus is characterized by sexual receptivity when the female will allow copulation. In the rat, estrus lasting 9 to 15 hours is characterized by behavioral changes, eg, increased running activity, quivering of the ears, and lordosis in the presence of another rat. In rodents the vaginal epithelium undergoes well-marked changes during the estrous cycle. Estrus is characterized by squamification and cornification of the cells and disappearance of leukocytes. During the 24 hours preceding the next ovulation, epithelial cells become abundant and their size and degree of pyknosis increase toward ovulation. The vaginal lavage or smear is thus an excellent indicator of the stage of the estrous cycle.

Bitches have long estrous cycles and ovulate spontaneously. During the active part of the cycle (proestrus, estrus) they are "in season" or "in heat." These periods usually occur at intervals of 4 to 18 months, the time and regularity of each interval being an individual and breed characteristic. Most bitches are in season in the USA during January to February and July to August.

OVULATION

Ovulation may occur spontaneously or as a reflex to a stimulus. Rabbits, ferrets and some other mustelids, the cat and the shrew do not ovulate spontaneously (Table 20–5). In the rabbit, ovulation occurs 10 to 13 hours after copulation or after some other stimulus, eg, injection of LH, salts or copper and cadmium, electrical stimulation of the head or of the lumbar region of the spinal cord, or orgasm induced by contact with other females. However, ovulation usually cannot be provoked by mechanical stimulation of the cervix, as it can in the cat. Ovulation is usually induced in the rabbit by an intravenous injection of 20 to 25 IU of human

Table 20–5. Species Differences in Type, Season, and Main Characteristics of Sexual Cycle in Laboratory Mammals

Ovulatory Mechanism		Species	Type and Season of Cycle	Length of Sexual Cycle (Days)	Duration of Heat	Time of Ovulation	Viability of Ova (Hours)
S P O N T A N E O U S	O V U L A T I O N	Dog	In estrus at 4–8 month interval in spring & fall depending on breed; mono-estrous	Proestrus, 9 days; estrus, 7–9 days	7–13 days	2nd–3rd day of estrus	A few days
		Gerbil	Polyestrous	4–6	12–18 hrs		
		Guinea Pig	Polyestrous	16–19	6–15 hrs	10 hrs from onset of estrus	20
		Hamster	Polyestrous	4	4–23 hrs	Early estrus	10
		Mouse	Polyestrous (anytime)	4–5	9–20 hrs	2–3 hrs from onset of estrus	10–12
		Opossum	Polyestrous (seasonal)	22–38	1–2 days	Early in heat	
		Rat	Polyestrous (anytime)	4–6 (pseudo-pregnancy lasts 13 days)	9–20 hrs	8–11 hrs from onset of estrus	8–12
R E F L E X	O V U L A T I O N	Cat	Polyestrous (seasonal in spring & fall)	15–28 (pseudo-pregnancy lasts 36 days)	4–10 days	24–36 hrs post-copulation	
		Ferret	Polyestrous (April–Aug.)		In absence of male, 5 mo.	30 hrs post-copulation	30
		Mink	Polyestrous	8–9		42–50 hrs postcopulation	
		Rabbit	Polyestrous (anytime)	Pseudo-pregnancy lasts 14–16 days	No clearly defined period	10.5 hrs postcopulation	8

(From the literature and from Fox, R. R. and C. W. Laird, 1970. Sexual cycles. In: Reproduction & Breeding Techniques for Laboratory Animals. E. S. E. Hafez [ed], Philadelphia, Lea & Febiger)

chorionic gonadotropin (HCG) in about 0.25 ml of sterile physiologic saline. In the hamster, ovulation occurs regularly every four days. The time of ovulation is estimated in relation to the time of onset of estrus, ie, the time at which the female first permits breeding by the male, and in relation to the photoperiod in the room.

EGG TRANSPORT

Species differences are noted in egg transport in the oviduct and rate of cleavage (Table 20–6). In the mouse and rat, the ova move in a few minutes through the first coil of the ampulla of the oviduct to reach the dilated portion of the ampulla. In the mouse, in which transport of ova is relatively fast, the late morula or early blastocyst is transported to the uterus on the morning of the third day after fertilization, and by the afternoon of this day, all the blastocysts are in the uterus. In the rat, transport of the ova takes a half-day longer than in the mouse.

In the rabbit, a concentric coat of oviductal secretion is formed around the fertilized or unfertilized eggs while in the oviduct.

PREGNANCY AND PSEUDOPREGNANCY

Progesterone is required for maintenance of pregnancy before and after implantation. Ovariectomy performed after implantation during the first third or half of pregnancy leads to fetal resorption or to abortion in most species. The ovary, the main source of progesterone, is essential throughout pregnancy in the rat, hamster, mouse and rabbit. In these species, pregnancy can be maintained after ovariectomy by injection of physiologic doses of exogenous progesterone. Dogs, cats and guinea pigs do not abort if ovariectomy is performed during mid- or late gestation, because the placenta produces sufficient progesterone to maintain pregnancy.

Pseudopregnancy in the rabbit may result from a sterile copulation, an injection of LH, or from the stimulation caused when one doe mounts another or when a doe mounts the young in her own litter. It lasts 16 to 17 days, during which time the doe does not conceive. At the end of pseudopregnancy, the doe may pull hair from her body and attempt to make a nest, but will fail to keep it clean. Corpora lutea secrete progesterone during pseudopreg-

Table 20–6.   Earliest Time at Which Cleavage Stages and Blastocysts Can Be Found with Some Regularity

Animal	Day Found				
	2-Cell Stage	4-Cell Stage	16-Cell Stage	Blastocyst	Implantation
Cat*	early 3	late 3	4	5–6	13–14
Ferret*	3	late 3	4–5	6–7	11–12
Guinea pig†	2	4	5	late 5	6
Gerbil†	2	3	late 4	5	6
Hamster†	2	3	early 4	4	5
Mink*	3	4	5–6	6–7	delayed
Mouse†	2	early 3	late 3	early 4	early 5
Opossum†	3	3	4	early 5	6
Rabbit*	2	late 2	early 3	late 3	7
Rat†	2 and 3	late 3	late 4	5	late 5

*Species with induced ovulation: day 1 is the first 24 hours following coitus.
†Species with spontaneous ovulation: day 1 is the day on which sperm are found in the vagina.
(From Enders, A.C., 1970. Fertilization, cleavage, and implantation. In: Reproduction and Breeding Techniques for Laboratory Animals. E.S.E. Hafez [ed], Philadelphia, Lea & Febiger)

nancy, causing the development of the uterus and mammary glands.

Species differences are noted in gestation period, litter size (Table 20–7), and rate of fetal differentiation and organogenesis (Table 20–8).

Table 20–7. Species Differences in Gestation Period and Litter Size

Animal Species	Gestation Period (days) Range	Mean	Number of Young/Litter*
Cat	56–65		2–8
Chinchilla	105–111		1–4
Dog	58–63		4–8
Gerbil (Mongolian)	24–26		1–7
Guinea pig	65–70		1–6
Mink	40–75	51	4–10
Mouse	19–20	19	4–8
Rabbit (Domestic)	26–36	31	3–10
Rat	21–23	22	6–9

*Litter size varies with the breed

Table 20–8. Differentiation and Organogenesis in the Rat and Dog Embryo

Days of Gestation Rat	Dog	Embryologic Characteristics
6	13–14	Blastocyst implanting, with trophoblast and inner cell mass; endoderm delaminating.
7	14–15	Chorionic villi simple and unbranched; embryonic disc well defined; differentiation into embryonic and extra-embryonic parts begins.
8	15–16	Rapid trophoblast differentiation; primitive streak, primitive knot, and head process; start of mesoderm formation; cephalocaudal elongation of embryonic area, and longitudinal axis of embryo established.
9	16	Occipital somites; neural plate and early neural folds appear.
10–10.5	17–18	Closure of neural tube, primary brain vesicles; cervical and upper thoracic somites; two visceral arches; formation of heart, optic, and otic primordia; endoderm differentiates into fore-, mid- and hindgut; liver primordium below heart.
11–11.5	20–21	Lower thoracic and lumbar somites, four visceral arches, tail bud organized; arm and leg buds; mesonephric tubules well developed; cephalic and caudal flexures present, embryo C-shaped; stomach, bile duct, and gallbladder distinct.
12–12.5	22–23	Secondary brain vesicles; sacral and upper caudal somites; mandibular, maxillary and frontonasal processes; differentiation of handplates; auricular hillocks first appear.
13–13.5	24–25	Caudal somites, round handplates and footplates, prominent facial processes and clefts.
14–14.5	26–27	Caudal somites, in rat and dog; body uncoils; auricular hillock distinct, pinna forming; precartilage in mandible, eyes shifting anteriorly.
15–15.5	28–29	Caudal somites, in rat only; facial clefts closed, complete diaphragm; eyes now facing anteriorly.
16	30–31	Last caudal somite, in rat only; head lifted off chest; ossification of skeleton begins.

(From Hendricks, A.G. and Houston, M.L., 1970. Gestation and prenatal development. In: Reproduction and Breeding Techniques for Laboratory Animals. E.S.E. Hafez [ed], Philadelphia, Lea & Febiger)

BREEDING SYSTEMS, SEXING AND SELECTION

Inbreeding, line breeding, and cross-breeding are used in commercial colonies of laboratory animals (Fig. 20–10). In-breeding and line breeding are used to reduce the number of ancestors and to retain the characteristics of an outstand-

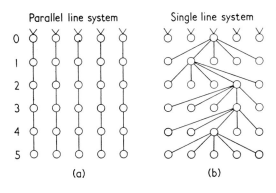

Fig. 20–10. **Pedigrees illustrating a parallel-line breeding system *(A)* and a single-line breeding system *(B)*. O is the foundation stock. *(Adapted from Falconer, 1962. Notes for Breeders of Common Laboratory Animals, Porter and Lane-Petter [eds.] Editor, E.S.E. Hafez. New York, Academic Press)***

ing buck or doe. In crossbreeding, pure-bred animals from two stocks are mated. The development of strains with good reproductive efficiency involves rigorous selection for conception rate, litter size, milk production, maternal behavior, growth rate and neonatal survival. Milk production can be evaluated indirectly by weighing the litter three to four weeks postpartum.

The following criteria should be considered in selection: general health, vigor, high productivity, good mothering ability, tameness and ease of handling. The mothering ability of the female can be judged from her performance with two or three litters. The true value of the female as a milk producer and mother cannot be assessed if her litters are fostered by another female. Furthermore, low milk production may be genetically fixed in a colony by repeatedly fostering the young of low milk producers, and by using these females for breeding stock.

Sexing of the neonate can be performed in most species. The young rabbit is re-

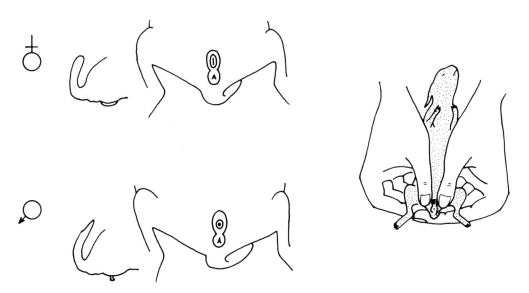

Fig. 20–11.   **Sexing young rabbits. The penis of the male is a rounded protrusion 1.2 mm dorsal to the anus; a pair of reddish-brown specks occur near the vent. The vulva of the female has a slit-like opening and is less than 1.2 mm from the anus; no specks are apparent. *(From Sanford, 1958. Reproduction and Breeding of Rabbits, E.S.E. Hafez [ed], York, England, Fur & Feathers)***

strained firmly and gently on its back with its head up; the left hand should cross the rabbit's chest and hold the front legs forward, next to the head (Fig. 20–11). The tail is held back with the index finger. The external sex organs are exposed for inspection by exerting slight pressure with the forefinger and thumb of the right hand. The vaginal mucosa of the female protrudes to form a slit anterior to the anus. The male organ has a rounded tip with two reddish brown specks. The distance between the anus and the sex organ is slightly longer in the male than in the female. Sex differences are not pronounced in very young animals.

The distance between the anus and prepuce is greater in newborn male kittens (12 mm) than between the anus and the vulva in females (7 mm). In the female the skin area between anus and vulva is virtually hairless, while in the male, hair covers the area between the anus and the penis which will develop into the scrotum.

REFERENCES

Gay, W.I. (1965). Methods of Animal Experimentation. Vols. I, II, New York, Academic Press.

Growth: Including Reproduction and Morphological Development (1962). Washington, D.C., Fed. Am. Soc. Exp. Biol.

Hafez, E.S.E. (ed) (1970). Reproduction and Breeding Techniques for Laboratory Animals. Philadelphia, Lea & Febiger.

Harvey, E.B., Yanagimachi, R. and Chang, M.C. (1961). Onset of estrus and ovulation in the golden hamster. J. Exp. Zool. *146*, 231.

Kent, Jr., G.C. (1968). Physiology of reproduction. Chap. 8. *In* The Golden Hamster: Its Biology and Use in Medical Research. R.A. Hoffman, P.F. Robinson and H. Magalhaes (eds), Ames, Iowa, Iowa State University Press, p. 119.

Laboratory Animals (1964). Animals for Research. Washington, D.C., National Research Council, Institute of Laboratory Animal Resources.

Lane-Petter, W., Worden, A.N., Hill, B.F., Paterson, J.S. and Vevers, G. (eds) (1970). The Care and Management of Laboratory Animals. 3d ed. London, Universities Federation for Animal Welfare.

The care of laboratory animals (1961). *In* Merck Veterinary Manual. 2nd ed., Rahway, N.J., Merck & Co.

The UFAW Handbook on the Care and Management of Laboratory Animals (1967). London, Livingstone.

21

Poultry

A.B. GILBERT

The bird differs from the mammal in that it has no well-defined estrous cycle and the breeding season is not divided into "estrus" and "pregnancy" phases. Instead, follicular development and oviductal function continue at the same time. Modern breeds also differ greatly from wild species; for example, "broodiness" has disappeared from many breeds of chicken and the modern hen is an "indeterminate layer," given the right conditions. Similar improvements are being made with other species of domestic poultry, although some retain their wild state and become photorefractory.

THE FEMALE

The Egg

Large by mammalian standards, the egg is a complex structure chemically and physically; its complexity is determined by the requirements to protect and to provide sustenance for the embryo living in a hostile environment. The egg is formed basically of three components: the "central yolk mass," which is equivalent to the mammalian egg, the "white" and the shell (Fig. 21–1), although no component is homogeneous either structurally or chemically (Table 21–1). Moreover, the physical characteristics of eggs are difficult to define precisely because they vary from species to species, from bird to bird, and even from egg to egg within a bird (Gilbert, 1971a, 1979).

Central Yolk Mass. The "central yolk mass" consists of the secondary oocyte in a nonfertile egg or the fertile zygote. It is bounded by the perivitelline membrane and weighs about 19 gm in the chicken, although its size is age-dependent; lower weights are associated with pullets. The living cytoplasmic part of the cell, the *blastodisc* (or the *blastoderm* if the egg is fertile), occupies a small fraction of the oocyte. It is a grayish-white spot, 3 mm in diameter, containing in the infertile egg the female haploid pronucleus, which produces either a male or female embryo; in the bird, unlike the mammal, the female is the heterogametic sex. In a fertile egg the area is covered by a double layer of cells at the time of laying.

The germinal disc floats on a cone of

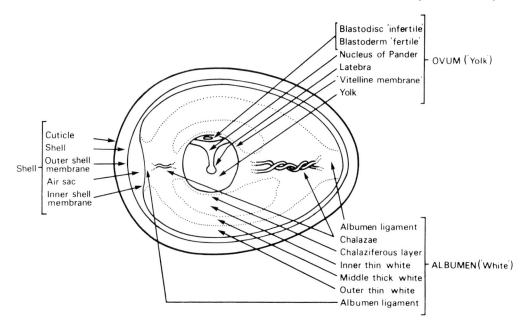

Fig. 21–1. Diagram of the components of the egg. *(From Gilbert, 1971a)*

Table 21–1.   Main Components of the Egg

Common Name	Specific Name	Components

Yolk	Oocyte	Blastodisc (infertile) or blastoderm (fertile)
		 Nucleus of Pander
		 Latebra (white yolk)
		 Yolk (yellow yolk)
		 Vitelline membranes:
		     (a) Primary membrane, the oocyte cell membrane
		     (b) Secondary membrane, the perivitelline membrane
		     (c) Two tertiary membranes, formed from the oviduct

White	Albumen	Albumen ligaments Chalazae Chalaziferous region Inner thin white Middle thick white Outer thin white
Shell	Shell membranes	Inner membrane (air space) Outer membrane
	Mineralized part	Organic matrix Inorganic crystals Pigment
	Cuticle	Organic part Pigment

(From Gilbert, 1971. In: Physiology and Biochemistry of the Domestic Fowl. D. J. Bell and B. M. Freeman [eds], New York, Academic Press)

light-colored yolk ("white yolk") that extends downward to end in a ball, the *latebra* (Fig. 21–1). The chemical composition of white yolk differs from that of the yellow: it contains a greater proportion of protein and its structure has certain peculiarities. The major part of the "central yolk mass" is composed of the familiar orange-yellow viscid fluid (yellow yolk), which has been described as an oil-water emulsion with the continuous phase as aqueous protein. Scattered throughout this are various droplets and yolk spheres (Bellairs, 1964).

Chemically, it is a heterogeneous mass containing proteins, lipids, pigments and a variety of minor organic and inorganic substances (Fig. 21–2). The low-density fraction contains nearly all the lipid fraction as lipoproteins and much phosphorus. The water-soluble fraction (livetin), which comprises 10% of the mass, contains about 30% of the protein mainly as plasma albumin, one of the plasma glycoproteins, and plasma γ-globulin. The granular fraction is formed of phosvi-

tin and includes most of the remaining phosphorus as well as the majority of the calcium and lipovitellins.

The function of the yolk is to provide the initial material for embryogenesis, but it is difficult to distinguish between substances essential for embryonic development and those that might be present as accidents of the mechanisms responsible for yolk formation.

Albumen ("White"). Surrounding the central yolk mass, and constituting about two-thirds of the egg in weight, is a layer of albumen that has at least seven major regions, which are possibly produced by the rotation of the egg in the oviduct. The albumen is almost pure aqueous protein and contains about 40 proteins (Table 21–2) (Feeney and Allison, 1969). Since the albumen surrounds the developing embryo, it forms an aqueous mantle preventing desiccation. It is also known to contribute material to the embryo during later development. Other functions have been ascribed to specific proteins, since many of the proteins *in vitro* have bac-

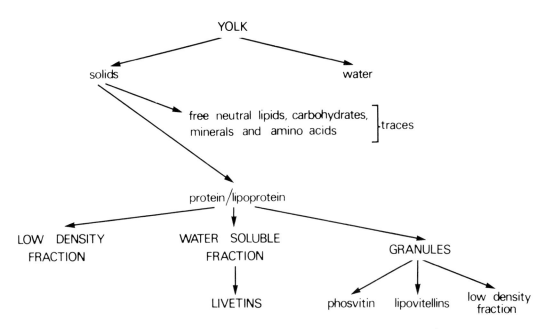

Fig. 21–2. Simplified breakdown of the major components of yolk. *(Redrawn from Gilbert, 1971a)*

Table 21–2. Summary of Characteristics of Egg White Proteins from the Hen

Component	Per Cent	Molecular Weight	Biologic Properties
Ovalbumin	54.0	46,000	
Ovotransferrin	12–13	76,600–86,000	Binds iron, copper, manganese, zinc; may inhibit bacteria
Ovomucoid	11.0	28,000	Inhibits trypsin
Globulins	8.0	Between 36,000 and 45,000	
Lysozyme	3.4–3.5	14,300–17,000	Splits specific β-(1.4)-D-glucosaminides; lyses bacteria
Ovomucin	1.5–2.9	?	Antiviral hemagglutination
Flavoprotein	0.8	32,000–36,000	Binds riboflavin
Ovomacroglobulin	0.5	760,000–900,000	
Ovoglycoprotein	0.5–1.0	24,400	
Ovoinhibitor	0.1–1.5	44,000–49,000	Inhibits proteases, including trypsin and chymotrypsin
Avidin	0.05	68,300	Binds biotin
Papain inhibitor	0.1	12,700	Inhibits proteases, including papain and ficin

tericidal properties, while others have enzymatic activity and some are enzyme inhibitors.

Shell. The shell is composed of three structures: the membranes, the mineralized part and the cuticle (Simkiss and Taylor, 1971). The two membranes are toughened sheets of fibrous protein, together about 70 μm thick (Simons and Wiertz, 1965); at one end they separate to form the air sac. Functionally the membranes provide the surface on which mineralization can occur; fibers from them penetrate outward to produce the organic matrix of the shell. They may reduce the speed of bacterial entry, allowing the bactericidal properties of egg white to act more effectively.

The mineral of the shell is almost pure calcite (calcium carbonate), with some magnesium, formed of radiating crystals (Becking, 1975). Running vertically through the shell are pores that allow gases to pass. The shell, apart from the pores, is impervious to most substances and forms a physical barrier to substances that might adversely affect the microenvironment of the embryo. It also provides mechanical strength and a rigid support to maintain the orientation of the heterogeneous internal components. Its strength is determined mainly by its curvature and

thickness, although other factors are involved (Carter, 1970). It also provides calcium for the developing embryo. The outer covering of the egg, the proteinaceous cuticle (Simons and Wiertz, 1970), reduces water loss and bacterial contamination.

Endocrinology

By comparison with mammals, avian endocrinology is poorly understood, and even less is known of the male than of the female. Nevertheless, certain generalities can be made. About nine hormones (or groups of hormones) are directly involved in the reproductive processes, but the specific physiology of each and their interrelationships with each other are largely unknown. Inferences may be drawn from mammalian studies, but this must be done with care. Also, the importance of a particular hormone may be hidden by "fashions" in research, and at present the wealth of research on the gonadotropins may lead the unwary to the conclusion that they are the main, or only, hormones involved.

In the female, reproduction can be divided into two endocrinologically distinct phases: the one in which the physiologic mechanisms lead to sexual maturation and subsequently to the maintenance of this state, and the other in which the processes result in the development of a single egg. Difficulties in understanding ovarian endocrinology arise because similar hormones are active in each case, and one phase is superimposed on the other.

Steroid hormones are produced by the left ovary in early embryogenesis (Guichard et al, 1977), and the estrogens present may be responsible for the suppression of the right ovary at this time. Without the presence of the right ovary, the right oviduct fails to develop further. After hatching, growth and development of the ovary and oviduct continue, but only slowly, the oviduct presumably in response to a steady output of steroids.

As birds approach sexual maturity they become photostimulatory (Murton and Westwood, 1977), and it is this change that is responsible for the future sexual development. In response to a changing daylength (not total daylight), the hypothalamus alters its output of the gonadotropin-releasing factors (RF) (Fig. 21–3). Doubt still exists regarding whether there is a specific follicular stimulating hormone releasing factor (FSH-RF) as well as the luteinizing hormone releasing factor (LHRF); the LHRF is probably similar to that of the mammal. These hormones, by way of the portal system, act on the pituitary and bring about a change in the pattern of gonadotropin output; little is known of this, although Sharp (1975) has shown that the luteinizing hormone (LH) level in plasma steadily rises until just before the first egg is laid, when it falls. Sharp (1975) suggested that this decrease is associated with a raised plasma steroid level, brought about by the development of the follicles.

Both follicular stimulating hormone (FSH) and LH have the ovary as their specific target, and they are responsible for the rapid development of the ovarian follicles and the increased steroidogenesis by the ovarian steroid-producing cells. It is often forgotten that the major hormones in reproduction are the estrogens, andro-

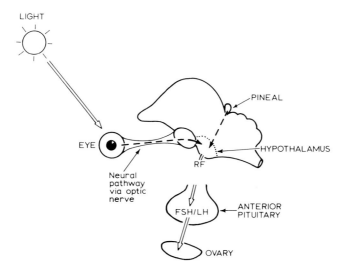

Fig. 21–3. Light may affect reproductive activity through at least three pathways: *(a)* through the eye, *(b)* through the pineal gland and *(c)* through its direct effect on the hypothalamus. The relative importance of these is not known, but it seems most likely that the main one is through the eye. The pineal, however, is an important organ. In response to the action of light, the hypothalamus produces a Releasing Factor (RF) which in turn regulates the pituitary production of the gonadotropins. *(Redrawn from Gilbert, 1971c)*

gens and progestogens, and that these steroid hormones bring about the major external and internal changes that are associated with an adult breeding animal.

Estrogens have many functions. They cause the oviductal occluding plate to break down, increase the uptake of calcium from the gut, act on the liver to produce the specific yolk lipoproteins, increase deposition of body fat, cause changes in the vascular system, stimulate growth of the oviduct, cause the pubic bones to "spread," play a role in the deposition of extra calcium in the bones, help to bring about the typical "female" behavior and, significantly, play a feedback role in the control of gonadotropin output.

Progesterone is more specific in its effects, but it is also involved in calcium metabolism, behavior and oviductal growth and function. In contrast to estrogen, it appears to have a positive feedback effect on gonadotropin production.

Androgens are also produced by the female and play a part in oviductal growth and differentiation and the development of the secondary sexual characteristics. They may affect pituitary output of the gonadotropins.

All three types of hormones are probably produced throughout the breeding season at their individual rates. It is their interaction that maintains the reproductive state: sometimes they are complementary, as in the regulation of oviductal growth by both estrogens and androgens, but in some instances they are antagonistic, as when androgens and estrogens act against each other in comb development.

Little is known about the cells responsible for steroid production in the ovary. It has been claimed that estrogens are produced by the interstitial cells of both the stroma and follicles, but the evidence is based mainly on the interpretation of the activity of the steroid-dehydrogenase enzymes that are present. The source of androgens is unknown; the only work

regarding their origin, which suggests that the stromal interstitial cells are responsible (Woods and Domm, 1966), has not been repeated. Progesterone seems to be a product of the granulosa cells of the preovulatory follicle but, unlike the mammal, the granulosa cells of the postovulatory follicles in birds may be incapable of producing it (Dick et al, 1978).

With the final maturation of the follicles, a daily cycle is initiated, which is superimposed on the general endocrinologic pattern. Light again plays a major role because the initial entrainment appears to be the transition from light to dark which, in some way, triggers the specific release of LH for a short time. The release of LH occurs about eight hours after the transition from light to dark, and it results in an increase in the concentration of LH in the blood, which rises to a peak value of 3 to 4 ng/ml, a level that is two to three times greater than the resting value (Wilson and Sharp, 1973). About four hours later, ovulation occurs.

There has been much speculation about the specific trigger for LH release, and estrogens, progesterone and androgens have all been invoked. As yet, the evidence only indicates that an increase in plasma LH occurs approximately at the same time as an increase of estrogens, androgens and progesterone occurs in the plasma (Senior and Cunningham, 1974; Shahabi et al, 1975; Etches and Cunningham, 1976).

A characteristic feature of avian reproduction is that each successive egg of a sequence is laid later ("lag") during the day (Fraps, 1961, 1965). This is brought about by LH release occurring later during the dark period and not by the length of time the egg spends in the oviduct (see p. 429 and Fig. 21–9).

Because LH release occurs later in the dark period with each successive member of the sequence, it is inevitable that one release will be scheduled to take place during the following light phase. This cannot occur because it appears that LH release occurs only during the hours of darkness, although the reasons are unknown. Since no LH release occurs, there can be no ovulation on that day and no subsequent egg production or oviposition 24 hours later (pause day). After the pause day, the following sequence is initiated by the next onset of darkness and the cycle is repeated.

After ovulation, egg formation appears to be largely outside direct hormonal control, though estrogens and progesterone are known to play important roles in the production of egg-white proteins (Palmitter, 1971) and calcium metabolism. It has been shown that prostaglandins can affect oviposition, though how important they are in the normal functioning of the hen has yet to be elucidated (Day and Nalbandov, 1977). Similarly, the oxytocic-like hormones can affect oviposition, but in the normal animal other mechanisms also may operate to bring about the expulsion of the egg (Gilbert, 1971b).

It is conceivable that many other hormones could indirectly affect reproductive performance. Body calcium is affected by both parathyroid hormone and calcitonin, and the availability of calcium for shell formation is an important factor (Simkiss and Taylor, 1971). Corticosteroids and thyroxine, regulated by their own pituitary hormones which, in turn, are controlled by the hypothalamic RF, may influence reproductive activity by their effect on general metabolic processes. In these terms, reproduction indeed becomes a "whole-body" response.

Egg Formation

The formation of the egg involves the transport of large quantities of material across numerous biologic membranes and the formation of many new substances, particularly specific proteins and lipids. The size and composition of the egg are affected by numerous genetic, environmental, and physiologic factors.

Ovary. In birds as in mammals, two ovaries and oviducts are formed during embryogenesis, but a characteristic feature of birds is the usual suppression of further development of these organs on the right side (Gilbert, 1979). The suppression is probably brought about by the early production of steroid hormones by the left ovary; removal of the left ovary results in development of the right gonad into an ovo-testis with both male and female tissue. The functional left ovary (Fig. 21–4) acts in the same way as mammalian ovaries; it produces the female gametes, releases them, and acts as an endocrine organ producing the steroid hormones.

The ovary consists of a medulla, which contains connective tissue, blood vessels and nerves, and a cortex. The cortex contains the oogonia, which give rise to the oocytes. The pear-shaped, immature ovary is about 15 mm long by 5 mm wide, lying in the body cavity, ventral to the aorta and cranial to the kidney, and close to the two adrenal glands. Although variable, the blood supply usually arises from the gonadorenal artery. Two veins drain blood from the ovary. It is extensively innervated from the sympathetic chain by way of the adrenal/ovarian plexus. For a more detailed description, see Gilbert (1979).

Follicular Development and Gametogenesis. At the onset of sexual maturity, ovarian weight increases. In the chicken this increase ranges from 0.5 gm to between 40 and 60 gm; most of this increase comes from the four to six developing follicles, the largest of which weighs about 20 gm, with a diameter of about 40 mm. The main mass of the ovarian tissue only increases to about 6 gm (Amin and Gilbert, 1970).

Of the thousands of oocytes present, usually only one starts its development at any given time in response to mainly FSH

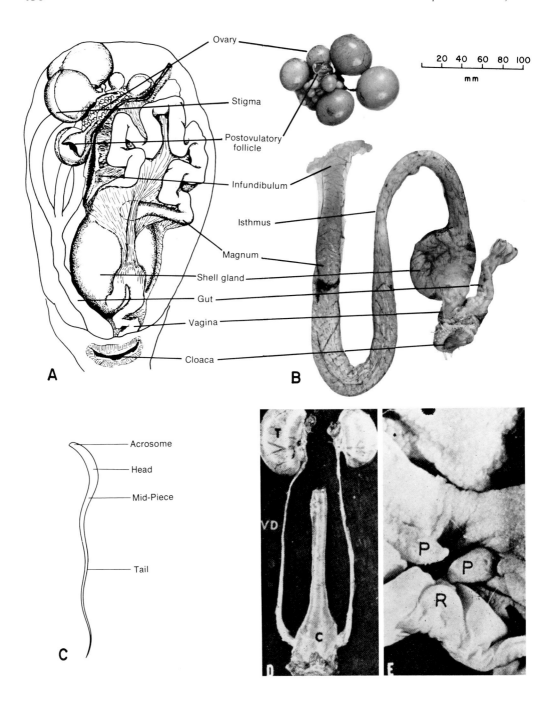

Fig. 21–4. *A*, The ovary of the domestic hen *in situ*. *B*, The ovary and oviduct of the domestic hen. *C*, Characteristic appearance of chicken spermatozoa. Acrosome *(A)*; head *(H)*; midpiece *(M)*; tail *(T)*. *D*, Reproductive organs of the male domestic chicken. The ductus deferens *(VD)* leads from the testis *(T)* to the cloaca *(C)* (0.48 ×). *E*, Copulatory apparatus located in the posterior ventral cloaca of the male chicken showing papillae *(P)* and the rudimentary copulatory organ *(R)* (5.6 ×).

but possibly also LH. The interval between the onset of development of successive follicles appears to be about 26 hours or multiples thereof. Maturation of the female pronucleus begins in the follicle and is completed in the oviduct. The first (reduction) division occurs two hours before ovulation, possibly under control of LH; the second (maturation) division occurs in the oviduct and may be initiated by the penetration of the sperm.

Development of the oocyte and yolk deposition can be considered in three phases (Marza and Marza, 1935), although other workers have separated the phases differently (Gilbert, 1979). The first, lasting for many months or years, is characterized by a slow deposition of material consisting mainly of neutral fat. This phase is superseded by the second period of several weeks, during which time the size of the follicle increases from about 1 mm to 6 mm. During the last stage (the rapid-growth phase) of seven to eight days, the main mass of yolk material is laid down and follicular weight increases almost linearly.

None of the material of the yolk is formed directly within the ovary, although some rearrangement and combination may occur inside the oocyte. The liver is the major source of yolk proteins and phospholipids (McIndoe, 1971). Since the ovarian follicle only accumulates yolk, the structure of the follicle must reflect this activity (Perry et al, 1978): in particular, the blood supply within the follicular wall is peculiarly complex, terminating in a capillary network adjacent to the basal lamina of the granulosa layer.

The accumulation of specific components of the yolk is important. Although a uniform amount of material is deposited each day, the proportions of the fractions vary from time to time. More is becoming known of this process (Gilbert, 1971c), but the functional significance of the differential accumulation is not understood. Since follicles at different stages of development are present in the ovary at the same time, variations in liver output are not the cause; the follicle or the oocyte itself must control the accumulation of material in this way.

Only two courses of development are available to the follicle: more commonly it continues until such time as ovulation occurs, with the release of the oocyte, but it may become atretic. The changes that occur in atresia are not known precisely, but the result is a general liquefaction of the yolk material and its resorption.

Ovulation. The events leading to ovulation are complex and not entirely understood, but the release of LH from the pituitary is involved and this may be controlled by feedback mechanisms of steroid hormones (Wilson and Sharp, 1973; Shahabi et al, 1975; Etches and Cunningham, 1976).

Ovulation occurs through rupture of the stigma (Figs. 21–4, 21–5), a specialized region of the follicle wall, in response to LH. In general the follicle must be mature enough to react to the stimulus (Fig. 21–6); however the follicle may be capable of reacting for several days, since premature (17–30 hours) ovulations can be induced, and some follicles ovulate 24 hours after they would normally be expected. The physical events that lead to the rupture of the stigma are unknown; none of the many suggestions is tenable in the light of modern research, but intrafollicular pressure is not involved.

Oviduct. The oviduct, about 700 mm long in the chicken, is suspended by the peritoneal dorsal ligament, which continues around it to form the ventral ligament (Aitken, 1971; Gilbert, 1979). It is highly vascular and the muscular layers are well supplied with nerves from the autonomic system.

The functions of the oviduct are production of the remaining formed elements of the egg (Fig. 21–1) and the movement of the developing egg along the oviduct.

The oviduct consists of a glandular folded mucosa and of muscle arranged into an outer longitudinal layer and an inner circular layer. The remainder of the formed elements of the egg are dissimilar and each is produced by a specialized region of the oviduct (Fig. 21–4); thus, five major regions can be distinguished, each with a characteristic distribution of mucosa and a muscularis (Fig. 21–7) (for more detailed descriptions, see Aitken, 1971; Gilbert, 1979).

Infundibulum. The infundibulum actively engulfs the shed ovum. In rare cases, unovulated follicles attached to the ovary may pass down the oviduct and subsequent ovulation may occur within the oviduct. The activity of the infundibulum is due to the fimbriae, projections of the cephalic surface of the infundibulum, which become engorged with

blood, and also to the activity of the smooth muscle of both the infundibulum and the ventral ligament.

After passing through the initial funnel-shaped region of the infundibulum, the ovum reaches the "chalaziferous region," a narrower, glandular portion (Fig. 21–4). This region may contribute to the formation of the chalazae, for albumen is added, and to the perivitelline membrane. However, the formation of the perivitelline membrane and the chalazae is a continuous process that starts in the infundibulum and ends in the distal oviduct.

Magnum. The albumen-secreting region is longer than any in the reproductive system, and it is distinguishable from the infundibulum and isthmus by its greater external diameter and its thicker walls (Fig. 21–4). The lining is thrown into

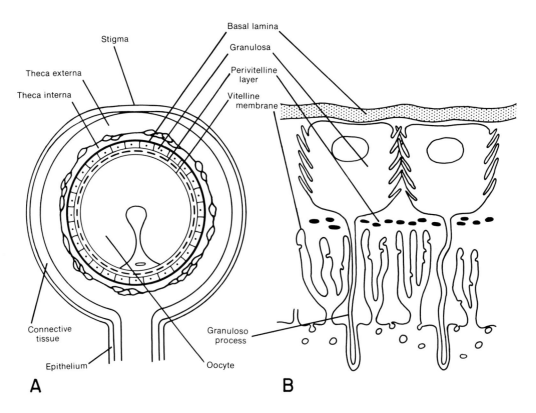

Fig. 21–5. *A*, Cross section of a maturing follicle. *(Redrawn from Gilbert, 1979) B*, Diagram of the oocyte surface and the granulosa layer. *(Redrawn from Perry et al., 1978)*

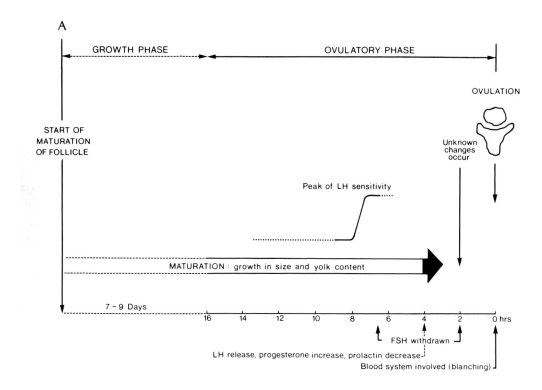

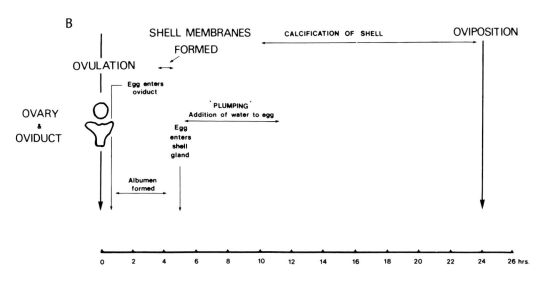

Fig. 21–6. *A*, Stages in the maturation of the follicle and ovulation. *(From Gilbert, 1971d)* *B*, Stages in the formation of the egg after ovulation. *(From Gilbert, 1971d)*

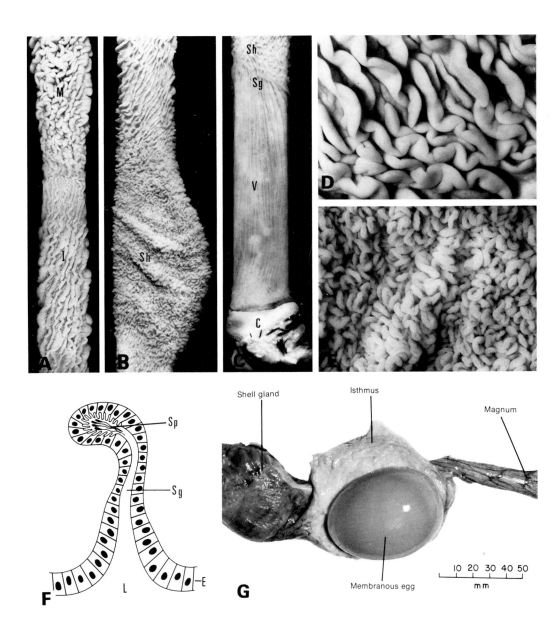

Fig. 21–7.   *A*, The luminal surface of the magnum and isthmus. *B*, The luminal surface of the shell gland and its junction with the isthmus. *C*, The luminal surface of the vagina. *D*, Higher power views of the mucosal folds of the magnum. *E*, Higher power view of the mucosal folds of the shell gland. *F*, "Sperm host" glands of the junction between the shell gland and the vagina. *G*, An egg *in situ* in the isthmus. Note the membranous covering only and the typical egg shape.

   C, Cloaca; *E*, epithelium; *I*, isthmus; *L*, lumen of oviduct; *M*, magnum; *Sg*, "sperm host glands"; *Sh*, shell gland; *Sp*, spermatozoa; *V*, vagina.

folds filled with secretory cells (Fig. 21–7).

Unlike yolk proteins, which are not formed in the ovary, the majority of the proteins of albumen are formed in the oviductal tissues; a possible exception is ovotransferrin (conalbumen), which appears to be similar to serum transferrin. However, the cells responsible for the formation of most of the 40 proteins are unknown, but avidin is formed in the epithelial goblet-cells, and ovalbumin and lysozyme are produced by the tubular glands. Ovotransferrin and ovomucoid may also be produced in the cells of the tubular glands (Palmitter, 1971).

The general processes involved in protein synthesis are well understood. Amino acids and ATP are activated to form a complex that reacts with transfer RNA (tRNA); this in turn combines with the complex ribosomal messenger RNA (mRNA) formed under the influence of DNA. The amino acids are then incorporated into the peptide chain according to the coding supplied by the mRNA. Most of the evidence suggests that similar processes occur in the cells of the chicken oviduct and that steroids control these processes (O'Malley et al, 1969; Schimke et al, 1977). Estrogens are involved, for their administration increases total oviductal proteins and specifically lysozyme, ovalbumin and avidin. In contrast, progesterone may be involved in the formation of only one protein, avidin, though it may have other actions also.

Release of the formed protein occurs as the developing egg passes down the magnum. On *a priori* grounds, three possibilities exist for the release of protein: (1) the developing egg, by mechanical or other means, causes release of the albumen, (2) hormonal mechanisms synchronize release, or (3) some neural coordinating mechanism is involved. At present, (1) seems most likely, but the others cannot be excluded.

Isthmus. The membrane-secreting region of the oviduct lies between the magnum and the shell gland and is narrower, with thinner walls, than either of the others (Fig. 21–4). Its structure is similar to that of the magnum, although the glands appear to contain material of a keratinous nature.

Little is known of the mechanisms involved in the formation of the protein or its release and deposition on the egg. However, partly membraned eggs can be obtained, and it has been observed that, while the albumenous egg is entering the isthmus, a membrane is deposited on those parts in contact with the glandular tissue. In the chicken, the typical shape of the egg is produced by the membranes and not by the calcified shell, which is laid down on an already formed shape (Fig. 21–7). At this stage the membranes are loosely applied to the egg, which has about 50% of its final mass.

Shell Gland. The shell gland is characterized by a pouch-like section joined to the isthmus by a short "neck" and extensive muscularization (Fig. 21–4). Microscopically, there are extensive mucosal folds studded with glandular cells.

The egg remains in the shell gland for approximately 20 hours (Fig. 21–6). During the first six hours or so, a watery fluid produced from the neck region passes into the egg ("plumping"), resulting in a twofold increase in the mass of egg white. However, plumping may continue throughout the time the egg stays in the shell gland. Thereafter, the main process is calcification. During its stay in the shell gland, rotation of the egg around the polar axis leads to the completion of the formation of the chalazae, which started in the infundibulum, and the stratification of the albumen.

Shell calcification probably starts in the isthmus, where small projections from the membrane, the mammillary cores, are formed (Simkiss and Taylor, 1971). In the shell gland, growth of the crystals continues at a constant rate of mineralization (about 300 mg calcium/hour). The oviduct does not store calcium, and about 20% of the calcium in the blood is removed as it passes through the shell gland. The cells responsible for transferring calcium into the lumen are not known for certain, nor is it known how the organic matrix is formed.

The final task of the shell gland is the formation of the cuticle and pigmentation; the latter consists of porphyrins deposited shortly before laying.

Vagina. The vagina acts as a passage from the shell gland to the cloaca and plays no part in the formation of the structure of the egg. However, it does play an important role in the processes involved in fertilization, for it contains the "sperm-storage glands" typical of birds (Fig. 21–7). It is a relatively short, S-shaped tube closely bound by connective tissue to the shell gland and similar in structure to other regions of the oviduct.

Transport of the Egg Through the Oviduct and Oviposition. The transport of the egg along the oviduct is similar in all domestic poultry. The developing egg spends about 15 minutes in the infundibulum. In the magnum its speed averages about 2 mm/minute and hence it takes about two to three hours to traverse this region (Fig. 21–6). After entering the isthmus the egg may be held for a short while before proceeding on its way; it takes about one to one and one-half hours to pass along the isthmus. About 20 hours of its total time in the oviduct (26 hours) is spent in the shell gland. Passage through the vagina takes a few seconds. The rate of transport is not uniform throughout the oviduct, so each region must have its own, unknown mechanism for coordination and control.

It is not known precisely how oviposition is initiated, but both hormonal and neural mechanisms are involved (Gilbert, 1971b): it may be related to the ovulation of the ovum 24 hours previously or events associated with it. Whatever the initial stimulus, contraction of the shell gland musculature forces the egg into the vaginal region. This may be controlled neurally, for stimulation of the central nervous system can affect it, but hormones are also involved. There is a depletion of the oxytocic hormones (mainly arginine vasotocin in birds) in the pituitary and an increase in their level in the blood. Blood oxytocinase decreases in activity and the shell gland musculature becomes increasingly sensitive to these hormones. Steroid hormones may play a role similar to that in mammals.

With the contractions of the shell gland musculature there occurs relaxation of the uterovaginal "sphincter," possibly analogous to the cervix of mammals. As this occurs the egg is pushed further into the vagina, and the general distention of the vaginal wall brings into operation the bearing-down reflex; this involves changes in respiration and stance, and contraction of the abdominal body muscles.

THE MALE

Semen: Its Physical and Chemical Properties

An ejaculate of a cock varies from 0.2 to 0.5 ml and has an average density of about 3×10^6 sperm/mm^3.

Morphology of the Spermatozoon. Although differences exist between domestic birds, the shape and size of spermatozoa are reasonably consistent, but sperm of avian species differ from those of domesticated mammals in being smaller, in having long, filamentous heads and in having no kinoplasmic droplet (Fig. 21–4). An acrosomal cap, head, and middle and principal tail pieces are present (Lake, 1971).

In the fowl, the sperm head, curved and 12 to 13 μm long, is surmounted by the acrosomal cap (2 μm long). The tail midpiece is about 4 μm long, and the remainder of the sperm's length of 100 μm is made up of the principal-piece of the tail. At its widest part, the sperm measures about 0.5 μm.

The spermatozoon is surrounded by the cytoplasmic membrane. The acrosome is simple with an inner spine surrounded by a conical cap. The head contains the nuclear material of the gamete. The midpiece contains the cylindric centriole surrounded by a sheath of about 30 curved, plate-like mitochondria; the latter are dissimilar to the elongated, curved cylinders of mammalian sperm and fewer in number. The typical outermost fibers of the mammalian posterior midpiece and the principal-piece of the tail appear to be missing. Throughout the tail runs the ring of nine doublet fibers surrounding two central ones.

Chemical Composition of the Spermatozoon and Seminal Plasma. Complex lipids and glycoproteins can be demonstrated around the sperm and their presence in the acrosomal cap may be related to sperm penetration of the female gamete. The tail contains phospholipid,

which may be metabolized as one source of energy.

The chemical composition of seminal plasma has been studied, and between 50 and 60 constituents are known (Lake, 1971). However, it is still not possible to distinguish those substances that are essential for spermatozoal function from those that are present as accidents of the reproductive system. Seminal plasma is essentially an aqueous solution of salts with some amino acids. When uncontaminated with "transparent fluid," fowl seminal plasma differs considerably from that of the mammal: it is almost completely lacking in fructose, citrate, ergothioneine, inositol, phosphoryl choline and glycerophosphoryl choline. Another interesting feature is the low chloride content, and glutamate is the chief anion.

Formation of Semen

The Testis and Spermatogenesis. The anatomy of the reproductive organs of the bird differs in several important respects from that of mammals (Figs. 21–4, 21–8)

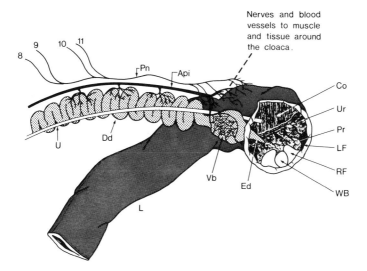

Fig. 21–8. Diagrammatic representation of the pelvic portion of the right male reproductive organs, including the distal part of the ductus deferens and the erectile phallic structures in the cloaca. *Dd*, ductus deferens; *Api*, artery, pudenda interna; *L*, large intestine; *Pr*, proctodaeum; *Ur*, urodaeum; *Co*, coprodaeum; *Lf*, lymph fold; *WB*, white body; *RF*, round fold; *Vb*, vascular body (gefassreicher Korper); *Ed*, ejaculatory duct (papilla). *(Redrawn from Lake, 1971)*

(Lake, 1971). The avian testes are situated in the abdomen and the ductus epididymidis is short by comparison. It is not known whether these differences are related to the peculiar reproductive phenomena in birds, in which sperm can survive for many days in the female oviduct.

In the avian male both gonads and ducts develop. The testes together weigh between 14 and 60 gm in the cock, depending on the breed, and they are suspended from the dorsal body wall just posterior to the lung and ventral to the kidney. Blood is supplied by way of the anterior renal artery and a variable testicular artery, but the arrangement of blood vessels within the testis is much simpler than that in mammals and there is no pampiniform plexus (Waites, 1970).

In contrast to the arrangement in the mammals, the seminiferous tubules are not grouped into evident lobules surrounded by connective tissue, but branch and anastomose freely within the tunica albuginea (Marvan, 1969); in the mature cock, branches of the latter penetrate between the tubules to act as a supporting skeleton. Interstitial tissue is negligible, but it contains the androgen-secreting Leydig cells. The seminiferous tubules of immature males are lined by a single-celled layer of Sertoli cells and stem spermatogonia, while mature males have irregularly shaped tubules lined by a multi-layered germinal epithelium. The spermatogonia give rise to the primary spermatocytes, secondary spermatocytes and spermatids; the latter eventually metamorphose into spermatozoa (Lake, 1956; McIntosh and Porter, 1967).

The time required for maturation of the sperm in the testis depends on the avian species. Usually spermatogonia multiply about the fifth week after hatching and primary spermatocytes appear about the sixth week. About the tenth week these cells multiply rapidly, secondary spermatocytes appear and the tubules increase in size. Spermatids first appear soon afterward and continued development occurs in the tubules until the 20th week. Thereafter the testis appears capable of producing spermatozoa in large quantities.

The Duct System. The fowl lacks the characteristic coiled and subdivided ductus epididymidis seen in most mammals. Moreover, spermatozoa are stored mainly in the ductus deferens, and there are no accessory reproductive organs, namely, seminal vesicles, prostate, Cowper's gland and urethral glands.

Ductules corresponding with the rete testis and ductuli efferentes of mammals are embedded within the connective tissue attachment of the testis to the body wall. The seminiferous tubules unite with those of the rete testis, and these unite with the ductuli efferentes. Thereafter the ductuli efferentes connect in several places with the short ductus epididymidis, which opens into the ductus deferens, an extensive convoluted tube running the whole length of the abdomen on the surface of the kidney (Fig. 21–8). At the distal end the ductus enlarges before passing through an erectile papilla extending into the cloaca; there is no comparable structure to the ampulla of mammals (Fig. 21–8).

The male has no penis, but an erectile phallus is present, and it is believed that this makes contact with the everted vagina during copulation (Fig. 21–4). Erection of the phallus is caused by its becoming engorged with blood, as are related cloacal structures (lymph folds). Presumably erection is controlled by nerves.

Reproductive Function in the Male

Though evidence is lacking, it is generally believed that FSH controls tubular growth and differentiation, and that LH affects the interstitial Leydig cells, which produce the steroid hormones.

Other hormones—prolactin, TSH and

ACTH—have been implicated in controlling mechanisms, but clear evidence for their normal role is yet to be found. Androgens may not have any effect on spermatogenesis, except perhaps with transformation of secondary spermatocytes to spermatids. Their main action is a negative feedback on gonadotropin output by the pituitary.

FERTILIZATION AND FERTILITY

"Fertilization," "fertility" and "fecundity" have been used ambiguously, although "fertilization," the term used for the penetration of the female gamete by that of the male and the fusion of the gametic pronuclei, is not now used synonymously with the word "insemination." On the other hand, "fertility" has been used to mean "producing many offspring." This has led to confusion because "poor fertility" has been used to describe flocks in which both fertilization and embryonic mortality are high and those in which true fertilization is low. In the industry, "fertility" has yet another meaning. In the present text its use is restricted to its association with fertilization. The factors affecting fertility are complex (Kamar and Hafez, 1975), and they involve both the male and the female. Only a general consideration can be given here.

Mating and Insemination

Natural Mating. The introduction of semen into the female is associated with complex courtship behavior in birds (Wood-Gush, 1971; Hafez, 1974).

A diurnal rhythm in mating frequency exists in males and is correlated with semen production. However, libido in the male is not necessarily correlated with high fertility and there is a tendency for the most frequent copulators to produce many aspermic ejaculates. Embryonic mortality may increase in flocks served by highly active roosters.

Best fertility is obtained with a ratio of males to females that prevents the formation of a "peck order" among them. Particularly troublesome is the "dominant male" (associated with aggression): he may not be the most active sexually, but he may prevent other males from mating successfully. Older males introduced into a society may not be able to express their full reproductive potential. Cocks tend to prefer mating with their own breed or strain; possibly this is related to early learning of visual and behavioral cues, for contact in early life with other strains enhances future mating. Hence heterosexual rearing is important.

Libido of the female is affected by her rearing environment—whether or not males were present—and by her social ranking. Receptivity (crouching) varies from hen to hen and this affects the chance of successful mating because the female probably determines whether or not the sexual advances of the male will lead to copulation.

Artificial Insemination. Factors to be considered in artificial insemination include: (a) laying intensity of female; (b) age of hen and/or advancing season; (c) position of egg in the oviduct; (d) placement of semen in the oviduct; (e) number of spermatozoa inseminated; (f) frequency of insemination; (g) quality of semen due to inherent factors and to its collection; and (h) dilution and storage of semen (Lake, 1967; Kamar and Hafez, 1975).

Semen Quality

Clearly the "quality" of the male ejaculate must play a major role in the fertilization processes. Unfortunately, semen characters are closely interrelated and it is difficult to study any one in isolation.

Of greatest importance are the characters of the spermatozoa themselves, since the seminal plasma, in the female, may soon be diluted and dispersed by oviductal secretions. Moreover, artificial diluents are used successfully in AI and the carrier medium can differ widely from the natural medium and still be effective. Hence seminal plasma may be essential only for the survival of spermatozoa within the male and to act as the carrier medium for their transfer to the female.

From AI studies it is clear that the concentration of spermatozoa in the inseminated dose is not as important as the total number deposited in the female. Fertility from intravaginal insemination of 1, 10, 100 and 1000 million spermatozoa resulted in a level of fertility of 21, 38, 95 and 97% respectively and a duration of fertility of 4, 7, 14 and 16 days respectively (Takeda, 1967). However, in natural mating, concentration is an important factor because, unlike AI, the quantity of semen in each ejaculate cannot be varied at will, and concentration determines the number of spermatozoa actually deposited in the female.

The total numbers inseminated are modified by the state of the spermatozoa. A large number of abnormal spermatozoa, as determined by morphologic and staining characters, leads to reduced fertility. "Dead" spermatozoa reduce the concentration, and hence total number, of viable spermatozoa introduced into the female. Motility is also correlated with fertilizing capacity.

Survival of the Spermatozoa in the Oviduct

The ejaculated semen is deposited at the entrance to the oviduct and contractions of the oviduct aid in moving the semen further into the vagina. Thereafter spermatozoa become distributed throughout the oviduct for a short while, but within 24 hours all have disappeared from the lumen. However, spermatozoa survive within the avian female for long periods (Lake, 1967, 1975): up to 32 days in the chicken and 70 days in the turkey, though the commonly known "fertile period" is usually shorter. Individuals and breeds vary considerably.

This longevity of spermatozoa is associated with at least one specialized region of the oviduct. At the uterovaginal junction, specialized tubular glands, the "sperm host glands (Fig. 21–7)," are located (Bobr et al, 1964). At present the factors concerned in the entry of sperm into the glands and their survival are unknown, and it is common experience that some hens fail to maintain sperm internally.

Whatever the cause, sperm rapidly enter the glands after insemination. Moreover, the glands may provide some barrier to the passage of defective sperm, since more embryonic deaths result when artificial insemination is intramagnal. Similar glands occur in the infundibular region of the oviduct and, following intramagnal insemination, sperm can be found there in large numbers. However, this is contrary to the findings following normal vaginal insemination, and the role of the infundibular glands has yet to be elucidated.

Fertilization

As fertilization occurs in the infundibular region, another requirement is the release of sufficient spermatozoa from the sperm host glands and their transport along the oviduct. Small, discrete "packets" of sperm may be released from the glands about the time of ovulation, for sperm can easily be collected from the oviductal lumen at this time. On the other hand, reappraisal of the evidence (Burke et al, 1969) suggests that release may be a continual process and sperm may be present, though difficult to recover, at all times.

Oviductal motility may be important for sperm to ascend the oviduct and waves of muscular contraction may pass along it in a direction opposite to that for the transport of the egg. Ciliary transport may also be important. Certainly, both dead sperm and inanimate particles can ascend the oviduct (Mimura, 1939). However, sperm motility could be significant, and secretions of the oviduct may aid transport because extracts from the magnum increased respiration and sustained motility of spermatozoa better than extracts from other regions.

Polyspermy is common but only one male nucleus fuses with the female pronucleus. However, about 150 spermatozoa are embedded in the vitelline membrane in a way similar to that in which a large number of sperm are present in the mammalian zona pellucida. It is not certain that this number is essential, although a reduction in numbers of spermatozoa at the anterior end of the oviduct leads to reduced fecundity. Alternatively an excessive number of spermatozoa at the infundibulum may lead to early disturbances of embryonic development: perhaps the vitelline membrane has no mechanism to prevent polyspermy, as it has in the mammal, and the rapid coating of the ovum with a thick layer of albumen may have made this development unnecessary under normal conditions. Capacitation is considerably less important than in mammals (Kamar and Hafez, 1975).

Lysosomal enzymes in the acrosome of mammalian sperm are involved in the dispersal of the corona radiata and penetration of the ovum. Lake (1971) suggested that the lipoglycoprotein associated with the acrosome of cock sperm may aid penetration, and others have demonstrated the presence of enzymes capable of initiating rupture of the vitelline membrane (Ho and Meizel, 1975, 1976).

REPRODUCTIVE PERFORMANCE IN THE FEMALE

Reproductive Cycles

In contrast to the mammal, a daily sequence of single ovulations typifies the bird, each followed by the completion of an egg and daily oviposition. This sequence is terminated by events about which we know little, though it may be the tactile stimulus of the eggs in the nest. In some species, if eggs are removed from the nest, further ovipositions occur until the required number remain to stimulate suspension of ovarian activity. Thereafter incubation behavior occurs; the ovary and oviduct regress to the juvenile condition, a situation presumably analogous to the anestrous condition of the seasonally breeding mammal.

In domestic birds, egg laying is not strictly a daily occurrence; a sequence of one or more successive daily eggs usually is interspersed with one day when no egg is laid (pause day). Therefore, it is difficult to equate the laying patterns of

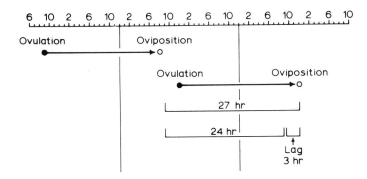

Fig. 21–9. The relationships between successive ovipositions and lag. *(From Gilbert and Wood-Gush, 1971)*

domestic and wild species. To avoid confusion, the term "clutch" should be restricted to the sequence of eggs of wild birds, usually of a fixed number within small limits; "sequence" and "cycle" are terms better used for those of domestic species.

Typically hens oviposit during the daylight hours. The first egg of a cycle is laid usually between 6 and 8 a.m. and each successive egg is laid later in the day than the previous one (Fig. 21–9). This difference in time from day to day is termed "lag" (Fraps, 1961, 1965); it depends on the length of the sequence and the position in the sequence of the two eggs concerned; in long sequences lag tends toward zero but may become negative, and lag between the first two members of a cycle tends to be greater than that between others. The terminal egg is laid toward late afternoon; the next egg follows about 36 hours afterward, ie, it is the first egg of a sequence laid early in the morning.

The cycle is terminated by failure of LH release, even though a mature follicle is present that can be ovulated in response to hormonal injection, and even though

the pituitary is capable of releasing LH. Whatever the reason for LH not being released, if a mature follicle is present it will ovulate when the next release of LH occurs. During this period more yolk material may be laid down. More commonly, however, a sequence is terminated without a mature follicle being present and the largest follicle will mature about 36 hours later. Often "gaps" can be seen within the hierarchy of developing follicles and they indicate the end of a subsequent sequence (Gilbert and Wood-Gush, 1971). Occasionally follicles become atretic and are unable to ovulate.

Total Egg Production

Much of the present discussion deals specifically with the chicken, though in general terms it is applicable to other domestic species (Table 21–3). After the first egg, production rises sharply and a peak of 90% or more is reached. Thereafter egg production gradually falls to a level of about 50% at the end of the first year of production. Since there is an early period of growth and development during

Table 21–3. Comparative Reproductive Performance* of Some Birds of Commercial Importance

Species	Incubation Period (Days)	Age at Sexual Maturity (Months)	Egg Weight (gm)	No. Eggs in First Laying Year	Fertility (%)	Hatchability of Fertile Eggs (%)
Chicken *(Gallus gallus)*						
Layer	21	5–6	58	300	97	90
Broiler	21	6	65	180	92	90
Turkey *(Meleagris gallopavo)*	28	7–8	85	90	83	84
Duck *(Anas platyrhynchos)*						
Layer	27–28	6–7	60	300	95	75–80
Meat type	28	6–7	65			
Goose *(Anser anser)*						
Small type	30	9–10	135	30–70	70	70
Large type	33	10–12	215			
Pheasant *(Phasianus colchicus)*	24–26	10–12	30	50–75	95	85
Guinea fowl *(Numida meleagris)*	27–28	10–12	40	80–200	90	95
Quail *(Coturnix coturnix)*	15–16	1.5–2	10	300	90	75–85

*Only general figures are given since values are greatly affected by breed, location and nutrition. In particular the values given in the last three columns depend greatly on management practices, and it is likely that fertility and hatchability will be the same for all species.

which reproduction does not occur, and since the hen's reproductive phase lasts for a limited period, the number of eggs that can be produced by a hen depends on when her sexual maturity starts, when it ends, and her rate of laying during this period.

Onset of Sexual Maturity. Reproductive potential of any animal depends on genetic makeup, which can be altered by selective breeding. In poultry, breeders have concentrated on selecting for early sexual maturity. The most important factor in bringing about sexual maturity is an increasing daylength, since the hypothalamus-pituitary axis is regulated by this (Fig. 21–3). However, light strength, above a very low level, and wavelength are unimportant. Hence there is a direct correlation between daylength during the rearing period and the age at onset of sexual maturity (Morris, 1967).

Many problems arise with egg quality during the early periods: among them are the production of defective eggs or soft-shelled eggs and 40% of the ova produced may fail to enter the oviduct. Hatchability is generally poor.

Main Period of Lay. After the first period, hens typically lay regular sequences, the length of the sequence being a characteristic of the hen. This period lasts about one year; though eggs may be produced longer, production levels are low (and uneconomical) and egg quality deteriorates. Commercially it is customary to kill birds at this time or to induce a cessation of lay for several months followed by a shorter second year's production, when egg quality and performance are improved over those at the end of the first year.

Total production depends on the length of the day period, but rate of lay is also important and this depends on the length of the sequence, eg, a hen laying three-egg sequences can only attain a maximum of 75% production, whereas one having sequences of 24 eggs can reach 96%.

Other factors are important (Gilbert, 1972). There appears to be an optimal egg size for production; large and small eggs are associated with poorer production figures. Since one important factor controlling egg size is the size of the yolk, yolk transport is related to total production.

Calcium is involved not only in shell formation but also in regulating total production. This is not surprising: the turnover of calcium in the formation of one shell is equivalent to 10% of the total body calcium, and few animals have such a consistent drain on their calcium resources as the bird (Simkis and Taylor, 1971). Also, mechanisms exist to protect the skeletal tissues; surprisingly, however, these do not reside within the shell gland, and while eggs are produced the shell gland forms a shell. The protective mechanisms appear to be associated with control of the production of oocytes through the pituitary gonadotropin output: hens on a calcium-deficient diet will produce eggs if given gonadotropin, but the birds die with severe skeletal defects. Hens that lay eggs without shells ovulate at a higher rate than normal hens and will do so even when on a severely calcium-deficient diet.

Physical environment, nutrition and disease affect reproductive performance, but little is known of the way in which they affect the bird's physiology (Gilbert and Wood-Gush, 1971). Below 10 hours of daylight and/or 10 lux at the food trough, reproductive performance is curtailed. Temperatures between 13°C and 21°C are recommended as optimal for egg production, though higher temperatures may lead to increased production figures. Altitude, gaseous environment, humidity and noise may also affect egg production.

Nutritional deficiencies affect egg production, but excessive amounts of certain substances may have deleterious effects (Scott et al, 1969). The increasing practice of feeding artifactual foodstuffs, such as drugs, may be beneficial in certain circumstances, but a balance must be reached between therapeutic and toxic effects. Disease causes decreased egg production and an increase in defective eggs.

End of the Laying Season. Some birds (eg, turkey) seem to become photorefractory, and hence, no longer respond to the light stimulus. Their egg production consequently decreases. On the other hand, the chicken pituitary is still capable of producing gonadotropins, the circulating plasma levels of these are high, yet reproductive efficiency decreases. This may indicate a change in relative hormone levels that adversely affects reproductive physiology. Alternatively, the ovary may become more refractory to stimulation by the gonadotropins. The deterioration in shell quality is, at least partly, a result of the steady increase in egg size with age.

REPRODUCTIVE PERFORMANCE IN THE MALE

As for the female, many interactions of management and husbandry of males influence their reproductive performance. There are wide differences in the onset of sexual maturation and in semen production between breeds (strains) and between individuals of one breed (Masliev and Davtyan, 1969). Some characteristics are inherited, such as volume of semen and sperm motility, concentration, vigor and number of sperm per ejaculate (Soller et al, 1965). However, fertility is not necessarily correlated with any of these and enormous variability is found both within and between different genetic lines. The expression of these traits can be altered by manipulation of the environmental factors, and the response depends on the genotype.

Photoperiod is a potent stimulator of sexual maturation (Siegel et al, 1969); cocks reared on about 13 hours of daylength produced semen earlier than those on daylengths less than eight hours. However, increasing the daily light period after 20 weeks of age was more effective than rearing on a constant daylength. The color of light is important, and red seems to be the most efficient. After sexual maturity, semen production appears to be stimulated by longer periods of daylight.

Often closely associated with light in the wild state, temperature is known to affect reproductive performance. However, as with lighting experiments, care is required in the interpretation of the results of temperature manipulation, because this may affect food intake which, by itself, will affect reproductive activity. In general, low ambient temperatures (8° C) retard sexual development of males, but may cause better semen production than high temperatures (30° C), when males are active. Most experiments indicate that for adults, an ambient temperature between 20° C and 25° C tends to result in optimal semen production, whereas extremes on either side are deleterious (Subhas and Huston, 1967).

Nutritional factors may affect semen production; males are usually expected to consume rations designed for hens, but if an adequate quantity of food is provided, it is likely that most specific requirements will be met. The high level of calcium in these rations appears to cause no harm. The level of dietary protein during growth is known to affect both the onset of sexual maturity and semen production afterward; a concentration of 9% or less delays the onset and 16% affects adult semen production. In the adult, dietary protein between 10% and 20% seems acceptable, though the type of protein may be important. Kamar (1964) and Gleichauf (1967) found that the inclusion of animal protein enhanced semen production. Adequate vitamins are important, but surprisingly, slight deficiencies in vitamin A may be beneficial (Lake, 1969). The required level of vitamin E may depend on the amount of linoleic acid in the diet. Of the essential fatty acids, this appears to be the most important, and a deficiency causes a decrease in the fertilizing capacity of spermatozoa (Lille and Menge, 1967).

REFERENCES

Aitken, R.N.C. (1971). The oviduct. *In* Physiology and Biochemistry of the Domestic Fowl. D.J. Bell and B.M. Freeman (eds), London, Academic Press, pp. 1237–1239.

Amin, S.A. and Gilbert, A.B. (1970). Cellular changes in the anterior pituitary of the domestic fowl during growth, sexual maturity and laying. Br. Poult. Sci. 11, 451.

Becking, J.H. (1975). The ultrastructure of the avian eggshell. Ibis 117, 143.

Bellairs, R. (1964). Biological aspects of the yolk of the hen's egg. *In* Advances in Morphogenesis. M. Abercrombie and J. Brachet (eds), New York, Academic Press, pp. 217–272.

Burke, W.H., Ogasawara, F.X. and Fuqua, C.L. (1969). Transport of spermatozoa to the site of fertilization in the absences of oviposition and ovulation in the chicken. Poult. Sci. 48, 602.

Bobr, L.W., Lorenz, F.W. and Ogasawara, F.X. (1964). Distribution of spermatozoa in the oviduct and fertility in domestic birds. I. Residence sites of spermatozoa in fowl oviducts. J. Reprod. Fertil. 8, 39.

Carter, T.C. (1970). The hen's egg: Some factors affecting deformation in statically loaded shells. Br. Poult. Sci. 11, 15.

Day, S.L. and Nalbandov, A.V. (1977). Presence of prostaglandin F (PGF) in hen follicles and its physiological role in ovulation and oviposition. Biol. Reprod. 16, 486.

Dick, H.R., Culbert, J., Wells, J.W., Gilbert, A.B. and Davidson, M.F. (1978). Steroid hormones in the postovulatory follicle of the domestic fowl *Gallus domesticus*. J. Reprod. Fertil. 53, 103.

Etches, R.J. and Cunningham, F.J. (1976). The interrelationships between progesterone and luteinizing hormone during the ovulation cycle of the hen *(Gallus domesticus)*. J. Endocrinol. 71, 51.

Feeney, R.E. and Allison, R.G. (1969). Evolutionary Biochemistry of Proteins. New York, J. Wiley & Sons.

Fraps, R.M. (1961). Ovulation in the domestic fowl. *In* Control of Ovulation. C.A. Villee (ed), London, Pergamon Press, pp. 133–162.

Fraps, R.M. (1965). Twenty-four hour periodicity in the mechanism of pituitary gonadotrophin release for follicle maturation and ovulation in the chicken. Endocrinology 77, 8.

Gilbert, A.B. (1971a). The egg: its physical and chemical aspects. *In* Physiology and Biochemistry of the Domestic Fowl. D.J. Bell and B.M. Freeman (eds), London, Academic Press, pp. 1379–1399.

Gilbert, A.B. (1971b). Egg albumen and its formation. *In* Physiology and Biochemistry of the Domestic Fowl. D.J. Bell and B.M. Freeman (eds), London, Academic Press, pp. 1291–1329.

Gilbert, A.B. (1971c). The ovary. *In* Physiology and Biochemistry of the Domestic Fowl. D.J. Bell and B.M. Freeman (eds), London, Academic Press, pp. 1163–1208.

Gilbert, A.B. (1971d). The female reproductive effort. *In* Physiology and Biochemistry of the Domestic Fowl. D.J. Bell and B.M. Freeman (eds), London, Academic Press, pp. 1153–1162.

Gilbert, A.B. (1972). The activity of the ovary in relation to egg production. *In* Egg Production and Formation. B.M. Freeman and P.E. Lake (eds), Edinburgh, British Poultry Science, pp. 3–21.

Gilbert, A.B. (1979). Female genital organs. *In* Form and Function in Birds. A.S. King and J. McLelland (eds), London, Academic Press, pp. 237–360.

Gilbert, A.B. and Wood-Gush, D.G.M. (1971). Ovulatory and ovipository cycles. *In* Physiology and Biochemistry of the Domestic Fowl. D.J. Bell and B.M. Freeman (eds), London, Academic Press, pp. 1353–1378.

Gleichauf, R. (1967). Der Einfluss von tierischem Eiweiss auf die Samenproduktion von weissen Leghorn-Hahnen. Arch Geflügelk. 31, 111.

Guichard, A., Scheibs, D., Haffen, K. and Cedard, L. (1977). Radioimmunoassay of steroid hormones produced by embryonic chick gonads during organ culture. J. Steroid. Biochem. 8, 599.

Hafez, E.S.E. (1974). The Behavior of Domestic Animals. 3rd ed. Philadelphia, Lea & Febiger.

Ho, J.J.L. and Meizel, S. (1975). Hydrolysis of the hen egg vitelline membrane by cock sperm acrosin and other enzymes. J. Exp. Zool. 194, 429.

Ho, J.J.L. and Meizel, S. (1976). Biochemical characterization of an avian spermatozoan acrosin and comparison of its properties to those of bovine trypsin and mammalian acrosins. Comp. Biochem. Physiol. 543, 213.

Kamar, G.A.R. (1964). The effect of amino acids and ascorbic acid on reproduction of chickens. 5th Int. Congr. Anim. Reprod. A.I. Trento 2, 168.

Kamar, G.A.R. and Hafez, E.S.E. (1975). Sperm maturation and fertility in poultry. Anim. Breed. Abst. 43, 99.

Lake, P.E. (1956). The structure of the germinal epithelium of the fowl testis with special reference to the presence of multinuclear cells. Q.J. Microscop. Sci. 97, 487.

Lake, P.E. (1967). The maintenance of spermatozoa in the oviduct of the domestic fowl. *In* Reproduction in the Female Mammal. E.G. Lamming and E.C. Amorosco (eds), London, Butterworths, pp. 254–266.

Lake, P.E. (1969). Factors affecting fertility. *In* The Fertility and Hatchability of the Hen's Egg. T.C. Carter and B.M. Freeman (eds), Edinburgh, Oliver and Boyd, pp. 3–29.

Lake, P.E. (1971). The male in reproduction. *In* Physiology and Biochemistry of the Domestic Fowl. D.J. Bell and B.M. Freeman, (eds), London, Academic Press, pp. 1411–1447.

Lake, P.E. (1975). Gamete production and the fertile

period with particular reference to domesticated birds. Proc. Zool. Soc., Lond. *35*, 225.

Lillie, R.N. and Menge, H. (1967). Effect of linoleic acid deficiency on the fertilizing capacity and semen fatty acid profile of the male chicken. J. Nutr. *95*, 311.

McIndoe, W.M. (1971). Yolk synthesis. In Physiology and Biochemistry of the Domestic Fowl. D.J. Bell and B.M. Freeman (eds), London, Academic Press, pp. 1209–1223.

McIntosh, J.R. and Porter, K.R. (1967). Microtubules in the spermatids of the domestic fowl. J. Cell. Biol. *35*, 153.

Marvan, F. (1969). Post-natal development of the male genital tract of *Gallus domesticus*. Anat. Anz. *124*, 443.

Marza, U.D. and Marza, R.V. (1935). The formation of the hen's egg. 1-IV. Q.J. Microsc. Sci. *78*, 134.

Masliev, I.T. and Davtyan, A.D. (1969). Influence of nutrition on the semen production of roosters. Wld's Poult. Sci. J. *25*, 315.

Mimura, H. (1939). On the mechanism of travel of spermatozoa through the oviduct in the domestic fowl with special reference to artificial insemination. Okajimas Folia Anat. Jpn. *17*, 459.

Morris, T.R. (1967). Light requirements of the fowl. In Environmental Control of Poultry Production. T.C. Carter (ed), Edinburgh, Oliver & Boyd, pp. 15–39.

Murton, R.K. and Westwood, N.J. (1977). Avian Breeding Cycles. Oxford, Clarendon Press.

O'Malley, B.W., McGuire, W.L., Kohler, P.O. and Korenman, S.G. (1969). Studies on the mechanism of steroid hormone regulation of synthesis of specific proteins. Recent Prog. Horm. Res. *25*, 105.

Palmitter, R.D. (1971). Interaction of estrogen, progesterone and testosterone in the regulation of protein synthesis in chick oviduct. Biochemistry *10*, 4399.

Perry, M.M., Gilbert, A.B. and Evans, A.J. (1978). Electron-microscope observations on the ovarian follicle of the domestic fowl during the rapid growth phase. J. Anat. *125*, 481.

Schimke, R.T., Pennequin, P., Robins, D. and McKnight, G.S. (1977). Hormonal regulation of egg white protein synthesis in chicken oviduct. In Hormones and Cell Regulation. J. Dumont and J. Nunez (eds), Amsterdam, North Holland Publishing Co., pp. 209–221.

Scott, M.L., Nesheim, M.C. and Young, R.J. (1969). Nutrition of the Chicken, Ithaca, N.Y., Scott.

Senior, B.E. and Cunningham, F.J. (1974). Oestradiol and luteinizing hormone during the ovulatory cycle of the hen. J. Endocrinol. *60*, 201.

Shahabi, N.A., Norton, H.W. and Nalbandov, A.V. (1975). Steroid levels in follicles and the plasma of hens during the ovulatory cycle. Endocrinology *96*, 962.

Sharp, P.J. (1975). A comparison of variations in plasma luteinizing hormone concentrations in male and female domestic chicken (*Gallus domesticus*) from hatch to sexual maturity. J. Endocrinol. *67*, 211.

Siegel, H.S., Siegel, P.B. and Beane, W.L. (1969). Semen characteristics and fertility of meat-type chickens given increasing daily photoperiods. Poult. Sci. *48*, 1009.

Simkiss, K. and Taylor, T.G. (1971). Shell formation. In Physiology and Biochemistry of the Domestic Fowl. D.J. Bell and B.M. Freeman (eds), London, Academic Press, pp. 1331–1343.

Simons, D.C.M. and Wiertz, G. (1965). Differences in measurement of membrane thickness in hen's egg shells. Br. Poult. Sci. *6*, 283.

Simons, D.C.M. and Wiertz, G. (1970). Notes on the structure of shell and membranes of the hen's egg; a study with the scanning electron microscope. Ann. Biol. Anim. Biochim. Biophys. *10*, 31.

Soller, M., Snapir, N. and Schindler, H. (1965). Heritability of semen quantity, concentration and motility in White Rock Roosters, and their genetic correlation with rate of gain. Poult. Sci. *44*, 1527.

Subhas, T. and Huston, T.M. (1967). The influence of different environmental temperatures on the plasma levels of FSH activity in female fowl. Poult. Sci. *46*, 1324.

Takeda, A. (1967). Behaviour of spermatozoa in the genital tract of the hen. IV. Persistence of spermatozoa in their transport in the hen's oviduct following artificial insemination. Jap. Poult. Sci. *4*, 62.

Waites, G.M.H. (1970). Temperature regulation and the testis. In The Testis. A.D. Johnson, W.R. Gomes and N.L. Van Demark (eds), New York, Academic Press, pp. 241–279.

Wilson S.C. and Sharp, P.J. (1973). Variations in plasma LH levels during ovulatory cycle of the hen *Gallus domesticus*. J. Reprod. Fertil. *35*, 561.

Wood-Gush, D.G.M. (1971). The Behaviour of the Domestic Fowl. London, Heinemann.

Woods, J.E. and Domm, L.V. (1966). A histochemical identification of the androgen-producing cells of the domestic fowl and albino rat. Gen. Comp. Endocrinol. *7*, 559.

IV. reproductive
failure

22

Reproductive Failure in Females

M.R. JAINUDEEN
and
E.S.E. HAFEZ

Sterility is some permanent factor preventing procreation, and infertility or temporary sterility is the inability to produce viable young within a stipulated time characteristic for each species. In this chapter we shall examine the phases of the reproductive process that are most vulnerable and show how hormonal imbalances or adverse environmental, genetic and hereditary factors exert their influences (Fig. 22–1). An important factor—reproductive infections—is discussed in a subsequent chapter.

OVARIAN DYSFUNCTION

The two main functions of the ovary, the production of ova and secretion of

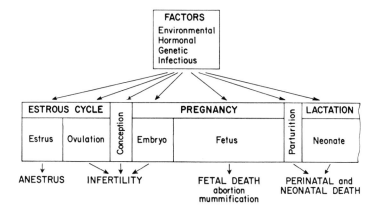

Fig. 22–1. Schematic representation of factors that adversely affect the reproductive process. Common types of reproductive failure are anestrus, infertility, fetal, perinatal and neonatal deaths. Note that infertility may result from fertilization failure or embryonic mortality.

449

ovarian hormones, are intimately related and directed toward successful reproduction.

Anestrus

Anestrus denotes a state of complete sexual inactivity with no manifestations of estrus. It is not a disease but a sign of a variety of conditions (Table 22–1). Although anestrus is observed during certain physiologic states, eg, before puberty, during pregnancy and lactation and in seasonal breeders, it is most often a sign of temporary or permanent depression of ovarian activity (true anestrus) caused by seasonal changes in the physical environment, nutritional deficiencies, lactation stress and aging (Fig. 22–2). Certain pathologic conditions of the ovaries or the uterus also suppress estrus.

Seasonal Anestrus. During seasonal anestrus there are no cyclic changes in the ovaries and reproductive tract. The extent of seasonal anestrus varies with the species, breed and physical environment, and is more pronounced in sheep and horses than in cattle, pigs and most laboratory mammals. Anestrus in the mare occurs during the winter and spring when daylight hours are short. The ovaries become small and hard, contain neither follicles nor corpora lutea, and there are low serum concentrations of LH, progesterone and estradiol (Oxender et al, 1977). Artificial lighting is an important management technique for renewing synthesis and release of gonadotropins in the seasonally anestrus mare, but the response to injections of gonadotropic hormones has been disappointing.

Anestrus During Lactation. In several species, ovulation and related reproductive activity are suppressed for a variable period after parturition and during lactation. The incidence and duration of anestrus varies greatly between different species and breeds, and is also influenced by the season of parturition, level of milk production, number of young being nursed and the degree of postpartum involution of the uterus. For example, during periods of high temperatures and on poor diets, Brahman cows that are nursing calves are particularly subject to anestrus. The duration of anestrus in cows nursing

Table 22–1. Abnormalities of Estrus

Species	Abnormality	Causes	Physiologic Mechanisms
Cattle	Anestrus	Pyometra, mummification	Maintenance of corpus luteum
		Lactation	Suckling stimulus inhibits gonadotropin release
		Cystic ovaries	Deficiency of LH and/or GnRH
		Ovarian hypoplasia and freemartinism	Failure to produce ovarian estrogens
		Nutritional and vitamin deficiencies	Gonadotropin production by anterior pituitary
	Subestrus, silent estrus (quiet ovulation)	High lactation	Endocrine imbalance
	Nymphomania	Cystic ovaries	
Sheep	Anestrus	Season, lactation	Effect of photoperiod on gonadotropin secretion
Swine	Anestrus	Lactation	As for cattle
Horse	Anestrus	Season, diet, ovarian hypoplasia	As for sheep
	Prolonged estrus	Early in breeding season; aging; undernutrition	
	Split estrus, silent estrus		

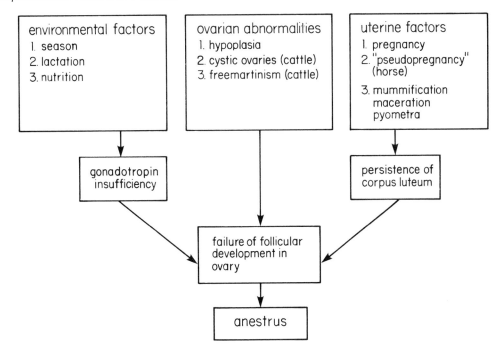

Fig. 22–2. Scheme illustrating the possible causes leading to a failure of follicular development in the ovary and anestrus in farm animals. Note that pregnancy is an important cause for an absence of estrus.

calves is longer than in similar cows milked twice daily; this suggests that nursing or frequency of milk removal may influence the pituitary gonadotropic activity. Ovarian activity, based on plasma progesterone levels, is resumed in less than 30% suckling cows by 60 days postpartum (Chupin et al, 1976). Two injections of GnRH, first given at 20 to 30 days after calving and the second approximately 10 days later, will reduce the anestrous period in suckled cows (Webb et al, 1977).

In sheep, lactational anestrus lasts five to seven weeks. Some ewes suckling lambs will come into estrus, but most ewes show estrus about two weeks after the lambs are weaned. Most foaling mares come into estrus within 5 to 15 days after foaling, but some nervous lactating mares may experience lactational anestrus due to psychologic disturbance rather than to the stress of lactation.

Physiologic interactions between lactation and depression of ovarian function have not been fully established, but may be related to pituitary dysfunction associated with lactation. The duration of anestrus is closely related to the length and intensity of lactation, and ovarian cysts are common during the early postpartum period.

Anestrus Due to Aging. Farm animals with the exception of the horse are rarely maintained into old age for economic reasons, and even more rarely given the opportunity to breed late in life. In rodents, the incidence of irregular or anovulatory cycles rises steadily with age, and even when the females are no longer fertile, mating frequently results in pseudopregnancy. Abnormal corpora lutea and ovaries lacking corpora lutea accounted for over 80% of cases of infertility in cows 14 to 15 years old (Erickson et al, 1976). This ovarian dysfunction may be related to one or all of the following: (1) failure of follicular cells to respond fully to hormonal stimuli; (2) change in the quantity and/or quality of hormonal secretion; (3) reduced stimulus. Regardless of the mechanism involved, anestrus due to aging probably alters the functional relationship of the hypothalamus-pituitary-ovarian axis, thereby leading to a decrease in gonadotropin secretion or a change in the ovarian response to these hormones.

Nutritional Deficiencies. Energy level has a significant effect on ovarian activity. Inadequate nutrition suppresses estrus in young growing females more than in adults. Low levels of energy lead to ovarian inactivity and anestrus in beef cows that are suckling calves and in sows after weaning.

Deficiencies of minerals or vitamins cause anestrus. Phosphorus deficiency in range cattle and sheep causes ovarian dysfunction, which in turn leads to delayed puberty, depressed signs of estrus and eventually cessation of estrus. Gilts or cows fed a manganese-deficient diet experience ovarian disturbances ranging from weak signs of estrus to anestrus. Vitamin A or E deficiencies may cause irregular estrous cycles or anestrus.

Abnormalities of the Ovary or Uterus. *Ovarian hypoplasia* occurs in Swedish mountain cattle. Affected animals have infantile reproductive tracts and never exhibit estrus. The morphology of the ovary differs from that of seasonal anestrus. Follicles of varying diameter up to the preovulatory size, which commonly are present in the ovaries of anestrous animals, are absent in ovarian hypoplasia. Ovarian hypoplasia tends to be associated with white coat color, being inherited as an autosomal recessive. Some mares with small inactive ovaries have abnormal sex chromosome complement (eg,XO) as well as low plasma estrogen and high LH levels (Hughes and Trommershausen-Smith, 1976).

Freemartins, or heifers born co-twin to bulls, have poorly developed ovaries and fail to show estrus. *Cystic ovaries* in cattle may lead to a prolonged period of anestrus.

Uterine distention in cattle and swine due to pathologic conditions (eg, pyometra, mucometra, fetal mummification or maceration) or *pseudopregnancy* (pregnancy failure) in the mare is associated with a retention of the corpus luteum and therefore a suppression of the estrous cycle. *Prolonged diestrus,* apparently unique to mares, results from spontaneous prolongation of the life of the cyclic corpus luteum beyond the normal 14 to 15 days, and is a major cause of anestrus during the natural breeding season. Persistence of the corpus luteum may be attributed to a failure of production and/or release of uterine luteolysin.

Atypical Estrus

Short estrus, prolonged estrus, "split" estrus, nymphomania and "silent" estrus are not uncommon (Table 22–1). Estrus may be of short duration and without well-marked signs. It may be undetected in young animals without the presence of a teaser male or it may occur during the night, particularly in cattle. Prolonged estrus, lasting from 10 to 40 days, without ovulation characterizes the transition from seasonal anestrus to resumption of cyclic activity in mares during the breeding season. "Split" estrus or behavioral estrus interrupted by one or two days of sexual nonreceptivity is also observed in mares, especially at the start of the breeding season.

Nymphomania occurs more frequently in dairy cattle than in beef cattle and horses. Nymphomania is one of the signs of cystic ovaries in cattle. Nymphomaniac cows show intense estrous behavior persistently or at frequent but irregular intervals, depressed milk production, a frequent copious discharge of clear mucus from the vulva, edema and relaxation of the sacrosciatic ligaments and a raised tail head. Nymphomaniac mares are excitable, vicious and intractable. They will not tolerate the approach of another horse nor will they stand for mating. The occurrence of ovarian cysts in the mare analogous to the condition in cows is doubtful, since in the mare ovariectomy has no effect on the abnormal behavioral pattern (nymphomania).

"Silent" estrus (quiet ovulation) or the occurrence of ovulation without overt estrus occurs in all farm animals, particularly young animals and those on a submaintenance ration. It is suspected when the interval between two consecutive estrous periods is double or triple the normal length. A high incidence of silent estrus occurs in sheep during the first estrous cycle of the breeding season, apparently related to the absence of a corpus luteum from a previous cycle, and at the end of the breeding season, probably due to estrogen deficiency. Several silent estruses occur in beef cows and ewes that suckle young and in dairy cows milked three times daily. Silent estrus is frequently encountered in maiden mares and in mares with a foal at foot.

Ovulatory Failure

Ovulatory failure may be due to failure of the follicle to ovulate during a normal cycle or to cystic ovaries.

Anovulatory estrus is more common in swine and horses than in cattle and sheep. The animal shows normal behavioral estrus and the ovarian follicle reaches preovulatory size but does not rupture. Anovulatory follicles become partly luteinized and then regress during the estrous cycle, as does a normal corpus luteum.

Cystic ovaries, common in dairy cattle and swine, are rarely encountered in beef cattle or other species. Cystic ovaries are frequently encountered in high-producing dairy cows during the first three

months of lactation. Affected animals are either nymphomaniacs or anestrous. One or both ovaries contain multiple small cysts or one or more large cysts. These are either follicular or luteal cysts. Follicular cysts undergo cyclic changes, ie, they alternately grow and regress but fail to ovulate. Luteal cysts contain a thin rim of luteal tissue, also fail to ovulate, but persist for a prolonged period. The cystic fluid has a high concentration of progesterone and a low concentration of estrogen, but there is no relationship between these hormonal concentrations in the cyst's fluid and behavioral patterns (nymphomania or anestrus).

In swine, cystic ovaries are an important cause of reproductive failure. Large multiple luteinized follicles are more common than small multiple cysts, and they contain progesterone. Estrous cycles are irregular with prolonged periods between cycles. Signs of estrus are pro-

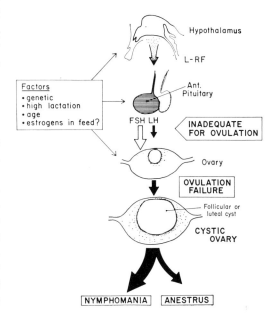

Fig. 22–3. Endocrine sequence, types of cysts and behavioral manifestations associated with cystic ovaries in the cow. Inadequate secretion of LH results in ovulatory failure and formation of follicular or luteal cysts. Note that affected cows are either nymphomaniac or in anestrus.

nounced but nymphomania does not occur. It is not certain whether cystic ovaries in cattle and swine result from a failure of the ovulatory mechanism, from adrenal cortex hyperfunction, or from a disturbance in the hypothalamo-pituitary axis that leads to premature release of LH or insufficient LH to cause ovulation. The development of cystic ovaries in cattle has been related to high milk production, seasonal changes, hereditary predisposition and pituitary dysfunction (Fig. 22–3). Administration of gonadotropin-releasing hormone (GnRH) or LH is an effective treatment for cows with ovarian follicular cysts. For treatment of luteal cysts, prostaglandin $F_{2\alpha}$ or its analogues have been used successfully.

DISORDERS OF FERTILIZATION

Disorders of fertilization include failure of fertilization and atypical fertilization.

Fertilization Failure

Fertilization failure may result from death of the egg before sperm entry, structural and functional abnormality in the egg or sperm, physical barriers in the female genital tract preventing gamete transport to the site of fertilization or ovulatory failure and cystic ovaries, as discussed previously.

Abnormal Eggs. Several types of morphologic and functional abnormalities have been observed in unfertilized eggs, eg, giant egg, oval-shaped egg, lentil-shaped egg and ruptured zona pellucida. Failure to undergo fertilization and normal embryonic development may be due to inherent abnormalities of the egg or to environmental factors. For example, fertilization is lower in animals exposed to elevated ambient temperature prior to breeding. In sheep, some of the conception failures at the beginning of the breeding season are associated with a high incidence of abnormal ova.

Abnormal Sperm. Semen usually contains a varying percentage of morphologically abnormal sperm. The physiologic significance of abnormal sperm in relation to fertilization failure has not been studied in animals other than cattle. Certain forms of male infertility are related to structural defects of the DNA protein complex.

Sperm aging and injury may cause alterations in the acrosomal cap. Acrosin, an enzyme within the acrosome, appears to be important in the penetration of the zona pellucida of the egg. It has been postulated that acrosin inhibitors present in seminal plasma combine with acrosin only in spermatozoa with acrosomal damage and prevent defective spermatozoa from fertilizing the egg. In bull, ram and boar, a good correlation exists between fertility and acrosomal integrity.

Sperm aging during prolonged storage has been attributed either to a leakage of vital intracellular constituents such as cyclic AMP or to the formation of lipid peroxides from sperm plasmalogen when sperm are stored under anerobic conditions (Mann, 1976). Although antigenic incompatibility between sperm and egg could lead to sperm rejection and fertilization failure, there is at present no convincing evidence that antibodies pass from serum into the lumen of the uterus or oviduct to agglutinate sperm.

Structural Barriers to Fertilization. Congenital or acquired defects of the female genital tract interfere with transport of the sperm and/or the ovum to the site of fertilization (Table 22–2). Congenital defects are the result of arrested development of the different segments of the müllerian duct (oviduct, uterus and cervix) or of an incomplete fusion of these ducts caudally. Acquired defects are caused by trauma or infection, particularly at time of parturition.

Several types of developmental anomalies affect the reproductive tracts of all species. Animals with anatomic de-

Table 22–2. Structural and Functional Causes of Fertilization Failure

Cause	Abnormality	Affected Species	Mechanism Interfered With
Structural Obstructions			
Congenital	Mesonephric cysts Uterus unicornis Double cervix	More common in swine, sheep and cattle than in horses	Sperm transport
Acquired	Tubal adhesions Hydrosalpinx Occluded uterine horns	All species, sheep and swine particularly	Egg pick-up, fertilization Egg transport
Functional			
Hormonal	Cystic ovaries	Cattle and swine	Ovulation
	Abnormal cervical and uterine secretions	Cattle; sheep on estrogenic pastures	Gamete transport
Management	Delayed insemination	In all species, horses and swine particularly	Death of egg
	Insemination too early	Cattle	Death of sperm

fects have cycles of normal length and normal estrus, and these abnormalities cannot be detected clinically. Anatomic abnormalities are common in swine and account for about half of the total cases of reproductive failure.

Common anatomic abnormalities are adhesions of the infundibulum to the ovary or uterine horns; this interferes with the pick-up of the egg or causes a mechanical obstruction of one part of the reproductive duct system. Bilateral or unilateral missing segments of the reproductive tract also cause anatomic sterility. Other congenital abnormalities, such as a double cervix, may interfere with reproduction but do not always cause infertility. An abnormal shape or position of the cervix, a narrow cervical canal preventing the transport of sperm to the oviducts, or a torn cervix may also cause reproductive failure.

Small unilateral or bilateral cysts found in the mesosalpinx or on the oviducts do not interfere with fertility; however, large cysts may occlude the lumen of the oviduct and prevent egg transport.

A classic congenital anomaly associated with the gene for white coat color is "white heifer disease" in cattle, in which the prenatal development of the müllerian ducts is arrested, and the vaginal canal is obstructed by the presence of an abnormally developed hymen. The degree and area of hypoplasia differ, so that various anomalies of the oviducts, uterus, cervix and vagina are formed. This congenital anomaly can be differentiated from the freemartin syndrome by the presence of normal ovaries, vulva and labia in the animals with white heifer disease.

Atypical Fertilization

The complex process of fertilization is subject to several aberrations, namely polyspermy, monospermic fertilization of an egg containing two female pronuclei, failure of pronucleus formation and gynogenesis or androgenesis. Atypical fertilization may occur spontaneously as a result of aging of the gametes or elevation of environmental temperature. It has also been induced experimentally by x-rays or the administration of certain toxic substances.

The aging of the ovum is gradual, during which various functions are successively lost (Table 22–3). An early effect of egg aging is that the resulting embryo is not viable and is resorbed before birth. Further aging leads to abnormalities in fertilization, particularly involving the pronuclei. The biophysical and biochemi-

Table 22–3. Aging of Gametes and Atypical Fertilization

Gamete	Mechanism	Abnormality
Sperm	Reduction or loss of DNA	Reduced viability of embryo
Egg	Incomplete maturation with failure to release the second polar body	Triploid embryo
	Inhibition of the block to polyspermy	Triploid or heteroploid embryo

cal reactions associated with sperm entry into the egg become slower, a condition leading to increased polyspermy (entry of more than one sperm).

Polyspermy occurs in several species of laboratory and farm animals. In swine, a delay in copulation or injection of progesterone given 24 to 36 hours before ovulation leads to some eggs having more than two pronuclei (Day and Polge, 1968). It is not clear whether these potential triploid embryos are caused by failure to extrude the second polar body or to polyspermy, which may result from failure of the block to polyspermy during ovum aging. The incidence of polyspermy increases when mating or insemination is delayed, resulting in triploid embryos that do not survive. This means that in horses and swine with a relatively long estrus, the timing of breeding in relation to ovulation is critical for normal fertilization and embryonic survival. Similarly, polyspermy could account for some returns to service by delayed insemination of cattle.

PRENATAL MORTALITY

Many interacting factors determine prenatal development, such as nutritional and genetic factors, which control the size and viability of the offspring, and uterine infections, which interfere with implantation, placentation or prenatal survival. Evidence is also accumulating in farm animals that incompatibility of the fetus and dam may lead to prenatal mortality.

A small percentage of prenatal loss is involved in the normal reproductive process and may be regarded as unavoidable. Prenatal mortality, responsible for approximately a third of all gestation failures, can be divided into *embryonic* and *fetal* mortality.

Embryonic Mortality

Approximately 25% to 40% of embryos in cattle, sheep and swine are lost between the time of sperm penetration of the ovum and the end of implantation (Table 22–4). Most losses occur before or immediately following implantation, resulting in complete resorption of the conceptus. It is also noted in large litters of swine and during multiple pregnancies in cattle and sheep. The time of mortality in cattle affects the resumption of estrus in two ways. In the first form, the fertilized egg develops to the morula or early blastocyst stage but degenerates before the middle of the estrous cycle. The corpus luteum regresses as in a normal cycle and the animal returns to estrus. In the second form, the blastocysts degenerate after midcycle but prior to, or immediately following, implantation. The regression of the corpus luteum is thus delayed for a period that is longer than the length of one estrous cycle.

The effects of embryonic mortality on the estrous cycle of swine are determined by the number of embryos that survive and the stage of pregnancy (Dunne, 1975). For example, if all embryos are lost by day 4 of gestation, the sow returns to estrus after a normal cycle length, but if one to four embryos survive beyond day 4, the pregnancy would still terminate but the next estrous period is delayed by six days. For pregnancy to continue beyond day 10 at least a total of four embryos must be present in both uterine horns, whereas for it to continue beyond 12 days, as few as

Table 22–4. Causes of Embryonic Mortality

| Species | Period of Maximal Mortality | | Possible Causes |
	Days of Gestation	Stage of Development	
Cattle	16–25	Rapid growth and differentiation of embryo and extraembryonic membranes	Progesterone deficiency; inbreeding; multiple pregnancy; blood group homozygosity; J-antigen in sera
Sheep	14–21	Transition from yolk sac to allantoic placentation	Estrogenic pastures; inbreeding; increasing maternal age; hemoglobin types; overfeeding; multiple pregnancy; high environmental temperature
Swine	16–25	Spacing of embryos, transuterine migration	Inbreeding; chromosomal aberrations; overcrowding; overfeeding; increasing maternal age; high environmental temperature; transferrins
Horses	30–60	Corpus luteum of pregnancy regresses and accessory corpora lutea are formed	Lactation; twinning; nutrition

one embryo is sufficient. Thus, a litter of four or fewer piglets born indicates death of embryos between days 12 and 30 of gestation.

Causes

Prenatal mortality can be due to maternal factors, embryonic factors or to fetal maternal interactions. Maternal failure tends to affect an entire litter, resulting in complete loss of pregnancy. In contrast, embryonic failure affects embryos individually, often leaving others in the litter unharmed. In other cases the maternal environment may be insufficient, allowing the support of only a few strong embryos. Heredity, nutrition and age of the dam, overcrowding in utero, hormonal imbalance and thermal stress all contribute to embryonic mortality.

Endocrine Factors. Accelerated or delayed transport of the egg, as a result of estrogen-progesterone imbalance, leads to preimplantation death. An abnormally undersized conceptus might not be able to counteract the uterine luteolytic effect, with consequent regression of the corpus

luteum and termination of pregnancy. In swine, as stated previously, at least four living blastocysts are needed by day 10 of pregnancy to counteract the uterine luteolytic effects.

A critical period of embryonic survival is the late blastocyst stage. Normally, the developing corpus luteum secretes progesterone, which acts on the female tract in close synchrony with the development of the embryos. Thus, failure of blastocyst implantation may result from delayed progestational changes in the endometrium at the appropriate time. For example, the pregnancy rate was increased in normal cows by the injection of 100 mg of progesterone one week after breeding.

Lactation. During lactation embryonic mortality occurs in cattle, sheep and horses, and is characterized by prolonged estrous cycles after breeding. Mating of mares at foal heat leads to early embryonic mortality, which has been attributed to reduced effectiveness of uterine defense mechanisms, stress of lactation and incomplete regeneration of the endometrium. In other species, the detrimental effects of lactation on embryonic development are not clear, but could disturb the uterine hormonal balance or environment, leading to death of embryos.

Chromosomal Aberrations. A relationship exists between chromosomal anomalies and embryonic mortality, particularly in swine. In addition, heteroploidy produced by delaying mating for 36 hours causes embryonic mortality.

Heredity. The frequency and repeatability of embryonic loss is partly determined by the genotype of the sire and dam and the breeding system. In cattle, embryonic mortality is higher in inbred than in outbred lines. In swine and sheep, inbreeding of the dam contributes more to a reduction in litter size than does inbreeding of the embryo.

Nutrition of the Dam. Caloric intake and specific nutritional deficiencies affect ovulation rate and fertilization rate, as well as cause prenatal death. In swine, high caloric intake or continuous unlimited feeding increases ovulation rate, thereby increasing the incidence of embryonic mortality prior to implantation. However, following implantation, fetal death is decreased by unlimited feeding. In sheep, full feeding before breeding also increases ovulation rate as well as embryonic death. The effect of caloric intake on prenatal death in cattle is presently controversial, although hypoglycemia induced by insulin or lower energy intake reduced fertility, probably due to embryonic mortality in lactating cows. In the mare, the critical period for embryonic resorption is between 25 and 31 days after ovulation. No resorption occurs if mares are maintained on an adequate plane of nutrition until 35 days after service.

Reproductive failure occurs more in sheep than in cattle grazing on plants that contain compounds with estrogenic activity, eg, subterranean clover (*Trifolium subterraneum*) and red clover (*Trifolium pratense*). The estrogenic activity is due to plant isoflavones and related substances with hydroxyl groups (Fig. 22–4). The

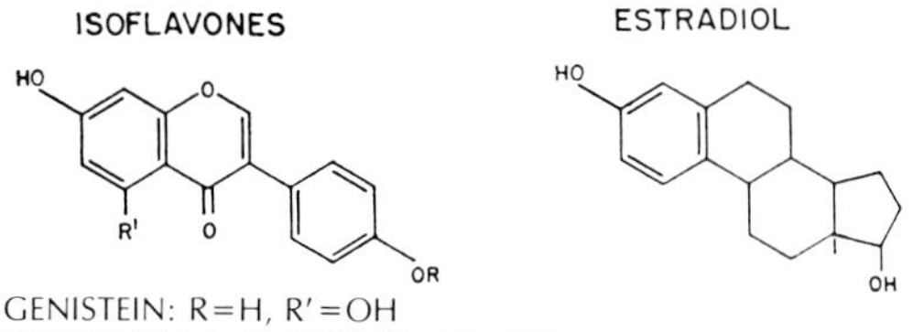

ISOFLAVONES   ESTRADIOL

GENISTEIN: R=H, R'=OH
BIOCHANIN-A: R=METHYL, R'=OH
DAIDZEIN: R=R'=H
FORMONENTIN: R=METHYL, R'=H

Fig. 22–4. The structures of plant estrogens (isoflavones) and estradiol. Isoflavones are found in many forage species of the family *Leguminosae*; formonetin has low estrogenic activity whereas biochanin-A, genistein and daidzein are metabolized in the rumen of the sheep to nonestrogenic compounds.

substance mainly responsible is the isoflavone formonetin, which is converted in the rumen of the sheep to equol, a weak estrogen. The other estrogenic isoflavones present in clovers (genistein and biochanin A) are converted to nonestrogenic metabolites in the rumen. Ewes grazing pastures with high levels of formonetin exhibit a permanent reduction in fertility associated with persistent glandular cysts affecting both cervix and endometrium. The infertility is largely due to a depressed fertilization rate as a result of a failure of sperm transport (Lightfoot et al, 1974). An amino acid (mimosine) extracted from the pasture legume (*Leucaena leucocephala*) caused both a lowered ovarian response to gonadotropins and an increase in embryonic death.

Several common plants possess estrogenic activity, eg, barley grain (*Hordeum vulgare*), oat grain (*Avena sativa*), the fruits of the apple (*Pyrus malus*) and cherry (*Prunus avium*), the tuber of the potato (*Solanum tuberosum*), and Bengal gram (*Cicer arietinum*).

Age of Dam. A higher incidence of embryonic mortality is observed in gilts and in sows after the fifth gestation. In the ewe, it is highest beyond six years. There is no evidence of the effects of age on embryonic deaths in cattle. Embryonic loss in old mares may be associated with uterine atony.

Overcrowding in Utero. As pregnancy advances, the embryo becomes increasingly dependent upon the placenta for its survival. Since the degree of placental development is primarily influenced by the availability of space and vascular supply within the uterus, increasing the number of implantations decreases the vascular supply to each site and restricts placental development. This results in a high embryonic and fetal death rate and probably explains the higher incidence of embryonic mortality in cattle and sheep following twin rather than single ovulations. However, it should be noted that uterine capacity does not limit the ability of the cow and ewe to carry twins provided they are located in separate uterine horns.

In cattle and sheep with multiple ovulations, the number of embryos surviving is reduced to a fairly constant number (2½ to 3 embryos per female) within the first three or four weeks of pregnancy, which implies that embryonic loss increases as the number of ova shed increases. This has been shown by transferring different numbers of sheep embryos to foster mothers. Embryonic mortality increased as the numbers increased. Most of the deaths due to crowding occur during the early stages of attachment, about the 14th day. Mortality does not seem to be due to a deficiency of progesterone. Overcrowded uteri in rodents also have a high percentage of fused placentae, leading invariably to embryonic or fetal loss. Transuterine migration of the embryos has special physiologic importance in certain species for equal distribution of conceptuses; for example, a high incidence of embryonic mortality occurs in pigs if transuterine migration is prevented.

Thermal Stress. Embryonic mortality increases in a number of species following exposure of the mother to elevated ambient temperatures, especially in tropical areas (Jainudeen, 1976). In early stages of development the embryo is directly affected by increased maternal body temperature due to thermal stress. The pig embryo is most susceptible to heat stress during the first two weeks of gestation, particularly during implantation. A greater incidence of embryonic deaths was noted among gilts exposed to high temperatures 8 to 16 days postbreeding than among those exposed during 0 to 8 days postbreeding (Omtvedt et al, 1971). On the contrary, the sheep embryo is most vulnerable to experimentally induced heat stress during the early cleavage stages while within the oviduct. Although acclimatization and diurnal temperature variations can modify these effects, continuous exposure to air temperatures causing a 1.5° C rise in rectal temperature will usually destroy the embryos (Thwaites, 1969). Cattle exposed to 32° C for 72 hours immediately after breeding failed to maintain pregnancy (Dunlap and Vincent, 1971). Also, interestrual periods of over 30 days following breeding suggest that embryonic loss could occur later.

The effects of thermal stress on the early embryo are not apparent until the later stages of its development. Fertilized eggs of sheep and cattle, when subjected to high temperatures either in *vitro* or in *vivo*, are damaged but continue to develop, only to die during the critical stages of implantation. This may explain some of the prolonged estrous cycles in cows during hot weather. Apparently heat stress causes alterations in metabolic regulating mechanisms (RNA synthesis), particularly within the sheep embryo, during the early cleavage stage (Ulberg and Sheean, 1973).

Semen. A part of the embryonic mortality can be attributed to the semen. Infertile matings by highly fertile bulls are primarily due to embryonic mortality, while those of bulls with low fertility are due to fertilization failure and embryonic deaths. These fertility differences between bulls in artificial insemination programs is attributable to genetic factors, which are not revealed in routine tests of semen quality. In swine, semen stored for three days before insemination produced zygotes much more susceptible to early embryonic death, presumably owing to the reduced DNA content in aged spermatozoa.

Incompatibility. Immunologic incompatibilities may block fertilization (prezygotic selection), or cause embryonic, fetal or neonatal death. In cattle, homozygosity for certain blood groups and certain substances related to transferrin (β-globulin) and J-antigen in sera have been associated with increased embryonic loss as well as decreased fertilization rate. Ewe fertility may be influenced by the combination of ram and ewe hemoglobin types varying, however, from year to year.

In swine, reproductive failure in *BB* ×*AB* transferrin matings is due to a higher embryonic mortality, rather than to fertilization failure.

Repeat Breeders

"Repeat-breeder" females return to service repeatedly after being bred to a fertile male. These females are not sterile because, if bred repeatedly, many of them ultimately conceive. Fertilization failure and early embryonic mortality are two major factors causing the reproductive failure.

The nature of the reproductive failure can be determined by the length of the interval between insemination and return to estrus. An interval equal to a normal estrous cycle length suggests failure of fertilization, particularly in animals with normal ovarian function. Poor sperm transport and/or survival, abnormal ova, adverse tubal environment or genital tract abnormalities are factors that could prevent fertilization. Depending on the stage at which embryonic deaths occur, this interval will either be normal, as in fertilization failure, or prolonged. It is generally believed that the majority of embryonic losses occur after day 16 of gestation, though there is some evidence that these deaths could occur before day 13 or perhaps earlier.

Fertilization failure, more common in sheep and swine than cattle, is due to genital tract abnormalities. The cause(s) for loss of approximately 50% of the embryos within three weeks of pregnancy in repeat-breeder cows are obscure, although infections, hormonal imbalances, hereditary and management factors have been suspected.

Abortion

Abortion refers to pregnancies that terminate with the expulsion of a fetus of recognizable size prior to the period of viability, which is arbitrarily defined as 260 days for cattle, 290 days for horses and 110 days for swine. Fetal death is not an essential prelude to abortion.

Abortions may be *spontaneous* or *induced*, *infectious* or *noninfectious*. Spontaneous abortion is more prevalent in cattle, particularly dairy cattle, than in sheep and horses. Noninfectious causes of spontaneous abortion may be genetic, chromosomal, hormonal or nutritional factors (Table 22–5). Spontaneous abortion may also occur in animals bred immediately after puberty or immediately after parturition. Mares seem to be endocrinologically susceptible to abortion between the fifth and tenth months of pregnancy.

Chromosomal abnormalities of the fetus are frequently associated with spontaneous abortion in man (Carr, 1970). Most chromosomal abnormalities lead to a grossly altered genetic content of the zygote, and among spontaneously aborted human embryos, perhaps as many as 40% are so afflicted. Chromosomal abnormalities are known to cause embryonic losses in swine, but their importance in abortion of farm animals is unknown.

Abortion due to hereditary factors results from abnormal development of some

Table 22–5. Noninfectious Causes of Abortion in Farm Animals

Causes	Cow	Mare	Sow	Ewe or Goat
Chemicals, drugs and poisonous plants	Nitrates Chlorinated naphthalenes Arsenic Perennial broom-weeds Pine needles	None Phenothiazine (?)	Dicoumarin Aflatoxin Wood preservatives Creosote Pentachlorophenols	Anthelmintics Phenothiazine Carbon tet-rachloride Lead, nitrate, locoweeds, lupines, sweet clover, onion grass, veratrum
Hormonal	High doses of estrogens or glucocorticoids Progesterone deficiency	High doses of estrogens or cortisone (?)	High doses of estrogens or glucocorticoids (?)	High doses of estrogens, cortisol or ACTH Progesterone deficiency
Nutritional	Starvation, malnutrition Deficiencies of vitamin A or iodine	Reduced energy intake	Deficiencies of vitamin A, iron and calcium	Lack of TDN or energy, deficiencies of vitamin A, copper, iodine and selenium
Genetic or chromosomal	Fetal anomalies	Fetal anomalies	Congenital or genetic lethal defects	Lethal genetic defects
Physical	Douching or insemination of pregnant uterus, stress (transport, fever, surgery)	Manual dilatation of cervix, natural service during pregnancy (?), rectal palpation of the very young blastodermic vesicle	Stress (transportation, fighting, injury)	Severe physical stress
Miscellaneous	Twinning, allergies, anaphylaxis	Twinning	Poor management	Twinning (?)

(Adapted from Roberts, 1971. *Veterinary Obstetrics and Genital Diseases,* published by the author, Ithaca, N.Y.)

vital organ or generally low viability of the fetus. Habitual abortions at 3 to 4½ months of pregnancy in Angora goats is due to a hereditary defect of the anterior pituitary gland, leading to a deficiency of luteotropic hormone secretion required to maintain the corpus luteum of pregnancy. Abortions are occasionally induced with high doses of estrogens, prostaglandin $F_{2\alpha}$ or glucocorticoids, particularly in young females bred at an early age and in meat-producing animals.

Twin pregnancy is the most common cause of abortion in mares; over two thirds of twin pregnancies terminate in abortions. The inability of a mare to successfully carry twin fetuses to term may be related to placental insufficiency arising from competition between the placentae. This may lead to the death of one fetus and eventually to the abortion of both fetuses.

Fetal Mummification

Fetal mummification is characterized by fetal death, failure of abortion, resorption of placental fluids, dehydration of the fetus and its membranes, and involution of the uterus. It is more common in cattle and swine than in sheep and horses.

Two types of mummification are known, the "hematic" type common to cattle and the "papyraceous" type of swine. In cattle, the gravid uterus becomes filled with gummy reddish brown material, and there is a massive intercotyledonary hemorrhage that causes the maternal and fetal cotyledons to separate. In some cases an "unsuccessful abortion" occurs; the fetus then undergoes gradual autolysis and maceration until only a compact mass of fetal bones is left. This latter form may also be associated with chronic mucopurulent vaginal discharge in cattle.

The syndrome occurs mainly from the fifth to seventh months of gestation in all breeds of cattle. Affected cows conceive normally in the subsequent breeding period. Occasionally, bovine mummified fetuses are aborted spontaneously, but in most cases they are carried many months beyond the gestation period.

Twin-bearing ewes may abort a mummified fetus during late gestation and maintain the other lamb to full term, or they may deliver a mummified fetus attached to the placenta of a viable offspring.

Swine embryos that die in the first six weeks of gestation are completely resorbed. The fetuses that die during later stages are retained and expelled as mummified fetuses along with normal piglets at farrowing. Mummified fetuses are more prevalent in large than in small litters, in older sows than in gilts, and in some breeds than in others.

Mummification may be due to interference with the fetal blood supply, deficiency in placentation, anomalies in the umbilical cord of the fetus or infection in the gravid uterus. The viruses of the SMEDI (stillbirth, mummification, embryonic death and infertility) group are an important cause of mummification in swine, particularly in susceptible gilts and young sows. Transplacental infection leads to the establishment of the virus in one or two fetuses. These affected fetuses subsequently die after the infection has been transmitted to adjacent fetuses, which also die later.

The mummification syndrome may also be genetically inherited, because it occurs in numerous pedigreed cows and in consecutive generations of cows bred to unrelated males. A high incidence of fetal mummification in the Jersey and Guernsey breeds also tends to support a hereditary influence. The retention of a mummified fetus within the uterus could be due to a fetal influence on the endometrium, through suppression of the uterine luteolytic mechanism, causing persistence of the CL.

PERINATAL AND NEONATAL MORTALITY

Perinatal mortality refers to death of the offspring shortly before, during or up to 24 hours after parturition at normal term. Nutrition and age of the dam and hereditary factors appear to be the major contributing factors. In cattle, the incidence of perinatal mortality ranges between 5% and 15% of all births, with a high incidence in primiparous animals, in male fetuses, and in calves sired by Friesian or Hereford bulls.

Stillborn piglets resemble live littermates but their lungs do not float in water. One to two piglets in approximately one third of all litters are dead at birth with advancing parity, in extremes of litter size, and in litters in which the gestation period is less than 110 days (Randall and Penny, 1970). Two types of stillbirth occur in swine. In the first type, usually due to infectious causes, fetuses die prepartum, whereas in the second type, due to noninfectious causes, piglets die during parturition. The presence of meconium on the skin, in the mouth and in the trachea of the piglet differentiates the latter from the former. A high incidence of piglet death is observed with uterine inertia, with prolongation of farrowing time or the interval between births of piglets, in litters that contain less than four or more than nine piglets, and among piglets in the last third of the litter (Sprecher et al,

1974). These intrapartum deaths may be due to the low tolerance of piglets to anoxia and are usually associated with rupture of the umbilical cord prior to delivery. Since piglets suffer irreversible brain damage within five minutes after umbilical rupture or impeded umbilical flow, delivery must be completed rapidly. One method of reducing intrapartum deaths that is being investigated is the intramuscular injection of $PGF_{2\alpha}$ to induce parturition.

In sheep, most losses between implantation and weaning occur during the perinatal period, due to starvation of the neonate or to dystocia among lambs born to maiden ewes, ewes on poor pasture, or ewes with "clover disease."

Neonatal mortality—death of the neonate during the first few weeks of life—is related to heredity, environmental factors, nutrition and infection. Several nutritional deficiencies may contribute to neonatal mortality (Table 22–6). *White muscle disease*, a myopathy of lambs and calves, results from a deficiency or metabolic disturbance of selenium. The hyperacute type, usually due to myocardial damage in younger animals, leads to death in a few hours. The subacute type, usually due to skeletal muscle damage in older animals, causes muscular weakness, difficulty in nursing and death in one to two weeks. *Hypomagnesemia*, a decrease in blood magnesium levels, is characterized in milkfed calves by irritability, nervousness and, in severe cases, by tetany.

Hemolytic icterus or *neonatal isoerythrolysis* occurs in neonates of horses, swine and cattle. The term isoerythrolysis implies lysis or destruction of erythrocytes by isoantibodies. If destruction of erythrocytes proceeds faster than their replacement, acute anemia ensues. Neonatal isoerythrolysis in horses is presumably restricted largely to Arabian breeds. The disease, rarely observed in foals from a primiparous mare, is due to passage of fetal erythrocyte antigens into the maternal circulation. The production of maternal antibodies becomes progressively elevated until the third or fourth parturition, when these antibodies are transmitted through the colostrum to affect the foal. Lethargy, weakness, jaundice of the mucous membranes and increased heart and respiratory rates are noted during the first two days of life. The disease in foals, apparently not due to an Rh-like factor as in man, can be treated by exchange transfusion with whole blood or packed erythrocytes washed with saline to remove the antibodies. Preventive measures include withholding from suckling the dam for 24 hours postpartum and feeding co-

Table 22–6. Neonatal Mortality Due to Nutritional Deficiencies

Disease	Species	Cause	Description
"White muscle" disease	Lambs and calves	Selenium deficiency	Acute and chronic forms, degeneration of cardiac and skeletal muscles, death
Hypomagnesemia	Milkfed calves	Decrease in serum Mg^{++}	Irritability, nervousness and tetany
Piglet anemia	Baby pigs during first week	Iron deficiency	Low hemoglobin levels in blood, weakness and prostration, inability to nurse
Goiter	Foals and lambs	Iodine deficiency	Enlargement of thyroids
Enzootic ataxia	Lambs	Copper deficiency	Locomotor incoordination leading to paralysis
Neonatal hypoglycemia	Newborn piglets	Low blood glucose	Loss of appetite, coma

lostrum from another mare. Hemolytic icterus, affecting a few piglets in a litter one to seven days after farrowing, results from vaccination of pregnant sows with hog cholera and swine fever vaccines. Rarely, neonatal isoerythrolysis in swine may be due to transplacental isoimmunization. In cattle, the causative immunogen has been identified as babesia vaccine in Australia and anaplasma vaccine in America, both vaccines being of bovine blood origin.

Neonatal mortality may also be a result of long labor, poor maternal nutrition, weakness of the mother or the young, bacterial infection of the young through the umbilical cord, poor maternal behavior, or delayed onset of lactation. Exposure of the newborn pig to low environmental temperature leads to hypothermia, hypoglycemia and death. Heat prostration and some deaths occur in newborn lambs exposed to high environmental temperature. Another source of danger to the neonate is mammalian or avian predators, such as the feral pig *(Suis scrofa)*, fox

(Vulpes vulpes), dingo *(Canis dingo)*, raven *(Corvus coronoides)*, wedge-tailed eagle *(Aquila audax)* and sea eagle *(Haliaetus leucogaster)*.

DISORDERS OF GESTATION, PARTURITION AND LACTATION

Dystocia

Dystocia, difficult or obstructed parturition, may be due to fetal, maternal or mechanical causes (Fig. 22–5). Fetal dystocia (Fig. 22–6) results from abnormalities in the presentation or position of the fetus and from postural irregularities of its head or limbs; it may be due to a relatively or absolutely oversized fetus, and to fetal monstrosities. Fetal dystocia is common in certain breeds of dairy cattle, in cattle and sheep with multiple pregnancies and in sows with small litters. It may also result from breeding females of smaller breeds to sires of larger breeds. An excessively large fetus in rela-

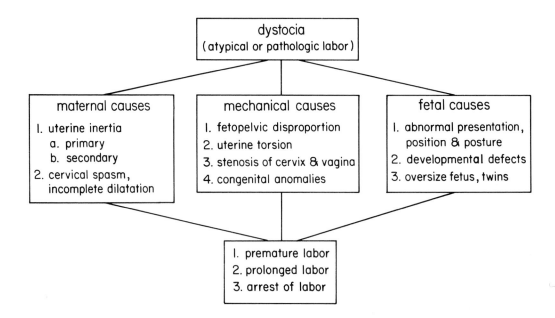

Fig. 22–5. Schematic representation of maternal and fetal causes that lead to various forms of dystocia. Fetomaternal mechanical causes, abnormal fetal presentation, position and posture, and maternal causes may lead to premature labor, prolonged labor and arrest of labor.

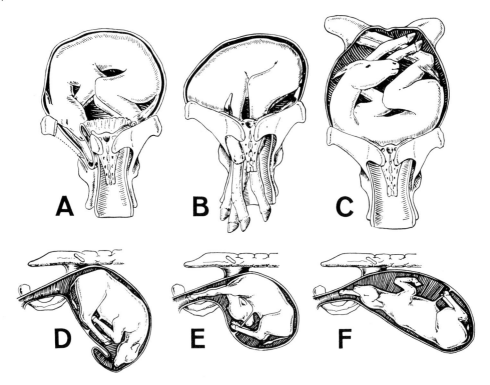

Fig. 22–6. Malpresentation in horses *(A, B, C)* and cattle *(D, E, F)*. *A*, Ventro-transverse presentation with ventral displacement of the uterus. *B*, Ventro-transverse presentation; uterine body gestation. *(A, B* and *C from Arthur, 1964. Wright's Veterinary Obstetrics. London, Bailliere, Tindall & Cox. D, E, and F redrawn from Diseases of Cattle. U.S.D.A., Special Report, 1942.)*

tion to the size of the birth canal of the dam presents difficulties even though the presentation is normal. Deviations of the head and flexion of the various joints in anterior presentation, flexion of both hindlimbs (breech) in posterior presentation, or twins may cause dystocia.

Maternal dystocia is more frequent in dairy cattle and sheep than in horses and swine. It occurs frequently in primiparous animals and in animals with multiple young. The absence of uterine contractions or inertia may be primary or secondary. Primary uterine inertia due to excessive stretching is common in multiple pregnancy in cattle and in large litters in swine. Secondary uterine inertia is due to exhaustion of the uterine muscle secondary to obstructive dystocia. Failure of the cervix to dilate properly leads to "spasm" of the cervix in cattle.

Anomalies of the soft parts of the reproductive passages or the bony pelvis are occasional causes of dystocia. One group of anomalies causes narrowing of the birth canal (eg, abnormalities or fractures of the pelvis and stenosis or obstruction of the cervix, vagina or vulva), while another group of abnormalities prevents entry of the fetus into the birth passages (eg, failure of the cervix to dilate or torsion of the uterus).

Metabolic Disorders of Late Gestation and Parturition

Metabolic disorders associated with reproductive failure fall into two groups. Neuromuscular disorders are associated with disturbances in the metabolism of calcium, magnesium and phosphorus. These syndromes are observed chiefly

Table 22–7.   Disorders of Gestation, Parturition and Lactation

Syndrome	Species	Causes	Description
Milk fever	Cattle	Hypocalcemia	At parturition, recumbency, circulatory collapse, loss of consciousness and death
Grass tetany	Cattle	Hypomagnesemia	Late gestation or lactation, convulsions and tetany
Ketosis	Cattle	Hypoglycemia	Disturbance of carbohydrate metabolism during first month after calving; ketones in blood and urine; rapid loss of weight, drop in milk production
Pregnancy toxemia	Sheep	Hypoglycemia	During late gestation in ewes carrying twins, disturbance of carbohydrate metabolism
Retained placenta	Cattle and sheep	Dystocia, infections	Failure to expel placenta after calving
Vaginal prolapse	Cattle and sheep	Excessive relaxation of pelvic ligaments, restricted exercise, twinning	Late gestation
Uterine prolapse	Cattle and sheep	Dystocia, placental retention	Follows parturition
Hydrops of fetal membranes	Cattle, sheep and horses	Fetal anomalies, placental dysfunction, fetal-maternal incompatibility	Excessive accumulation of fluids within the amniotic or allantoic cavity
Twinning	Cattle, sheep and horses	Spontaneous or induced	Abortion
Prolonged gestation	Cattle	Genetic and fetal abnormalities	Fetal death
	Sheep	Ingestion of teratogens in early pregnancy	Fetal death
	Swine	Genetic, in certain inbred lines	Fetal death

prior to and at the time of parturition, eg, milk fever in cattle, lactation tetany in cattle, and lambing sickness. Disturbances in carbohydrate metabolism are most common in sheep and cattle (Table 22–7).

Hypocalcemia or Milk Fever. This is one of the most common disorders associated with parturition in dairy cows, characterized by recumbency, circulatory collapse and loss of consciousness. Milk fever in cattle is due to a sharp fall in plasma calcium and inorganic phosphorus levels. If the disease is not treated with calcium borogluconate, it often terminates in death.

Grass Tetany. This is a metabolic disease of cows, in late gestation or lactation, grazing on lush pasture; it is characterized by hypomagnesemia, convulsions and tetany. It resembles milk fever.

Ketosis in Cattle. Ketosis, which occurs one to six weeks after calving, is characterized by lack of appetite, rapid loss of weight and decreased milk production, nervous signs, hypoglycemia, ketonemia and ketonuria, and depletion of the alkali reserves.

Pregnancy Toxemia, "Twin-Lamb" Disease, Ketosis of Pregnancy or Prepartum Paralysis. This syndrome occurs in sheep during the last two weeks of pregnancy and is caused by hypoglycemia, which may stem from twin pregnancy, sudden changes in caloric intake or sudden restriction of exercise, such as occurs with heavy snowfall. Ewes overfed in early pregnancy or underfed in late pregnancy are highly susceptible to pregnancy toxemia, resulting in their death or the loss of lambs due to lack of vigor at birth. Ewes often recover if fetuses are aborted.

Retained Placenta

Failure of the placenta to be expelled during the third stage of labor is a common complication in ruminants, primarily due to failure of the fetal villi to detach themselves from the maternal crypts. There is an increase in progesterone, decrease in estradiol-17β and possible decrease in prolactin levels in plasma of dairy cows with retained placentae (Chew et al, 1977). This may indicate asynchrony of endocrine mechanisms controlling the second and third stages of labor. Retention of the placenta beyond 12 hours in cattle is considered pathologic and is associated with abortions due to *Brucella abortus* or *Vibrio fetus*, dystocia, uterine inertia, twinning and induction of parturition with dexamethasone. It is more common in dairy than beef breeds and adversely affects milk production and fertility owing to delayed uterine involution. Although extensive putrefactive changes occur if the placenta is retained for several days, antibiotic therapy is more effective than manual removal of the placenta.

Lactation Failure

Lactation failure following parturition is due to physiologic as well as physical factors. Milk letdown is inhibited by fright or painful conditions of the udder, particularly in heifers. Ewes underfed during late pregnancy often fail to produce colostrum during the first few hours of the neonatal period. Colostrum may also be inaccessible to the lamb as a result of damage to the teats, presence of wool and dirt around the udder or abnormal maternal behavior.

A relationship exists between lactation and physiologic stresses associated with high environmental temperatures. The lactation failure in these instances is probably related to depression of secretions from the anterior pituitary and other hormones essential for lactation.

In cattle, inflammation of the mammary glands or mastitis results in complete failure or a reduction in the production of milk, depending on whether the inflammation is acute or chronic.

Lactation failure or the metritis-mastitis-agalactia (MMA) syndrome is a frequently encountered, noninfectious condition in the gilt or sow that occurs within two or three days after parturition. It is characterized by a decrease in or an absence of milk secretion leading to starvation of the piglets. The cause is unknown, but infectious agents, notably *Escherichia coli* and Mycoplasma spp., endotoxins, poor management and feeding practices and endocrine disturbances have been incriminated. Milk secretion in affected sows can be stimulated by oxytocin injections.

Prolapse of the Vagina or Uterus

Prolapse of the vagina or protrusion of the vagina through the vulva occurs during the latter half of pregnancy. It is common in multiparous cows and in ewes on restricted exercise, grazing estrogenic pastures or carrying twins, and it is due to an excessive relaxation of the pelvic ligaments.

Prolapse of the uterus is more common in cows and ewes than in mares and sows, and it occurs a few hours after parturition or may be delayed by several hours. It is a sequela to dystocia or pathologic retention of the placenta or hyperestrogenism in ewes fed on pastures rich in estrogenic compounds.

Hydramnios and Hydrallantois

Hydramnios, the excessive accumulation of amniotic fluid, is less common than hydrallantois, the accumulation of allantoic fluid. Hydramnios, observed more often in cattle than sheep or swine, is associated with certain cranial abnormalities of the fetus. In these defective fetuses, swallowing is impaired, causing amniotic fluid to accumulate as gestation progresses. Fetuses of the Guernsey and Jersey breeds in prolonged gestation have hydramnios.

Hydrallantois occurs in cattle especially in twin pregnancies, and it is characterized externally by an enormous enlargement of the abdomen after the sixth month of gestation. It has also been reported in horses after the seventh month of gestation and was associated with fetal abnormalities (Vandeplassche et al, 1976). In cattle the syndrome has been attributed to fetal-maternal incompatibility and placental dysfunction (Arthur, 1975), but in horses it may be related to fetal abnormalities.

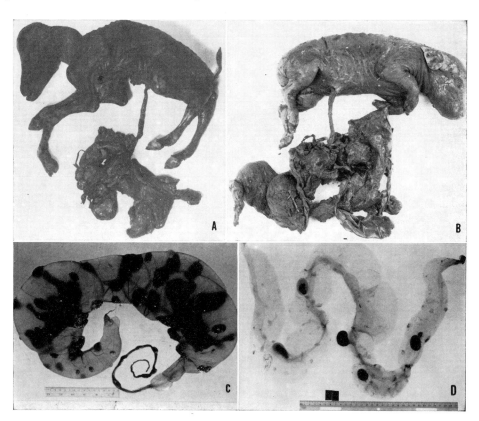

Fig. 22–7. *A.* "Hematic" mummification of a seven-month-old bovine fetus, removed by Cesarean section from a Guernsey cow approximately one month after expected date of calving. *(Photograph by S. J. Roberts.)*

B, "Papyraceous" mummification of a pig fetus expelled at farrowing together with viable young. The fetus was surrounded by fetal envelopes that resemble parchment; note torsion of the umbilical cord.

C, Conceptus of beef cow at 105 days of pregnancy showing two viable fetuses and five degenerative remnants of placentae and of the chorioallantois. The cow was previously treated with pregnant mare serum and human chorionic gonadotropins to induce multiple pregnancy.

D, Conceptus of beef cow at 105 days of gestation showing degeneration of the placenta and five fetuses. The cow was previously treated with gonadotropins to induce multiple pregnancy.

Multiple Pregnancy

In cattle, horses, sheep and goats, the frequency of multiple pregnancies is higher than that of multiple births, owing to the high incidence of abortion and fetal resorption (Fig. 22–7). In the cow, the sequelae of twinning include shortened gestation period, abortion, stillbirth, dystocia and retained placenta. Economic losses are related to decreased fertility, neonatal mortality, decrease in birth weights of calves, longer calving intervals, and lower butter fat production. In addition, over 90% of the females born co-twin to a male are sterile (freemartins). Neonatal mortality in sheep is greater among twins than among singles. Ewes carrying twins are more susceptible to pregnancy toxemia (twin-lamb disease). In mares a high percentage of twin fetuses are aborted.

Prolonged Gestation

Abnormally long gestations, due to fetal abnormalities in cattle, sheep and swine, result from genetic and nongenetic factors.

There are two general types of the syndrome in cattle, and each is governed by a single autosomal recessive gene. In the type seen in Holstein and Ayrshire breeds, the large fetuses have no facial abnormalities, and when delivered surgically they are weak, unable to nurse and die in six to eight hours from severe hypoglycemia. There is hypoplasia of the adenohypophysis and the adrenal cortex. The plasma progesterone level in a cow carrying an affected calf does not fall before parturition as it does in a normal cow. In the second type of syndrome, seen in the Guernsey and Jersey breeds, the fetuses are small, many exhibit facial abnormalities and hydramnios, they lack an adenohypophysis and survive *in utero* for long periods past term but live for only a few minutes when delivered surgically.

Prolonged gestation occurred in an in-bred line of swine and in ewes that consumed the teratogenic plant, *Veratrum californicum*, 14 days after conception. Since the fetal hypothalamo-hypophyseal-adrenal axis initiates parturition in sheep and possibly in other species, the abnormally long gestation periods associated with functional defects of the fetal pituitary and adrenal glands may well be due to a breakdown of the fetal mechanism that initiates parturition.

REFERENCES

Arthur, G.H. (1975). Veterinary Reproduction and Obstetrics. Philadelphia, Lea & Febiger.

Carr, D.H. (1970). Chromosome abnormalities and spontaneous abortions. In Human Population Cytogenetics, Edinburgh, University Edinburgh Press.

Chew, B.P., Keller, H.F., Erb, R.E. and Malven, P.V. (1977). Periparturient concentrations of prolactin, progesterone and the estrogens in blood plasma of cows retaining and not retaining fetal membranes. J. Anim. Sci. 44, 1055.

Chupin, D., Pelot, J., Alonso de Miguel, M. and Thimonier, J. (1976). Progesterone assay for study of ovarian activity during postpartum anoestrus in the cow. Proc. 8th. Internat. Congr. Anim. Reprod. & Artif. Insem. Krakow, Poland, 3, 346.

Day, B.N. and Polge, C. (1968). Effects of progesterone on fertilization and egg transport in the pig. J. Reprod. Fertil. 17, 227.

Dunlap, S.E. and Vincent, C.K. (1971). Influence of postbreeding thermal stress on conception rate in beef cattle. J. Anim. Sci. 32, 1216.

Dunne, H.W. (1975). Abortion, stillbirth, fetal death, infections and infertility. In Disease of Swine. H.W. Dunne and A.D. Leman (eds). Ames, Iowa State University Press.

Erickson, B.H., Reynolds, R.A. and Murphree, R.L. (1976). Ovarian characteristics and reproductive performance of the aged cow. Biol. Reprod. 15, 555.

Hughes, J.P. and Trommershausen-Smith, A. (1976). Karyotype anomalies in reproductive failure of the equine. Proc. 8th Internat. Congr. Anim. Reprod. & Artif. Insem. Krakow, Poland, 4, 586.

Jainudeen, M.R. (1976). Effects of climate on reproduction among female farm animals in the tropics. Proc. 8th Internat. Congr. Anim. Reprod. & Artif. Insem. Krakow, Poland, Plenary Sess. 29–38.

Lightfoot, R.J., Smith, J.F., Cumming, I.A., Marshall, T., Wroth, R.H. and Hearnshaw, H. (1974). Infertility in ewes caused by prolonged grazing on oestrogenic pastures: Oestrus, fertilization and cervical mucus. Aust. J. Biol. Sci., 27, 409.

Mann, T. (1976). Progress in the biology of the gametes I. Problems in the male. Proc. 8th Internat. Congr. Anim. Reprod. & Artif. Insem. Krakow, Poland, *Plenary Sess. 39.*

Omtvedt, I.T., Nelson, R.E., Edwards, R.L., Stephens, D.F. and Turman, E.J. (1971). Influence of heat stress during early, mid and late pregnancy of gilts. J. Anim. Sci. *32*, 312.

Oxender, W.D., Noden, P.A. and Hafs, H.D. (1977). Estrus, ovulation, and serum progesterone, estradiol, and LH concentrations in mares after an increased photoperiod during winter. Am. J. Vet. Res. *38*, 203.

Randall, G.C.B. and Penny, R.H.C. (1970). Stillbirths in the pig: An analysis of the breeding records of five herds. Br. Vet. J. *126*, 593.

Sprecher, D.J., Leman, A.D., Dziuk, P.D., Cropper, M. and DeDecker, M. (1974). Causes and control of swine stillbirths. J. Am. Vet. Med. Assoc. *165*, 698.

Thwaites, C.J. (1969). Embryo mortality in the heat stressed ewe. II. Application of hot room results to field conditions. J. Reprod. Fertil. *19*, 255.

Ulberg, L.C. and Sheean, L.A. (1973). Early development of mammalian embryos in elevated ambient temperatures. J. Reprod. Fertil. Suppl. *19*, 153.

Vandeplassche, M., Bouters, R., Spincemaille, J. and Bonte, P. (1976). Dropsy of the fetal sacs in mares: Induced and spontaneous abortion. Vet. Rec. *99*, 67.

Webb, R., Lamming, G.E., Haynes, N.B., Hafs, H.D. and Manns, J.G. (1977). Response of cyclic and post-partum suckled cows to synthetic injections of LH-RH. J. Reprod. Fertil. *50*, 203.

23

Reproductive Failure in Males

M.R. JAINUDEEN
and
E.S.E. HAFEZ

The fertility of a male is related to several phenomena: (1) sperm production; (2) viability and fertilizing capacity of the ejaculated sperm; (3) sexual desire; and (4) the ability to mate. The sterile male is readily identified, but the male with reduced fertility poses serious problems and causes economic losses to breeders and the artificial insemination (AI) industry.

AI has made many valuable contributions to understanding the factors that influence male reproductive functions.

The purpose of this chapter is to review the functional aspects of male reproductive failure resulting from anatomic, physiologic, endocrinologic, environmental, nutritional, genetic, psychogenic or pathologic factors (Table 23–1).

CONGENITAL MALFORMATIONS

Segmental Aplasia of the Wolffian Ducts. In this defect, small or large segments of one or both wolffian ducts (eg,

Table 23–1. Causes of Reproductive Failure in Male Farm Animals

Cause	Reproductive Failure Mechanisms
Developmental	Testicular hypoplasia
	Segmental aplasia of wolffian duct system
	Cryptorchidism
	Testicular hypoplasia
Neuroendocrine and psychologic	Disorders of erection and ejaculation, impotence
Semen factors	Disorders in semen concentration, sperm morphology and motility
Pathologic	Testicular degeneration
	Orchitis
	Seminal vesiculitis
	Epididymitis
Immunologic	Sperm agglutination
Chromosomal	Klinefelter syndrome
	Chromosomal translocations, constrictions

epididymis, vas deferens or ampulla) are missing. Males with unilateral tubal deficiencies or occlusions often have normal fertility, but those with the bilateral condition are sterile. It is more common among the offspring of certain bulls that also exhibit this condition. It is characterized in cattle by a total or partial absence of one or both epididymides, but more often the right epididymis. Segmental aplasia of the epididymis is commonly associated with a localized accumulation of spermatozoa within an occluded epididymis, which is known as a spermatocele.

Cryptorchidism. The descent of the testes involves the abdominal migration to the internal inguinal ring, passage through the inguinal canal and finally migration within the scrotum. In cryptorchidism, one or both testes fail to descend from the abdominal cavity into the scrotum.

Testicular descent in mammals results from swelling and subsequent regression of the gubernaculum (Wensing, 1973). Early in this process the gubernaculum extends from the caudal pole of the testis to the external inguinal ring (Fig. 23–1). Traction that develops from swelling of the extra-abdominal portion of the gubernaculum draws the testis into the inguinal canal. Subsequent regression of the gubernaculum enables the testis to descend further into the scrotal position. Abnormal gubernacular development has been associated with cryptorchidism in swine.

Factors controlling gubernacular development and testicular descent are poorly understood. It has been postulated that gonadotropin deficiency might be responsible for failure of testicular descent. The administration of human chorionic gonadotropin (HCG) or luteinizing hormone-releasing hormone (LH-RH) can promote testicular descent in man but not in swine (Colenbrander et al, 1978). The normal development of the gubernaculum and the scrotal position of the testis in decapitated pig fetuses and normal serum LH concentrations in unilateral cryptorchid pigs (Colenbrander et al, 1977) suggest that these two processes are independent of the hypothalamo-hypophyseal-gonadal axis.

The incidence of cryptorchidism is higher in swine and horses than in other farm animals. It is probably a hereditary defect transmitted by the male; it is dominant in the horse and recessive in other species. One or both testes may be located in the abdominal cavity or, more commonly, in the inguinal canal. The left testis is affected more often than the right testis.

Bilaterally cryptorchid animals are sterile owing to thermal suppression of spermatogenesis, whereas unilaterally cryptorchid animals have normal spermatogenesis in the scrotal testis. Unilaterally cryptorchid animals are usually fertile but have reduced sperm concentrations; they display normal secondary sexual characteristics because their testes secrete testosterone at nearly normal levels.

The presence or absence of an undescended testis can be determined in the horse by measuring the testosterone concentration in peripheral blood 30 to 90 minutes after a single intravenous injection of 6000 IU of HCG. On the basis of plasma testosterone concentration in the sample, it is possible to distinguish between castrated horses (less than 40 pg/ml) or cryptorchid horses (more than 100 pg/ml) (Cox, 1975). Despite the ability of a unilaterally cryptorchid male to reproduce, it should not be used for breeding because the trait can be transmitted to its offspring.

Testicular Hypoplasia. Hypoplasia of the testes, a congenital defect in which the potential for development of the spermatogenic epithelium is lacking, occurs in all farm animals, particularly in bulls of several breeds.

Inherited testicular hypoplasia is best

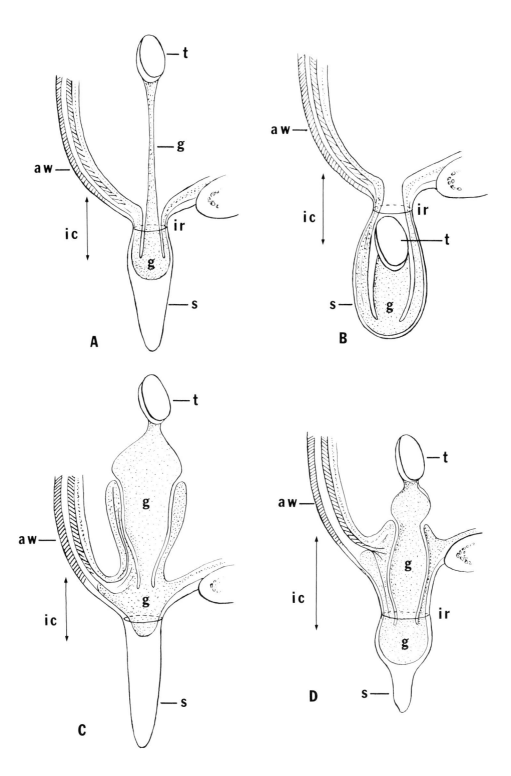

Fig. 23-1. *A, B.* Schematic representation of testicular descent. The increase in size of the extraabdominal part of the gubernaculum *(g)* causes the testis *(t)* to move toward the inguinal canal *(ic)*. The major cause of cryptorchidism in pigs is an aberrant development of the gubernaculum. *C.* Swelling of the gubernaculum that extends intraabdominally prevents the descent of the testis. *D.* Swelling of the gubernaculum that occurs partly within the inguinal canal causes the testis to be located closer to the internal inguinal ring. *aw,* Abdominal wall; *s,* scrotum; *ir,* inguinal ring. *(Adapted from Wensing, C.J.G., 1973. Kon. Ned. Akad. Wetensch. Proc. C, 76, 373–381)*

known in Swedish Highland cattle and is caused by a recessive autosomal gene with incomplete (about 50%) penetrance. Testicular hypoplasia also occurs in other breeds of cattle, but a genetic basis has not been well-documented, although a famil-ial distribution has been noted (Galloway and Norman, 1976). The testicular hypo-plasia observed in beef cattle with exces-sive scrotal fat may be due to thermal degeneration of the spermatogenic epithelium.

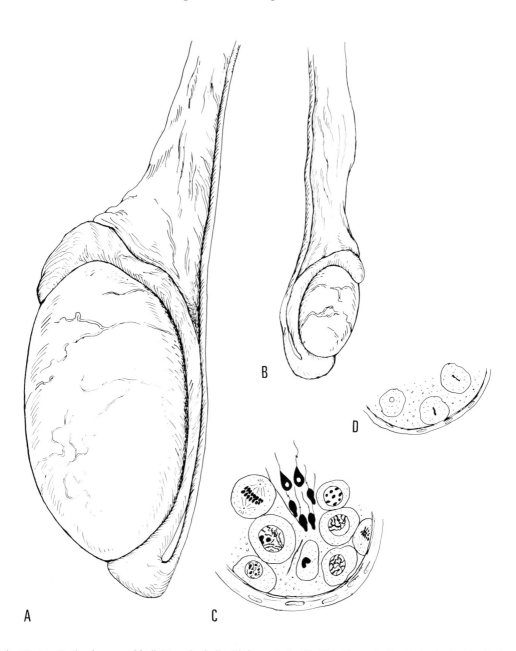

Fig. 23–2. Testis of a normal bull *(A)* and a bull with hypoplasia *(B)*. Note the reduction in testicular size in the bull with testicular hypoplasia. Transverse section through seminiferous tubules of a normal testis *(C)* and a testis with hypoplasia *(D)*. In testicular hypoplasia, the seminiferous tubules are lined almost entirely by Sertoli cells.

Testicular hypoplasia is suspected only at puberty or later because of reduced fertility or sterility. One or both testes may be hypoplastic. In Swedish Highland cattle, the hypoplasia is confined to the left testis. In sterile bulls, the semen is watery and contains few or no spermatozoa. In less severe forms, semen, libido and the ability to serve are not affected, but sperm numbers may be reduced. Histologically the seminiferous tubules are characterized by a lack of germinal elements, predominance of Sertoli cells and a failure of spermatogenesis (Fig. 23–2).

A hypoplastic testis is reduced in size. Although severe cases of testicular hypoplasia may be diagnosed by scrotal and testicular measurements (Fig. 23–3), the less obvious cases are difficult to diagnose (Almquist et al, 1976). Karyotype analysis may aid diagnosis, since a high incidence of chromosomal secondary constrictions· is present in leukocyte cultures from the blood of bulls with testicular hypoplasia (Galloway and Norman, 1976).

As in the bull, testicular hypoplasia in boars and rams is characterized by small testes and semen with low sperm concentration (boar) or with a high percentage of abnormal spermatozoa (ram).

EJACULATORY DISTURBANCES

Ejaculatory disturbances are of two types: lack of sex drive or libido and failure to copulate, which encompasses disturbances in erection, mounting, intromission, or ejaculation.

Lack of Libido

Lack of libido or sexual desire *(impotentia coeundi)* may be hereditary or may originate from psychogenic disturbances, endocrine imbalance or environmental factors. Even though seminal characteristics may be satisfactory, fertility may be adversely affected due to poor libido.

Bull. Lack of sexual desire is more fre-

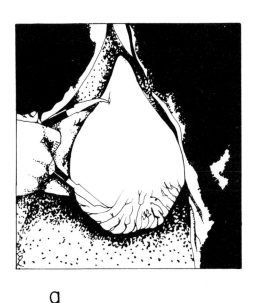

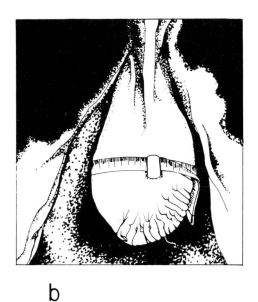

a   b

Fig. 23–3. *A*, Measurement of testicular circumference; *B*, measurement of testicular length, width and depth. The size or volume of the testis can be estimated from these measurements. A high correlation exists between testicular size and sperm production in bulls and rams, and testicular hypoplasia can be diagnosed.

quent in some strains and breeds of cattle than in others. In the Swedish Highland breed, bulls with poor libido had significantly more offspring with the same defect than normal bulls. Sexual drive shows similarities within twin pairs and differences among pairs. The similarity within pairs may partly reflect an hereditary effect.

Some bulls become apprehensive about sudden changes in the environment, such as changing the farm, the barn, the herdsman or the locality of semen collection. Since fear and apprehension are inimical to sexual expression, the intensity of sexual behavior declines until the bull becomes accustomed to the new situation. Inhibition may develop as a result of repeated frustration, faulty management, wrong techniques during semen collection, distraction during coitus and too-rapid withdrawal of the teaser animal after copulation. Inhibition is characterized by refusal to copulate, incomplete erection, or incomplete ejaculation (Hafez and Bouissou, 1975).

Bulls exhibit considerable differences in semen characteristics and libido, often without any signs of illness, weakness or abnormal condition. Poor libido is believed to be due to a deficiency in circulating androgens, but in Holstein bulls the concentration of circulating testosterone is unrelated to libido or semen characteristics (Foote et al, 1976).

Stallion. Abnormal mating behavior in stallions is most often due to mismanagement at time of breeding. Overuse, rough treatment at service or too-frequent ejaculation during winter may exert a detrimental effect on the behavior of young stallions. Pain resulting from injury at copulation or associated with mounting attempts is also a common cause of impotence (Pickett and Voss, 1975). Seasonal variations in libido and the secretory and gametogenic activity of the stallion reproductive tract are mediated, at least in part, by the pattern of testosterone secre-

tion. The greatest sperm output in stallions occurs during July, two months after the seasonal peak in plasma testosterone levels (Berndtson et al, 1974).

Rams. Despite the production of normal numbers of fertile sperm, rams may have low fertility because of their inability to breed sufficient numbers of ewes. This low-service frequency results from a lack of libido, poor dexterity or interference from other rams.

Seasonal factors, such as daylight and temperature, influence the sexual performance of rams of different breeds under a wide variety of both natural and controlled experimental conditions. A decline in the hours of light generally appears to favor enhanced sexual performance, but evidence of such a relationship is conflicting. Ram fertility is also adversely affected during periods of high temperature.

Boar. Low libido in the boar is associated with obesity, heat stress or too-high a plane of nutrition. Libido also may be seriously impaired by mismanagement of young boars during service.

Inability to Copulate

Physical disabilities may impede or prevent mating by causing failure in copulatory behavior, ie, mounting, intromission or ejaculation.

Failure to Mount. Inability to mount is a common disorder encountered in older bulls and boars. It is associated with locomotor dysfunction arising from dislocations, fractures, sprains and osteoarthritic lesions of the hindlimbs and vertebrae. Degenerative changes in the articular surface of the stifle and hock joints and exostoses of the thoracolumbar vertebrae interfere with mobility and ability to mount, particularly in older bulls.

Failure to Achieve Intromission. This is a condition in which the penis fails to enter the vagina. It may result from insufficient protrusion of the penis from the sheath or deviation of the penis.

"Phimosis," or stenosis of the preputial orifice due to congenital, traumatic or infectious causes, may prevent the normal protrusion of the penis. The pendulous prolapse that occurs in *Bos indicus* breeds (Santa Gertrudis and Brahman), or the inherent tendency to preputial eversion that occurs in some polled *Bos taurus* breeds (Hereford and Aberdeen Angus, may lead to trauma, inflammatory changes and eventually to preputial prolapse and phimosis (Fig. 23–4). At service, affected bulls are unable to protrude the penis more than two or three inches, or not even through the preputial orifice in more severe cases. The condition may be corrected by surgical amputation of the prolapsed preputial mucosa. Selective breeding and culling of *B. taurus* bulls with a predisposition to preputial prolapse may help to reduce the incidence.

Another serious cause of inability to protrude the penis is hematoma of the penis due to rupture of the corpus cavernosum of the penis (Fig. 23–4). It commonly occurs in bulls during coitus, where the penis is thrust against the perineum of the cow. A hematoma develops at the sigmoid flexure and causes swelling, which may be palpated anterior to the scrotum. Preputial prolapse often accompanies the condition. A few days later, the hematoma is well circumscribed and reduced in size. At this time the animal is willing to serve, but unable to fully extend the penis owing to adhesions.

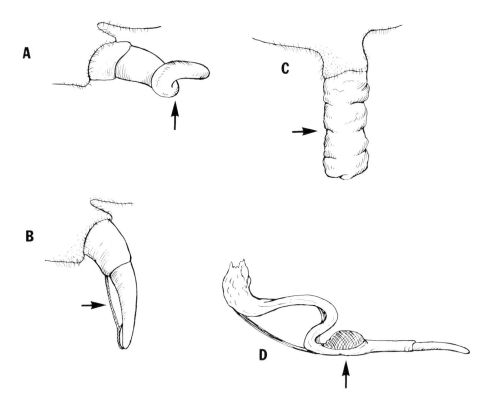

Fig. 23–4. Abnormalities of the penis and prepuce that interfere with intromission in the bull. *A*, "Corkscrew" penis or spiralling of the glans penis prior to intromission. *B*, Persistence of the frenulum, causing a downward deviation of the penis. *C*, Prolapse of prepuce causing phimosis. *D*, Hematoma of the penis anterior to the sigmoid flexure.

Spontaneous regression of the hematoma may occur after a prolonged period of sexual rest, but in most cases surgical evacuation of the hematoma and repair of the damaged tunica albuginea have been successful.

Tumors of the glans penis may occasionally prevent protrusion of the penis. Fibropapillomas of viral origin are frequently noted on the glans penis of two-year-old bulls. Affected bulls are reluctant to serve or are incapable of achieving intromission. Although spontaneous regression of the tumor can occur, surgical extirpation or vaccination with a tissue vaccine is employed to control the condition.

Congenital deformities of the penis or prepuce may render intromission difficult or impossible. One such deformity is the "persistence of the frenulum" commonly encountered in beef Shorthorn and Aberdeen Angus bulls (Fig. 23–4). In this condition, the frenulum attaches the ventral aspect of the glans penis to the preputial mucosa. At coitus, the deformity is noted as a ventral or downward deviation of the penis. Rarely, intromission may be accomplished. The deformity can be corrected by ligating and cutting the band of tissue.

Three distinct types of congenital deviations of the penis are encountered in bulls. A spiral deviation of the penis occurs in most normal bulls after intromission. A similar spiralling occurs with the "corkscrew" type of penile deviation (Fig. 23–4) where the spiralling precedes intromission and prevents coitus. The condition is prevalent among polled Hereford bulls and can be corrected by surgical fixation of the dorsal ligament of the penis to the tunica albuginea. Less common types of deviations of the penis are the ventral or "rainbow" deviation and the mild S-shaped deviation.

In the boar, abnormalities of the penis, eg, persistent frenulum, penile hypoplasia and enlargement of the preputial diverticulum, frequently result in a failure to achieve intromission and are the major causes of poor mating performance. With these defects the boar is unable to erect his penis, to penetrate the vagina or to lock it in the cervix.

Failure to Ejaculate. This condition is occasionally observed with bulls even when accompanied by vigorous thrust at intromission. Poor semen collection techniques, eg, improper temperature or pressure within the artificial vagina, often cause failure of ejaculation in bulls used for AI purposes.

In the stallion, ejaculatory disorders, ranging from intromission without ejaculation to an abnormal copulatory pattern with or without occasional ejaculation, are frequently encountered. These disorders are probably caused by a functional disturbance of the nervous mechanisms that regulate the ejaculatory process. Unfamiliar surroundings, fatness, poor condition or exhaustion due to frequent services may exert a detrimental effect upon these nervous mechanisms (Rasbech, 1975).

FERTILIZATION FAILURE

Fertilization failure is an important cause of infertility in males that have normal libido and are capable of mating and ejaculating. This capacity or reduced capacity is related to defective semen characteristics or to errors in breeding techniques.

Evaluation of Semen Quality

The ideal method of evaluating the fertility of a breeding male other than his ability to produce pregnancy is by the examination of his semen.

The evaluation of the ejaculate is an important part of the physical examination of an infertile male. A semen analysis is relatively simple to perform and several important conclusions can be obtained from the results. The various laboratory methods of evaluating semen, performed routinely in artificial insemination centers, are described in a subsequent chapter. Briefly, the semen evaluation procedures include recording volume, color and density of the ejaculate, and estimating sperm concentration, forward progressive motility of spermatozoa (%), and examining smears for sperm abnormalities and the presence of leukocytes or abnormal cells. Average standards for any species could be established and any deviations from these "normal" standards can be recognized and correlated with fertility. Certain terms are used to express these deviations in semen characteristics (Table 23–2).

Semen evaluation in relation to fertility is conducted more in bulls than in other species. In general, the minimal standards for a classification of "probably fertile" specimen of bull semen are: over 500 million spermatozoa per milliliter are present, more than 50% of motile sperm make forward progression, and more than 80% of the spermatozoa conform to normal morphology. If any one of these criteria is not met, particularly with samples of three or more ejaculates, the bull is suspected of being infertile. However, only when motile spermatozoa are totally absent and the reproductive system has been carefully examined for disease can it be stated that the bull is sterile. Similar

criteria have not been well defined for other species.

Presently there is no single biochemical criterion for differentiating potentially fertile and infertile males. In man, biochemical analysis of seminal plasma is a useful diagnostic tool to evaluate the secretory functions of the male accessory glands (Polakoski and Zaneveld, 1977), but similar analyses are not performed routinely for farm animals.

Sperm Abnormalities. The recognition of defectively formed spermatozoa in stained smears under the microscope has provided useful information. However, with the advent of phase contrast and differential interference contrast microscopy on unfixed preparations as well as electron microscopy of testicular tissue and spermatozoa, opportunities of recognizing sperm abnormalities have increased.

Acquired Sperm Abnormalities. Morphologic abnormalities of spermatozoa can be *primary, secondary* or *tertiary*. Primary abnormalities are due to failure of spermatogenesis, whereas secondary abnormalities occur during the passage of spermatozoa through the epididymis. Damage to spermatozoa during or after ejaculation or handling for AI are designated as tertiary abnormalities. In general the potential fertility is low when a high percentage of spermatozoa have primary and secondary abnormalities.

Table 23–2. Semen Analysis Nomenclature

Parameter	Evaluation Criterion	Nomenclature
Volume	None	Aspermia
	Reduced	Hypospermia
	Increased	Hyperspermia
Sperm concentration	Zero	Azoospermia
	Reduced	Oligozoospermia
	Normal	Normozoospermia
	Increased	Polyzoospermia
Sperm motility	Decreased	Asthenozoospermia
Sperm viability	All dead	Necrozoospermia
Abnormal spermatozoa	High percentage	Teratozoospermia

For additional information see Andrologia (1974) 6, 100; Biol. Reprod. (1973), *8/1*, iii.

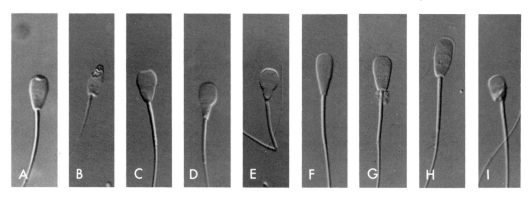

Fig. 23–5. Head abnormalities in ejaculated bull sperm as shown under differential interference contrast microscopy (1000×). *A*, Knobbed acrosome; *B*, ruffled acrosome; *C*, incomplete acrosome; *D*, asymmetric pyriform; *E*, pyriform; *F*, tapered; *G*, *H*, cratered; *I*, small head with abaxial implantation. *(From Mitchell, J.R., Hanson R.D., Fleming, W.N., 1978. Proc. VIIth Tech. Conf. A.I. and Reprod. NAAB Inc., Columbia, Missouri)*

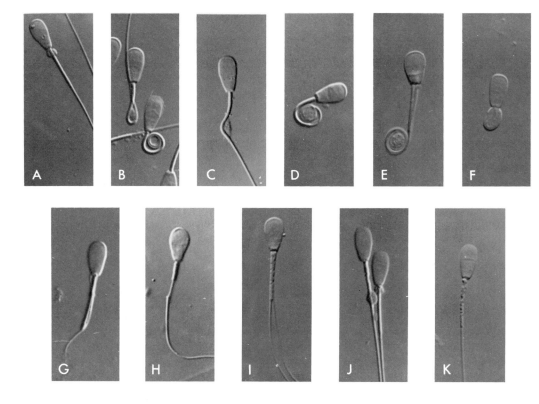

Fig. 23–6. Abnormalities in ejaculated bull sperm as shown under differential interference contrast microscopy (1000 ×). *A*, Proximal and distal protoplasmic droplets; *B*, translocating protoplasmic droplets; *C*, tail opening following droplet translocating; *D*, coiled tail with droplet; *E*, coiled double tail; *F*, "dag" defect; *G*, folded tail; *H*, filamentous tail; *I*, double tail; *J*, "corkscrew" midpiece with droplet; *K*, "corkscrew" midpiece. *(From Mitchell, J.R., Hanson, R.D., and Fleming, W.N., 1978. Proc. VIIth Tech. Conf. A.I. and Reprod. NAAB Inc., Columbia, Missouri)*

For routine examination in AI centers, the abnormal spermatozoa are classified as follows: abnormal or detached heads, cytoplasmic droplets attached to the anterior, middle or distal part of the midpiece, coiled or bent tails, and other abnormalities (Figs. 23–5, 23–6).

Inherited Sperm Defects. A variety of structural defects of spermatozoa can cause sterility or subfertility of bulls and boars by interfering with fertilization (Blom, 1973; Bishop, 1972; Saacke et al, 1968). Some of these sperm defects are inherited; however, the mode of inheritance is not clear, because of the small number of animals or relatives of the affected animals that have been investigated. Of the two defects of the sperm head, the *diadem defect* or nuclear pouch formation defect is responsible for low-

ered fertility of bulls from Norway, Sweden and Denmark (Fig. 23–7). The defect, confined to the anterior border of the postnuclear cap, results from invaginations of the nuclear membrane and is a sign of disturbed spermatogenesis. The *knobbed-sperm defect*, found mostly in Friesian bulls, takes the form of an eccentrically placed thickening of the acrosome; it affects all spermatozoa in the ejaculate. Affected bulls are sterile. It is an inherited character and is due to an autosomal recessive gene. A similar *knobbed sperm* defect was noted in the semen of large White boars. After insemination, affected spermatozoa failed to fertilize eggs.

The *decapitate defect* is an abnormality of the sperm tail reported in the semen of sterile Guernsey bulls. It is associated

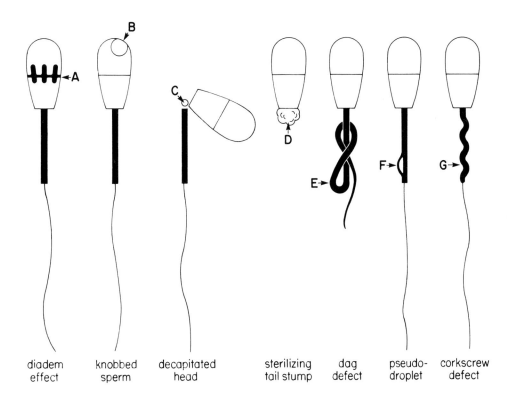

diadem effect knobbed sperm decapitated head sterilizing tail stump dag defect pseudo-droplet corkscrew defect

Fig. 23–7. Some inherited sperm abnormalities associated with reproductive failure in bulls. *A*, Diadem effect; *B*, knobbed sperm; *C*, decapitated head; *D*, sterilizing tail stump; *E*, dag defect; *F*, pseudodroplet; *G*, corkscrew defect.

with an ultrastructural abnormality in the neck or implantation region of the spermatozoon. These defective spermatozoa disintegrate into isolated heads and tails in the caput epididymis. In the *dag defect*, noted in the semen of three infertile Danish Jersey bulls, the main piece is strongly coiled and folded over the midpiece, giving the impression of a short tail. It is due to elevated levels of zinc in spermatozoa as well as in the seminal plasma (Blom and Wolstrup, 1976). A *sterilizing tail stump defect* was noted in a sterile Danish Holstein-Friesian bull (Blom, 1976). Over 60% of the sperm heads had short tail stumps, 2 to 3 μm long. Another sperm defect from five sterile Friesian bulls is the *pseudodroplet defect*. It is characterized by a rounding or elongated thickening of the midpiece. A rare sperm-tail defect noted in aging bulls, especially in the Red Dane breed, is a midpiece shaped like a *corkscrew* in 15 to 50% of spermatozoa; these sperm were dead.

In view of the inherited nature of the structural defects of spermatozoa that cause sterility or subfertility, a detailed microscopic examination of spermatozoa must be made on all young breeding bulls before they are used for AI.

Pathology of Testes and Accessory Glands

Pathologic conditions of the testes, epididymis and seminal vesicles (Table 23–3) may interfere with fertilization by disturbing spermatogenesis or sperm maturation, causing abnormal semen characteristics, or preventing the passage of spermatozoa from the testes to the urethra.

Testicular Degeneration. In testicular degeneration, the seminiferous tubules lose their ability to produce physiologically normal spermatozoa. It is the most common cause of acquired infertility and lowered semen quality. Testicular degeneration is probably the most common abnormality in the testes of bulls that are discarded for reduced reproductive efficiency. Testicular degeneration has been attributed to a variety of causes (Table 23–3). A high percentage of morphologically abnormal spermatozoa is present in the semen, which is thin and watery due to a reduction in sperm concentration.

Orchitis. Inflammation of the testes may result from direct injury with subsequent infection, but in most cases it is caused by bacterial infection from infected accessory glands that reach the testes along the vas deferens. In the acute phase, the affected testicles are swollen and hot, followed either by abscess formation or by chronic fibrosis. Spermatogenesis in affected animals is reduced considerably or even absent due to irreversible damage to the seminiferous tubules. If only one testicle is affected, immediate removal should be considered, so that degenerative changes in the unaffected testis may be avoided.

Epididymitis. Maturation and storage of spermatozoa occur in the epididymis, and any inflammation may adversely affect semen quality and fertility. Epididymitis is caused primarily or secondarily by the organisms responsible for orchitis. Prognosis in such cases is poor, as obstructions occur that prevent the passage of spermatozoa from the testis to the vas deferens.

Epididymitis is a common lesion of the reproductive tract of the ram and is caused by *Brucella ovis* (Watt, 1972). The organism may be disseminated in the ejaculate and is transmitted from ram to ram through copulation with ewes previously served by infected rams. Chronic epididymitis in rams is characterized by enlargement of the tail of the epididymis, associated with adhesions of the tunica vaginalis and fibrosis. Diagnosis is based on palpation of testes, bacterial examination of semen and the complement fixation test of serum.

Table 23–3. Summary of Diseases of Male Reproductive Organs with Infertile Semen

Disease	Species Affected	Causes	Lesion	Seminal Changes
Testicular degeneration	Bull, ram	Thermal, localized or systemic infections; nutrition (vitamin A); vascular lesions; aging; obstructive lesions of the head of epididymis; noxious agents; hormonal factors	Testicular size reduced; fibrosis; disturbances in spermatogenesis; seminiferous tubules destroyed in advanced cases	Increase in immature and abnormal sperm with normal motility; later ejaculate is thin and watery due to reduction in sperm concentration; giant cells; azoospermia or necrozoospermia in severe cases
Orchitis	Bull, ram, boar	Brucellosis, tuberculosis	Inflammatory changes in testis leading to degeneration of seminiferous tubules	Asthenozoospermia; oligozoospermia; teratozoospermia; giant cells; erythrocytes and leukocytes; normal semen volume
Epididymitis	Bull, ram	Brucellosis; viral infections	Inflammation of epididymis; infiltration of lymphocytes and neutrophils; dead sperm and giant cells	Poor semen characteristics; semen contaminated by inflammatory exudate
Seminal vesiculitis	Bull	Brucellosis	Unilateral inflammation of seminal vesicles; glands enlarged and fibrosed	Purulent exudate in semen, normozoospermia, asthenozoospermia; lowered fructose content

(Adapted from Jubb and Kennedy, 1970. Pathology of Domestic Animals, New York Academic Press; Laing, 1970. Fertility and Infertility in Domestic Animals, London, Bailliere Tindall and Cassell)

Heat Stress

Temperature is one of the important environmental factors modifying reproduction. Elevated body temperatures, during periods of high ambient temperature or pyrexia from disease, lead to testicular degeneration and reduce the percentage of normal and fertile spermatozoa in the ejaculate.

In several species, there are seasonal variations in the quality and fertility of semen. Bulls subjected to high environmental temperatures have reduced semen quality. Rams may retain a satisfactory level of fertility throughout the whole year, but in many instances, fertility is depressed when matings occur during the hot months of the year. Conception failure in ewes mated to heat-stressed rams is related more to failure of fertilization than to embryonic mortality.

When the scrotal contents of rams are heated to approximately 40°C for 1½ to 2 hours, a sharp increase in the proportion of morphologically abnormal spermatozoa occurs in the ejaculate 14 to 16 days later (Braden and Mattner, 1970). Spermatozoa that were developing in the testis at the time of heating showed damage (eg, dead and tailless spermatozoa), whereas epididymal spermatozoa were unaffected. Acrosomal damage is characterized by swelling, vesiculation and eventual disintegration (Williamson, 1974).

Volume of semen and total sperm per ejaculation from boars are greater during cool weather. Boars exposed to elevated ambient temperatures show decreases in sperm concentration and motility, and increases in sperm abnormalities and acrosomal changes (Wettermann et al, 1976). Also, infectious diseases resulting in a febrile reaction are likely to affect the subsequent fertility of the boar for a five- to six-week period. The decrease in conception rate in artificially inseminated gilts maintained under elevated environmental temperatures is apparently due to embryonic mortality.

Breeding Techniques

Fertilization failure attributed to the male may result from poor breeding management or from faulty techniques in AI. Also, synchronization of estrus in cattle and sheep with progestational compounds, the ingestion of estrogenic pasture grass by sheep, or the imposition of stress during insemination may interfere with sperm transport and cause fertilization failure.

Breeding Management. Under natural mating programs, the frequency of service and the ratio of females assigned to each breeding male (Table 23–4) depend on the species, age, libido, fertility and nutrition of the male, the duration of the mating season, the system of management and the size of pasture or range.

Spermatogenesis is a continuous pro-

Table 23–4. A Guide to Breeding Frequency and Number of Breeding Females for Each Breeding Male

| Species | Hand Breeding | | Pasture Breeding |
	Services/Wk	No. Females/Season (Year)	No. Females/Season (Year)
Stallion	3–12 (2–5)*	30–120 (3–12)	
Bull	4–12 (2–4)	80–120 (20–60)	10–25 (10–15)
Boar	4–10 (2–4)	30–60 (10–40)	20–40 (10–20)
Ram	6–24 (6–12)	40–80 (30–40)	40–80 (20–30)

*Figures given in parenthesis are for immature males.

(Adapted from Roberts (1971). Veterinary Obstetrics and Genital Diseases, published by the author, Ithaca, N.Y.)

cess, but frequent and repeated ejaculations adversely affect male libido and semen characteristics. Although libido returns to normal after a week of sexual rest, semen characteristics are not restored to normal for six weeks. Similarly, after a period of prolonged inactivity, semen characteristics and fertility remain low for the first few services.

Seasonal variations are especially important in seasonally breeding species such as stallions and rams; changes in the ratio of daylight to darkness are reflected in the quality and quantity of semen.

Infertility and Artificial Insemination. The male makes several contributions to reproductive failure in an AI program, eg, defective semen, improper insemination techniques or failure of sperm transport in the female tract. These and other factors affecting fertility in AI are considered in Chapter 26. Changes in the fertility of frozen semen during storage are important in the efficient use and design of AI programs.

Failure of fertilization is a major cause of lowered fertility following long-term storage of semen. With normal aging or injury, the acrosomal cap of bovine spermatozoa undergoes a distinct sequence of alterations, which leads to the formation of the equatorial segment and loss of the anterior acrosome (Fig. 23–8). The rate of alteration depends upon the bulls, the ejaculates and the procedures for handling semen and it has a good correlation with conception rates.

Similar acrosomal changes occur

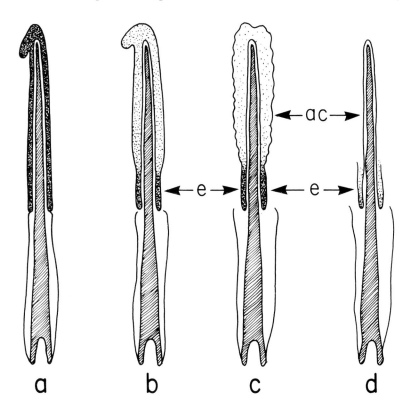

Fig. 23–8. Schematic representations of sagittal section through the bovine sperm head under the electron microscope showing sequential changes of the acrosome *(ac)* due to sperm aging or injury. *a*, Intact acrosomal cap with a distinct apical ridge; *b, c,* gradual loss of apical ridge and swelling of anterior acrosomal cap with formation of equatorial segment, *e; d*, loss of anterior acrosomal cap, leaving the equatorial segment. *(Redrawn from Saacke et al., 1968. J. Anim. Sci., 27, 1391)*

within spermatozoa during *in vitro* storage, which subsequently decrease fertility and increase embryonic mortality in cattle, swine and sheep (Fig. 23–9). These changes in fertility with *in vitro* aging of spermatozoa occur at all storage temperatures and at rates that are positively related to the storage temperature. Earlier observations with bull semen that had been frozen and stored for two years in liquid nitrogen showed that optimal fertility was achieved after storage for four months, but by the sixth month fertility declined (Salisbury, 1968). Subsequent investigations have shown no decrease in the fertility of semen stored at −196°C for several years from the horse (Merkt et al, 1975), the bull (Foote, 1972; Lee et al, 1977) and the ram (Salamon and Visser, 1974), provided there are no fluctuations in storage temperature, a common problem in field liquid nitrogen containers.

The poor fertility of frozen ram semen may be attributed to acrosomal damage and impaired viability and transport of spermatozoa to the oviducts. A much higher proportion of spermatozoa is damaged in ram semen than in bull semen during freezing and thawing (Watson and Martin, 1972).

Differences in fertilizing capacity of bulls can be determined by inseminating cows with mixed semen, so that sperm from two bulls compete directly at the level of fertilization. The number of calves born to each bull provides a "heterospermic index of fertility," and may differentiate sires according to the ability of semen from some sires to survive freezing far better than that from others. Fertility differences between bulls also arise through differences in the survival of spermatozoa after insemination.

Immunologic Factors

The presence of antibodies in the serum of the male or female may cause sperm agglutination or immobilization. These antibodies can cause a reduction in fertility by their interaction with spermatozoa in the semen.

In man, depending on the antibodies present, different types of sperm agglutination have been recorded. IgM antibodies reacted largely to the acrosomal area and tip of the tail and caused head-to-head and tip of tail-to-tip of tail agglutination patterns (Fig. 23–10), whereas IgG reacted with the equatorial segment

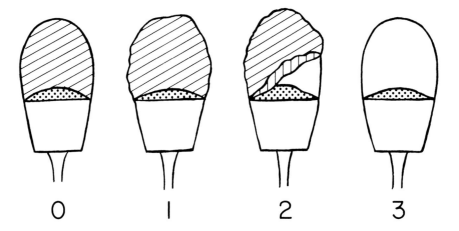

Fig. 23–9. Drawings of ram spermatozoa stained with Giemsa showing various stages in the disruption and loss of the acrosome during chilling and deep-freezing (1250 ×). Score *0*, "normal" acrosome; scores *1* and *2*, acrosome loosened or damaged; score *3*, acrosome completely lost. *(Redrawn from Watson, P.F., 1975. Vet. Rec. 97, 12)*

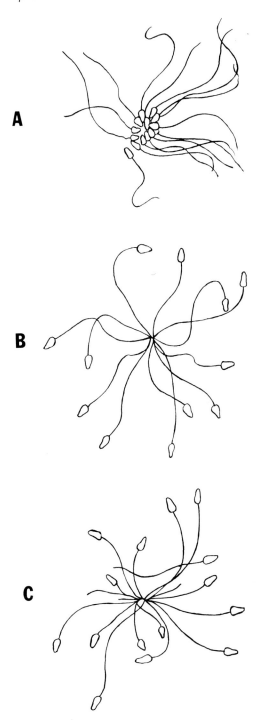

and the main tail piece, and caused head-to-tail and tail-to-tail types of agglutination (Menge, 1976).

The seminal antigens associated with infertility in cattle reside in the sperm components, and the infertility is due to inhibition of fertilization and/or embryonic mortality (Menge, 1969).

The role of immunologic factors in reproductive failure in farm animals is not well documented. Even in cattle, a species in which isoimmunization against sperm is good, the semen must be administered in conjunction with immunologic adjuvants to stimulate the formation of effective antibodies.

Egg-yolk and milk used in semen extenders may also act as antigens. Antibodies against egg-yolk antigens have been detected in uterine mucus and tissue from cows that had been inseminated repeatedly (Griffin et al, 1971). When cows were inseminated with extenders containing egg-yolk, the fertility rate was lower in cows showing uterine titers to egg-yolk antigens than in cows not showing uterine titers.

NUTRITION AND MALE INFERTILITY

The effects of nutritional restrictions on fertility are more notable in the female than in the male. Nutritional deficiencies delay the onset of puberty and depress production and characteristics of semen in the male. The young and growing animal is much more susceptible to nutritional stress than the mature animal. In addition, nutrition affects the endocrine rather than the spermatogenic function of the testis. Common nutritional factors include caloric, protein and vitamin deficiencies, but minerals or toxic agents may also be important.

Underfeeding. Despite the ability of a mature male to maintain sperm production and testosterone secretion under low levels of nutrition, the young male shows retarded sexual development and delayed

Fig. 23–10. Typical sperm agglutination patterns caused by human serum agglutinins. *A*, head-to-head; *B*, tip of tail-to-tip of tail; *C*, tail-to-tail. *(Menge, A.C., 1977. In: Techniques of Human Andrology. E.S.E. Hafez* [ed.], *New York, Elsevier/North-Holland)*

puberty. This is due to suppression of endocrine activity of the testes and consequently to retardation of growth and secretory function of the male organs of reproduction. When mature bulls, rams and boars are fed low-energy rations for prolonged periods, libido and testosterone production are affected much earlier than semen characteristics (Parker and Thwaites, 1972; Braden et al, 1974; Mann, 1974). The effects of undernutrition may be corrected in mature animals, whereas it is less successful in young animals because of the permanent damage caused to the germinal epithelium of the testis.

Obesity and overfeeding reduce libido and sexual activity in rams, boars and bulls, particularly during hot weather.

Protein deficiency affects the young more than the mature male. Young bulls on a protein-deficient diet show decreases in libido and semen characteristics, whereas mature bulls, rams and boars are rarely affected. Diets high in protein are not essential for optimal sperm production in the ram (Braden et al, 1974).

Vitamin Deficiencies. Dietary vitamin A or carotene deficiency leads to testicular degeneration in all farm animals. The effect of vitamin A on the testes is probably indirect and due to suppression of the release of pituitary gonadotropins. Injections of gonadotropic hormones or vitamin A will restore spermatogenesis, except in cases in. which the damage to the testis is permanent. While bull calves maintained on a low vitamin A diet show degenerative changes in the germinal epithelium of the testis and azoospermia, mature bulls show no adverse effects in spermatogenesis.

Cattle are more resistant to vitamin A deficiency than swine. For example, night-blindness and incoordination of movement precede recognizable reduction in fertility of mature bulls, whereas testicular degeneration is one of the earliest signs of avitaminosis A in the mature

boar. Vitamin E (tocopherol or wheat germ oil) is important for normal reproduction, but its role in the fertility of male farm animals is obscure.

Mineral Deficiencies. There is a paucity of information concerning the effects of trace mineral deficiencies upon male reproductive functions. Iodine deficiency has been suspected as a cause of poor libido and semen characteristics in bulls. Also, improvement in sperm production and fertility have been noted following supplementary feeding of copper, cobalt, zinc and manganese.

Toxic Agent. Plant estrogens exert adverse effects on male accessory organs (Mann, 1974), but infertility of sheep and cattle grazing on estrogenic pastures are related to changes in cervical mucus and to a failure of sperm transport in the female tract. Many chemicals, rare earth salts and ionizing radiations interfere with spermatogenesis in a variety of mammalian species, but their contribution to male infertility remains to be established.

INFERTILITY AND CHROMOSOME ABERRATIONS

Chromosomal aberrations play an important role in human reproductive failure. From a breeding point of view, it is important to eliminate males that are affected by chromosomal aberrations, particularly those resulting in decreased fertility (Table 23–5). Chromosomal aberrations involve either the sex chromosomes or the autosomes.

Klinefelter's Syndrome. Cytogenetic studies in man have revealed sex chromosome abnormalities with a number of phenotypic defects, including those of the reproductive system. The XXY syndrome in man (Klinefelter's syndrome) is associated with testicular atrophy, aspermatogenesis and infertility. Sex chromosome aneuploidy is now being uncovered in domestic animals. For example the

Table 23–5. Male Infertility and Chromosomal Aberrations

Aberration	Animal	Karyotype	Description	Reference
Klinefelter's syndrome	Ram, bull, boar, stallion	XXY	Testicular hypoplasia, azoospermia, aspermatogenesis	Bishop (1972)
Chimerism	Bull	XX/XY	Bulls born co-twin to heifers	Stafford (1972); Fechheimer (1973)
Mosaics	Stallion	XX/XXY and XX/XY/XO/XXY	Male pseudohermaphrodite with azoospermia	Bouters et al, (1972); Basrur et al (1969)
Robertsonian translocation	Bull	1/29 translocation	Infertility in Swedish Red and White cattle and Canadian Guernseys but not in other breeds	Gustavsson (1969); Bongso and Basrur (1976); Pollock and Bowman (1974)
	Ram	$53xyt_1$, $53xyt_2$ and $52xyt_3t_3$	Normal fertility in affected males and their progeny	Bruere (1975)
	Boar	$(13q^-$; $14q^+)$ and $(11p^+$; $15q^-)$	Low semen quality and reduced litter size	Hageltorn et al (1976)
Autosomal secondary constrictions	Bull	XY	Infertility with or without testicular hypoplasia	Bongso and Basrur (1976); Galloway and Norman (1976)

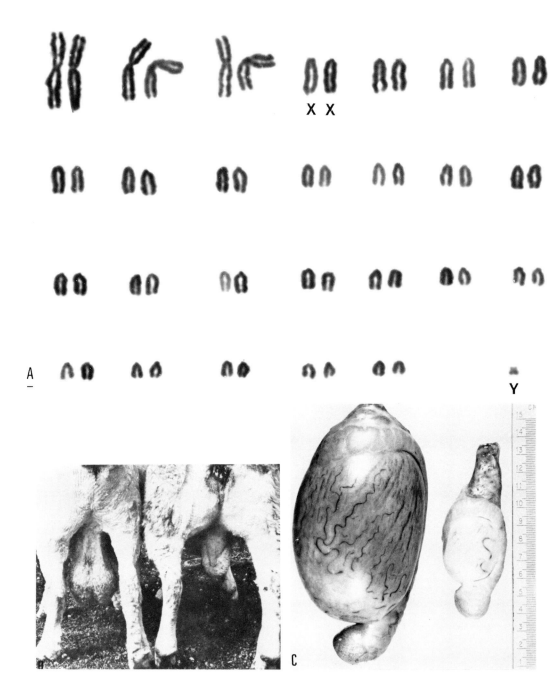

Fig. 23–11. Klinefelter syndrome in sheep. *A*, Karyotype from a 55-XXY chromatin-positive Klinefelter ram. *B, C,* Photographs illustrating the difference in scrotal size between a normal ram *(left)* and a chromatin-positive Klinefelter ram *(right)*. *(Bruere, A.N., 1974. Vet. Rec. 95, 436)*

XXY syndrome has been reported in infertile bulls, rams (Fig. 23–11) and boars (Bishop, 1972).

XX/XY Chimerism. Infertile bulls born co-twin to heifers are XX/XY chimeras (Stafford, 1972), due to a common allantoic circulation between the twins. During a cytogenetic survey of 743 young bulls in three AI centers, 13 animals were chimeric (Fechheimer, 1973), but three of the chimeric bulls were singletons. In the latter cases, the female co-twin probably died and was resorbed. That chimerism develops from other mechanisms in singletons cannot be entirely eliminated.

Mosaics. Culture of peripheral blood leukocytes in a bilaterally cryptorchid male pseudohermaphrodite horse revealed 64XX/65XXY mosaicism (Bouters et al, 1972). The animal showed strong male behavior and had a small penis which, on erection, released a watery fluid devoid of spermatozoa. In another pseudohermaphrodite horse, four cell types with XX, XY, XO and XXY sex chromosomes were noted (Basrur et al, 1969).

Robertsonian Translocation. Structural rearrangements of chromosomes may be a cause of infertility. Breakage and reunion of two acrocentric or telocentric chromosomes results in a robertsonian translocation. The 1/29 translocation (Fig. 23–12) has been recognized in several breeds of cattle with (Bongso and Basrur, 1976) or without deleterious effects on fertility (Pollock and Bowman, 1974). In the Swedish Red and White breed of cattle, the translocation reduced the nonreturn rate (fertility) of bulls (Gustavsson, 1969) as well as increased the incidence of delayed returns to service (embryonic mortality) among the daughters of carrier bulls (Gustavsson, 1969). Robertsonian translocations in rams (53 xyt$_1$; 53 xyt$_2$ and 52 xyt$_3$t$_3$ do not affect fertility of affected sires or their daughters (Bruere, 1975), whereas in boars, translocations (13q$^-$; 14q$^+$ and 11p$^+$; 15q$^-$) are associated

with low semen quality and small litter size (Hageltorn et al, 1976).

Autosomal Constrictions. Cytogenic investigations of Guernsey bulls used for AI revealed the presence of chromatid breaks or secondary constrictions in large and medium-sized autosomes. This aberration was associated with low fertility (Bongso and Basrur, 1976). Bulls with testicular hypoplasia were also shown to have similar secondary constrictions in

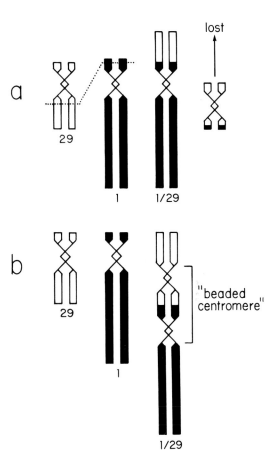

Fig. 23–12. Proposed schemes for the origin of 1/29 translocation in the bull. *a*, Chromosome 29, with a break in the long arms, and chromosome 1, with a break close to the centromere on the short arms, reunite so that the distal arms of chromosome 29 attach to the short arms of chromosome 1 with a loss of a chromosome fragment during subsequent divisions. *b*, Reunion of chromosomes 29 and 1 so that the long arms of chromosome 29 attach to the short arms of chromosome 1 with no loss of chromosome segment. *(Courtesy of Dr. T.A. Bongso)*

both large and medium sized chromosomes (Galloway and Norman, 1976).

REFERENCES

Almquist, J.O., Branas, R.J. and Barber, K.A. (1976). Post-puberal changes in semen production of Charolais bulls ejaculated at high frequency and the relation between testicular measurements and sperm output. J. Anim. Sci. 42, 670.

Basrur, P.K., Kanagawa, H. and Gilman, J.P.W. (1969). An equine intersex with unilateral gonadal agenesis. Can. J. Comp. Med. 33, 297.

Berndtson, W.E., Pickett, B.W. and Nett, T.M. (1974). Reproductive physiology of the Stallion IV. Seasonal changes in the testosterone concentration of peripheral plasma. J. Reprod. Fertil. 39, 115.

Bishop, M.W.H. (1972). Genetically determined abnormalities of the reproductive system. J. Reprod. Fertil. Suppl. 15, 51.

Blom, E. (1973). The ultrastructure of some characteristic sperm defects and a proposal for a new classification of the bull spermiogram. Nord. Vet. Med. 25, 383.

Blom, E. (1976). A sterilizing tail stump sperm defect in a Holstein-Friesian bull. Nord. Vet. Med. 28, 295.

Blom, E. and Wolstrup, C. (1976). Zinc as a possible causal factor in the sterilizing sperm tail defect, the 'Dag-defect' in Jersey bulls. Nord. Vet. Med. 28, 515.

Bongso, A. and Basrur, P.K. (1976). Chromosome anomalies in Canadian Guernsey bulls. Cornell Vet. 66, 476.

Bouters, R., Vandeplassche, M. and de Moor, A. (1972). An intersex (male pseudohermaphrodite) horse with 64 XX/65XXY mosaicism. Equine Vet. J. 4, 150.

Braden, A.W.H. and Mattner, P.E. (1970). The effects of scrotal heating in the ram on semen characteristics, fecundity and embryonic mortality. Aust. J. Agric. Res. 21, 509.

Braden, A.W.H., Turnbull, K.E., Mattner, P.E. and Moule, G.R. (1974). Effect of protein and energy content of the diet on the rate of sperm production in rams. Aust. J. Biol. Sci. 27, 67.

Bruere, A.N. (1975). Further evidence of normal fertility and the formation of balanced gametes with one or more different Robertsonian translocations. J. Reprod. Fertil. 45, 323.

Colenbrander, B., Kruip, T.A.M., Dielemen, S.J. and Wensing, C.J.G. (1977). Changes in serum LH concentrations during normal and abnormal sexual development in the pig. Biol. Reprod. 17, 506.

Colenbrander, B., van Straaten, H.W.M. and Wensing, C.J.G. (1978). Gonadotrophic hormones and testicular descent. Maturitus, 1, 131.

Cox, J.E. (1975). Experience with a diagnostic test for equine cryptorchidism. Equine Vet. J. 7, 179.

Fechheimer, N.S. (1973). A cytogenic survey of young bulls in the USA. Vet. Rec. 93, 535.

Foote, R.H. (1972). Aging of spermatozoa during storage in liquid nitrogen. Proc. 4th Tech. Conf. A.I. and Reprod., 28.

Foote, R.H., Munkenbeck, N. and Greene, W.A. (1976). Testosterone and libido in Holstein bulls of various ages. J. Dairy Sci. 59, 2011.

Galloway, D.B. and Norman, J.R. (1976). Testicular hypoplasia and autosomal secondary constrictions in bulls. 8th. Internat. Congr. Anim. Reprod. and AI. IV, 710.

Griffin, J.F.T., Nunn, W.R. and Hartigan, P.J. (1971). An immune response to egg-yolk semen diluent in dairy cows. J. Reprod. Fertil. 25, 193.

Gsutavsson, I. (1969). Cytogenetics, distribution and phenotypic effects of a translocation in Swedish cattle. Hereditas, 63, 68.

Hafez, E.S.E. and Bouissou, M.F. (1975). The behaviour of cattle. In The Behaviour of Domestic Animals. 3rd ed. E.S.E. Hafez (ed), Philadelphia, Lea & Febiger.

Hageltorn, M., Gustavsson, I. and Zech, L. (1976). Detailed analysis of a reciprocal translocation (13q⁻; 14q⁺) in the domestic pig by G- and Q-staining techniques. Hereditas, 83, 262.

Lee, A.J., Salisbury, G.W., Boyd, L.J. and Ingallis, W. (1977). In vitro aging of frozen semen. J. Dairy Sci. 60, 89.

Mann, T. (1974). Effects of nutrition on male accessory organs. In Male Accessory Sex Organs. D. Brandes (ed), New York, Academic Press.

Menge, A.C. (1969). Early embryo mortality in heifers isoimmunized with semen and conceptus. J. Reprod. Fertil. 18, 67.

Menge, A.C. (1976). Immunology. In Techniques of Human Andrology. E.S.E. Hafez (ed), New York, North-Holland Publishing.

Merkt, H., Klug, E., Krause, D. and Bader (1975). Results of long-term storage of stallion semen frozen by the pellet method. J. Reprod. Fertil. Suppl. 23, 105.

Parker, G.V. and Thwaites, C.J. (1972). The effects of undernutrition on libido and semen quality in adult Merino rams. Aust. J. Agric. Res. 23, 109.

Polakoski, K.L. and Zaneveld, L.J.D. (1977). Biochemical examination of the human ejaculate. In Techniques of Human Andrology. E.S.E. Hafez (ed.). New York, North-Holland Publishing Co.

Pickett, B.W. and Voss, J.L. (1975). Abnormalities of mating behaviour in domestic stallions. J. Reprod. Fertil. Suppl. 23, 129.

Pollock, D.L. and Bowman, J.C. (1974). A Robertsonian translocation in British Friesian cattle. J. Reprod. Fertil. 40, 423.

Rasbech, N.O. (1975). Ejaculatory disorders of the stallion. J. Reprod. Fertil. Suppl. 23, 123.

Saacke, R.G., Amman, R.R. and Marshall, C.E. (1968). Acrosomal cap abnormalities of sperm from subfertile bulls. J. Anim. Sci. 27, 1391.

Salamon, S. and Visser, D. (1974). Fertility of ram spermatozoa frozen-stored for 5 years. J. Reprod. Fertil. 37, 433.

Salisbury, G.W. (1968). Fertilizing ability and biological aspects of sperm storage in vitro. 6th Int. Congr. Anim. Reprod. & A.I. Paris. II: 1189.

Stafford, M.R. (1972). The fertility of bulls born co-twin to heifers. Vet. Rec. *90*, 146.

Watson, P.F. and Martin, I.C.A. (1972). A comparison of changes in the acrosomes of deep-frozen ram and bull spermatozoa. J. Reprod. Fertil. *28*, 99.

Watt, D.A. (1972). Testicular abnormalities and spermatogenesis of the ovine and other species. Vet. Bull. *42*, 181.

Wensing, C.J.G. (1973). Abnormalities of testicular descent. Proc. Kon. Ned. Akad. v. Wettensch. C *76*, 373.

Wettermann, R.P., Wells, M.E., Omtvedt, L.T., Pope, C.E. and Turman, E.J. (1976). Influence of elevated ambient temperatures on reproductive performance of boars. J. Anim. Sci. *42*, 664.

Williamson, P. (1974). The fine structure of ejaculated ram spermatozoa following scrotal heating. J. Reprod. Fertil. *40*, 191.

24

Intersexuality

R.A. McFEELY
and
L.R. KRESSLY

Intersexuality, of interest to man since antiquity, continues to attract the attention of scientists and breeders alike. To the scientist, intersex animals provide clues to understanding the mechanisms of sex determination, while the breeder is more concerned with the economic drawbacks of raising animals destined to be infertile. Modern techniques of cytogenetics have provided investigators with new tools to help unravel the mysteries of normal and abnormal sexual development. Since the demonstration of sexual dimorphism in some somatic cell nuclei and the development of methods for the display of chromosomes, an increasing number of reports have appeared in the literature describing animals in which the diagnosis of sex is uncertain. By studying the chromosomes of an intersex animal, in most cases the sex of the animal can be established, providing a base from which one can investigate the altered development. Although a solid understanding of sex determination and development has yet to be established and many theories

remain unproven, sound advice can frequently be passed on to the breeder.

NORMAL SEXUAL DEVELOPMENT

In domestic mammals, all normal individuals produce either male or female gametes, each containing one set of autosomes and one sex chromosome. Female mammals produce ova containing an X chromosome, while male mammals produce spermatozoa that may contain either an X chromosome or a Y chromosome. Thus, at the most important instant in an individual's life when fertilization occurs, the sex of that individual is determined when the X-bearing ovum is united with either an X-bearing sperm, producing a genetic female, or a Y-bearing sperm, producing a genetic male.

Once the sex has been established, the process of differentiation of the reproductive organs begins. Gonadogenesis, or the development of the gonad, is initiated early in organogenesis. Although the controlling mechanisms of gonadogenesis are not fully understood, it is apparent that this process is influenced by the genetic sex. Initially, the sex of the primitive gonad, which develops from the genital ridge, cannot be distinguished. Epithelial and primordial germ cells migrate into the indifferent gonad. If the gonad is destined to become a testis, the migrating cells give rise to the seminiferous tubules. If the gonad is to become an ovary, there occurs additional migration and proliferation of these cells, giving rise to the ovarian cortex with the medullary or testicular portion remaining vestigial (Fig. 24–1).

The development of the tubular ductal system is closely related to the development of the gonad. Initially two ducts are present. In the normal individual, one duct persists while the other regresses. In the male, fetal androgens cause persistence of the mesonephric or Wolffian duct. A second substance, which is not an androgen, causes regression of the paramesonephric or Müllerian duct. In the absence of these substances the Wolffian ducts regress and the Müllerian ducts develop into the oviducts, uterus, cervix and anterior vagina. Good experimental evidence shows that the control of development of tubular genital organs is largely a local event and not the result of circulating hormones.

The posterior portion of the genital system is formed by the development of the urogenital sinus and the genital tubercle, influenced by the presence or absence of

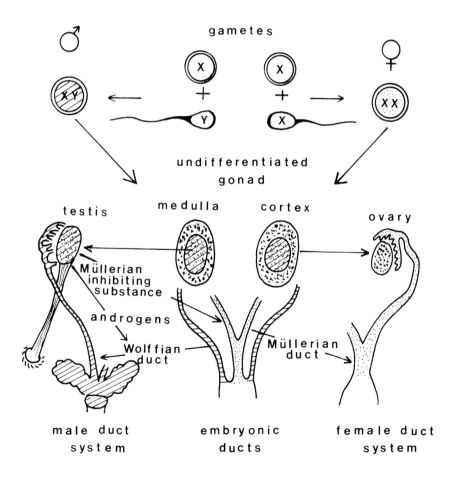

Fig. 24–1. Diagram of normal sex determination in mammals.

substances secreted by the fetal gonad. The development of the external genitalia is greatly influenced by circulating androgens, and a female fetus can be masculinized by hormones at a critical stage of development.

In normal development, the genetic sex, the gonadal sex and the sex of the accessory genital organs are identical, and each one depends upon events that have occurred in the preceding stages. It is remarkable, however, that at the indifferent stage of gonadogenesis all individuals have the potential to develop, sometimes almost completely, in the opposite direction of their genetic sex. A variety of factors can disturb normal sexual development, producing a broad spectrum of genital anomalies or intersexuality.

ABERRATIONS OF GENETIC SEX

Genetic sex is either XX or XY in the normal animal. Aberrations can result from structural abnormalities in either one or both of the sex chromosomes, by increases or decreases in the number of sex chromosomes, or by the presence of both male and female cells in the same individuals.

Although structural abnormalities may be difficult to detect, a number of new techniques for chromosome identification (Evans et al, 1973) make it possible to identify minor changes in structure. In man and in the mouse, a variety of abnormalities of the X chromosome have been reported, usually associated with some anomaly of development (Fig. 24–2).

Cases of abnormal reproductive development or function have been described in cattle (McFeely and Kanagawa, 1974) and horses (Payne et al, 1968), associated with structural abnormalities of one of the sex chromosomes.

Monosomy of the X chromosome (XO) has been reported in a large number of mares (Blue et al, 1978). These animals are phenotypic females with a history of anes-

trus or irregular estrus cycle activity. In general, they possess small uteri and small inactive ovaries devoid of follicular activity. A similar abnormality has been described in a pig (Gustavsson, 1973). In many respects this syndrome is similar to monosomy X (Turner's syndrome) described in women, in whom the combination is often lethal to the fetus and results in a spontaneous abortion.

Extra sex chromosomes appear to be more compatible with life. The XXY syndrome in man (Klinefelter's syndrome) is associated with testicular pathology and infertility, and in pure cases of XXY sex chromosome constitution in the bull, stallion, ram and pig, similar pathology seems to prevail (Bishop, 1972). The XXY was first identified in animals in the rare tricolor male cat.

A mare with normal phenotype but with small ovaries and hypoplastic endometrium and possessing three X chromosomes (Chandley et al, 1975) and a similar case involving a heifer (Norberg et al, 1976) have been reported. A number of reports of individual cases with various combinations of sex chromosomes detail similar developmental anomalies.

Animals which possess a mixture of male and female cells are referred to as chimeras and are covered in a separate section on freemartins.

ABNORMAL GONADAL DEVELOPMENT

Of particular interest are cases where the gonad develops in a manner which is inconsistent with the genetic sex. Obviously, in cases of gonadal hypoplasia in both sexes or cryptorchidism in the male, there are developmental defects in the gonad but because they do not usually confuse the diagnosis of sex, they are not considered in intersexuality.

True hermaphroditism with various combinations of ovaries, testes and ovotestes has been described in the

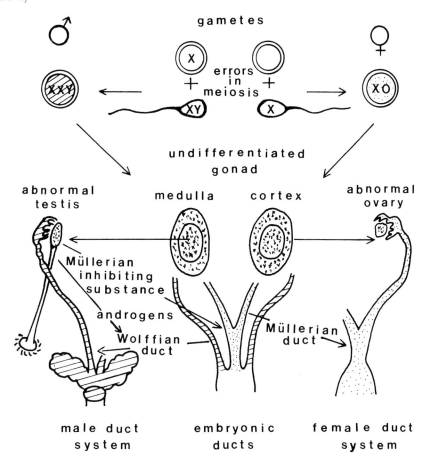

gametes

errors
in
meiosis

undifferentiated
gonad

abnormal
testis

medulla

cortex

abnormal
ovary

Müllerian
inhibiting
substance

androgens

Wolffian
duct

Müllerian
duct

male duct
system

embryonic
ducts

female duct
system

Fig. 24–2. Diagram of aberrations of genetic sex.

horse, goat, cow and pig (McFeely and Kanagawa, 1974). In some cases, this is associated with chimerism of male and female cells. These cases could be readily classified in the previous category and the development of the gonad attributed to the local effect of either male or female cells in the primordial gonad. However, there is no evidence at present to prove or disprove this theory.

Other hermaphrodites with both ovarian and testicular tissue have been described in which only one genetic sex has been demonstrated. While it is impossible to rule out undetected chimerism in these cases, there are other possible explanations for this phenomenon. They may have a similar origin to a sizeable number

of cases described in the goat, pig (Okamoto and Masuda, 1977) and dog that have male gonadal sex but are genetic females (Bishop, 1972). Although the testicles in these animals are not entirely normal, usually because they are undescended, there is little doubt that they are basically testicular in nature. The accessory sex organs vary in these cases, presumably depending upon the hormone production by the embryonic gonad. Several theories have been advanced to explain the contradiction between genetic and gonadal sex. It has been postulated that the Y chromosome, or part of it, could be translocated to one of the X chromosomes or an autosome and thereby remain undetected. However, these animals would then have an XXY constitution and a different type of abnormality would be expected.

Another hypothesis elaborates on the theory that the Y chromosome has developed by successive deletion of the X chromosome. If this is the case, the X chromosome may carry genes for maleness which are normally not expressed. In these cases of genetic females with testes, the maleness genes on the X chromosome may find a way to expression. This theory also applies to other types of intersexuality, such as the cow described previously with a structural abnormality of the sex chromosomes. In fact, the theory can be stretched further to ascribe the function of the Y chromosome as somehow permitting expression of maleness genes on the X chromosome. A somewhat similar theory attributes a hormone-like role to the H-Y antigen, which is disseminated by XY cells and thereby influences XX cells to engage in testicular organization (Ohno et al, 1976). However attractive they may seem, these theories remain to be proven.

ABNORMAL DEVELOPMENT OF THE ACCESSORY GENITAL ORGANS

Aberrations in the accessory genital organs are often the most easily detected as they frequently confuse or distort the phenotypic sex. In these cases the genetic and gonadal sex develop sequentially, but there are discrepancies in the development of the derivatives of the Wolffian or Müllerian ducts, or the urogenital sinus and the external genitalia.

Male Pseudohermaphroditism. As the development of the accessory genital organs appears to depend upon substances produced by the gonad in the male, male pseudohermaphrodites could result from either a failure of normal production by the testes or by lack of response to these substances by the target organs.

Male pseudohermaphroditism has been described in the testicular feminizing syndrome of man. In these cases genetic and gonadal sex are male, but phenotypic, behavioral and legal sex are female. The developmental abnormality in these cases is the result of insensitivity of the target organs to androgens. The testes produce testosterone and yet the Wolffian elements do not develop. However, the substance causing regression of the Müllerian elements must also be produced because they are not present in the postnatal individual. These individuals often have adequate breast development.

A few cases similar to testicular feminization in man have been described in cattle (Bishop, 1972) and swine (Lojda, 1975). As there is a suggestion that this syndrome may have a hereditary basis, it is interesting to speculate upon the role of some autosomal gene in the development of this syndrome. Further research is required to clarify this abnormality.

Female Pseudohermaphroditism. Some cases of female pseudohermaphroditism in man occur in individuals with a deficiency of an enzyme in the metabolic pathway in the adrenal gland, causing a block in the production of cortisone and a resultant shunting of the pathway, which increases the production of androgens. The adrenogenital syndrome is not well-documented in farm animals.

Increased circulatory androgens from a functional ovarian or adrenal tumor in the dam, causing masculinization of a female fetus, have been suggested as a possible cause of female pseudohermaphroditism, but convincing evidence is lacking. However, animals have been partially masculinized by the administration of drugs with androgenic properties to pregnant females. This has been done experimentally and clinically, when these drugs were used therapeutically. Androgens injected experimentally into pregnant ewes did not affect the gonadal development of the female lambs, which had normal Müllerian derivatives but also possessed Wolffian duct derivatives and masculinized external genitalia (Alifakiotis et al, 1976).

FREEMARTINISM

Freemartinism, which occurs primarily in cattle, is the sexual modification of a female twin by *in-utero* exchange of blood from a male fetus (Fig. 24–3). A freemartin-like syndrome has also been reported in sheep, goats and pigs. As in other forms of intersexuality, the freemartin syndrome was known in ancient times. Although the derivation of the word freemartin is uncertain, it undoubtedly is derived from the abnormal reproductive condition of these animals. Marcum (1974) has presented an excellent review of this condition.

Aberrations of Reproductive Organs. The gonads of the freemartin are intra-abdominal and rarely descend through the inguinal canal. The anatomic structure ranges widely from modified ovaries to structures resembling testes. Conclusive evidence of spermatogenesis in proven freemartins is lacking. Testosterone appears to be the major steroid produced by the freemartin gonad.

The anatomic and histologic structure of the reproductive tract also varies from individual to individual. In general, the Wolffian and Müllerian ducts are poorly developed, which is apparent early in embryonic life. Often a rudimentary epididymis and ductus deferens are observed. The presence of seminal vesicles appears to be a constant finding. The external genitalia, with few exceptions, resemble those of the normal female; however, they lead into a much shorter vagina. Frequently the clitoris is enlarged and the tuft of hair at the ventral commissure of the vulva is elongated. The mammary glands are underdeveloped and the teats shorter than those in three- to four-month-old normal heifers.

In experimental studies, Jost et al (1972) were able to determine that the initial changes in sexual development began after the 48th day of gestation and signs of masculinization were noticed after the 60th day.

Etiology of the Syndrome. During multiple pregnancy in cattle, the chorioallantoic membranes of adjacent embryos fuse. Even when two chorionic sacs develop in separate uterine horns, they eventually

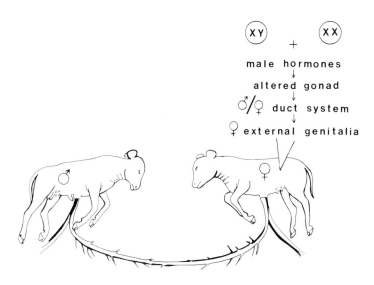

Fig. 24–3. Schematic illustration of the etiology of the bovine freemartin.

meet and unite. An early and intimate vascular connection is likely to be established between fetuses in the same horn, whereas a connection between fetuses in different horns would probably develop later. The vascular connection permits the exchange of blood elements between the twin fetuses early in gestation.

Freemartinism has been ascribed by early scientists to the male twin's hormones that reach the female through vascular anastomoses between the fused placentae. The dominance of the male twin was explained on histologic evidence; namely the appearance of interstitial cells in the fetal testis much earlier than in the fetal ovary, suggesting that the testis is active endocrinologically long before the ovary. This "hormonal" theory played a crucial role in the establishment of the endocrinology discipline. However, recent knowledge raises questions about the hormone theory. There is no evidence to support the claim that androgens can alter a mammalian ovary into a testis; likewise, androgens cannot cause regression of the Müllerian ducts, which is a common feature of the freemartin. The external genitalia, a sensitive area for masculinization in the female, remain female. Attempts to experimentally induce the freemartin syndrome have failed. Although intersexuality is produced by the administration of androgens to the pregnant dam, the changes are not characteristic of the freemartin.

In some true hermaphrodites it is possible to find an ovary or ovotestis on one side and a testicle on the other. The Wolffian and Müllerian derivatives tend to follow the lead of the gonad on that side. It is difficult to accept the hormone theory as the basis of the freemartin syndrome when the male gonad of a true hermaphrodite does not produce a masculinizing effect on the contralateral side in the same individual, much less in a twin.

The fusion of the chorioallantoic membranes is so intimate that extensive arterial and venous anastomoses occur, and blood-forming cells and other cells are exchanged between the fetuses. As a result of this reciprocal exchange between dizygotic twins, identical erythrocyte antigen types occur in both twins and sex chromosome chimerism appears in peripheral blood mononuclear leukocytes. Each twin contains two genetic populations of cells, one corresponding to its own genotype and one produced by the cells acquired from the co-twin. Early studies demonstrated the ability of freemartins to tolerate skin grafts from their dizygotic twins as a result of this cellular exchange.

In view of the recent advances in immunology and cytogenetics, some people feel that the presence of the male cells, and most particularly, the Y chromosomes or perhaps the H-Y antigen produced by the male cells, provide the key to the altered development. Although it was reported initially that a correlation existed between the degree of chimerism and the extent of genital tract modification, it is now clear that the degree of masculinization in the freemartin cannot be predicted from the proportion of circulating XY lymphocytes. It is interesting that the ratio of XX to XY lymphocytes is similar in both the freemartin and her co-twin, and that this ratio appears to remain relatively constant over a prolonged period (Marcum, 1974). Between groups of twins there is a wide variation in the ratio of male to female cells. Attempts to alter the genital tract of the freemartin postnatally with common sex steroid hormones generally fail, suggesting that these animals may lack the components necessary for stimulation by gonadal hormones (Greene et al, 1977).

There is conflicting evidence for the cellular theory as well. In the marmoset, it is well-documented that sex chromosome chimerism occurs frequently without apparent detriment to the reproductive organs. The time during organogenesis at which chimerism develops may be a critical factor. However, in chimeras produced experimentally by fusing mouse blastocysts, intersexuality is not a common occurrence.

Clearly the factors responsible for the development of this fascinating syndrome remain to be defined. It appears that something alters gonadogenesis in the female twin, and the altered gonad somehow influences the development of the remainder of the genital system.

Incidence of Freemartinism. In general the incidence of twinning in cattle is low and varies somewhat with breed and perhaps with cow families. Combined statistics indicate that approximately 92% of heterosexual twin females are freemartins (Marcum, 1974). In the other cases, chorioallantoic vascular anastomoses either fail to develop or occur after the critical stage in organogenesis. Multiple births with more than two fetuses can also result in the freemartin condition, as long as vascular anastomoses occur and both sexes are represented.

Occasionally, a single-born animal is a chimera (Wijeratne, et al, 1977). It is possible that under certain conditions involving heterosexual twins, one of the twins degenerates during early gestation and the surviving twin exhibits chimerism. The fusion of zygotes early in development presents another plausible explanation. A particularly interesting and unique case of a diploid-triploid true hermaphroditic cow was postulated to have arisen as a result of double fertilization and fusion of the fertilized meiotic products (Dunn et al, 1970).

Although the male co-twin appears to be anatomically normal, there is increasing evidence that the fertility of these animals is reduced and their reproductive lifetime is shortened.

Diagnosis of Freemartinism. The diagnosis of freemartinism is based upon the following: (1) clinical genital abnormalities; (2) presence of sex chromatin bodies in circulating leukocytes in male co-twin; (3) presence of sex chromosome chimerism (XX/XY) of hemopoietic cells; (4) blood typing; and (5) skin grafting, which provides indirect evidence for fetal vascular anastomosis.

Rectal palpation and vaginoscopy with a glass speculum are usually used for clinical diagnosis. Freemartins often have only a rudimentary uterus and a small vagina. Although chromosomal analysis is fairly expensive, it may be economically feasible in cases in which clinical diagnostic procedures are uncertain. The cost of raising an animal destined to be sterile may outweigh the cost of the karyotyping.

Freemartinism in Other Species. Although the freemartin syndrome generally refers to cattle, similar syndromes have been described in other farm animals (Marcum, 1974).

In goats, the freemartin-like condition has been observed infrequently, although twinning is not rare and intersexes are common. Some intersexes have XX/XY lymphocyte and bone marrow chimerism, and the anatomic findings are somewhat consistent with the freemartin syndrome, although not appreciably different from other types of intersexuality in goats. There is extensive evidence demonstrating that intersexuality in goats is frequently associated with the gene for polledness. Abnormal sexual development appears to be as common in births of twin females as in heterosexual twins, and the role of chimerism remains in doubt.

In sheep, fraternal twins are born more frequently than in cattle; however, sex chromosome chimerism occurs in a small percentage of sheep twins. Several cases have been described, in which developmental abnormalities of the reproductive tract, similar to those seen in bovine freemartins, have been associated with sex chromosome chimerism.

In the pig, a few cases of a freemartin-like syndrome have been reported. Side-to-side fusion of fetal sacs is common, but end-to-end chorioallantoic vascular fusion seems to be rare. Such vascular fusion must occur very early in these rare cases, since in later development chorionic overlapping is the rule, without parabiosis. As in the goat, intersexuality occurs frequently in the pig in the absence of sex chromosome chimerism. Although the etiology of intersexuality in the pig is not as well understood as it is for the goat, the incidence can be increased by selective breeding, which suggests a genetic basis for at least some forms of hermaphroditism. In some instances, the condition may be identical to the bovine freemartin condition.

In the horse, vascular anastomoses between the allantochorions in twin pregnancies occur frequently, but developmental anomalies are rare despite sex chromosome chimerism. Fusion of the vessels and exchange of blood elements probably occur after the critical stages of gonadogenesis.

The freemartin syndrome holds many clues to complete understanding of sex determination. Like other forms of intersexuality, it will continue to attract the interest of scholars and breeders alike.

REFERENCES

Alifakiotis, T.A., Call, J.W. and Foote, W.C. (1976). Androgen induction of intersexuality in sheep. Theriogenology 6, 21.

Bishop, M.W.H., (1972). Genetically determined abnormalities of the reproductive system. J. Reprod. Fertil. Suppl. 15, 51.

Blue, M.G., Bruere, A.N. and Dewes, H.F. (1978). The significance of the XO syndrome in infertility of the mare. N.Z. Vet. J. 26, 137.

Chandley, A.C., Fletcher, J., Rossdale, P.D., Peace, C.K., Ricketts, S.W., McEnery, R.J., Thorne, J.P., Short, R.V. and Allen, W.R. (1975). Chromosome abnormalities as a cause of infertility in mares. J. Reprod. Fertil. 23, 377.

Dunn, H.O., McEntee, K. and Hansel, W. (1970). Diploid-triploid chimerism in a bovine true hermaphrodite. Cytogenetics 9, 245.

Evans, J.J., Buckland, R.A. and Sumner, A.T. (1973). Chromosome homology and heterochromatin in goat, sheep and ox studied by banding techniques. Chromosoma (Berl) 42, 383.

Greene, W.A., Dunn, H.O. and Foote, R.H. (1977). Sex-chromosome ratio in cattle and their relationship to reproductive development in freemartins. Cytogenet. Cell Genet. 18, 97.

Gustavsson, I., (1973). Chromosomal errors in the reproduction of the domestic pig. In Les accidents chromosomiques de la reproduction. A. Bove and C. Thibault (eds), pp. 154–164, Paris, Inserm.

Jost, A., Vigier, B. and Prepin, J. (1972). Freemartin in cattle: The first steps of sexual organogenesis. J. Reprod. Fertil. 29, 349.

Lojda, L. (1975). The cytogenetic pattern in pigs with hereditary intersexuality similar to the syndrome of testicular feminization in man. Documenta Veterinaria 8, 71.

Marcum, J.B. (1974). The freemartin syndrome. Animal Breeding Abstracts 42, 227.

McFeely, R.A. and Kanagawa, H. (1974). Intersexuality. In Reproduction in Farm Animals, 3rd ed. E.S.E. Hafez (ed), Philadelphia, Lea & Febiger.

Norberg, H.S., Refsdal, A.O., Garm, O.N. and Nes, N. (1976). A case report on X-trisomy in cattle. Hereditas 82, 69.

Ohno, S., Christian, L.C., Wachta, S.S. and Koo, G.C. (1976). Hormone-like role of H-Y antigen in bovine freemartin gonad. Nature 261, 597.

Okamoto, A. and Masuda, H. (1977). Cytogenetic studies of intersex swine. Proc. Japan Acad. 53, 276.

Payne, H.W., Ellsworth, K. and de Groot, A. (1968). Aneuploidy in an infertile mare. J. Am. Vet. Med. Assoc. 153:1293.

Wijeratne, W.V.S., Munro, I.B. and Wilkes, P.R. (1977). Heifer sterility associated with single-birth freemartinism. Vet. Res. 100, 333.

25

Reproductive Infections

J.W. KENDRICK
and
J.A. HOWARTH

Infectious agents cause diseases that affect reproduction in farm animals. Infectious agents that cause epidemics of abortion or a substantial number of abortions are described in the text. Other agents that are associated sporadically with abortion are listed in Table 25–1.

VIRUS DISEASES

Granular Venereal Disease (Granular Vaginitis)

This is a widespread condition observed in cattle that masquerades under a misnomer. It is, in fact, hyperplasia of the

Table 25–1. Causes of Sporadic Abortion

Cattle	Sheep	Horses	Swine
Parainfluenza$_3$ virus	Tick-borne fever	Equine viral arteritis	Foot and mouth disease
Bluetongue virus	Wesselsbron virus	Equine infectious anemia	Picorna viruses
Malignant catarrhal fever	Rift Valley fever	Dourine	Influenza
Foot and mouth disease	Nairobi sheep disease	Piroplasmosis	Japanese B virus
Rinderpest	Rinderpest		Hemagglutinating virus
Tick-borne fever	Foot and mouth disease		African swine fever
Bovine petechial fever	*Coxiella burnetti* (Q fever)		Bovine viral diarrhea
Anaplasmosis	Bovine viral diarrhea		Infectious bovine rhino-
Piroplasmosis			tracheitis
Trypanosomiasis			
Toxoplasmosis			
Globidiosis			

503

lymphatic follicles that are normally found in the subepithelial tissues of the vulvar mucosa and the integument of the penis. It is a sequel to the viral disease, infectious pustular vulvovaginitis (IPV), but it may be caused by other conditions that stimulate lymphoid hyperplasia. Affected animals have varying numbers of nodules that protrude above the surrounding membranes; these are 1 to 2 mm in diameter. The condition is probably not related to fertility, although one survey found that more severely affected animals had approximately a 10% reduction in fertility. Cattle of all ages are affected, but it is seen most frequently in heifers, in which it occurs in the most severe form. It does not interfere with breeding, and one should not withhold an animal from breeding because of its presence. Bulls with severe granular venereal disease may refuse to serve.

Treatment. Spontaneous recovery is the rule, and in most cases treatment is unnecessary. If treatment is necessary, mild antiseptics in the form of ointments, douches and powders may be used. An exception to this rule is the case of a bull in which the condition may prevent service. Mild antiseptics and antibiotics should be used in treatment. In severe or refractory cases, cauterization of the individual lesions with silver nitrate has been done. If this is done, special care must be taken to prevent adhesions by using antibiotic or antiseptic ointments and frequently extending the penis during the healing period.

Bovine Viral Diarrhea

This disease is found in all parts of the United States and is distributed widely throughout the world. Four clinical forms of this infection in cattle are recognized. Most frequently, infection with this virus is subclinical. There are many herds, or groups of herds, in which the infection rate, as determined by the presence of serum antibody, is more than 50% of the animals, and the history of clinical signs referrable to this infection are absent. The acute infection is characterized by a high temperature, nasal discharge, erosions in the oral cavity and digestive tract, and diarrhea. The acute form has been observed as a herd outbreak in which almost all of the animals were affected, as well as cases in which only an individual animal was affected. The latter is the more common situation and, in all probability, is a reflection of the fact that most other animals in the herd are already immune from a previous exposure.

Chronic viral diarrhea is characterized by poor appetite, emaciation, and slow rate of growth. The onset may not be accompanied by any of the acute signs but may appear insidiously, being recognized only by the loss of condition in the affected animal. There may be diarrhea. The course of the disease is two to 6 months, and if death does not occur, those animals that survive represent a continuing economic loss.

Mucosal disease is another form of this condition, in which all the symptoms of acute viral diarrhea occur, except that the lesions progress in severity so that the animals die approximately 14 days after onset. Mucosal disease usually affects animals between 8 and 18 months of age, and it differs from the other forms of BVD virus infection because the animal is unable to develop immunity to it.

The etiologic agent of bovine viral diarrhea and mucosal disease is the bovine viral diarrhea (BVD) virus. It is spread by contact among animals housed or pastured together. Infection confers a long-lasting immunity that is recognized by the presence of antibody in the serum.

This disease affects the reproductive performance of cattle by causing abortion, mummification of the fetus and birth defects. Contrary to earlier opinion, the incidence of abortion caused by this virus is low. The diagnosis of BVD abortion is difficult, and there is no satisfactory means of confirming a clinical diagnosis. In those cases in which abortion has been ascribed to this virus, such circumstantial evidence as the concurrent appearance of febrile disease or the appearance of antibody in the serum following abortion has been interpreted as indicating that BVD virus caused the abortion.

Using the foregoing criteria, abortion due to BVD has been diagnosed at all stages of gestation; however, most abortions caused by BVD occur during the first three or four months. Experimental inoculation of pregnant, susceptible cattle has resulted in abortion only during the first three months of pregnancy. Abortion at this time, when the conceptus is small, often goes unrecognized in the average livestock operation, and infertility, rather than abortion, is diagnosed.

Viral infection and fetal damage during the middle third of gestation, when the body organs are in the developmental stages, results in the birth of calves with defective brain, eye and hair development. Birth defects referrable to the brain and eye have occurred most frequently. Brain defects result in inability to stand, walk or nurse. Eye lesions result in blindness and opacity of the cornea. However, fetuses may experience BVD infection during the last third of gestation without recognizable sequelae. All fetuses surviving intrauterine infection are actively immune at the time of parturition.

BVD virus is apparently the cause of Hairy Shaker disease of newborn lambs. This occurs when the infected ewe transmits the disease through the placenta to the fetus. These lambs are smaller than normal, have a coarse tremor, have difficulty in standing and nursing, and have an abnormally pigmented fleece with varying amounts of long, hairy fibers. The lambs are immune to BVD virus at birth, indicating an intrauterine infection, and dams of lambs with this disease are also immune to the BVD virus. The basic le-

sion of hypomyelinogenesis is not reversible. These lambs do poorly and rarely survive. Experimental infection of ewes with BVD virus has produced placentitis and abortion.

BVD virus rarely infects pigs. The disease causes poor condition, abortion, small litters and poor growth in young pigs.

Control. The prevention of this disease in cattle, and thus the prevention of abortion and birth defects, is accomplished by vaccination with a modified live virus vaccine. The best time to administer this vaccine is to replacement animals when approximately one year of age. It creates a life-long immunity; however, booster vaccinations are often given on an annual basis. Pregnant animals should not be vaccinated (Kahrs et al, 1970; Kendrick, 1971, 1976).

Infectious Bovine Rhinotracheitis and Infectious Pustular Vulvovaginitis

This disease is caused by a bovine herpesvirus that affects cattle and rarely affects other species. Infections are frequently observed in the United States and the distribution is worldwide. In addition to respiratory and genital disease, infectious bovine rhinotracheitis (IBR) virus causes conjunctivitis, encephalitis, acute gastrointestinal disease of newborn calves and abortion. The diseases caused by this virus ordinarily occur singly; ie, an outbreak of vulvovaginitis usually is not accompanied by the upper respiratory infection, or vice versa. The reason for this unique characteristic is unknown. Some investigators believe that different strains of the virus exist; however, these have not been defined. The characteristics of transmission, virulence, immunity and immunization are similar for all diseases. There is no danger of human infection from this bovine strain of herpesvirus.

The virus is transmitted easily by contact, and the application of the virus to any mucous membrane of a susceptible animal results in localized disease in that area, systemic infection and systemic immunity. Infection, regardless of the type of disease, results in resistance to reinfection and is characterized by the presence of specific viral antibody in the serum. This immunity is life-long and serves to protect the animal, even at low levels. Recovered animals rarely shed the virus; exceptions are some cows that might have recrudescence of infection at the time of parturition, and the bull that might harbor the virus in the sheath for several months after the acute infection.

Clinical Signs. Infectious pustular vulvovaginitis is characterized by pustule formation in the vulva. The condition is painful and the animal stands with the back arched and tail elevated. Irritation of the vulva is indicated by straining, frequent urination, stamping of the feet, restlessness and switching of the tail. There is a small-to-moderate amount of tenacious yellow discharge. The vulva is swollen, and in the case of white-skinned animals, a slight reddening can be observed. Small white pustules, approximately 2 mm in diameter, occur on the vulvar mucosa. The pustules may become confluent and form a sheet of fibrinous, necrotic exudate over the affected area. In some cases, the lesions extend into the vagina, where they occur as a more diffuse necrosis. Examination reveals sheets of tissue peeling from the wall of the vagina.

Recovery is usually uncomplicated and occurs in seven to ten days. Within two or three weeks after recovery, the characteristic nodules of granular venereal disease appear in the vulva. Usually, vulvar and vaginal infection do not extend into the uterus. Contamination of insemination pipettes or of semen with the virus will result in infection of the uterus and endometritis.

Breeding should be suspended during the acute phase of the disease; however, it may be resumed afterward because normal fertility following recovery is to be expected. Endometritis is a cause of infertility and recovery of the uterus results in a return to normal fertility.

Infectious pustular balanoposthitis is the corresponding condition in the bull. The same lesions occur, but because of the movement of the penis, the exudate is often scrubbed off and the lesions appear as hemorrhagic ulcers. Edema of the penis and prepuce may be sufficient to cause phimosis or paraphimosis. In severe cases, adhesions of the penis and prepuce have occurred. Spontaneous recovery of these lesions also occurs, as in the cow; however, to avoid adhesions and secondary infection, oily solutions of an antibiotic should be placed in the sheath.

IBR virus is a common cause of abortion in cattle. Abortion occurs following natural infection, but it frequently results from improper use of the modified live virus vaccine for prevention of this disease. This is a highly effective vaccine, and when properly used, it creates a long-lasting immunity with excellent protection against the disease. However, it causes abortion in approximately 50% of pregnant animals that are vaccinated during the last half of the gestation period.

The period between infection and abortion is 21 to 90 days. The aborted fetus is autolyzed. Some cows have retained placenta; however, metritis is usually mild or absent. Recovery is the rule, and except in those cases where secondary infection of the uterus occurs, fertility in the subsequent breeding season is usually normal.

Diagnosis. The disease in mature animals is usually diagnosed by the clinical signs and confirmed by the isolation of the virus from swabbings of the lesions. Serologic examination may give supporting evidence of the infection; however, it often falls short as a diagnostic tool because the presence of antibody in the serum may be the result of infection that took place several months, or even several years, prior to the present condition. Abortion from the IBR virus is suspected when there is a history of other forms of the disease, a storm of abortions during the last half of pregnancy, or use of the modified live vaccine during pregnancy. Necropsy of the fetus reveals autolysis that is nonspecific. Diagnosis is confirmed by finding microscopic focal necrosis, which is commonly present in the liver but may occur in many other organs of the body. Frequently the virus can be isolated from the fetus, although in some instances the autolysis precludes this isolation.

Control. The best method of control for this disease is the routine immunization of all potential replacement animals when they are six months to one year of age. Such a regimen selects calves that have lost their maternally conferred immunity at approximately six months and before breeding age. A single vaccination should be sufficient; however, booster inoculations may be given when the animal is not pregnant (Kendrick, Gillespie and McEntee, 1958; Kendrick and Straub, 1967; McKercher and Wada, 1964).

Epizootic Bovine Abortion (EBA, Foothill Abortion)

Epizootic bovine abortion refers to a well-defined clinical condition that occurs principally in the foothills and mountain areas surrounding the Central Valleys of California. It occurs most commonly in beef cattle because these animals are most frequently pastured in the endemic areas. The highest incidence is in heifers, but older animals moved into endemic areas for the first time may subsequently abort. Usually a cow aborts only once, makes a complete recovery, and fertility is normal in subsequent gestations. The abortion rate may exceed 80% when large numbers of animals are exposed for the first time.

The cause is suspected to be a virus. The disease can be experimentally produced by feeding the tick, *Ornithodoros coriaceus*, on susceptible cattle during the second trimester of gestation. The ticks used for these transmission studies are gathered in areas where abortion has occurred. The period from inoculation to abortion is approximately three months.

Clinical Signs. Most abortions occur in the last third of pregnancy and, because most abortions are in beef cattle that are on a seasonal breeding program, there is a seasonal incidence of abortion. One of the most striking characteristics of this abortion disease is the freshness of fetal tissues, indicating that death of the fetus occurs approximately at the time of expulsion from the uterus. This is in marked contrast to other abortion diseases, in which the fetus is badly autolyzed before being aborted.

Diagnosis. At the present time, the only satisfactory diagnostic criteria are pathologic lesions of the fetus; these include petechial and ecchymotic hemorrhages in the conjunctiva and oral mucosa. The muzzle of such fetuses frequently is bright red, and dermatitis is present. Subcutaneous edema and straw-colored pleural and peritoneal effusion are usually present. A swollen nodular liver is a dramatic but irregularly occurring finding. Petechial hemorrhages are found throughout the body and the lymph nodes are enlarged and moist.

The most specific lesions are microscopic. These consist of general reticuloendothelial hyperplasia, seen most prominently in the liver, lymph nodes and spleen. This lesion may also be present in the brain and heart.

A significant finding is the depletion of lymphocytes and reticuloendothelial hyperplasia of the thymus. There may be some focal necrosis and inflammatory changes in the liver and spleen. The lesions described are not necessarily specific for a single etiologic agent, but rather indicate an infectious agent that causes a chronic disease of the fetus. The mechanism for abortion under these circumstances is probably stress, which causes adrenal cortical hyperplasia which, in turn, releases cortical hormones to initiate parturition.

Control. No vaccine is available since the etiologic agent has not been identified. Avoiding the foothill and mountain areas prevents the disease; however, this is a vast rangeland area and must be used for beef cattle production. Adjusting the calving season so that the middle trimester of pregnancy does not coincide with the tick season has reduced the incidence of abortion. The tick season in the foothills begins in May and continues to October. By breeding to calve in August, all cows are past the susceptible midtrimester by the time the ticks become active. Animals that have experienced the disease will not abort again, and these should be maintained in the herd. There is no evidence that this disease spreads from cow to cow (Kendrick, 1976).

Bluetongue

Bluetongue is an arthropod-borne viral disease of sheep, goats, cattle and wild ruminants. The infection of sheep is characterized by elevated temperature, ranging from 104°F to 108°F, severe depression and anorexia. The name of the disease comes from the cyanotic appearance of the mucosa of the mouth and tongue. Erosions occur on the lip, cheek, tongue, in the mouth, and in the nostrils. Eventually there is a mucopurulent discharge and encrustations on the lip and muzzle. Lameness, due to laminitis and inflammation of the coronary band, may occur. The course of the acute disease is 6 to 14 days, and mortality is 10% to 40%.

The disease influences reproductive performance primarily by affecting the fetus. Virus of the naturally occurring disease and the modified live virus vaccine used to protect sheep against the disease may cross the placenta and produce fetal disease, characterized by abnormal brain development. Although lambs are born at full term, some are stillborn, others are spastic and lie struggling until death. Others, called "dummy lambs," are unable to stand or may be uncoordinated and fail to nurse. The period of greatest danger from disease or vaccination in sheep is between the fourth and eighth weeks of pregnancy, but infection at a later date may produce mild but significant birth defects.

The disease is spread by a small biting insect, the *Culicoides* gnat.

Control. Of the total of 17 international strains of bluetongue virus currently recognized, four have been identified in the United States and only one has been modified for vaccine use in the United States. The commonly used egg-adapted bluetongue vaccine was withdrawn from commercial use because of excessive virulence. A tissue cultured vaccine is available and it should not be used within three weeks of breeding or during pregnancy.

Ulcerative Dermatosis

Ulcerative dermatosis (lip and leg ulceration and venereal disease) is a viral infection of sheep characterized by circumscribed ulceration of the skin of the lips, legs, feet and external genital organs. Transmission under natural conditions occurs when the virus enters through a break in the skin. Venereal transmission produces lesions of the penis, prepuce and vulva. The disease is characterized by localized ulceration in the areas described. These ulcers become covered with scabs, and sometimes may undergo a secondary infection. The genital lesions interfere with breeding. This condition must be differentiated from a noncontagious balanoposthitis, commonly called "pizzle-rot," which is probably caused by dietary-induced changes in the urine. No vaccine for ulcerative dermatosis is available. Infection produces low-grade immunity and an animal can be reinfected within five months.

Equine Rhinopneumonitis (Viral Abortion, Equine Herpesvirus I Infection)

The equine herpesvirus I infection is primarily a disease of the upper respiratory tract. The horse first experiences exposure to this virus as a foal in the fall near weaning time. Horse farms are familiar with the mild upper respiratory infection that affects most weanlings each year. The clinical signs include a temperature of 102°F to 105°F, serous nasal discharge and congestion of the nasal mucosa. After several days the nasal discharge becomes purulent, and there is a mild-to-moderate cough. Appetite is affected only in the more severely involved animals. The course of the disease is two to four weeks, and spontaneous recovery is the rule. This epizootic disease in foals produces a massive exposure of mares that are in mid-pregnancy.

Few, if any, horses in the United States escape this infection during the first years of life. Infection confers immunity, as indicated by the presence of antibody in the serum. However, unlike most viral immunity, soon after infection it begins a relatively rapid decline and, when the titer of antibody in the serum reaches a level of less than 1:100, the horse is again susceptible to respiratory infection by the virus. In this way, individual horses may be repeatedly infected during their lifetime. The rate of abortion in a band of mares depends upon their susceptibility, but abortion rates as high as 80% have been reported.

Clinical Signs. Abortion usually occurs after the eighth month of pregnancy and, because of seasonal breeding, most abortions occur between January and April. There are no premonitory signs indicating that abortion is about to occur. The placenta is not retained, and the mare undergoes a prompt recovery without aftereffects. Fertility in subsequent years is not affected. The fetuses are not decomposed, indicating that death occurs approximately at the time of expulsion from the uterus.

Diagnosis. By far the most frequent cause of epizootics of abortion in mares is the equine herpesvirus I, and this infection should be suspected until otherwise eliminated. A positive diagnosis is based on gross and microscopic lesions in the aborted fetus and on virus isolation and identification. Gross examination reveals a straw-colored fluid in the body cavities and edema of the lung, which are present in 80% to 90% of aborted fetuses. Small foci of necrosis, varying in size from minute up to 5 mm in diameter, occur in the liver in about 50% of aborted fetuses. Hemorrhages and edema may occur in other parts of the body. The diagnosis is confirmed by the presence of intranuclear inclusion bodies in fetal tissues and by viral isolation.

Control. The equine herpesvirus I is highly contagious and probably all susceptible animals are infected at the time of the fall outbreak of upper respiratory infection in the weanlings. It is seldom possible to prevent abortion in a band of mares by isolating the nonaffected from those that have aborted. However, if possible, such isolation should be attempted. Abortion from this infection rarely occurs in successive years, and in the absence of other control methods it may not occur in a single band of mares for several years.

Immunization is the best method of prevention; two methods are available. A hamster-adapted virus is used in a "controlled infection" program for the prevention of this disease. This system has been used successfully on farms where the disease has been diagnosed. It should be initiated immediately after an outbreak of abortion when immunity is at a high level.

Vaccination of every horse on the farm is done yearly in July and October. In mares without recent exposure to the virus, the vaccine may cause abortion. Therefore, this immunization method should be used only following diagnosis of this disease. A modified live virus vaccine has recently been placed on the market. All animals over three months of age should be immunized with two doses given four to eight weeks apart. Semiannual revaccination is recommended (Doll and Bryans, 1963; Purdy et al, 1978).

Hog Cholera

Hog cholera as an intrauterine infection assumes major importance, both as a cause of reproductive failure, and as an obstacle, in the completion of the national hog cholera eradication program. It has been demonstrated that injection of attenuated hog cholera vaccine, or natural hog cholera infection, can produce fetal death with absorption or mummification, or can arrest fetal growth, resulting in anomalies and death in the newborn.

Viral invasion during the first 10 days of gestation results in absorption of fetuses, while infection at 15 to 30 days produces fetal abnormalities. Infection after 30 days of gestation also produces fetal abnormalities and is a cause of stillborn and weak pigs. Pigs that survive intrauterine hog cholera infection may be carriers and shed the virus for several weeks after birth.

The use of hog cholera vaccines is no longer allowed in the United States, and under the eradication program all infected herds are being eliminated. Therefore, it should be emphasized that abortion, stillbirth, weak piglets, piglets that die shortly after birth, "shaky or jittery" piglets, blindness, hairlessness or high death rate from birth to weaning may be indicative of hog cholera (Dunne and Clark, 1968).

Pseudorabies (Aujeszky's Disease)

Pseudorabies is an acute, highly fatal, viral disease of cattle, sheep, dogs and cats that is characterized by an intolerable itching at the site of the virus entrance, convulsions and death. Pseudorabies is a highly contagious infection of swine, with the nasal secretion of pigs being the principal source of virus for swine and other animals.

Mortality rates range from nearly 100% in newborn piglets to 3% in fat hogs; it is rare in sows. Most infections in swine are mild and unrecognized but result in many carrier animals. Clinically, diseased pigs may have a temperature up to 106°F, uncoordination progressing to complete paralysis, and convulsions. Sows may exhibit clinical sickness with fever and loss of appetite. When sows are affected early in pregnancy, some or all of the fetuses may die and abortion results.

The drastic increase in pseudorabies in swine in the United States during the 1970s is of major concern in the hog cholera eradication program, because both diseases have many features in common.

Control. Swine should not be housed with other animal species because virus spread by apparently normal pigs can cause a uniformly fatal disease in such animals. An accurate serologic test is available for the detection of infected swine herds. A live virus vaccine for pigs is used in some European countries but it has not been approved for use in the United States (Baskerville et al, 1973; Kluge and Mare, 1974).

SMEDI Viruses

Enteroviruses and parvoviruses are widespread among pig populations throughout the world. These viruses seldom cause clinical disease in adult, growing or suckling pigs, but they are highly contagious and spread rapidly in susceptible herds. When the first exposure of females to these viruses occurs during pregnancy, there is transplacental infection that results in fetal disease. The name "SMEDI" was derived from the first disease conditions recognized in litters when pregnant sows were infected with enteroviruses: stillbirth, mummification, embryonic death and infertility. Infections with SMEDI viruses react differently from infections with hog cholera and pseudorabies viruses, because sows do not sicken and newborn pigs are seldom affected.

Field observations indicate that SMEDI viruses spread through a breeding herd after the introduction of a carrier boar or sow. Sows infected before breeding usually become immune and farrow normally. Stillbirth, mummification and infertility result when infection occurs during the critical first 30 days of gestation. Sows that have abnormal litters generally farrow normally thereafter, unless a new strain of virus is introduced into the herd.

As there is no vaccine available, it is generally recommended that a closed breeding herd be maintained. All members of the breeding herd should be held in close contact for at least a month before breeding to permit cross-transmission of SMEDI viruses present in the herd (Dunne and Leman, 1975; Wrathall, 1975).

PROTOZOAN DISEASES

Bovine Trichomoniasis

Trichomoniasis is a contagious, venereal disease of cattle characterized by infertility, pyometra and abortion. It is caused by the protozoan parasite *Trichomonas fetus*. Abortion and pyometra are often the first signs of trichomoniasis that are noticed in a herd, but they occur in relatively few animals. Infertility characterized by repeat breeding and irregularly long or short estrous cycles is the most

constant symptom and occurs in a high percentage of animals in a recently infected herd. Heifers are usually considered to be most susceptible; however, immunity does not increase with age but is created through exposure. Infection is confined to the genital tract, and with rare exceptions, is transmitted only during the breeding act.

Clinical Signs. The introduction of trichomonad infection into a clean herd passes unnoticed for a considerable period of time. Infection of the female results in a mild vaginitis, which usually is not observed. The first recognition of the disease is the occurrence of pyometra, abortion, or decreased fertility. The infection is first established in the vagina, and then passes to the uterus within 20 days. It causes endometritis, which results in an inhospitable environment for the developing zygote, or infection of the zygote, which results in its death and resorption. Inflammation of the uterus may cause shortening of the estrous cycle. Long estrous cycles occur when the embryo survives beyond 14 days of age and the estrous cycle is interrupted. Later, when the embryo dies and is expelled or resorbed, estrous cycles are again initiated. The pyometra resulting from this infection consists of macerated fetal tissue. It persists for a long period of time unless recognized and treated. Abortion from trichomoniasis always occurs prior to the fifth month of pregnancy and often the membranes are expelled intact with the fetus.

Diagnosis. The appearance of the previously mentioned symptoms, usually following the introduction of new animals into the herd, is suggestive of trichomoniasis or vibriosis. The diagnosis of trichomoniasis is confirmed by finding the organism in at least one animal in the herd. No other tests, such as serologic tests, are satisfactory for the detection of trichomoniasis. The organisms may be found in large numbers in the tissues, stomach contents and placental fluids of aborted fetuses. They are present in the uterus for several days after abortion. Trichomonads are found in large numbers in pyometric fluids, particularly if the cervix has remained sealed and bacterial contamination of the uterus has not occurred. Sometimes, however, bacterial contamination supersedes the trichomonad infection. Trichomonads may be found in aspirated vaginal fluid, particularly during the two or three days just prior to the occurrence of estrus. Since the bull is the source of infection and remains infected over a long period of time, he is often selected for herd diagnosis. Smegma from the penis is collected by pipet or sponge and examined either by culture or direct microscopic examination.

Treatment and Control. Most cows recover spontaneously and individual treatment is unnecessary, except in the case of pyometra or secondary infection associated with abortion. The usual recommendation is 90 days of sexual rest to allow immunity to eliminate the infection. In herds in which artificial insemination can be used and there is no danger of animal-to-animal transmission, breeding has not been interrupted and the recovery rate in the female has been satisfactory. Bulls must be treated and those worth less than twice their salvage value should be replaced. The cost of treatment involves not only the treatment itself, but the time and effort necessary to test the bull to be certain that treatment has been successful. A satisfactory treatment for bulls is the combined use of sodium iodide, acriflavin, and Bovoflavin, ointment. Dimetridazole is an effective systemic treatment for bulls when administered orally at a dose of 50 mg/kg for five days.

The best method for controlling trichomoniasis is to use artificial insemination. The herd should be examined to find animals with pyometra and other recognizable genital diseases. These are treated or eliminated. The use of artificial insemination in a herd for two years will eliminate the infection in most instances. When replacement animals are added to the herd, they should come from herds with high fertility rates.

Toxoplasmosis

Ovine abortion caused by *Toxoplasma gondii* is of major importance in New Zealand, where it is not associated with clinical disease in the ewe. Abortion occurs during the last month of gestation, and weak, full-term lambs may be observed. Specific gross lesions are not seen in the fetus, but placental lesions of focal necrosis of the cotyledon up to 2 mm in diameter occur. In some cases the lesions may be found only by microscopic examination. Histologically, foci of necrosis and gliosis are observed in the brain. Diagnosis is confirmed by the recognition of white necrotic areas in the cotyledon, in which the toxoplasma organism is microscopically visible. Impression smears of cotyledons may also reveal toxoplasma. The agent can be isolated by injecting placental or fetal tissue intraperitoneally in mice and observing the organism in peritoneal exudate one to two weeks later (Watson and Beverley, 1971).

BACTERIAL DISEASES

Vibriosis

Bovine vibriosis is a contagious venereal disease of cattle characterized by infertility and abortion. This disease is caused by the bacterium *Campylobacter fetus*, which was formerly called *Vibrio fetus*. It is worldwide in distribution and occurs throughout the United States. It is seldom seen in dairy cattle because the semen sources for artificial insemination are essentially free of this agent. On the other hand, natural breeding still predominates in beef cattle, and the disease prevails in many herds. The clinical effects of this infection are limited to the reproductive system and spread is only by breeding.

Clinical Signs. Infertility or delayed conception occurs in almost all infected animals and is economically by far the most damaging aspect of this disease. Some cycles following infection are of normal length. Frequently one or more cycles are abnormally long because conception occurs and the developing embryo delays estrus until it dies and is resorbed. Inflammation of the uterus may cause an abnormally short cycle. Physical signs of the disease are slight in the female, and are characterized by endometritis and the transient presence of small amounts of pus in the vagina. The disease is self-limiting and recovery occurs in approximately two months in 75% of the cases. Of the remaining 25%, almost all recover in a period of three or four months, although occasionally carrier animals can be identified in which the infection persists through pregnancy and for periods in excess of one year.

The organism inhabits the penis and the prepuce of the bull in a symbiotic relationship that does not cause any physical changes, nor does it affect semen quality and libido. Bulls may be mechanical carriers of the agent at any age, but establishment of the carrier state usually occurs only in bulls three years or older.

While abortion is associated with vibriosis of cattle, it should be emphasized that this occurs in a small number of cases. Ovine vibriosis is characterized by abortion in late gestation, which may affect up to 80% of the flock.

Diagnosis. The disease is suspected in herds in which there is an increase in the number of services per conception, and in beef herds in which the breeding and calving seasons are prolonged. The presence of the disease must be confirmed by one or more laboratory tests. The vaginal mucus agglutination test may be positive within 60 days after the infection, and

may return to negative within seven months. The best stage of the cycle for sample collection is during diestrus. False-positive results occur immediately after estrus; false-negatives, during estrus.

Because of the variables mentioned, the standard procedure for testing is to select infertile cows that were bred at least 60 days prior to the sampling date. The presence of positive titers is indicative of herd infection. *Campylobacter fetus* may be cultured from vaginal and cervical mucus. The sample is collected aseptically in a pipet, refrigerated, and taken to a laboratory for culture. Fluorescent antibody is used to identify organisms in washings from the bull's sheath. All diagnostic procedures are subject to considerable variation in accuracy. Therefore, they are used primarily for identifying the presence of infection in the herd. Thereafter, the infected herd is treated as a unit. Abortion in sheep is diagnosed by direct microscopic observation or culture of *Campylobacter fetus* from the placenta or the aborted fetus.

Treatment and Prevention. Females with normal genitalia eliminate the disease spontaneously in most cases. Bulls may be treated by systemic administration of streptomycin, local application of streptomycin or a combination of both. The disease may be eliminated from a herd by removing animals with physically abnormal genitalia and using artificial insemination with noninfected semen for a period of two years. A bovine vaccine is available that should be used on females approximately two months prior to the beginning of the breeding season. An ovine vaccine is effective in preventing abortion when used prior to breeding or in early gestation (Bryner, 1976).

Listeriosis

This disease primarily affects the nervous system but may be responsible for outbreaks of abortion in cattle and sheep.

Abortion occurs in late pregnancy and is accompanied by retained placenta, metritis and death of the dam. Mortality can be prevented by local and systemic antibiotic treatment. The disease is associated with silage feeding and stress. No vaccine is available.

Brucellosis

This infectious disease of animals and man is caused by one of the *Brucella* species (*B. abortus, B. melitensis, B. suis, B. ovis, B. canis*); it is characterized by genital and mammary gland infections in animals and by undulant fever in man. In cattle the disease is also known as Bang's disease and contagious abortion.

Brucellosis is a worldwide problem of both public health and economic importance. Loss to the livestock producer occurs from abortion, premature birth, infertility and decreased milk yield.

Man may be infected by a food product of raw milk origin. Contact with the organism in vaginal discharges, fetuses, placentas, urine, manure and carcasses causes a large proportion of human cases. Transmission of brucellosis in animals occurs through the ingestion of feedstuffs contaminated by uterine discharges and milk. Infected male animals may transmit *Brucella* during breeding.

In young animals not yet sexually mature, the infection causes no ill effects and is eliminated within a few weeks. In adult, nonpregnant female animals the organism localizes in the mammary gland and later spreads to the uterus when pregnancy occurs. In pregnant animals, *Brucella* organisms invade the uterus, placenta and fetus, the final result being the death and premature expulsion of the fetus.

Clinical Signs. In *B. abortus* infection in cattle, abortion after the fifth month of pregnancy is the most prominent clinical sign. In subsequent pregnancies calving may be normal, although second or third abortions may occur with the same cow.

Retention of the placenta and uterine infection are common sequelae following abortion. Severe uterine infections may cause sterility. In the bull, infection of the testicle may occur, and if both testicles are involved sterility will result.

In *B. suis* infection of swine, the most common symptom is abortion or birth of weak pigs. Sows that have aborted once usually farrow normal litters thereafter. A persistent but scanty discharge from the uterus may follow abortion. Sows so affected are either temporarily or permanently sterile, depending on the persistence of the uterine infection. The testicles of boars may be infected, with sterility resulting.

Infection of sheep with *B. ovis* is characterized by infertility in rams due to epididymitis. Occasionally abortion or the birth of weak lambs occurs. Infection of dogs with *B. canis* is characterized by abortion, failure to whelp, epididymitis and sterility.

Diagnosis. Laboratory procedures used in the diagnosis of brucellosis include isolation of the organism and tests for the presence of specific antibodies in serum and milk. *Brucella* organisms can be isolated from aborted fetuses, the placenta, the uterine discharges, milk and semen.

The milk ring test for *Brucella* antibodies is the most practical method for locating infected dairy herds and for surveillance of brucellosis-free herds; it is performed on bulk milk samples. In eradication programs, herds that show a positive ring test are then examined by serologic tests to identify the infected individuals. The serum agglutination test is the most commonly used procedure for detecting brucellosis in individual bovine animals.

Control. The control of bovine brucellosis is based on hygienic measures, vaccination with strain-19 *B. abortus*, and testing and disposal of infected animals. The living attenuated strain-19 vaccine used in the control of bovine brucellosis affords good protection and does not spread from animal to animal when properly used. Heifer calves vaccinated at four to six months of age are resistant to infection for seven years or more and do not react to serologic tests used in eradication programs. Effective vaccination with strain 19 can reduce infection rates to a point where procedures based on testing and the elimination of the reactors are economically feasible.

The control of brucellosis in swine is based on detection and elimination of infected herds. As yet there is no vaccine available for use in swine. The control of *B. ovis* infection in rams can be achieved by the elimination of infected rams and vaccination of immature rams to serve as replacements. (Meyer, 1966; Joint FAO/WHO Expert Committee on Brucellosis, 1971).

Enzootic Abortion of Ewes (EAE)

The disease occurs in many parts of the world and has been diagnosed in the western United States. The rate of abortion in a newly infected flock may be as high as 30%. In flocks experiencing a re-infection, the rate is less than 5%. Ewes abort during the last month of pregnancy. Abortions continue until normal lambing time and some fetuses are expelled alive at term but are diseased. Fetuses are usually fresh and have no specific lesions. Diagnostically significant lesions do occur in the placenta. Cotyledons are necrotic and have a dry, dull-yellowish, or grey color. The allantochorion is brownish and thickened, often described as "leather-like." The etiologic agent of EAE is a member of the *Chlamydia* group. This disease must be differentiated from vibriosis.

Diagnosis and Control. Diagnosis of EAE depends upon finding the elementary bodies of the infectious agent in smears from the cotyledon or from placental exudate.

The disease produces immunity. A single infection produces life-time immunity and the disease can be prevented by vaccination (Marsh, 1965; Parker, Hawkins and Brenner, 1966).

Leptospirosis

Leptospirosis of cattle, horses and swine is caused by a variety of leptospiral serotypes widely distributed throughout the United States and the rest of the world. Both domestic and wildlife animals are sources of leptospiral infection for man. Only a small portion of the animals infected with *Leptospira* have obvious sickness, but a high percentage may develop persistent kidney infections and spread the organism. Cattle recovering from leptospirosis remain kidney carriers for about three months, but swine, wild rodents and skunks may become permanent sources of infection.

Transmission of leptospirosis most commonly occurs when an infected animal contaminates pastures, drinking water and feed with infected urine. Infection may also be spread by aborted fetuses and contaminated semen.

Clinical Signs. *Leptospira pomona* has been the most common cause of bovine leptospiral infections. This serotype and *L. grippotyphosa* may cause an acute hemolytic anemia with hemoglobin in the urine, particularly in younger cattle. Abortion occurring during the last third of pregnancy and a characteristic drop in milk yield (agalactia) are the more common clinical signs.

Clinical illness in cattle is not as well defined with infections due to *L. canicola* and *L. icterohemorrhagiae*, but each is capable of causing abortion. In herd outbreaks of *L. hardjo* infection, a few abortions may occur, but the most common feature observed is failure to conceive, which may be a persistent problem lasting many months.

Infection with *Leptospira* may cause

abortion in pigs, sheep, goats and horses. In horses an additional sequela is a disease of the eye called periodic ophthalmia.

Diagnosis. Leptospiral organisms are not readily isolated from animals with clinical signs; this is particularly true with aborted bovine fetuses. Isolation of *Leptospira* from the urine denotes chronic infection and thus does not really assist in making a diagnosis.

The most reliable and most commonly used diagnostic procedures are the tube and plate agglutination tests. Serum samples from animals suspected of having leptospirosis should be tested with each of several serotypes during the sickness and again when such animals are convalescent.

Control. Streptomycin and tetracyclines are used in the treatment of acutely ill animals and to eliminate the kidney carrier state.

Vaccination of cattle, horses and swine with leptospiral vaccines has been effective in controlling clinical signs of the disease. Because immunity wanes quickly, animals should be vaccinated at least once yearly.

Currently, a vaccine containing *L. pomona* and another containing the five commonly occurring leptospiral serotypes are available commercially (Amatredjo and Campbell, 1975; Szatalowicz et al, 1969).

REFERENCES

Afshar, A. (1965). Virus diseases associated with bovine abortion and infertility. Vet. Bull. *35*, 165.

Amatredjo, A. and Campbell, R.S.F. (1975). Bovine leptospirosis. Vet. Bull. *45*, (12), 875.

Baskerville A., McFerran, J.B. and Dow, C. (1973). Aujeszky's disease in pigs. Vet. Bull. *43*, (9), 465.

Blood, D.C., Henderson, J.A. and Radostitz, O.M. Veterinary Medicine. 5th ed. Philadelphia, Lea & Febiger.

Bryner, J.H. (1976). Vibrio fetus induced abortion. Theriogenology *5*, 128.

Doll, E.R. and Bryans, J.T. (1963). Epizootiology of equine viral rhinopneumonitis. J. Am. Vet. Med. Assoc. *142*, 31.

Doll, E.R. and Bryans, J.T. (1963). A planned infection program for immunizing mares against viral rhinopneumonitis. Cornell Vet. *53*, 249.

Dunne, H.W. and Clark, C.D. (1968). Embryonic death, fetal mummification, stillbirth, and neonatal death in pigs of gilts vaccinated with attenuated live-virus hog cholera vaccine. Am. J. Vet. Res. *29*, 787.

Dunne, J.W. and Leman, A.D. (1975). Diseases of Swine. 4th Ed., Iowa, Iowa State University Press.

Hanson, L.E., Tripathy, D.N. and Killinger, A.H. (1972). Current status of Leptospirosis immunization in swine and cattle. J. Am. Vet. Med. Assoc. *161*, 1235.

Howarth, J.A. (1960). The differential diagnosis of bovine abortion. Proc. U.S. Livestock Sanit. Assoc. *64*, 401.

Joint FAO/WHO Expert Committee on Brucellosis; 5th Report (1971). Who Tech. Rep. Ser., No. 464, 76 p.

Jubb, K. and Kennedy, P. (1970). Pathology of Domestic Animals. 2nd ed., New York, Academic Press.

Kahrs, R.F., Scott, F.W. and deLahunta, A. (1970). Bovine viral diarrhea-mucosal disease, abortion and congenital cerebellar hypoplasia in a dairy herd. J. Am. Vet. Med. Assoc. *156*, 851.

Kendrick, J.W., Gillespie, J.H. and McEntee, K. (1958). Infectious pustular vulvovaginitis of cattle. Cornell Vet. *48*, 458.

Kendrick, J.W. and Straub, O.C. (1967). Infectious bovine rhinotracheitis-infectious pustular vulvovaginitis virus infection in pregnant cows. Am. J. Vet. Res. *28*, 1269.

Kendrick, J.W. (1971). Bovine viral diarrhea-mucosal disease virus infection in pregnant cows. Am. J. Vet. Res. *32*, 533.

Kendrick, J.W. (1976). Bovine viral diarrhea virus-induced abortion. Theriogenology *5*, 91.

Kendrick, J.W. (1976). Epizootic bovine abortion. Theriogenology *5*, 99.

Kluge, J.P. and Mare, C.J. (1974). Swine pseudorabies: Abortion, clinical disease and lesions in pregnant gilts infected with pseudorabies virus. (Aujeszky's disease). Am. J. Vet. Res. *35*, 911.

Marsh, H. (1965). Newson's Sheep Diseases. 3rd ed., Baltimore, Williams & Wilkins.

McKercher, D.G. and Wada, E.M. (1964). The virus of infectious bovine rhinotracheitis as a cause of abortion in cattle. J. Am. Vet. Med. Assoc. *144*, 136.

Meyer, M.C. (1966). Host-parasite relationships in Brucellosis. I. Reservoirs of infection and interhost transmissibility of the parasite. U.S. Livestock Sanitary Assoc. Proc. *70*, 129.

Parker, H.D., Hawkins, W.W. and Brenner, E. (1966). Epizootiologic studies of ovine virus abortion. Am. J. Vet. Res. *27*, 869.

Purdy, C.W., Ford, S.J. and Porter, R.C. (1978).

Equine rhinopneumonitis vaccine: Immunogenicity and safety in adult horses, including pregnant mares. Am. J. Vet. Res. *39*, 377.

Roberts, S.J. (1971). Veterinary Obstetrics and Genital Diseases (Theriogenology). Ithaca, N.Y., S.J. Roberts (distributed by Edwards Bros., Ann Arbor, Mich.)

Szatalowicz, F.T., Griffin, T.P. and Stunkard, J.A. (1969). The international dimensions of Leptospirosis. J. Am. Vet. Assoc. *155*, 2122.

Watson, W.A. and Beverley, J.K.A. (1971). Epizootics of toxoplasmosis causing ovine abortion. Vet. Rec. *88*, 120.

Wrathall, A.E. (1975). Reproductive Disorders in Pigs. Review Series No. 11 of the Commonwealth Bureau of Animal Health. Farnham Royal, Slough, United Kingdom.

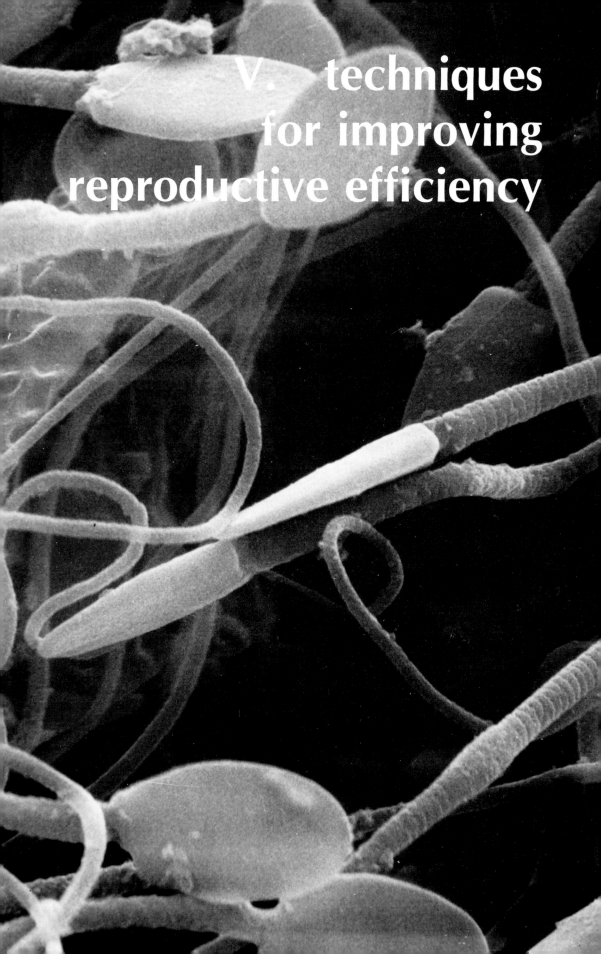

V. techniques
for improving
reproductive efficiency

26

Artificial Insemination

R.H. FOOTE

Artificial insemination (AI) is the most important single technique ever devised for the genetic improvement of animals. This is possible because a few highly selected males produce enough spermatozoa to inseminate thousands of females per year, whereas only relatively few progeny per selected female can be produced per year even by embryo transfer. Methods have been developed for inseminating cattle, sheep, goats, swine, horses, dogs, cats, poultry and a variety of laboratory animals and insects. The earliest carefully documented use of AI was in 1780 when Spallanzani, an Italian physiologist, obtained pups by this method. Other scattered reports appeared in the 19th century, but it was not until 1900 that extensive studies with farm animals began in Russia and shortly thereafter in Japan (Nishikawa, 1962).

Major advantages of AI are as follows: (1) genetic improvement, (2) control of venereal diseases, (3) availability of accurate breeding records necessary for good herd management, (4) economic service and (5) safety through elimination of dangerous males on the farm. AI is practically essential in conjunction with synchronization of estrus programs, and it has been proposed as a means of sex control through separation of spermatozoa containing X- and Y-chromosomes.

When properly done, disadvantages are few. However, it is necessary to have sufficiently well-trained personnel to provide proper service, and to have appropriate arrangements for corralling females for detection of heat and insemination, particularly under range conditions.

It is important that carefully selected young dairy bulls be sampled as soon as possible and rigidly culled after the progeny test. Only a few outstanding sires need be selected to breed a large cow population, provided proper procedures of sexual preparation, semen collection and processing are used to harvest and preserve the maximal number of viable sperm from each sire. Similar types of performance and progeny test programs are important to make maximal genetic progress with meat-type farm animals. AI facilitates cross-breeding, requiring that only one breed be maintained on the farm.

The technique has been used most widely for breeding cattle. Today about 60% of the dairy cows in the USA and nearly all cattle in Denmark and Japan are inseminated artificially. The proportion of cattle bred by AI is also high in most European countries, whereas it is low in Africa, South America and parts of Asia.

Development of AI in beef cattle has been slower in the USA and many other countries because of the difficulty of heat detection and insemination under range conditions, restrictive regulations of the breed associations, and the considerable genetic improvement that can be made for highly heritable traits, using performance-tested bulls, without artificial insemination. In Japan and other countries where beef cattle are reared under confined conditions, a high percentage are inseminated artificially.

Swine are inseminated commercially in several countries, particularly in Europe. AI in sheep is extensive in Russia (more than 35,000,000), other countries of Eastern Europe and parts of South America. Commercial programs for insemination of goats are available in several countries. AI in horses has been actively researched, but it is not widely used commercially, partly due to certain breed society restrictions; such service is available in Japan and several European countries.

When energy intake is restricted, the rate of growth is decreased, testis growth is retarded, age at puberty is increased and sperm output may be reduced. Prolonged deficiencies of essential nutrients can cause infertility. On the other hand, excessive feeding is wasteful and may produce fat, sluggish males.

Testis size is important because of the high correlation between testis size and sperm-producing potential (Amann, 1970; Coulter and Foote, 1979). Testicular development and associated sperm output under good conditions for Holstein bulls is shown in Figure 26–1. Testis size increases rapidly as puberty approaches, and by one year of age bulls have reached approximately 50% of their mature potential. Testis size also is a highly heritable trait (Coulter et al, 1976) that is easily monitored by measuring scrotal circumference. It should be evaluated in all males considered for breeding (Foote, 1975b; Coulter and Foote, 1979).

Housing for males should be convenient, comfortable for the animals and safe for the handlers. Exercise is presumed to be important for the general health of the animal, and bull studs frequently put bulls on mechanical exercisers. This has not been proven to affect fertility. Pens and paddocks provide ade-

MANAGEMENT OF MALES AND SEMEN COLLECTION

The production of high-quality semen depends upon males that have been kept under good conditions. When young males are properly fed and managed, semen can be collected successfully at the following approximate ages: bulls, 12 months; rams, goats and boars, 7 to 8 months; stallions, 24 months.

Physical Condition of Males

Feeding has a notable effect on rate of sexual development in all farm animals.

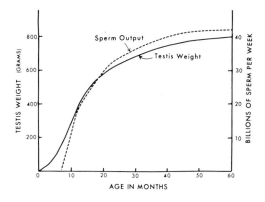

Fig. 26–1. Testicular size and sperm output of growing Holstein bulls ejaculated frequently. *(Adapted from Hahn et al., 1969. J. Anim. Sci. 29, 41.)*

quate exercise for bulls, rams, bucks, boars and stallions.

Semen Collection

Of major importance to an AI program is the correct collection of semen. This involves scheduling males for semen collection at optimal intervals, sexual preparation and correct techniques of semen collection (Corteel, 1977; Pickett and Back, 1973; Salisbury et al, 1978).

Mounts and Teasing Procedures. Live mounts, such as a teaser female, another male or a castrated male, have proven to be the most successful techniques for routine semen collection. Some males, especially boars, can be trained to mount dummies equally well (Fig. 26–2). An estrogen-treated female may provide added incentive during the training period. Dummies may be constructed to hold the artificial vagina. Dummies have an advantage over live mounts in providing stability, and over teaser females particularly in permitting disease control. In either case, a strong mount providing adequate support should be provided. Live mounts should be restrained to minimize lateral as well as forward movement and still provide easy access for semen collection. Good footing is necessary. Convenient posts should be included for restraining bulls during sexual preparation.

Sexual preparation prior to semen collection from bulls increases the number of sperm cells obtained by as much as 100% (Amann, 1970; Almquist, 1978). False mounting a bull several times and/or intensive teasing for five or ten minutes without false mounting is effective. Fluids from the accessory sex glands secreted during this preparatory period may flush out contaminating material from the urethra. Stimuli that are effective in increasing the sexual response of bulls include: changing the teaser, changing the location of the teaser, bringing a new bull into the collection area and false mount-

ing. Various combinations should be tried with sluggish males to keep the intensity of the sexual stimuli high. Beef bulls generally exhibit less libido than dairy bulls and require more ingenuity to provide proper sexual stimuli prior to semen collection.

Methods of sexually preparing boars, rams, bucks and stallions have not been studied in as much detail. However, teasing stallions does not increase the number of spermatozoa ejaculated (Pickett and Voss, 1973).

Frequency of Semen Collection. Increasing the frequency of semen collection decreases the number of sperm per collection, but increases the number of sperm obtained per unit of time. Thus, frequent ejaculation of superior sires provides more spermatozoa to inseminate more females. Billions of sperm can be obtained per normal male per week from any farm animal if properly managed at the time of semen collection (Table 26–1). Ranges are given because of variability among animals (Foote, 1975b) and in semen collection procedures. Large numbers of spermatozoa usually are obtained initially after periods of sexual rest until the epididymal reserves are partially depleted.

Bulls. Under practical conditions artificial breeding organizations often prefer to collect semen from bulls twice a day two days per week to harvest more sperm to freeze at one time. With this procedure, a majority of the sperm produced can be collected. With liquid semen programs, bulls can be ejaculated three times per week and semen made continuously available from each bull by using it over a period of two to three days. Bulls can be ejaculated daily without reducing fertility, but the number of sperm per ejaculate is reduced and more stimuli often are required for proper sexual preparation.

Rams. Rams can be ejaculated many times a day for several weeks before severely depleting epididymal reserves of sperm. This is because of the small ejacu-

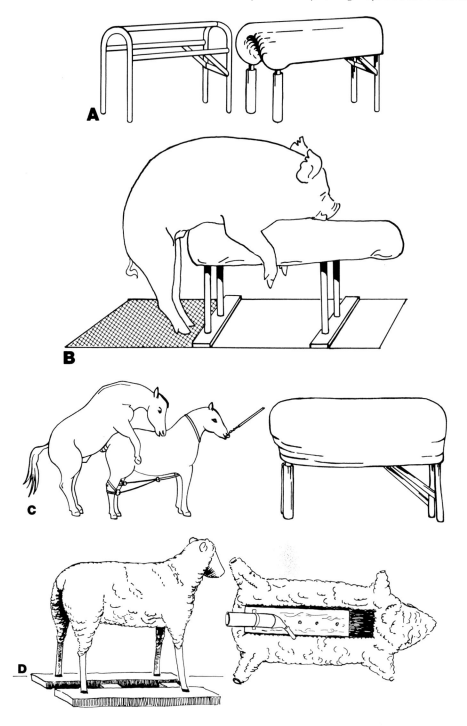

Fig. 26–2. Dummies and mounts used for collecting semen from farm animals. *A, Left to right,* dummy cow constructed from steel pipe. Completed dummy covered and well padded. *B,* Boar mounting padded dummy. A mat to give secure footing is provided. *C, Left to right,* mare hobbled, restrained and tail-bandaged before using as a mount; padded dummy. *D, Left to right,* side view of a portable dummy for rams. Bottom view showing an artificial vagina secured for semen collection.

Table 26–1. **Semen Characteristics and Sperm Output of Normal Mature Farm Animals Ejaculated at Different Frequencies***

Item	Dairy Cattle	Beef Cattle	Sheep	Goats	Swine	Horses
Number of semen collections per week	2–6	2–6	7–25	7–20	2–5	2–6
Characteristics of average ejaculates†						
Volume (ml)	5–10	4–8	.8–1.2	.5–1.5	150–300†	30–100†
Sperm concentration (million/ml)	1000–2000	800–1500	2000–3000	3000–6000	200–300	150–300
Total sperm/ejac. (billion)	5–15	5–10	1.6–3.6	1.5–6.0	30–60	5–15
Total sperm/week (billion)	15–40	10–35	20–40	25–35	100–150	15–40
Progressively motile sperm (%)	50–75	40–75	60–80	60–80	50–80	40–75
Morphologically normal sperm (%)	70–95	65–90	80–95	80–95	70–90	60–90

*Semen characteristics are influenced by age, size, breed and frequency of collection. The low volumes, concentrations and sperm/ejaculation and the high sperm/week correspond to the high frequency of semen collection. Season also affects sheep, goats and horses.
†Gel-free volume.

lates (Table 26–1) and the large epididymal reserves. Rams often mate or are ejaculated many times per day during the breeding season. Bucks have similar semen characteristics but are not ejaculated as frequently as rams.

Boars and Stallions. These animals expel large numbers of sperm in each ejaculate and deplete their epididymal reserves more quickly. It is best not to attempt regular semen collections more often than every other day. If daily ejaculations are required for several days, a short sexual rest for two or three days is recommended.

Artificial Vagina. The best procedure for collecting semen is with the artificial vagina (AV). The AV is simple in construction and simulates natural copulation. The basic design is illustrated with an AV for bulls (Fig. 26–3). The unit provides suitable temperature, pressure and lubrication to evoke ejaculation, and a calibrated tube is attached to collect the semen. Sanitation and the technical skill of both the semen collector and handler of the male are important. Such skill should lead to obtaining semen of high quality and minimizing the possibility of injury during collection. A separate sterile AV

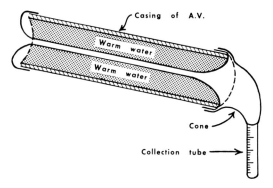

Fig. 26–3. Artificial vagina for bulls shown in longitudinal section to illustrate construction.

should be used for collecting each semen sample.

Bull. For the bull, temperature of the AV is more important than the pressure it exerts on the penis. Water may be used to control both, but final pressure may be adjusted by pumping in air. The temperature inside the vagina should be near 45°C at the time of collection, although temperatures from 38°C to 55°C have been employed. An excellent system is to place the assembled vaginas in a conveniently located incubator (Fig. 26–4, *A*) set at 45°C or slightly higher if cooling will

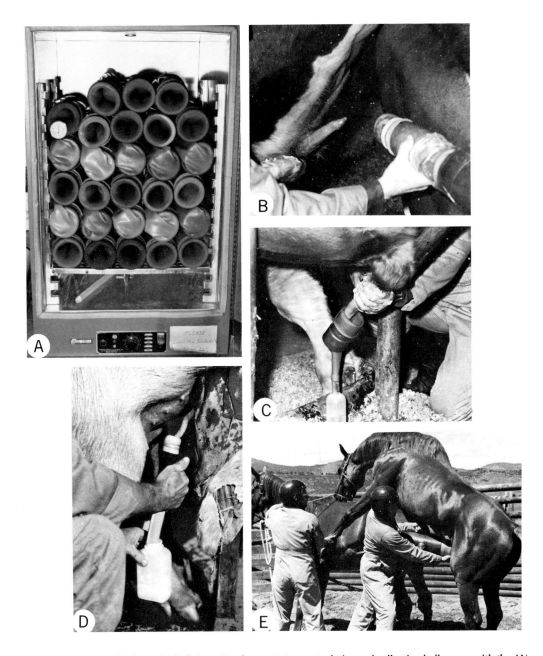

Fig. 26–4. *A*, Assembled AVs for bulls in an incubator. *B*, Proper technique of collecting bull semen with the AV. *C*, Achieving an erection and semen collection by electroejaculation in the bull. The semen collection tube is protected by a jacket of warm water. *D*, Collecting boar semen by applying manual pressure to a short AV. *E*, Collecting stallion semen. The handler of the stallion is holding the foreleg to prevent possible striking of the collector. *(Courtesy of B.W. Pickett).*

occur prior to semen collection. At the same time, the semen collection tube should be maintained near body temperature to prevent sperm damage upon ejaculation due either to overheating or to cold shock.

The AV should be lubricated carefully with sterile lubricant. During collection the artificial vagina should be held parallel and close to the cow slanted in line with the expected path of the bull's penis. The penis should be guided into the artificial vagina by grasping the sheath with the hand immediately behind the preputial orifice (Fig. 26–4, B).

Insertion of the penis is usually best accomplished on the upward movement as soon as the bull mounts. Precise timing is required. As the bull thrusts forward for ejaculation, the operator allows the artificial vagina to move forward with the thrust and aligned with the penis. Then he gently tilts the artificial vagina down so the semen can run into the collecting tube. The experienced operator will alter the conditions of the AV and the semen collection procedure slightly according to the sexual behavior of individual males.

Rams. The recommended temperature of the smaller-sized artificial vagina for the ram is identical to that described for the bull. The technique of collection is similar. The forward thrust of the ram is less vigorous, but rapid. The semen collector must coordinate movements swiftly with those of the ram (Salamon, 1976). An AV may be attached to a dummy (Fig. 26–2, D).

Goat. Buck libido varies, partly with season, as in sheep. Bucks also mount and ejaculate quickly. The technique of semen collection is similar for bucks and rams.

Boar. Pressure is especially important for collecting semen from the boar. The boar ejaculates when the curled tip of the penis is firmly engaged in the sow's cervix, the AV or the operator's hand (Fig. 26–4, D). When using the AV or the gloved hand (King and Macpherson,

1973), pressure should be exerted on the coiled distal end of the penis throughout ejaculation. Once ejaculation has started, the boar usually will remain quiet for the several minutes required for ejaculation.

The ejaculate from the boar consists of three fractions: the presperm fraction, the sperm-rich fraction and the postsperm fraction. Both pre- and postsperm fractions contain mostly seminal fluids with gelatinous, pellet-like material from the Cowper's glands. This material tends to seal the cervix of the sow during mating, preventing loss of semen. The presperm fraction can be discarded before the sperm-rich fraction is collected, or the gel can be filtered out with a filter similar to one illustrated for the horse (Fig. 26–5).

Stallion. Prior to ejaculation the penis should be washed with warm soapy water and rinsed with clean water to remove smegma and other debris on the surface of the penis (Pickett and Back, 1973). The mare should be hobbled (Fig. 26–2, C). The AV for the stallion must be larger than for other farm animals in order to accommodate the stallion's erect vascu-

Fig. 26–5. *Top,* Artificial vagina (modified Japanese model) for stallions. *Bottom,* Enlarged collection bottle fitted with a filter to remove the gel. *(Adapted from Komarek et al., 1965. J. Reprod. Fertil. 10, 337).*

lar-muscular penis. Many types of AVs have been developed, one of which is shown in Figure 26–5. A handle to maintain a firm grip on the AV is necessary because of its size and to cope with the vigorous thrusting of a stallion. The AV is partially filled with water to give an internal temperature of 45°C to 50°C. Air space connected to an expansion valve is provided to permit expansion of the liner when the penis engages the AV and increases the pressure. This pressure and friction stimulates the stallion to ejaculate. When pulsations begin at the base of the penis, which is characteristic of ejaculation, the end of the AV with the collection bottle should be lowered sufficiently to allow semen to flow into the collection vessel (Fig. 26–4, E). Ejaculation is completed in about 25 seconds.

Electroejaculation. Electrical stimulation is the preferred method, only when males cannot be trained or refuse to serve the AV (often under range conditions); also, injuries and infirmities may make mounting impossible. This method can be used successfully in the bull, ram and buck, and it is possible to obtain a semen sample of reduced volume from the boar. It is not desirable to collect and use semen from males unable to serve because of a probable genetic defect. Portable electroejaculators that run on 110 volts as well as on the 12-volt car ignition system are available with different probe sizes for different species.

Bull. A rectal probe with either ring or straight electrodes or finger electrodes can be used to provide the electrical stimulation. The penis usually erects and the semen is collected without the possibility of contamination in the prepuce (Fig. 26–4, C). Sine waves are equal to or better than pulse waves for stimulation (Furman et al, 1975).

Excess fecal material is removed from the rectum and the lubricated probe or gloved hand is inserted into the rectum. Voltage is gradually increased, with re-peated rhythmic stimulation periods alternated with short rest periods. Experience is necessary to achieve the proper combination for erection followed by ejaculation. Secretion from the accessory sex glands takes place at lower voltages and ejaculation at higher voltages. Semen samples obtained with an electroejaculator are usually of larger volume with a lower concentration of sperm, but the fertility and total sperm numbers are equivalent to samples obtained with an artificial vagina.

Bulls have been ejaculated by this method for a period of several years with no apparent ill effects. However, there usually occurs some stimulation of motor nerves with limb extension, so a restraining rack with good footing should be used (Fig. 26–4, C).

Rams. The ram responds exceptionally well to electrical stimulation, and the response is more rapid than in the bull. It is recommended that stimuli be applied every seven seconds with increments of one volt (Cameron, 1977). Ejaculation usually occurs with four to seven stimuli. Semen can be collected with the ram in a standing or recumbent position on a table. The glans penis should be lightly secured with sterile gauze so that the filiform appendage and urethra are directed into the collection tube prior to ejaculation to minimize loss of semen. Ejaculate volume is slightly larger than for samples collected with the AV, and sperm concentration is correspondingly lower. Male *goats* can be ejaculated in a similar fashion.

Boar. An electrostimulator designed for use in the ram also can be used in the boar. Because of the fat insulation it may be necessary to apply 10 volts or more to initiate ejaculation. This results in discomfort and causes the animal to strain; therefore, an anesthetic or analgesic is recommended to immobilize the boar. A small volume of semen usually is obtained. The procedure cannot be recommended for use in the boar.

Massage Method. If an electroejaculator is not available, massage by way of the rectum of the vesicular glands and ampullae of the vas deferens can induce semen flow in the bull. Massage of the sigmoid flexure also may be desirable to cause protrusion of the penis so that an assistant can collect the semen with a minimum of bacterial contamination. Semen so collected usually has a lower sperm concentration than ejaculates obtained with the AV.

EVALUATING SEMEN

Semen evaluation should be rapid and effective so that carefully collected samples can be processed to preserve initial quality and fertility. Inferior samples should be rejected. No single test has been developed that is an accurate predictor of the fertility of individual ejaculates, but. when several tests are carefully combined (Foote, 1975b; Graham, 1978; Salisbury et al, 1978), ejaculates can be selected for use that have a higher fertility potential than those discarded. Several characteristics of semen that are usually evaluated are listed in Table 26–1.

Appearance and Volume

Semen that has been properly protected to avoid cold shock during collection should arrive at the laboratory slightly below body temperature. The tube should be properly labeled to identify the source of semen. It should have a relatively uniform opaque appearance indicative of high sperm cell concentration. Translucent samples contain few sperm. The sample should be free from hair, dirt and other contaminants. Semen with a curdy appearance, containing chunks of material (other than the gel in boar and stallion semen), should not be used; this indicates infection of the reproductive system. Some bulls consistently produce yellow semen owing to the presence of a harmless pigment, riboflavin. This should not be confused with urine, which has a distinctive odor.

The volume of the semen can be determined from the calibrated collection tube. Volumes of individual ejaculates may range more widely than the averages given previously (Table 26–1). In general, young animals and those of smaller size within a species produce smaller volumes of semen. Frequent ejaculation results in lower average volume and, when two ejaculates are obtained consecutively, the second usually has the lower volume. Small volume is not harmful, but if accompanied by a low sperm concentration, the number of sperm available is limited.

Motility and Live Spermatozoa

The percentage of progressively motile cells should be estimated on a microscope stage incubator at 37°C to 40°C under high power (400×). The fresh semen should be prepared as a thin film on a microscope slide, diluted with sufficient physiologic solution (such as saline) that individual cells are visible. At 400×, the percentage of motile cells, their rate of motility and their gross morphology can be scrutinized. Often television equipment is used to display the sperm cells. The motility can be recorded to the nearest 10% or in units on a scale from 0 to 10. Most semen will have less than 80% progressively motile sperm cells, except dog semen, which often has 80% forward-moving cells. The rate of movement can be estimated on an arbitrary scale by microscopic observation. Circular or reverse motion is often a sign of cold shock or of a medium that is not isotonic with semen. Oscillatory motion frequently occurs in aged and dying spermatozoa.

The proportion of live to dead cells can be estimated by supravital staining with a stain mixture such as nigrosin-eosin. The cells that were alive when the stain was applied exclude the stain, and the dead cells stain red with eosin against the dark nigrosin background. The results are highly correlated with visual estimates of progressively motile cells, but the latter averages are lower than the percentage of unstained spermatozoa.

Photoelectric and electronic methods of observing sperm cell velocity, swimming patterns and the proportion of moving spermatozoa as they pass a phototube have been developed (Liu and Warme, 1977), and these yield values correlated with fertility. However, the instrumentation is expensive and not available for general use.

Other techniques used to evaluate motility include impedance change measurements and microscopic-electronic estimation of swirling. These criteria are affected by both sperm cell motility and concentration.

Concentration

Accurate determination of the number of spermatozoa per milliliter of semen is extremely important; it is a highly variable semen characteristic. When combined with the volume of the ejaculate, this quantity of spermatozoa determines how many females can be inseminated, each with the optimal number of sperm cells.

The simplest way of routinely estimating sperm cell concentration is to estimate the optical density (turbidity) of the sample (Foote, 1978) with a nephelometer or photoelectric colorimeter (Fig. 26–6). The procedure is applicable to semen of all farm animals provided the gel has been filtered out. The semen sample is diluted carefully. Calibration is done by making sperm counts with a hemocytometer and comparing these with optical densities of the samples determined photometrically. Thereafter the calibrated photometer can be used to estimate sperm concentrations quickly and accurately. Deibel et al (1978) have described in detail a rapid electronic

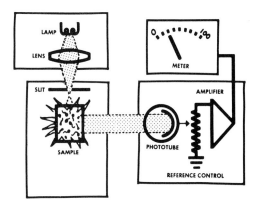

Fig. 26–6. Diagram of a photometer that can be used to determine accurately the concentration of sperm cells in a sample by reflected or transmitted light. The portion of light reaching the phototube is converted to an electrical signal, which deflects the needle on the meter scale.

method of counting spermatozoa that requires expensive equipment, but permits counting sperm cells in the extender.

Morphology

Semen from most males contains some abnormally formed spermatozoa (Table 26–1). This usually is not associated with lower fertility rates until the proportion of abnormals exceeds about 20%. Even then, certain types of abnormalities may not be associated with infertility. Samples with large numbers of abnormal sperm can be detected when estimating the percentage of motile cells, as described previously. Phase contrast, particularly Nomarski interference optics, are helpful. A more precise estimate of the types of abnormal sperm and their incidence can be gained from slides prepared with an India-ink background or with the supravital stain. Many sperm (200) should be counted. Skill is required in classifying the cell types.

Special attention should be given to the acrosome (Saacke, 1972), as this plays an important role in fertilization. The apical ridge of the acrosome of bull or boar sperm deteriorates with aging or injury of the cell and the acrosin enzyme may be lost (Church and Graves, 1976). Eventually the acrosome may loosen and be lost. This can only be seen with appropriate phase or interference microscopes or with specially stained preparations.

Improper freezing may damage the acrosome and cell membranes. Loss of enzymes from the cell, such as glutamic oxaloacetic transaminase (GOT), and the ability of spermatozoa to swell in hypotonic media indicate the effectiveness of cryoprotection by the medium and freezing procedures used in preserving the integrity of sperm cells (Foote, 1975b; Graham, 1978). Viability of the sperm, when incubated at 37°C after freezing and thawing, gives some indication of how long sperm will survive in the female following insemination.

Other Criteria

Various biochemical measurements of the DNA and associated protein of the sperm head have been researched as possible indicators of semen quality. Also, metabolic tests, such as oxygen utilization, fructolysis, lactic acid production, and methylene blue or resazurin color reduction time, have been used. These primarily reflect the combined effects of sperm cell motility and concentration. The pH of semen and shifts in pH also were used at one time. The complex functions of the sperm cell that contribute to its fertilizing capacity can best be monitored by a combination of carefully conducted tests.

PRESERVATION OF SEMEN

Ejaculated sperm do not survive for long periods unless various agents are added. The agents that comprise good extending media have the following functions: (1) provide nutrients as a source of energy, (2) protect against the harmful effect of rapid cooling, (3) provide a buffer

to prevent harmful shifts in pH as lactic acid is formed, (4) maintain the proper osmotic pressure and electrolyte balance, (5) inhibit bacterial growth, (6) increase the volume of the semen so that it can be used for multiple inseminations and (7) protect the sperm cells during freezing (Pickett and Berndtson, 1974; Foote, 1975b; Graham, 1978; Berndtson and Pickett, 1978; Salisbury et al, 1978).

Semen Extender

Pure substances and clean equipment should be used to exclude toxic materials from the sperm environment. Extenders should be prepared aseptically and stored for less than a week unless frozen. A simple carbohydrate, such as glucose, usually is added as a source of energy for the sperm. Both egg yolk and milk are used to protect against cold shock of the sperm cells as they are cooled from body temperature to 5°C. These substances also contain nutrients utilized by sperm. A variety of buffers may be used to maintain a nearly neutral pH and an osmotic pressure of approximately 300 milliosmoles, which is equivalent to that of semen, blood plasma and milk. To inhibit the growth of microorganisms in semen, penicillin, streptomycin, polymyxin B or other combinations of antibiotics that cover a broad bacterial spectrum are added.

Glycerol usually is added to protect sperm against the otherwise lethal effects of freezing. Dimethylsulfoxide (DMSO) and sugars such as lactose and raffinose also may be beneficial, as they serve as dehydrating agents.

Composition of Extenders. Practically all extenders for liquid or frozen semen have either egg yolk or heated milk or a combination of the two as basic ingredients. Egg yolk simply combined with sodium citrate or organic buffers and heated milk or skim milk have been used widely for bull semen, and with modifications for ram, buck, boar and stallion semen.

For many years the emphasis in AI programs was on the use of unfrozen semen. Numerous extender formulations were recommended (Salisbury et al, 1978). The ones used most widely were egg yolk buffered with sodium citrate or tris, or heated milk extenders for bull semen. These were adapted for use with other species. With the remarkable discovery by Polge and co-workers of the protective effect of glycerol during freezing, the emphasis shifted to the use of frozen semen.

Even with the best freezing techniques, more spermatozoa must be put into each breeding unit than with liquid (unfrozen) semen, because of the loss of some viable cells during freezing. Thus, unfrozen semen still is used partially in places such as New Zealand, where a large proportion of the cows are inseminated during a short breeding season. Many cows can be inseminated with semen at high extension rates (fewer sperm cells per insemination) with an extender called Caprogen (see Salisbury et al, 1978). Caprogen consists of sodium citrate, glucose, glycine, glycerol, caproic acid and antibacterial agents; it is gassed with nitrogen before mixing with semen. The quantity of egg yolk in this extender has been reduced to as little as 2% by volume.

In areas where little refrigeration is available, semen may be stored at ambient temperatures. A carbonated egg-yolk extender called Illinois Variable Temperature extender (IVT), and coconut milk have given satisfactory fertility when semen was stored for up to a few days at moderate ambient temperatures.

Most cattle are inseminated with frozen semen. This offers the user a wide variety of choice. It permits semen to be collected at one time and place and to be used anywhere, even after long periods of storage, provided that it is stored continuously at −196°C with a good supply of liquid nitrogen. Thus there need be no wastage of this semen. Examples of extenders for frozen semen are given in Table 26–2.

The egg yolk-citrate and egg yolk-tris extenders listed were originally developed for unfrozen semen. Heated skim milk, homogenized milk or reconstituted dry milk, with 7% to 10% glycerol included, are used widely. Enzymes such as amylase and β-glucuronidase may be added, but fertility has not been improved consistently by their addition.

Pelleting extended bull semen and freezing it on solid carbon dioxide (dry ice) is practiced in a few countries. A sugar such as raffinose (Table 26–2) or 11% lactose may be used. Pellets offer an inexpensive way of preserving spermatozoa, but they are difficult to properly identify when large numbers of bulls are involved. However, in some species in which successful freezing of sperm cells is difficult, the pellet method has been the most successful.

Goat semen can be frozen in skim milk with about 9 gm of glucose per liter and 7% glycerol by volume (Corteel, 1977). An equally satisfactory egg yolk-tris extender (Table 26–2), based on one developed for bull semen, has been tested by Fougner (1976).

Ram semen frozen in the extender listed in Table 26–2 (Salamon, 1976) or in heated milk can be used successfully. However, it is not as fertile as when used undiluted or used unfrozen at low dilutions. Much of the semen used in countries where AI of sheep is practiced is not frozen. Boar semen has been difficult to freeze satisfactorily. However, a variety of extenders have been developed (Larsson, 1978); one of the more successful extenders is given in Table 26–2 (Pursel and Johnson, 1975). Glycerol is detrimental to boar fertility (Wilmut and Polge, 1977), and 2% by volume or less is included in the extended semen during freezing. Some frozen boar semen is available commercially. Much of the boar semen used for insemination in Europe is unfrozen, preserved for up to three days in a glucose-bicarbonate-yolk-extender gassed with CO_2.

Stallion semen also can be pelleted or frozen in straws. Also egg yolk-tris and cream-gelatin extenders have been used (Pickett et al, 1975). The cream-gelatin extender developed by Hughes and Loy has given high fertility rates with unfrozen semen. Glycerol depresses the fertility of stallion semen; semen from some stallions freezes poorly, so the methods of freezing stallion semen are less than optimal (Sullivan, 1978). Nevertheless, successful commercial AI with frozen stallion semen has been reported (Nishikawa et al, 1972).

Antibacterial agents are especially useful in controlling microorganisms present in extended semen stored at higher temperatures (Salisbury et al, 1978). It is important that proper antibiotics be used to control different types of organisms that may be encountered in different species. Glycerol may partially interfere with the inhibitory action of antibiotics on microorganisms.

Vegetable dyes may be included in the extender at sufficient concentrations to distinctly color it. This will not harm the sperm and facilitates identification of semen from different males or breeds.

Semen Processing

The processing of semen through cooling it to 5°C is similar whether it is to be used frozen or unfrozen. Semen is collected at body temperature. Following collection, it should be kept warm (30°C) prior to extension to avoid cold shock. This can be done by placing semen and extender in a water bath kept at 30°C. An aliquot of semen should be removed for

Table 26–2. Composition of Extenders with Egg Yolk for Frozen Semen, as Compiled from the Literature

| Ingredients* | For Frozen Semen (−196°C) | | | | | | |
| | Ampules or Straws | | | Pellet Freezing | | | |
	Bull	Bull	Buck	Bull	Ram	Boar	Stallion†
Sodium citrate dihydrate (gm)	23.2	—	—	—	—	—	—
Tes-N-Tris (gm)	—	—	—	—	—	12	—
Tris (gm)	—	24.2	24.2	—	36.3	2	—
Citric acid monohydrate (gm)	—	13.4	13.4	—	19.9	—	—
Glucose or fructose (gm)	—	10.0	10.0	—	5.0	32	50
Lactose (gm)	—	—	—	—	—	—	3
Raffinose (gm)	—	—	—	139	—	—	3
Casein (gm)	—	—	—	—	—	—	—
Penicillin (units/ml)	1000	1000	1000	1000	1000	1000	1000
Streptomycin (μg/ml)	1000	1000	1000	1000	1000	1000	1000
Polymyxin B (units/ml)	500	500	—	500	—	—	—
Glycerol (ml)	70	70	80	47	50	10	50
Egg yolk (ml)	200	200	100	200	150	200	50
Orvus ES paste (ml)	—	—	—	—	—	5	—
Distilled water to final volume	1000	1000	1000	1000	1000	1000	1000

*Ingredients usually are dissolved in distilled water and then glycerol, more water and egg yolk are added to bring the final volume to one liter. Two buffers and extenders often are prepared, as the glycerol level in the initial medium mixed with sperm usually has a lower glycerol concentration than the extender in which sperm are frozen. Antibiotic levels also vary. See text for details.

†Also frozen in straws. Up to 0.5 gm each of sodium phosphate and sodium-potassium tartrate is included in the medium, according to Japanese workers (Nishikawa et al, 1972). However, there is no completely satisfactory extender for freezing stallion semen.

sample evaluation, and the remainder can be mixed with three to four parts of extender at 30°C. It is recommended that semen be held for 30 minutes at 30°C to increase the antibiotic action of the extender. The mixture is cooled gradually to 5°C for all species, except unfrozen boar semen, which usually is held at 15°C. Buck semen frequently is centrifuged first to prevent possible coagulation (Nishikawa, 1962; Corteel, 1977). Cooling should be slow, taking at least one hour to cool the mixture from 30°C to 5°C. Cooling usually is done with a surrounding water jacket to prevent cold shock.

Extension of Semen. Semen is extended at specified rates so that the volume of semen inseminated will contain sufficient spermatozoa to give high fertility without wasting many cells. The approximate extension rates possible with average ejaculates, appropriate sperm numbers to in-seminate and other relevant data under good management conditions are summarized in Table 26–3. The number of sperm cells required for insemination varies depending on the site of deposition. When extended semen can be placed through the cervix, the number of spermatozoa required is less than that for intracervical or vaginal insemination. When this variation and the range of sperm cell concentrations that occur in fresh semen are taken into consideration, the actual extension rates may be more extreme than those listed in the table.

Extension rates are higher with unfrozen semen than with frozen semen. Unfrozen bull semen can be extended 200 to 300 times with fewer than 5 million motile cells per insemination required for high fertility. Processing liquid semen after cooling to 5°C (or 15°C for boar semen) is simple. Tubes of extended

Table 26–3. Extension, Storage and Insemination Requirements with Frozen Semen (Based upon Average Conditions)

Item	All Frozen Semen				
	Cattle	Sheep	Goats*	Swine	Horses*
Storage temperature (°)	−196	−196	−196	−196	−196
Storage time (years)	>1	>1	>1	>1	>1
Extension rate of 1 ml semen (ml)	10–75	5–10	10–25	4	2
Insemination dose					
Volume (ml)	.2–1	.05–.2†	.5	50	20–50
Motile sperm (10^6)	15	200	200	5000	1500‡
Best time to inseminate during estrus	9 hr after onset to end of estrus	10–20 hr after onset of estrus	12–36 hr after onset of estrus	15–30 hr after onset of estrus	Every second day, starting on day 2 of estrus
Site of semen deposition	uterus and cervix	uterus if possible	uterus if possible	cervix into uterus	uterus
No. of breeding units per male:					
Per ejaculate	300	15	15	10	5
Per week	1000	150	150	30	15
Conception on first insemination (% pregnant)	60	50	60	65	30–50

*Seasonal breeders; refers to AI during the normal breeding season.
†Small volumes with a high concentration of spermatozoa are preferable; sperm in extended semen should be reconcentrated by centrifugation.
‡Based on three inseminations per estrus with 500 million motile sperm each insemination. Fewer sperm have been used with frozen semen, but it is doubtful if the number of sperm for best results is less than that required with fresh semen.

semen should be nearly full to avoid excess air and agitation in shipment. They should be packaged in a manner that maintains the temperature constant and avoids exposure to light. With unfrozen bull, buck, ram, boar and especially stallion semen, fertility declines within a few days of collection. It is recommended that semen be used the day of collection or the next day.

Glycerol Addition for Freezing Bull Semen. Glycerol is used almost universally as the cryoprotective agent for freezing semen. The amounts and methods of adding glycerol vary, depending upon the extenders, freezing methods and species (Table 26–2), and the recipes developed by individual laboratories (Foote, 1975b; Graham, 1978; Berndtson and Pickett, 1978; Saacke, 1978; Salisbury et al, 1978).

Glycerol usually is added to semen after cooling to 5°C; however, it affords just as much protection when added just before freezing. The final amount varies from less than 5% in some yolk-sugar media to 10% in milk. Some add glycerol slowly by dripping or by adding small amounts over a period of one hour; others recommend a one-step addition. Extended semen normally is held for several hours at 5°C before freezing. Tris buffers and the sugar buffers used with pellet freezing offer the advantage that glycerol can be included in the initial media used for cooling sperm. In most other extenders, sperm are damaged morphologically by adding glycerol at room temperature. With glycerol added to the initial extender, several laboratories using tris-buffered extender process semen prior to freezing in ambient temperature, avoiding the need for large cold rooms.

The semen-extender mixture is held for several hours before freezing to allow sperm cells to equilibrate with the extender (usually at 5°C). About four to six hours are optimal, depending upon the medium used. Originally this was thought to be partly necessary for equilibration of

spermatozoa with glycerol. However, it is clear that little time, if any, is required for glycerol equilibration, and it can be added shortly before freezing. Other changes take place at 5°C that enhance sperm survival during freezing.

Bull spermatozoa are packaged in three ways: (1) polyvinyl chloride straws containing 0.25 to 0.5 ml of extended semen; (2) glass ampules containing 0.5 to 1 ml; (3) pellets containing approximately 0.1 to 0.2 ml. When the smaller volumes are frozen as a unit, the sperm concentration per milliliter is increased correspondingly, so that the total sperm per insemination dose is maintained. For example, semen frozen in 0.1-ml packages should have ten times as many sperm per unit volume as semen frozen in 1.0-ml packages to contain the same total number of spermatozoa.

As the technology for freezing cattle semen was developed, glass ampules were used nearly exclusively. Ampules provide a sterile container that can be automatically labeled, filled and sealed. The latter prevents any cross-contamination. Each ampule contains sufficient sperm for a single insemination. Six to eight ampules are attached to a metal cane (Fig. 26–7, *A*) which also carries the bull's identification.

Subsequently, polyvinyl chloride straws were developed by Cassou, patterned after Danish and Japanese straws used in earlier days of artificial insemination with liquid semen. Methods for handling various types of straws have been described (Berndtson and Pickett, 1978; Pickett and Berndtson, 1974; Salisbury et al, 1978). The straw requires less storage space than the ampule, has slightly better freezing characteristics, and can be labeled, filled and sealed automatically (Fig. 26–7, *A,B*). Also, sperm can be transferred to the cow at the time of insemination with minimal loss of cells.

Pelleted semen is prepared by pipetting about 0.1-ml drops of extended semen

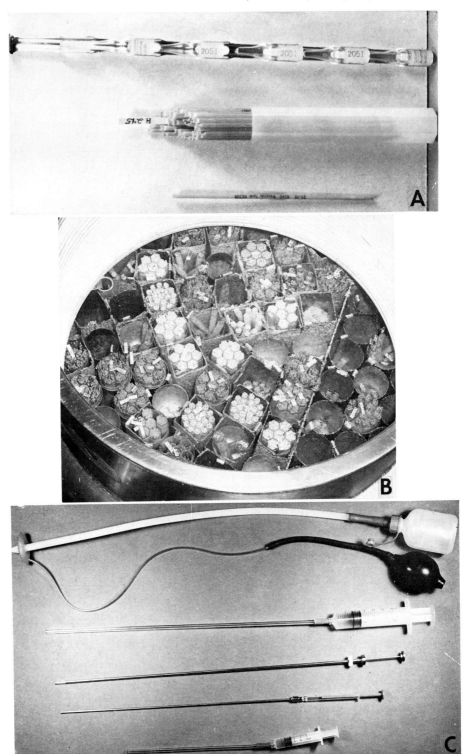

Fig. 26–7. *A,* A 0.5-ml straw and goblet holding 36 straws in comparison with a 1.0-ml ampule on a six-ampule cane. *B,* Storage of straws under liquid N_2 at $-196°$ C. *C,* Insemination equipment for sows, mares, cows, ewes and bitches (top to bottom).

into hemispherical depressions made in a block of dry ice. Sperm survival is good following freezing. Pellets take little space when they are stored in bulk; they offer the cheapest storage method. The main disadvantage is the difficulty of placing bull identification on each pellet, although this has been done by incorporating a small printed paper disc during freezing. Cross-contamination by other spermatozoa or by microorganisms also is possible. Pellets can be placed in bulk containers that are labeled and then repackaged at low temperatures into individually labeled containers prior to field distribution. Another packaging technique is to place sperm directly into catheters used eventually for insemination, and to freeze these as breeding units.

Freezing Bull Semen. Mechanical freezers and freezers using dry ice, liquid air, liquid O_2 and liquid N_2 have all been tried successfully. Liquid N_2 has increased in popularity because it is also the refrigerant of choice for low-temperature, long-time storage of semen. Extended semen frozen as pellets, in straws or in ampules is held at about 5° C prior to freezing. Straws usually are frozen in nitrogen vapor and stored at −196° C. Because of the large surface area of the straw and its thin wall, heat transfer is rapid and semen freezes rapidly, usually within a few minutes. Ampules often are frozen at about 3° C per minute to −15° C. At this point the rate of freezing is increased until −150° C is reached. The ampules on canes then are transferred to liquid N_2 at −196° C. Pellets start to freeze within a few seconds after being placed on a block of dry ice. In a few minutes they reach −79° C and are transferred quickly into containers immersed in liquid nitrogen at −196° C.

The exact nature of the freezing process, which allows spermatozoa under one set of conditions to be frozen rapidly (pellets) or moderately rapidly (straws) and under another set of conditions requires a slower rate (ampules), is not understood. Freez-

ing too rapidly may cause thermal shock and internal ice formation. Slow freezing causes salt concentrations to increase as water freezes out. This increase in osmotic pressure over a prolonged period of slow freezing may damage the proteins and lipoproteins of the sperm cells and the acrosome. The rate of freezing can affect ice crystal formation and this should be taken into account in selecting the thawing rate.

Storage of Frozen Semen. Frozen semen placed initially in dry ice and more recently in liquid nitrogen has been used successfully after more than 20 years of storage. With the lower temperature of liquid N_2, it should be possible to store frozen semen indefinitely. However, it is cheaper and genetically advantageous to use semen from outstanding sires promptly, rather than to store it for long periods.

A variety of efficient vacuum-sealed liquid N_2 refrigerators are available for storing frozen semen. These range in size from central units with a storage capacity of several hundred thousand 0.25-ml "ministraws" or 0.5-ml "midistraws" with a N_2 reserve that lasts about six months, to the common field units that hold up to several thousand straws and a N_2 reservoir that lasts for up to six weeks. Ampules require more space; pellets stored in bulk occupy the least space. The large central storage units can hold up to 750,000 0.1-ml pellets, and possibly provide economical banking of semen from young bulls in sampling programs. In some countries, after a large number of "breeding units" are frozen, the bulls are slaughtered, thereby decreasing total costs.

It is extremely important to check the liquid nitrogen refrigerator periodically to see that the nitrogen level is maintained. Loss of all liquid nitrogen, permitting the temperature to rise considerably, can result in killing the spermatozoa, even when the semen still appears to be frozen.

Thawing Bull Semen. Frozen semen should be held continuously at low temperatures until used. After thawing, frozen spermatozoa do not survive as long as unfrozen sperm, and they refreeze poorly. Therefore, one must be certain that the semen is going to be used soon once it has been thawed. Straws have been successfully thawed at temperatures ranging from that of ice water to 65° C or higher (Berndtson and Pickett, 1978; Gilbert and Almquist, 1978; Graham, 1978; Saacke, 1978). At the higher thaw temperatures, the retention of normal acrosomes and the proportion of progressively motile sperm cells was higher. Thaw time must be controlled carefully to avoid killing the cells by overheating. It is recommended that under field conditions ampules be thawed in ice water; this takes about eight minutes. Higher temperatures (37° C) may be superior (Pace and Sullivan, 1978) for straws, but may depend upon the extender. Pellets are best thawed in a liquid medium at 40° C, but under practical field conditions an ice-water thawing bath is easier to maintain and is satisfactory.

Frozen Semen of Other Farm Animals. The same general principles described for cattle semen appear to apply to handling frozen semen of other species. However, AI with frozen semen has not been developed extensively on a commercial scale for these species. Either the semen has been more difficult to freeze, resulting in lower fertility, or problems with the management of the females, including detection of estrus, have not made AI programs with frozen semen attractive or economical on an extensive basis.

Ram semen can be pelleted (Table 26–2) or frozen in straws with a milk, yolk-lactose or yolk-tris extender. Semen collected during the normal breeding season freezes well. In the pellet method, the semen is cooled to 5° C, glycerol added as necessary, and the semen held for about two hours before freezing as 0.1-ml to 0.4-ml pellets on dry ice (Salamon, 1976). Storage is in liquid nitrogen at −196° C. Pellets are thawed in a solution similar to

the freezing extender, but with glycerol and egg yolk omitted. Spermatozoa should be concentrated by centrifugation and a small volume containing the desired number of sperm cells (Table 26–3) inseminated, if possible, through the cervix. Fertility is considerably higher and fewer spermatozoa are required with intrauterine insemination.

Buck (goat) spermatozoa also can be frozen with good survival when semen is collected during the breeding season. The semen is pre-extended, centrifuged and resuspended in heated milk (Corteel, 1977) or yolk-tris media (Fougner, 1976). The centrifugation removes seminal plasma, which contains an enzyme that can cause coagulation (Nishikawa, 1962). Following cooling, the semen is frozen in straws and stored at −196° C. Thawing should be done quickly, and intrauterine insemination is highly desirable, if possible.

Boar semen has been difficult to freeze, but there are several successful methods (Pursel and Johnson, 1975; Larsson, 1978). Sperm survival is improved by holding the raw semen for two hours to expose sperm cells to the seminal plasma. Then the cell concentration is increased by centrifugation, extender is added, and the extended semen is cooled to 5° C. About 2% by volume of glycerol is added, and the semen pelleted as described previously. Following storage at −196° C, enough pellets are rapidly thawed in a solution such as that given in Table 26–4

Table 26–4. Thawing Solution for Boar Spermatozoa

Ingredients	Amount
Dextrose, anhydrous (gm)	37
Sodium citrate, dihydrate (gm)	6
Sodium bicarbonate (gm)	1.25
Disodium ethylenediamine tetraacetate (gm)	1.25
Potassium chloride (gm)	0.75
Double distilled water to volume (ml)	1000

Adapted from Pursel and Johnson, 1975.

to provide the necessary sperm for insemination.

The success with which stallion semen can be frozen varies greatly. The Nagase type of yolk-lactose-glycerol or yolk-raffinose-glycerol extender (similar to the one given for bulls in Table 26–2), or a more complex extender (Nishikawa et al, 1972) can be used for straws as well. Only the sperm-rich fraction is collected, or much of the seminal plasma is removed by centrifugation. Semen is extended and cooled to 5° C. Stallion spermatozoa are sensitive to glycerol and this compound may be added just prior to freezing. Storage is at −196° C. The fertility of stallion semen is decreased by freezing.

INSEMINATION OF THE FEMALE

High fertility with artificial insemination depends upon (1) high quality semen, (2) proper technique of thawing and inseminating the semen, (3) healthy females in sound breeding condition, and (4) insemination at the proper time of the estrous cycle. The latter is extremely important. Viable sperm must be in the vicinity of the egg and capacitated (in most species at least) shortly after ovulation. Because ovulation is difficult to detect, insemination should be timed relative to estrus.

Detection of Estrus

The best indication of estrus is when the female stands when mounted by the male or other females. However, this test often is not practical for AI. Cows, does, sows and mares cycle about every 20 to 21 days, and ewes, about every 16 to 17 days. The following are procedures for detecting estrus.

Cattle. Cows in loose housing should be observed early in the morning and in the evening every day for standing when mounted by other cows. Nonpregnant cows in stanchion barns should be turned out morning and evening and watched for 20 to 30 minutes each time for standing heat. Restlessness, bawling and attempting to mount may be signs of coming into estrus. Beef cattle should be similarly checked. Also, there are many aids for detecting estrus (Foote, 1975a; Britt, 1977). These include pressure-sensitive devices placed on the backs of animals which change color when an animal stands for mounting, the use of surgically sterilized bulls carrying devices for marking females that stand, electronic probes (Foote et al, 1979) and the discharge of clear mucus from the vagina. Reddish mucus on the tail is an indication that estrus occurred one or two days previously. Also, cows in estrus are more active and this activity can be monitored by pedometers placed on the legs. Odors can be detected by specially trained dogs.

Because of the difficulty of detecting estrus in large herds, particularly in cold climates, a program of estrous cycle regulation could be advantageous. Several techniques that produce a highly synchronized and fertile estrus have been developed; however, not all the drugs used in these techniques have been approved by the Food and Drug Administration for use in cattle in the USA.

Sheep. Estrus is difficult to detect in ewes. Therefore, vasectomized rams with crayon or colored grease applied to the brisket or contained in a special harness should be used. Ewes in heat and rams will seek each other out. By frequently changing rams and/or the colors used, the time of onset of heat can be judged.

Goats. Detection of estrus in does also is facilitated by using a vasectomized buck with marking crayon or grease paint. Also, exposure of does to a cloth with buck urine on it often causes does in estrus to wag their tails and exhibit other symptoms of estrus.

Swine. The sound, sight and smell of a vasectomized or intact boar in an adjacent pen are helpful. Sows in heat tend to seek out the boar and assume a rigid stance (lordosis) with ears erect when mounted, or similarly when hands are placed firmly on their back. The vulva swells and reddens as blood flow increases. Sows normally come into estrus about three to eight days after weaning their litter. Weaning time can be used as a method of synchronizing estrus.

Horses. Mares should be teased daily with a stallion in special teaser paddocks. Indications of acceptance of the stallion are elevation of the tail, spreading of the legs, standing, and frequent urination and contractions of the vulva ("winking").

Optimum Insemination Time

Animals should not be inseminated after parturition until the uterus is fully involuted and the females are cycling normally. Then insemination near the time of ovulation is important. Because the length of estrus and time of ovulation are variable, it is necessary to give a range of insemination times that yield "optimal" results (Table 26–3). Cows should not be inseminated before 50 days after calving for best conception rates; otherwise, more inseminations per conception are required. Conception rate also is lower when cows are inseminated during the first part of estrus (Foote, 1979; Robbins et al, 1978). Bull sperm pass rather quickly through the reproductive tract and capacitation time probably is short. A practical procedure is to check for estrus twice daily and inseminate on the same day all cows first seen at the morning check. Cows first seen in estrus in the evening should be inseminated the next morning. There is no precisely defined "best" time (Gwazdauskas, 1978).

In the ewe, insemination should take place at the middle or during the second half of estrus. Double inseminations during estrus, particularly with frozen semen, increase fertility. Goats can best be inseminated about 12 hours after the onset of estrus, and should be inseminated again the next day if they are still in estrus. Sows come into estrus about three to five days after farrowing, but they do not ovulate at this time and should not be bred. They come into heat three to eight days after weaning their pigs and may be inseminated at this time. Since sows ovulate about 30 to 36 hours after the beginning of estrus, with rapid loss of fertility after ovulation, it is best to inseminate either late on the first day or early on the second day of estrus. An advantage of about 10% is gained by inseminating on both the first and second days of estrus.

Inseminating a mare during the first heat (foal heat), about nine days after foaling, is not advised because the uterus is not fully involuted and the conception rate is lower. Mares may be inseminated at the next estrus about 30 days after foaling. Because mares have a long and variable estrus, it is best to inseminate at least every second day during estrus, starting on the second day. When mares are palpated daily, insemination can be done to coincide with ovulation, which precedes the end of estrus by one or two days.

Insemination Procedure

Identification of the female is the first step. The breeding record should be checked for date of parturition; previous breedings are noted and the current insemination is recorded. Basic breeding records are important for all species.

Cattle. The dairy cow is usually inseminated while standing in a stanchion or stall. Beef cows in estrus should be penned and a squeeze chute provided to restrain the animal during insemination. Stress should be avoided.

Semen can be deposited in the cervix with the aid of a speculum. However, the rectovaginal technique is more effective and more widely used (Salisbury et al, 1978). By manipulation of the cervix with the hand in the rectum, the inseminating catheter is passed just through the annular rings of the cow's cervix (Fig. 26–8, *A*). Part of the semen is deposited through the cervix into the body of the uterus and the remainder in the cervix as the catheter is withdrawn. With the straw inseminating gun most spermatozoa are transferred to the cow. With the catheter, care must be taken to expel the semen slowly, to avoid excessive sperm losses in the catheter. A common error is to penetrate beyond the body of the uterus into the uterine horns. The number of sperm cells required for optimal conception rate for frozen semen

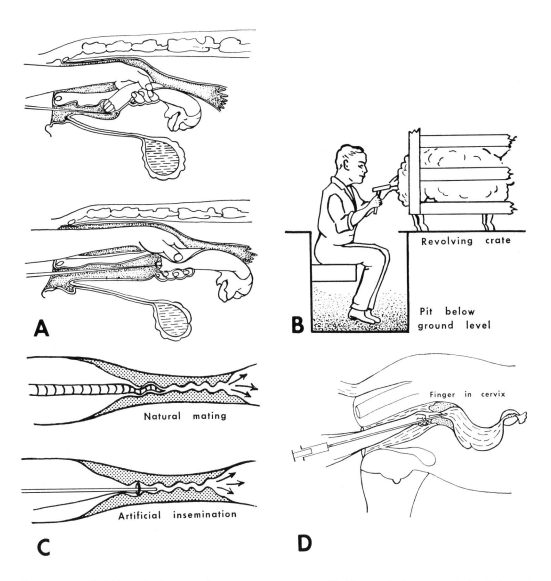

Fig. 26–8. Artificial insemination procedures. *A, Top,* wrong way of holding cervix for rectovaginal technique of inseminating cows. *Bottom,* by using correct procedure, it is relatively easy to deposit semen through the cervix. *(Redrawn from Bonadonna, 1957. Nozioni Di Fisiopathologia Della Reproduzione E Di Fecondazione Artificiale Degli Animali Domestici, Milan, courtesy of T. Bonadonna). B,* Pit and restraining crate facilitating ewe insemination. *C,* Comparison of cervical semen deposition in natural mating and artificial insemination in swine. Arrows indicate flow of semen into the uterus. *D,* For insemination of mares the index finger in the cervix assists in guiding the catheter held by the other hand into the uterus.

(Table 26–3) is higher than that for liquid semen. Pellets require the same insemination equipment as ampules.

Pregnancy is a possibility in a previously inseminated cow, and the catheter should not be forced into the uterus; approximately 3% to 5% of pregnant cows show signs of estrus. The use of dye in extender when training technicians to inseminate reproductive tracts and intact cows at the specified location appears to be an extremely useful method to improve the accuracy of semen placement and the efficiency of technicians.

Sheep. The ewe can be held securely by putting the hind legs over a rail or placing her in an elevated crate during insemination (Salamon, 1976). A rotating platform arrangement (Fig. 26–8, *B*) with the inseminator in a pit can be used to inseminate more than 100 ewes per hour. This permits one person to put a ewe on the platform, another to inseminate, and a third one to release the ewe.

Insemination with the aid of a speculum and light permits semen to be deposited into or through the cervix rather than the vagina. A small volume of concentrated sperm cells (Table 26–3) is most effective in producing pregnancy. Frozen extended semen should be reconcentrated so that about 200 million motile sperm cells are inseminated. With unfrozen semen, 50 million motile sperm are sufficient. With progestogen treatment of ewes to induce estrus at a synchronized time, sperm number requirements may be as great as 1500 million cells. Double inseminations 12 hours apart increase fertility.

Goats. Insemination procedures are similar to those used for sheep. Insemination with the aid of a lighted speculum increases the likelihood that spermatozoa will be deposited into the uterus, where fertility normally is higher than for vaginal or cervical deposition.

Swine. The sow is best inseminated without being restrained to avoid possible loss of spermatozoa. With some rubbing and pressure on the back, the sow in estrus usually stands calmly during insemination. The insemination tube is easily guided into the cervix because the vagina tapers directly into the cervix, which itself tapers. It is not possible to pass the catheter into the uterus, but with the inflated cuff (Fig. 26–7, *C*) most of the 50 ml suggested for insemination will be forced into the uterus. Otherwise much semen may be lost. Large semen volume and high sperm numbers (Table 26–3) are required in sows for maximal fertility. Why the large volume is necessary is unknown. Frozen boar spermatozoa appear to be removed rapidly by the uterus. Even when many billions are inseminated intracervically, few reach the oviducts. Surgical deposition of a few million sperm in .05 to .10 ml of fluid directly into the oviducts gives good fertility, but it is not practical.

Horses. The mare should be restrained by hobbles, backed against baled hay or a board wall or put in a breeding chute to protect the inseminator. The technique of insemination is illustrated in Figure 26–8, *D*. The area around the vulva should be scrubbed before insemination to minimize contamination. The arm in a plastic sleeve, lightly lubricated, is inserted into the vagina and the index finger inserted into the cervix. The inseminating catheter then is easily guided into the uterus, where 20 to 50 ml of raw or extended semen is deposited. Better results with fresh or stored unfrozen semen are obtained when 500 million motile sperm are inseminated. The minimal number of frozen sperm required for high conception rates has not been established.

FACTORS AFFECTING CONCEPTION RATE IN AI

Accurate measurement of fertility is an important part of any organized artificial insemination program. The major factors determining fertility in artificial breeding are: (1) the fertility of the males used to produce the semen, (2) the care with which the semen is collected, processed and stored, (3) the skill of the inseminating technician and (4) management of the females. Males should be carefully selected, isolated and tested before joining a stud. They should have completely normal testes (Coulter and Foote, 1979), produce high quality semen and be free from disease. Those failing to meet standards should be culled. Regular health checks should be repeated periodically and all health codes followed.

The underline of males, particularly around the prepuce, should be clean at the time of semen collection. All equipment used for collecting, processing and inseminating semen should be clean and sterile. Traces of washing solutions should be removed by proper rinsing. Thorough rinsing of liners with boiling distilled water, followed by isopropyl alcohol, should produce clean, sterile liners that dry readily before they are used. All glassware should be washed carefully, rinsed thoroughly in distilled water and sterilized in an oven at 150° C for at least one hour.

Special sanitary precautions are needed for semen collection. The collector should wear disposable plastic gloves. The mount should be covered or disinfected between collections.

Only high quality semen should be processed for insemination. Antibiotics should be added to the semen. Frozen semen should be examined after freezing and inferior samples discarded. Semen can be stored successfully for a long time if stored continuously under liquid nitrogen at −196° C. However, lack of liquid nitrogen for even a few hours may result in complete destruction of a sperm bank.

Improper handling of semen during storage (Pickett and Berndtson, 1974; Foote, 1975b; Saacke, 1978) also decreases fertility. Placing sperm through the cervix

or flushing cells through it (swine) is important. Skill and experience are required. Penetrating the cervix is particularly difficult in sheep and goats. The female should be in optimal breeding condition, and proper estrus detection is of extreme importance. Nearly all experiments show that more than one insemination per estrus increases conception rates. Presumably this results from the difficulty of predicting ovulation and appropriately timing each insemination.

Commercial AI organizations should monitor fertility and evaluate all contributing components possible. Ideally this would be to gather complete information on young born, but this is not practical. A useful report, when all services are reported, is the 60- to 90-day report tabulated on a monthly basis. Two months after each month when the inseminations were performed (which is also about 90 days after the beginning of that month), the records of all inseminated cows are checked to determine what proportion were reported for reinsemination. With computers it would be feasible to introduce a more precise time, such as a 60- or 75-day interval. The relationship between various measurements is illustrated in Figure 26–9. The advantages and limi-

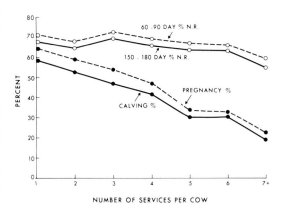

Fig. 26–9. The relationship between fertility estimated by nonreturn, pregnancy and calving rates following varying numbers of services. *(From Foote, 1978. Search 8(2), 18.)*

tations of the nonreturn report have been described (Oltenacu and Foote, 1976).

Pregnancy information may be obtained by palpation of the reproductive organs per rectum in cattle and horses, by milk progesterone assays in cattle and goats, by running an immunologic blood test for mares and by ultrasonic measurements in sheep and swine. Information on litter size in swine also is important.

Realizing the potential of artificial insemination depends upon the combined skill and cooperation of people—the organization producing and distributing semen, the inseminator and the manager of the herd or flock.

REFERENCES

Almquist, J.O. (1978). Bull semen collection procedures to maximize output of sperm. NAAB Proc. 7th Tech. Conf. Artif. Insem. Reprod., 33–36.

Amann, R.P. (1970). Sperm production rates. *In* The Testis. A.D. Johnson, W.R. Gomes and N.L. Van Demark (eds), New York, Academic Press, Vol. I, pp. 443–482.

Berndtson, W.E. and Pickett, B.W. (1978). Techniques for the cryopreservation and field handling of bovine spermatozoa. *In* The Integrity of Frozen Spermatozoa. Washington, D.C., Conf. Natl. Acad. Sci., 53–57.

Britt, J.H. (1977). Strategies for managing reproduction and controlling health problems in groups of cows. J. Dairy Sci. 60, 1345.

Cameron, R.D.A. (1977). Semen collection and evaluation in the ram. The effect of method of stimulation on response to electroejaculation. Aust. Vet. J., 53, 380.

Church, K.E. and Graves, C.N. (1976). Loss of acrosin from bovine spermatozoa following cold shock: protective effects of seminal plasma. Cryobiology 13, 341.

Corteel, J.M. (1977). Production, storage and insemination of goat semen. Proc. Symposium Management of Reproduction in Sheep and Goats. Am. Soc. of Anim. Sci., 41–57.

Coulter, G.H. and Foote, R.H. (1979). Bovine testicular measurements as indicators of reproductive performance and their relationship to productive traits in cattle: A review. Theriogenology 11, 297.

Coulter, G.H., Rounsaville, T.R. and Foote, R.H. (1976). Heritability of testicular size and consistency of Holstein bulls. J. Anim. Sci. 43, 9.

Deibel, F.C., Graham, E.F. and Evensen, B. (1978). Technique and application of electronic counting and sizing of spermatozoa. NAAB Proc. 7th Tech. Conf. Artif. Insem. Reprod., 45–52.

Foote, R.H. (1975a). Estrus detection and estrus detection aids. J. Dairy Sci. 58, 248.

Foote, R.H. (1975b). Semen quality from the bull to the freezer: An assessment. Theriogenology 3, 219.

Foote, R.H. (1978). Principles and procedures for photometric measurement of sperm cell concentration. NAAB Proc. 7th Tech. Conf. Artif. Insem. Reprod., 55.

Foote, R.H. (1979). Time of artificial insemination and fertility in dairy cattle. J. Dairy Sci. 62, 355.

Foote, R.H., Oltenacu, E.A.B., Mellinger, J., Scott, N.R. and Marshall, R.A. (1979). Pregnancy rate in dairy cows inseminated on the basis of electronic probe measurements. J. Dairy Sci. 62, 69.

Fougner, J.A. (1976). Uterine insemination with frozen semen in goats. VIIIth Internat. Congr. Anim. Reprod. Artif. Insem., Krakow, 4, 987.

Furman, J.W., Ball, L. and Seidel, G.E., Jr. (1975). Electroejaculation of bulls using pulse waves of variable frequency and length. J. Anim. Sci. 40, 665.

Gilbert, G.R. and Almquist, J.O. (1978). Effects of processing procedures on post-thaw acrosomal retention and motility of bovine spermatozoa packaged in .3 ml straws at room temperature. J. Anim. Sci. 46, 225.

Graham, E.F. (1978). Fundamentals of the preservation of spermatozoa. *In* The Integrity of Frozen Spermatozoa. Proc. Conf. Natl. Acad. Sci., Washington, D.C., 4–44.

Gwazdauskas, F.C. (1978). The effect of time of insemination on conception rates in dairy herds. NAAB Proc. 7th Tech. Conf. Artif. Insem. Reprod., 92–96.

King, G.J. and Macpherson, J.W. (1973). Comparison of two methods for boar semen collection. J. Anim. Sci. 36, 563.

Larsson, K. (1978). Deep-freezing of boar semen. Cryobiology 15, 352.

Liu, Y.T. and Warme, P.K. (1977). Computerized evaluation of sperm cell motility. Comput. Biomed. Res. 10, 1.

Nishikawa, Y. (1962). Fifty years of artificial insemination of farm animals in Japan. English Bul. 2, Dept. Anim. Sci., Kyoto University, Japan, 1–43.

Nishikawa, Y., Iritani, A., and Shinomiya, S. (1972). Studies on the protective effects of egg yolk and glycerol on the freezability of horse spermatozoa. VIIth Internat. Congr. Anim. Reprod. Artif. Insem., Munich, 2, 1545.

Oltenacu, E.A.B. and Foote, R.H. (1976). Monitoring fertility of A.I. programs: Can nonreturn rate do the job? NAAB Proc. 6th Tech. Conf. Artif. Insem. Reprod., 61–68.

Pace, M.M. and Sullivan, J.J. (1975). Effect of timing of insemination, numbers of spermatozoa and extender components on the pregnancy rate in mares inseminated with frozen stallion semen. J. Reprod. Fertil. Suppl., 23, 115.

Pace, M.M. and Sullivan, J.J. (1978). A biological comparison of the .5 ml ampule and .5 ml French straw systems for packaging bovine spermatozoa. NAAB Proc. 7th Tech. Conf. Artif. Insem. Reprod., 22–32.

Pickett, B.W. and Back, D.G. (1973). Procedures for preparation, collection, evaluation and insemination of stallion semen. Ft. Collins, Col., Colorado State University, General Series 935, 26 pp.

Pickett, B.W. and Voss, J.L. (1973). Reproductive management of the stallion. Ft. Collins, Col., Colorado State University, General Series 934, 501–531.

Pickett, B.W. and Berndtson, W.E. (1974). Preservation of bovine spermatozoa by freezing in straws. A review. J. Dairy Sci. 57, 1287.

Pickett, B.W., Burwash, L.D., Voss, J.L. and Back, D.G. (1975). Effect of seminal extenders on equine fertility. J. Anim. Sci. 40, 1136.

Pursel, V.G. and Johnson, L.A. (1975). Freezing of boar spermatozoa: fertilizing capacity with concentrated semen and new thawing procedure. J. Anim. Sci. 40, 99.

Robbins, R.K., Sullivan, J.J., Pace, M.M., Elliott, F.I., Bartlett, D.E. and Press, P.J. (1978). Timing of insemination of beef cattle. Theriogenology, 10, 247.

Saacke, R.G. (1972). Semen quality tests and their relationship to fertility. NAAB Proc. 4th Tech. Conf. Anim. Reprod. Artif. Insem., 22–28.

Saacke, R.G. (1978). Factors affecting spermatozoan viability from collection to use. NAAB Proc. 7th Tech. Conf. Artif. Insem. Reprod., 3–9.

Salamon, S. (1976). Artificial Insemination of Sheep. University of Sidney, N.S.W., Australia, 104 pp.

Salisbury, G.W., VanDemark, N.L. and Lodge, J.R. (1978). Physiology of Reproduction and Artificial Insemination of Cattle, 2nd ed. San Francisco, W.H. Freeman & Co.

Sullivan, J.J. (1978). Characteristics and cryopreservation of stallion spermatozoa. Cryobiology 15, 355.

Wilmut, I. and Polge, C. (1977). The low temperature preservation of boar spermatozoa. 3. The fertilizing capacity of frozen and thawed semen. Cryobiology, 14, 483.

27

Induction and Synchronization of Ovulation

J.H. BRITT
and
J.F. ROCHE

The possibility of inducing estrus and ovulation in acyclic females and of synchronizing estrus and ovulation in groups of females offers an opportunity to increase the efficiency of animal production and to increase the use of artificial insemination. Efficiency of production would increase because initial conception would occur at an earlier age and intervals between successive pregnancies would be reduced. Broader use of AI would result in faster improvements in traits of economic importance because of increased male selection.

During the last two decades, tremendous progress has been made in unravelling the complex endocrine relationships that control ovarian function in major domestic species (Stabenfeldt et al, 1978). This chapter deals with induction of estrus and ovulation in prepubertal females and in females during lactational and/or seasonal anestrus, and with synchronization of estrus and ovulation in groups of

females. Mechanisms for controlling estrus and ovulation will be discussed and current methods for inducing and synchronizing estrus and ovulation will be presented.

CONTROL OF OVULATION

Rationale for Induction of Ovulation

The interval between birth and conception and intervals between successive pregnancies in farm animals are influenced by both genetic and environmental factors. For example, ewe lambs born in temperate climates must reach puberty by eight to ten months of age in order to be bred during the first naturally occurring breeding season after birth. Otherwise, first conception will not normally occur until ewes are 20 to 22 months old because of seasonal anestrus. Seasonal anestrus also prolongs the interval from birth to conception in goats and horses that do not reach puberty by the first breeding season after birth.

Breeding can occur year-round in swine and in cattle, so that a delay in onset of puberty does not affect lifetime productivity as dramatically as in other species. Nevertheless, if gilts could be bred at an earlier age, productivity would be increased relative to that now achieved. Seasonal breeding is common in beef cows and in dairy cows in certain countries, and hence age of onset of puberty is an important management consideration.

In females nursing their young, lactational anestrus prevents rebreeding during the first few weeks or months after parturition. In sheep and goats, lactational anestrus also normally coincides with seasonal anestrus. A similar situation exists in horses, except mares exhibit a foal estrus about 5 to 15 days after parturition. Duration of lactational anestrus depends on several factors, including nutritional status, parity and suckling intensity. In general, lactational anestrus

persists longer if energy intake by the dam is limited, if the dam is primiparous, and if the number of young being nursed is greater than normal for a given age and parity of dam.

In the situations just described, productivity would be increased if the age at puberty could be reduced and if breedings could occur during lactational and/or seasonal anestrus. It is possible to induce estrus and ovulation during these acyclic periods in domestic females by administering exogenous hormones or by manipulating environmental factors (eg, photoperiod, suckling intensity) that suppress ovarian activity (Fig. 27–1). In order to be considered useful on a practical basis, a method of inducing estrus and ovulation should result in a high response rate and fertility at the induced estrus should be acceptable.

Rationale for Synchronization of Ovulation

In a group of randomly cycling females, the time of occurrence of estrus cannot be predicted with any certainty for an individual animal. Running a male of the species with a group of females is the most efficient method for detecting and breeding females that are in estrus. However, where hand or supervised mating is to be practiced (generally in the case of sows and mares) or where AI is to be used, it is necessary to know when a female is in estrus. Detection of estrus is time-consuming, laborious, and subject to human error. If estrous cycles of randomly cycling females could be altered so that all females in a group would be in estrus and ovulate within a predictable time-span, then the necessity for detection of estrus would be obviated. It is possible to synchronize the time of estrus and ovulation in a group of females by manipulating their estrous cycles through administration of exogenous hormones (Fig. 27–1). The objectives of estrous synchronization

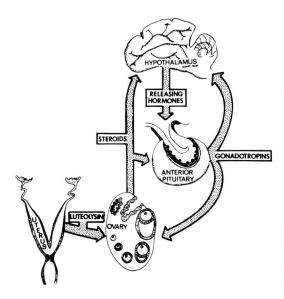

Fig. 27–1. Possible pathways for controlling ovarian function in domestic females. Exogenous releasing hormones, gonadotropins, steroids or luteolytic agents can be used aione or in combination to induce or synchronize ovulation in groups of females.

programs are the following: (1) to be able to predict in advance the day and time a group of females is ready to be bred, and (2) to obtain normal fertility at this controlled estrus.

METHODS OF INDUCING OVULATION

In order to induce ovulation in anestrous or acyclic females, a single follicle or group of follicles must be stimulated to develop to a state of maturity so that a surge of luteinizing hormone (LH) or a hormone with LH-like properties (eg, human chorionic gonadotropin, HCG) will cause ovulation. This can be accomplished by several methods (Table 27–1).

Gonadotropin Treatments

Follicle growth and ovulation in acyclic domestic animals can be stimulated by administering hormones that have gonadotropic activity. Either follicle stimulating hormone (FSH), in presence of LH, or pregnant mare serum gonadotropin (PMSG) will stimulate follicular growth. However, PMSG is preferred because it has a longer metabolic half-life and is more readily available. Follicle growth can be stimulated with a single dose of PMSG, whereas multiple doses of FSH are usually required to obtain similar responses. Once follicle growth has been stimulated with either FSH or PMSG, it may be necessary to administer LH or HCG about 48 to 96 hours later in order to induce ovulation. In some cases this is not required because developing follicles secrete estrogen in sufficient quantities to induce an endogenous LH surge, which then results in ovulation.

The response following a given treatment regime with exogenous gonadotropins varies, ranging from no ovulatory response to a response that exceeds the normal ovulation rate by several-fold. It is not known why the ovulatory response is so variable, but it is likely due to the

Table 27–1. Some Practical Methods for Inducing Estrus and Ovulation in Acyclic Females

Species	Reproductive State	Treatment Regime	Response
Cattle	Prepuberal	5 mg estradiol valerate, day 1; 3 mg Norgestomet, day 1; 6 mg Norgestomet ear implant, days 1 to 9	Majority in estrus within 4 days after treatment
	Postpartum-lactational anestrus	Dairy cows: 100 μg GnRH, day 14 postpartum	Majority ovulate on the day following treatment
		Suckled cows: 100–500 μg GnRH on two occasions, 10 to 14 days apart	Majority begin estrous cycles after second treatment
		Suckled cows: 5 mg estradiol, day 1; Progestogen for 7 to 12 days after estrogen; PMSG, 500–800 IU on last day of progestogen (optional)	Majority cycle after treatment; response is enhanced after PMSG
Sheep	Prepuberal or seasonal anestrus	Progestogen (progesterone, MAP, FGA) for 12 to 14 days; PMSG, 500–750 IU on last day of progestogen	Majority cycle after treatment; PMSG is necessary for good response
Goat	Prepuberal or seasonal anestrus	Progestogen (FGA) for 18 to 21 days; PMSG, 400–800 IU 1 to 2 days before end of progestogen	Majority cycle after treatment
Swine	Prepuberal or anestrus	PMSG, 250–1500 IU; HCG, 500–1000 IU given 96 hours after PMSG	Majority cycle after treatment
		PMSG, 200–300 IU + HCG, 200–300 IU given simultaneously	Majority cycle after treatment
Horse	Seasonal anestrus	Artificially lengthen photoperiod beginning in late fall or early winter	Majority begin cycling before mares exposed to natural photoperiod

number of developing follicles that are able to respond at the time of treatment.

In addition to treatment regimes that employ exogenous gonadotropins, it has been possible to stimulate endogenous FSH and LH release and subsequent ovulation by administering gonadotropin-releasing hormone (GnRH). GnRH does not normally induce additional follicular growth, but it will cause a mature follicle to ovulate. The lack of follicular stimulation following GnRH administration is presumably related to the short half-life of endogenously released LH and FSH.

Cattle. Mechanisms involved in normal initiation of the first ovulatory cycle appear similar for pubertal heifers and postpartum cows. There is an increase in the frequency of episodic LH releases from the anterior pituitary followed by a transient increase in progesterone. The source of this progesterone may be either the ovary or adrenal gland, but the site is not known with certainty. When progesterone decreases, there is a surge of LH that is followed by ovulation (Gonzalez-Padilla et al, 1975a; Humphrey et al, 1976).

GnRH releases a surge of LH in postpartum cattle depending on the endocrine state of animals at time of treatment. Britt et al (1974) induced ovulation with a single injection of GnRH on day 14 postpartum in a high proportion of dairy cows (Fig. 27–2). GnRH induces a peak of LH equivalent to the normal preovulatory surge after 20 days postpartum in dairy cows (Kesler et al, 1977) and after 30 days postpartum in beef cows (Webb et al, 1977). However, a single injection of GnRH or two injections 10 to 14 days apart to induce ovulation have led to contradictory results. Some researchers have reported that either treatment induces ovulation and shortens the postpartum interval, while others have reported no effect on the postpartum interval. Postpartum interval, breed of animal, age, season and plane of nutrition all influence the response to GnRH (Fig. 27–3).

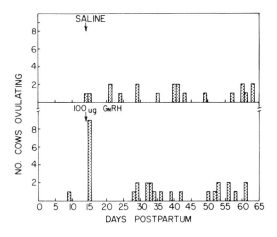

Fig. 27–2. Ovulatory response in Holstein cows given saline or gonadotropin-releasing hormone (GnRH) on day 14 after parturition. Ten cows were treated in each group and ovarian activity was monitored until day 65 postpartum. *(From Britt et al, 1974. Reprinted by permission from J. Anim. Sci.)*

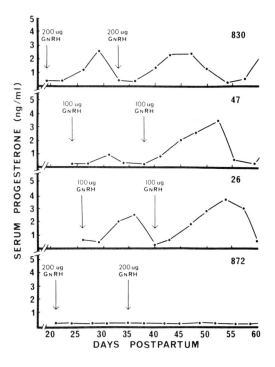

Fig. 27–3. Serum progesterone in four suckled cows given gonadotropin-releasing hormone (GnRH) at approximately 3 and 5 weeks after parturition. Progesterone concentrations of >1 ng/ml were taken as evidence that ovulation had occurred. *(From Fonseca, Britt and Ritchie, unpublished results)*

Sheep. The seasonal breeding pattern of the ewe can be altered by varying the photoperiod, which presumably stimulates endogenous gonadotropin release. By reducing daylight hours, ewes in seasonal anestrus can be induced to ovulate. Recent research (Legan et al, 1977) suggests that seasonal change in sensitivity to negative feedback of estradiol in response to varying photoperiod is one of the main endocrine causes of anestrus in sheep.

Despite the effect of photoperiod on seasonal breeding and the fact that estrus can be induced in anestrous ewes by changing photoperiod, little practical use has been made of this method. However, this procedure for inducing estrus and ovulation during the anestrous period may have merit in intense sheep production systems in which photoperiod can be regulated by confining ewes to indoor housing.

Swine. Ovulation can be induced in the prepubertal gilt by giving a single injection of 250 to 1000 IU of PMSG, followed 48 to 96 hours later by an injection of HCG, 500 IU. Inseminations are performed 24 hours after the HCG injection. The number of ovulations following such treatments increase linearly with the dose of PMSG; generally, with doses of 500 IU or less, the ovulation rate is near that which would be expected at the first natural ovulation, but larger doses result in a superovulatory response. The ovulatory response is variable and difficult to predict with any degree of precision.

Ovulation can also be successfully induced in gilts by giving a single injection of a combination of PMSG and HCG, presumably because the HCG promotes greater estrogen secretion from developing follicles than do higher doses of PMSG given alone (Baker and Downey, 1975). As little as 200 IU of each gonadotropin given in combination give a response similar to that when 500 IU of each gonadotropin are given as separate injections 48 to 96 hours apart. When the PMSG/HCG combination is given, ovulation occurs about 110 to 120 hours later, compared to about 40 hours after HCG when the two gonadotropins are given separately. Thus, inseminations after the combination treatment should be given about 96 hours after the treatment, compared to 24 hours after HCG when the double treatment is used.

GnRH or its potent analogues can be used to induce ovulation in gilts when the GnRH is given 60 to 72 hours after an injection of PMSG. However, administering GnRH concurrently with PMSG does not result in as good an ovulatory response as does the PMSG/HCG combination.

Even though ovulation can be induced easily in prepubertal gilts and the percentage of ova that are fertilized is high, pregnancy rates at 25 days after breeding are disappointingly low (Shaw et al, 1971). Reduced pregnancy rates in gilts result from the fact that the prepubertal gilt's normal luteotropic mechanism is not sufficiently developed to provide sustained support for the newly formed corpora lutea. Pregnancies can be maintained by administering HCG at regular intervals to provide the needed luteotropic support or by feeding a progestogen to provide the needed progestational activity; however, neither of these approaches is practical. If gilts are near the age at which puberty normally occurs when ovulation induction regimes are given, then a higher percentage of the females will maintain pregnancies to term (Schilling and Cerne, 1972).

Mares. The mare is normally anestrous during that portion of a year when daylength is shortest, and estrous cycles begin on a regular basis as the photoperiod increases. Estrus and ovulation can be induced during seasonal anestrus in mares by extending the daily photoperiod with artificial light (Oxender et al, 1977). Mares are confined to small areas such as indoor stalls or small paddocks where photoperiod can be easily regulated.

Ovulation has also been induced in seasonally anestrous mares by twice daily injections of equine pituitary extracts for 14 days. However, this method appears to be impractical on a commercial basis. More recent work using a combination of extended photoperiod and twice daily injections of GnRH for eight days has shown more promise.

Steroid Treatments

In sheep, cattle and goats, gonadotropin treatments do not normally stimulate behavioral estrus even though follicle growth and ovulation occurs. If females are to be mated naturally following an ovulation induction procedure, they must exhibit estrus. Apparently cattle, sheep and goats require a period of exposure to elevated progestogen levels to prime higher brain centers for behavioral estrus to be manifested prior to ovulation. Thus, an ovulation induction regime in these species should include a period of progestogen treatment if mating is to occur. In swine, estrus is normally manifested following gonadotropin treatment even though the female has not been primed previously by progestogen treatment.

Gonadal steroids or their synthetic counterparts suppress the release of gonadotropins from the anterior pituitary (Fig. 27–1). Following withdrawal of this suppression, an endogenous surge of gonadotropins ensues. Good results in inducing estrus and ovulation have been obtained when progestogen is given for a few days followed by a single dose of estrogen. Estrogen promotes release of FSH and LH that have accumulated in the anterior pituitary during the period of progestogen treatment. Similar responses have been obtained when progestogen treatment is followed by an injection of PMSG. Presumably PMSG stimulates estrogen synthesis in developing follicles, and increasing levels of circulating estrogens stimulate endogenous FSH and LH

release. Unlike the situation with gonadotropin treatments, ovulation rates following steroid treatments are usually normal unless gonadotropins are included in the treatment regime.

Cattle. Progesterone treatment alone does not always induce ovulation. Gonzalez-Padilla et al (1975b) administered 20 mg progesterone, followed 48 hours later by 2 mg estradiol benzoate, in pubertal heifers; they observed an ovulatory surge of LH. Ovulation occurred in these heifers following the LH surge.

Progesterone treatment for 7 to 12 days in combination with gonadotropin or estrogen is probably the most successful method that has been employed to induce estrus and ovulation in acyclic cattle (Gonzalez-Padilla et al, 1975c; Chupin et al, 1975). Progesterone appears to prime the reproductive system to allow it to respond to gonadotropins. PMSG has been the most widely used gonadotropin in such a regime because a single injection of PMSG at the end of the progesterone treatment is effective. The dose required depends on the season of the year, breed of animal and level of nutrition. About 400 to 800 IU of PMSG appear to give best results; however, there is considerable variation in ovarian response to a single dose.

Sheep. The methods for induction and synchronization of ovulation in sheep are based on the use of progesterone or progestogens to simulate a normal luteal cycle followed by injection of gonadotropin to stimulate follicular growth and ovulation. Progestogens (progesterone; medroxyprogesterone acetate, MAP; fluorogestone acetate, FGA Cronolone) are administered by means of intravaginal pessary (MAP or Cronolone) or subcutaneous implants (progesterone) for 12 to 14 days (Table 27–1).

To induce ovulation in pubertal ewe lambs or anestrous ewes, an injection of PMSG, 500 to 750 IU (Thimonier et al, 1968; Gordon, 1975a,b), is given. The

PMSG initiates follicular growth and ovulation, and such treatments will advance the breeding season in ewes by four to eight weeks (Gordon, 1975a,b). In lactating ewes during the anestrous season, ovulation occurs over a greater spread of time, necessitating two fixed-time inseminations at 48 and 60 hours after treatment when AI is being used. This treatment is used commercially in anestrous ewes in Europe.

Goats. During the nonbreeding season, estrus and ovulation can be induced in goats by using intravaginal sponges containing 45 mg FGA for 18 to 21 days followed by an injection of PMSG two days before removal of the sponge. The dose of PMSG varies from 400 to 800 IU, depending on age, season and parity (Corteel, 1975). Estrus occurs in the majority of does beginning 12 hours after sponge removal, and kidding rate varies from 40% to 80%. Fertility following such induction procedures is higher during the normal breeding season than during the anestrous period, but this can be partially overcome by increasing the number of sperm inseminated (Corteel, 1975).

Altered Suckling

Treatment regimes utilizing gonadotropins or steroids will stimulate follicle growth and ovulation during the prepubertal period, during seasonal anestrus and during lactational anestrus. However, lactational anestrus occurs because frequent nursing inhibits normal gonadotropin release. In both swine and cattle, ovulation occurs sooner after parturition when the amount of time that the young are allowed to nurse each day is restricted. If the young are weaned early, estrus occurs soon after weaning. Although this procedure does not allow one to predict with great accuracy when an individual female will ovulate, it is a useful procedure for shortening the duration of lactational anestrus.

Cattle. The longer postpartum interval in beef cattle compared to dairy cattle is due in part to the inhibitory effects of suckling on reproductive activity. Suckling can depress LH and the frequency of LH episodic releases (Carruthers et al, 1977). A new approach to ovulation induction in beef cows has been the use of a nine-day progestogen treatment combined with removal of the calf for 48 hours. Contrary to the expected depression on fertility due to this stress, it appears that calf removal can increase the number of cows showing estrus and consequently the pregnancy rate (Mares et al, 1977). Apparently the LH level in the serum of cows increases faster when calves have been removed (Smith et al, 1977), but whether this is the mechanism of action remains to be elucidated.

Sows. The responses reported following attempts to induce ovulation during lactational anestrus in swine are variable and in some cases contradictory. Ovulation has been induced during lactational anestrus in sows by administering PMSG, 1500 IU, followed 96 hours later by HCG. Sows are inseminated about 24 and 36 hours after HCG without regard to estrous detection, and conception rates have been reported to be 60% to 80% (Kinney et al, 1977). If PMSG is given alone, the percentage of sows that show estrus during early lactation is low but increases as lactation progresses (Martinat-Botte, 1975). Reducing the number of pigs allowed to nurse enhances the response to PMSG treatments.

In sows that are anestrus after weaning, a single injection of a combination of PMSG/HCG causes estrus in over 80% of the treated animals within 10 days, and over 90% of those in estrus conceive (Schilling and Cerne, 1972).

Sows normally begin estrus four to five days after weaning, and this interval can be shortened by one to two days by administering PMSG at weaning. However, this offers no real advantage unless anestrus persists for a longer-than-normal period of time, which frequently happens when sows are housed in total confinement.

SYNCHRONIZATION OF OVULATION

The time of estrus and ovulation in cyclic domestic animals is controlled mainly by the secretion of progresterone from the corpus luteum, presumably through a negative feedback effect on gonadotropin secretion (Fig. 27–1). Thus, the regulation of the cycle in domestic animals really means controlling the lifespan of the corpus luteum. The lifespan of the corpus luteum can be regulated by several procedures (Table 27–2).

Long-term Progestogen Administration

The administration of progesterone for the length of a cycle will control time of ovulation in cows, ewes, goats and mares,

but in pigs there is an increased incidence of cystic ovaries. Fertility in cattle is reduced in matings at this controlled estrus. In the ewe and goat fertility is normal provided sufficient numbers of spermatozoa are used. Fertility following such treatments in the mare is not known.

Cattle. The classic method for controlling the time of ovulation has been to administer progesterone or a synthetic progestogen to animals for the length of an estrous cycle (Fig. 27–4). Progesterone has little effect on the lifespan of the corpus luteum unless treatment begins within four days after estrus. Since the stage of the cycle when progesterone treatment begins is seldom known, it is necessary to administer the treatment for approximately the length of the luteal phase of a cycle. This ensures that all animals are in the follicular stage of the cycle at the end of treatment, and the majority of animals should show estrus

Table 27–2. Some Practical Methods for Synchronizing Estrus in Farm Animals

Species	Method	Treatment Regime	Response
Cattle	Long-term progestogen	Progestogen (oral or implant) for 14 to 21 days	Synchrony of estrus is good but fertility is low
	Prostaglandin	One luteolytic dose of PGF$_{2\alpha}$ or its analogue to animals between days 5 and 18 of an estrous cycle	Majority in estrus 2 to 5 days after treatment; fertility is good
		Two luteolytic doses of PGF$_{2\alpha}$ or its analogue given 11 to 12 days apart to a group of animals	Majority in estrus 2 to 5 days after second treatment; fertility is acceptable
	Estrogen plus short-term progestogen	5 mg estradiol, day 1; progestogen (implant) for 9 to 12 days beginning on day 1	Majority in estrus 2 to 4 days after implant removal; fertility is good
	Progestogen plus prostaglandin	Progestogen (implant or oral) for 5 to 7 days; luteolytic dose of PGF$_{2\alpha}$ or its analogue given on the last day of progestogen treatment	Majority in estrus 2 to 5 days later; fertility is good
Sheep	Progestogen + PMSG	Progestogen pessary for 12 to 14 days; PMSG, 500–750 IU given at time of pessary removal	Estrus is synchronized in cyclic and acyclic females; fertility is good provided rams are managed properly
Goat	Progestogen + PMSG	Progestogen for 18 to 21 days; PMSG, 400–800 IU, given 1 to 2 days before end of progestogen treatment	Synchronization of estrus and fertility is good
Swine	Progestogen	Progestogen (oral) for 18 days	Synchronization of estrus and fertility is good provided dose of progestogen is high enough to prevent ovarian cyst formation
Horse	Prostaglandin	One luteolytic dose of PGF$_{2\alpha}$ or its analogue to mares in diestrus	Estrus begins 2 to 5 days later and fertility is good

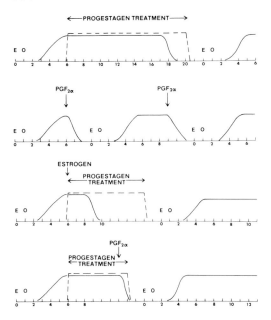

Fig. 27–4. Schematic representation of methods for synchronizing estrus in cattle. The abscissa represents days of an estrous cycle. The solid line represents corpus luteum function and the broken line represents level of progestogen administered. E, Estrus; O, ovulation.

within two to five days after the end of treatment. This method has been widely tested in cows, but while synchronization of estrus is good, fertility to natural or artificial insemination is low (Hansel, 1967; Jochle, 1972). The poor fertility may be due to imbalance in hormones at the synchronized estrus (Hansel et al, 1975), delayed cleavage of embryos (Wishart and Young, 1974), or to other unknown factors.

Ewes. To synchronize estrus, progesterone or progestogens can be administered by implant or intravaginal pessary for 12 to 14 days. Ewes will show estrus 36 to 60 hours after treatment. Quinlivan and Robinson (1969) have shown that sperm transport is reduced in such ewes. To overcome this, attention must be given to mating management of rams (Jennings and Crowley, 1972; Gordon, 1975a). This can be by either hand-mating each ewe once or twice to highly fertile rams (Jen-

nings and Crowley, 1972), or by delaying the introduction of rams to the flock until 48 hours after pessary withdrawal (Gordon, 1975a). A ratio of one ram per 10 ewes is preferred and large numbers of ewes should not be run in a group during the mating period. Attention to mating management will overcome the depressed transport of sperm that occurs in ewes synchronized in this manner (Quinlivan and Robinson, 1969). If AI is to be used, 250 to 500 × 10⁶ motile sperm must be inseminated, between 44 and 58 hours after pessary removal (Colas, 1975).

Goats. In cyclic females, progestogen-impregnated vaginal sponges are inserted for 18 to 21 days, instead of 12 to 14 days as in sheep, and an injection of PMSG, 400 to 600 IU, is given one to two days before sponge withdrawal. Fertility is normal to artificial or natural mating at this estrus (Corteel, 1975).

Sows. Long-term progestogen treatment for synchronizing estrus caused a large increase in the occurrence of cystic ovaries in sows (Webel, 1978), indicating poor control of gonadotropin secretion during treatment. A new progestogen administered at proper dose levels may overcome this problem (Webel, 1978).

Mares. Long-term progestogen treatment has not been widely tested for ovulation control. Daily injections of 50 mg progesterone for 18 days were effective for controlling ovulation in pony mares when an injection of HCG was given six days after the last progesterone injection (Holtan et al, 1977).

Use of Luteolytic Agents

Various factors are known to affect the lifespan of the corpus luteum of domestic animals. These include injections of oxytocin during the early luteal phase, estrogen treatments, presence of foreign bodies in uterus, irrigation of uterus with irritants such as iodine solutions or benzyl alcohol, administration of LH antibodies and administration of prostaglandins. Few of these methods, however, have been used to synchronize estrus, with the notable exception of prostaglandins (Hafs et al, 1974).

Cattle. A luteolytic dose of prostaglandin $F_{2\alpha}$ or one of its potent synthetic analogues causes regression of the bovine corpus luteum and a precipitous decline in progesterone leading to follicular growth, estrus and ovulation within two to four days (Fig. 27–5). The endocrine changes following prostaglandin indicate that the fall in progesterone, the subsequent rise in estrogen and the LH surge are similar to that found during normal luteolysis (Louis et al, 1973).

The fact that prostaglandin is effective only in causing the mature corpus luteum to regress means that single injections of prostaglandin will synchronize estrus in only a portion of a randomly cyclic herd. To overcome this fact, various management strategies have been developed to ensure that only animals with a responsive corpus luteum are treated. These are the following:

1. Double injections of prostaglandin $F_{2\alpha}$ or one of its synthetic analogues 11 or 12 days apart (Fig. 27–4). Following the first injection, only about 66% of randomly cyclic animals will respond (Fig. 27–6). However, within 11 or 12 days, the

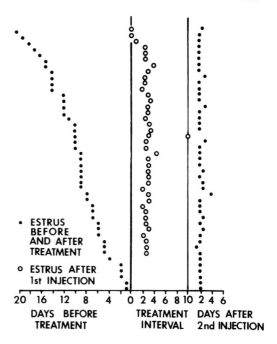

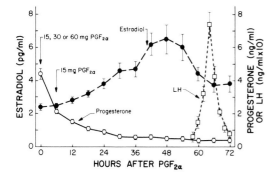

Fig. 27–6. Estrus before and after double injection of the prostaglandin analog Cloprostenol (ICI 80, 996) into heifers. A group of 45 randomly cycling heifers were given 500 μg dosages of Cloprostenol on two occasions 10 days apart. *(Redrawn by permission from Cooper, 1974. Vet. Rec. 95, 200)*

Fig. 27–5. Serum estradiol, progesterone and LH after 15(2×), 30 or 60 mg prostaglandin $F_{2\alpha}$ in cattle. *(From Hafs, VIII Int. Cong. Anim. Reprod. A.I., pp. 17–26. Reprinted by permission from author)*

majority of animals should be in the luteal phase of an estrous cycle and hence respond to the second injection. Animals are given one insemination (80 hours) or two inseminations (72 and 96 hours) after the second $PGF_{2\alpha}$ injection, or they are observed for estrus and bred about 12 hours after estrus begins (MacMillan et al, 1977).

2. Animals observed to be in estrus after first $PGF_{2\alpha}$ injection can be inseminated, and those not bred are given a second injection 11 or 12 days later. Cows in estrus after this injection are inseminated (MacMillen et al, 1977). This is a convenient way to group heat detection of cows into two periods lasting about four days each.

3. Following rectal examination, animals with a palpable corpus luteum can be given a single injection of prostaglandin and bred at a fixed time or following

observation of estrus. The efficacy of this method depends on the accuracy of the method of determining presence of a corpus luteum. The onset of estrus may be spread over a period of time.

4. Following 5 to 7 days of heat detection, all animals not bred are given a single injection of prostaglandin and bred at a fixed time or following observation of estrus. This is a convenient way to overcome the fact that cows between days 0 and 5 of the cycle are not responsive to prostaglandins.

5. Progestogen can be given for five to seven days to prevent estrus in cattle that are in late diestrus or proestrus at the start of treatment. Then $PGF_{2\alpha}$ is given on the last day of progestogen treatment to regress corpora lutea in animals that are in early or mid-diestrus (Fig. 27–4). Inseminations are given at a fixed time after $PGF_{2\alpha}$ treatment or at estrus.

Fertility at the controlled estrus is normal following the use of prostaglandins. It is possible to administer GnRH or estradiol benzoate after prostaglandin treatment to control more precisely the time of ovulation. GnRH will reduce the spread in time of ovulation, but it does not reduce the number of inseminations required for normal fertility.

The injection of 400 to 500 μg of estradiol benzoate to synchronize the LH peak after prostaglandin treatment has yielded conflicting results. Some workers have reported no beneficial effect, while others have found that fertility is improved. Perhaps dose, time of administration and type of animal treated influence the response. In general, no dramatic benefits have accrued from the further use of GnRH or estrogen to synchronize ovulation in cattle.

Ewes. When a mature corpus luteum is present, $PGF_{2\alpha}$ or its synthetic analogues will cause its regression in the ewe. Using an eight-day interval between two successive $PGF_{2\alpha}$ injections is effective for synchronizing estrus, but fertility is low at the induced estrus. However, when a 14-day interval between injections is used, fertility appears normal at the controlled estrus. Of course, $PGF_{2\alpha}$ treatments cannot be used to induce estrus in prepubertal or anestrous ewes, because they do not have a functional corpus luteum. Therefore, it is unlikely that prostaglandin treatment regimes will replace progestogen treatments for synchronizing estrus in ewes, because the progestogens can be used effectively for both cyclic and acyclic individuals and, during early parts of the normal breeding season in sheep, many ewes are likely to be in the late stages of seasonal anestrus and ewe lambs may still be prepubertal.

Sows. The porcine corpus luteum is not responsive to $PGF_{2\alpha}$ until day 12 after estrus. Thus, only a small portion of treated animals respond to a single injection of $PGF_{2\alpha}$. Lifespan of corpora lutea in gilts can be prolonged by administering a luteotropin such as estrogen, and when gilts treated in this manner are given $PGF_{2\alpha}$, their estrous cycles are synchronized (Guthrie, 1975). However, this has not proven to be a practical method for synchronizing estrus, because the estrogen has to be injected daily for several days. Thus, prostaglandin offers little prospect as a practical method to control ovulation in groups of swine.

Mares. Mares are generally treated as individuals and not in groups. Prostaglandin $F_{2\alpha}$ will induce premature luteal regression and a fertile estrus and ovulation. It is useful for shortening prolonged luteal phases, which often occur after the foal estrus in certain mares. Thus, it has a valuable clinical role in mares. Estrus has been synchronized successfully in groups of mares by administering a treatment regime involving two injections of $PGF_{2\alpha}$ and two injections of HCG (Palmer and Jousset, 1975). In this regime, $PGF_{2\alpha}$ is injected on day 0, HCG on day 6, $PGF_{2\alpha}$ on day 14 and HCG again on day 20. The first $PGF_{2\alpha}$ and HCG injections are given to assure that all mares have a functional corpus luteum at the time of the second

$PGF_{2\alpha}$ treatment. The second HCG injection synchronizes the time of ovulation following the second $PGF_{2\alpha}$ treatment, and results obtained following this four-injection regime indicate that about 70% to 80% of treated mares will ovulate in a four-day interval. This procedure may prove useful for regulating time of estrus for mares during a breeding season so that optimal fertility can be obtained and stallions can be used to their fullest extent.

Use of Progesterone Combined with Luteolytic Agent

Cattle. Wiltbank and Kasson (1968) showed that the key to achieving normal fertility in cattle when using progesterone or progestogen is to limit the period of administration to between 9 and 12 days. Others have confirmed this observation (Mauleon, 1974; Roche, 1974; Wishart and Young, 1974; Sreenan, 1975). This necessitates the use of a luteolytic agent in combination with the progesterone, because exogenously administered progesterone has little effect on the lifespan of the mature corpus luteum. Estrogen has been the most widely used luteolytic agent, but estrogens are at best only partially luteolytic in cattle and in fact can be luteotropic at certain stages of the cycle (Wiltbank, 1966; Lemon, 1975).

For effective synchronization of estrus, only animals between day 0 and days 8 or 9 of an estrous cycle need have the lifespan of the corpus luteum reduced, as the remaining animals will all be in the follicular stage of a cycle after a 9- or 12-day progesterone treatment. Estrogens are not totally effective in causing luteolysis when given during early stages of a cycle, and generally only about 90% of cows will be well-synchronized. The remainder will show delayed onset of estrus after treatment.

Fertility is normal at the synchronized estrus, and provided that a large proportion of animals are in heat within 24 and 36 hours of each other, insemination at a prearranged time is effective. The following treatments can be used:

1. Nine-day ear implant of synthetic progestogen SC-21009 (Norgestomet). Animals are injected with 5 mg of estradiol valerate and 3 mg of Norgestomet at the time of implant insertion to cause regression of the corpus luteum. All animals are inseminated 48 and 72 hours after removal of implants (Wishart and Young, 1974).

2. Ten-day treatment with intravaginal pessary containing 3 gm progesterone. Retention rate of pessaries is 98% in heifers and 90% in cows. Animals are injected with 5 mg of estradiol valerate and 250 mg of progesterone at the start of treatment, and inseminated 48 and 72 hours after pessary removal (Sreenan, 1975).

3. Twelve-day treatment with progesterone-releasing intravaginal devices containing 6.6% progesterone. Retention rate of coils with a diameter of 4.6 cm is 97% in dairy or beef animals. Cows are bred once at 56 hours after treatment, but about 10% of dairy cows show estrus four to seven days after removal of coils; thus, these animals should be watched for about four days after the fixed-time AI and rebred at estrus (Roche, 1976).

Fertility is normal following these various treatments, provided normal cyclic animals are treated and semen of good quality is used. Dairy cows should be at least five-weeks postpartum and beef cows, six-weeks postpartum before being treated with a short-term progesterone regime. In the case of prostaglandin, dairy cows should be six-weeks and beef cows, eight-weeks postpartum before initiation of treatment.

Management Factors Affecting Responses in Cattle

The variation in fertility from farm to farm following estrous synchronization procedures is large. Estrous synchronization works well when nutrition and management are good, but it is not a substitute for inadequate management. Estrous synchronization will not affect the calving pattern dramatically in a herd following its use in any one year, although the average interval from calving to conception may be reduced by five to ten days. The calving pattern prior to the use of estrous synchronization procedures dictates the number of cows that can be treated at any given time during a breeding season.

The need for estrus detection is not completely eliminated when estrous synchronization is used. In dairy cows in particular, a certain portion of animals following progesterone or prostaglandin treatments will be in estrus one to six days after fixed-time AI. These cows have little chance of conceiving unless they are rebred. In addition, synchronized cows that do not conceive at fixed-time AI generally show a repeat estrus 16 to 26 days later. Thus, if AI is to be used for the repeat services, cattle have to be observed to determine when estrus occurs. The major advantage of estrous synchronization in cattle is to allow use of AI on a large portion of the breeding herd at one time. This should result in more widespread use of genetically superior sires. The economic effects of using estrous synchronization procedures have not been fully evaluated to date.

REFERENCES

Baker, R.D. and Downey, B.R. (1975). Induction of estrus, ovulation and fertility in prepuberal gilts. Ann. Biol. Anim. Biochim. Biophys. *15*, 375.

Britt, J.H., Kittok, R.J. and Harrison, D.S. (1974). Ovulation, estrus and endocrine response after GnRH in early postpartum cows. J. Anim. Sci. *39*, 915.

Carruthers, T.D., Kosugiyama, M. and Hafs, H.D. (1977). Effects of suckling on interval to first postpartum ovulation and on serum luteinizing hormone and prolactin in Holsteins. In Abstracts, 69th Annual Meeting, American Society of Animal Science, p. 142.

Chupin, D., Pelot, J. and Thimonier, J. (1975). The control of reproduction in the nursing cow with a progestogen short-term treatment. Ann. Biol. Anim. Biochim. Biophys. *15*, 263.

Colas, G. (1975). The use of progestogen SC9880 as an aid for artificial insemination in ewes. Ann. Biol. Anim. Biochim. Biophys. *15*, 317.

Corteel, J.M. (1975). The use of progestogens to control the oestrous cycle of the dairy goat. Ann. Biol. Anim. Biochim. Biophys. *15*, 353.

Gonzalez-Padilla, E., Wiltbank, J.N. and Niswender, G.D. (1975a). Puberty in beef heifers. I. The interrelationship between pituitary, hypothalamic and ovarian hormones. J. Anim. Sci. *40*, 1091.

Gonzalez-Padilla, E., Niswender, G.D. and Wiltbank, J.N. (1975b). Puberty in beef heifers. II. Effect of injections of progesterone and estradiol-17β on serum LH, FSH and ovarian activity. J. Anim. Sci. *40*, 1105.

Gonzalez-Padilla, E., Ruiz, R., LeFever, D., Denham, A. and Wiltbank, J.N. (1975c). Puberty in beef heifers. III. Induction of fertile estrus. J. Anim. Sci. *40*, 1110.

Gordon, I. (1975a). The use of progestogens in sheep bred by natural and artificial insemination. Ann. Biol. Anim. Biochim. Biophys. *15*, 303.

Gordon, I. (1975b). Hormonal control of reproduction in sheep. Proc. Br. Soc. Anim. Prod. *4*, 79.

Guthrie, H.D. (1975). Estrous synchronization and fertility in gilts treated with estradiol-benzoate and prostaglandin $F_{2\alpha}$. Theriogenology *4*, 69.

Hafs, H.D., Louis, T.M., Noden, P.A. and Oxender, W.D. (1974). Control of the estrous cycle with prostaglandin $F_{2\alpha}$ in cattle and horses. J. Anim. Sci. *38* (Suppl. 1), 10.

Hansel, W. (1967). Control of the ovarian cycle in cattle. In Reproduction in the Female Mammal. G.E. Lamming and E.C. Amoroso (eds), London, Butterworths.

Hansel, W., Schechter, R.J., Malven, P.V., Simmons, K.R., Black, D.L., Hackett, A.J. and Saatman, R.R. (1975). Plasma hormone levels in 6-methyl-17-acetoxyprogesterone and estradiol benzoate treated heifers. J. Anim. Sci. *40*, 671.

Holtan, D.W., Douglas, R.H. and Ginther, O.J. (1977). Estrus, ovulation and conception following synchronization with progesterone, prostaglandin $F_{2\alpha}$ and human chorionic gonadotropin in pony mares. J. Anim. Sci. *44*, 431.

Humphrey, W.D., Koritnik, D.R., Kaltenbach, C.C., Dunn, T.G. and Niswender, G.D. (1976). Progesterone and LH in postpartum suckled beef cows. J. Anim. Sci. *43*, 290 (Abstr.).

Jennings, J.J. and Crowley, J.P. (1972). The influence of mating management on fertility in ewes following progesterone-PMS treatment. Vet. Rec. *90*, 495.

Jochle, W. (1972). Pharmacological aspects of the control of the cycle in domestic animals. Proc. VII Intern. Cong. Anim. Reprod. AI *1*, 97.

Kesler, D.J., Garverick, H.A., Youngquist, R.S., Elmore, R.G. and Bierschwal, C.J. (1977). Effect of days postpartum and endogenous reproductive hormones on GnRH-induced LH release in dairy cows. J. Anim. Sci. *46*, 797.

Kinney, T.J., Hausler, C.L., Hodson, H.H., Jr. and Snyder, R.A. (1977). Induced follicular development, ovulation and conception in lactating sows. *In* Abstracts, 69th Annual Meeting, American Society of Animal Science, p. 178.

Legan, S.J., Karsch, F.J., and Foster, D.L. (1977). The endocrine control of seasonal reproductive function in the ewe: A marked change in response to the negative feedback action of estradiol on luteinizing hormone secretion. Endocrinology *101*, 818.

Lemon, M. (1975). The effect of oestrogens alone or in association with progestogens on the formation and regression of the corpus luteum in the cyclic cow. Ann. Biol. Anim. Biochim. Biophys. *15*, 243.

Louis, T.M., Hafs, H.D. and Seguin, B.E. (1973). Progesterone, LH, estrus and ovulation after prostaglandin F$_{2\alpha}$ in heifers. Proc. Soc. Exp. Biol. Med. *143*, 152.

MacMillan, K.L., Curnow, R.J. and Morris, G.R. (1977). Oestrus synchronization with a prostaglandin analogue: I. Systems in lactating dairy cattle. N.Z. Vet. J. *25*, 366.

Mares, S.E., Peterson, L.A., Henderson, E.A. and Davenport, M.E. (1977). Fertility of beef herds inseminated by estrus or by time following Syncro-Mate-B (SMB) treatment. *In* Abstracts, 69th Annual Meeting, American Society of Animal Science, p. 185.

Martinat-Botte, F. (1975). Induction of gestation during lactation in the sow. Ann. Biol. Anim. Biochim. Biophys. *15*, 369.

Mauleon, P. (1974). New trends in the control of reproduction in the bovine. Livestock Prod. Sci. *1*, 117.

Oxender, W.D., Noden, P.A. and Hafs, H.D. (1977). Estrus, ovulation, and serum progesterone, estradiol, and LH concentrations in mares after an increased photoperiod during winter. Am. J. Vet. Res. *38*, 203.

Palmer, E. and Jousset, B. (1975). Synchronization of oestrus and ovulation in the mare with a two PG-HCG sequences treatment. Ann. Biol. Anim. Biochim. Biophys. *15*, 471.

Quinlivan, T.D. and Robinson, T.J. (1969). Numbers of spermatozoa in the genital tract after artificial insemination of progestogen treated ewes. J. Reprod. Fertil. *19*, 73.

Roche, J.F. (1974). Effect of short-term progesterone treatment on oestrous response and fertility in heifers. J. Reprod. Fertil. *40*, 433.

Roche, J.F. (1976). Synchronization of oestrus in cattle. World Rev. Anim. Prod. *12*, 79.

Schilling, E. and Cerne, F. (1972). Induction and synchronization of oestrus in prepuberal gilts and anoestrous sows by a PMS/HCG-compound. Vet. Rec. *91*, 471.

Shaw, G.A., McDonald, B.E. and Baker, R.D. (1971). Fetal mortality in the prepuberal gilt. Can. J. Anim. Sci. *51*, 233.

Smith, M.F., Walters, D.L., Harms, P.G. and Wiltbank, J.N. (1977). LH levels after steroids and/or 48 hr calf removal in anestrous cows. *In* Abstracts, 69th Annual Meeting, American Society of Animal Science, p. 209.

Sreenan, J.M. (1975). Effect of long- and short-term intravaginal progestogen treatments on synchronization of oestrus and fertility in heifers. J. Reprod. Fertil. *45*, 479.

Stabenfeldt, G.H., Edquist, L.E., Kindahl, H., Gustafsson, B., and Bane, A. (1978). Practical implications of recent physiologic findings for reproductive efficiency in cows, mares, sows, and ewes. J. Am. Vet. Med. Assoc. *172*, 667.

Thimonier, J., Mauleon, P., Cognie, Y. and Ortavant, R. (1968). Declenchement de l'oestrus et obtention precoce de gestations chez des agnelles a l'aide d'eponges vaginales impregnees d'acetate de fluorogestone. Ann. Zootech. *17*, 275.

Webb, R., Lamming, G.E., Haynes, N.B., Hafs, H.D. and Manns, J.G. (1977). Response of cyclic and post-partum suckled cows to injections of synthetic LH-RH. J. Reprod. Fertil. *50*, 203.

Webel, S.K. (1978). Ovulation control in the pig. *In* Control of Ovulation. D.B. Crighton, G.R. Foxcroft, N.B. Haynes and G.E. Lamming (eds), London, Butterworths.

Wiltbank, J.N. (1966). Modification of ovarian activity in the bovine following injection of oestrogen and gonadotrophin. J. Reprod. Fertil. Suppl. *1*, 1.

Wiltbank, J.N. and Kasson, C.W. (1968). Synchronization of estrus in cattle with an oral progestational agent and an injection of an estrogen. J. Anim. Sci. *27*, 113.

Wishart, D.F. and Young, I.M. (1974). Artificial insemination of progestin (SC 21009) treated cattle at predetermined times. Vet. Rec. *95*, 503.

28

Pregnancy Diagnosis

M.R. JAINUDEEN
and
E.S.E. HAFEZ

The main purposes of pregnancy diagnosis are as follows: to enable early identification of the nonpregnant animal so that production time lost as the result of infertility may be reduced by appropriate treatment or culling; to certify pregnancy for sale or insurance purposes and in breeding programs that utilize expensive hormonal techniques; economic management of animal production. Clinical and laboratory methods are available for the diagnosis of pregnancy. The choice of method depends on the species, stage of gestation, cost, accuracy and speed of diagnosis (Table 28–1).

CLINICAL METHODS OF PREGNANCY DIAGNOSIS

Clinical methods include rectal examination, radiography and ultrasonic techniques.

Rectal Examination

Rectal examination is the method of choice for the diagnosis of pregnancy in cows and mares. In this technique, an operator inserts his arm into the rectum of the animal and manually explores the uterus to detect the uterine enlargement occurring during pregnancy, the fetus or fetal membranes (Fig. 28–1, Table 28–2). This technique is accurate, can be performed at an early stage of gestation and the result is known immediately. The ewe and sow are not suitable for such manual exploration of uterine contents for anatomic reasons. However, the middle uterine arteries, particularly in parous sows, are readily accessible for palpation through the rectal wall. Detecting a "fremitus" in one or both uterine arteries is a rapid and simple clinical test for pregnancy in the sow from 28 days onward (Meredith, 1976).

Radiography

Radiography may be employed for the detection of pregnancy in sheep and swine; it is based on the identification of the fetal skeleton on an x-ray plate. Among the disadvantages of this method

560

Table 28–1. Some Commonly Used Methods of Pregnancy in Farm Animals

Species	Diagnostic Method	Sample	Procedure	Time in Gestation When Applicable (Days)	Comments
Cattle	Rectal palpation	—	Manual	45 to term	Accurate and rapid
	Hormonal	Milk	Progesterone assay	21 till end of lactation	A test for nonpregnancy
Sheep	Vaginal biopsy	Vaginal mucosa	Histologic	40 to term	Time-consuming
	Ultrasonic	—	—	Beyond 100 days	Too late in gestation
Swine	Vaginal biopsy	Vaginal mucosa	Histologic	21 to term	Time-consuming
	Ultrasonic	—	—	30 to 90	Accurate and yields immediate results
	Rectal palpation	Middle uterine arteries	Manual	28 to term	Limited to parous sows
Horse	Rectal palpation	—	Manual	35 to term	Accurate and rapid
	Ultrasonic	—	—	56 to term	Provides proof of fetal viability
	Hormonal	Serum	Hemagglutination inhibition or bioassay for PMSG	50 to 100	Individual variations in level of PMSG can affect result
	Hormonal	Urine	Chemical test for estrogen	150 to 250	Easy to perform and accurate during late gestation

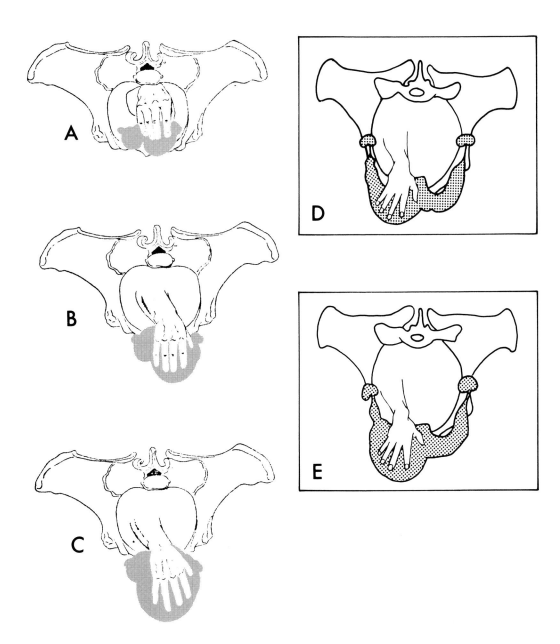

Fig. 28–1. *Left,* Pregnancy diagnosis in the cow by rectal palpation at Day 70 *(A)*, Day 90 *(B)*, Day 110 *(C)*. *(Redrawn from Arthur, 1964, In: Wright's Veterinary Obstetrics, London, Bailliere, Tindall & Cox) Right,* Pregnancy diagnosis in the mare at Day 40 *(D)* and Day 60 *(E)*. Note the size and position of the pregnant uterine horn.

Table 28-2. Diagnosis of Pregnancy in the Cow and Mare by Rectal Palpation

Animal	Month of Gestation	Major Findings
Cow	First	Quiescent uterus and a fully developed corpus luteum in one ovary.
	Second*	Enlargement and dorsal bulging of pregnant horn due to fetal fluids; on application of digital pressure, gives a resilient sensation. "Fetal membrane slip" is present (slipping of the chorioallantois when pregnant horn is gently lifted with fingers); amnionic vesicle present anterior to the intercornual ligament.
	Third*	Descent of uterus commences; fetal membrane slip is present and fetus is palpable.
	Fourth to Seventh	Uterus on abdominal floor, fetus difficult is palpate; cotyledons, 2 to 5 cm in diameter, are palpated as circumscribed areas in uterine wall; middle uterine arteries hypertrophy and pulse changes to a distinct fremitus.
	Seventh to term	Cotyledons, fremitus and fetal parts are palpable.
Mare	First	Contracted and firm cervix; turgid uterine horns.
	Second*	Chorioallantoic sac (size of an orange) bulges ventrally in lower third of uterine horn; uterine horns are turgid.
	Third	Chorioallantoic sac grows rapidly and descends into uterine body, changes from spherical to oval, and tenseness is gradually lost. Uterus begins to descend.
	Fourth	Dorsal surface of the uterus is felt as a distended dome between the stretched broad ligaments. Fetus and/or fetal parts palpable.
	Fifth to Seventh	Uterus lies deep in abdominal cavity. Usually possible to palpate fetus, except in large pluriparous mares.
	Seventh to term	Fetus easier to palpate. Uterus begins to ascend.

*Period when pregnancy can be accurately diagnosed.

are that it can be applied only during the last third of gestation, it is costly, it necessitates restraint, and it poses a radiation hazard to the health of the operator.

Ultrasonic Techniques

Sound waves striking a moving object are reflected to the transmitting source at a slightly altered frequency. This Doppler phenomenon can be used in the sow and ewe to detect fetal movements, pulse, and uterine arterial circulation. The ultrasonic detector consists of an amplifier and a transducer, which is applied as a probe to the animal's abdomen or inserted into the rectum in order to locate fetal movement, fetal heart sounds and the swishing of the umbilical vessels. This technique is accurate in ewes at least 100-days pregnant (Richardson, 1972) and in sows beyond 30 days of gestation (Pierce et al, 1976).

The Doppler ultrasound and a special rectal probe may be used for the detection of the fetal circulation in the mare and the cow (Mitchell, 1973). In the mare, a positive diagnosis can be made from 56 days of gestation, and it is the only diagnostic test that provides proof that the fetus is alive.

Another ultrasonic technique for pregnancy diagnosis in swine is based on the detection of the difference in acoustic impedance between the contents in the gravid uterus and the abdominal viscera and ingesta (Lindahl et al, 1975). The equipment consists of an amplitude-depth ultrasonic analyzer used in conjunction with a 2-MHz transducer. The tip of the transducer is applied to the right lower flank of the standing sow just lateral to the nipple line about 5 cm posterior to the umbilicus (Fig. 28-2). When the ultrasound contacts tissues of varying acoustic impedance, some of the energy is reflected to the transducer and converted into a signal display (echoes) on a cathode ray oscilloscope. A band of echoes is obtained from a depth of 15 to 20 cm in pregnant animals, as compared to only about 5 cm in nonpregnant animals. A 95% accuracy in pregnancy diagnosis can

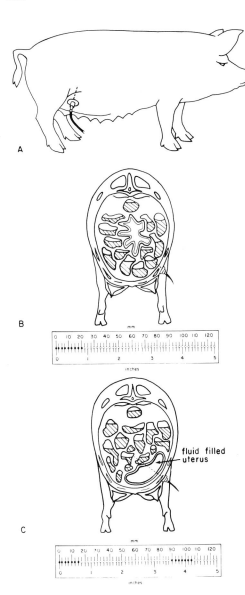

be achieved in sows and gilts between 30 and 90 days of gestation. This ultrasonic technique requires little restraint of the animal, yields immediate results and can be adapted for field use.

LABORATORY METHODS

Laboratory methods include vaginal biopsy and hormonal assays.

Vaginal Biopsy

Histologic changes that occur in the vaginal epithelium during pregnancy form the basis of a pregnancy test in the ewe (Richardson, 1972) and the sow (Walker, 1972). In the ewe, the diagnosis depends on the reduction of cell layers in the vaginal epithelium from about 12 in nonpregnancy to 5 in pregnancy, and on the uniformity of the epithelial cell type (cuboidal) in pregnancy. The technique is 95% accurate for pregnancies beyond 40 days. In the sow, the vaginal epithelium during estrus consists of about 16 cells in height; it is reduced to five or four during diestrus (Fig. 28–3). If pregnancy ensues, the vaginal epithelium is reduced to two or three layers of cells arranged parallel to the basement membrane. The test is reliable in 95% of sows not less than 21 days pregnant.

Numerous vaginal biopsy instruments have been designed, but the preferred one operates on the principle of a rotary-action coaxial rod and has a tube with cutting edges around a slot at one extremity. It should be at least 24 cm long for the sow. Biopsy samples are subjected to routine histologic procedures, such as formalin fixation, paraffin-wax embedding and hematoxylin and eosin staining. Since the method involves sampling, processing and microscopic examination, which are time-consuming and costly, the vaginal biopsy technique has limited practical applicability.

Fig. 28–2. **Pregnancy diagnosis in swine by ultrasonic amplitude-depth analysis.** *A,* Placement of transducer on the lower flank of the standing sow about 5 cm posterior to the umbilicus and just lateral to the nipple line. *B,* Display pattern of a sow or gilt that is not bred or is less than 30 days pregnant. *C,* Display pattern of a sow or gilt between 30 and 90 days of pregnancy. The first set of lights (0–15) represents the different layers of the abdominal wall, the gap in light pattern is due to fetal fluids, and the second set of lights (90–105) are the echoes from the distal wall of the uterus and adjacent intestines. Note that the gap in light pattern and the second set of lights are absent in a nonpregnant animal *(Scanoprobe, Ithaco. Inc., Ithaca, N.Y.)*

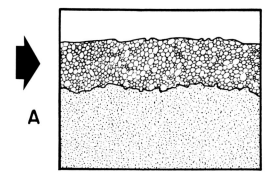

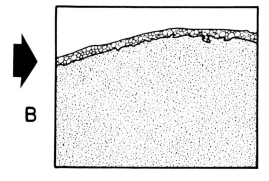

Fig. 28–3. Vertical section through the vaginal epithelium of the sow (250×). *A,* During diestrus in a nonpregnant sow the vaginal epithelium is several layers thick. *B,* At day 30 of pregnancy, the depth of the epithelium is only two to three cells.

Hormonal Assays

Biologic as well as chemical methods are employed for the detection of pregnancy-dependent hormones in body fluids. Since high levels of PMSG and estrogens appear in the body fluids during pregnancy (Fig. 28–4), laboratory diagnosis of pregnancy was limited mainly to the mare. With the development of highly sensitive radioimmunoassay and competitive protein-binding techniques for the detection of plasma progesterone, pregnancy can now be diagnosed in farm animals at a much earlier stage with an accuracy of over 90% more than was possible with clinical methods.

Progesterone. Whereas the plasma progesterone content declines as the corpus luteum regresses in the nonpregnant animal, in the pregnant animal it persists or rises. This difference in progesterone levels is the basis for an early pregnancy test in the cow, ewe and sow (Robertson and Sarda, 1971).

A close relationship exists between the progesterone levels in plasma and in whole milk of lactating dairy cattle (Pope et al, 1976). Progesterone levels in whole

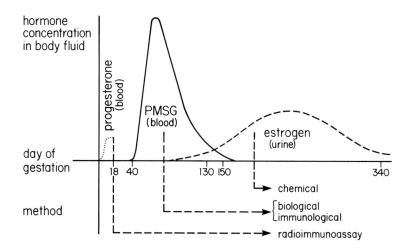

Fig. 28–4. The laboratory methods of pregnancy diagnosis at various stages of gestation in the mare. *(Redrawn from Walker, 1977. Vet. Rec. 100, 396)*

milk are approximately twice those in plasma and reflect the secretory activity of the corpus luteum in the normal cycle and early pregnancy. Thus the possibility of using milk progesterone levels as an aid to pregnancy diagnosis in lactating dairy cattle has been studied extensively (Bishop et al, 1976; Dobson and Fitzpatrick, 1976; Heap et al, 1976; Pennington et al, 1976; Pope et al, 1976).

A milk sample is obtained at 21 days after service, a time at which progesterone levels are low (<2 ng/ml) if the animal is not pregnant. The main difficulty with this test, however, is deciding the exact progesterone level upon which to base a positive diagnosis of pregnancy (Fig. 28–5). It varies among laboratories, the time of sampling after insemination, the fat percentage of the milk, the preservation of the milk prior to analysis and the assay technique. For these reasons, the milk progesterone test of pregnancy has been developed to determine relative rather than absolute concentrations of progesterone for the correct identification of pregnancy in pregnant and nonpregnant animals. The success rate of this test from a single specimen of whole milk sampled at 21, 24, or 42 days after insemi-

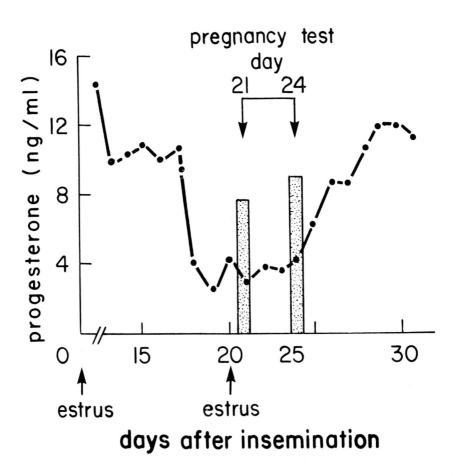

Fig. 28–5. The progesterone-in-milk as a method of pregnancy diagnosis in the lactating cow. The milk progesterone concentrations in a nonpregnant cow before and after estrus (●—●). The height of vertical bars represents the upper limit of the progesterone concentration for a diagnosis of nonpregnancy at 21 and 24 days after inseminations. *(Adapted from the data of Dobson and Fitzpatrick, 1976. Br. Vet. J. 132, 538; Heap et al, 1976. Br. Vet. J. 132, 445)*

nation ranges from 77.5% to 85.8% for pregnant cows and from 85.7% to 100% for nonpregnant cows (Heap et al, 1976).

Several factors affect the accuracy of pregnancy diagnosis by the milk progesterone test performed at 21 to 24 days as compared with rectal diagnosis at 45 to 60 days after breeding (eg, early embryonic mortality, abnormal estrous cycle lengths, and insemination not coinciding with ovulatory estrus due to inaccurate detection of estrus). Thus the application of the milk progesterone test, as a method of pregnancy diagnosis, is reliable only in determining nonpregnancy.

Gonadotropin. PMSG appears in the blood of mares as early as 40 days following conception, and its detection has been regarded as evidence of pregnancy. Peak levels are found between 50 and 120 days;

thereafter, there is a gradual decline. Both immunologic and biologic methods have been employed for its detection.

Immunologic diagnosis of pregnancy in mares is based on the principle that PMSG, when present in the blood specimen to be tested, prevents agglutination of sensitized sheep red cells by anti-PMSG (hemagglutination-inhibition test, or HI). Agglutination of red cells means a negative result (ie, no pregnancy), and failure of agglutination, a positive result (Fig. 28–6). This immunologic test is most accurate between 50 and 100 days of gestation. Errors are due mostly to individual variations in PMSG levels. An immunologic gel diffusion test, also based on the presence of PMSG in the tested blood, is considered unsuitable for routine pregnancy diagnosis (Walker, 1977).

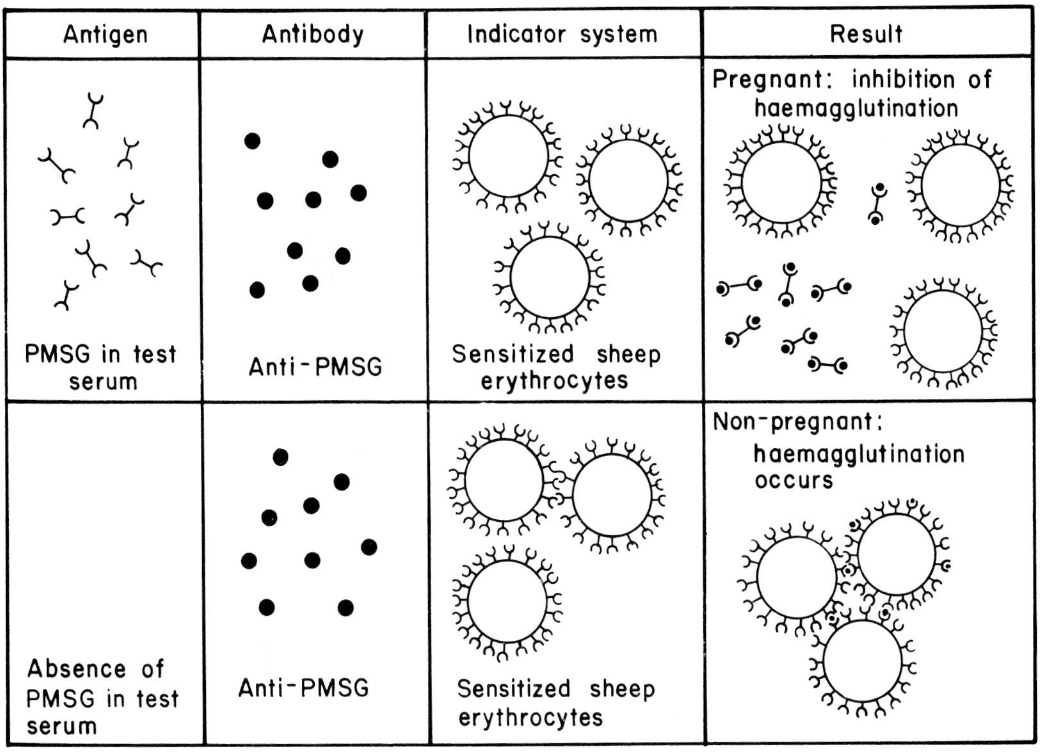

Antigen	Antibody	Indicator system	Result
PMSG in test serum	Anti-PMSG	Sensitized sheep erythrocytes	Pregnant: inhibition of haemagglutination
Absence of PMSG in test serum	Anti-PMSG	Sensitized sheep erythrocytes	Non-pregnant: haemagglutination occurs

Fig. 28–6. The hemagglutination inhibition test for the detection of pregnant mare serum gonadotropin (PMSG). Note that inhibition of hemagglutination is positive for pregnancy and occurrence of hemagglutination is negative. *(Redrawn from Jeffcott et al, 1969. Vet. Rec. 84, 552)*

The bioassay using immature female rats *(Ascheim-Zondek test)* is based on the FSH activity of PMSG. Immature rats are injected with whole blood or serum and killed 72 hours later. Hemorrhagic ovulation spots and an edematous uterus are evidence that pregnancy exists; this test is most accurate between 50 and 80 days of gestation. A more rapid and accurate bioassay is the 24-hour utero-ovarian response in the immature female mouse (Walker, 1977).

Estrogens. The detection of high levels of estrogen in the urine of pregnant mares from approximately day 120 is a laboratory test for pregnancy. The Cuboni chemical test relies on the development of green fluorescence produced when an extract of hydrolyzed urine is heated with sulfuric acid. A simple method of performing the test using readily available laboratory apparatus (Cox, 1971) consists of three basic steps: (1) the estrogens, which are normally present in the conjugated form, are set free by acid hydrolysis or sulfuric acid added to the urine; (2) the estrogens are extracted from the hydrolyzed urine by the addition of benzene; and (3) green fluorescence develops with the addition of concentrated sulfuric acid.

The urine test for pregnancy in the mare is reliable after day 150; it has limited applications because it becomes reliable relatively late in gestation. The presence of high concentrations of urinary estrogens in the mare is not diagnostic of pregnancy, since high levels are also present in other conditions such as ovarian cysts and nymphomania.

REFERENCES

Bishop, C.A., Bond, C.P. and Roberts, C. (1976). Early diagnosis of nonpregnancy in cattle. The first eighteen months of a commercial service. Br. Vet. J. *132*, 529.

Cox, J.E. (1971). Urine tests for pregnancy in the mare. Vet. Rec. *89*, 606.

Dobson, H. and Fitzpatrick, R.J. (1976). Clinical application of the progesterone-in-milk test. Br. Vet. J. *132*, 538.

Heap, R.B., Holdsworth, R.J., Gadsby, J.E., Laing, J.A. and Walters, D.E. (1976). Pregnancy diagnosis in the cow from milk progesterone. Br. Vet. J. *132*, 445.

Lindahl, I.L., Totsch, J.P., Martin, P.A. and Dziuk, P.J. (1975). Early diagnosis of pregnancy in sows by ultrasonic amplitude-depth analysis. J. Anim. Sci. *40*, 220.

Meredith, M.J. (1976). Pregnancy diagnosis in the sow by examination of the uterine arteries. Proc. Int. Pig Vet. Soc., Ames, Iowa, D.5.

Mitchell, D. (1973). Detection of fetal circulation in the mare and cow by the Doppler ultrasound. Vet. Rec. *93*, 365.

Pennington, J.A., Spahr, S.L. and Lodge, J.R. (1976). Factors affecting progesterone in milk for pregnancy diagnosis in dairy cattle. Br. Vet. J. *132*, 487.

Pierce, J.E., Middleton, C.C. and Phillips, J.M. (1976). Early pregnancy diagnosis in swine using Doppler ultrasound. Proc. Int. Pig Vet. Soc., Ames, Iowa, D.3.

Pope, G.S., Majzlik, I., Ball, P.J.H. and Leaver, J.D. (1976). Use of milk progesterone concentrations in plasma and milk in the diagnosis of pregnancy in domestic cattle. Br. Vet. J. *132*, 497.

Richardson, C. (1972). Pregnancy diagnosis in the ewe: a review. Vet. Rec. *90*, 264.

Robertson, H.A. and Sarda, I.R. (1971). A very early pregnancy test for mammals: its application to the cow, ewe and sow. J. Endocrinol. *49*, 407.

Walker, D. (1972). Pregnancy diagnosis in pigs. Vet. Rec. *90*, 139.

Walker, D. (1977). Laboratory methods of equine pregnancy diagnosis. Vet. Rec. *100*, 396.

29

Embryo Transfer

T. SUGIE, G.E. SEIDEL, JR.,
and
E.S.E. HAFEZ

Although the mammalian ovary contains hundreds of thousands of oocytes (Erickson, 1966), the number of progeny a female produces is small. In farm animals, the number of times a female can become pregnant is severely limited by the extended duration of gestation. Furthermore, only one or two ova are usually shed per estrous cycle in nonlitter-bearing species. The number of offspring that a female can produce in her lifetime can be greatly increased by repeatedly allowing her to become temporarily pregnant, recovering the embryos in early pregnancy, and transferring them to the reproductive tracts of other females to complete gestation. The process can be further amplified if the donor superovulates.

The first successful embryo transfer was reported by Heape (1890) with rabbits. The first successful embryo transfers in farm animals were reported several decades ago: for sheep and goats (Warwick and Berry, 1949), pigs (Kvansnickii, 1951), and cattle (Willett et al, 1951). These and similar historic events are summarized in Table 29–1. Recently commercial companies for embryo transfer in farm animals have been established in Australia, Argentina, Canada, New Zealand, the United States and several countries in Europe (Church and Shea, 1977; Seidel and Seidel, 1978).

This chapter considers the methodology of embryo transfer in farm animals, with particular emphasis on superovulation, collection of ova, examination of ova for fertilization and normality, handling and storage of embryos, *in-vitro* fertilization, and techniques and applications of embryo transfer.

SUPEROVULATION

The objective of superovulation is to increase the yield of viable ova.

Methods

Usually subcutaneous or intramuscular injections of pregnant mare's serum gonadotropin (PMSG) or follicle-stimu-

Table 29–1. Summary of Historical Development of Embryo Transfer and Related Techniques

Author	Event	Species
Heape, 1890	First successful embryo transfer	Rabbit
Beidl et al, 1922	Successful embryo transfer	Rabbit
Nicholas, 1933	Successful embryo transfer	Rat
Warwick and Berry, 1949	Successful embryo transfer	Sheep and goats
Kvansnickii, 1951	Successful embryo transfer	Pig
Willett et al, 1951	Successful embryo transfer	Cattle
Marden and Chang, 1952	First intercontinental shipment of embryos stored at 10°C	Rabbit
Alberta Livestock Transplants, Ltd., 1971	First commercial company formed for embryo transfer in farm animals	Cattle
Whittingham et al, 1972	Offspring produced from long-term frozen embryos	Mouse
Wilmut and Rowson, 1973	Offspring produced from frozen embryos	Cattle
1974	International Embryo Transfer Society formed	—
Steptoe and Edwards, 1978	Baby girl born after embryo transfer	Man

lating hormone (FSH) are given to stimulate additional follicular growth. This treatment is often followed by intravenous administration of luteinizing hormone (LH) or human chorionic gonadotropin (HCG) several days later to induce ovulation of the follicles, although exogenous LH or HCG is not required for adult cows, sheep and goats. Methods of superovulation are summarized in Table 29–2. Some researchers inject 3 mg of estradiol-17β on day 19 and again on day 20, when cows are treated with PMSG on day 16 of the estrous cycle (Boland et al, 1978).

With current procedures, superovulation increases the yield of normal embryos about five-fold in the cow, goat, sheep and rabbit, but only slightly in pigs and horses. With all species, there are tremendous individual variations in response. Many donors do not produce any normal embryos, while a few produce large numbers. For example, about half of the embryos recovered from a large group of superovulated donors are typically produced by one-fourth of the donors. Unfortunately, it is not possible to predict how a particular donor will respond; this is a major obstacle to successful application.

The single greatest advance in superovulation methodology in the last decade has been the use of prostaglandin F$_{2\alpha}$ or PGF$_{2\alpha}$ analogues such as cloprostenol (Table 29–2), because superovulatory treatment can be initiated anytime between day 6 of the estrous cycle and natural corpus luteum regression. However, the optimal time for treatment is between days 8 and 12 of the cycle in cattle (Betteridge, 1977). Not only does PGF$_{2\alpha}$ increase the flexibility of timing superovulation, it is also an excellent treatment for producing large numbers of normal embryos. Most donors are in estrus two to three days after prostaglandin injection.

Insemination

Superovulated donors are usually inseminated more often and with more sperm per inseminate than other donors. Even so, fertilization rates of ova from superovulated donors are usually considerably below those of ova from untreated donors (Elsden et al, 1976). This may be due to suboptimal sperm transport, ovulation over a period of time, defective oocytes, or other causes.

Repeated Superovulation

Experiments with repeated superovulation are sometimes difficult to interpret because they are often confounded by the

Table 29–2. Doses of Gonadotropins for Superovulation

Animal	Day of estrous cycle	Gonadotropin for Follicular Growth			Gonadotropin for Ovulation	
		PMSG[b] (IU)	or	FSH[c] (mg)	HCG[d] (IU)	LH[d] (mg)
Cow	15–16[a]	1500–3000		20–50	1500–2000	75–100
Calf	—	1000–2000		20–50	1000–1500	50–75
Goat	16–17[a]	1000–1500		12–20	1000–1500	50–75
Kid	—	1000–1200		10–15	1000–1500	50–75
Ewe	12–14[a]	1000–2000		12–20	1000–1500	50–75
Lamb	—	1000–1200		10–15	1000–1500	50–75
Pig	15–16	750–1500		10–20	500–1000	25–50
Rabbit	—	25–75		2–3	25–75	2–3

[a]Alternatively, treatment may be initiated on days 6 to 15 of the cycle in the goat and cow and days 6 to 13 in the ewe if followed with a luteolytic dose of prostaglandin $F_{2\alpha}$ two to three days later with the cow and one day later with the ewe and goat.

[b]PMSG is given as a single dose intramuscularly or subcutaneously.

[c]FSH (NIH-FSH-S1 equivalents) is divided among daily or twice daily subcutaneous injections for four to five days in the bovine and three days in the other species.

[d]LH (NIH-LH-S1 equivalents) or HCG is given intravenously five days after initiation of superovulation in the calf; three days after initiation in the rabbit, and at the beginning of estrus in the other species. The gonadotropin injection for ovulation in the cow, ewe and goat may be omitted because sufficient endogenous LH is usually released following follicular growth.

sequelae of repeated surgery (Maurer and Foote, 1971). On the average, donors respond similarly to first, second and third superovulatory treatments. The response to subsequent treatments, however, is less in some individuals, probably due to the production of antibodies against gonadotropins. This may be avoided to some extent by increasing the interval between hormone treatments. Antibody production may be minimized by using gonadotropins derived from the same species as the one being treated.

COLLECTION OF OVA

Embryos can be collected from the oviducts or uteri after slaughter or excision of the reproductive tract, or either surgically or nonsurgically removed from the intact animal (Table 29–3). Ova representing 40% to 80% of the corpora lutea can usually be recovered from superovulated, intact animals. Recovery rates are slightly higher from excised tracts. Why the number of ova recovered is often considerably lower than the number of corpora lutea is unknown; in some cases,

some of the oocytes may not be picked up by the fimbria because of a greatly enlarged, superovulated ovary. Other possibilities include the loss of ova from the reproductive tract due to altered steroid levels, the formation of corpora lutea without ovulation, and the failure to recover all ova present.

Surgical techniques have been used to collect ova from laboratory animals (Daniel, 1971) and farm animals (Dziuk, 1971; Murray, 1978). Ova reside in the oviduct for three to four days after estrus and then migrate to the uterus. Flushing the oviduct one to three days after estrus yields more ova than flushing the uterus five or more days after estrus. However, uterine embryos are often collected because they result in higher pregnancy rates (Newcomb and Rowson, 1975) and can be frozen more successfully with some species than younger embryos.

Surgical Methods

Oviductal ova can be flushed either from the fimbria toward the uterus or from the uterotubal junction toward the fim-

Table 29–3. Techniques of Ova Collection

Technique	Species
Collection From Isolated Reproductive Tract:	
Remove reproductive organs and trim adjacent adipose tissue and ligaments	All species
Flush each oviduct and uterine horn, avoiding contamination of flushings with blood	All species
Surgical Methods:	
Flush 2–20 ml medium through the oviduct from the upper part of uterine horn toward the fimbria using a syringe and blunt needle (Fig. 29–1A); collect flushings through a small glass tube inserted into the infundibulum.	Cattle, sheep, goat, rabbit
Flush 15–20 ml medium through the oviduct from the infundibulum, through the uterotubal junction, and into the upper part of the uterine horn using a small glass tube attached to a syringe (Fig. 29–1B); collect flushings through a blunt needle or fine glass tube inserted into the uterine lumen through a puncture wound.	Pig, mare, some uses in other species
Flush the uterus from the base of the uterine horn toward the uterotubal junction or in the reverse direction (Fig. 29–1C), using 10 to 100 ml of medium, depending on the size of the uterus; collect flushings through a blunt needle or small glass tube inserted into the uterine lumen through a puncture wound.	All species
Nonsurgical Methods:	
Dilate cervix (heifers) with expander; insert Foley catheter, Sugie's instrument, or similar device into uterine horn by manual guidance per rectum (Fig. 29–3); inflate balloon with air; irrigate uterine horn with 100–800 ml flushing medium; for mares, inflate balloon in cervix and flush both uterine horns simultaneously.	Cattle, horses

bria. In species other than the pig and horse, flushing toward the fimbria (Fig. 29–1, A) is the preferred method because a high percentage of ova can be recovered with little damage to the reproductive tract, although periovarian adhesions occur in some animals.

Due to the valve-like structure of the uterotubal junction in the sow and the mare, the flushing medium is introduced into the ampulla of the oviduct through the ostium of the fimbria with a small glass tube connected to a syringe (Fig. 29–1, B). The medium is flushed toward the tip of the uterine horn and collected through a blunt needle or a fire-polished, fine glass tube punctured through the uterine wall. A disadvantage of this method is the frequent formation of post-operative adhesions of the uterus or oviduct.

Ova may be recovered from the uterine horns after they have left the oviducts, usually five days after estrus and later (Fig. 29–1C). The flushing medium is introduced into the base of the uterine horn and flushed toward the tip where the medium is collected through a blunt syringe needle or a small glass tube inserted into the lumen through a puncture wound in the uterine wall. The procedure may also be carried out in the reverse direction. Fewer ova are recovered with these procedures than by flushing the oviducts, and serious damage may result from the formation of uterine adhesions. A volume of 2 ml to 20 ml is used to flush the oviducts, whereas 10 ml to 100 ml are used to flush the uterus, depending on its size.

Nonsurgical Methods

For many applications, nonsurgical techniques for collection of ova are desirable because all surgical techniques may

Fig. 29–1. Surgical techniques of ova recovery. *A*, Flushing oviduct toward fimbria; *B*, flushing oviduct toward uterotubal junction; *C*, flushing uterus toward base of the uterine horn. →

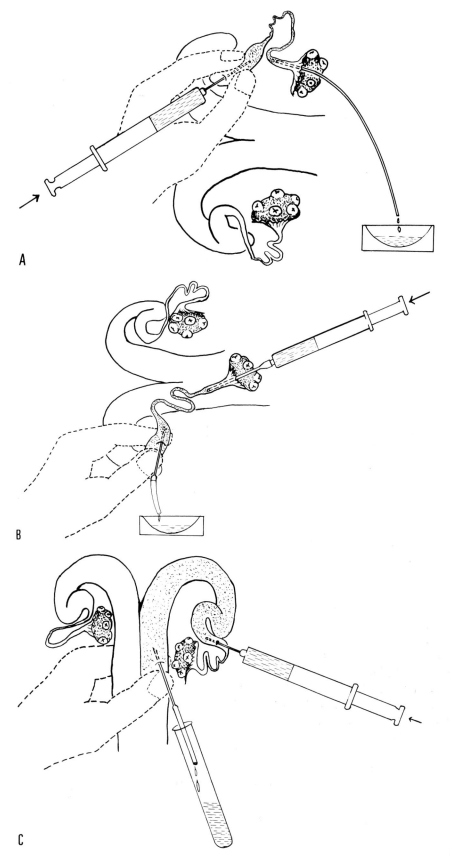

A

B

C

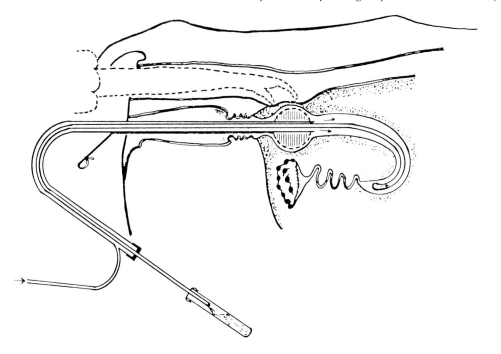

Fig. 29–2.   A nonsurgical technique for ova collection from the bovine uterus developed by T. Sugie.

lead to the formation of adhesions and because there is less risk to the life and health of the donor. Since the first work by Rowson and Dowling (1949), many attempts to recover bovine and equine ova nonsurgically have been reported. Sugie devised an apparatus (Fig. 29–2) that led to a remarkable improvement in the non-surgical collection of bovine ova from the uterus.

Currently, Foley catheters, three-way Stewart catheters, and similar devices are used for nonsurgical embryo recovery in cattle and horses (Fig. 29–3). The cervix of the heifer is often dilated with an ex-pander to facilitate the insertion of the collection device (Fig. 29–3). Each uterine horn is filled with 30 to 60 ml of medium, which is then allowed to flow into the collection vessel while the uterus is gently massaged per rectum. This is re-peated until 300 to 800 ml of medium have been used. The Foley catheter is then inserted into the other uterine horn and the process repeated. The same principle

is used to recover ova from mares, except that the balloon is inflated in the cervix and both horns are flushed simulta-neously.

Successful nonsurgical methods have made repeated recovery from the same donor practical. One consequence is that ova are now often recovered from mares and cows without superovulation (Oguri and Tsutsumi, 1974; Elsden et al, 1976).

SELECTION OF OVA FOR TRANSFER

After collection from the donor, ova are isolated under a microscope with a mag-nification of $10\times$ to $15\times$. A fine pipette (Fig. 29–4) is used to move them into fresh culture medium for morphologic examination, usually with a compound microscope at 40 to 200 magnifications. Embryos should be kept in a container that prevents evaporation of the culture medium and provides a 5% CO_2 atmo-sphere if the medium requires one. Some types of containers are illustrated in Fig-

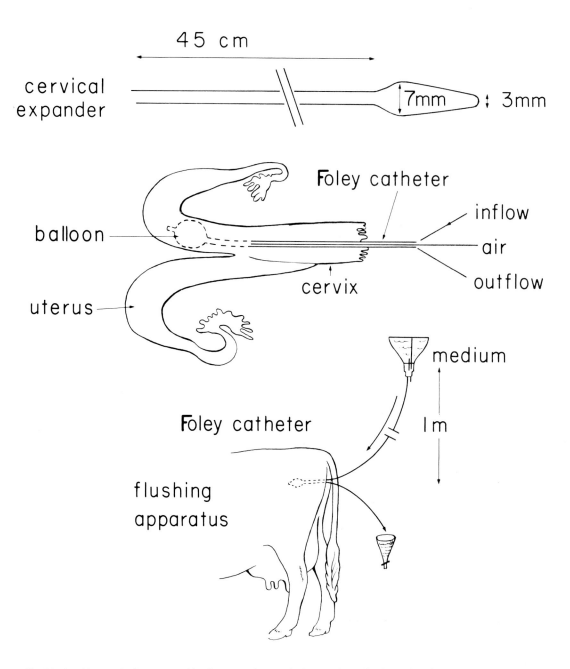

Fig. 29–3. Nonsurgical recovery of bovine ova using cervical expander and Foley catheter. *(Redrawn from Elsden et al., 1976)*

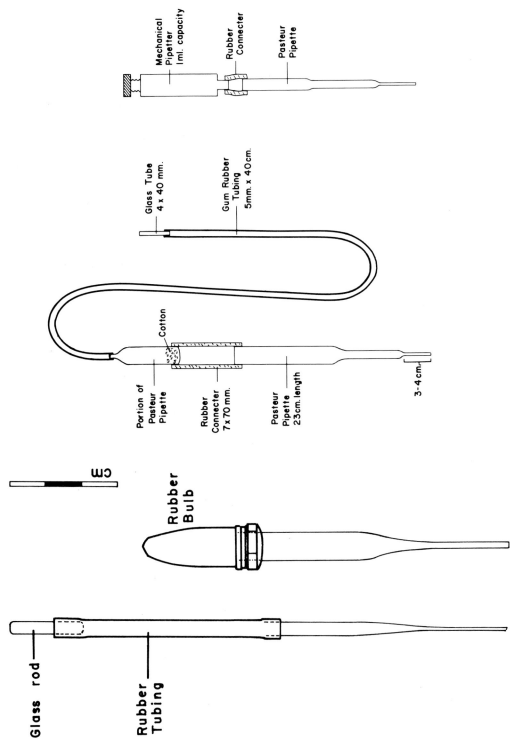

Fig. 29–4. Examples of pipettes for manipulating ova.

ure 29–5. Paraffin oil is used frequently to cover the medium to prevent evaporation and contamination with microorganisms.

Usually, only morphologically normal ova are transferred (Fig. 29–6); however, a few that appear morphologically abnormal may develop into normal young (Shea et al, 1976a; Elsden et al, 1978) (Fig. 29–7). Ova that show gross structural ab-

normalities, ova that are unfertilized, degenerating or fragmenting, ova that have empty or nearly empty zonae pellucidae (Figs. 29–7, 29–8), and embryos that are retarded in development by two or more days rarely develop normally and should be discarded. Stages of embryonic development normally found at various times after ovulation are presented in

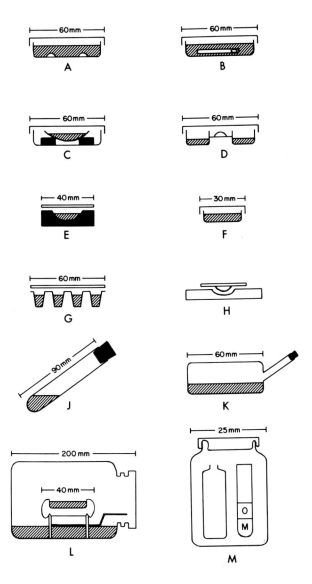

Fig. 29–5. Containers for ova. *A,* Microdroplets under paraffin oil; *B,* capillary tube of medium under paraffin oil; *H,* Hanging drop method.

A to H, Various vials containing culture media in equilibrium with the atmosphere of the incubator; *J to M,* Vials containing culture media which are tightly sealed. *(From Brinster, 1976)*

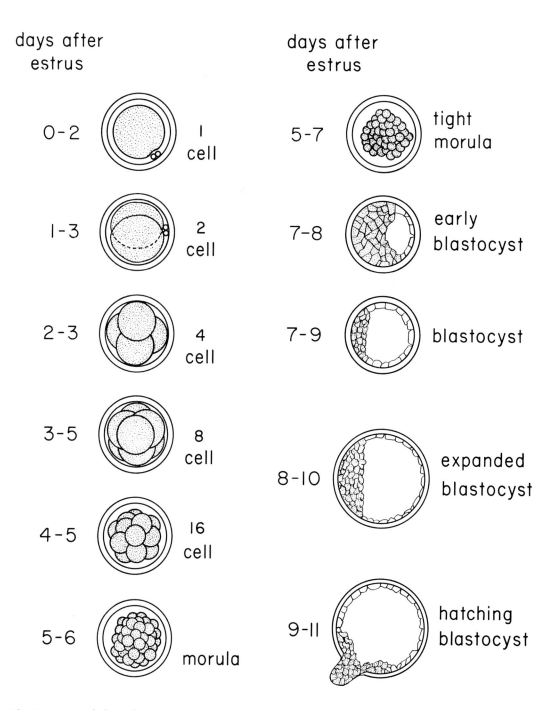

days after
estrus

days after
estrus

0-2   1 cell

1-3   2 cell

2-3   4 cell

3-5   8 cell

4-5   16 cell

5-6   morula

5-7   tight morula

7-8   early blastocyst

7-9   blastocyst

8-10   expanded blastocyst

9-11   hatching blastocyst

Fig. 29–6. Morphologically normal bovine embryos recovered at various stages of development.

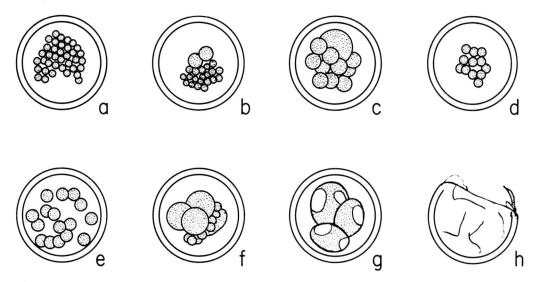

Fig. 29–7. Morphologically abnormal embryos. *a,* Tight morula with oval zona; *b,* morula with excluded blastomeres; *c,* irregular blastomeres; *d,* morula with debris; *e,* loose blastomeres; *f,* irregular cell mass; *g,* vacuoles in cytoplasm; *h,* cracked, empty zona pellucida.

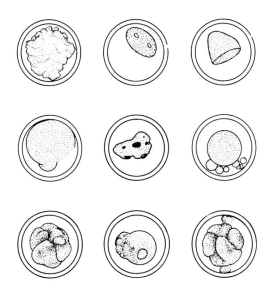

Fig. 29–8. Degenerating one-cell ova which should not be used for routine embryo transfer.

Table 29–4. Ova at any stage, from one-cell to the hatched blastocyst, can develop to term following transfer to a suitable environment, but success rates may be lower with very early and very late stages. Under most conditions, embryos between the eight-cell (four-cell in pigs) and blastocyst stage result in the highest pregnancy rates. Older embryos may tolerate *in-vitro* handling better than younger embryos; in most species, older embryos are best for freezing.

TRANSFER OF OVA

Site of Transfer

Generally ova of fewer than eight cells should be transferred to the oviduct. Young embryos are especially susceptible to damage in the uterus, possibly because uterine secretions are toxic to them. Ova with more than eight cells should be transferred to the uterine horn. The pig and rabbit are exceptions; porcine ova in

Table 29–4.   Stages of Embryonic Development at Various Times After Ovulation

Stage of Embryonic Development*	Days after Ovulation, by Species					
	Cow	Mare	Ewe	Goat	Pig	Rabbit
1-cell	0–1	0–1	0–1	0–1	0–1	0–1
2-cell	0–2	0–2	0–1	0–1	0–1	0–1
4-cell	1–2	1–2	1–2	1–2	2–3	1
8-cell	2–4	2–3	2–3	2–3	3–4	1
Early morula	3–5	2–4	2–4	2–4	3–4	2
Compacted morula	4–6	4–5	4–5	4–5	3–5	2
Early blastocyst	6–7	5–6	5–6	5–6	4–5	3
Blastocyst	6–8	6–7	6–7	6–7	5–6	3
Expanded blastocyst	7–9	7–8	7–8	7–8	5–7	4–6
Hatching blastocyst	8–10	8–9	8–9	7–9	6–8	7

* Embryos move from the oviduct to the uterus two to three days after ovulation in the rabbit and pig, and three to four days after ovulation in other species. From Betteridge, 1977, and unpublished observations.

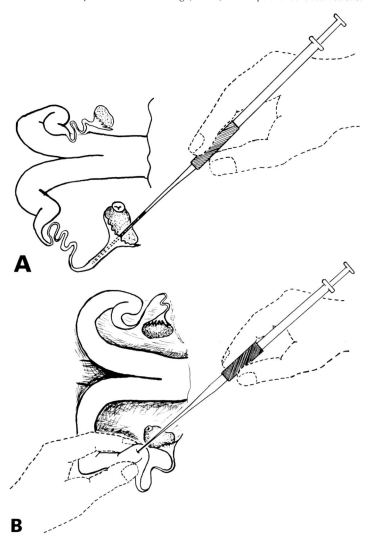

Fig. 29–9.   Surgical techniques. *A,* Embryo transfer to the oviduct; *B,* embryo transfer to the uterine horn.

the late four-cell stage should be transferred to the uterus, and 16-cell rabbit ova should be transferred to the oviduct. Similar results are obtained when ova are transferred to the upper, middle, and lower portion of the uterine horn (Sreenan, 1978; Hafez, 1962).

Surgical Transfer

Embryos are usually transferred surgically. Midventral laparotomy under general anesthesia is the most common method, although flank and lumbar approaches are often used with cattle and rabbits. The reproductive tract is exposed as for ova collection. For transfer into the oviduct, the tip of a capillary pipette containing the embryos is inserted into the infundibulum and ampulla of the oviduct, where embryos are then deposited in a drop or two of medium (Fig. 29–9, *A*). When transfer is made to the uterus, the wall of the uterine horn is punctured with a blunt needle and the embryos expelled from the tip of the capillary pipette inserted into the uterine lumen (Fig. 29–9, *B*).

For bovine embryo transfer, laparotomy may be performed under local anesthesia. The tip of the uterine horn is exposed through an incision in the flank. The ovum is then deposited in the uterine lumen as previously described. Flank incisions also work well in rabbits (Fig. 29–10).

Nonsurgical Transfer

Results of nonsurgical embryo transfer have been summarized by Foote and Onuma (1970). With the technique developed by Sugie (1965) (Fig. 29–11), the cervix is bypassed. Currently, a more common method is to deposit ova in the uterus through the cervix with an artificial insemination straw gun 6 to 12 days after estrus. Many calves have been produced by this method in Europe (Sreenan, 1978; Trounson et al, 1978a).

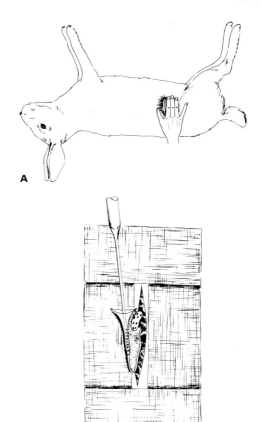

A

B DORSAL

Fig. 29–10. Embryo transfer by flank laparotomy in the rabbit. *A,* The recommended incision site for flank laparotomy and embryo transfer into the oviduct. In New Zealand rabbits, the distance between the thigh and the incision site is three fingers.

** *B,* Method of transferring embryos into the oviduct. The mesosalpinx attached to the oviduct is exposed through the incision and the glass pipette containing the eggs is inserted into the infundibulum.**

SYNCHRONIZATION OF ESTRUS BETWEEN DONOR AND RECIPIENT

For successful embryo transfer, synchronization between the stage of ovum and the reproductive tract of the recipient is necessary (rabbit: Chang, 1950; sheep: Moore and Shelton, 1964; pig: Webel et al, 1970; cow: Rowson et al, 1972). This is usually accomplished by selecting recipients that were in estrus at the same time as the donor, either naturally or as a result

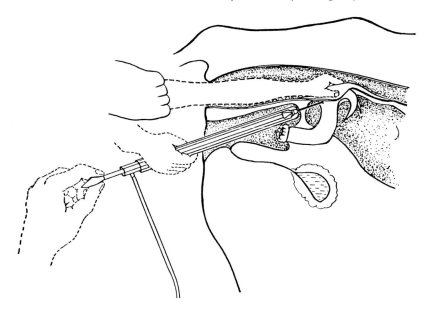

Fig. 29–11. Nonsurgical technique for ova transfer in the cow, in which the cervix is bypassed. *(From Sugie, 1965)*

of estrus synchronization (Sreenan et al, 1975). For optimal results, the recipient should be in estrus within 12 hours of the donor. Pregnancy rates decline drastically if the difference is greater than 24 hours in cows and 48 hours in sheep and goats. Recipients for embryos that have been frozen or stored at low temperature should be selected to be in physiologic synchrony with the stage of development of the embryo. Thus, if embryos are frozen for two days, they should be transferred to recipients that were in estrus two days after the donor.

Pregnancy losses from asynchronous transfers are probably due to placement of embryos in an adverse environment, or to an inability of asynchronous embryos to exert a luteotropic action on the corpus luteum of the recipient.

COMPOSITION OF MEDIA FOR EMBRYOS

The compositions of media frequently used for embryos from farm animals are presented in Table 29–5. Usually 50 mg streptomycin sulfate and 100,000 IU of potassium penicillin G are added per liter, but other concentrations, other antibiotics and antifungal agents are frequently used as well. Media should be forced through a filter of .45 μ or smaller to remove bacteria.

Media should contain either bovine serum albumin (BSA) or blood serum (often from fetal calves) that has been inactivated by being held at 56°C for 30 minutes. BSA is usually added at .3% to 1%, but concentrations from .1% to 3.2% have been used. Blood serum is usually added at 5% to 20%, but 1% to 50% concentrations have been used. All media except the modified phosphate-buffered saline are bicarbonate-buffered, and therefore, require an atmosphere of 5% CO_2 to maintain proper pH. This is accomplished with a mixture of 5% CO_2 in air, or better, 5% CO_2, 5% O_2, and 90% N_2. A CO_2-incubator, stoppered and gassed test tubes, or small gassed and air-tight containers may be used. When media must be kept in an

Table 29–5.　Compositions of Media Frequently Used for Embryos From Farm Animals

Ingredients (mg/l)	Brinster's mouse ova culture medium-3 (GIBCO)	Menezo's B-2 medium (Menezo, 1976)	Eagle's minimum essential medium* (GIBCO)	Modified Dulbecco's PBS (Gwatkin, 1972)	Ham's nutrient mixture F-10 (GIBCO)	Synthetic oviduct fluid (Tervit, et al, 1972)	Tissue-culture medium 199† (GIBCO)	Whitten's medium (Gwatkin, 1972)
Inorganic salts								
NaCl	5546	5250	6800	8000	7400	6300	8000	5140
KCl	356	800	400	200	285	533	400	356
$CaCl_2$	189	—	200	100	33	190	140	—
$MgCl_2 \cdot 6H_2O$	—	—	—	100	—	100	—	—
$MgSO_4 \cdot 7H_2O$	294	200	200	—	153	—	200	294
$NaHCO_3$	2106	2500	2200	—	1200	2106	350	1900
Na_2HPO_4	—	61	—	1150	154	—	48	—
$NaH_2PO_4 \cdot H_2O$	—	—	140	—	—	—	—	—
KH_2PO_4	162	60	—	200	83	162	60	162
Carbohydrates								
glucose	1000	1200	1000	1000	1100	270	1000	1000
Na pyruvate	56	250	—	36	110	36	—	36
Na lactate (DL)	2253	—	—	—	—	370	—	2416
Ca (lactate)$_2 \cdot 5H_2O$	—	644	—	—	—	—	—	527
ribose	—	—	—	—	—	—	.5	—
deoxyribose	—	—	—	—	—	—	.5	—
Amino acids	none	contains 23	contains 13	none	contains 20	none	contains 21	none
Vitamins	none	contains 1	contains 8	none	contains 10	none	contains 16	none
Nucleic acids and precursors	none	none	none	none	contains 2	none	contains 8	none
Trace elements	none	none	none	none	contains 3	none	contains 1	none
Other components								
bovine serum albumin	5000	10000	varies	varies	varies	varies	varies	3000
cholesterol	—	125	—	—	—	—	.2	—
Na acetate	—	50	—	—	—	—	50	—
lipoic acid	—	—	—	—	.2	—	—	—
Tween 80	—	50	—	—	—	—	20	—
glutathione	—	—	—	—	—	—	.05	—
α tocopherol PO$_4$ (Na)	—	—	—	—	—	—	.01	—

* The formulation with Earle's salts.
† The formulation with Hank's salts.
GIBCO catalog (1976–1977), Grand Island Biological Company, Grand Island, NY.

air environment (without CO_2) for long periods, 25 mM HEPES buffer is usually added and NaCl decreased to maintain proper osmolality.

Phenol red, 1 to 20 mg/L, is often added as a pH indicator. The pH of media may range from 7 to 8, but best results are obtained between 7.2 and 7.6. The osmolality may be adjusted from 250 to 320 mOsm/kg by varying the NaCl concentration. Osmolalities between 270 and 300 mOsm/kg are most commonly used for embryos.

Water is the principal ingredient and purity is important. Double distillation or glass distillation of deionized water is usually adequate. Gwatkin (1972) has summarized further information about preparation of media for embryo culture.

STORAGE OF EMBRYOS

Culture and Storage Between 0° and 37°C

For experimental manipulations and storage between recovery and transfer, embryos are often kept in culture medium at 37°C. In most studies, development of embryos *in vitro* is slowed, often to about two-thirds the normal *in-vivo* rate. However, embryos frequently continue to develop for two to three days or more, although pregnancy rates are usually reduced if they are transferred after more than 24 hours *in vitro* (Davis and Day, 1978). Culture of bovine embryos from the one- to two-cell stage to the four-cell stage and from morulae to blastocysts has been successful. However, attempts to culture them from the early cleavage stages through the blastocyst stage have been disappointing (Seidel, 1977). Recent studies with sheep (Wright et al, 1976; Tervit and Goold, 1978) and swine (Lindner and Wright, 1978; Davis and Day, 1978) have shown that development of embryos of four or more cells from these species proceeds very well *in vitro* in relatively simple media.

Embryos are often held successfully for several hours to a day between collection and transfer at ambient temperature (15° to 25°C). However, if they are cooled to 0° to 10°C or transferred to the ligated oviduct of a rabbit, they can be stored for several days with little reduction in viability (Hafez, 1971; Lawson et al, 1972; Boland et al, 1978). Porcine embryos are an exception and do not survive cooling below 15°C (Maurer, 1978). Development of the embryos continues at a normal rate in the rabbit oviduct, but is arrested during storage at 0° to 10°C.

For long-term storage and easy transportation, embryos in the late morula or early blastocyst stage may also be frozen to liquid nitrogen temperature. The major disadvantage of frozen embryos is that pregnancy rates are only about half of those obtained when nonfrozen embryos are transferred (Maurer, 1978; Trounson et al, 1978b). However, the technology of freezing embryos is improving rapidly.

Principles of Freezing Embryos

The principles that apply to freezing any living cells also apply to freezing embryos (Maurer, 1978). Embryos may be damaged during freezing or thawing either by the formation of large intracellular ice crystals or by the increased intracellular concentration of solutes and accompanying changes that result from the dehydration of cells during freezing (solution effects). Whereas fast freezing minimizes damage from solution effects, it leads to the formation of large ice crystals that cause severe mechanical damage. On the other hand, while slow freezing prevents large ice crystal formation, it leads to increased damage from solution effects. Therefore, the optimal freezing rate for a given tissue depends on its relative tolerances to damage from ice crystals and toxicity from solution effects.

The addition to the freezing medium of cryoprotectants such as glycerol or dimethyl sulfoxide results in freezing at lower temperatures. This probably retards dehydration of cells and the resultant harmful solution effects; thus, embryos may be cooled slowly enough to prevent the formation of large ice crystals. Because cryoprotectants drastically increase the osmolality of the medium and may be toxic, they often are added and removed in steps.

Procedures for Freezing Embryos

The procedures currently used for freezing embryos from farm animals were developed largely by workers in England (Table 29–6). The usual medium is modified Dulbecco's phosphate-buffered saline (Table 29–5), supplemented with bovine serum albumin or serum. Cryoprotectant is added in steps either at 0° or 20°C (Maurer, 1978; Trounson et al, 1978b). Embryos are cooled rapidly to 0°C and at a rate of 1°C/minute to −7°C, at which point freezing is initiated by adding a small crystal of ice to the medium (seeding). Seeding minimizes temperature fluctuations resulting from heat of fusion. Embryos are then cooled very slowly (less than .3°C/minute) to about −33°C, and finally, placed into liquid nitrogen.

To prevent the enlargement of small, harmless ice crystals, embryos are thawed rapidly in a 25°C water bath; the cryoprotectant is then removed by stepwise dilution at room temperature. Pregnancy rates may be improved to some extent if embryos are cultured for several hours before transfer and any morphologically abnormal ones discarded.

MAINTENANCE OF PREGNANCY AFTER EMBRYO TRANSFER

Pregnancy rates, the percentage of recipients pregnant, should not be confused with the percentage of embryos surviving. Under certain ideal conditions, up to 80% of the embryos survive to term following transfer to a synchronous recipient. Under field conditions, 50% to 60% survival to term is more common (Shea et al, 1976a). The highest pregnancy rates in sheep, cattle, and goats are obtained with the transfer of one embryo into each uterine horn of the recipient. This frequently results in twins. In pigs and rabbits, six to ten embryos should be transferred into each side to obtain a normal-sized litter, since only about one-half of the embryos transferred are represented by viable young at birth. Transfer of 40 to 60 embryos per recipient does not result in larger litters than normal at term (Hafez, 1964).

Pregnancy and embryo survival rates that have been determined in early pregnancy will usually be only slightly inflated relative to term pregnancy rates. Some of the higher pregnancy rates reported for embryo transfer to farm animals are 75% for sheep (Moore, 1968); 91% for cattle (Rowson et al, 1969, 1972); 70% to 83% for goats (Sugie and Soma, 1970); and 100% for pigs (Pope et al, 1972).

IN-VITRO MATURATION OF OOCYTES

Graafian follicles contain primary oocytes that normally complete morphologic development in the following sequence in response to the preovulatory LH surge. First, the nuclear membrane, also known as the germinal vesicle, breaks down. Second, the secondary oocyte is formed by completion of the first meiotic division and extrusion of the first polar body. The metaphase plate for the second meiotic division is formed shortly thereafter. Third, ovulation occurs. In the horse and certain carnivores, ovulation probably occurs before formation of the secondary oocyte.

Years ago, Pincus and Enzmann (1935) discovered that oocytes removed from graafian follicles of rabbits prior to the LH surge and placed in suitable culture media would mature spontaneously. In recent years, this has been confirmed for a

Table 29–6. Development to Term Following Transfer of Frozen-Thawed Embryos

Species	Cryoprotectant		Developmental Stage of Ova	Number of Ova Transferred	Number of Recipients	Number Pregnant	Number of Liveborn	References
Cattle	2.0 M	DMSO	Blastocyst	21	11	1	1	Wilmut and Rowson, 1973
	1.5 M	DMSO	Morula and blastocyst	23	17	6*	—	Bilton and Moore, 1976b
	1.5 M	DMSO	Morula and blastocyst	8	5	1	1	Bilton and Moore, 1977
	1.0 M	Glycerol		8	6	4	4	
	1.5 M	DMSO	Blastocyst	23	11	8	10	Willadsen et al, 1978
				9	9	6*	—	
	1.5 M	DMSO	Blastocyst	17	16	7*	—	Lehn-Jenson and Greve, 1978
	1.5 M	DMSO	Blastocyst	42	42	17*	—	Trounson et al, 1978b
Sheep	1.5 M	DMSO	Blastocyst	20	11	7	8	Willadsen et al, 1976
	1.5 M	DMSO	Morula and blastocyst	15	8	3	5	Smorag, 1977
Goats	1.0 M	Glycerol	Blastocyst	5	2	2	3	Bilton and Moore, 1976a

* Determined by palpation per rectum

Table 29–7. Fertilization of Oocytes Matured *In Vitro*

Species	Medium for Maturation	Hours for Maturation	Method of Fertilization	Criterion of Fertilization	Per cent Fertilization	References
Cattle	Follicular fluid or growth medium	27–32	Oviduct of ewe	Cleavage	15	Sreenan, 1970
	Tyrode's plus follicular fluid	24–26	Oviduct of cow	Pronuclei plus swollen sperm or cleavage	52	Hunter et al, 1972
	Ham's F-10 plus fetal calf serum	28	Oviduct of pig	Sperm penetration	17	Shea et al, 1976b
	Modified Krebs-Ringer bicarbonate	20–24	*In vitro*	Pronuclei	16	Iritani and Niwa, 1977
	Fetal calf serum	24	Oviduct of cow	Sperm penetration or pronuclei	49	Trounson et al, 1977b
	Ham's F-10 plus serum	22–24	Oviduct of cow	Development to 8 or more cells; two calves born	23	Newcomb et al, 1978
Sheep	Growth medium	24–28	Oviduct of ewe	Pronuclei or cleavage	17	Quirke and Gordon, 1971
	Modified Medium 199	24	Oviduct of ewe	Morulae or blastocysts; 24 lambs born	46	Moor and Trounson, 1977
Pig	Modified Medium 199	20–27	Oviduct of sow	Sperm penetration	80	Motlík and Fulka, 1974
	Modified Krebs-Ringer bicarbonate	Until meta-phase II	*In vitro*	Pronuclei	36	Iritani et al, 1978

number of species. However, a high incidence of chromosomal abnormalities (McGaughey and Polge, 1971) and cytoplasmic deficiencies (Moor and Trounson, 1977) indicates that the process is not always normal.

Table 29–7 is a summary of experiments with farm animals in which oocytes were matured or partially matured *in vitro* and then fertilized, usually *in vivo*. In some of these studies, fertilization was clearly abnormal, and may have been abnormal in most of them. However, normal offspring resulted from two of the studies. Best results were obtained when the entire follicle was cultured with appropriate hormones (Moor and Trounson, 1977). This method apparently does not produce the abnormalities associated with *in-vitro* maturation of isolated oocytes.

IN-VITRO FERTILIZATION

Fertilization is much easier to study *in vitro* than in the oviduct, where it occurs *in vivo*. To date, *in-vitro* fertilization has been used primarily for research, but in the near future it will also have applications in animal husbandry in such areas as testing male fertility and fertilizing oocytes from slaughterhouse ovaries.

Definition of Fertilization

Fertilization may be defined as the integration of paternal and maternal DNA into the embryo. This implies more than just the penetration of the oocyte by a sperm. For many years it was thought that syngamy, the fusion of the male and female pronuclei, was the appropriate event to mark the completion of fertilization. However, studies (Longo, 1973) have shown that the male and female pronuclei never fuse in mammals and that the maternal and paternal genetic material first come together just before the two-cell stage. Thus, sperm capacitation, the acrosome reaction, penetration of the oocyte by the sperm, activation of the oocyte,

formation of pronuclei, breakdown of the pronuclear membranes, and completion of the first mitosis may all be considered as part of the fertilization process.

Capacitation

Most early experiments with *in-vitro* fertilization were unsuccessful because the sperm had not been capacitated, a term that refers to a modification of ejaculated sperm in the female reproductive tract, making them capable of fertilizing oocytes. Ejaculated sperm from every mammalian species studied thus far require capacitation under normal *in-vivo* conditions. Capacitation times range from less than 30 minutes in the mouse and cat to six to ten hours or longer in the rabbit.

The molecular mechanism of capacitation is not known. One theory is that the membranes of the uncapacitated sperm cell are specifically stabilized to protect against damage in the female reproductive tract and that capacitation is the reversal of this stabilization. Another is that capacitation consists of the removal of inhibitors of acrosomal enzymes. In each species, capacitation is completed by the time most of the sperm come into contact with the oocytes. At this time, the cell membrane over the acrosome must be destabilized to permit the acrosome reaction. If the acrosome reaction occurs too early, the enzymes in the acrosome are lost. Such sperm are probably incapable of fertilization.

A probable explanation of capacitation invokes decapacitation factor, which is thought to be produced in the epididymis or accessory sex glands of the male and bound to the sperm. Capacitation, then, would simply be the removal of the factor from the cell membrane of the sperm. In the classic experiment leading to this hypothesis, capacitated sperm (defined as those capable of fertilization) were decapacitated (made incapable of fertilization) by adding seminal plasma, and recapacitated by a second exposure to the female reproductive tract (Chang, 1957). The concept of decapacitation factor is supported by the demonstration that molecules originating in seminal plasma are removed from the sperm cell membrane during capacitation (Oliphant and Brackett, 1973b).

A number of different molecules that decapacitate sperm have been found in semen but none has been purified completely. Aonuma et al (1978) have shown that zinc inhibits fertilization *in vitro*. It is possible that several different compounds, working independently or synergistically, cause decapacitation.

Obtaining Capacitated Sperm

Capacitated sperm are often obtained from the reproductive tract of a mated female. A common method is to flush medium through the uterus (Seidel et al, 1976). The oviducts also contain capacitated sperm, albeit in very low numbers.

For some species, capacitation may be accomplished *in vitro*, although the mechanism as well as the duration of the process may differ from the *in-vivo* phenomenon. Media of high ionic strength seem to promote *in-vitro* capacitation (Oliphant and Brackett, 1973a; Brackett and Oliphant, 1975).

Obtaining Oocytes

Oocytes for *in-vitro* fertilization are usually obtained from the oviducts shortly after ovulation, but they may also be obtained from follicles or the surface of the ovary. While sperm can penetrate oocytes ranging in age from the immature primary stage prior to the breakdown of the nuclear membrane to aged oocytes recovered more than a day after ovulation, normal embryonic development does not result from the fertilization of oocytes at either extreme. With immature oocytes, the chromatin of the sperm fails to decondense properly and the block to polyspermy is defective (Moore and Bedford, 1978). In aged oocytes, the incidence of polyspermy is also high. If the oocyte is too old, embryonic development is abnormal or fertilization may not take place at all. The fertilizable life of ovulated oocytes from several species is presented in Chapter 11.

Unlike the sperm, the oocyte requires no exposure to the reproductive tract following release from the gonad in order to be fertile. However, under some conditions, *in-vitro* fertilization is enhanced by the presence of cells of the corona radiata and cumulus oophorus (Seidel et al, 1976). In other systems, *in-vitro* fertilization proceeds normally when these cells have been removed mechanically or enzymatically with hyaluronidase.

In comparison to *in-vivo* fertilization, the percentage of oocytes fertilized *in vitro* is usually lower. In addition, fertilization may take longer to complete *in vitro* and the incidence of polyspermy may be increased.

Media

Modified, balanced salt solutions such as Krebs-Ringer bicarbonate support *in-vitro* fertilization. The acrosome reaction seems to occur much more readily in media containing serum albumin in the form of either heat-treated serum or bovine serum albumin (Bavister and Yanagimachi, 1977). It is also critical to provide an energy source, usually glucose or pyruvate, to support spermatozoal motility and the metabolism of the oocytes. In addition, proper levels of zinc, calcium and magnesium ions are required (Aonuma et al, 1978; Reyes et al, 1978).

Criteria of Successful In-Vitro Fertilization

Proof of *in-vitro* fertilization is the birth of normal young, genetically marked by means of the male parent, resulting from the transfer of the fertilized ova to a recipient female. Due to the great cost in time and effort of transferring embryos, however, other criteria are often used, such as penetration of sperm into the ooplasm, swelling of the sperm head, pronuclear formation, morphologically normal cleavage, blastocyst formation, demonstration of a Y chromosome in the embryo, breakdown of cortical granules, and evidence of a sperm tail in the ooplasm. None of these alone is sufficient proof of normal fertilization. Parthenogenetic embryos, for example, also exhibit some of these traits.

Because information from farm animal species is limited, it has been necessary to extrapolate from laboratory species. There is circumstantial evidence for *in-vitro* fertilization in swine, sheep and cattle (Table 29–8), but conclusive evidence obtained by embryo transfer is not yet available for any farm animal species.

POTENTIAL USES AND LIMITATIONS OF EMBRYO TRANSFER AND RELATED TECHNIQUES

Embryo transfer is a useful experimental technique; it also has many applications in animal production. It allows critical experimental approaches to problems in genetics, cytology, animal breeding, immunology, evolution, and the physiology and biochemistry of reproduction. For example, the technique can be used to evaluate the relative contributions of the aging oocyte and the aging reproductive tract to decreased reproduction in older animals (Maurer and Foote, 1971).

Success Rates in Animal Production

Successful embryo transfer usually depends on proper execution of the following: (1) superovulation, (2) estrus detection in the donor and recipients, (3) insemination of the donor, (4) recovery of embryos, (5) short-term storage of embryos *in vitro*, (6) embryo transfer, (7) proper management of the recipient through parturition, and (8) keeping the progeny healthy until useable or saleable. If even one of these steps is done poorly, the entire process may fail. Information on success rates is now becoming available from commercial bovine embryo transfer (Shea et al, 1976a; Church and Shea, 1977; Elsden et al, 1978). When one considers all treated donors, the average number of calves produced per superovulation is between three and four with the best current technology. The median number is two and the mode, zero. Occasionally, litters of more than 20 calves are produced, but this occurs much less often than once in 100 attempts. The major contribution to this variability is the unpredictable response to superovulation; the second greatest problem is that many ova are unfertilized or abnormal. Since it is possi-

Table 29–8. *In Vitro* Fertilization of Ova From Cattle, Sheep and Swine

Species	Source of Oocytes	Method of Capacitation	Results	References
Cattle	Matured *in vitro*	In uterus or oviduct of slaughtered cow or rabbit	17 of 103 with male pronucleus or sperm remnants	Iritani and Niwa, 1977
Cattle	Follicle or oviduct	*In vitro* in high ionic strength medium	14 of 25 ova fertilized; 10 of the 14 cleaved	Brackett et al, 1978
Swine	Matured *in vitro*	Epididymal sperm in uterus or oviduct of slaughtered sow	23 of 64 ova with pronuclei and sperm tail	Iritani et al, 1978
Sheep	Follicle or oviduct	In reproductive tract of ewe	4 of 78 with male pronuclei and second polar bodies	Dauzier and Thibault, 1959

ble to induce superovulation in cows four or five times a year, more than ten calves can be obtained per cow per year on the average. Information from individual cows that have superovulated more than four or five times is still fragmentary.

Unfortunately, information is not yet available from large-scale commercial embryo transfer in species other than cattle. However, each species has its advantages and problems, such as seasonal breeding in sheep and horses.

Applications

Genetic improvement through the ovum has been greatly hampered by the inability to harvest and use the ova that are not ovulated and therefore, undergo atresia. Less than .1% of the oocytes produced by females of farm animal species are ovulated, and many of these do not develop into young. In cattle, embryo transfer can be used to increase rare blood lines rapidly, obtain more offspring from valuable cows, and accelerate genetic progress by facilitating progeny testing of females and reducing the generation interval. However, embryo transfer methods are not nearly as effective as artificial insemination for making genetic progress (Land, 1977). One problem is that reproductive rates of donors are increased at the expense of decreased reproductive rates of the recipients, because the females in the recipient pool frequently remain nonpregnant for prolonged periods until an embryo is transferred to them (Seidel and Seidel, 1978). Therefore, the total reproductive performance of the population is usually greatly reduced relative to conventional breeding.

The transport of frozen embryos over long distances can be an inexpensive means of exporting livestock. Frozen bovine embryos were first transported successfully from New Zealand to Australia (Bilton and Moor, 1977). Another common use of embryo transfer is to obtain additional progeny from infertile cows (Bowen et al, 1978), although the number of progeny produced per donor is much lower than with cattle of normal fertility.

It may be possible to increase the ratio of beef produced to feed consumed by reliable induction of twinning (Anderson, 1978). Transferring either a single embryo into each uterine horn of an unmated cow or a second ovum to the horn contralateral to the corpus luteum of a recipient cow that conceived a few days earlier is more effective than mild superovulation for production of twins in cattle (Rowson et al, 1971; Anderson, 1978).

In pig breeding, embryo transfer is of limited value. The transfer of supersized litters to unmated females or additional embryos to previously bred recipients does not appreciably increase litter size. One important use of embryo transfer in swine, however, is the introduction of new genetic material into specific pathogen-free herds.

Superovulation is not very effective in horses. However, a single ovum can usually be recovered nonsurgically during each estrous cycle six or seven days after ovulation. Further, nonsurgical transfer works well with mares (Oguri and Tsutsumi, 1974). It appears that embryo transfer may be especially useful for obtaining additional progeny from old, infertile brood mares.

Future Developments

The future of embryo transfer will be exciting. Some progress has already been made with superovulation of prepuberal animals (Onuma et al, 1970; Wright et al, 1976; Trounson et al, 1977a), although pregnancy rates from transferring calf embryos are low (Seidel et al, 1971). Another exciting area is sexing embryos. Although it is expensive and time-consuming, about two-thirds of older bovine embryos can be sexed successfully prior to transfer (Hare and Betteridge, 1978). The production of exact genetic copies of outstanding animals by cloning may also be possible before the end of this century.

Many problems remain to be solved before embryo transfer techniques reach a practical scale comparable to artificial insemination. Methods for superovulation, freezing embryos, and nonsurgical transfer are not completely reliable, and there is no simple nonsurgical technique for collecting embryos from sheep, goats, pigs and rabbits. Some of these problems are being resolved and embryo transfer, just as artificial insemination, may some day play a significant role in animal reproduction.

REFERENCES

Anderson, G.B. (1978). Methods for producing twins in cattle. Theriogenology 9, 3.

Aonuma, S., Okabe, M. and Kawaguchi, M. (1978). The effect of zinc ions on fertilization of mouse ova in vitro. J. Reprod. Fertil. 53, 179.

Bavister, B.D. and Yanagimachi, R. (1977). The effects of sperm extracts and energy sources on the motility and acrosome reaction of hamster spermatozoa in vitro. Biol. Reprod. 16, 228.

Beidl, A., Peters, H. and Hofstattler, R. (1922). Experimentelle studien über die einnistung und weiter entwicklung des eies im uterus. Z. Geburtsh. Gynäk. 84, 60.

Betteridge, K.J. (ed.) (1977). Embryo Transfer in Farm Animals. Ottawa, Canada Dept. Agric. Monograph 16. 92 pp.

Bilton, R.J. and Moore, N.W. (1976a). In vitro culture, storage, and transfer of goat embryos. Aust. J. Biol. Sci. 29, 125.

Bilton, R.J. and Moore, N.W. (1976b). Storage of cattle embryos. J. Reprod. Fertil. 46, 537. (Abstr.)

Bilton, R.J. and Moore, N.W. (1977). Successful transport of frozen cattle embryos from New Zealand to Australia. J. Reprod. Fertil. 50, 363.

Boland, M.P., Crosby, T.F. and Gordon, I. (1978). Morphological normality of cattle embryos following superovulation using PMSG. Theriogenology 10, 175.

Bowen, R.A., Elsden, R.P. and Seidel, G.E., Jr. (1978). Embryo transfer for cows with reproductive problems. J. Am. Vet. Med. Assoc. 172, 1303.

Brackett, B.G., Oh, Y.K., Evans, J.F. and Donawick, W.J. (1978). In vitro fertilization of cow ova. Theriogenology 9, 89. (Abstr.)

Brackett, B.G. and Oliphant, G. (1975). Capacitation of rabbit spermatozoa in vitro. Biol. Reprod. 12, 260.

Chang, M.C. (1950). Development and fate of transferred rabbit ova or blastocyst in relation to the ovulation time of recipients. J. Exp. Zool. 114, 197.

Chang, M.C. (1957). A detrimental effect of seminal plasma on the fertilizing capacity of sperm. Nature 179, 258.

Church, R.B. and Shea, B.F. (1977). The role of embryo transfer in cattle improvement programs. Can. J. Anim. Sci. 57, 33.

Daniel, J.C., Jr. (ed) (1971). Methods in Mammalian Embryology. San Francisco, W.H. Freeman, 532 pp.

Dauzier, L. and Thibault, C. (1959). Données nouvelles sur la fécondation in vitro de l'oeuf de la lapine et de la brebis. C.R. Acad. Sci. 248, 2655.

Davis, D.L. and Day, B.N. (1978). Cleavage and blastocyst formation by pig eggs in vitro. J. Anim. Sci. 46, 1043.

Dziuk, P.J. (1971). Obtaining eggs and embryos from sheep and pigs. In Methods in Mammalian Embryology. J.C. Daniel, Jr. (ed), San Francisco, W.H. Freeman, pp. 76–85.

Elsden, R.P., Hasler, J.F. and Seidel, G.E., Jr. (1976). Non-surgical recovery of bovine eggs. Theriogenology 6, 523.

Elsden, R.P., Nelson, L.D. and Seidel, G.E., Jr. (1978). Superovulating cows with follicle stimulating hormone and pregnant mare's serum gonadotrophin. Theriogenology 9, 17.

Erickson, B.H. (1966). Development and senescence of the postnatal bovine ovary. J. Anim. Sci. 25, 800.

Foote, R.H. and Onuma, H. (1970). Superovulation, ovum collection, culture, and transfer: a review. J. Dairy Sci. 53, 1681.

Gwatkin, R.B.L. (1972). Chemically defined media for mammalian eggs and early embryos. In Vitro, 8, 59.

Hafez, E.S.E. (1962). Effect of progestational stage of the endometrium on implantation, fetal survival, and fetal size in the rabbit, Oryctolagus cuniculus. J. Exp. Zool. 151, 217.

Hafez, E.S.E. (1964). Effects of overcrowding in utero on implantation and fetal development in the rabbit. J. Exp. Zool. 156, 269.

Hafez, E.S.E. (1971). Egg storage. In Methods in Mammalian Embryology. J.C. Daniel, Jr. (ed), San Francisco, W.H. Freeman, pp. 117–132.

Hare, W.C.D. and Betteridge, K.J. (1978). Relationship of embryo sexing to other methods of prenatal sex determination in farm animals: a review. Theriogenology 9, 27.

Heape, W. (1890). Preliminary note on the transplantation and growth of mammalian ova within a uterine foster mother. Proc. Roy. Soc. (Lond.) 48, 457.

Hunter, R.H.F., Lawson, R.A.S. and Rowson, L.E.A. (1972). Maturation, transplantation, and fertilization of ovarian oocytes in cattle. J. Reprod. Fertil. 30, 325.

Iritani, A. and Niwa, K. (1977). Capacitation of bull spermatozoa and fertilization in vitro of cattle follicular oocytes matured in culture. J. Reprod. Fertil. 50, 119.

Iritani, A., Niwa, K. and Imai, H. (1978). In vitro fertilization of pig follicular oocytes matured in culture. Biol. Reprod. 18 (Suppl. 1), 18A (Proc. XI Ann. Mtg. Soc. Study Reprod.)

Kvansnickii, A.V. (1951). Inter-breed ova transplantation. Sovetsk. Zootech. 1, 36.

Land, R.B. (1977). The genetics of breed improvement. In Embryo Transfer in Farm Animals. K.J. Betteridge (ed), Ottawa, Canada Dept. Agric. Monograph 16, pp. 57–59.

Lawson, R.A.S., Rowson, L.E.A. and Adams, C.E. (1972). The development of cow eggs in the rabbit oviduct and their viability after re-transfer to heifers. J. Reprod. Fertil. 28, 313.

Lehn-Jensen, H. and Greve, T. (1978). Low temperature preservation of cattle blastocysts. Theriogenology 9, 313.

Lindner, G.M. and Wright, R.W., Jr. (1978). Morphological and quantitative aspects of the development of swine embryos in vitro. J. Anim. Sci. 46, 711.

Longo, F.J. (1973). Fertilization. A comparative ultrastructural review. Biol. Reprod. 9, 149.

Marden, W.G.R. and Chang, M.C. (1952). Aerial transport of mammalian ova for transplantation. Science 115, 705.

Maurer, R.R. (1978). Freezing mammalian embryos: a review of the techniques. Theriogenology 9, 45.

Maurer, R.R. and Foote, R.H. (1971) Maternal aging and embryonic mortality in the rabbit. I. Repeated superovulation, embryo culture, and transfer. J. Reprod. Fertil. 25, 329–341.

McGaughey, R.W. and Polge, C. (1971). Cytogenetic analysis of pig oocytes matured in vitro. J. Exp. Zool. 176, 383.

Menezo, Y. (1976). Milieu synthétique pour la survie et la maturation des gamétes et pour la culture de l'oeuf fécondé. C.R. Acad. Sci. 282, 1967.

Moor, R.M. and Trounson, A.O. (1977). Hormonal and follicular factors affecting maturation of sheep oocytes in vitro and their subsequent developmental capacity. J. Reprod. Fertil. 49, 101.

Moore, H.D.M. and Bedford, J.M. (1978). Ultrastructure of the equatorial segment of hamster spermatozoa during penetration of oocytes. J. Ultrastruct. Res. 62, 110.

Moore, N.W. (1968). The survival and development of fertilized eggs transferred between Border Leicester and Merino ewes. Aust. J. Agric. Res. 19, 295.

Moore, N.W. and Shelton, J.N. (1964). Egg transfer in sheep: effect of degree of synchronization between donor and recipient, age of egg, and site of transfer on the survival of transferred eggs. J. Reprod. Fertil. 7, 145.

Motlík, J. and Fulka, J. (1974). Fertilization of pig follicular oocytes cultivated in vitro. J. Reprod. Fertil. 36, 235.

Murray, F.A. (1978). Embryo transfer in large domestic mammals. In Methods in Mammalian Reproduction. J.C. Daniel, Jr. (ed), New York, Academic Press, pp. 285–305.

Newcomb, R., Christie, W.B. and Rowson, L.E.A. (1978). Birth of calves after in vivo fertilization of oocytes removed from follicles and matured in vitro. Vet. Rec. 102, 461.

Newcomb, R. and Rowson, L.E.A. (1975). Conception rate after uterine transfer of cow eggs in relation to synchronization of estrus and age of eggs. J. Reprod. Fertil. 43, 539.

Nicholas, J.S. (1933). Development of transplanted rat eggs. Proc. Soc. Exp. Biol. Med. 30, 1111.

Oguri, N. and Tsutsumi, Y. (1974). Non-surgical egg transfer in mares. J. Reprod. Fertil. 41, 313.

Oliphant, G. and Brackett, B.G. (1973a). Capacitation of mouse spermatozoa in media with elevated ionic strength and reversible decapacitation with epididymal extracts. Fertil. Steril. 24, 948.

Oliphant, G. and Brackett, B.G. (1973b). Immunological assessment of surface changes of rabbit sperm undergoing capacitation. Biol. Reprod. 9, 404.

Onuma, H., Hahn, J., and Foote, R.H. (1970). Factors affecting superovulation, fertilization, and recovery of superovulated ova in prepuberal cattle. J. Reprod. Fertil. 21, 119.

Pincus, G.W. and Enzmann, E.V. (1935). The comparative behavior of mammalian eggs in vivo and in vitro. I. The activation of ovarian eggs. J. Exp. Med. 62, 665.

Pope, C.E., Christenson, R.K., Zimmerman-Pope, V.A. and Day, B.N. (1972). Effect of number of embryos on embryonic survival in recipient gilts. J. Anim. Sci. 35, 805.

Quirke, J.F. and Gordon, I. (1971). Culture and fertilization of sheep ovarian oocytes. III. Evidence of fertilization in the sheep oviduct based on pronucleate and cleaved eggs. J. Agric. Sci. 76, 375.

Reyes, A., Goicocchea, B. and Rosado, A. (1978). Calcium ion requirement for rabbit spermatozoal capacitation and enhancement of fertilizing ability by ionophore A23187 and cyclic adenosine 3':5'-monophosphate. Fertil. Steril. 29, 451.

Rowson, L.E.A. and Dowling, D.F. (1949). An apparatus for the extraction of fertilized eggs from the living cow. Vet. Rec. 61, 191.

Rowson, L.E.A., Lawson, R.A.S. and Moor, R.M. (1971). Production of twins in cattle by egg transfer. J. Reprod. Fertil. 25, 261.

Rowson, L.E.A., Lawson, R.A.S., Moor, R.M. and Baker, A.A. (1972). Egg transfer in the cow: synchronization requirements. J. Reprod. Fertil. 28, 427.

Rowson, L.E.A., Moor, R.M. and Lawson, R.A.S. (1969). Fertility following egg transfer in the cow: effect of method, medium, and synchronization of estrus. J. Reprod. Fertil. 18, 517.

Seidel, G.E., Jr. (1977). Short-term maintenance and culture of embryos. In Embryo Transfer in Farm Animals. K.J. Betteridge (ed), Ottawa, Canada Dept. Agric. Monograph 16, pp. 20–24.

Seidel, G.E., Jr., Bowen, R.A. and Kane, M.T. (1976). In vitro fertilization, culture, and transfer of rabbit ova. Fertil. Steril. 27, 861.

Seidel, G.E., Jr., Larson, L.L., Spilman, C.H., Hahn, J. and Foote, R.H. (1971). Culture and transfer of calf ova. J. Dairy Sci. 54, 923.

Seidel, G.E., Jr. and Seidel, S.M. (1978). Bovine

embryo transfer: costs and success rates. Adv. Anim. Breeder, *26*, 6.

Shea, B.F., Hines, D.J., Lightfoot, D.E., Ollis, G.W., and Olson, S.M. (1976a). The transfer of bovine embryos. *In* Egg Transfer in Cattle. L.E.A. Rowson (ed), Commission of the European Communities, EUR 5491. Luxembourg, pp. 145–152.

Shea, B.F., Latour, J.P.A., Bedirian, K.N. and Baker, R.D. (1976b). Maturation in vitro and subsequent penetrability of bovine follicular oocytes. J. Anim. Sci. *43*, 809.

Smorag, Z. (1977). (Discussion of freezing embryos from farm animals) *In* The Freezing of Mammalian Embryos. K. Elliott and J. Whelan (eds), Ciba Foundation Symposium 52 (New Series). Amsterdam, Elsevier, Excerpta Medica, pp. 211–212.

Sreenan, J.M. (1970). In vitro maturation and attempted fertilization of cattle follicular oocytes. J. Agric. Sci. *75*, 393.

Sreenan, J.M. (1978). Non-surgical embryo transfer in the cow. Theriogenology *9*, 69.

Sreenan, J.M., Beehan, D. and Mulvehill, P. (1975). Egg transfer in the cow: factors affecting pregnancy and twinning rates following bilateral transfer. J. Reprod. Fertil. *44*, 77.

Sugie, T. (1965). Successful transfer of a fertilized bovine egg by non-surgical techniques. J. Reprod. Fertil. *10*, 197.

Sugie, T. and Soma, T. (1970). The survival and development of fertilized ova transferred reciprocally between Saanen and Japanese native goats. Bull. Nat. Inst. Anim. Ind. Jpn. *23*, 7.

Tervit, H.R. and Goold, P.G. (1978). The culture of sheep embryos in either a bicarbonate-buffered medium or a phosphate-buffered medium enriched with serum. Theriogenology *9*, 251.

Tervit, H.R., Whittingham, D.G. and Rowson, L.E.A. (1972). Successful culture in vitro of sheep and cattle ova. J. Reprod. Fertil. *30*, 493.

Trounson, A.O., Rowson, L.E.A., and Willadsen, S.M. (1978a). Non-surgical transfer of bovine embryos. Vet. Rec. *102*, 74.

Trounson, A.O., Shea, B.F., Ollis, G.W. and Jacobson, M.E. (1978b). Frozen storage and transfer of bovine embryos. J. Anim. Sci. *47*, 677.

Trounson, A.O., Willadsen, S.M. and Moor, R.M. (1977a). Reproductive function in prepubertal lambs: ovulation, embryo development, and ovarian steroidogenesis. J. Reprod. Fertil. *49*, 69.

Trounson, A.O., Willadsen, S.M. and Rowson, L.E.A. (1977b). Fertilization and development capability of bovine follicular oocytes matured in vitro and in vivo and transferred to the oviducts of rabbits and cows. J. Reprod. Fertil. *51*, 321.

Warwick, B.L. and Berry, R.O. (1949). Inter-generic and intra-specific embryo transfer. J. Hered. *40*, 297.

Webel, S.K., Peters, J.B. and Anderson, L.L. (1970). Synchronous and asynchronous transfer of embryos in the pig. J. Anim. Sci. *30*, 565.

Whittingham, D.G., Leibo, S.P. and Mazur, P. (1972). Survival of mouse embryos frozen to $-196°C$ and $-269°C$. Science *178*, 411.

Willadsen, S.M., Polge, C. and Rowson, L.E.A. (1978). The viability of deep-frozen cow embryos. J. Reprod. Fertil. *52*, 391.

Willadsen, S.M., Polge, C., Rowson, L.E.A. and Moor, R.M. (1976). Deep freezing of sheep embryos. J. Reprod. Fertil. *46*, 151.

Willett, E.L., Black, W.G., Casida, L.E., Stone, W.H. and Buckner, P.J. (1951). Successful transplantation of a fertilized bovine ovum. Science *113*, 247.

Wilmut, I. and Rowson, L.E.A. (1973). Experiments on the low-temperature preservation of cow embryos. Vet. Rec. *92*, 686.

Wright, R.W., Jr., Anderson, G.B., Cupps, P.T., Drost, M. and Bradford, G.E. (1976). In vitro culture of embryos from adult and prepuberal ewes. J. Anim. Sci. *42*, 912.

Appendix I

Chromosome Numbers of Bovinae, Equinae, and Caprinae Species

Common Name	Scientific Name	Chromosome Number (2N)	Fundamental Number
Domestic cattle	*Bos taurus*	60	62
Banteng	*Bos banteng*	60	62
Zebu	*Bos indicus*	60	62
Yak	*Bos grunniens*	60	62
European bison	*Bison bonasus*	60	62
American bison	*Bison bison*	60	62
Gaur	*Bos gaurus*	58	62
Nyala	*Tragelaphus angasi*	55	58
Congo buffalo	*Syncerus caffer nanus*	54	60
African buffalo	*Syncerus caffer caffer*	52	60
Asiatic buffalo	*Bubalus bubalis*	48	58
Anoa	*Anoa depressicornis*	48	60
Nilgai	*Boselaphus tragocamelus*	46	60
Four-horned antelope	*Tetracerus quadricornis*	38	38
Sitatunga	*Tragelaphus spekei*	30	58
Mongolian wild horse	*Equus przewalskii*	66	94
Domestic horse	*Equus caballus*	64	94
Donkey	*Equus asinus*	62	104
Nubian ass	*Equus asinus africans*	62	104
Mongolian wild ass	*Equus hemionus*	56	104
Tibetan wild ass	*Equus kiang*	56	104
Persian wild ass	*Equus onager*	56	104
Grevy zebra	*Equus grevyi*	46	78
African zebra	*Equus burchelli*	44	82
Grant zebra	*Equus burchelli boehmi*	44	82
Mountain zebra	*Equus zebra*	34 (?)	60
Domestic goat	*Capra hircus*	60	60
Ibex	*Capra ibex*	60	60
Markhor	*Capra falconeri*	60	60
Saiga antelope	*Saiga tatarica*	60	60
Aoudad	*Ammotragus lervia*	58	60
Afghanistan sheep	*Ovis ammon cycloceros*	58	60
Kara-Tau sheep	*Ovis ammon nigimontana*	56	60
Domestic sheep	*Ovis aries*	54	60
Mouflon	*Ovis musimon*	54	60
Red sheep	*Ovis orientalis*	54	60 (?)
Bighorn sheep	*Ovis canadensis*	54	60
Laristan sheep	*Ovis ammon laristanica*	54	60
Musk ox	*Ovibos moschatus*	48	60
Himalayan tahr	*Hemitragus jemlahias*	48	60
Rocky Mountain goat	*Oreamnos americanus*	42	60

Appendix II

Chromosome Numbers and Reproductive Ability in Equine, Bovine and Caprine Hybrids

Species and Chromosome Number (2N)		Hybrids Chromosome Number (2N)	Reproductive Ability
Sire	**Dam**		
Mongolian wild horse, 66 (*E. przewalskii*)	Domestic horse, 64 (*E. caballus*)	65	Fertility (?)
Donkey, 62 (*E. asinus*)	Domestic horse, 64 (*E. caballus*)	63 (Mule)	Sterile
Domestic horse, 64 (*E. caballus*)	Donkey, 62 (*E. asinus*)	63 (Hinny)	Males are sterile, females are fertile, only in very exceptional cases
Nubian ass, 62 (*E. asinus africanus*)	Donkey, 62 (*E. asinus*)	62	Fertile
Mongolian wild ass, 56 (*E. hemionus*)	Donkey, 62 (*E. asinus*)	59	Fertile (?)
Grevy zebra, 46 (*E. grevyi*)	Domestic horse, 64 (*E. caballus*)	55 (Zebroid)	Sterile
African zebra, 44 (*E. burchelli*)	Donkey, 62 (*E. asinus*)	53 (Zebronkey)	Sterile
Donkey, 62 (*E. asinus*)	Mountain zebra, 34 (?) (*E. zebra*)	48	Sterile
American bison, 60 (*Bison bison*)	Zebu, 60 (*Bos indicus*)	60	Females are fertile
American bison, 60 (*Bison bison*)	Domestic cattle, 60 (*Bos taurus*)	60 (Cattalo)	Male F_1 are sterile
Domestic cattle, 60 (*Bos taurus*)	American bison, 60 (*Bison bison*)	60 (Cattalo)	Male F_1 are sterile
Domestic goat, 60 (*Capra hircus*)	Barbary sheep, 58 (*Ammotragus lorvia*)	59 (?)	Full-term fetuses, but no live hybrid
Domestic goat, 60 (*Capra hircus*)	Domestic sheep, 54 (*Ovis aries*)	57	Embryos are resorbed or aborted at six weeks pregnancy
Domestic sheep, 54 (*Ovis aries*)	Mouflon, 54 (*Ovis musimon*)	54	Fertile in both sexes
Bighorn sheep, 54 (*Ovis canadensis*)	Domestic sheep, 54 (*Ovis aries*)	54	Reduced fertility

Appendix III

Reproductive Diseases of Viral, Protozoan and Bacterial Origin

| Disease | Species | Etiology | Diagnosis | | Control |
			Clinical	Other	
Epizootic bovine abortion	Cattle	Unknown	Abortion in late pregnancy, lymphoid tissue lesions in fetus	Microscopic lesions of chronic inflammation	Change breeding season
Granular veneral disease	Cattle	Lymphoid reaction	Nodular vulvitis, balanitis	—	None
Infectious pustular vulvovaginitis (IPV)	Cattle	Bovine herpesvirus	Fibrinonecrotic vulvitis and vaginitis	Isolation of virus, serum neutralization test	Cessation of breeding, vaccination
Infectious bovine rhinotracheitis (IBR)	Cattle	Bovine herpesvirus	Respiratory disease, abortion	Isolation of virus, serum neutralization test	Vaccination
Bovine viral diarrhea (BVD)	Cattle	Virus	Abortion in early pregnancy, birth defects	Serum neutralization test	Vaccination
Catarrhal vaginitis	Cattle	Virus	Catarrhal vaginitis	Isolation of virus	None
Ulcerative dermatitis	Sheep	Virus	Ulceration of lips, legs, vulva and sheath	Lamb inoculation	Inspection of sale rams or rams purchased before breeding
Hog cholera	Swine	Virulent infection or modified live virus vaccination	Stillborn pigs, edematous dead pigs, weak pigs	History of pregnant sow vaccination or infection	Quarantine and slaughter of infected herds

Appendix III (cont'd)

Reproductive Diseases of Viral, Protozoan and Bacterial Origin

Disease	Species	Etiology	Diagnosis		Control
			Clinical	Other	
SMEDI	Swine	Virus	Stillbirth mummified fetus, embryonic death, infertility	Virus isolation	Allow exposure before breeding, maintain closed herd
African swine fever	Swine	Virus	Disease resembling hog cholera, abortion in pregnant sows	Exposure to warthogs or other infected swine	Quarantine and slaughter of infected herds
Equine rhino-pneumonitis	Horses	Equine herpesvirus I	Abortion in late pregnancy, respiratory disease in young	Focal necrosis of liver and edema of lungs in fetus, inclusion bodies, isolation of virus	Vaccination
Equine viral arteritis	Horses	Virus	Respiratory infection, cellulitis, abortion	Isolation of virus and serum neutralization test	Isolation of infected herds
Coital vesicular exanthema	Horses	Virus	Pustules on vulva, vagina, sheath and penis	None	Isolation, cessation of breeding
Enzootic abortion	Sheep	Chlamydia	Abortion, fresh (not autolyzed)	Staining elementary bodies in placenta, complement-fixation test	Vaccination
Trichomoniasis	Cattle	*Trichomonas fetus*	Infertility, pyometra and abortion in cows	Culture preputial cavity for trichomonads	Breeding rest, artificial insemination and treatment of bulls
Toxoplasmosis	Sheep Swine	*Toxoplasma gondii*	Encephalitis, abortion	Histopathology, dye test for antibodies	Isolation
Listeriosis	Cattle Sheep	*Listeria monocytogenes*	Nervous signs, circling, abortion	Isolation of bacterium	Avoid stress and feeding silage
Vibriosis	Cattle	*Campylobacter fetus* var. *venerealis*	Infertility	Mucus agglutination test, isolation, fluorescent antibody	Artificial insemination, vaccination
	Sheep	*Campylobacter fetus* var. *intestinalis*	Abortion	Isolation of bacterium	Vaccination

| Disease | Species | Etiology | Diagnosis | | Control |
			Clinical	Other	
Leptospirosis	Cattle	*Leptospira pomona* *Leptospira hardjo*	Hemolytic anemia, abortion in late pregnancy, agalactia	Agglutination test	Vaccination, elimination of carriers with antibiotic treatment
	Swine	*Leptospira pomona* *Leptospira grippotyphosa* *Leptospira canicola*	Abortion in late pregnancy, birth of weak pigs	Isolation of *Leptospira*	
	Horses	*Leptospira pomona* *Leptospira grippotyphosa* *Leptospira ictero- hemorrhagiae*	Abortion in late pregnancy, periodic ophthalmia		
	Sheep	*Leptospira pomona*	Hemolytic anemia, abortion in late pregnancy		
Brucellosis	Cattle	*Brucella abortus*	Abortion in late pregnancy, sterility in bulls	Isolation of bacterium, serum and milk agglutination tests	Vaccination, test and slaughter
	Swine	*Brucella suis*	Abortion, weak pigs, sterility in boars		
	Sheep Goat	*Brucella melitensis*	Abortion		
	Sheep	*Brucella ovis*	Epididymitis in rams, abortion		
	Dog	*Brucella canis*	Abortion		

(Afshar, 1965. Vet. Bull. 35, 165; Blood and Henderson, 1963, Veterinary Medicine, 2nd ed., Baltimore, Williams & Wilkins; Howarth, 1960, Proc. U. S. Livestock Sanit. Assoc. 64, 401.)

Index

Page numbers in *italics* indicate figures; numbers followed by "t" indicate tables.